# ANALYTICAL DETERMINATION OF NICOTINE AND RELATED COMPOUNDS AND THEIR METABOLITES

*edited by*

John W. Gorrod

*Toxicology Unit, John Tabor Laboratories, University of Essex, Colchester CO4 3SQ, UK*

and

Peyton Jacob, III

*University of California at San Francisco, Division of Clinical Pharmacology, San Francisco, CA 94110, USA*

ELSEVIER
Amsterdam – Boston – London – New York – Oxford – Paris
San Diego – San Francisco – Singapore – Sydney – Tokyo

ELSEVIER SCIENCE B.V.
Sara Burgerhartstraat 25
P.O. Box 211, 1000 AE Amsterdam, The Netherlands

First edition 1999
Second impression 2002

Library of Congress Cataloging in Publication Data

```
Analytical determination of nicotine and related compounds and their
   metabolites / edited by J. Gorrod and Peyton Jacob III.
       p.   cm.
    Includes bibliographical references and idex.
    ISBN 0-444-50095-2 (alk. paper)
    1. Nicotine--Analysis.  2. Nicotine--Metabolism.  I. Gorrod, J.
 W. (John W.)   II. Peyton, Jacob.
 QP801.N48A53 1999
 615'.785--dc21                                        99-38876
                                                          CIP
```

ISBN: 0-444-50095-2

# *Foreword*

Nicotine is a drug that has had enormous scientific, cultural, industrial, and health impact on our society. In science, nicotine was the chemical used to probe and conceptualize the receptor [1]. The idea that drugs and hormones act on receptors has become, of course, a basic tenet of modern pharmacology. More recently, nicotine has been used to probe a variety of nicotine cholinergic receptor subtypes in the human nervous system [2]. Characterization of the structure, regulation, and function of nicotinic receptor subtypes will undoubtedly lead to better understanding of the mechanisms of addiction, mood control, cognition, and pain control and most likely will result in the development of novel nicotine-related medications to treat a variety of disorders.

The cultural impact of nicotine was seen hundreds of years ago in South American Indians, where shamans used nicotine in rituals to induce death-like states [3]. The recovery of the shaman from the nicotine-induced delirium/coma reinforced the appearance of his supernatural powers. Nicotine is the addictive principle of tobacco and, as such, has served to spread the use of tobacco throughout the world [4]. At one time in the United States and currently in many countries of the world, 50% or more of men were or are cigarette smokers.

The industrial implications of nicotine have primarily been related to the fact that nicotine maintains the use of tobacco and, therefore, is crucial to the viability of the tobacco industry. Nicotine has also been used as an insecticide for many types of crops, and most recently has become important for the manufacturing of nicotine-containing medications.

Finally, the importance of nicotine in the area of public health is tremendous. Nicotine itself in high doses can be toxic, but serious direct toxicity is a rare event. Direct toxicity occurs primarily in accidents with exposure to insecticides or occasionally with intentional human poisonings. The major health effect of nicotine is by mediating tobacco use, which results in millions of premature deaths around the world yearly.

Regulatory strategies to control the tobacco-induced disease epidemic are at present very much focused on nicotine. At various times, nicotine-enriched cigarettes, aimed at reducing smoke intake, or nicotine-reduced cigarettes, aimed at gradually lowering the level of nicotine dependence, have been advocated as ways to reduce disease from tobacco [5,6]. Taxation of cigarettes based on nicotine content has been proposed. Labeling of nicotine delivery from cigarettes and/or nicotine content of tobacco is required in some countries or states, and it is likely to be required more widely in the future. Recently, the American Medical Association in the United States endorsed a public health strategy in which nicotine levels of tobacco would be mandatorily reduced, while nicotine-containing medications are made more easily available to smokers as they are weaned from their tobacco use [7].

Whether the issue is science, industrial processes, or public health, when the focus is nicotine there is a critical need for accurate and reliable measurement. The sources and extent of human exposure to nicotine need to be determined. Nicotine and metabolite levels need to be measured, usually in low concentrations, in biological fluids and in body tissues. Industrial processes require the quality control of nicotine levels in various products. Public health plans require sensitive methods for measuring exposure of individuals to nicotine and other constituents of tobacco products.

This book, 'Analytical Determination of Nicotine and Related Compounds and Their Metabolites', edited by John Gorrod and Peyton Jacob, provides for the first time a single comprehensive source of information on the analytical chemistry of nicotine and related alkaloids. The editors attempted a task that has not been achieved with much success in the past — that is, to get scientists from academia and the tobacco industry to work together to describe the state of the art of the chemistry and analytical methods for measurement of nicotine. Both the scope and the detail of this book are impressive. Chapters describe the history, pharmacology and toxicology of nicotine, the biosynthesis of nicotine and other alkaloids in the tobacco plant, the general chemistry of nicotine, and the analytical methodologies that have been used to measure nicotine and related alkaloids in biological specimens, in tobacco and pharmaceutical products, and in tobacco smoke. There is also a comprehensive review of the chemistry and toxicology of nicotine-derived nitrosamines, an important class of tobacco carcinogens.

The result of this effort is a unique monograph which will be an invaluable resource for tobacco chemists, neuroscientists studying nicotine, clinical researchers, policy makers, as well as tobacco and pharmaceutical industry scientists. This book joins a number of other recent monographs on the topics of nicotine psychopharmacology [8], nicotine disposition and metabolism [9], nicotine addiction [10], nicotine safety and toxicity [11], and nicotinic receptors [12] as part of a core library for nicotine and tobacco scientists.

# REFERENCES

1 Langley JN. On the reaction of cells and of nerve-endings to certain poisons, chiefly as regards the reaction of striated muscle to nicotine and to curare. *J. Physiol., Lond.*, 1905: **33**; 374–413.

2 McGehee DS, Role LW. Physiological diversity of nicotinic acetylcholine receptors expressed by vertebrate neurons. *Annu. Rev. Physiol.*, 1995: **57**; 521–546.

3 Wilbert J. *Tobacco and Shamanism in South America*. Yale University Press: New Haven, 1987.

4 Benowitz NL. Pharmacologic aspects of cigarette smoking and nicotine addiction. *N. Engl. J. Med.*, 1988: **319**; 1318–1330.

5 Russell MAH. Low-tar medium nicotine cigarettes: A new approach to safer smoking. *Brit. Med. J.*, 1976: **1**; 1430–1433.

6 Benowitz NL, Henningfield JE. Establishing a nicotine threshold for addiction. *N. Engl. J. Med.*, 1994: **331**; 123–125.

7 Henningfield JE, Benowitz NL, Slade J, Houston TP, Davis RM, Deitchman SD. Reducing the addictiveness of cigarettes. *Tob. Control*, 1998: **7**; 281–293.

8 *Nicotine Psychopharmacology. Molecular, Cellular, and Behavioral Aspects*. Wonnacott S, Russell MAH, Stolerman IP; Oxford University Press: Oxford, 1990.

9 *Nicotine and Related Alkaloids: Absorption, Distribution, Metabolism andExcretion*. Gorrod JW, Wahren J; Chapman & Hall: London, 1993.

10 *Nicotine Addiction: Principles and Management*. Orleans CT, Slade J; Oxford University Press: New York, 1993.

11 *Nicotine Safety and Toxicity*. Benowitz NL; Oxford University Press: New York, 1998.

12 *Neuronal Nicotinic Receptors: Pharmacology and Therapeutic Opportunities*. Arneric SP, Brioni JD; Wiley-Liss, Inc.: New York, 1998.

Neal L. Benowitz
*University of California, San Francisco*

# List of Contributors

Joseph L. Banyasz
Philip Morris, Research Center, PO Box 26583, Richmond, VA 23261-6583, USA

Neal L. Benowitz
University of California at San Francisco, San Francisco, CA 94110, USA

Michael F. Borgerding
R.J. Reynolds Tobacco Company, R&D Analytical Division, Winston-Salem, NC 28102, USA

Harold Burton
Department of Agronomy, University of Kentucky, Lexington, KY 40546-0091, USA

Lowell Bush
Department of Agronomy, University of Kentucky, Lexington, KY 40546-0091, USA

Gary D. Byrd
Biological Chemistry Division, R.J. Reynolds Tobacco Company, PO Box 1236, Winston-Salem, NC 27102-1236, USA

Peter A. Crooks
Division of Pharmaceutical Sciences, University of Kentucky, Lexington, KY 40536-0091, USA

Margareta Curvall
Swedish Match, Snuff Division, Research and Analysis, SE-118 85 Stockholm, Sweden

Riley A. Davis
Research and Development, Bowman Gray Technical Center, R.J. Reynolds Tobacco Company, PO Box 1236, Winston-Salem, NC 27102-1236, USA

Edward F. Domino
Department of Pharmacology, University of Michigan, Ann Arbor, MI 48109-0632, USA

Hilda B. Gjika
Department of Biochemistry, Brandeis University, Waltham, MA 02254, USA

John W. Gorrod
Toxicology Unit, John Tabor Laboratories, University of Essex, Wivenhoe Park, Colchester, Essex CO4 3SQ, UK

Stephen S. Hecht
University of Minnesota Cancer Center, 420 Delaware Street S.E., Minneapolis, MN 55455, USA

Walter P. Hempfling
Philip Morris USA, Research, Development and Engineering, Research Center, PO Box 26583, Richmond, VA 23261-6583, USA

Peyton Jacob, III
Division of Clinical Pharmacology, University of California at San Francisco, Bldg. 100, Room 235, 1001 Potrero Avenue, San Francisco, CA 94110, USA

Robert A. Jenkins
Chemical and Analytical Science Division, Oak Ridge National Laboratory, Building 4500S, PO Box 2008, Mail Stop 6120, Oak Ridge, TN 37831-6120, USA

John J. Langone
Center for Devices and Radiological Health, Food and Drug Administration, Rockville, MD 20852, USA

Peter N. Lee
P.N. Lee Statistics and Computing Ltd., Hamilton House, 17 Cedar Road, Sutton, Surrey SM2 5DA, UK

Erich Leitner
Department of Food Science, Institute of Biochemistry and Food Science, Technical University of Graz, A-8010 Graz, Austria

Donald E. Leyden
Worldwide Scientific Affairs, Philip Morris International, CH-2003 Neuchâtel, Switzerland

Michael W. Ogden
R.J. Reynolds Tobacco Company, Research & Development, PO Box 1236, Winston-Salem, NC 27102-1236, USA

Thomas A. Perfetti
R.J. Reynolds Tobacco Company, R&D Analytical Division, Winston-Salem, NC 28102, USA

Jean-Jacques Piadé
Philip Morris Europe, Fabriques de Tabac Réunies, CH-2003 Neuchâtel, Switzerland

Suresh Ralapati
BATF/National Laboratory Center, Rockville, MD 20850-3188, USA

Klaus Rustemeier
INBIFO Institut für biologische Forschung GmbH, Fuggerstrasse 3,
D-51149 Cologne, Germany

G. Schepers
INBIFO Institut für biologische Forschung GmbH, Fuggerstrasse 3,
D-51149 Cologne, Germany

Barbara Siegmund
Department of Food Science, Institute of Biochemistry and Food Science,
Technical University of Graz, A-8010 Graz, Austria

Anthony R. Tricker
Philip Morris Europe, CH-2003 Neuchâtel, Switzerland

Mui C. Tsai
Toxicology Unit, John Tabor Laboratories, University of Essex, Wivenhoe
Park, Colchester, Essex CO4 3SQ, UK

Helen Van Vunakis
Department of Biochemistry, Brandeis University, Waltham, MA 02254,
USA

# Contents

*Chapter 1*

# Pharmacological significance of nicotine

Edward F. Domino

*Department of Pharmacology, University of Michigan, Ann Arbor, MI 48109-0632, USA*

## I. INTRODUCTION

It is generally acknowledged that nicotine is the principal alkaloid that accounts for the widespread human use of tobacco products throughout the world. In addition to its importance in tobacco products, nicotine is used as an insecticide for agricultural and horticultural purposes. Nicotine alone as a chemical has a commercial and environmental impact: in addition, its chemical and pharmacological significance is of fundamental biological importance.

## II. HISTORICAL ASPECTS

Jean Nicot, a French ambassador 449 years ago, in 1550 sent tobacco and seeds to Paris from Portugal and, thus, introduced the French to tobacco from the New World. *Nicotiana tabacum* was indigenous to tropical America; the dried leaves were smoked by native Indians at the time Columbus discovered the New World. The dried leaves were called tábacum which was then used by the Spanish and Portuguese, and subsequently by the French and Italians. The English obtained their tobacco directly from America. It did not take long for the rest of the world to become involved with tobacco use and abuse. Crude aqueous extracts of tobacco were used by native tropical American populations, but by 1571 such extracts became more generally available.

The history of the chemical identification of nicotine has been well described (see [1,2]). Cerioli in 1807 and Vauquelin in 1808 isolated the 'essential oil' or 'essence of tobacco'. In 1828, Posselt and Reimann isolated nicotine from tobacco and in 1843 Melsens described its chemical empirical formula. Pictet and Crepieux synthesized nicotine in 1893 and in 1904 Pictet and Rotschy described the chemical isomerism of nicotine. Only 21 years ago, in 1978, Pitner et al. identified the spacial orientation of natural (S)-nicotine, herein called nicotine.

The experimental use of nicotine contributed substantially to our knowledge of physiology. Langley and Dickenson [3,4] noted that large concentrations of nicotine blocked autonomic ganglia, proving that autonomic nerves synapse in ganglia. Langley's continued use of nicotine as a pharmacological tool greatly enlarged our knowledge of neuronal cholinergic synapses [5]. Heymans and colleagues showed that nicotine stimulates respiration via chemoreceptors in the carotid and aortic bodies associated with the carotid sinus and aortic arch [6]. The only central nervous system synapse in the spinal cord shown to be activated via the nicotinic actions of acetylcholine is the Golgi recurrent collateral-Renshaw cell synapse [7–11].

## III. OCCURRENCE IN NATURE (see also Chapter 2)

Nicotine and related alkaloids are found in genus *Nicotiana* plants. *Nicotiana tabacum* is cultivated throughout the world for preparation of cigars, cigarettes, pipe and chewing tobacco. Nicotine usually constitutes about 2–8% of the dry weight of the cured leaf, although a much larger range exists in some *Nicotiana* plants. Other related alkaloids, including anabasine, anatabine, and nornicotine are also present, usually in much lower amounts. There are different relative amounts of all of these alkaloids in various *Nicotiana* species. The leaf content of nornicotine in some species is as high as 15–20% of the nicotine content. The chemical reviews on the alkaloids of tobacco by Jackson [12] and Pailer [13] are especially useful sources of more information. In other species such as *N. glauca* (tree tobacco), the main alkaloid is anabasine rather than nicotine. Interestingly, anabasine has been found in the marine worm *Amphiporus* and the ant *Aphaenogaster*. Other species of plants besides *Nicotiana* contain nicotine including *Duboisia, Equisetum* (horsetails), *Lycopersicum* (tomatoes), *Lycopodium* (club mosses), *Sedum* (succulent plants) and *Solanum* (potatoes). Therefore, it is not unexpected that trace amounts of nicotine are present in some edible human foods [14–18]. This is of interest in establishing threshold levels in human fluids of second hand tobacco smoke exposure. Surprisingly, this area of research has not been pursued, in part because of a lack of interest by granting agencies. Commercial nicotine is a byproduct of the tobacco industry. By adding lime or caustic soda to a filtered, concentrated aqueous extract of tobacco plant parts, the alkaloid can be isolated with either an organic solvent or by steam distillation. The addition of a chemical drying agent such as potash and subsequent fractional distillation further purifies nicotine.

## IV. CHEMISTRY (see also Chapters 4 and 5)

The chemical characteristics of nicotine are well known [19]. Pure nicotine is a colorless liquid with a characteristic acrimonious odor. Nicotine boils at 246–247°C. On exposure to air and light, or even on standing in the dark in a sealed bottle, over time the colorless or pale yellow oily liquid becomes the brownish color of stored nicotine. Brown colored nicotine is as toxic as pure colorless or pale yellow nicotine. Nicotine is known chemically as 3-(1-methyl-2-pyrrolidinyl)pyridine; 1-methyl-2-(3-pyridyl)pyrrolidine; or β-pyridyl-α-N-methylpyrrolidine. Its empirical formula is $C_{10}H_{14}N_2$. Its molecular weight is 162.23; C 74.03%, H 8.70%, N 17.27%. Its density $d_4^{20}$ is 1.0097. Pure nicotine has a specific rotation $[\alpha]_D^{25} = -169°$. When (S)-nicotine is boiled at 250°C with potassium tertiary butoxide, it racemizes. Therefore, nicotine can exist either as the (S)- or (R)-form. In this chapter, the natural (S) form is referred to as nicotine unless otherwise noted. After the absolute configuration of hydroxyproline and related compounds was determined, Hudson and Neuberger suggested that natural nicotine be designated as Ls-nicotine because the COOH group of N-methyl-Ls-proline can be replaced with a pyridine group [20]. The two isomers of nicotine are important in understanding its molecular pharmacology; they were used in the historical accumulation of evidence of cholinergic receptive substances, i.e., cholinergic nicotinic receptors.

A key chemical concept that determines nicotine's absorption, excretion, pharmacology and toxicology is that its charged and uncharged forms are pH dependent. Nicotine is dibasic because of its pyrrolidine (pKa=7.84) and pyridine nitrogens (pKa=3.04) at 15°C. At a pH of 7.4 and a temperature of 37°C about 69% of the pyrrolidine nitrogen is ionized or positively charged, whereas the pyridine nitrogen exists in an unionized form [21]. Thus, the pH of a nicotine solution dramatically alters its protonated state. Uncharged organic bases are lipophilic, whereas charged organic bases are hydrophilic. Nicotine at pH 7.4 exists in both forms; one diffuses through lipoprotein membranes, the other does not. Nicotine, like other alkaloids with similar pKas, exists with a ratio of charged to uncharged forms of about 2 to 1 at pH 7.4. This chemical fact permits nicotine to have important biological actions.

## V. TOXICOLOGY

The symptoms of nicotine poisoning are similar to those of tobacco poisoning. The naive tobacco smoker who overindulges experiences a strange feeling which soon progresses to increased salivation, nausea, and vomiting. Cardiac palpitation is noted with a rapid, pounding pulse and an increase in blood pressure. Nicotine poisoning involves central and peripheral nervous system effects consisting of initial stimulation followed by depression. Very rapid light headedness, severe nausea, vomiting, skeletal muscle weakness, prostration, tremor and convulsions can occur. Death is due to peripheral skeletal neuromuscular blockade. Hence, artificial ventilation is the most important therapeutic measure in treating a person with a potentially lethal nicotine overdose. Self-intoxication with *Nicotiana tabacum* and *rustica* historically was used by shamans and others for inducing stupor or narcosis [22]. *Nicotiana tabacum* was native to Pre-Conquest Middle and South America and the West Indies. *Nicotiana rustica* was native to Mexico and North America prior to the arrival of Europeans and was probably the form used by the more northern Indian tribes. After the Conquest, Europeans introduced *Nicotiana tabacum* to North America including the territory of the United States and Canada [23]. Large amounts of *Nicotiana* were used in native religious ceremonies to communicate with gods in which the celebrant was dangerously close to death. Similarly, the acute toxic effects of aqueous nicotine solutions have been known for centuries. Holmstedt [2] summarized the literature on the toxicity of nicotine and related compounds. In 1828, when Possett and Reimann isolated nicotine, they studied its toxicity in rabbits, dogs, and themselves, in a classic approach used by some scientists throughout history. The use of a nicotine solution as an insecticide has been known before the conquest of the New World. It is a very efficient agent to kill softbodied insects such as plant lice. It is now sold as a 40% nicotine sulfate solution to prevent volatilization. The role of acetylcholine as a neurotransmitter in organisms higher on the phylogenetic scale results in (S)-nicotine being more toxic than [R]-nicotine. Aqueous solutions of nicotine sulfate continue to be used widely throughout the world as relatively cheap and environmentally friendly insecticides. The threshold limit for commercial exposure to nicotine is a concentration of 0.5 mg/m$^3$ [24].

## VI. PHARMACOLOGY

As described above, the fact that at pH 7.4 nicotine exists in both charged and uncharged forms means that the uncharged form easily penetrates lipoprotein membranes and the charged form does not. Schmiterlöw and colleagues [25–27] showed dramatically the tissue distribution of nicotine and metabolites with autoradiographs of [$^{14}$C]-nicotine given i.v. to animals. It distributes rapidly into many tissues of the body including fetuses of pregnant mice. Initially, the brain and spinal cord contain a high concentration of nicotine; within 15 minutes there is only a little left, while its concentration in the liver increases. Within 30 minutes after injection, most of the nicotine is present in the liver, the adrenals, the stomach, and bone marrow, but very little in the central nervous system. Why is nicotine in the bone marrow? We know bone marrow is relatively fatty, but is this the only reason nicotine is there?

About 85 years ago, Dale [28] studied the actions of various synthetic and endogenous choline derivatives, including acetylcholine, together with those of the plant alkaloids muscarine and nicotine. The effects of muscarine were similar to some actions of acetylcholine, as well as parasympathetic nerve stimulation. On the other hand, the effects of nicotine were similar to those of acetylcholine after the muscarinic antagonist atropine was given. In addition, nicotine had actions similar to those of sympathetic nerve stimulation. Dale postulated that acetylcholine was an autonomic nervous system neurotransmitter; it had dual actions, muscarinic as well as nicotinic [28]. Both nicotine and acetylcholine exist in very similar molecular forms. Although acetylcholine is a very flexible molecule compared to nicotine, it can easily configure to resemble nicotine. The pyridine nitrogen of nicotine is an electron donor which is similar to the keto oxygen of the acetyl group of acetylcholine. The positive charge of the quaternary nitrogen of acetylcholine is similar to the positive charge of the pyrolidine nitrogen of nicotine. These facts have been previously emphasized [21,30]. Figures 1 and 2 illustrate the remarkable similarity of both molecules.

## Nicotine binding sites throughout the body

There is a great deal of evidence for multiple nicotine binding sites throughout the body, especially in the nervous system. The number of types of binding sites varies from one to as many as five. Much of this early research has been reviewed previously [29-32]. It is now known that there are at least five muscarinic receptor subtypes ($M_1$-$M_5$) with transduction mechanisms to alter second messengers such as a decrease in cyclic AMP, an increase in PI turnover, or an increase in potassium ion conductance. The primary composition of nicotinic cholinergic receptors has been described. There are specific genes for each of the five subunits of several nicotinic

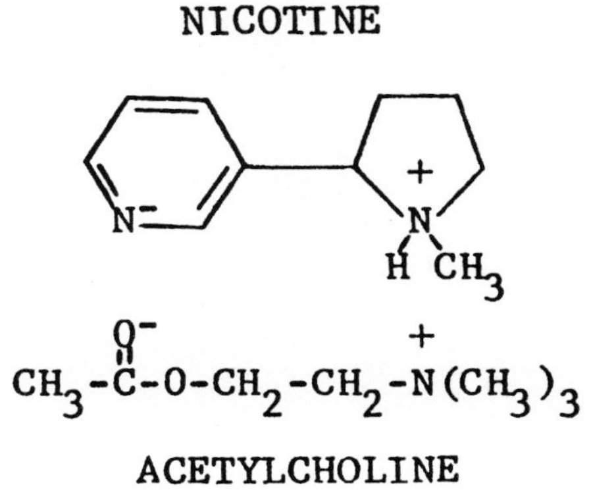

**NICOTINE**

**ACETYLCHOLINE**

Fig. 1 Structural similarity of nicotine to the neurotransmitter acetylcholine using conventional chemical symbols. Note that the positive and negative charges on both molecules are at the same distance from each other.

cholinergic receptors (nAChRs) which encode a variety of protein subunits similar to the skeletal muscle receptor. To date, at least eight neuronal genes have been isolated for *alpha* subunits (*alpha 2–alpha* 9). There are five *beta*, and one each of *delta, epsilon,* and *gamma* subunits. The subunit mRNA distribution for each subunit varies considerably throughout the body. There are at least three major subfamilies of nAChRs as summarized in Table 1.

## Pharmacological and toxicological actions of the optical isomers of nicotine

Ever since the synthesis of nicotine by Pictet and Rotschy in 1904, there has been continued interest in the action of the two enantiomers of nicotine. The natural (S)-nicotine is usually, but not always, more active than (R)-nicotine (see [2]). Over the years, the absolute purity of the two isomers has been a problem. When the degree of mixed contamination is known, (S)-nicotine usually is clearly more potent than (R)-nicotine [33]. However, the latter is not inert. As described above these facts are important in the

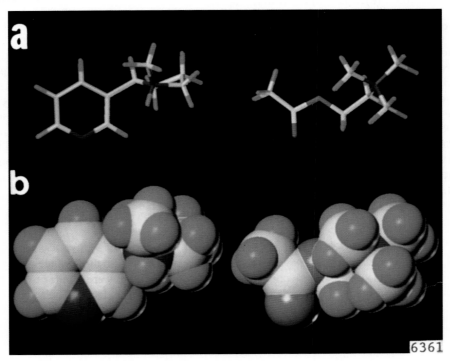

Fig. 2 Stereochemistry of nicotine (a) and acetylcholine (b) using two different molecular models. Both molecules are color coded by atom type. (Kindly provided by Charles Spivak, Ph.D., NIDA Addiction Research Center, Baltimore, MD. From reference 30, page 6 with permission of NPP Books.)

pharmacological actions of nicotine at the molecular level and provide evidence that nAChRs have stereoselective preference.

The pharmacological effects of nicotine are very diffuse and affect all major organs of the body wherever there are nicotinic cholinergic receptors. Abood and colleagues have pointed out that there are actions of nicotine unrelated to cholinergic receptors [34]. Nevertheless, most actions of nicotine involve the latter. These are summarized in Table 2 into major categories involving multiple systems.

## CONCLUSIONS

Even a quick perusal of the remarkably diverse effects of nicotine impresses one with their complexity. Nicotinic cholinergic receptors are the major targets of nicotine's action. As multiple ligand-gated ion channels, these receptors are very complex, involving many presynaptic and postsynaptic sites, especially in the central nervous system. Receptor depolarization leads to an influx of $Ca^{++}$ and $Na^+$ intracellularly providing an ionic basis for the release of multiple chemical messengers. Repeated nicotine exposure leads to tachyphylaxis and tolerance due to receptor desensitization. Does this

TABLE 1

Subfamilies of nicotinic cholinergic receptors

| Property | Skeletal Muscle | Autonomic ganglionic | Central nervous system |
| --- | --- | --- | --- |
| Subunits | $\alpha1,\beta1,\varepsilon,\delta,\gamma$ | $\alpha3,\alpha4,\alpha5,\alpha7,\beta2,\beta4$ | $\alpha2,\alpha3,\alpha4,\alpha5,\alpha6,$ $\alpha7,\alpha8,\alpha9,\beta2,\beta3,\beta4$ |
| Examples of $(\alpha4)_2(\beta2)_3,(\alpha3)_x(\beta2)_y$ subunit composition as hetero- or homo-oligomers | $(\alpha1)_2B1\delta\varepsilon$ (adult) $(\alpha1)_2B1\delta\gamma$ (fetal) | $(\alpha4)_x7)_y,(\alpha4)_x(\beta2)_y$ | $(\alpha7)_5,(\alpha8)_5,(\alpha9)_5$ |
| Effector | int. $Na^+/K^+/Ca^{++}$ (cond. ps large) | int. $Na^+/K^+/Ca^{++}$ (cond. small to large) | int. $Na^+/K^+/Ca^{++}$ (cond. small to large, high $Ca^{2+}$ perm.) |
| Selective ligands | $[^3H]$- or $I^{[125]}$ $\alpha$-bungarotoxin | $[^3H]$ cysteine $[^3H]$ methylcarbamyl-choline $[^3H]$ nicotine | $[^3H]$- or $I^{[125]}$ $\alpha$-bungarotoxin $[^3H]$ $\kappa$-bungarotoxin |
| Channel blockers | decamethonium gallamine | chlorisondomine hexamethonium | chlorisondomine mecamylamine |

abbreviations: cond.=conductance; perm.=permeability; ps=picosiemens; x,y=variable numbers not identified.

*Analytical Determination of Nicotine*

TABLE 2

Dose-dependent pharmacological actions of nicotine

I.  Central Nervous System
    A. Stimulation of brainstem activating system producing an arousal or "wake-up" effect.
    B. Positive cognitive effects increasing memory, teaching behavior, etc.
    C. Relaxation, especially in stressful situations.
    D. Mixed stimulant and depressant actions.
    E. Increased orienting response.
    F. Increases barrier to irrelevant and improved gating of relevant stimuli.
    G. Decreased monosynaptic reflexes such as the patellar reflex.
    H. Reinforcing self-administration behavior.
    I. Dependence, addiction, withdrawal.
    J. Stimulation of the chemoreceptor trigger zones in area postrema to produce nausea and vomiting.
    K. Tremor, convulsions and death.
II. Peripheral Nervous System
    A. Stimulation and blockade of both sympathetic and parasympathetic autonomic ganglia leading to multiple autonomic effects on the cardiovascular, gastrointestinal, etc. systems.
    B. Peripheral afferent stimulation involving taste and smell.
    C. Aortic and carotid body chemoreceptor stimulation.
    D. Stimulation of pulmonary, gastrointestinal and skeletal muscle spindle afferents.
    E. Direct and indirect modulation of skeletal muscle tone to increase and decrease tension in different muscles.
III. Cardiovascular System
    A. Increase in heart rate in small concentrations and decreases with poisoning.
    B. Significant bradycardia after overnight withdrawal on chronic use of tobacco.
    C. Increases in systolic and diastolic blood pressure.
    D. Increases in cardiac stroke volume and output.
    E. Cutaneous vasoconstriction resulting in decreased blood flow and skin temperature.
    F. Increased skeletal muscle blood flow.
    G. Increase in platelet reactivity.
    H. Modulation of COX-2 expression in vascular cells.
    I. Enhancement of cardiac ischemia.
    J. Increased circulating thromboxane and prostacyclin levels.
IV. Gastrointestinal System
    A. Complex effects usually reducing activity including decreased gastric emptying, reduction of gastric acid and pancreatic secretion.
    B. Therapeutic benefit in ulcerative colitis.
V.  Endocrine System
    A. Release of ACTH, growth hormone, β-endorphin, cortisol, vasopressin, prolactin, catecholamines, serotonin, etc.
    B. Increase in circulating free fatty acids, glycerol, lactate, glucose.
    C. Antiestrogen effects
VI. Metabolic Effects
    A. Increased energy expenditure.
    B. Decreased energy consumption.
    D. Decreased body weight.

TABLE 2 Continued

---

VII. Hepatic, Renal, and Other Organ Interactions
    A. Most are secondary to the diffuse and complex actions of nicotine to release various chemical messengers (neurotransmitters, modulators, hormones, and autocoids) which affect all major organ systems.
    B. Biotransformation of nicotine involves multiple CYP450 microsomal enzymes. Current emphasis is on the role of the polymorphism of the *CYP2A6* gene but many other CYP450s are involved.
    C. Curing of tobacco forms NNK, a procarcinogen (see Chapter 11).
VIII. Therapeutic Potential of Nicotine as a Lead Compound for New Potential Therapeutic Agents
    A. Substitute for treating tobacco addiction and dependence.
    B. Parkinson's disease.
    C. Alzheimer's disease.
    D. Ulcerative colitis.
    E. New class of analgesic agents.
    F. Insight into the pathology and genetics of the gating deficit in prepulse inhibition in schizophrenic patients and some relatives.

---

mean receptor desensitization to acetylcholine as well? In all probability, yes. Such a mechanism accounts for the mixed stimulant/depressant actions of nicotine. The diffuse pharmacological actions of nicotine are dramatic evidence of the functional significance of the nicotinic cholinergic system in all living organisms possessing such receptors.

## ACKNOWLEDGEMENTS

Supported in part by NIDA grant DA-10992.

## REFERENCES

1 Koenig P (Ed.). *Die Entdeckung des reinen Nicotins im Jahre 1828 an der Universität Heidelberg durch Reimann und Posselt: mit einer Beschreibung Ihrer Vorläufer und mit Abbildungen.* Bremen: Arthur Geist, 1940.
2 Holmstedt B. Toxicity of nicotine and related compounds. In: *The pharmacology of nicotine.* Rand MJ, Thurau K (Eds.), McLean VA: IRL Press, ICSU Symposium Series 1988: **9:** 61–68.
3 Langley JN, Dickinson WL. Nicotine. *J. Physiol.*, 1890: **11**; 123, 265, 509.
4 Langley JN, Dickinson WL. On the local paralysis of peripheral ganglia, and on the connection of different classes of nerve fibers with them. *Proc. Roy. Soc.*, 1889: **46:** 423–431.
5 Langley NJ. On the stimulation and paralysis of nerve cells and nerve endings. Part II. Paralysis by curari, strychnine and brucine and its antagonism by nicotine. *J. Physiol.*, 1918: **52:** 247–266.

6 Heymans C, Bouckaert JJ, Dautrebande L. Sinus carotidien et réflexes res-
   piratories. III. Sensibilité des sinus carotidiens aux substances chimiques. Action
   stimulante respiratorie réflexe du sulfure de sodium, du cyanure de potassium, de
   la nicotine et de la lobéline. *Arch internat de pharmacodyn et de thérap.*, 1931: **40**;
   54–91.

7 Eccles JC, Fatt P, Koketsu K. Cholinergic and inhibitory synapses in a pathway
   from motor-axon collaterals to motoneurons. *J. Physiol. London*, 1954: **126**;
   524-562.

8 Eccles JC, Eccles RM, Fatt P. Pharmacological investigations on a central synapse
   operated by acetylcholine. *J. Physiol. London*, 1956: **131**; 154–169.

9 Curtis DR, Eccles RM. The excitation of Renshaw cells by pharmacological agents
   applied electrophoretically. *J. Physiol. London*, 1958: **141**; 435–445.

10 Curtis DR, Eccles JC, Eccles RM. Pharmacological studies on spinal reflexes.
   *J. Physiol. London*, 1957: **136**; 420–434.

11 Ueki S, Koketsu K, Domino EF. Effects of mecamylamine on the Golgi recurrent
   collateral-Renshaw cell synapse in the spinal cord. *Exper. Neurol.*, 1961: **3**;
   141–148.

12 Jackson KE. Alkaloids of tobacco. *Chem. Rev.*, 1941: **29**; 123–197.

13 Pailer M. Chemistry of nicotine and related alkaloids (including biosynthetic
   aspects). In: *Tobacco alkaloids and related compounds*, Von Euler US (Ed.), Proc.
   IVth Wenner-Gren International Symposium, New York: Pergamon Press/
   Macmillan Co, 1965; 15–36.

14 Castro A, Monji N. Dietary nicotine and its significance in studies on tobacco
   smoking. *Biochem. Arch.*, 1986: **2**; 91–97.

15 Sheen SJ. Detection of nicotine in foods and plant materials. *J. Food Sci.*, 1988: **53:**
   1572–1573.

16 Davis RA, Stiles MF, deBethizy JD, Reynolds JH. Dietary nicotine: a source of
   urinary cotinine. *J. Food Chem. Toxicol.*, 1991: **29**; 821–827.

17 Domino EF, Hornbach E, Demana T. To the Editor. The nicotine content of common
   vegetables. *New Engl. J. Med.*, 1993: **369**; 437.

18 Domino EF, Hornbach E, Demana T. Relevance of nicotine content of common
   vegetables to the identification of passive tobacco smokers. *Med. Sci. Res.*, 1993: **21**;
   571–572.

19 The Merck Index, 11th Edition, Compound #6434. *Nicotine*, Budavari S, O'Neil MJ,
   Smith A, Heckelman PE (Eds.), Rahway NJ, Merck and Co. Inc., 1989, p 1030.

20 Hudson CS, Neuberger A. The stereochemical formulas of the hydroxyproline and
   allohydroxypyroline enantiomorphs and some related compounds. *J. Org. Chem.*,
   1950: **15**; 24–34.

21 Domino EF. Behavioral, electrophysiological, endocrine, and skeletal muscle actions
   of nicotine and tobacco smoking. In: *Electrophysiological effects of nicotine*, Remond
   A, Izard C (Eds.), Amsterdam: Elsevier/North Holland Biomedical Press, 1979,
   133–146.

22 Cooper JM. Stimulants and narcotics. In: Handbook of South American Indians.
   Bull No 143, *Bur. Am. Ethnol.*, 1949: **5**; 525.

23 Goodspeed TH. *The genus Nicotiana*. Waltham MA: Botanica Co, 1954.

24 Baselt RC. *Disposition of toxic drugs and chemicals in man*. Second Edition. Davis
   CA: Biomedical Publications, 1982; 549–555.

25 Appelgren LE, Hansson E, Schmiterlöw CG. The accumulation and metabolism of $^{14}$C-labeled nicotine in the brain of mice and cats. *Acta Physiol. Scand.*, 1962: **56**; 249–257.

26 Hansson E, Schmiterlöw CG. Physiological disposition and fate of $C^{14}$ labelled nicotine in mice and rats. *J. Pharmacol. Exp. Ther.* 1962: **137**; 91–102.

27 Schmiterlöw CG, Hansson E, Appelgren LE, Hoffman PC. Physiological disposition and biotransformation of $C^{14}$-labelled nicotine. In: *Isotopies in experimental pharmacology*, Roth LJ (Ed.), Chicago: University of Chicago Press, 1965, 75–90.

28 Dale HH. The action of certain esters and ethers of choline and their relation to muscarine. *J. Pharmacol. Exp. Ther.*, 1914: **6**; 147–190.

29 *US Surgeon General's Report. The health consequences of smoking-nicotine addiction*, Washington DC: Superintendent of Documents, US Government Printing Office, 1988, 1–639.

30 Domino EF. Significance of multiple nicotinic cholinergic receptors for tobacco smoking. In: *Brain imaging of nicotine and tobacco smoking*, Domino EF (Ed.), Ann Arbor: NPP Books, PO Box 1491, 48106, 1995, 1–12.

31 Mesulam MM. Chapter 12. Structure and function of cholinergic pathways in the cerebral cortex, limbic system, basal ganglia, and thalamus of the human brain. In: *Psychopharmacology: the fourth generation of progress,* Bloom FE, Kupfer DJ (Eds.), New York: Raven Press, 1994; 135–146 .

32 Reiner PB, Fibiger HC. Functional heterogeneity of central cholinergic systems. In: *Psychopharmacology: the fourth generation of progress*, Bloom FE, Kupfer DJ (Eds.), New York: Raven Press, 1994; pp 147–153.

33 Domino EF. Some comparative pharmacological actions of ( − )-nicotine, its optical isomer, and related compounds. In: *Tobacco alkaloids and related compounds.* Von Euler US (Ed.), Proc IVth Wenner-Gren International Symposium, New York: Pergamon Press/Macmillan Co., 1965, 303–313.

34 Abood LG, Reynolds DT, Bidlack JM. Stereospecific [$^3$H]-nicotine binding to intact and solubilized rat brain membranes and evidence for its noncholinergic nature. *Life Sci.*, 1980: **27**; 1307–1314.

*Chapter 2*

# Biosynthesis of nicotine and related compounds

Lowell Bush[1], Walter P. Hempfling[2] and Harold Burton[1]

[1] *Department of Agronomy, University of Kentucky, Lexington, KY 40546-0091, USA*
[2] *Philip Morris USA, Research, Development and Engineering, Research Center, P.O. Box 26583, Richmond, VA 23261-6583, USA*

## I. INTRODUCTION

The pyridine alkaloids are characteristic compounds in the chemistry of *Nicotiana* spp. Nicotine is the principal alkaloid, usually accounting for greater than 90% of the alkaloid fraction, in commercial tobacco, *N. tabacum*. Nornicotine, anatabine and anabasine are the other predominant alkaloids in tobacco, but seldom do they individually accumulate to more than 5% of the total alkaloid fraction. Other pyridine alkaloids have been reported to be present in very small amounts in tobacco and are not discussed in this chapter. Although, alkaloids are actively metabolized in plants their evolutionary function in tobacco is uncertain and much detail of their metabolism is still unknown. Alkaloid accumulation and biosynthesis have been reviewed with emphasis on accumulation and utilization (see [1–3]). In this chapter the emphasis will be on recent biochemical advancements in alkaloid biosynthesis and on the biological aspects of nicotine biosynthesis.

Total alkaloid content of commercial tobacco is primarily controlled by genes at two nonlinked loci, *Nic1* and *Nic2* [4]. The low alkaloid alleles are recessive and the dosage effect of *nic1* on leaf nicotine levels is ~2.4 times greater than *nic2* [5]. The *Nic* genes are expressed mainly in the plant roots and the result of the expression is mainly measured as leaf alkaloid content. The generally accepted biosynthetic pathways will be presented with emphasis on perturbations of the enzymes with molecular techniques. The biology of alkaloid biosynthesis will evaluate alkaloid biosynthesis in transformed root tissues and plant growth regulator effects on alkaloid biosynthesis.

13

## II. NICOTINE BIOSYNTHESIS

The precursors for the pyridine, pyrrolidine and piperidine rings of nicotine, nornicotine, anatabine and anabasine have been determined and bio-synthetic schemes proposed that are generally consistent with the obtained results. Even though the general schemes are known, the mechanisms in many of the reactions are unknown and continue to receive research attention. Knowledge of the details of the reactions within a biosynthetic pathway will not only be helpful in understanding the chemistry of the biosynthesis, but aid in development of biosynthesis of potentially new alkaloids within the tobacco plant system and in understanding the in vivo regulation of alkaloid biosynthesis. The four principal tobacco alkaloids and their derivatives all have the pyridine ring in common (Fig. 1). General biosynthetic pathways for the principal alkaloids of tobacco are depicted in Fig. 2. The pyridine ring is derived from nicotinic acid via the pyridine nucleotide cycle. The most rate defining steps in this portion of the scheme are quinolinic acid phosphoribosyltransferase (QPT) (Fig. 2, step 1) and nicotinic acid mononucleotide glycohydrolase (Fig. 2, step 2) [6]. QPT activity is found in roots of tobacco and activity increases in roots of plants that have been decapitated to induce nicotine accumulation. Root QPT activity is proportional to leaf nicotine levels and is considered to be the main regulatory enzyme for nicotinic acid production in nicotine biosynthesis [6–8]. Enzymes in the nicotinic acid nucleotide cycle seem to have less regulatory control for nicotine biosynthesis, because of the importance of nicotinamide adenine dinucleotide (NAD) in other cell metabolic events. Wagner et al. [6] demonstrated that nicotinic acid was formed directly from nicotinic acid mononucleotide, the product of QPT, by nicotinic acid

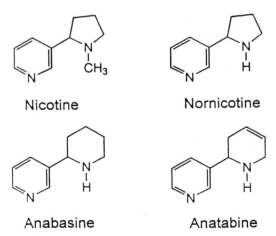

Nicotine                           Nornicotine

Anabasine                          Anatabine

Fig. 1  Principal alkaloids of tobacco.

mononucleotide glycohydrolase. The $K_m$ (4 mmol $l^{-1}$) for this reaction is high and probably ensures, in rapidly nicotine synthesizing tissue, that nicotinic acid mononucleotide is available for NAD biosynthesis for other cellular metabolism. However, in other studies on nicotinic acid mononucleotide

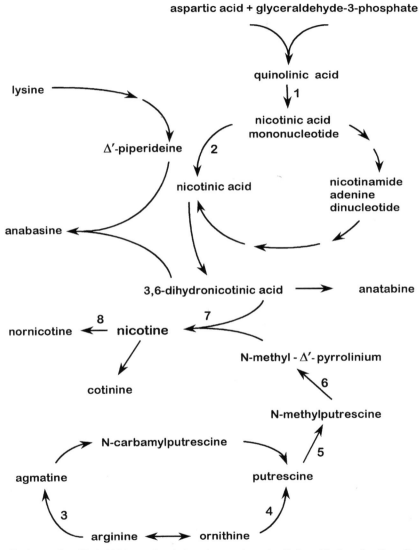

Fig. 2 Pathway for alkaloid biosynthesis in tobacco. 1=quinolinic acid phosphoribosyltransferase; 2=nicotinic acid mononucleotide glycohydrolase; 3=arginine decarboxylase; 4=ornithine decarboxylase; 5=putrescine methyltransferase; 6=methylputrescine oxidase; 7=nicotine synthase; 8=nicotine demethylase.

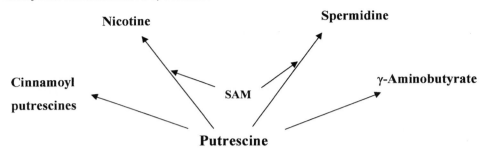

Fig. 3 Metabolic utilization of putrescine in tobacco.

glycohydrolase no apparent regulatory function in nicotine biosynthesis has been shown [9]. The other steps from nicotinic acid to the dihydronicotinic acid are less well understood, but dihydronicotinic acid is the intermediate utilized for synthesis of the pyridine ring of the tobacco alkaloids.

The pyrrolidine ring moiety of nicotine is derived from putrescine that is formed from ornithine or perhaps from arginine. Ornithine decarboxylase (ODC) (Fig. 2, step 4) catalyzes the decarboxylation of ornithine to putrescine. Utilization of putrescine has been proposed as a regulatory step for nicotine biosynthesis as at least four competing reactions are known to diverge in tobacco (Fig. 3). Cinnamoylputrescines constitute as much as 6% of the dry weight of cell cultures [10] and γ-aminobutyrate may accumulate to as much as 80 mg per plant [11]. Perhaps the two most important pathways are the methylation and the *N*-aminopropylation of putrescine utilizing SAM (S-adenosyl-methionine) for nicotine and polyamine formation, respectively. Many reports of attempts to supply additional putrescine to root cultures or cell cultures to increase nicotine content have had limited success [12,13]. However, a specific inhibitor of ODC activity, 2-difluor-omethylornithine (DFMO), when supplied to tobacco callus cultures did not decrease nicotine content [14].

Tobacco alkaloids are primarily synthesized in the roots and transformation of tobacco root cultures by *Agrobacterium rhizogenes* causes a condition called hairy root (for details see in this chapter subsection on *A. rhizogenes* infection and nicotine biosynthesis). These cultures produce greater amounts of nicotine than control root cultures. An ODC gene from yeast (*Saccharomyces cerevisiae*) was integrated into hairy root cultures [15]. The presence of the yeast gene increased ODC activity in selected lines 3-fold. Free putrescine increased up to 2-fold in some genotypes, but there was no effect on the polyamines, spermidine and spermine, or putrescine conjugates. Nicotine content in 3 of 6 lines was increased approximately 2-fold greater than the mean of the controls. There was not a precise correlation between ODC level and nicotine levels in individual lines but the correlation was positive, $r=0.66$ and significant. These results suggest that enzymes other than ODC are to some degree limiting alkaloid production. Arginine decarboxylase (ADC) (Fig. 2, step 3) which catalyzes the decarboxylation of arginine to agmatine may also initiate the formation of putrescine. There is conflicting evidence in the literature on the importance of ADC and ODC in the formation of putrescine in tobacco (see [16,17]). In studies using specific inhibitors for ADC and ODC, it was only when inhibitors for both enzymes were used in combination that alkaloid level was reduced in callus cultures, indicating that both enzymes are important in tobacco [14]. Transgenic tobacco overexpressing ADC had 10 to 20-fold more agmatine than the controls, but no increase in levels of putrescine or the polyamines [18]. No diversion of polyamine metabolism into conjugates or nicotine was detected. The other two enzymes in the pathway to putrescine from arginine, agmatine iminohydrolase and *N*-carbamoylputrescine amidohydrolase, did not increase with the ADC insertion and further studies with a coordinated increase in all three enzymes seems warranted. Also, in this study the effect of ADC was measured in leaf tissue and not ADC expression in the root where nicotine is synthesized. However, cumulative effects of the ADC should have been observed in the leaf if they did occur in the root because of the length of time (>80 days) of the experiment.

Following putrescine formation the next step in nicotine biosynthesis is the methylation of putrescine by putrescine methyltransferase (PMT) (Fig. 2, step 5). PMT activity in tobacco roots was proportional to leaf nicotine in several related and unrelated high and low alkaloid genotypes [19,20]. Wagner et al. [7] concluded that PMT, along with QPT, are the two most rate-limiting enzymatic steps in nicotine biosynthesis. PMT has been purified and sequenced by two groups [21,22]. Hibi et al. [21] found levels of PMT mRNA to decrease in the same rank order as nicotine content of the tobacco plants decreased. With decreased PMT activity a concomitant increase in polyamine content was measured indicating that putrescine formation from arginine and ornithine is much less affected than nicotine biosynthesis by the *Nic* genes, suggesting the *Nic* genes may be regulatory genes that act specifically or primarily on nicotine biosynthesis [21] It is interesting to note that the PMT amino acid sequence is highly homologous to the sequences of spermidine synthase, however PMT and spermidine synthase catalyze different reactions with high specificity for substrate. From a constructed phylogenetic tree based on amino acid sequences, Hibi et al. [21] concluded that in tobacco PMT evolved from spermidine synthase. For this evolution to occur, utilization of SAM and not decarboxylatedSAM (dSAM) would be required and the binding would have to be such that for PMT the methyl group of SAM was close to an amino group of putrescine rather that the propyl group of dSAM. When a portion of the PMT sequence is inserted in the antisense orientation, nicotine content of the tobacco was reduced [22]. All the above data confirm the earlier conclusions that PMT is a very significant regulatory enzymatic step in nicotine biosynthesis.

The enzyme $N$-methylputrescine oxidase (MPO) catalyses the conversion of $N$-methyputrescine to the $N$-methyl-$\Delta'$-pyrrolinum salt (Fig. 2, step 6). Activity of the enzyme in root tissue is not proportional to nicotine accumulation in tobacco [7,19]. The enzyme has been purified to homogeneity from roots transformed with *Agrobacterium rhizogenes* but it has not been sequenced nor DNA fragments identified as the gene [23]. MPO has a relatively high $K_m$, 0.1 mM for $N$-methylputrescine. The enzyme will also utilize putrescine and cadaverine as substrate, but has a preference for $N$-methylputrescine [24,25]. MPO does not seem to be a regulatory site for nicotine biosynthesis in tobacco.

Nicotine synthase catalyzes the last step in nicotine biosynthesis, the decarboxylation and condensation of 3,6-dihydronicotinic acid with $N$-methyl-$\Delta'$-pyrrolinum salt (Fig. 2, step 7). This enzyme has reportedly been isolated from plants but others have not been able to successfully repeat the experiments [26]. The isolated enzyme required $O_2$, was stimulated by $Mg^{2+}$, and was inhibited by NADH, FAD and ATP. Specificity of nicotine synthase for $N$-methyl-$\Delta'$-pyrrolinum is probably significant because with MPO oxidizing $N$-methylputrescine, putrescine and cadaverine without this specificity nicotine, nornicotine and anabasine could be formed readily by the same alkaloid synthase. However, there may be regulatory control of nicotine biosynthesis at this step as many 5-substituted nicotinic acids supplied to root tissue yield the 5-substituted nicotines [27]. Substitution at the 2 and 6 positions inhibit, and most substitutions at the 4 position appeared to interfere with, nicotinic acid incorporation into nicotine. Much less recent research has been done on the specificity of the pyrrolidine moiety for nicotine synthase [28], but $N$-ethylputrescine is readily incorporated into $N'$-ethyl-S-nornicotine [29]. In somatic hybrids between tobacco (*N. tabacum*) and *Duboisia leichhardtii* the enzymatic pathways for alkaloid biosynthesis from both species have been shown to be present when appropriate precursors were supplied [30]. Tropane alkaloids of *Duboisia* and nicotine both require $N$-methyl-$\Delta'$-pyrrolinium ion for synthesis, but in small plants of the somatic hybrids tropane alkaloid production was suppressed and nicotine was the predominant alkaloid accumulated. This suggests a preferential allocation of the $N$-methyl-$\Delta'$-pyrrolinium toward nicotine synthesis at the first divergent step in nicotine and tropane biosynthetic pathways. This apparent greater competitiveness of nicotine synthase compared to other tobacco alkaloid synthases for substrate also seems likely to occur within tobacco genotypes (see [16]). Certainly, additional work on the in vivo and in vitro activity of nicotine synthase must be completed before the entire regulatory control of nicotine biosynthesis is understood.

Nicotine is synthesized mainly as $(-)$-2′S-nicotine. The $(+)$-2′R-nicotine enantiomer is present as 0.1 to 0.6% of the total nicotine in raw and processed tobacco [31]. Both enantiomers of nicotine are pharmacologically active, although there are conflicting reports as to the effects and potency of the (+)-2′R enantiomer. A discussion of biological activity of nicotine is beyond the scope of this chapter and readers are referred to papers referenced in Armstrong et al. [31] and Chapter 1.

## III. NORNICOTINE BIOSYNTHESIS

The metabolism and degradation of nicotine during senescence and curing of tobacco leaf revolve around the ability of the carbon atoms α to the pyrrolidine nitrogen atom to be oxidized (Fig. 4). Oxidation of the 2′-carbon will result in the formation of 2′-hydroxynicotine (pseudooxynicotine [PON]), a key intermediate for formation of 4-(N-methylnitrosamino)-1-(3-pyridyl)-1-butanone (NNK). Oxidation at the 5′-carbon atom and subsequent hydration and oxidation will lead to the formation of cotinine, whereas oxidation of the methyl group of nicotine and hydration of the methyl iminium ion will result in either demethylation to give nornicotine or oxidation of the hydroxymethyl group to yield formylnornicotine. All three of these oxidation products of the pyrrolidine moiety of nicotine accumulate in tobacco [32–34]. Analogous oxidation reactions occur in animal tissues [35]. The amounts of PON, cotinine or nornicotine formed is dependent upon the carbon atom oxidized as the 1′, 2′; 1′, 5′; and 1′, 6′ nicotine iminium ions are not in equilibrium [33].

High levels of nornicotine yield a tobacco product that has undesirable smoking quality and is the principal, if not sole amine precursor, for N′-nitrosonornicotine (NNN) formation. Nornicotine is formed mainly by the demethylation of nicotine by nicotine demethylase (Fig. 2, step 8). Many

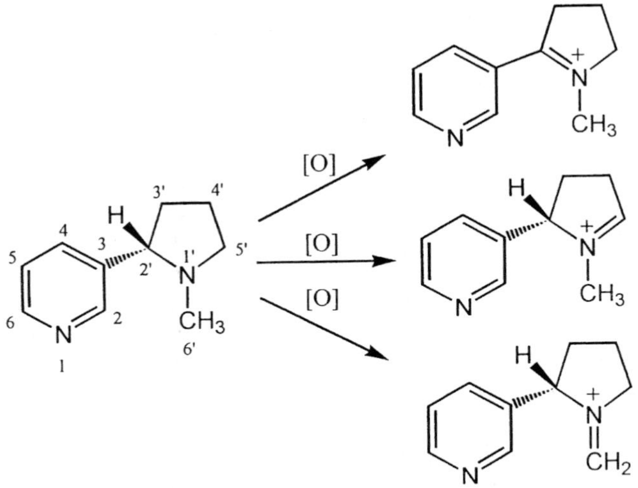

Fig. 4 Primary oxidation of nicotine in plants.

tobacco tissues contain nicotine demethylase, but the enzyme is only expressed when nicotine is available as substrate. Depending upon the genotype the enzyme is expressed both during plant growth and senescence. Partial characterization of nicotine demethylase in vivo and in vitro has been reported [36,37]. The enzyme is associated with the microsomal fraction as well as membrane vesicles or spherules of cell disruption and separation of a purified protein has proven difficult. Demethylation may be associated with the enzyme complex cytochrome P-450 [37] and supports the concept of demethylation as an oxidative reaction. More recent reports suggest that the oxidative demethylation is a monooxygenase, mixed function oxidase [38,39]. If a cytochrome P-450 dependent enzyme is involved, a mechanism of demethylation could be hydroxylation of the *N*-methyl group of nicotine and as this hydroxyl derivative would be unstable, it is non-enzymatically decomposed to nornicotine and formaldehyde. *N'*-Formylnornicotine has been detected in tobacco cells rapidly demethylating nicotine [40] and the authors have preliminary data to support directly the formation of formaldehyde in the demethylation process.

Also, recent studies have demonstrated that the demethylation is not a transmethylation mechanism as proposed in earlier literature reports [39]. Other studies have concluded that oxidation does not occur at C-2' and that nicotine-1'-*N*-oxide is not an intermediate in the demethylation. More recently direct evidence has been reported that oxidation at the C-5' is not involved in nornicotine formation [41]. These results support the conclusion that demethylation takes place as a result of the oxidation of the *N*-methyl group of nicotine [34].

Under optimal conditions tobacco tissues or tobacco cell cultures will convert nicotine to nornicotine at nearly 100% efficiency [40,42]. The nutritional status of the biologic material influences the level of conversion as heterotrophic, photoheterotrophic and mixotrophic cell cultures converted 96, 90 and 66%, respectively, of the nicotine to nornicotine during a 14 day culture period [40]. Nicotine present in tobacco is >95% (−)-2'S-nicotine, whereas the nornicotine is most frequently a racemic mixture (see 2). (−)-2'S -nicotine demethylation to nornicotine by cell cultures yielded only (−)-nornicotine (38). This result provides strong evidence that the demethylation in cell cultures unlikely involves opening of the pyrrolidine ring of nicotine and supports the oxidation of the *N*-methyl group of nicotine for demethylation. In cell cultures the demethylation most likely occurs intracellularly and nicotine must be taken from the medium, demethylated and the nornicotine released into the medium as well as remaining intracellularly [40]. *N. glauca* and *N. otophora* fed (−)-2'S -nicotine formed only the (−)-2'S −nornicotine [43]. However, in longer term experiments partial racemization of the nornicotine was measured. It was suggested that availability of endogenous pyridoxal may be responsible for the slow partial in vivo racemization of nornicotine in the plant. In raw cured tobacco and processed tobacco, (−)-nornicotine ranged from 57 to 96% of the total nornicotine [44,45]. A mechanism for the wide differences in enantiomer formation is presently not available, but probably involves the physiological status of the tobacco tissue at the time of demethylation.

## IV. ANATABINE AND ANABASINE

The pyridine rings of anatabine are both derived from nicotinic acid. This suggests that anatabine formation is entirely dependent upon QPT, the pyridine nucleotide cycle and an 'anatabine synthase'. Assuming QPT to be the main regulatory control for precursor formation the competitiveness of anatabine synthesis compared to nicotine synthase for 3,6-dihydronicotinic acid would determine the relative amount of each alkaloid formed. In genotypes for lowered alkaloid content anatabine levels are reduced to a greater extent than nicotine, suggesting a lower $K_m$ of nicotine synthase for the substrate (see [16]). The decarboxylation condensation reactions are similar to those of nicotine formation. 3,6-Dihydronicotinic acid is decarboxylated to form 1,2-dihydropyridine and 2,5-dihydropyridine. It is proposed that the 1,2-dihydropyridine reacts with 2,5-dihydropyridine to yield 3,6-dihydroanatabine which is aromatized to (−)-2′S-anatabine [46]. This proposed formation of (−)-S is consistent with recent data that the (−)-S and (+)-R enantiomers are approximately 85 to 90% and 10 to 15%, respectively, of the total anatabine in raw and processed tobacco [44,45].

Anabasine biosynthesis may be even more similar to nicotine formation than anatabine. The generally accepted pathway is the decarboxylation condensation of 3,6-dihydronicotinic acid with Δ′-piperideine derived from lysine (Fig. 2). QPT activity would be the most rate-limiting step for the nicotinic acid moiety as in nicotine. Lysine decarboxylase activity, to yield cadaverine, may be the rate-limiting step for the Δ′-piperideine formation. A bacterial lysine decarboxylase expressed in hairy root cultures, tobacco roots transformed with *A. rhizogenes*, had a higher enzyme activity and resulted in accumulation of up to 0.7% cadaverine on a dry weight basis in root tissue [47]. However, in both wildtype (low lysine decarboxylase activity) and enhanced lysine decarboxylase activity genotypes, supplying lysine increased the cadaverine accumulation indicating that even low levels of lysine decarboxylase are able to decarboxylate large amounts of lysine provided sufficient substrate is available. Utilization of the cadaverine for anabasine biosynthesis appears to be independent of the site of formation, cytoplasm or plastid and low in vivo activity may be dependent upon the available substrate. However, a positive correlation between lysine decarboxylase and levels of cadaverine and anabasine was measured [47]. The best transformed genotype had increased expression or accumulation of lysine decarboxylase, cadaverine and anabasine of 20-, 10- and 20-fold, respectively. This superior line also accumulated hydroxycinnamoylcadaverines, at least 100-fold more than the controls and more of added $^{14}$C-lysine was in the hydroxycinnamoylcadaverines than anabasine after four days [48]. Hydroxycinnamoylcadaverines are not natural constituents of tobacco and this result demonstrates the potential for such genetic transformation for study of secondary metabolic processes in plants and perhaps more importantly for production of new metabolic products in transgenic plants.

## V. BIOLOGICAL ASPECTS OF NICOTINE SYNTHESIS

### *Agrobacterium rhizogenes* infection and nicotine biosynthesis

*The 'hairy root' phenomenon*
Neoplasias among plants are rare. However, infections of wounded dicotyledonous plants by virulent forms of *Agrobacterium tumefaciens* and *A. rhizogenes* (see [49]) give rise to unorganized tumor-like growths (teratomas) known as 'crown gall' and 'hairy root', respectively. In both cases, the bacteria are capable of colonizing those teratomas as the result of genetic transformations of the host cells at the sites of infection. Phenotypic changes in such transformed cells include distinctive morphological alterations, rapid cell division, modified phytohormone (auxin, cytokinins) levels, and the production of large amounts of opines, novel carbohydrate-amino acid conjugates. The opine produced by the transformed host, not further metabolized by the plant, is characteristic of the infecting strain of *Agrobacterium*, which uses it as source of carbon, nitrogen and energy.

Virulence depends upon the presence, within the infecting bacterium, of a large plasmid, a circular double-stranded DNA molecule that replicates independently of the bacterial chromosome. The Ti (tumor-inducing) or Ri (root-inducing) plasmids of *A. tumefaciens* and *A. rhizogenes*, respectively, bear so-called 'virulence' (*vir*) and other necessary genes, notably those in T (transferred)-DNA. Virulence genes enable wound invasion and propagation of infection (see [50]). It is of interest that the *vir* genes of both Ti and Ri plasmids are highly conserved, while the homology between Ti-T-DNA and Ri-T-DNA is much less.

Transformed host cells are genetically modified owing to the incorporation, into host genomic DNA, of T-DNA [51]. Those cells, taken from crown gall or hairy root teratomas, can be cultured in vivo, and plants that bear the genes encoding the transformed traits can be regenerated from such cultures. Plant genetic engineering is highly dependent upon this gene delivery system to craft chimeric plants boasting novel and desirable traits.

Ri-T-DNA is divided into a 'left' ($T_L$) and a 'right' ($T_R$) portion, separated by a nonintegrated sequence of plasmid DNA, each half containing characteristic genes [52]. $T_R$-DNA bears genes encoding opine biosynthesis and genes homologous with the auxin biosynthesis genes of the Ti plasmid, while the

$T_L$-DNA (size 21.1 kilobases) contains, among 18 open reading frames (ORF) [53], three genes ('root locus' *rolA, rolB,* and *rolC*) that are necessary for the transformation of infected cells to the hairy root phenotype. Indeed, only $T_L$-DNA need be incorporated in order to develop the hairy root teratoma, although opines are not produced therein. Transformation of cells with single *rol* genes brings about, to different extents, incomplete development of hairy roots; *rolA, B* and *C* interact synergistically to produce the complete teratoma, with the notable exception of opine production.

A useful review of the hairy root process, with regard to the activities of *rol* genes, has recently appeared [54].

## Genes active in Ri-$T_L$-DNA

The various functions of the *rolA, rolB,* and *rolC* 'oncogenes', and of other ORFs in $T_L$-DNA, are not fully understood. Each of these can exert powerful developmental effects upon transformed plants in which they are overexpressed.

## rolA (ORF 10)

Transformation of *Nicotiana tabacum* cv. Xanthi with *rolA* alone manifests severe wrinkling of leaves in regenerated plants and certain other developmental irregularities [55]. Vilaine et al. [56] reported that the wrinkled leaf/dwarfism phenotype consequent upon transformation of tobacco with a construct *rolA::gus (Escherichia coli* β-glucuronidase gene) was associated with the production of a fusion protein. The fusion protein was found in the microsomal cellular fraction, suggesting that the *rolA* protein product (in theory, a small basic protein of size 11.4 kDa) was on or within the plasmalemma membrane. Transformation of plants with the *gus* gene alone expressed β-glucuronidase (gus) in the cytosolic fraction.

## rolB (ORF 11)

The *rolB* gene product appears to interact with auxin perception, in that transformed tobacco cells respond to much lower concentrations (ca. three to four orders of magnitude) of auxin (naphthalene acetic acid, indole acetic acid) than do untransformed cells [57,58]. The interaction was revealed by measuring the levels of auxin necessary to elicit a transmembrane electrical potential in mesophyll protoplasts, mediated by the auxin-transduction pathway of the plasma membrane (see [59,60]). The auxin-binding activity of membrane fragments from tobacco cells transformed with *rolB* was also enhanced as compared to that of untransformed cells [61].

Evidence of further involvement of *rolB* with auxin is found in the report that auxin regulates expression of the *rolB* gene through its native promoter, as indicated by the use of *a rolB::gus* reporter gene constructed by linking the *rolB* promoter to the *gus* gene. Gus activity was increased by 20–100-fold in transformed tobacco mesophyll protoplasts by supplying auxin [62]. Certain α-1,4-linked oligogalacturonides appear to inhibit the auxin-induced expression of the *rolB* gene, thereby diminishing the formation of hairy root teratomas on leaf explants transformed with *rolB* under control of its own auxin-activated promoter. No such inhibition was noted when the transformation took place with *rolB* under control of a tetracycline-dependent promoter [63].

Filippini et al. [64] obtained evidence supporting the view that *rolB*, expressed in *Escherichia coli*, gave rise to a membrane-bound protein with phosphatase activity, possibly directed toward phosphotyrosine. This role for the *rolB* protein would be consistent with activity in a kinase/phosphatase cascade that may function in the signal transduction associated with auxin binding.

### *rolC (ORF 12)*

Estruch et al. [65] reported that the *rolC* gene product, expressed in *E. coli* as a fusion product, was a protein with β-glucosidase activity. Moreover, such an enzyme was immunopurified from leaves of tobacco that had been transformed with the *rolC* gene controlled by the CAMV35S promoter. The protein catalyzed the release of cytokinins from cytokinin glucosides, and it was suggested that the enhancement of free cytokinin levels was responsible, in part, for the alterations brought about by *rolC* expression in transgenic plants. However, estimation of the levels of various phytohormones in *rolC*-transformed tobacco plants indicated that the level of the cytokinin isopentenyladenosine (ipa) in leaves was only about 20% of that found in wild-type plants [66]. The gibberellin $GA_{19}$ was found in amounts some 5- to 6-fold higher than that in the leaves of untransformed plants, while auxins did not significantly differ in the two types of plants. The finding that ipa was significantly reduced consequent upon expression of *rolC* is at variance with the suggestion that a cytokinin glucosidase activity is responsible for the development of the *rolC* phenotype.

### *Non-rol genes in $T_L$-DNA*

Lemcke and Schmülling [67] recently described a study designed to identify possible additional genes active in the hairy root process, from among the remaining ORFs residing in $T_L$-DNA. The individual ORFs were placed under control of a tetracycline-dependent 35S promoter, and used to transform *N. tabacum*. After plant regeneration, indications of morphogenetic differences or modification of response to phytohormones (growth of transgenic seedlings) were sought. Three loci manifested such effects.

ORF3n-transformed plants underwent morphological changes in leaves and internode shortening, and were delayed in flowering. The same transformation gave rise to responses of seedlings to auxin and cytokinin different from those of wild-type seedlings, probably by suppressing dedifferentiation of tissues consequent upon growth in the presence of the phytohormones. ORF3n transformation also brought about marked necrosis at the tips of otherwise normal leaves.

Transformation with ORF13 produced dwarfism and irregular internode distances. Root growth of seedlings showed evidence of reduced graviotropism. Leaves were uneven, with irregular vascular ramification.

ORF8 also modified responses of transformed seedlings to auxin, in that they were able to grow on media containing auxin concentrations that completely inhibit growth of wild-type seedlings. Some considerable homology was found between ORF8 and the *iaaM* gene of *A. tumefaciens*, which encodes a tryptophan monooxygenase active in oxidizing tryptophan to indole-3-acetamide, an intermediate in auxin synthesis. It was suggested that the gene product of ORF8 is indeed tryptophan monooxygenase.

The authors proposed that these ". . . genes play mainly a negative regulatory role in pathogenesis", and that they were significant in establishing infections of host plants.

*Hairy root and nicotine production*
The following is a brief review of the hairy root transformation of *Nicotiana* spp., related to what is so far known about how it affects nicotine biosynthesis.

Seven *Nicotiana* species were transformed with the Ri plasmid, giving rise to hairy root teratomas, from which transformed cells were taken and cultured in phytohormone-free media [68,69]. Both $T_L$- and $T_R$-DNA were incorporated, with more frequent detection of $T_L$-DNA than of $T_R$-DNA. Integration events ranged from one in *N. cavicola* and *N. africana* to as many as six in *N. rustica*. The alkaloid contents of cultures were measured at suitable times, and compared with the alkaloids of untransformed plants of the same species. The number of integrations of T-DNA did not proportionately affect alkaloid contents.

Hairy root cultures of all seven transformed species contained significantly greater amounts (fresh weight basis) of nicotine, nornicotine, anatabine and anabasine than did roots of untransformed plants, grown aseptically on agar or in soil. The spectra of alkaloids were similar to those described by Saitoh et al. [70].

With one exception, some 65%–95% of the total alkaloids of soil-grown plants was found in shoots. This was consistent with the proportion of alkaloids transported to the medium in the various hairy root cultures. The noted exception was *N. umbratica*, which retained about 93% of its alkaloids in the roots of aseptically grown plants. On the other hand, the medium of *N. umbratica* hairy root cultures contained large amounts of alkaloids. Apparently, shoots and culture medium are not always equivalent sinks for alkaloid transport from the root or cells in culture, nor are the transport mechanisms involved necessarily the same.

The various teratomas induced by transforming *N. tabacum* with the Ri plasmid, the Ti plasmid, and Ti plasmids from which genes encoding auxin (*aux*) or cytokinin (*cyt*) synthesis had been modified to eliminate their expression, were compared by Saito et al. [71]. Neither crown gall (wild-type Ti plasmid) nor 'shooty' teratoma (*aux*- Ti plasmid) cells, in separate culture, produced alkaloids, while hairy root cells in culture accumulated nicotine. However, when shooty teratoma cells were co-cultivated with hairy root cells, nicotine was transferred through the medium to the former from the latter, where it was accumulated and converted to nornicotine. Plants regenerated from cultures transformed with wild-type Ri plasmid accumulated about twice as much nicotine, and many-fold more nornicotine, than did leaves of untransformed plants.

*Nicotiana tabacum* cv. Xanthi leaf sections were transformed with individual *rol* genes from $T_L$-Ri-DNA, through the agency of *A. tumefaciens* and a cloned vector. *rolA* and *rolB* were incorporated under the control of their own promoters, producing hairy root teratomas at the site of infection. *rolC* had to be introduced linked to the CaMV35S RNA promoter to yield hairy roots. Finally, still other leaves were transformed with *rolABC*, using only native promoters, and with wild-type RiA4-T-DNA from an agropine (opine) strain of *A. rhizogenes*. Portions of the teratomas, and of untransformed root, were transferred to phytohormone-free culture medium, and cultures were established [72].

Growth rates of the various untransformed and transformed cultures varied over a 33-fold range. The slowest increase of root biomass occurred in the untransformed root culture, which grew without branching. Hairy root cultures transformed with *rolA* alone grew twice as fast, *rolB*-transformed roots three times as fast, and *rolC*-transformed roots seven times more rapidly. The hairy root transformed by *rolABC* accumulated biomass 16-fold more rapidly, and that transformed by RiA4-T-DNA some 33-fold more rapidly than untransformed root. Roots transformed with *rolC* and *rolABC* resembled those transformed with RiA4-T-DNA, in that they were highly branched. Those transformed with *rolA* and *rolB* manifested less branching.

Nicotine production was determined in four cell lines derived from each transformation. In all cases nearly all nicotine produced was liberated into the culture medium. With one exception, the rank of nicotine productivity by the various cell lines was largely in line with their growth rates. RiA4-T-DNA- and *rolABC*- transformed cells produced the largest amounts of nicotine (up to 38-fold more nicotine than the untransformed root), while *rolC* transformation yielded roots with as much as 17-fold more nicotine than the untransformed organ. *rolA* roots accumulated up to five-fold more nicotine than untransformed roots, yet grew only twice as fast. The authors suggested that the increase of nicotine synthesis in the transformant may be due to more than just increased growth rate, pointing out that putrescine (the major precursor of the pyrrolidine moiety of nicotine) was reported to be relatively increased in *rolA*-transformed roots, (2-3-fold higher than in roots transformed with *rolB*, *rolC* or *rolABC*) [73].

Mothes et al. [74] reported that only the actively growing root is the source of nicotine in tobacco. Two lines of evidence were developed that supported this notion.

The shoot of the long-day plant, *Nicotiana sylvestris*, grew and flowered in days of 20 hr duration, but the root mass of the same plants did not increase, over a period of some 40 days. The total nicotine contents of such plants were maintained nearly constant during that time, declining after flowering. During short (12 hr daylight), the shoot grew less (rosette stage) than in long days, but the root more than doubled in mass over the same 40-day period. Total plant nicotine increased about 2.4-fold. The day-neutral plants *N. tabacum* (Virginia Gold) and *N. rustica* manifested no such dependency upon day length.

Leaf cuttings (*Blattstecklinge*) of *N. rustica*, which had developed roots in culture, were subjected to repeated excision of portions of their root tissue, and their nicotine contents were compared to those of uninjured rooted cuttings of similar age. Root tissue that had been removed was replaced by active growth. The lamina of rooted cuttings, whose roots had been cut four times over a period of nearly six months, contained nearly three times as much nicotine as did lamina of undisturbed rooted cuttings. This difference in nicotine contents did not depend upon greater uptake of nitrogen from the culture medium by the former cuttings; nitrogen uptake was about the same in the two kinds of rooted cuttings. Indeed, the rooted cuttings from which root tissue had been removed contained only about one-third the amount of soluble amide-*N* as did those that had not been so treated. Nicotine had been formed at the expense of the soluble amide-*N* pool. The pools of ammonia-*N* and nitrate-*N* were about the same size in both types of cuttings.

The authors asserted that "the production of nicotine follows the growth of the root" ("Die Nikotinproduktion folgt dem Wachstum der Wurzel"). Consistent with this conclusion is the relationship between the accumulation of nicotine and the active accumulation of hairy root tissue, and as well with Baldwin's [75] observation that pot-bound *N. sylvestris* does not respond to leaf damage by producing additional nicotine. In the four decades since their publication, these very interesting observations have not received the experimental attention they deserve. Further investigation of the relationship between transformation with the *rolC* gene (CAMV 35S promoter), *rolABC* genes together, and nicotine biosynthesis in *N. tabacum* cv. Xanthi was conducted by Palazón et al. [76]. Transformed cells in culture and plants regenerated therefrom were examined.

Roots transformed with the *rolABC* genes together grew some 22-fold, and those transformed with *rolC* alone grew 14-fold more rapidly, than did untransformed roots. Total mean nicotine production in the *rolABC* root cell lines was about 107-fold, and in the *rolC* root lines some 21-fold, greater than that in cultures of untransformed roots.

Regenerated *rolABC*-plants, *rolC*-plants, and untransformed plants were rooted and grown in agar containing phytohormone-free culture medium. Both transformed plants manifested the hairy root syndrome. After six weeks in culture, in comparison to untransformed plants, this condition included higher rooting rate, more rapid root growth (ca. 5-fold more in *rolABC*-plants, 3-fold more in *rolC*-plants), shorter internodal distances and altered leaf morphology (about 5-fold more leaf biomass in *rolABC*-plants, 2.5-fold more in *rolC*-plants.

The *rolABC*-plants contained 11-fold more total nicotine in leaves and 5-fold more in roots, than did untransformed plants. Plants transformed with *rolC* contained over three times more nicotine in leaves and twice as much nicotine in roots than the untransformed plant. Among untransformed *rolABC*- and *rolC*-plants the specific amount (dry weight basis) of nicotine in roots did not greatly vary (0.8–1.2 mg nicotine $g^{-1}$), but that function tended to increase in leaves from 1.6 mg $g^{-1}$ (untransformed), to 2 mg $g^{-1}$ (*rolC*), up to 3.8 mg $g^{-1}$ (*rolABC*), as the ratio of root mass to leaf mass increased (0.73 in untransformed, 0.83 in *rolABC*-plants).

The amounts of the putative product of the *rolC* gene, a protein of size 20.1 kDa was estimated in root cell lines and roots and leaves of plants, using a polyclonal antibody preparation raised against the protein expressed by subcloning in *Escherichia coli* the *rolC* ORF. The relative amounts of this protein in transformed cultured cell lines were ". . . positively correlated with the capacity of the transgenic roots to grow and produce nicotine." Plants transformed with the *rolC* gene alone, under control of the strong unregulated CAMV35S promoter, expressed the highest levels of the *rolC* protein product, yet the regenerated plants did not contain the highest amounts of nicotine, nor attain the greatest root mass. On the other hand, plants transformed with *rolABC*, under the control of native promoters, reached the highest root mass and nicotine levels. This difference probably arises from regulatory interactions among *rolC* expression, the coexpression of *rolA* and *rolB*, and plant developmental processes.

Information is not available on the possible changes of the activities of enzymes of nicotine biosynthesis [17,77] subsequent to *A. rhizogenes* infection or transformation of tobacco with the various $T_i$-DNA constructs. This would seem to be essential knowledge needed toward testing the hypothesis that it is meristematic root tissue that is the site of nicotine biosynthesis, and toward defining the mechanisms by which nicotine production is coupled to root growth.

## Nicotine: an inducible, mobile chemical defense

*The alkaloidal response to herbivory*
It is customary practice to remove the flowering parts ('topping') of commercial *Nicotiana tabacum* before harvest, thereby removing the principle source of auxins and breaking apical dominance. Among the consequences of this act is the desirable increase of leaf mass; however, alkaloid levels in leaves also rise. The increase of alkaloids in the aerial parts of the plant, resulting from 'topping', can in part be attributed to the lessening of auxins flowing from the apex to the root, the site of nicotine biosynthesis. Auxins have been demonstrated to suppress nicotine synthesis in cultured tobacco callus (see [78,79]), and to suppress enzymatic activities in the nicotine biosynthetic pathway [77], see [80]. Alkaloids have also been shown to accumulate in the aerial portions of certain 'wild' *Nicotiana* spp., not as the result of removing apical dominance, but rather as the result of leaf damage. Such increases are thought to confer a survival advantage on the plant against herbivory. Evidence has appeared that nicotine production is probably only part of a portfolio of defensive responses by the genus *Nicotiana*, comprising (at least) ". . . alkaloids, phenolics, proteinase inhibitors, PR (pathogenesis response) proteins, and sesquiterpenoid phytoalexins" (see [81]). An engaging account of these and related induced responses in other plants is given in 'Induced Responses to Herbivory' [82].

*Nicotiana sylvestris*, a species native to South America, responds to leaf damage (insect larvae, mechanical cutting or puncturing) by accumulating, over about a week, as much as four-fold more shoot alkaloids than do comparable undamaged control plants [83]. The alkaloid pool comprises about 95% nicotine, with the balance being primarily nornicotine [84]. The damage-elicited increment of alkaloids appears in the aerial portion of the plant, with little change of root alkaloids. The accumulation of alkaloids in the shoot is not due to enhanced transpiration or reduced alkaloid degradation in the leaf [84]. Pot-bound plants respond to leaf damage in an attenuated manner. However, upon transplanting ('releasing') such plants to larger pots, the nicotine response to leaf damage was manifested [75].

The signal elicited in the leaf is a chemical rather than electrical one, and is a positive (initiation of production), rather than a negative (cessation of production), cue [85]. It is transmitted from damaged parts via the phloem to the root, as indicated by stem-girdling experiments: damaging leaves above the girdle produces no additional alkaloid production; damaging leaves below the girdle results in an increased alkaloid delivery to the entire plant through the xylem stream. Girdling, without at the same time damaging leaves, produces only a small alkaloidal response, indicating that the interruption of auxin flow in the phloem is not responsible for additional alkaloid production. Another line of evidence supporting the nonparticipation of auxin in the alkaloidal response is the observation that application of ethylene or other inhibitors of auxin transport did not elicit alkaloid production [84]. However, the application of auxin (indoleacetic acid, naphthalene acetic acid) to the wound site on the leaf does eliminate or suppress the alkaloidal response to that damage [84].

Removal of equal amounts of leaf tissue by *Manduca sexta* larvae or by cutting with scissors over similar time periods led to increases of whole-plant alkaloids of 2.2-fold in the former and 4-fold in the latter case [83]. This difference in damage response was traced at least in part to the two different kinds of injuries inflicted. The larva prefers to consume intervenous tissue before consuming veins, while scissors damage all structures. Subsequent experiments employed a 'pattern-tracing wheel', which placed multiple small punctures in the leaf without damaging veins, a better simulation of folivory than cutting the leaf. The alkaloidal response was shown to be proportional to the number of punctures so inflicted [86].

Detaching leaves from the stalk at the petiole without first damaging them produced a modest alkaloidal response, but damaging leaves first and then removing them 16 hr after such damage yielded a somewhat larger alkaloidal response [86].

Hanounik and Osborne [87] infested two cultivars of *N. tabacum* (nematode-resistant NC95 and nematode-susceptible McNair 30) with a root-knot nematode, *Meloidogyne incognita*, and determined the effects of infestation upon contents of nicotine in leaf (positions 7–9) and root. Seedlings in pots were infested with various amounts of eggs, and were then placed in the field for three months. At the end of this period, at the highest level of infestation, the amount of nicotine in roots was some 77% (dry weight basis) higher (NC95) and 56% higher (McNair 30) than in roots of the uninfested controls. Nicotine in leaves of infested plants reached 1.5% (NC95) and 0.7% (McNair 30), while control plants contained 0.9% and 0.8% nicotine, respectively. As a result of infestation, root masses decreased from the control value of 22 g (dry weight) to 15 g (NC95) and from 24 g to 0.8 g (McNair 30). The authors rationalized the small decrease of nicotine in leaves of infested McNair 30 as the consequence of extensive root damage in that susceptible plant, leading to faulty xylem transport. No information was provided regarding the possible effect of leaf damage upon root or leaf nicotine contents in the pot-bound plants.

The damage-elicited alkaloidal response was also shown to occur in other *Nicotiana* spp. growing in a native habitat (southwestern Utah, USA). *N. attenuata* (nicotine as principal leaf alkaloid), *N. repanda* and *N. trigonophylla* (nornicotine), and *N. glauca* (anabasine) all manifested significant accumulations of their principal alkaloids in undamaged leaves following damage to other leaves [88]. *N. attenuata* responded to the cutting of leaves by doubling the basal rate of nicotine accumulation as found in comparable undamaged plants; 'simulated browsing' (removing the flowering stalk above the basal rosette of leaves) quadrupled that basal rate, reaching 5.1% of dry weight by 18 days after damage. The incremental increase of nicotine was maintained over the growing season.

In the review following, the various notions of chemical ecology relating to "Darwinian fitness" tested by Baldwin and his coworkers, using the alkaloidal response as experimental vehicle, are not included as they are beyond our scope (see [89–91]).

### The alkaloidal response is protective

The efficacy of elevated foliar nicotine in reducing folivory by larvae of the alkaloid-tolerant tobacco specialist *Manduca sexta* (tobacco hornworm) was demonstrated [83] using leaves taken from undamaged plants (containing 0.5% of dry weight as alkaloids), damaged pot-bound plants (0.8% alkaloids), and damaged released plants (1.1% alkaloids). Hornworm larvae were reared on cuttings from these plants. The larval biomasses resulting were greatest when feeding on leaves from undamaged plants, least when feeding on leaves from damaged, released plants, and intermediate when leaves from damaged, pot-bound plants were used. In a parallel experiment, leaves from undamaged plants were detached and stem-fed with nicotine solutions, providing a series of leaves that contained nicotine over the range 0.2% to 1.7% dry weight. Hornworm larvae were reared on these leaves, with biomass accruing to 67 mg at the lowest level, and to 26 mg at highest level, of nicotine content. In both experiments, lesser amounts of leaf material was consumed at higher leaf alkaloid levels. Evidence was also obtained that larval survival diminished as alkaloid levels increased. These findings are consistent with the notion that nicotine, at least, confers protection against herbivory in a quantitative manner.

### The alkaloidal response is prompt and the product persists

The time course of the alkaloidal response was characterized using hydroponically grown *N. sylvestris* [92]. $^{15}N$-Nitrate at two different concentrations was added to the nutrient solution five hours before damaging leaves by multiple punctures with a pattern-wheel device. By two days after damage, the damaged plants contained 4.8-fold more nicotine than did undamaged plants, even though the rates of removal of $^{15}N$-nitrate from solution did not differ. $^{15}N$-label was detected in nicotine isolated from damaged plants, in significantly greater amounts than in nicotine isolated from controls, within 21 hr after damage. Five days after damage, $^{15}N$ incorporation into nicotine reached a plateau at the lower $^{15}N$-nitrate concentration, which had been depleted at three days. When a four-fold greater amount of $^{15}N$-nitrate was used (depleted by eight days after damage), $^{15}N$ incorporation into nicotine in damaged plants persisted for at least eight days. In an earlier publication, accumulation of nicotine under similar conditions was shown to persist for up to 10–12 days after leaf damage [93]. These findings indicate that de novo synthesis of nicotine is characteristic of the alkaloidal response.

Unlike earlier reports that exogenously supplied nicotine in the leaf was subject to rapid degradation [94,95], Baldwin et al. [92] found no evidence for metabolism of $^{15}N$-nicotine synthesized in vivo in either undamaged or damaged plants at eight days or in undamaged plants at 41 days. Some loss of $^{15}N$ from labeled nicotine was observed in damaged plants after 41 days, but the rate of loss over that period was less than 4% of the rate of constitutive nicotine synthesis, perhaps due to volatilization through the damaged areas. It was estimated that, over the life of the plant, nicotine turnover amounted to no more than about 33%. The authors contrasted their method, in which labeled nicotine newly synthesized from normal precursors was supplied to the intact plant, with the findings of Tso and Jeffrey [94], who supplied labeled nicotine to whole plants through the root, and with the results of Yoshida [95], who injected nicotine through the midrib. They suggested that such rapid metabolism perhaps reflected ". . . nicotine detoxification or other types of salvage metabolism, rather than normal nicotine turnover." Gershenzon [96] has asserted that all attempts to measure turnover of alkaloids by directly administering the compound to the plant, rather than supplying normal precursors, are ". . . potentially flawed."

Consistent with this view of the relative stability of endogenously synthesized nicotine was the report that the alkaloid, supplied by feeding through the root, did not support growth in nitrate-deprived *N. sylvestris* [97]. As compared to controls, nicotine pools doubled in size after nicotine feeding, but 28 days later they had become equal to the nicotine pools of comparable controls. The exogenously supplied excess nicotine was almost quantitatively converted to nornicotine, and much lesser amounts of myosmine. The authors concluded that "investment of nitrogen in nicotine is a one-time investment for *N. sylvestris*." Moreover, "these plants are homeostatic with regard to their nicotine pools." It would be of interest to determine whether endogenously biosynthesized nicotine, and exogenously supplied nicotine, are similarly distributed in the various symplastic and apoplastic compartments of the plant, and how that distribution might change with time. More recently, in hydroponically grown *N. attenuata*, Lynds and Baldwin [91] have confirmed the long-term stability of $^{15}N$-labeled nicotine produced by the plant.

Of the various antiherbivore defense compounds considered by Gershenzon [96], including terpenoids, phenolics, alkaloids, proteinase inhibitors and cyanogenic glycosides, nicotine is the most expensive (3.62 g glucose g$^{-1}$ nicotine) for a plant to synthesize. The value cited, calculated from known biosynthetic pathways, does not include the energetic costs of storage or transport. That total cost increases, of course, with increasing nicotine concentration, the hallmark of the alkaloidal response.

## *The alkaloidal response depends upon jasmonic acid production by damaged leaves*

The octadecanoid metabolite jasmonic acid (JA) and its methyl ester, arising from linolenic acid derived in part from degraded cellular membrane materials, have in recent years been recognized as capable of eliciting manifold physiological responses in plants (see [98]). Originally thought to be causally involved in leaf senescence, instead the jasmonates can bring about major accumulations of defense proteins, with senescence being probably an indirect outcome. Jasmonates are particularly associated with responses to stresses, including drought stress, the action of fungal elicitors, herbivory and leaf damage [99].

Hydroponically growing *N. sylvestris*, the roots of which were exposed to at least 90 μg methyl jasmonate (MJ), incorporated $^{15}N$ from $^{15}N$-nitrate into newly synthesized nicotine [85]. After five days, MJ-stimulated, undamaged plants accumulated two-to-three-fold more nicotine than did plants that were subjected to leaf damage alone. Measurement of JA pools in shoots and roots of leaf-damaged plants demonstrated that their levels were affected by damage. Thirty minutes after damage, JA in shoots had increased 1.7-fold, decaying to levels found in undamaged plants by 10 hours. In roots, no change in JA was manifested in 30 min after damage, but significantly increased after a period of two hours. A later study reported that JA peaked (10-fold increase) in the shoot 90 min post-wounding, and reached a level three-fold greater than in undamaged controls 180 min post-wounding [100], decaying to baseline levels six hours after wounding. The magnitudes of JA pools in the shoot at 90 min and nicotine at five days post-damage were positively correlated with the number of punctures inflicted upon inter-venous areas of the leaf. Nicotine content was also positively correlated with sizes of JA pools in damaged leaves [100,101]. These findings are consistent with production of JA in the damaged leaf, transport via the phloem to root, and induction of increased nicotine biosynthesis in the root as detected some hours thereafter [92].

The possibility that the rise of JA in the root owes to de novo synthesis, or release from JA-conjugates, at that site, and not to transport from the leaf, was tested by Zhang and Baldwin [99]. 2-$^{14}C$-JA was applied in a lanolin paste to an uninjured single leaf in an amount sufficient to elicit the alkaloidal response. The distribution of 2-$^{14}C$-JA and radioactivity was traced within the plants by eight harvests over 300 min after applying $^{14}C$-JA, examining treated and untreated leaves and roots. Some 15%–20% of the radioactivity was lost within the first hour. Within three hours after application, radioactive material was found in the treated leaf, root and leaves younger than the treated leaf, but in not older leaves. Five hours after application, some 80% of radioactivity remained in the treated leaf, roots contained 0.25% and untreated leaves 0.13% of the initial radioactivity. Even though the amount of radioactivity in the root was small compared to the total JA applied, it was yet commensurate with the increment of root JA observed after wounding. JA-conjugate pools, measured using several different hydrolysis procedures, amounted to only about 10% of the free JA pool, insufficient to account for the ten-fold increase of JA in the shoot 90 min post-wounding [100]. Again, the hypothesis of synthesis of JA in the damaged leaf, followed by transport to the root, was consistent with these findings. Wound-induced root JA pools are thought to be governed by the relative rates of JA synthesis in damaged leaf and metabolism of JA in the root. Subsequent work, however, called into question the model of JA transport from the leaf to the root, owing to the very rapid metabolism of JA applied at the wound site [102]. The possible involvement of other stress signals, such as the octadecapeptide systemin, in contributing to the alkaloidal response is not known.

Early in the course of these investigations, Baldwin [83] reported that the alkaloidal response in *N. sylvestris* is smaller when leaves are damaged by the hornworm caterpillar than when damage is mechanically inflicted with scissors. This relationship persists even when veins are not cut. The difference is apparently due to a partial decoupling between JA production at the wound site and the alkaloidal response [103]. Cutting with microscissors brought about a two-fold rise in JA in the damaged leaf, but a comparable amount of tissue damage by *M. sexta* larvae results in a three-fold increase, peaking 30–135 min ". . . after first bite." Compared with leaves damaged by microscissors, leaves damaged by caterpillars produced twice as much JA per cm of cut edge in the damaged area. Despite the difference in JA production, amounts of nicotine in plants damaged by the two methods did not differ. Moreover, and in line with the nicotine findings, although JA in roots was increased by both kinds of damage, its level did not differ between the two treatments.

The role played by caterpillar regurgitant (insect herbivore regurgitants are known to trigger release of volatile compounds when applied to damaged plant surfaces) in the JA and alkaloidal responses was tested by applying distilled water or a regurgitant preparation to leaves on separate plants that had been damaged in a standardized manner using a fabric pattern wheel. Wounded, water-treated leaves contained JA levels at 90 min post-wounding that had increased by about 93% as compared to unwounded controls; wounded, regurgitant-treated leaves contained 13-fold more JA than unwounded controls. Again, nicotine pools, greater than in unwounded plants, did not differ between water- and regurgitant-treated, wounded plants. The results suggest that *M. sexta* larvae can mitigate the alkaloidal response as compared to alkaloidal response from mechanical damage, and that such mitigation is mediated by larval regurgitant. Karban and Baldwin [104] characterize this phenomenon as a 'stealth' measure on the part of the insect.

Repeated application of a single dose of MJ to the roots of hydroponically growing *N. sylvestris* plants tends to facilitate the alkaloidal response following the later stimulations [105]. Over a period of 18 days, plants were induced one, two or three times, with intervals of six days to allow the alkaloidal response to subside before restimulation. Nicotine contents and rates of $^{15}N$-nicotine synthesized de novo from $^{15}N$-nitrate were assessed after each stimulation. Plants that had two induction periods prior to the third stimulation, reached their maximum nicotine levels about two days sooner than did previously unstimulated plants, or plants that had been stimulated once. Nicotine tended to accumulate with repeated stimulations, but the maximum amounts of net $^{15}N$ in nicotine post-stimulation did not so increase. The latter observation was attributed to the use of the same amount of MJ for each stimulation. The authors suggested that their findings were akin to a kind of "immunological 'memory'". It is not clear whether the more rapid alkaloidal response occurring after prior stimulations might confer additional survival advantages against herbivory. An explanation for basal nicotine formation occurring in the absence of elicitation is not available. It may be that the kind of regulation that governs nicotine biosynthesis does not allow for a 'fully off' mode.

*The alkaloidal response is reduced by inhibitors of jasmonate synthesis and auxin*

Certain inhibitors of mammalian cyclooxygenase and lipoxygenase, applied in lanolin paste to the wounded area (pattern-tracing wheel) of damaged *N. sylvestris* leaves, can inhibit or arrest the alkaloidal response that would otherwise take place [106]. Causing no visible damage to leaves upon application, *n*-propylgallate (*n*-PG) and acetylsalicylic acid brought about nearly complete inhibition of the alkaloidal response, while partial inhibition followed application of salicylhydroxamic acid, methyl salicylate (MSA) and antipyrine. When MJ was applied to undamaged leaves, nicotine content at five days thereafter exceeded that of damaged leaves. Mixing *n*-PG or MSA with MJ and applying the mixture to undamaged leaves did not alter the alkaloidal response, consistent with the notion that the action of the inhibitors was rooted in inhibition of JA production from linolenic acid. No increases in nicotine were noted upon application to unwounded leaves of lanolin pastes containing abscisic acid, linolenic acid, linoleic acid or traumatic acid, an octadecanoid metabolite active in leaf abscission in cotton.

When applied to the site of the wound of damaged leaves, the auxin indole-3-acetic acid in lanolin suppressed the rise in shoot JA and the alkaloidal response, suggesting that the auxin interfered in JA biosynthesis [100].

Imanishi et al. [107] recently reported that exposure of cultured *N. tabacum* BY-2 cell to MJ brought about enhanced synthesis of nicotine, as well as increases in levels of the precursor of its *N*-methylpyrrolidine moiety, *N*-methylputrescine. Evidence was obtained indicating that MJ elicits sequential expression of several genes in the nicotine pathway, probably involving more than one regulatory mechanism. The cellular response to jasmonate was characterized first by elevated expression of the mRNA species of ODC (ornithine decarboxylase), followed by mRNAs of PMT (putrescine methyltransferase) and of S-adenosylmethionine synthase (SAMS). No information was provided regarding MPO (*N*-methylputrescine oxidase), the remaining essential enzyme of the pathway leading to *N*-methyl-Δ'-pyrrolinium ion.

The complexity of the regulatory systems controlling the induction of nicotine biosynthesis was revealed by the finding that the inductions of PMT and SAMS mRNAs, but not of ODC mRNA, were prevented in the presence of cycloheximide, an inhibitor of cytosolic translation. This suggests that synthesis of other proteins intervenes between the jasmonate signal and expression of at least some genes. As with intact *N. sylvestris*, auxin appeared to interfere with the MJ-dependent production of nicotine through suppressing induction of ODC, PMT, and SAMS.

Jasmonate-dependent elicitation of nicotine synthesis appears to depend upon ODC as the means of providing putrescine, since the mRNA of ADC was not increased by MJ treatment. Moreover, the developmentally important polyamines spermidine and spermine, which also arise from putrescine, were not augmented by jasmonate owing to the fact that the necessary enzyme, S-adenosylmethionine decarboxylase, was not induced.

The absence of information about the role of shoot wounding toward affecting foliar nicotine contents in commercial tobaccos requires attention. Tobacco breeding strategies tend strongly to yield cultivars with consistent foliar chemical compositions. Due to such selections, it may be that the dramatic rises of alkaloid contents observed in 'wild' *Nicotiana* spp. consequent upon folivory have been diminished or even eliminated from commercially developed lines of *N. tabacum*.

*Root versus shoot as the site of nicotine biosynthesis*

Why does the tobacco plant produce nicotine in the root and then translocate it to the shoot? Why, instead, is nicotine not produced in the aerial part of the plant, where it acts in defense against folivory? Karban and Baldwin [108] provide a simple and persuasive rationale for this apparent paradox.

Located underground, the root is naturally protected from the kind of herbivory that endangers the aerial reproductive parts of the plant. If the leaf were the site of nicotine production, progressive removal of foliar tissue would, in proportion, diminish the alkaloidal response to damage. On the contrary, sequestration of nicotine biosynthesis in the root ensures that the agent of defense will be available even if the major portion of leaf tissue has been grazed away. Moreover, as the aerial sink size diminishes, the concentration of nicotine in the remaining leaves will attain even higher levels, assuming that nicotine synthesis in the roots persists. Baldwin and Schmelz [86] demonstrated that as much as 88% of leaf tissue could be removed from *N. sylvestris* without significantly reducing either the rate of nicotine production or its upward transport. Mothes et al. [74] showed that after removal of the floral parts and all but one leaf from *N. sylvestris,* and allowing sufficient time to elapse, the single leaf remaining accrued up to 20% nicotine (dry weight basis), a relative increase of 250-fold compared to its starting concentration.

In the beginning of this chapter, statements were made indicating that much of the detail of nicotine metabolism in the tobacco plant remains to be elucidated. Hopefully, as this chapter on biosynthesis and metabolism of the nicotinoids was presented, it has been obvious to the reader that there is much work to be done to complete the understanding of their biosynthesis and metabolism in the plant. As better analytical instruments and techniques are developed it will be possible to better understand the biochemical processes involved. For example, with the development of chiral chromatographic columns it is much easier to determine the R and S isomers of these alkaloids, but we still do not know the mechanism by which racemic nornicotine is formed from (−)-2′S-nicotine in the plant. Isolation and characterization of both the nicotine synthase and nicotine demethylase enzymes are critical for understanding the synthesis and metabolism of nicotine. Nicotine demethylase is of particular importance because it is the major enzyme for nicotine degradation in tobacco. Genes for the enzymes in the nicotine biosynthetic pathway are being identified and manipulated by molecular biological techniques. These powerful biochemical tools are beginning to provide valuable information for the alteration of naturally formed alkaloids in tobacco and the potential for development and 'green plant' manufacture of new, more useful, components in transgenic tobacco plants. Even though we know that plant growth regulators dramatically affect nicotine biosynthesis and accumulation, the roles of growth regulators on nicotine biosynthesis is not understood fully. Both the direct effects and the interaction effects of growth regulators on nicotine biosynthesis remain to be determined. Nicotine has been one of the most extensively studied constituents of plants and there is an immense catalog of information on parameters that alter nicotine accumulation during vegetative growth of the plant. However, there is much less known about specific effects of these parameters on biosynthesis and accumulation. With the rapid advancement of biochemical and molecular biological techniques we are on the verge of a much more complete understanding of the chemical and biochemical reactions of nicotine biosynthesis in the tobacco plant.

# REFERENCES

1 Tso TC. Organic metabolism – Alkaloids, Chapter 21. *Production, physiology, and biochemistry of tobacco plant.* Ideals, Inc, Beltsville, MD 1990.

2 Bush LP, Fannin FF, Chelvarajan RL, Burton HR. Biosynthesis and metabolism of nicotine and related alkaloids. In: *Nicotine and related alkaloids: absorption, distribution, metabolism and excretion,* Gorrod JW, Wahren J (Eds.), Chapman & Hall, London. 1993, pp. 1–30.

3 Bush LP. Alkaloid biosynthesis. In: *Tobacco – Production, chemistry and technology,* Davis DL, Nielsen MT (Eds.), Blackwell Science, Oxford, UK, 1999, pp. 812–834.

4 Legg PD, Chaplin JF, Collins GB. Inheritance of percent total alkaloids in *Nicotiana tabacum* L. *J. Hered.,* 1969: **60**; 213–217.

5 Legg PD, Collins GB. Inheritance of percent total alkaloid in *Nicotiana tabacum* L. II. Genetic effect of two loci in Burley 21 × LA Burley 21 populations. *Can. J. Genet. Cytol.,* 1971: **13**; 287–291.

6 Wagner R, Feth F, Wagner KG. Regulation in tobacco callus of enzyme activities of the nicotine pathway. II. The pyridine-nucleotide cycle. *Planta,* 1986: **168**; 408–413.

7 Wagner R, Feth F, Wagner KG. The regulation of enzyme activities of the nicotine pathway in tobacco. *Physiol. Plant,* 1986: **68**; 667–672.

8 Pudliner HJ. *Nicotine accumulation in tobaco and tobacco callus.* Master Science Thesis. Univ Kentucky. Lexington, KY. 1980.

9 Fannin FF, Bush LP. Enzymes for alkaloid biosynthesis in *N. tabacum, N. glauca* and their hybrid. *Tob. Chem. Res. Conf.,* 1988: **42**; 20.

10 Schiel O, Jarchow-Redecker K, Gerd-Walter P, Increased formation of cinnamoyl putrescines by fedbatch fermentation of cell suspension cultures of *Nicotiana tabacum. Plant Cell Reports,* 1984: **3**; 18–20.

11 Tso TC, McMurtrey JE. Mineral deficiency and organic constituents in tobacco plants. II Amino acids. *Plant Physiol.,* 1960: **35**; 865–870.

12 Fannin FF, Bush LP. Manipulation of alkaloid biosynthesis in detached tobacco roots by diamine and nicotinic acid feeding. *Tob. Chem. Res. Conf.,* 1982: **46**; 32.

13 Miller RD, Collins GB, Davis DL. Effects on nicotine precursors on nicotine content in callus cultures of burley tobacco alkaloid lines. *Crop Sci.,* 1983: **23**; 561–565.

14 Phillips GC, Eggett BC, Collins GB. Effects of polyamine metabolic inhibitors on growth and alkaloid accumulation in tobacco callus cultures. *Tob. Sci.,* 1992: **36**; 1–5.

15 Hamill JD, Robins RJ, Parr AJ, Evans DM, Furze JM, Rhodes MJC. Over-expressing a yeast ornithine decarboxylase gene in transgenic roots of *Nicotiana rustica* can lead to enhanced nicotine accumulation. *Plant Mol. Biol.,* 1990: **15**; 27–38.

16 Bush LP. Physiology and biochemistry of tobacco alkaloids. *Recent Adv. Tob. Sci.,* 1981: **7**; 75–106.

17 Tiburcio AF, Galston AW. Arginine decarboxylase as the source of putrescine for tobacco alkaloids. *Phytochemistry*, 1986: **25**; 107–110.

18 Burtin D, Michael AJ. Overexpression of arginine decarboxylase in transgenic plants. *Biochem. J.*, 1997: **325**; 331–337.

19 Saunders JW, Bush LP. Nicotine biosynthetic enzyme activities in *Nicotiana tabacum* L. genotypes with different alkaloid levels. *Plant Physiol.*, 1979: **64**; 236–240.

20 Yoshida D. Mechanisms of alkaloid accumulations in the varieties of tobacco plants containing lower alkaloid. *Bull. Hatano Tob. Expt. Stn.*, 1973: **73**; 245–257.

21 Hibi N, Higashiguchi S, Hashimoto T, Yamada Y. Gene expression in tobacco low-nicotine mutants. *Plant Cell*, 1994: **6**; 723–735.

22 Nakatani HY, Malik VS. Purified tobacco protein involved in nicotine synthesis, DNA encoding, and use of sense and antisense DNAs corresponding thereto to affect nicotine content in tobacco. *US Patent*, No. 5,684,241, 1997.

23 McLauchlan WR, McKee RA, Evans DM. The purification and immunocharacterisation of *N*-methylputrescine oxidase from transformed root cultures of *Nicotiana tabacum* L. cv SC58. *Planta*, 1993: **191**; 440–445.

24 Saunders JW, Pudliner HJ, Bush LP. Nicotine accumulation in callus and small plants of tobacco (*Nicotiana tabacum* L.) grown in media supplemented with nicotine. *Plant Sci. Lett.*, 1981: **23**; 315–319.

25 Hashimoto T, Mitani A, Yamada Y. Diamine oxidase from cultured roots of *Hyoscyamus niger*. *Plant Physiol.*, 1990: **93**; 216–221.

26 Friesen JB, Leete E. Nicotine synthase – An enzyme from Nicotiana species which catalyses the formation of (S)-nicotine from nicotinic acid and 1-methyl-Δ'-pyrrolinium chloride. *Tetrahedron Lett.*, 1990: **31**; 6295–6298.

27 Fannin FF, Wei X, Bush LP, Burton HR. Aberrant alkaloid biosynthesis in tobacco from nicotinic acid analogs. *Tob. Chem. Res. Conf.*, 1994: **48**; 49.

28 Rueppel ML, Rapoport H. Aberrant alkaloid biosynthesis. Formation of nicotine analogs from unnatural precursors in *Nicotiana glutinosa*. *J. Am. Chem. Soc.*, 1971: **93**; 7021–7028.

29 Boswell HD, Watson AB, Walton NJ, Robins DJ. Formation of N-ethyl-S-nornicotine by transformed root cultures of *Nicotiana rustica*. *Phytochemistry*, 1993: **34**; 153–155.

30 Endo T, Hamaguchi N, Eriksson T, Yamada Y. Alkaloid biosynthesis in somatic hybrids of *Duboisia leichhaardtii* F. Muell. and *Nicotiana tabacum* L. *Planta*, 1991: **183**; 505–510.

31 Armstrong DW, Wang X, Ercal N. Enantiomeric composition of nicotine in smokeless tobacco, medicinal products, and commercial reagents. *Chirality*, 1998: **10**; 587–591.

32 Burton HR, Andersen RA, Fleming PD. Changes in chemical composition of burley tobacco during senescence and curing. 2. Acylated pyridine alkaloids. *J. Agric. Food Chem.*, 1988: **36**; 579–584.

33 Burton HR, Wei X, Caldwell WS. Identification and quantification of tobacco-specific nitrosamine precursors in air-cured tobacco. *Tob. Chem. Res. Conf.*, 1993: **47**; 39–40.

34 Burton HR, Wei X. Distribution of tobacco constituents within the leaf tissue. 2. Oxidized nicotine derivatives. *Tob. Chem. Res. Conf.*, 1994: **48**; 42.

35 Castagnoli N,Jr, Rimoldi JM, Bloomquist J, Castagnoli KP. Potential metabolic bioactivation pathways involving cyclic tertiary amines and azaarenes. *Chem. Res. Toxicol.*, 1997: **10**; 924–940.

36 Fannin FF, Bush LP. Nicotine demethylation in *Nicotiana*. *Med. Sci. Res.*, 1992: **20**; 867–868.

37 Chelvarajan RL, Fannin FF, Bush LP. Study of nicotine demethylation in *Nicotiana otophora*. *J. Agric. Food Chem.*, 1993: **41**; 858–862.

38 Hao DY, Yeoman MM. Mechanism of nicotine *N*-demethylation in tobacco cell suspension cultures. *Phytochemistry*, 1996: **41**; 477–482.

39 Hao DY, Yeoman MM. Evidence in favour of an oxidative *N*-demethylation of nicotine to nornicotine in tobacco cell cultures. *J. Plant Physiol.*, 1998: **152**; 420–426.

40 Hobbs MC, Yeoman MM. Biotransformation of nicotine to nornicotine by cell suspensions of *Nicotiana tabacum* L. cv. Wisconsin–38. *New Phytol.*, 1991: **119**; 477–482.

41 Botte M, Mabon F, Mouillour ML, Robins RJ. Biosynthesis of nornicotine in root cultures of *Nicotiana alata* does not involve oxidation at C–5' of nicotine. *Phytochemistry*, 1997: **117**; 117–222.

42 Fannin FF, Bush LP. Nornicotine accumulation in nicotine-fed detached leaves of *N. glauca*, *N. glauca/N. tabacum* grafted plants and *N. tabacum* X *N. glauca* hybrids. *Tob. Chem. Res. Conf.*, 1989: **43**; 28.

43 Fannin FF, Wei X, Bush LP. Mechanism for partial racemization of nornicotine in tobacco. *Tob. Chem. Res. Conf.*, 1996: **50**; 39.

44 Perfetti TA, Coleman WM, III. Chiral-gas chromatography-selected ion monitoring-mass selective detection analysis of secondary alkaloids in tobacco and tobacco smoke. *Beit. Tabakforsch. Intl.*, 1998: **18**; 35–42.

45 Armstrong DW, Wang X, Lee JT, Liu YS. Enantiomeric composition of nornicotine, anatabine, and anabasine in tobacco. *Chirality*, 1999: **11**; 82–84.

46 Leete E, Muller ME. Biomimetic synthesis of anatabine from 2,5-dihydropyridine produced by the oxidative decarboxylation of baikiain. *J. Am. Chem. Soc.*, 1982: **104**; 6440–6444.

47 Herminghaus S, Tholl D, Rugenhagen C, Fecker LF, Leuschner C, Berlin J. Improved metabolic action of a bacterial lysine decarboxylase gene in tobacco hairy root cultures by its fusion to *rbsS* transit peptide coding sequence. *Transgenic Res.*, 1996: **5**; 193–201.

48 Berlin J, Mollenschott, C, Herminghaus S, Fecker LF. Lysine decarboxylase transgenic tobacco root cultures biosynthesize novel hydroxycinnamoylcadaverines. *Phytochemistry*, 1998: **48**; 79–84.

49 Riker AJ, Banfield WM, Wright WH, Keitt GW, Sagen HE. Studies on infectious hairy root of nursery apple trees. *J. Agric. Res.*, 1930: **41**; 507–540.

50 Binns AN, Thomashow MF. Cell biology of *Agrobacterium* infection and transformation of plants. *Annu. Rev. Microbiol.*, 1988: **42**; 575–606 .

51 Chilton M-D, Tepfer D, Petit A, David C, Casse-Delbart F, Tempé J. *Agrobacterium rhizogenes* inserts T-DNA into the genomes of host plant root cells. *Nature*, 1982: **295**; 432–434.

52 White FF, Taylor BH, Huffman GA, Gordon MP, Nester EW. Molecular and genetic analysis of the transferred DNA regions of the root-inducing plasmid of *Agrobacterium rhizogenes*. *J. Bacteriol.*, 1985: **164**; 33–44.

53 Slightom JL, Durand-Tardif M, Jouanin L, Tepfer D. Nucleotide sequence analysis of TL-DNA of *Agrobacterium rhizogenes* agropine type plasmid. *J. Biol. Chem.*, 1986: **261**; 108–121.

54 Nilsson O, Olsson O. Getting to the root: The role of the *Agrobacterium rhizogenes rol* genes in the formation of hairy roots. *Physiol. Plant.*, 1996: **100**; 463–473.

55 Sinkar VP, Pythoud F, White FF, Nester EW, Gordon MP. *rolA* locus of the Ri plasmid directs developmental abnormalities in transgenic tobacco plants. *Genes Dev.*, 1988: **2**; 688–697.

56 Vilaine F, Rembur J, Chriqui D, Tepfer D. Modified development in transgenic tobacco plants expressing a *rolA*; GUS translational fusion and subcellular localization of the fusion protein. *Mol. Plant Microbe Interact.*, 1998: **11**; 855–859.

57 Spanò L, Mariotti D, Cardarelli M, Branca C, Costantino P. Morphogenesis and auxin sensitivity of transgenic tobacco with different complements of Ri T-DNA. *Plant Physiol.*, 1988: **87**; 479–483.

58 Shen WH, Davioud E, David C, Barbier-Brygoo H, Tempé J, Guern J. High sensitivity to auxin is a common feature of hairy root. *Plant Physiol.*, 1990: **94**; 554–560.

59 Maurel C., Barbier-Brygoo H., Spena A., Tempé J. and Guern J. Single *rol* genes from the *Agrobacterium rhizogenes* T$_L$-DNA alter some of the cellular responses to auxin in *Nicotiana tabacum*. *Plant Physiol.*, 1991: **97**; 212–216.

60 Maurel C, Leblanc N, Barbier-Brygoo H, Perrot-Rechenmann C, Bouvier-Durand M, Guern, J. Alterations of auxin perception in *rolB*-transformed tobacco protoplasts. Time course of *rolB* mRNA expression and increase in auxin sensitivity reveal multiple control by auxin. *Plant Physiol.*, 1994: **105**; 1209–1215.

61 Filippini F., Lo Schiavo F., Terzi M., Costantino P. and Trovato M. The plant oncogene *rolB* alters binding of auxin to plant cell membranes. *Plant Cell Physiol.*, 1994: **35**; 767–771.

62 Maurel C, Brevet J, Barbier-Brygoo H, Guern J, Tempé J. Auxin regulates the promoter of the root-inducing *rolB* gene of *Agrobacterium rhizogenes* in transgenic tobacco. *Mol. Gen. Genet.*, 1990: **223**; 58–64.

63 Bellincampi D, Cardarelli M, Zaghi D, Serino G, Salvi G, Gatz C et al. Oligogalacturonides prevent rhizogenesis in *rolB*-transformed tobacco explants by inhibiting auxin-induced expression of the *rolB* gene. *Plant Cell*, 1996: **8**; 477–487.

64 Filippini F, Rossi V, Marin O, Trovato M, Costantino P, Downey P, Lo Schiavo F, Terzi M. A plant oncogene as a phosphatase. *Nature*, 1996: **379**; 499–500 .

65 Estruch JJ, Chriqui D, Grossmann K, Schell J, Spena A. The plant oncogene *rolC* is responsible for the release of cytokinins from glucoside conjugates. *EMBO J.*, 1991: **10**; 2889–2895 .

66 Nilsson O., Moritz T., Imbault N., Sandberg G. and Olsson O. Hormonal characterization of transgenic tobacco plants expressing the *rolC* gene of *Agrobacterium rhizogenes* T$_L$-DNA. *Plant Physiol.*, 1993: **102**; 363–371.

67 Lemcke K, Schmülling, T. Gain of function assays identify non-*rol* genes from *Agrobacterium rhizogenes* TL-DNA that alter plant morphogenesis or hormone sensitivity. *Plant J.*, 1998: **15**; 423–433.

68 Parr AJ, Hamill JD. Relationship between *Agrobacterium rhizogenes* transformed hairy roots and intact, uninfected *Nicotiana* plants. *Phytochemistry*, 1987: **26**; 3241–3245.

69 Hamill JD, Parr AJ, Robins RJ, Rhodes MJC. Secondary product formation by cultures of *Beta vulgaris* and *Nicotiana rustica* transformed with *Agrobacterium rhizogenes*. *Plant Cell Rep.*, 1986: **5**; 111–114.

70 Saitoh F, Noma M, Kawashima N. The alkaloid contents of sixty *Nicotiana* species. *Phytochemistry*, 1985: **24**; 477–480.

71 Saito K, Yamazaki M, Kawaguchi A, Murakoshi I. Metabolism of solanaceous alkaloids in transgenic plant teratomas integrated with genetically engineered genes. *Tetrahedron*, 1991: **47**; 5955–5968.

72 Palazón J, Cusidó RM, Roig C, Piñol MT. Effect of *rol* genes from *Agrobacterium rhizogenes* TL DNA on nicotine production in tobacco root cultures. *Plant Physiol. Biochem.*, 1997: **35**; 155–162.

73 Altabella T, Angel E, Biondi S, Palazón J, Bagni N, Piñol MT. Effect of the *rol* genes from *Agrobacterium rhizogenes* on polyamine metabolism in tobacco roots. *Physiol. Plant*, 1995: **95**; 479–485 .

74 Mothes K, Engelbrecht L, Tschoepe K-H, Hutschenreuter-Trefftz G. Wurzelaktivität und Nikotinbildung. *Flora*, 1957: **144**; 518–523.

75 Baldwin IT. Damage-induced alkaloids in tobacco: pot-bound plants are not inducible. *J. Chem. Ecol.*, 1988: **4**; 1113–1120 .

76 Palazón J, Cusidó RM, Roig C, Piñol M. Expression of the *rolC* gene and nicotine production in transgenic roots and their regenerated plants. *Plant Cell Rep.*, 1998: **17**; 384–390.

77 Mizusaki S., Tanabe Y., Roguchi M. and Tamaki E. Changes in the activities of ornithine decarboxylase, putrescine *N*-methyltransferase and *N*-methylputrescine oxidase in tobacco roots in relation to nicotine biosynthesis. *Plant Cell Physiol.*, 1973: **14**; 103–110.

78 Takahashi M, Yamada Y. Regulation of nicotine production by auxins in tobacco cultured cells in vitro. *Agric. Biol. Chem.*, 1973: **37**; 1755–1757.

79 Tiburcio AF, Kaur-Sawhney R, Ingersoll RB, Galston AW. Correlation between polyamines and pyrrolidine alkaloids in developing tobacco callus. *Plant Physiol.*, 1985: **78**; 323–326.

80 Hashimoto T, Yamada Y. Regulation of tobacco alkaloid biosynthesis. In: *Phytochemicals and Health*, Gustine DL, Flores HE (Ed.), Current Topics Plant Physiol 1995: **15**; 130–144.

81 Baldwin IT. Chemical changes rapidly induced by folivory. In: *Insect-Plant Interactions*, Bernays EA (Ed.), Boca Raton: CRC Press, 1994: **5**; 1–23.

82 Karban R, Baldwin IT. *Induced Responses to Herbivory*. Chicago: Univ. Chicago Press, 1997.

83 Baldwin IT. The alkaloidal responses of wild tobacco in real and simulated herbivory. *Oecologia*, 1988: **77**; 378–381.

84 Baldwin IT. Mechanism of damage-induced alkaloid production in wild tobacco. *J. Chem. Ecol.*, 1989: **15**; 1661–1680.

85 Baldwin IT, Schmelz EA, Ohnmeiss TE. Wound-induced changes in root and shoot jasmonic acid pools correlate with induced nicotine synthesis in *Nicotiana sylvestris* Spegazzini and Comes. *J. Chem. Ecol.*, 1994: **20**; 2139–2158.

86 Baldwin IT, Schmelz EA. Constraints on an induced defense: the role of leaf area. *Oecologia*, 1994: **97**; 424–430.

87 Hanounik SB, Osborne WW. The relationships between population density and *Meloidogyne incognita* and nicotine content of tobacco. *Nematologica*, 1977: **23**; 147–152.

88 Baldwin IT, Ohnmeiss TE. Alkaloidal responses to damage in *Nicotiana* native to North America. *J. Chem. Ecol.*, 1993: **19**; 1141–1151.

89 Baldwin IT. Jasmonate-induced responses are costly but benefit plants under attack in native populations. *Proc. Nat. Acad. Sci. USA*, 1998: **95**; 8113–8118.

90 Baldwin IT, Gorham D, Schmelz EA, Lewandowski CA, Lynds GY. Allocation of nitrogen to an inducible defense and seed production in *Nicotiana attenuata*. *Oecologia*, 1998: **115**; 541–552.

91 Lynds GY, Baldwin IT. Fire, nitrogen, and defensive plasticity in *Nicotiana attenuata*. *Oecologia*, 1998: **115**; 531–540.

92 Baldwin IT, Karb MJ, Ohnmeiss TE. Allocation of $^{15}N$ from nitrate to nicotine: production and turnover of a damage-induced mobile defense. *Ecology*, 1994: **75**; 1703–1713.

93 Ohnmeiss TE, Baldwin IT. The allometry of nitrogen allocation to growth and an inducible defense under nitrogen-limited growth. *Ecology*, 1994: **75**; 995–1002.

94 Tso TC, Jeffrey RN. Biochemical studies on tobacco alkaloids. IV. The dynamic state of nicotine supplied to *N. rustica*. *Arch. Biochem. Biophys.*, 1961: **92**; 253–256.

95 Yoshida D. Degradation and translocation of $N^{15}$-labeled nicotine injected into intact tobacco leaves. *Plant Cell Physiol.*, 1962: **3**; 391–395.

96 Gershenzon J. The cost of plant chemical defense against herbivory: a biochemical perspective. In: *Insect-Plant Interactions*, Bernays EA (Ed.), Boca Raton: CRC Press, 1993: **5**; 105–173.

97 Baldwin IT, Ohnmeiss TE. Swords into plowshares? *Nicotiana sylvestris* does not use nicotine as a nitrogen source under nitrogen-limited growth. *Oecologia*, 1994: **98**; 385–392.

98 Sembdner G, Parthier B. The biochemistry and physiological and molecular actions of jasmonates. *Annu. Rev. Plant Physiol. Plant Mol. Biol.*, 1993: **44**; 569–589.

99 Zhang Z-P, Baldwin IT. Transport of [2-$^{14}$C]-jasmonic acid from leaves to roots mimics wound-induced changes in endogenous jasmonic acid pools in *Nicotiana sylvestris*. *Planta*, 1997: **203**; 436–441.

100 Baldwin IT, Zhang Z-P, Diab N, Ohnmeiss TE, McCloud ES, Lynds GY, Schmelz EA. Quantification, correlations and manipulations of wound-induced changes in jasmonic acid and nicotine in *Nicotiana sylvestris*. *Planta*, 1997: **201**; 397–404.

101 Ohnmeiss TE, McCloud ES, Lynds GY, Baldwin IT. Within-plant relationships among wounding, jasmonic acid and nicotine: implications for defense in *Nicotiana sylvestris*. *New Phytol.*, 1997: **137**; 441–452.

102 Karban R, Baldwin IT. *Induced Responses to Herbivory*. Chicago: Univ. Chicago Press, p. 88, 1997.

103 McCloud ES, Baldwin IT. Herbivory and caterpillar regurgitants amplify the wound-induced increases in jasmonic acid but not nicotine in *Nicotiana sylvestris*. *Planta*, 1997: **203**; 430–435.

104 Karban R, Baldwin IT. *Induced Responses to Herbivory*. Chicago: Univ. Chicago Press, 1997, p. 89.

105 Baldwin IT, Schmelz EA. Immunological "memory" in the induced accumulation of nicotine in wild tobacco. *Ecology*, 1996: **77**; 236–246.

106 Baldwin IT, Schmelz EA, Zhang Z-P. Effects of octadecanoid metabolites and inhibitors on induced nicotine accumulation in *Nicotiana sylvestris*. *J. Chem. Ecol.*, 1996: **22**; 61–74.

107 Imanishi S, Hashizume K, Nakakita M, Kojima H, Matsubayashi Y, Hashimoto T, et al. Differential induction by methyl jasmonate of genes encoding ornithine decarboxylase and other enzymes involved in nicotine biosynthesis in tobacco cell cultures. *Plant Mol. Biol.*, 1998: **38**; 1101–1111.

108 Karban R, Baldwin IT. *Induced Responses to Herbivory*. Chicago: Univ. Chicago Press, 1997, 90–91.

*Chapter 3*

# Biotransformation of nicotine in mammalian systems

John W. Gorrod[1] and Georg Schepers[2]

[1] *Toxicology Unit, John Tabor Laboratories, The University of Essex, Wivenhoe Park, Colchester, Essex CO4 3SQ, UK*
[2] *INBIFO Institut für biologische Forschung GmbH, Fuggerstrasse 3, D-51149 Cologne, Germany*

## I. INTRODUCTION

Despite the apparent simple chemical structure of nicotine, its biotransformation is extremely complex [1] and involves novel intermediates and diverse enzyme systems. Nicotine contains both aromatic and aliphatic carbon and nitrogen atoms within its structure, providing numerous sites for metabolic oxidation (phase 1) and conjugation (phase 2) reactions; in some cases metabolites arising from nicotine can undergo reductive reactions. This chapter is not intended to be an exhaustive comprehensive review of all aspects of the metabolism and enzymology of nicotine biotransformation and the factors that influence these processes, and the reader is referred to major reviews on this subject [2–7]. The present chapter will outline the major metabolic pathways, describe more recent findings, discuss some intermediates of interest and draw attention to where anomalies and deficiencies in our knowledge occur.

## II. PHASE 1 OXIDATION OF NICOTINE

### Oxidation at alicyclic carbon

Quantitatively the most important metabolic reaction of nicotine in humans is the formation of cotinine which involves the incorporation of oxygen at the 5′ position of nicotine and thereby converts the basic pyrrolidine moiety into a cyclic amide (Fig. 1); it has been calculated that more than 70% of nicotine metabolism in humans proceeds via the formation of cotinine [8]. Cotinine formation, unlike the oxidation of most xenobiotics, does not incorporate molecular oxygen into the substrate nicotine, but rather derives its oxygen from water [9]. It was recognised early that when nicotine was metabolised in vitro in the presence of cyanide a new compound was formed with a concomitant decrease in cotinine formation [10]. However it was not until the elegant studies by Murphy [9] that a key intermediate in the pathway was recognised. Murphy proposed that the initial metabolic reaction of nicotine involves electron and hydrogen abstraction to produce nicotine-$\Delta^{1'(5')}$-iminium ion which reacts with water to yield cotinine (Fig. 1). More recent work has indicated that nicotine can form all three isomeric iminium ions leading to additional metabolic products [11,12] (Fig. 1). From the above it can be seen that the formation of cotinine is more complex than originally thought and probably involves electron abstraction to form an aminium

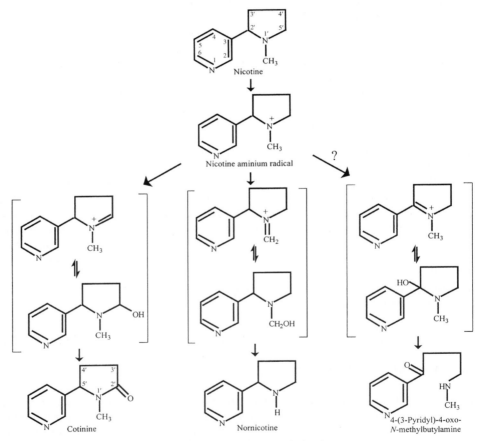

Fig. 1  Biological oxidation of nicotine.

radical [13] which is further oxidised to isomeric iminium ions. These reactions are catalysed by certain isoforms of cytochrome P450 (CYP) [14]. Whilst these initial electron abstractions may be carried out by a single CYP isozyme, the oxidation to isomeric iminium ions may involve more than one isozyme; the precise factors determining which iminium ion (and hence metabolic pathway) is formed, are at present but poorly understood. The formation of nicotine-$\Delta^{1'(5')}$-iminium ion appears to be mediated by CYP2A6 in humans [15] which is the same enzyme responsible for the 3'-hydroxylation of cotinine [16].

The involvement of nicotine-$\Delta^{1'(2')}$-iminium ion as an intermediate in nicotine metabolism is more problematical. The corresponding 2'-cyanonicotine has been detected in urine of smokers [17] although this may have been formed during work-up and analytical procedures [18]. It is clear that the failure to detect a compound as an excretory product does not prove that it is not formed in vivo. 2'-Cyanonicotine has been detected in tissue preparations during the metabolism of nicotine in the presence of cyanide [19] which supports the formation of the $\Delta^{1'(2')}$-iminium ion as an intermediate in nicotine metabolism. In this case however, neither 4-(3-pyridyl)-4-oxo-*N*-methylbutylamine nor the corresponding carbinol 4-(3-pyridyl)-4-hydroxy-*N*-methylbutylamine or 2'-hydroxynicotine or its dehydration product nicotine-$\Delta^{2'(3')}$-enamine have been detected as nicotine metabolites in mammalian systems to date. This suggests that this iminium ion is at best a minor intermediate in nicotine metabolism (Fig. 1).

## Metabolism of isomeric nicotine iminium ions

In the case of nicotine-$\Delta^{1'(5')}$-iminium ion it is clear that its oxidation is mediated by a cytosolic molybdenum containing aldehyde oxidase [22,23] to yield cotinine. The evidence is less clear as to the enzymes involved in the oxidation of the nicotine-$\Delta^{1'(2')}$- and $\Delta^{1'(6')}$-iminium ions, although the latter process is known to yield the *N*-demethylated alkaloid, nornicotine [11] (Fig. 1).

## Further oxidation of primary metabolites

*Nornicotine*
The metabolism of this alkaloid has been studied in vivo [24,25] and in vitro [25-26] and the lactam norcotinine recognised as the major metabolite (Fig. 2). The corresponding acid 4-(3-pyridyl)-4-aminobutyric acid was also isolated from the urine of dogs treated with nornicotine [24]. It was considered that in the course of this oxidation process a reactive intermediate is formed analogous to the $\Delta^{1'(5')}$-iminium ion in cotinine formation. In an attempt to examine this possibility, nornicotine was incubated in the presence of cyanide [27]. Whilst cyanide was incorporated into the product

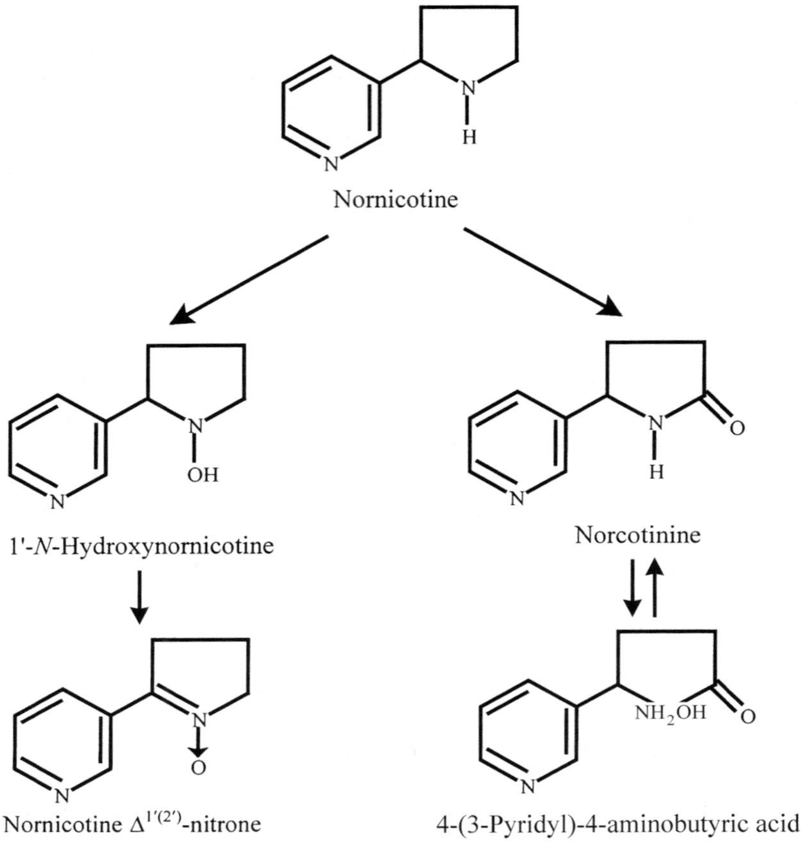

Fig. 2 Biological oxidation of nornicotine.

formed, it was shown that this product was identical to one of the cyano compounds formed from nicotine viz. 1'-cyanomethylnornicotine, and that the methylene group in this compound was derived from formaldehyde, itself probably produced metabolically from the tissue preparation or from glycerol used as a preservative. Nornicotine is also sequentially oxidised at the pyrrolidine nitrogen to give nornicotine $\Delta^{1',(2')}$-nitrone via the proposed intermediate 1'-N-hydroxynornicotine [26]. An additional metabolite has also been detected but not as yet characterised [26].

*Cotinine*
Cotinine undoubtedly plays a major role in the overall picture of nicotine metabolism. It can be excreted unchanged in most species, and is vulnerable to oxidation at both carbon and nitrogen centres (Fig. 3). Hydroxylation at the 3'-position gives both *cis-* and *trans-*3'-hydroxycotinine with the latter

form predominating over the *cis* by far and represents the major urinary nicotine metabolite in humans [28–31]. Hydroxylation also occurs at the 5'-position to give 5'-hydroxycotinine [32], which predominates under physiological conditions over its open chain tautomer 4-(3-pyridyl)-4-oxo-*N*-methylbutyramide [33]. Cotinine and 5'-hydroxycotinine are both converted to the corresponding *N*-oxides by oxidation of the pyridyl nitrogen [34,35] (Fig. 3).

*N*-Demethylation of cotinine to give norcotinine has been observed following the administration of cotinine to various species [36–38] but not man [39]. Norcotinine has less frequently been recognised as a cotinine metabolite in vitro. In a thorough study using hepatic microsomal preparations from hamster, guinea pig, rabbit, rat, mouse or dog [40,41] norcotinine was not detected when cotinine was incubated with microsomes from

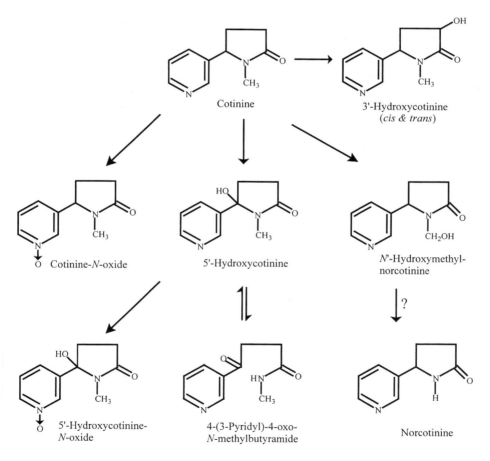

Fig. 3 Biological oxidation of cotinine.

49

untreated control animals or with those obtained following pre-treatment of animals with a variety of enzyme inducing agents. During the course of this study a novel metabolite was observed which was formed during the incubation of cotinine in guinea pig and rabbit preparations in which hydroxylation occurred at the *N*-methyl group to form 1'-*N*-hydroxyme-thylnorcotinine [42]. This compound appeared to be metabolically stable and did not give rise to norcotinine in vitro. The above results suggest that the major route to norcotinine is by oxidation of nornicotine [24,26]. It should be noted that norcotinine has been detected in the urine of smokers [43,44] and nicotine treated rats [20,21], where it occurs to about 1–2% of the total nicotine metabolites.

*Norcotinine*
Irrespective of its route of formation norcotinine would be expected to be metabolised, at least to some extent, prior to excretion. Norcotinine is converted to 4-(3-pyridyl)-4-oxo-butyramide by fortified rat microsomal preparations [45] (Fig. 4). This is of considerable interest as this metabolite is the open chain form of 5'-hydroxynorcotinine, a compound found in the urine of rats treated with (±)-nicotine [20] but not in the urine of rats treated with (−)-nicotine [21]. In human urine the corresponding dehydrated

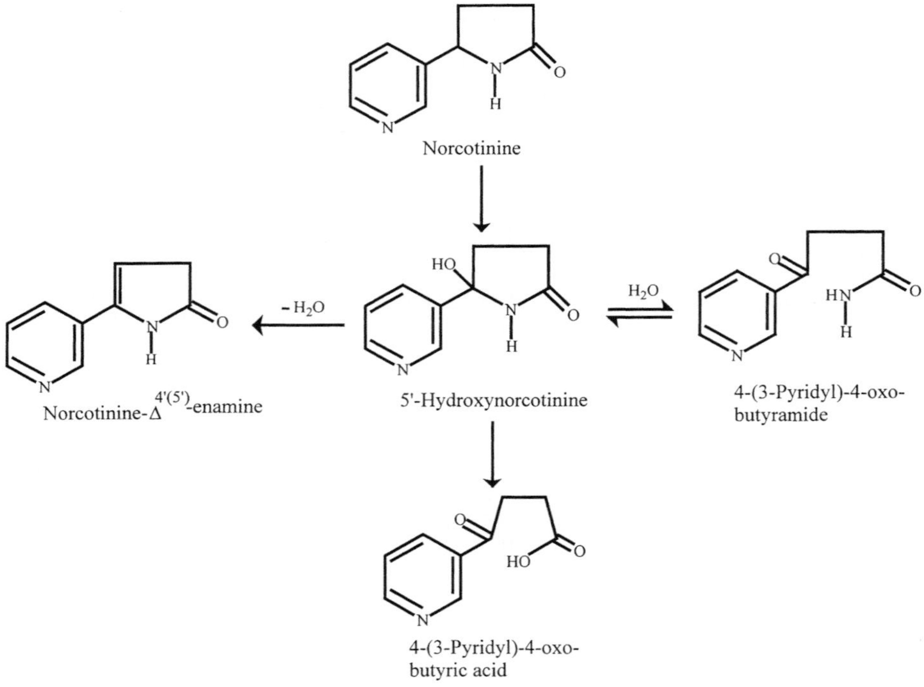

Fig. 4 Biological oxidation of norcotinine.

product, demethylcotinine-$\Delta^{4'(5')}$-enamine was found with a long half life of excretion [46]. However it cannot be precluded that the formation of the enamine results from the MS analytical procedures [47].

*Pyridylbutyric acid derivatives*
From the above it can be seen that both cotinine and norcotinine give rise to metabolites possessing a pyridyl moiety attached to a modified butyric acid residue. In the case of the one derived from cotinine, 4-(3-pyridyl)-4-oxo-*N*-methylbutyramide, it is converted to 4-(3-pyridyl)-4-oxobutyric acid (Fig. 5) by a single enzymic hydrolytic reaction rather than by *N*-demethylation followed by hydrolysis of the amide [40]. The metabolite derived from

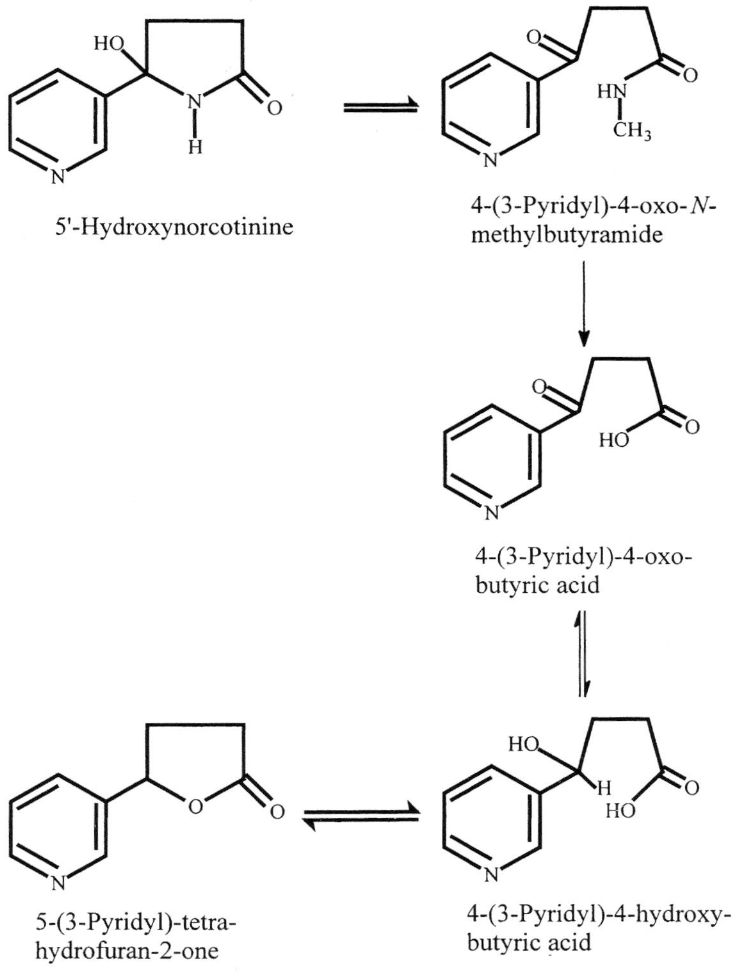

5'-Hydroxynorcotinine

4-(3-Pyridyl)-4-oxo-*N*-methylbutyramide

4-(3-Pyridyl)-4-oxo-butyric acid

5-(3-Pyridyl)-tetra-hydrofuran-2-one

4-(3-Pyridyl)-4-hydroxy-butyric acid

Fig. 5 Metabolism of 5'-hydroxycotinine to 4-(3-pyridyl)-4-oxo-butyric acid and its derivatives.

nornicotine, 4-(3-pyridyl)-4-oxo-butyramide, is also hydrolysed to 4-(3-pyr-idyl)-4-oxobutyric acid [45], a compound that is a common metabolite of the two metabolic pathways (Figs 4 and 5).

The further metabolism of 4-(3-pyridyl)-4-oxobutyric acid has been studied in vitro [48] and shown to be converted by reduction to the corresponding carbinol 4-(3-pyridyl)-4-hydroxybutyric acid which cyclises to 5-(3-pyridyl)-tetrahydrofuran-2-one (Fig. 5). These three compounds are interconvertable and incubation of any component with hepatic fractions gives rise to the other two, however in differing ratios, depending upon the tissue fraction used as enzyme source. In this study no further degradation products could be detected. 4-(3-Pyridyl)-4-hydroxybutyric acid has also been detected in the urine of (S)-(−)-nicotine treated rats [21] and in the urine of smokers where it was present as 5% of the total nicotine metabolites excreted (Schepers et al. unpublished observations). Recently it has been shown that this metabolic pathway be more important than previously thought, as the keto-acid and hydroxy-acid together account for about 14% of a nicotine dose (Hecht et al. 1999, *Chem. Res. Toxicol.* 12; 172–179).

Administration of 5-(3-pyridyl)-tetrahydrofuran-2-one to rats gave five Koenig positive metabolites in urine [49]. Two of these compounds were shown to be 4-(3-pyridyl)-4-hydroxybutyric acid and 3-pyridylacetic acid

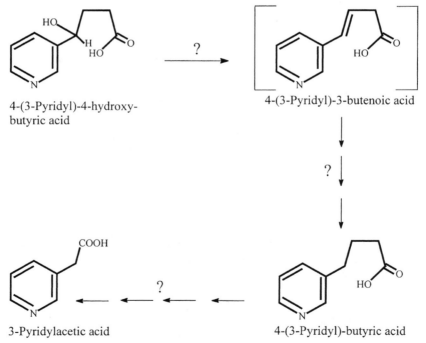

4-(3-Pyridyl)-4-hydroxy-butyric acid

4-(3-Pyridyl)-3-butenoic acid

3-Pyridylacetic acid

4-(3-Pyridyl)-butyric acid

Fig. 6 Proposed pathways of 4-(3-pyridyl)-4-hydroxybutyric acid to 3-pyridylacetic acid.

(Fig. 6). This latter compound has also been detected in the urine of animals or man following cotinine [36,50,51] nicotine [20,21,50] and 4-(3-pyridyl)-4-oxobutyric acid administration [52]. Thus it is now generally accepted that 3-pyridylacetic acid is one end product of nicotine phase 1 metabolism in mammalian systems. It is thought that 3-pyridylacetic acid is derived from 4-(3-pyridyl)-4-hydroxybutyric acid by dehydration to 4-(3-pyridyl)-3-butenoic acid followed by reduction to 4-(3-pyridyl)-butyric acid and fatty acid degradation analogous to that for phenylbutyric acid [51]. Little evidence for the presence of these intermediates or their coenzyme A derivatives has been presented.

Interestingly it was suggested that 4-(3-pyridyl)-butyric acid was present in the urine of rats receiving 5-(3-pyridyl)tetrahydrofuran-2-one [49] and more recently at low levels in the urine of rats receiving nicotine [21]. 4-(3-Pyridyl) butyric acid has also been detected in the urine of animals after administration of 4-(3-pyridyl)-4-oxo-butyric acid [52,53] however this biotransformation has not been achieved in vitro [48]. Many of the early ideas regarding the degradation of nicotine to pyridylacetic acid have been summarised by McKennis [54].

## Oxidation at alicyclic nitrogen

Nicotine has two nitrogen centres which can be oxidised chemically to produce isomeric nicotine-*N*-oxides, however in biological systems it is only the alicyclic pyrrolidine nitrogen which is susceptible to metabolic oxidation [55,56]. The oxidation of nicotine at the 1'-*N*-position can occur at two different stereochemical positions, i.e. the 1'R- and 1'S-configuration, thus leading for each nicotine enantiomer to two diastereoisomers: in the case of the natural S-(−)-nicotine to 1'R,2'S- and 1'S,2'S- nicotine-1'-*N*-oxide with the *N*-methyl group in the *cis*- and *trans*-position to the pyridine ring, respectively (Fig. 7). A large inter-species variation in the *cis/trans* ratio has been observed [57,58] with the *trans*-stereoisomer being almost exclusively produced in man [61–63]. In vivo *N*-oxidation of S-(−)-nicotine in the rat leads predominantly to the urinary excretion of the *cis*-diastereoisomer [21] whereas for R-(+)-nicotine the *trans*-diastereoisomer predominates (Schepers et al., unpublished results). Thus 1'-*N*-oxidation of both nicotine enantiomers in the rat produces predominantly the same absolute stereoconfiguration at the pyrrolidine nitrogen, i.e. the 1'R configuration.

This oxidation is mediated by microsomal flavin containing monooxygenases [58–60]; in humans this oxidation is carried out by FMO3 [61–63]. Nicotine-1'-*N*-oxide is reduced back to nicotine by hepatic and extrahepatic systems [64] and in man by the bacteria of the gastrointestinal tract [65].

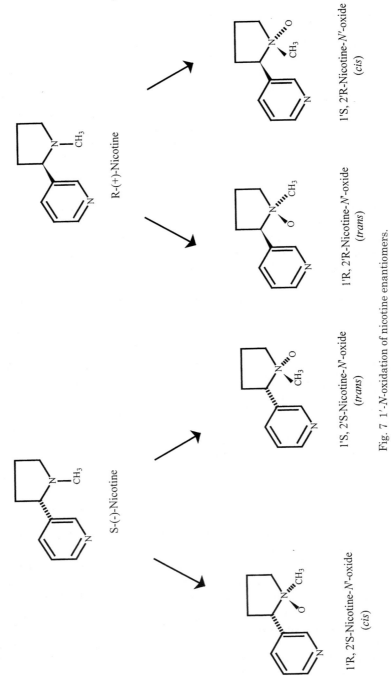

Fig. 7  1′-*N*-oxidation of nicotine enantiomers.

## III. PHASE 2 METABOLISM OF NICOTINE

Xenobiotics or metabolites thereof can be further metabolised by enzymatic transfer to a molecule or molecular substructure from endogenous compounds thus forming metabolites with higher molecular weight. These type of metabolites are called phase 2 metabolites.

These metabolites are often highly polar, water-soluble compounds and thus aid urinary excretion of foreign compounds and their phase 1 metabolites. Enzymes often involved in the formation of phase 2 metabolites are methyltransferases yielding methylated products, glucuronosyltransferases which form glucuronides, sulfotransferases which lead to sulfuric acid conjugates, glutathione transferases forming conjugates with glutathione which can themselves be further metabolised to mercapturic acids, acyl transferases leading, e.g. to conjugates between carboxylic acids and amino acids, acetyltransferases which transfers acetyl groups predominantly to amino groups, but also to hydroxyl and sulphydryl groups.

*N*-Methylation, catalysed by methyltransferases and the formation of glucuronides are especially important phase 2 pathways in nicotine metabolism and will be discussed in more detail later in this review. Beside these, other phase 2 pathways have also been reported to be involved in nicotine metabolism, i.e. the formation of conjugates of nicotine metabolites with amino acids like glycine [66,67] and phenylalanine [68]. Nucleotide analogues have been found in tissues of rabbits injected with nicotine or cotinine [69], and a conjugate with glutathione [70] has been found which arises via reaction with nicotine-$\Delta^{1'(5')}$-iminium ion. Conjugation with sulphuric acid has also been proposed [71]. These types of phase 2 metabolites will not be further discussed in this review.

In addition, there are indications of enzymatic dependent covalent binding of reactive nicotine metabolite intermediates (e.g. iminium species) with tissue macromolecules [72,73], and also non-enzymic dependant covalent binding of such nicotine metabolites to liver microsomes and sulphydryl containing polyamino acids has been described [74].

### Methylation

That methylation reactions are involved in nicotine metabolism had early been recognised by McKennis and co-workers [75]. From the urine of dogs treated intravenously with S-nicotine they isolated a quaternary ammonium compound derived from nicotine and identified this metabolite as *N*-methylnicotinium ion, methylated at the pyridine nitrogen (Fig. 8). In addition, they isolated *N*-methylcotininium ion, also methylated at the pyridine nitrogen, from the urine of dogs, as well as from the urine of a non-smoking man, who had orally received large quantities of S-cotinine.

More recently, after developing a selective HPLC method with electrochemical detection for analysing the *N*-methylnicotinium ion [76], and a

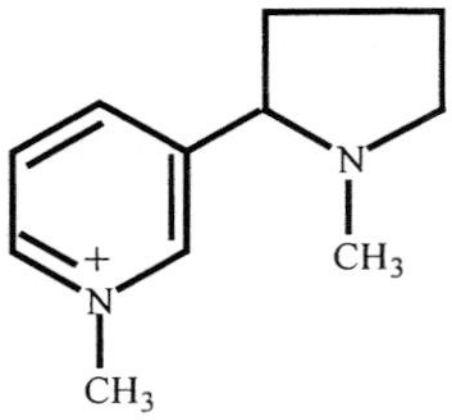

N-Methylnicotinium ion 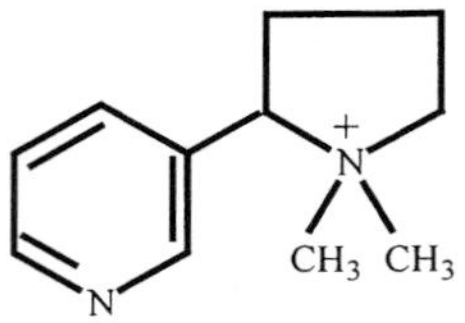      1'-N-Methylnicotinium ion 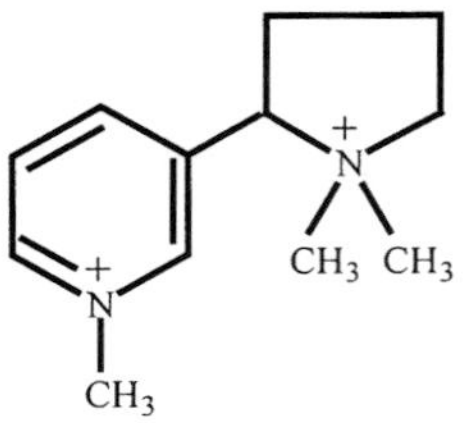      1,1'-N-Dimethylnicotinium ion

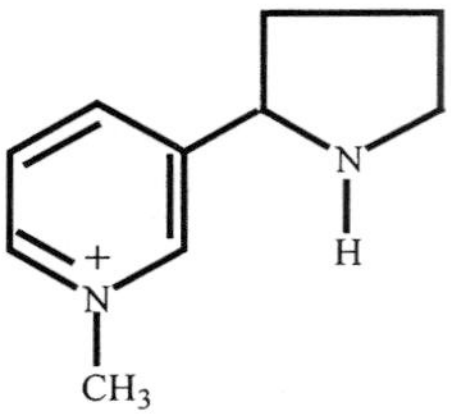

N-Methylnornicotinium ion 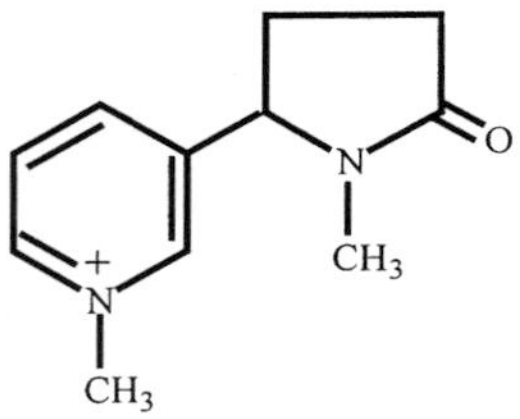      N-Methylcotininium ion 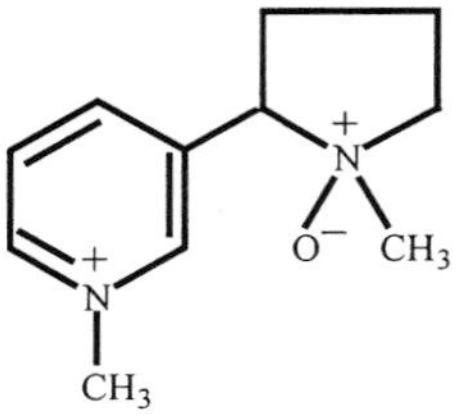      N-Methyl-1'-N-oxo-nicotinium ion

Fig. 8 *N*-Methylated nicotine metabolites.

cation-exchange HPLC for analysing seven potential metabolites of nicotine including several *N*-methylated nicotine derivatives [77], Crooks and co-workers carried out a comprehensive study of the *N*-methylation pathway of nicotine in the guinea pig, a species which is known to be very proficient at *N*-methylation. They found that metabolic *N*-methylation is, for this species, a stereo- as well as regiospecific pathway which involves only the R-(+)-enantiomer of nicotine and only methylation at the pyridine nitrogen. *N*-Methylated nicotine metabolites account for 15 to 20% of the total urinary metabolites in this species [74]. The following compounds were found and identified in urine from guinea pigs after R-nicotine administration: *N*-methylnicotinium ion [78–83], *N*-methylcotininium ion [78,79,81–83], *N*-methylnornicotinium ion [78,80,81,83,79] as well as the two diastereoisomers of *N*-methyl-1'-*N*-oxonicotinium ion [81]. *N*-Methylcotininium ion, *N*-methylnornicotinium ion as well as *N*-methyl-1'-*N*-oxonicotinium ion were also found as metabolites in the urine of guinea pigs receiving R-*N*-methylnicotinium acetate [81,84].

No *N*-methylated metabolites of S-nicotine were found in the guinea pig, neither in vivo nor in vitro, but interestingly, S-nicotine was found to be an inhibitor of S-adenosylmethionine (SAM) dependent cytosolic azaheterocycle *N*-methyltransferase which catalyses the *N*-methylation of R-nicotine [85]. This enzyme is widely distributed in guinea pig tissue and its activity was found to differ in the homogenates of various organs: spleen>brain>

lungs>liver [86]. Its activity was also found in guinea pig liver cytosol [87], but not in isolated hepatocytes [88].

For other species, no systematic studies like those for the guinea pig have been performed, and only isolated and sometimes conflicting findings for the N-methylation pathway of nicotine have been reported. For example, whereas Nwosu and Crooks [82] failed to find any N-methylated nicotine metabolite in the urine of rats, hamsters or rabbits treated with R- or S-nicotine or in vitro after incubation of S- or R-nicotine with rat liver cytosol [87], Kyerematen and co-workers [20,89] found that N-methylnicotinium ion in the urine of rats accounted for up to 3% of the administered racemic nicotine. This metabolite was not found in rats treated with S-nicotine [21], however in this latter study only very small amounts of 1'-N-methylnicotinium ion, 1,1'-N-dimethylnicotinium ion, and N-methyl-1'-N-oxonicotinium ion were found, accounting in total for less than 1% of the administered nicotine dose.

Nicotine N-methylation has also been demonstrated in vitro by incubation of either S- or R-nicotine with two purified SAM-dependent amine N-methyltransferases isolated from rabbit liver [86]. Using either enzyme alone, N-methylnicotinium ion was found as a metabolite of R- but not of S-nicotine; interestingly however, incubation with a mixture of both enzymes yielded N-methylnicotinium ion as a metabolite of S-nicotine, and both N-methyl- as well as 1'-N-methylnicotinium ions as metabolites of R-nicotine. N-Methylcotininium ion, was found in the same system when S- or racemic cotinine served as a substrate [91]. However, N-methylcotininium ion was not found as a urinary metabolite after ip. administration of S-cotinine to guinea pigs [92]. N-Methylnicotinium ion was found as a human nicotine metabolite in vivo in the urine of smokers [28] as well as in vitro by incubation of either R- or S-nicotine with human liver cytosol [87].

In the context of the role of the N-methylation pathway in nicotine metabolism discussed above, an early finding of Axelrod [93] remains remarkable, namely that nicotine itself is a metabolic product when nornicotine is incubated with SAM enriched rabbit lung cytosol.

## Glucuronidation

The first evidence for the existence of glucuronic acid conjugates in nicotine metabolism was presented at scientific meetings almost ten years ago [94,95]: concentrations of nicotine as well as of its metabolites cotinine and *trans*-3'-hydroxycotinine in urine samples from smokers and snuff users were found to increase on average by a factor of 2, 3, and 1.4, respectively, when the urine samples were treated with β-glucuronidase. This was interpreted to mean that the increase in nicotine and metabolite concentrations resulted from the enzymatic cleavage of the respective glucuronides (Fig. 9). The glucuronides of cotinine and *trans*-3'-hydroxycotinine were also found in the urine of non-smokers after oral administration of cotinine [94].

The observation that glucuronides of nicotine, cotinine, and *trans*-3'-hydroxycotinine are major human nicotine metabolites accounting for up to 40% of the metabolites excreted in urine was confirmed in the following years [96–100]. High interindividual variation in the urinary metabolite pattern was observed [98,101–103]. The glucuronides of nicotine and 3'-hydroxycotinine were also found as the major biliary metabolites of nicotine in the rat [104].

In 1992, the chemical synthesis of the *N*-β-D-glucuronide of cotinine as well as of *cis*- and *trans*-3'-hydroxycotinine conjugated at the pyridine nitrogen were described [105,106]. One glucuronide found in smoker's urine possessed an identical retention time in HPLC and showed, after partial purification, the same molecular masses of conjugate and aglycone as the chemically synthesised *N*-β-D-glucuronides of cotinine. From this, the metabolically formed conjugate of cotinine was postulated to be *N*-β-D-glucopyranuronosyl-S-(−)-cotininium inner salt [105]. The respective *N*-glucuronides of *cis*- and *trans*-3'-hydroxycotinine were not found in urine of smokers [103] The glucuronide of *trans*-3'-hydroxycotinine was isolated from smokers' urine and was identified by mass spectrometry and NMR spectroscopy as *trans*-3'-hydroxycotinine-*O*-β-D-glucuronide [71] (Fig. 9). In

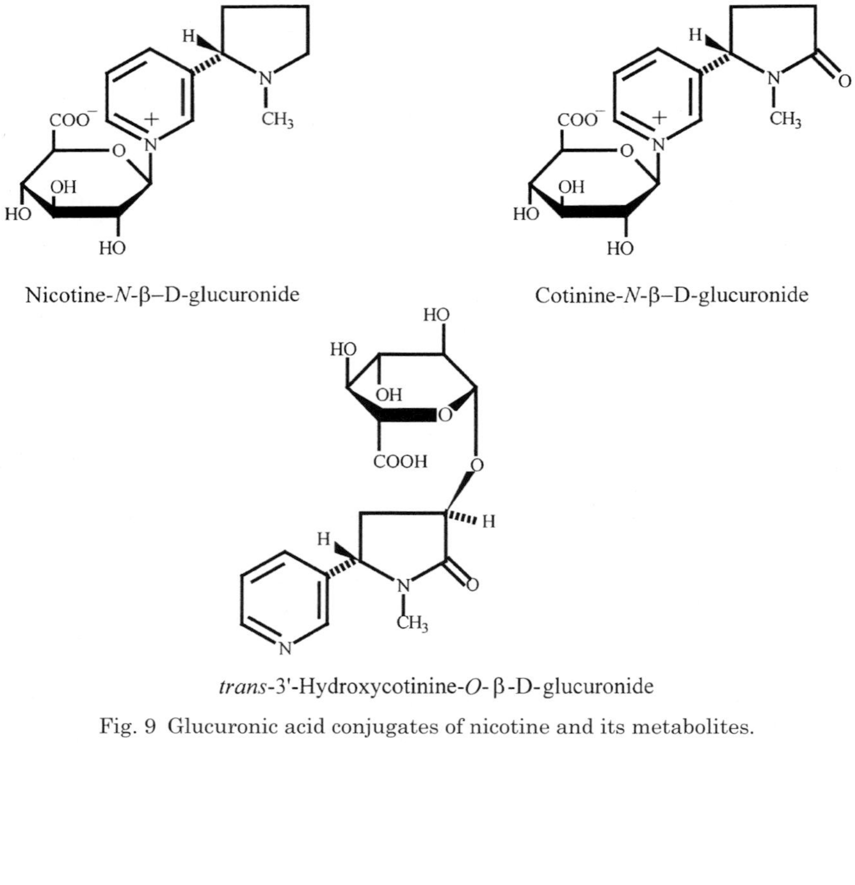

Nicotine-*N*-β–D-glucuronide

Cotinine-*N*-β–D-glucuronide

*trans*-3'-Hydroxycotinine-*O*-β-D-glucuronide

Fig. 9 Glucuronic acid conjugates of nicotine and its metabolites.

1993, the chemical synthesis of the *N*-β-D-glucuronide of nicotine conjugated at the pyridine nitrogen was described [107]. In the HPLC analysis of bile from rats treated with nicotine they found a peak having the same retention time as that of the synthesised glucuronide of nicotine. After hydrolysis of the compound associated with this peak, nicotine was found. From this finding the authors postulated that the structure of the metabolically formed glucuronide of nicotine was identical with that of the synthetic material. The glucuronides of nicotine and cotinine have been isolated from smoker's urine using various purification steps including liquid extraction, solid phase extraction, as well as four different HPLC methods, each of them applied repetitively [108]. The purified conjugates were identified by the daughter ion mass spectra of the molecule mass $[M+H]^+$ 339 Dalton for nicotine and 353 Dalton for cotinine glucuronide, using FAB tandem mass spectrometry. Final identification of the structures were performed by NMR. It was found that the structures of the metabolically produced conjugates of nicotine and cotinine were identical with those of the previously synthesised *N*-β-D-glucuronide, thus confirming the postulated structures (Fig. 9). The *N*-glucuronide of nicotine has also been identified as a metabolite by incubating nicotine with human hepatocytes [109].

The availability of authentic standard reference compounds was important for the development of a direct quantitative analytical method, i.e. by using LC/MS, as described for both the *N*-glucuronide of cotinine [105] and nicotine [110]. Whilst the *N*-β-D-glucuronides of nicotine and cotinine are available by chemical synthesis, the synthesis of *trans*-3'-hydroxycotinine-*O*-β-D-glucuronide has not presently been successful. Because of the difficulties in obtaining synthetic standard glucuronides the analysis of glucuronides is generally performed by an indirect method, i.e. by analysis of the aglycone after deconjugation. It was found that hot strong acid treatment for prolonged time (1M HCl, 16h, 70°C) partially hydrolyses the glucuronides of nicotine and *trans*-3'-hydroxycotinine, but not cotinine [71]. However, no hydrolysis of nicotine-*N*-glucuronide was observed even after stronger acid treatment (2.8M HCl, 18h, 70°C) [111]. The *N*-glucuronides of nicotine and cotinine can be deconjugated completely by short strong alkali treatment (1.1M NaOH, 35min, 75°C) [102]. Nicotine-*N*-glucuronide was also completely hydrolysed to nicotine by prolonged mild alkali treatment (0.015M NaOH, 18h, 70°C) [111], but not with prolonged, strong alkaline treatment [71,111]. Furthermore, we have observed that some batches of the same commercially available β-glucuronidase preparation (EC 3.2.1.31, from *Helix pomatia),* did not only cleave the glucuronides, but also attacked nicotine, thereby resulting in a false low or even negative nicotine glucuronide concentration. Therefore, appropriate controls are necessary when using chemical or enzymatic hydrolysis.

Glucuronide formation is catalysed by UDP-glucuronosyltransferases (UGTs, EC 2.4.1.17), of which at least two families comprising at least forty individual isozymes are known [112]. The isozymes which are involved in *N*-glucuronidation of nicotine and cotinine and *O*-glucuronidation of *trans*-3'-hydroxycotinine have not yet been identified. Due to high inter-individual variation in glucuronide formation and the observation that the extent of conjugation of nicotine and cotinine is highly correlated, but neither is correlated with the conjugation of *trans*-3'-hydroxycotinine, it was suggested that the same UGT isozyme might be involved in the *N*-glucuronide formation of nicotine and cotinine, but a different isozyme being utilised in the *O*-glucuronidation of *trans*-3'-hydroxycotinine [102].

Recently, the formation of nicotine-*N*-glucuronide, but not cotinine-*N*-glucuronide, has been observed in marmoset microsomal preparations [111] strongly indicating that nicotine and cotinine are not, in this species, common substrates for a single isozyme.

In this context it seems remarkable that, in general, the *N*-glucuronide of cotinine was found as the most abundant urinary nicotine phase 2 metabolite in studies performed in Northern America, whereas in European studies the formation of the *O*-glucuronide of *trans*-3'-hydroxycotinine seems to predominate. However, there are also some subjects from the American studies who did not show any significant *N*-glucuronide formation [98,102] These observations, however, need to be substantiated by investigating a much larger number of individuals.

## CONCLUSIONS

From the above it can be seen that the metabolism of nicotine is extremely complex involving multiple enzyme systems to handle the diverse range of intermediates involved in both phase 1 and phase 2 biotransformations. These enzymes are influenced by a host of physiological and pharmacological factors which control the levels of any constituent of a metabolic pathway giving rise to large interindividual variation. Additionally, many proposed intermediates in the pathways of nicotine metabolism have not been proven and further work will be needed to unravel the full metabolic picture of this ubiquitous molecule.

## REFERENCES

1 Gorrod JW. Nicotine: A case study in metabolic complexity. *Eur. J. Drug Metab. Pharmacokinet.*, 1995; Special Issue: 5–9.

2 Gorrod JW, Jenner P. The metabolism of tobacco alkaloids. *Essays in Toxicology*, 1975: **6**; 35–78.

3 Castagnoli N, Shigenaga M, Carlson T, Trager WF, Trevor A. The in vitro metabolic fate of (S)-nicotine. In: *Effects of Nicotine on Biological Systems*, Adlkofer F, Thurau K (Eds.), Birkhauser Verlag, Basel, 1991, 25–34.

4 Benowitz NL, Porchet H, Jacob III P. Pharmacokinetics, metabolism and pharmacodynamics of nicotine. In: *Nicotine Psychopharmacology: Molecular, Cellular and Behavioural Aspects*, Wonnacott S, Russell MAH, Stolerman IP (Eds.), Oxford University Press, New York, 1990, 112–157.

5 Kyerematen GA, Vessell ES. Metabolism of nicotine *Drug Metab. Revs.*, 1991: **23**; 3–41.

6 Gorrod JW, Wahren J (Eds.). *Nicotine and Related Alkaloids: absorption, distribution, metabolism and excretion,* Chapman and Hall, London, 1993.

7 Seaton M, Vesell ES. Variables affecting nicotine metabolism. *Pharmac. Ther.*, 1993: **60**; 461–500.

8 Benowitz NL, Jacob III P. Metabolism of nicotine to cotinine studied by a dual stable isotope method. *Clin. Pharmacol. Therap.*, 1994: **56**; 483–493.

9 Murphy PJ. Enzymatic oxidation of nicotine to nicotine-$\Delta^{1'(5')}$-iminium ion. A newly discovered intermediate in the metabolism of nicotine. *J. Biol. Chem.*, 1973: **248**; 2796–2800.

10 Booth J, Boyland E. Enzymic oxidation of $(-)$-nicotine by guinea pig tissue in vitro. *Biochem Pharmacol.*, 1971: **20**; 407–415.

11 Nguyen T-L, Gruenke LD, Castagnoli N. Metabolic *N*-demethylation of nicotine. Trapping of a reactive iminium species with cyanide ion. *J. Med. Chem.*, 1976: **19**; 1168–1169.

12 Nguyen T-L, Gruenke LD, Castagnoli N. Metabolic oxidation of nicotine to chemically reactive intermediates. *J. Med. Chem.*, 1979: **22**; 259–263.

13 Peterson LA, Castagnoli N. Regio- and stereochemical studies on the α-carbon oxidation of (S)-nicotine by cytochrome P-450 model systems. *J. Med. Chem.*, 1988: **31**; 637–640.

14 Nakayama H, Okuda H, Nakashima T, Imaoka S, Funae Y. Nicotine metabolism by rat hepatic cytochrome P-450. *Biochem. Pharmacol.*, 1993: **45**; 2554–2556.

15 Nakajima M, Yamamoto T, Nunoya K-I, Yokoi T, Nagashima K, Inoue K et al. Role of human cytochrome P4502A6 in C-oxidation of nicotine. *Drug Metab. Dispos.*, 1996: **24**; 1212–1217.

16 Nakajima M, Yamamoto T, Nunoya K-I, Yokoi T, Nagashima K, Inoue K et al. Characterisation of CYP2A6 involved in 3′-hydroxylation of cotinine in human liver microsomes. *J. Pharmacol. Exptl. Ther.*, 1996: **277**; 1010–1015.

17 Neurath GB, Dünger M, Orth D. Detection and determination of tautomers of 5′-hydroxynicotine and 2′-hydroxynicotine in smokers' urine. *Med. Sci. Res.*, 1992: **20**; 853–858.

18 Byrd GD, Dull GM, Dobson GP, Caldwell WS, Determination of nicotine-1′,2′- and 1′,5′-iminium ions in aqueous solutions. Proc. 47th Tobacco Chemists Research Conference, Gatlinburg 1993: **47**; 38–39.

19 Gorrod JW, Aislaitner G, Hindmarsh J, Jenner P. Evidence for the presence of nicotine iminium ions in areas of rat brain using the cyanide trapping technique after intraperitoneal administration of nicotine. Poster 68 presented at the 1st Annual SRNT meeting San Diego, 1995.

20 Kyerematen GA, Taylor LH, deBethizy JD, Vesell ES. Pharmacokinetics of nicotine and 12 metabolites in the rat. Application of a new radiometric high performance liquid chromatography assay. *Drug Metab. Dispos.*, 1988: **16**; 125–129.

21 Schepers G, Rustemeier K, Walk R-A, Hackenberg, U. Metabolism of S-nicotine in

noninduced and Aroclor-induced rats. *Eur. J. Drug Metab. Pharmacokinet.*, 1993: **18**; 187–197.

22 Brandänge S, Lindblom L. The enzyme 'aldehyde oxidase' is an iminium oxidase. Reaction with nicotine-$\Delta^{1'(5')}$-iminium ion. *Biochem. Biophys. Res. Comm.*, 1979: **91**; 991–996.

23 Gorrod JW, Hibberd AR. The metabolism of nicotine-$\Delta^{1',(5')}$-iminium ion, in vivo and in vitro. *Eur. J. Drug Metab. Pharmacokinet.*, 1982: **7**; 293–298.

24 Wada E, Bowman ER, Turnbull LB, McKennis H. Norcotinine (desmethylcotinine) as a urinary metabolite of nornicotine. *J. Med. Pharm. Chem.*, 1961: **4**; 21–30.

25 Papadopolous NM. Formation of nornicotine and other metabolites from nicotine in vitro and in vivo. *Can. J. Biochem.*, 1964: **42**; 435–442.

26 Aislaitner G, Li Y, Gorrod JW. In vitro metabolic studies on (−)-(S)-nornicotine. *Med. Sci. Res.*, 1992: **20**; 897–889.

27 Gorrod JW, Sai Y. Recognition of novel artifacts produced during the microsomal incubation of secondary alicyclic amines in the presence of cyanide. *Xenobiotica*, 1997: **27**; 389–399.

28 Neurath GB, Dünger M, Orth D, Pein FG. *Trans*-3′-hydroxycotinine as a main metabolite in urine of smokers. *Int. Arch. Occup. Environ. Health*, 1987: **59**; 199–201.

29 Voncken P, Schepers G, Schäfer K-H. Capillary gas chromatographic determination of *trans*-3′-hydroxycotinine simultaneously with nicotine and cotinine in urine and blood samples. *J. Chromatogr.*, 1989: **479**; 410–418.

30 Voncken P, Rustemeier K, Schepers G. Identification of *cis*-3′-hydroxycotinine as a urinary nicotine metabolite. *Xenobiotica*, 1990: **20**; 1353–1356.

31 Jacob III P, Shulgin AT, Benowitz NL. Synthesis of (3′R,5′S)-*trans*-3′-hydroxycotinine, a major metabolite of nicotine, Metabolic formation of 3′-hydroxycotinine in humans is highly stereoselective. *J. Med. Chem.*, 1990: **33**; 1888–1891.

32 Dagne E, Gruenke L, Castagnoli N. Deuterium isotope effects in the in vivo metabolism of cotinine. *J. Med. Chem.*, 1974: **17**; 1330–1333.

33 Nguyen TL, Dagne E, Gruenke L, Bhargava H, Castagnoli N. The tautomeric structures of 5′-hydroxycotinine, a secondary mammalian metabolite of nicotine. *J. Org. Chem.*, 1981: **46**; 758–760.

34 Dagne E, Castagnoli N. Cotinine-*N*-oxide a new metabolite of nicotine, *J. Med. Chem.*, 1972: **15**; 840–841.

35 Schepers G, Demetriou D, Stabbert R, Diehl B, Seeman JI. 5′-Hydroxycotinine-*N*-oxide, a new nicotine metabolite isolated from rat urine. *Xenobiotica*, 1999: **29**; 793–801.

36 Bowman ER, Hansson E, Turnbull LB, McKennis H, Schmiterlöw CG. Disposition and fate of (−)-cotinine-$H^3$ in the mouse. *J. Pharmacol. Exptl. Ther.*, 1964: **143**; 301–308.

37 Morselli PL, Ong HH, Bowman ER., McKennis H. Metabolism of (±)-cotinine–2-$^{14}$C in the rat. *J. Med. Chem.*, 1967: **10**; 1033–1036.

38 Harke H-P, Schüller D, Fram B, Mauch A. Demethylation of nicotine and cotinine in pigs. *Res. Commun. Chem. Path. Pharmacol.*, 1974: **9**; 595–599.

39 Bowman ER, McKennis H. Studies on the metabolism of (−)-cotinine in the human, *J. Pharmacol. Exptl. Ther.*, 1962: **135**; 306–311.

40 Li Y. The in vitro metabolism of cotinine and related compounds. Thesis presented to London University for the award of PhD. 1995.

41 Li Y, Li NY, Aislaitner G, Gorrod JW. In vitro metabolism of cotinine. *Med. Sci. Res.*, 1992: **20**; 903–904.

42 Li Y, Gorrod JW. *N*-Hydroxymethylnorcotinine, a new primary in vitro metabolite of cotinine. *Xenobiotica*, 1994: **24**; 409–415.

43 McManus KT, deBethizy JD, Garteiz DA, Kyerematen GA, Vessell ES. A new quantitative thermospray LC-MS method for nicotine and its metabolites in biological fluids. *J. Chrom. Sci.*, 1990: **28**; 510–516.

44 Byrd GD, Chang K-M, Greene JM, deBethizy JD. Evidence for urinary excretion of glucuronide conjugates of nicotine, cotinine and *trans*-3′-hydroxycotinine in smokers. *Drug Metab. Dispos.*, 1992: **20**; 192–197.

45 Eldirdiri N, Ulgen M, Jacob III P, Gorrod JW. The in vitro metabolism of norcotinine and related biotransformation products by microsomal preparations. *Eur. J. Drug Metab. Pharmacokinet.*, 1997: **22**; 385–390.

46 Kyerematen GA, Morgan ML, Chattopadhyay B, deBethizy JD, Vessell ES. Disposition of nicotine and eight metabolites in smokers and non-smokers: Identification in smokers of two metabolites that are longer lived than cotinine. *Clin. Pharmacol. Ther.*, 1990: **48**; 641–651.

47 Gorrod JW, Li Y, Castagnoli N. Is 5′-hydroxynorcotinine a metabolite of nicotine? 1995 Poster 11 presented at the 1st Annual SNRT meeting San Diego.

48 Sai Y, Gorrod JW. Metabolic interconversion of acid metabolites of cotinine in vitro. *Drug Metab Drug Interact.*, 1995: **12**; 93–103.

49 Bowman ER. Studies on the excretion of 5-(3-pyridyl)-tetrahydrofuranone-2 and its intermediary role in the metabolism of nicotine. *Virginia J. Science*, 1968: **19**; 115–121.

50 McKennis H, Bowman ER, Turnbull LB. Mammalian degradation of (−)-nicotine to 3-pyridylacetic acid and other compounds. *Proc. Soc. Exp. Biol. Med.*, 1961: **107**; 145–148.

51 McKennis H, Schwartz SL, Bowman ER. Alternate routes in the metabolic degradation of the pyrrolidine ring of nicotine. *J. Biol. Chem.*, 1964: **239**; 3990–3996.

52 McKennis H, Schwartz SL, Turnbull LB, Tamaki E, Bowman ER. The metabolic formation of γ-(3-pyridyl)-γ-hydroxybutyric acid and its possible intermediary role in mammalian metabolism of nicotine. *J. Biol. Chem.*, 1964: **239**; 3981–3989.

53 Turnbull LB, Bowman ER, McKennis H. The metabolism of 4-(3-pyridyl)-4-oxo-butyric acid. *Fed. Proc. Fed. Amer. Soc. Expl. Biol. Med.*, 1958: **17**; 325.

54 McKennis H. Disposition and fate of nicotine in animals. In: *Tobacco Alkaloids and Related Compounds*, von Euler US (Ed.), Pergamon Press. Oxford, 1965, 53–74.

55 Beckett AH, Gorrod JW, Jenner P. The analysis of nicotine-1′-*N*-oxide, in the presence of nicotine and cotinine, and its application to the study of in vivo nicotine metabolism in man. *J. Pharm. Pharmacol.*, 1971: **23**; 55S–61S.

56 Papadopolous NM. Nicotine-1′-oxide: A metabolite of nicotine in animal tissues. *Arch. Biochem. Biophys.*, 1964: **106**; 182–185.

57 Booth J, Boyland E. The metabolism of nicotine into two optically-active stereoisomers of nicotine-1′-oxide by animal tissues in vitro and by cigarette smokers. *Biochem. Pharmacol.*, 1970: **19**; 733–742.

58 Jenner P, Gorrod JW, Beckett AH. Species variation in the metabolism of R-(−)- and S-(+)-nicotine by α-*C*- and *N*-oxidation in vitro. *Xenobiotica*, 1973: **3**; 573–580.

59 Hibberd AR, Gorrod JW. Comparative *N*-oxidation of nicotine and cotinine by hepatic microsomes. In: *Biological Oxidation of Nitrogen in Organic Molecules; Chemistry, Toxicology and Pharmacology*. Gorrod JW, Damani LA (Eds.), Ellis Horwood, Chichester, 1985, pp. 246–250.

60 Damani LA, Pool WF, Crooks PA, Kaderlik RK, Ziegler DM. Stereoselectivity in the N'-oxidation of nicotine isomers by flavin-containing monooxygenase. *Mol. Pharmacol.*, 1988: **33**; 702–705.

61 Hines RN, Cashman JR, Philpot RM, Williams DE, Ziegler DM. The mammalian flavin-containing monooxygenases: Molecular characterization and regulation of expression. *Toxicol. Appl. Pharmacol.*, 1994: **125**; 1–6.

62 Cashman JR, Park SB, Yang Z-C, Wrighton SA, Jacob III P, Benowitz NL. Metabolism of nicotine by human liver microsomes: Stereoselective formation of *trans*-nicotine-*N'*-oxide. *Chem. Res. Toxicol.*, 1992: **5**; 639–646.

63 Park SB, Jacob III P, Benowitz NL, Cashman JR. Stereoselective metabolism of (S)-(−)-nicotine in humans: Formation of *trans*-(S)-(−)-nicotine-*N*-1'-oxide. *Chem. Res. Toxicol.*, 1993: **6**; 880–888.

64 Dajani RM. Gorrod JW, Beckett AH. In vitro hepatic and extra-hepatic reduction of (−)-nicotine-1'-*N*-oxide in rats. *Biochem. Pharmacol.*, 1975: **24**; 109–117.

65 Jenner P, Gorrod JW, Beckett AH. The absorption of nicotine-1'-*N*-oxide and its reduction in the gastrointestinal tract in man. *Xenobiotica*, 1973: **3**; 341 349.

66 Bowman ER, Chang RSL, Sprouse CT, McKennis H. *N*-3-Pyridylacetylglycine as a nicotine metabolite. Abstract 50; 27th Tobacco Chemists' Research Conference, Medical College of Virginia, 1973.

67 McKennis H. Experimental considerations of some alternate routes in the mammalian metabolism of nicotine. *Beitr. Tabakforsch. Int.*, 1976: **8**; 288–289.

68 Truhaut R, de Clercq M. Sur le métabolisme de la nicotine. Elimination de ses dérivés de dégradation chez le lapin et chez le rat. *Bull. Soc. Chim. Biol.*, 1959: **41**; 1693–1705.

69 Shen W-C, Greene KM, van Vunakis H. Detection by radioimmunoassay of nicotinamide nucliotide analogues in tissues of rabbits injected with nicotine and cotinine. *Biochem. Pharmacol.*, 1977: **26**; 1841–1846.

70 Hibberd AR, Gorrod JW. Enzymology of the metabolic pathway from nicotine to cotinine, in vitro. *Europ. J. Drug Metab. Pharmacokinet.*, 1983: **8**; 151–162.

71 Schepers G, Demetriou D, Rustemeier K, Voncken P, Diehl B. Nicotine phase 2 metabolites in human urine – structure of metabolically formed *trans*-3'-hydroxy-cotinine glucuronide. *Med. Sci. Res.*, 1992: **20**; 863–865.

72 Shigenaga MK, Trevor AJ, Castagnoli N. Metabolism-dependent covalent binding of (S)-[5-$^3$H]Nicotine to liver and lung microsomal macromolecules. *Drug Metab. Dispos.*, 1988: **16**; 397–402.

73 Kim BH, Shigenaga MK. Metabolism-dependent covalent binding of S-(levo)-tritiated nicotine to lung microsomes in vitro, *Arch. Pharmacal. Res. (Seoul)*, 1993: **16**; 89–93.

74 Obach RS, van Vunakis H. Non-metabolic covalent binding of nicotine-$\Delta^{1'(5')}$-iminium ion to liver microsomes and sulfhydryl-containing polyamino acids. *Biochem. Pharmacol.*, 1988: **37**; 4601–4604.

75 McKennis H, Turnbull LB, Bowman ER. *N*-Methylation of nicotine and cotinine in vivo. *J. Biol. Chem.*, 1963: **238**; 719–723.

76 Mousa S, van Loon GR, Houdi AA, Crooks PA. High-performance liquid chromatography with electrochemical detection for the determination of nicotine and *N*-methylnicotinium ion. *J. Chromatogr.*, 1985: **347**; 405–410.

77 Cundy KC, Crooks PA. High-performance liquid chromatographic method for the determination of *N*-methylated metabolites of nicotine. *J. Chromatogr.*, 1984: **306**; 291–301.

78 Nwosu CG, Godin CS, Houdi AA, Damani LA, Crooks PA. Enantioselective metabolism during continuous administration of S-(−)- and R-(+)-nicotine isomers to guinea-pigs. *J. Pharm. Pharmacol.*, 1988: **40**; 862–869.

79 Cundy KC, Sato M, Crooks PA. Stereospecific in vivo N-methylation of nicotine in the guinea pig. *Drug Metab. Dispos.*, 1985: **13**; 175–185.

80 Sato M, Crooks PA. *N*-Methylnornicotinium ion, a new in vivo metabolite of R-(+)-nicotine. *Drug Metab. Dispos.*, 1985: **13**; 348–353.

81 Pool WF, Houdi AA, Damani LA, Layton WJ, Crooks PA. Isolation and characterization of *N*-methyl-*N'*-oxonicotinium ion, a new urinary metabolite of R-(+)-nicotine in the guinea pig. *Drug Metab. Dispos.*, 1986: **14**; 574–579.

82 Nwosu CG, Crooks PA. Species variation and stereoselectivity in the metabolism of nicotine enantiomers. *Xenobiotica*, 1988: **18**; 1361–1372.

83 Houdi AA, Godin CS, Crooks PA. Effect of continous adminstration of nicotine on urinary histamine and *N*-methylhistamine levels in the guinea pig. *Toxicol. Lett.*, 1988: **44**; 161–166.

84 Pool WF, Crooks PA. Biotransformation of primary nicotine metabolites. I. In vivo metabolism of R-(+)-[$^{14}$C-NCH$_3$]*N*-methylnicotinium ion in the guinea pig. *Drug Metab. Dispos.*, 1985: **13**; 578–581.

85 Cundy KC, Crooks PA, Godin CS. Remarkable substrate-inhibitor properties of nicotine enantiomers towards a guinea pig lung aromatic azaheterocycle *N*-methyltransferase. *Biochem. Biophys. Res. Commun.*, 1985: **128**; 312–316.

86 Cundy KC, Godin CS, Crooks PA. Stereospecific in vitro *N*-methylation of nicotine in guinea pig tissues by an S-adenosylmethionine-dependent *N*-methyltransferase. *Biochem. Pharmacol.*, 1985: **34**; 281–284.

87 Crooks PA, Godin CS. *N*-Methylation of nicotine enantiomers by human liver cytosol. *J. Pharm. Pharmacol.*, 1988: **40**; 153–154.

88 Crooks PA, Sherratt AJ, Godin CS. Oxidative metabolism of nicotine enantiomers in isolated guinea pig hepatocytes. *Med. Sci. Res.*, 1992: **20**; 909–911.

89 Kyerematen GA, Owens GF, Chattopadhyay B, deBethizy JD, Vessell ES. Sexual dimorphism of nicotine metabolism and distribution in the rat. Studies in vivo and in vitro. *Drug Metab. Disp.*, 1988: **16**; 823–828.

90 Damani LA, Shaker MS, Godin CS, Crooks PA, Ansher SS, Jakoby WB. The ability of amine *N*-methytransferases from rabbit liver to *N*-methylate azaheterocycles. *J. Pharm. Pharmacol.*, 1986: **38**; 547–550.

91 Crooks PA, Godin CS, Damani LA, Ansher SS, Jakoby WB. Formation of quaternary amines by *N*-methylation of azaheterocycles with homogeneous amine *N*-methytransferases. *Biochem. Pharmacol.*, 1988: **37**; 1673–1677.

92 Cundy KC, Crooks PA. Biotransformation of primary nicotine metabolites II. Metabolism of [$^3$H]-S-(−)-cotinine in the guinea pig: Determination of in vivo urinary metabolites by high-performance liquid-chromatography. *Xenobiotica*, 1987: **17**; 785–792.

93  Axelrod J. Enzymatic formation of morphine and nicotine in a mammal. *Life Sci.*, 1962: **1**; 29–30.

94  Curvall M, Kazemi Vala E, Englund G, Enzell CR. Excretion of nicotine and some major metabolites in urine of smokers and snuff-users. German Research Council on Smoking and Health, 6th Scientific Symposium, Titisee, Germany, February 26–28, 1989.

95  Curvall M, Kazemi Vala E, Englund G, Enzell CR. Urinary excretion of nicotine and its major metabolites. Abstract 54; 43rd Tobacco Chemists' Research Conference, Richmond, VA, USA, September 29–October 3,1989.

96  Kyerematen GA, Morgan ML, Chattopadhyay B, deBethizy JD, Vesell ES. Disposition of nicotine and eight metabolites in smokers and nonsmokers: Identification in smokers of two metabolites that are longer lived than cotinine. *Clin. Pharmacol. Ther.*, 1990: **48**; 641–651.

97  Curvall M, Kazemi Vala E, Englund G. Conjugation pathways in nicotine metabolism. In: *Effects of Nicotine on Biological Systems*. Adlkofer F, Thurau K (Eds.), Birkhauser Verlag, Basel, 1991, pp. 69–75.

98  Byrd GD, Chang K-M, Greene JM, deBethizy JD. Evidence for urinary excretion of glucuronide conjugates of nicotine, cotinine, and *trans*-3′-hydroxycotinine in smokers. *Drug Metab. Dispos.*, 1992: **20**; 192–197.

99  Rustemeier K, Demetriou D, Schepers G, Voncken P. High-performance liquid chromatographic determination of nicotine and its urinary metabolites via their 1,3-diethyl-2-thiobarbituric acid derivatives. *J. Chromatogr*, 1993: **613**; 95–103.

100  Andersson G, Björnberg G, Curvall M. Oral mucosal changes and nicotine disposition in users of Swedish smokeless tobacco products: A comparative study. *J. Oral Pathol. Med.*, 1994: **23**; 161–167.

101  Mariner DC, Moore J, Cornelissen K, Sinclair NM. Urinary recovery of nicotine and five metabolites after smoking conventional cigarettes. *Med. Sci. Res.*, 1992: **20**; 861–862.

102  Benowitz NL, Jacob III P., Fong I, Gupta S. Nicotine metabolic profile in man: Comparison of cigarette smoking and transdermal nicotine. *J. Pharmacol. Exp. Therap.*, 1994: **268**; 296–303.

103  Byrd GD, Uhrig MS, deBethizy JD, Caldwell WS, Crooks PA, Ravard A, Riggs RM. Direct determination of cotinine-*N*-glucuronide in urine using thermospray liquid chromatography/mass spectrometry. *Biol. Mass Spectrom.*, 1994: **23**; 103–107.

104  Seaton MJ, Kyerematen GA, Vesell ES. Rates of excretion of cotinine, nicotine glucuronide, and 3-hydroxycotinine glucuronide in rat bile. *Drug Metab. Dispos.*, 1993: **21**; 927–932.

105  Caldwell WS, Greene JM, Byrd GD, Chang KM, Uhrig MS, deBethizy JD et al. Characterization of the glucuronide conjugate of cotinine: A previously unidentified major metabolite of nicotine in smokers' urine. *Chem. Res. Toxicol.*, 1992: **5**; 280–285.

106  Crooks PA, Bhatti BS, Ravard A, Riggs RM, Caldwell WS. Synthesis of *N*-glucuronic acid conjugates of cotinine, *cis*-3-hydroxycotinine and *trans*-3-hydroxycotinine. *Med. Sci. Res.*, 1992: **20**; 881–883.

107  Seaton MJ, Vesell ES, Luo H, Hawes EM. Identification of radiolabeled metabolites of nicotine in rat bile. Synthesis of S-(−)-nicotine *N*-glucuronide and direct separation of nicotine-derived conjugates using high-performance liquid chromatography. *J. Chromatogr.*, 1993: **621**; 49–53.

108 Schepers G, Demetriou D, Stabbert R., Diehl B. Isolation and identification of nicotine phase 2 metabolites from smokers' urine. Abstract A33; Society for Research on Nicotine and Tobacco. Second Annual Scientific Conference. Washington, DC, March 15–17,1996.

109 Byrd GD, Caldwell WS, Beck DJ, Kaminski DL, Li AP. Profile of nicotine metabolites from human hepatocytes. Abstract 173; ISSX Proceedings. Volume 6. Sixth North American ISSX Meeting Raleigh, NC. October 23–27,1994.

110 Byrd GD, Caldwell WS, Crooks PA, Ravard A, Bhatti BS. Direct determination of nicotine-*N*-glucuronide in human biological samples. Abstract P131; In: *International Symposium on Nicotine: The Effects of Nicotine on Biological Systems II.* The abstracts, Clarke PBS, Quik M, Thurau K, Adlkofer F (Eds.). Birkhauser Verlag, Basel, 1994.

111 Tsai M-C, Gorrod JW. Evidence for the biosynthesis of a glucuronide conjugate of (S)-( − )-nicotine, but not cotinine or *trans*-3′-hydroxycotinine by marmoset hepatic microsomes. *Drug Metab. Drug Interact.*, 1999; **15**.

112 Mackenzie PI, Owens IS, Burchell B, Bock KW, Bairoch A, Bélanger A et al. The UDP-glycosyltransferase gene superfamily: recommended nomenclature update based on evolutionary divergence. *Pharmacogenetics*, 1997: **7**; 255–269.

*Chapter 4*

# Chemical properties of nicotine and other tobacco-related compounds

Peter A. Crooks

*Division of Pharmaceutical Sciences, University of Kentucky, Lexington, KY 40536-0091, USA*

## I. SCOPE

In this chapter, the chemical properties of 15 of the most important constituents of tobacco will be discussed. The structural relationships between these alkaloids, their stereochemistry, and the occurrence of enantiomeric forms in nature will also be addressed. Information on the physico-chemical properties of the alkaloids, i.e. basicity (pKa), and log P values the alkaloids will be provided, and the significance of these properties with regard to biological activity will be discussed. A section on the chemical properties of the alkaloids will deal with electronic structure, oxidation and reduction, *N*-alkylation and dealkylation reactions, *N*-acylation reactions, reactions, pyridine ring substitution reactions, pyrrolidine ring substitution reactions, ring opening reactions, resolution and racemization procedures, and other miscellaneous reactions. A final section will deal with the total synthesis of nicotine and related alkaloids via enantioselective pathways, and the preparation of their isotopic forms.

## II. STRUCTURAL CHEMISTRY OF THE MAJOR TOBACCO ALKALOIDS

### A. The major tobacco alkaloids

The plant genus *Nicotiana*, of which there are over 60 known species, is rich in alkaloids, nicotine being the principal alkaloidal constituent in well over half of these species. These alkaloids are generally 3-pyridyl compounds, as a consequence of their biogenesis from nicotinic acid (see Chapter 2).

Analysis of leaf alkaloidal extracts within the species indicates that the major alkaloids are nicotine, nornicotine, anatabine, and anabasine. It is important to note that the *N*-demethylated alkaloid, nornicotine is the major alkaloid in 30 to 40% of *Nicotiana* species [1]. Most of the other alkaloids are present in the plant in minor amounts (less than 50 μg/g dry weight) and many are present in nanogram quantities. The chemical structures of the 3-pyridyl alkaloids present in *Nicotiana* species are illustrated in Fig. 1.

Nicotine (**1**), is the most abundant tobacco alkaloid, and exists in the S-(−)-enantiomeric form in nature, although recent studies indicate that very small amounts (0.1–0.4%) of the R-(+)-isomer may be present in *Nicotiana*

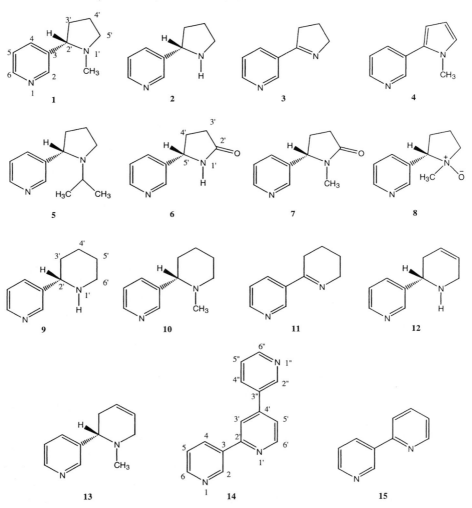

Fig. 1 Major 3-pyridyl alkaloids in tobacco.

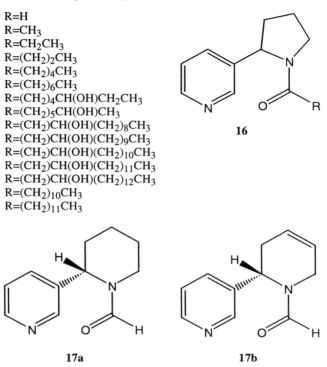

R=H
R=CH$_3$
R=CH$_2$CH$_3$
R=(CH$_2$)$_2$CH$_3$
R=(CH$_2$)$_4$CH$_3$
R=(CH$_2$)$_6$CH$_3$
R=(CH$_2$)$_4$CH(OH)CH$_2$CH$_3$
R=(CH$_2$)$_5$CH(OH)CH$_3$
R=(CH$_2$)CH(OH)(CH$_2$)$_8$CH$_3$
R=(CH$_2$)CH(OH)(CH$_2$)$_9$CH$_3$
R=(CH$_2$)CH(OH)(CH$_2$)$_{10}$CH$_3$
R=(CH$_2$)CH(OH)(CH$_2$)$_{11}$CH$_3$
R=(CH$_2$)CH(OH)(CH$_2$)$_{12}$CH$_3$
R=(CH$_2$)$_{10}$CH$_3$
R=(CH$_2$)$_{11}$CH$_3$

**16**

**17a**                    **17b**

Fig. 1 Continued.

plant species (see Chapter 2). In this respect, it is important to note that smokers are exposed to both enantiomers of nicotine, due to pyrolytic racemization of the S-(−)-isomer during smoking. As much as 11% of R-(+)-nicotine has been found in smoke condensate from cigarettes [2]. Nicotine is a 2-(3-pyridyl)-substituted analog of N-methylpyrrolidine and is related both structurally and biosynthetically to nornicotine (**2**), myosmine (**3**), β-nicotyrine (**4**), N′-isopropylnornicotine (**5**), norcotinine (**6**) and cotinine (**7**). The 5-membered heterocyclic ring in these alkaloids is derived from either L-ornithine or L-arginine in the plant. Unlike nicotine, nornicotine, the second most abundant tobacco alkaloid, exists in tobacco as a mixture of S-(−)- and R-(+)- isomers in nearly equal amounts, with the R-enantiomer usually predominating [3,4]. The presence of both enantiomeric forms results from the fact that nornicotine is biosynthesized in the plant from nicotine via an N-demethylase enzyme; which catalyses the formation of the optical isomers of nornicotine. The mechanism of demethylation has not been elucidated (Chapter 2). However, biosynthesis studies with 2′-[³H]-nicotine show that the 2′-label is lost during formation of R-(+)-nornicotine, but is retained in the formation of the S-(−)-isomer [4], suggesting the formation of a planar intermediate in the formation of the former isomer, and the possible existence of two distinct *N*-demethylase enzymes. Nicotine-*N*-1'-oxide (**8**) is a constituent of tobacco products, but it is generally believed that it originates mainly from aerial oxidation of S-(−)-nicotine during the drying and/or curing of tobacco leaf. Nicotine *N*-1'-oxide can exist in two diastereomeric forms, i.e. a 1'R,2'S-isomer and a 1'S,2'S-isomer, due to the chiral nature of the N'-atom in this molecule

The alkaloid S-(−)-anabasine (**9**), is representative of a second group of tobacco alkaloids, which includes S-(−)-*N*-methylanabasine (**10**), anabaseine (**11**), S-(−)-anatabine (**12**), S-(−)-*N*-methylanatabine (**13**), nicotelline (**14**), and 2,3'-bipyridyl (**15**). In these alkaloids, both 6-membered heterocyclic rings are derived biosynthetically from nicotinic acid in the plant. Similar to nicotine and nornicotine, anabasine, *N*-methylanabasine, anabaseine, and anatabine, all possess neuronal nicotinic receptor agonist activity [5], and anabaseine is a potent neurotoxin [6]. Several other minor alkaloids are worthy of mention; a wide variety of *N*-acyl analogs of nornicotine (**16**) have been detected in tobacco, ranging from acetyl to dodecanoyl, and are presumably formed through conjugation of nornicotine with fatty acids. A number of nor-alkaloids also exist in small amounts as their *N*-formylated derivatives (**17a** and **17b**) [1].

## B. Crystal and solution structure

In all but one (nicotelline) of the alkaloid structures illustrated in Fig. 1, the compounds consist of two heterocyclic rings connected by a 3-(pyridyl),2'-bond (or in the case of the cyclic lactam alkaloids **6** and **7**, a 3-(pyridyl)-5'-bond. In those compounds that are fully aromatic (i.e. β-nicotyrine and 2,3'-bipyridyl), the rings are planar, and the only degree of conformational freedom in the molecule is the rotation about the bond that connects the two aromatic heterocycles. However, in the majority of alkaloids, one heterocyclic ring is aromatic, and the other one is saturated, or contains one double bond. In these compounds, in addition to rotameric freedom between the two heterocyclic rings, there is the additional conformational flexibility within the saturated ring.

The nicotine molecule in its non-protonated form can exist in a *cis-trans* isomerism of the *N*-methyl and 2'-pyridyl moieties about the N–1'-C-2' bond axis, due to the pyrimidal inversion of the pyrrolidine nitrogen (see Fig. 2). X-ray crystallographic studies on the dihydriodide salt [7], indicate that in the solid diprotonated form, nicotine exists in the *transoid* configuration, while molecular modeling studies [8,9] conclude that nicotine free base exists predominantly as the *trans*-invertomer in the gas phase (*trans/cis* ratio 10 : 1). Chynoweth et al. [11] reported that the preferred form of nicotine in solution was the *cis*-invertomer, based upon a proton magnetic resonance

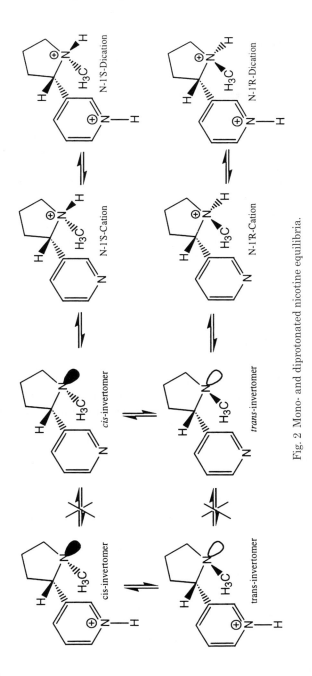

Fig. 2 Mono- and diprotonated nicotine equilibria.

nuclear Overhauser effect observed for the $N$-methyl group and the pyridyl $C_2$-H and $C_4$-H. However, this study did not take into account the rate of nitrogen inversion, which is relatively much faster than $1/T_1$, where $T_1$ is the proton relaxation time, thus compromising the interpretation of the experimental data. In subsequent NMR studies, Seeman et al. (for a comprehensive treatise on this topic see reference [14]) have established that the *trans*-invertomer of nicotine predominates in solution to the extent of greater than 90 percent. Information from circular dichroism studies [10], and from molecular orbital calculations [8], suggest that the conformer most favored, is the one where the C-2′–H-2 bond and the pyridine ring are coplanar, making the two rings perpendicular. NMR paramagnetic shift measurements suggest that pyrrolidine ring adopts an envelope conformation when nicotine is complexed with shift reagents [12]. With the aid of variously deuterated nicotines, Pitner et al. [13] have measured the vicinal coupling constants of the pyrrolidine ring protons of nicotine in solution. They have determined that an envelope conformation of the five-membered ring predominates, in which the methyl and pyridine moieties are equitorial, and the pyrrolidine and pyridine rings are juxtaposed in a perpendicular spatial orientation. Seeman [14] has concluded that nicotine exists in solution primarily in a conformation in which the $N'$-methyl group is *trans* to the pyridine ring, the pyrrolidine ring is in an envelope conformation, and the relative orientation of the pyridine and pyrrolidine rings is essentially orthoganol (see Chapter 5 for additional discussion of the conformation of nicotine).

In the case of the cyclic imino alkaloids myosmine (**18a**) and anabaseine (**18b**), the five- and six-membered cylic imino moieties in these compounds hydrolyze in aqueous solution, to afford the corresponding open-chain amino ketone **19a** and **19b**, respectively [15]. Zoltewicz et al. have shown that both anabaseine and myosmine exist in solution over the pH range 2–10 as a rapidly equilibrating set of four species (Scheme 1), which include the free base (**18**), the conjugate iminium ion acid (**20**), the open chain keto alkylammoniun ion (**19**), and the pyridinium-ammonium dication (**21**). In the case of anabaseine, at low pH, the dication is the predominant species, while at high pH the free base is the major component; at neutral or physiological pH, the two mono cations are essentially the only species present in solution, with the open chain cation slightly predominating [15]. In the case of myosmine, the conjugate iminium ion species (**20a**) is more acidic than its anabaseine counterpart (**20b**) by 1.5 pKa units, which is surprising when one considers the similarity of the pKas of the free bases. This has a significant effect on the solution equilibria, and at pH 7, about 80% of myosmine exists in solution as the cyclic imine free base (**18a**), with 7% as the conjugate iminium species (**20a**), and 13% as the keto alkylammonium ion (**19a**). Thus, much more of the cyclic free base of myosmine will be present at physiological pH compared to that of anabaseine.

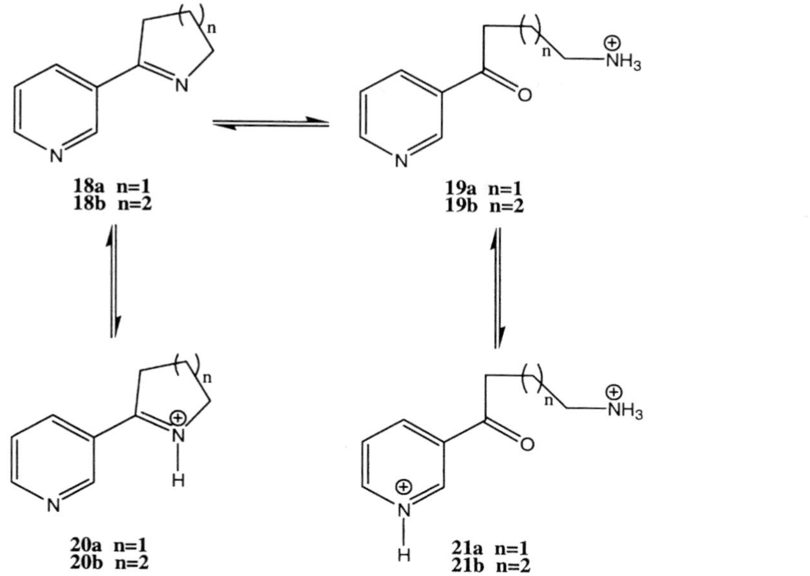

18a n=1
18b n=2

19a n=1
19b n=2

20a n=1
20b n=2

21a n=1
21b n=2

Scheme 1. Solution equilibria of cyclic and open-chain forms of myosmine (**18a**) and anabaseine (**18b**).

## III. PHYSICO-CHEMICAL PROPERTIES OF NICOTINE AND RELATED COMPOUNDS

### A. Basicity

The tobacco alkaloids are classified as weak organic bases. Nicotine contains two tertiary amino centers in its molecule, which have reported relative basicities (pKas) of approximately 3.0–4.0 (pyridino $N$) and 8.0–9.0 (pyrrolidino $N$) [16] (see Chapter 5 for a table of pKa values that have been reported in the literature). The relatively lower pKa value for the aromatic pyridino $N$ is due to the effect of $sp^2$ hybridization on base strength, since as the 's' character of an orbital increases, the electrons in that orbital are bound more tightly to the nucleus. Table 1 provides calculated [17–19] values for the pKas and log Ps of the tobacco alkaloids. Generally, the calculated values are in line with experimentally determined values (see Chapter 5). The two pKa values calculated for nicotine, nornicotine, anabasine, $N$-methylanabasine, anatabine, $N$-methylanatabine, and $N$-isopropylnornicotine are comparable, with the nor-alkaloid exhibiting a slightly higher pKa, compared to its corresponding $N$-alkylated analogue, for the non-aromatic amino group. However, as conjugation in the molecule increases, the basicities of the alkaloids decrease. For example, for both myosmine and anabaseine, the pyridino pKa is only 1.13, and the imino pKa is reduced to 7.81 and 7.46, respectively. The fully aromatic alkaloids nicotelline and 2,3′-bipyridine are

75

very weak bases (pKas 1.23 and 3.25, respectively) and would not be protonated at physiological pH. In the β-nicotyrine molecule, an *N*-methylpyrrole ring has replaced the usual *N*-methylpyrrolidino moiety, resulting in loss of basicity in this structure. This is due to the aromatic nature of the pyrrole ring, since the nitrogen lone pair in this structure is no longer available for proton sequestration, as it is incorporated into the 6-electron aromatic orbital of the pyrrole ring. In the alkaloids norcotinine and cotinine, the pyrrolidine ring has also lost its basicity due to conversion of the cyclic amino group into a *gamma*-lactam structure. The *gamma*-lactam nitrogen in cotinine is a neutral moiety, thus cotinine has a calculated pKa of 4.72 (pyridino), and is a much weaker base than nicotine. However, in norcotinine, the secondary *gamma*-lactam group is weakly acidic, and the *N'*-proton can be removed by strong bases.

TABLE 1

Calculated pKa and log P values of tobacco alkaloids

| Alkaloid | pKa | log P |
|---|---|---|
| Nicotine | 9.13 | 0.93 |
|  | 4.23 |  |
| Nornicotine | 9.83 | 0.50 |
|  | 4.23 |  |
| Myosmine | 7.81 | 0.70 |
|  | 1.13 |  |
| β-Nicotyrine | ND | 1.00 |
| *N*-isopropylnor-nicotine | 9.27 | 1.59 |
|  | 4.23 |  |
| Norcotinine | 15.27 | 0.20 |
|  | 4.72 |  |
|  | 0.06 |  |
| Cotinine | 4.72 | 0.04 |
|  | −0.41 |  |
| Nicotine *N*-1'-oxide | 4.63 | ND |
|  | 4.23 |  |
| Anabasine | 9.86 | 0.96 |
|  | 4.21 |  |
| *N*-Methylanabasine | 8.53 | 1.35 |
|  | 4.21 |  |
| Anabaseine | 7.46 | 15 |
|  | 1.13 |  |
| Anatabine | 8.77 | 0.93 |
|  | 4.13 |  |
| *N*-Methylanatabine | 7.67 | 1.17 |
|  | 4.13 |  |
| Nicotelline | 1.23 | 1.63 |
| 2,3'-bipyridyl | 3.25 | 1.31 |
|  | ND |  |

Nicotine and other tobacco alkaloids that possess a basic pyrrolidino nitrogen readily form stable salts with both mineral and organic acids. The hydrochloride, dihydrochloride, perchlorate and hemisulfate salts of nicotine have been described, and all are deliquescent crystalline compounds exhibiting high solubility in polar solvents such as water and ethanol and insolubility in solvents of low polarity, such as diethyl ether. Several organic acid salts of nicotine e.g. the bitartrate and salicylate salts, are also stable and less hygroscopic than the mineral salts, and *para*-toluoyl tartrate salts have been utilized in the resolution of racemic nicotine into its pure enantiomeric forms (see Section III G). Perfetti [20] has carried out a detailed structural study of nicotine salts of organic acids, and has shown that three basic types of salts can be formed, which have acid:base ratios of either 1:1, 2:1, or 3:1. The type of salt obtained is dependent upon the structure of the acid (i.e. whether aliphatic or aromatic) and its functionality. Salts with a 1:1 ratio are usually formed from amino acids or benzoic-type acids, and salt formation occurs at the pyrrolidino nitrogen. Salts with a 2:1 acid:base ratio are formed from formic acid, aliphatic dicarboxylic acids, and/ or nitroaromatic acids; salt formation occurs at both the pyrrolidino and the pyridino nitrogen. Nicotine forms 3:1 salts with aliphatic monocarboxylic acids, one acid equivalent is bound to the pyrrolidino nitrogen while the other two acid molecules dimerize and bind to the nitrogen of the pyridine ring. Nicotine bitartrate is generally the salt of choice when nicotine is used in analytical or pharmacological studies, since it is an easily manipulatable free-flowing solid, obtainable in high purity, has a high melting point, has good solubility in water and other protic solvents, and is not deliquescent in nature.

Relatively fewer salts of nornicotine have been reported. As expected, mineral acid salts are extremely deliquescent, however, the diperchlorate, camsylate, and fumarate salts appear to be stable forms of the alkaloid. The HCl, HBr, HI, and acetate salts of anabasine are all hygroscopic; the preparation of nitrate, oxalate, malonate, and naphthalene 2-sulfonate salts are known. *N*-methylanabasine forms stable perchlorate and methiodide salts.

Since nicotine and nicotine analogs that interact with nicotinic acetyl choline receptors do so in their *N*-1'-protonated forms, pKa values of tobacco alkaloids are important criteria, since they determine the relative proportions of free base and protonated species at physiological pH. Nicotine, nornicotine, anabasine, *N*-methylanabasine, anatabine, and *N*-methylanatabine all possess significant agonist activity at nicotinic receptors in functional assays [21–23, 5] and all exist predominantly in their *N*-1'-protonated forms at physiological pH. On the other hand, cotinine is only a weak agonist at these receptors [24], and is predicted not to be protonated at physiological pH. Fig. 2 illustrates the essential equilibria in the protonation of the S-(−)-nicotine molecule. The *N*-1-protonated forms are forbidden, due to the considerably lower basicity of the pyridyl nitrogen compared to the pyrrolidine nitrogen, while the *N*-1, *N*-1'-diprotonated species are formed predominantly in acidic media (pH < 4.0). It is important to note that protonation of the pyrrolidino-*N* atom can occur either at the top face of the nitrogen atom or at the bottom face, via protonation of the respective nitrogen invertomers, resulting in two different configurations at the protonated nitrogen center, and thus two diastereomeric forms of the protonated S-(−)-nicotine molecule. It is also worth pointing out that protonation of nicotine with strong acids is well known to be a diffusion-controlled process, and is thus faster than nitrogen inversion. Determining which one of the two possible *N*-1'-protonated nicotine species is involved in nicotinic receptor binding, as well as a consideration of their solution conformations, will provide valuable data for mapping the topography of the nicotinic receptor, and for accurate pharmacophore development in pharmacological and toxicological studies.

*Partition coefficients*

Table 1 provides information on the log P values of the tobacco alkaloids.

Values range from 1.63 for the very lipophilic nicotelline, to −0.20 for the polar norcotinine molecule. As might be expected, increasing ring size from 5 to 6 carbon atoms, and increasing the size of the *N*-1'-alkyl group, generally leads to an increase in log P. Removal of the pyrrolidino-*N* substituent significantly reduces log P, whereas introduction of a 5'-, or 6'-oxo atom into the molecule dramatically reduces lipophilicity (norcotinine and cotinine). Imino alkaloids generally have higher log P values than the parent nor-alkaloid (e.g. myosmine versus nornicotine, and anabaseine versus anabasine), but have lower values than the corresponding *N*-1'-methyl alkaloid (e.g. myosmine versus nicotine, and anabaseine versus *N*-methylanabasine). Also, increasing the aromatic character in the alkaloid structure generally increases log P.

## IV. CHEMICAL PROPERTIES OF NICOTINE AND RELATED COMPOUNDS

### A. Oxidation reactions

Nicotine free base is relatively unstable both in solution and as the neat liquid, a yellow coloration developing with time, and especially in the presence of light and air; heating solutions of nicotine under these conditions accelerates the discoloration. Aged samples of nicotine free base take on a dark brown coloration and become more viscous in nature. These changes have been attributed to aerial oxidation processes that most likely involve free radical reactions. No studies have been reported on the detailed mechanism of these oxidation processes. However, when an aqueous solution of nicotine is exposed to air/oxygen and sunlight, methylamine, myosmine, and nicotine-*N*-1'-oxide have been identified as degradation products [25,26].

It has also been observed that when solutions containing a mixture of nicotine and cotinine are left to stand for several weeks, the relative amount of cotinine in the sample increases as a function of time [27]. These results suggest that nicotine degradation may be mediated, at least in part, via *N*-1′-oxidation, and oxidation of the pyrrolidino ring carbon(s) *alpha* to the *N*-1′-atom. This instability of nicotine free base may be of some significance in work where very dilute solutions of nicotine are used, e.g. in pharmacological work, or in methodologies involving radiolabeled forms of nicotine. Care should be exercised in making sure that nicotine solutions are freshly prepared, or that previously prepared solutions are checked for homogeneity. The use of nicotine salts is recommended, since this form of nicotine is much more stable to oxidative degradation processes.

Chemical oxidation of nicotine and other tobacco alkaloids, with strong oxidizing agents. such as chromium oxide, potassium permanganate, sodium hypochlorite, nitric acid, or sulfuric acid containing either mercuric sulfate or selenium, results in the formation of products in which oxidative carbon-carbon bond cleavage in the pyrrolidine ring has occurred [25–37]. Nicotinic acid is usually a major product, and this oxidation step was utilized in earlier classical degradation studies to establish that tobacco alkaloids were 3-pyridino compounds. Treatment of nicotine with concentrated nitric acid affords, in addition to nicotinic acid, 48% of a product identified as 3-(5-nitro-1(2)H-pyrazol-3-yl)pyridine (**22**) [38–40].

**22**          **23**          **24**

**25**          **26**

*N-Oxidation.* Oxidation with milder agents, such as hydrogen peroxide, affords *N*-oxide products. Reaction of nicotine with aqueous $H_2O_2$ or *meta*-chloroperoxybenzoic acid affords a mixture of the diastereomeric 1'S,2'S and 1'R,2'S nicotine-*N*-1'-oxides, **23** and **24**, respectively, which can be separated by silica gel chromatography [41–46]. Brandange et al. [47] have reported the synthesis of almost pure samples of these diastereomeric forms (~2% contamination with the other diastereomer) utilizing peroxymolybdate and peroxytungstate reagents. Treatment of nicotine with aqueous $H_2O_2$ and acetic acid forms mainly the *N*-1'oxides when the reaction is run at ambient temperature [43]; however, when the reaction is carried out at 80–90°C, the major product is the *N*-1-, *N*-1'-dioxide (**25**) [43]. Nicotine *N*-1, *N*-1'-dioxide is best prepared by treatment of nicotine with either 30% $H_2O_2$ or perlauric acid [43]; the latter procedure is particularly convenient, since the lauric acid by-product can easily be removed from the reaction mixture by filtration. Selective removal of the *N*-1'-oxo group can be achieved by treatment with sulfur dioxide [43] or titanium chloride [48]; using this procedure, Taylor and Boyer prepared nicotine-*N*-1-oxide (**26**) via selective *N*-deoxygenation of the *N*-1, *N*-1'-dioxide [43]. Recently, Schmidt and Neitmeyer [49] have reported the preparation of nicotine-*N*-1-oxide by treating nicotine with $H_2O_2$/glacial acetic acid and aqueous sodium bisulfite/acetic acid for 30 min. at 0°C, and a 32% yield of nicotine-*N*-1-oxide is afforded when nicotine-*N*-1, *N*-1'-dioxide is incubated with baker's yeast; nicotine (6%) is also obtained as a byproduct [50].

Treatment of S-(−)-cotinine with 35% aqueous $H_2O_2$ in acetic acid at 70°C affords a 90% yield of S-(−)-cotinine-*N*-1-oxide (**27**) [51]. Kisaki et al. have reported the formation of myosmine-*N*-oxide (**28**) from the treatment of nornicotine with aqueous hydrogen peroxide [52]. The *N*-1'-mono (**29**) and *N*-1, *N*-1'-dioxides (**30**) of 2,3'-bipyridyl (ratio 26:1, respectively) have been prepared via reaction with $H_2O_2$/acetic acid [53]; *meta*-chloroperoxybenzoic acid oxidation of 2,3'-bipyridyl affords only the *N*-1'-monooxide **29** in 78% yield [54].

*Pyrrolidino-C oxidation.* Treatment of nicotine with bromine in acetic acid affords good yields of 3',3'-dibromocotinine (**31**) [55–58]. S-(−)-Nicotine can be converted to cotinine in good yield and with retention of configuration by reaction with bromine in acetic acid followed by treatment with aqueous HCl/zinc [58]. Treatment of nicotine with aqueous iodosobenzene, also affords cotinine, but in low yield [59], and when cotinine is reacted with iodosobenzene in the presence of Mn(III)TppCl as catalyst, a mixture of 3'-hydroxycotinine (**32**) and 5'-hydroxycotinine (**33**) is formed in low yield [60]. A mixture of *cis*- and *trans*-3-hydroxycotinines (**34** and **35**, respectively) is obtained when cotinine is oxidized with LDA/oxydiperoxymolybdenum(MoOH) reagent [61], and Lin et al. have reported the preparation of *trans*-3'-hydroxycotinine from the reaction of S-(−)-cotinine with LDA/

(+)-(camphorylsulfonyl)oxaziridine [62]. When nicotine is reacted with sodium periodate and ruthenium oxide in $CCl_4$/water, a mixture of cotinine (25%), nornicotine (10%), and 5′-methoxymyosmine (**36**) (12%) is formed [63], and a mixture of 3′-oxonicotine, 4′-oxonicotine, and cotinine (**37**, **38**, and **39**, respectively) is formed, when nicotine is treated with zinc powder/ pyridine/acetic acid, with iron as catalyst [64]. Wenkert has reported the oxidation of *N*-methylanabasine to *N*-methyl-6-oxoanabasine (**40**) in 93% yield with $Hg^{2+}$/EDTA in dichloromethane [65].

*Pyridino-C-oxidation.* Examples of direct C-oxidation reactions in the pyridine ring of nicotine are relatively rare. Oxygen is usually introduced via intermediate halogenopyridine or diazopyridine derivatives (see later). Oxygen can be introduced directly into the pyridine ring by the action of microorganisms (see below), and the yield of the oxidized product may be quite respectable.

*Dehydrogenation reactions.* High temperature reactions of nicotine in the presence of quartz or zirconium oxide have been reported to afford dehydrogenated products [26,66–73]. Myosmine (**18a**) has been prepared by heating nicotine with quartz at 570°C [69,71]; other by-products in this reaction are β-nicotyrine (**4**), 3-vinylpyridine, and 3-ethylpyridine. When nicotine is reacted with hydrous zirconium oxide at 300°C in the gaseous phase, myosmine and β-nicotyrine are formed [73], and according to Shibagaki et al., when the reaction is conducted at 500°C, a 72.5% yield of β-nicotyrine is obtained [73]. β-Nicotyrine is also formed when nicotine is treated with platinum-asbestos catalyst at 320°C [72], with silver oxide/$H_2O$ [74], with $V_2O_5$ in air [70], or with palladium black/xylene in the presence of 1-methoxy-4-propenylbenzene [66]. Reaction of nicotine with $V_2O_5$ at 450°C in air afforded 3-cyanopyridine [67–69,71]. Johnson has reported that when nicotine is treated with palladium-asbestos at 230-280°C, β-nicotyrine and the ring-opened product, methyl-*bis*-4-(3-pyridylbutyl)amine (**41**) is formed [42]. Oxidation of anabasine or anatabine with silver acetate in acetic acid, zinc powder, or palladium, affords 2,3'-bipyridyl [75–78]. There are several reports on the formation of a dimeric oxidation product from the reaction of nicotine with sulfur. Earlier studies identified the product as 3,3'-(1,1'-dimethyl-5,5'-sulfanediyl-dipyrrol-2-yl)*bis*-pyridine (**42**) [79–81]. However, later studies have shown that this product is the 3,3'-sulfanediyl dimer **43** [82].

*Oxidation by microorganisms.*    There are several reports on the oxidation of tobacco alkaloids by microorganisms. Incubation of an aqueous solution of S-(−)-nicotine with the fungus *Cunninghamella echinulata* at 28°C for 5 days affords a mixture of S-(−)-nornicotine and myosmine (**44**) [83]. Several *Arthrobacter* bacterial species biotransform S-(−)-nicotine into a mixture of S-(−)-2-oxonicotine (**45**) and S-(−)-6-oxonicotine (**46**) [84–86], and incubations of nicotine with extracts of a soil bacterium [85,87] also afford these oxidation products. *Pseudomonas nicotinopharga* biotransforms nicotine into the ring-opened product, 4-(6-hydroxy-3-pyridyl)-4-oxobutyric acid (**47**) [88,89]. Soil bacteria of type B have been shown to convert nornicotine into 6-oxomyosmine (**48**) [90], and incubation of nornicotine and anabasine with *pseudomonas artene* affords 4-(6-hydroxy-3-pyridyl)-4-oxobutyric acid (**47**) and 6-oxoanabaseine (**49**), respectively [90,91].

## B. Reduction reactions

Relatively few reduction reactions of nicotine have been documented. Complete reduction of nicotine to an epimeric mixture of 3S- and 3R-(1-methylpyrrolidin-2S-yl)piperidine (**50** and **51**, respectively) with sodium metal in alcohols, and with hydrogenation over platinum, or palladium-on-charcoal as catalyst, has been reported [92–95]. In the latter reduction, the ring opened product, *N*-methyl-4-(3-piperidinyl)butylamine (**52**) is also formed [96,97]. Catalytic hydrogenation of anabasine with platinum/aqueous HCl, Raney nickel/aqueous NaOH, or copper-nickel catalyst at 200°C, afforded the fully saturated 2S,3'R-dodecahydro-2,3-*bis*-pyridyl product, **53** [98]. This product was also formed from the reaction of anabasine with sodium/butanol [99,100], and from electrochemical reduction of anabasine in aqueous $H_2SO_4$ [100]. Anatabine can be converted into anabasine by

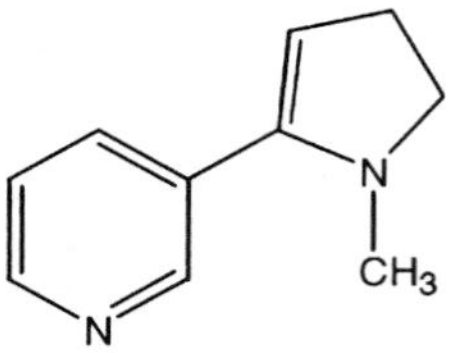

44 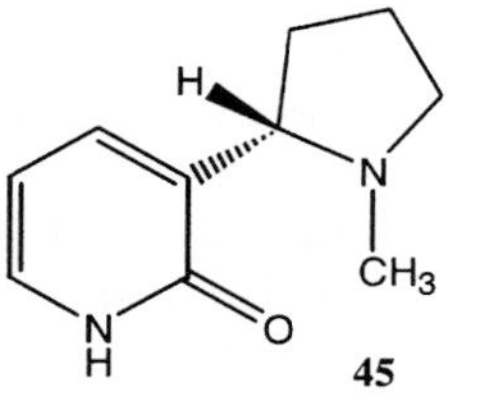   45 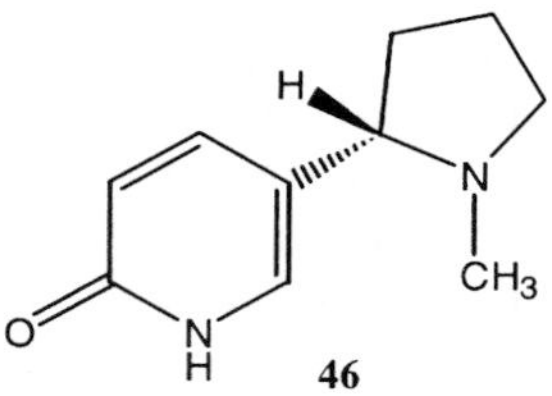   46

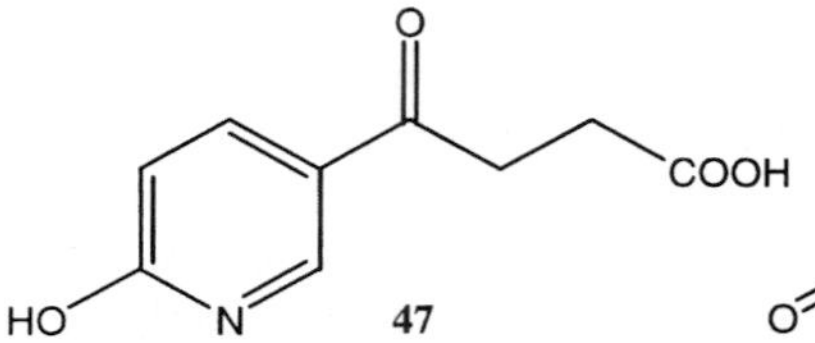

47 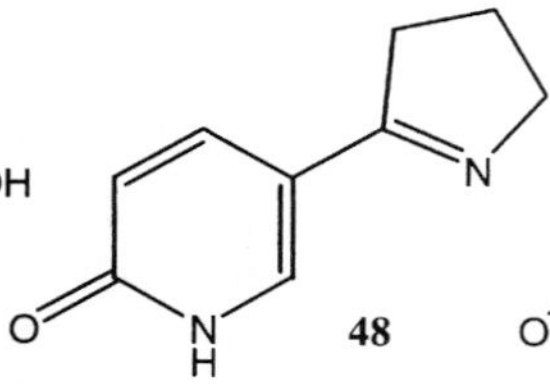   48 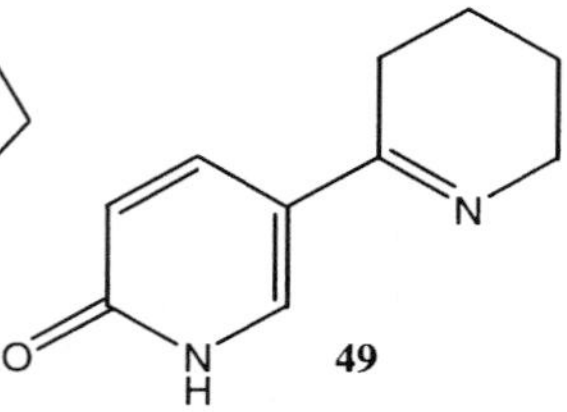   49

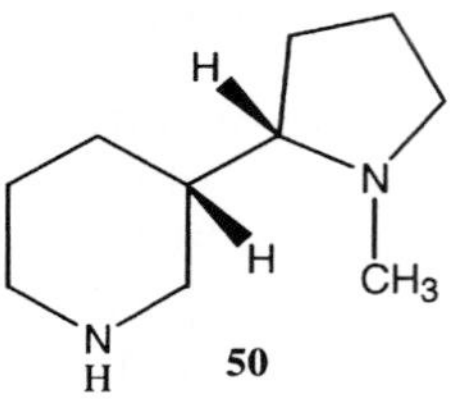

50 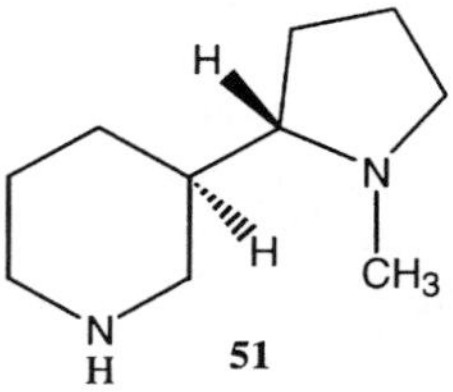   51 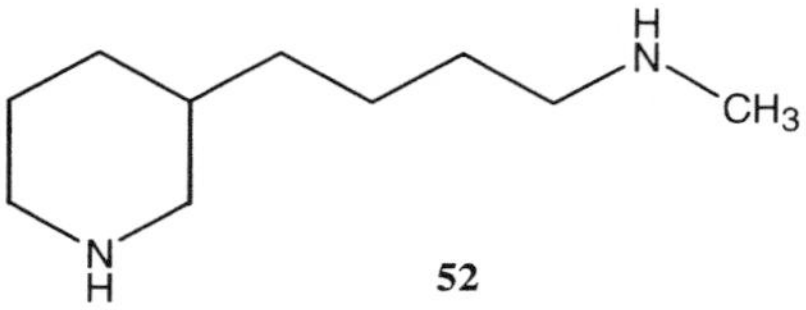   52

hydrogenation with palladium-on-carbon in ethanol [101,102]; this procedure has been used to prepare deuterium-labelled forms of anabasine [102]
(see section V). Myosmine can be reduced to nornicotine with sodium
borohydride [103,104], and reduction of anabaseine with sodium cyanoborohydride in methanol, or by hydrogenation over palladium-on-charcoal
affords high yields of anabasine [101,103]. 2,3′-Bipyridyl is exhaustively
hydrogenated to dodecahydro-2,3-*bis*-pyridyls **53** and **54** with platinum/
acetic acid at 50°C [105]; however, reduction with Ni-Al alloy in 1 M
KOH/methanol, or with tin in aqueous HCl affords the 1′,2′,3′,4′-hexahydro-
2,3′-*bis*-pyridine product **55** [106,107]. Reduction of 2,3′-bipyridyl with
*bis*-(1,4-dihydro-1-pyridyl)zinc *bis*-pyridine complex in THF affords the
partially reduced product **56** [108]. β-Nicotyrine can also be converted into
nornicotine by hydrogenation over palladium-on-charcoal catalyst [109],
although if the hydrogenation is carried out with platinum in acetic acid,
complete reduction of only the pyridine ring occurs to give **57**, and the ring-
opened compound, *N*-methyl-4-(3-piperidinyl)butylamine (**52**) is also formed
[110]. An early report describes the formation of 3-(*N*-methyl-2,5-dihy-
dropyrrol-2-yl)pyridine (**58**) from the reduction of β-nicotyrine with zinc in
aqueous HCl [111].

Reduction of S-(−)-cotinine to S-(−)-nicotine can be achieved with lithium aluminum hydride in THF or diethyl ether [56,112,113]; this reduction procedure retains the configuration of the 2′-carbon, and is a useful procedure for introducing tritium or deuterium into the 5′-position of the nicotine molecule [114] (see section V). In a similar manner, S-(−)-norcotinine can be converted into S-(−)-nornicotine by treatment with either lithium aluminum hydride in THF, or with 1.0 M BH$_3$-THF complex [112,113,115]; the latter procedure affords superior yields of product. Treatment of S-(−)-cotinine with sodium dihydro-*bis*-(2-methoxyethoxy)aluminate in diethyl ether or sodium aluminum hydride in THF [116] is reported to afford 5′-hydroxynicotine (**59**) as the reduction product.

## C. *N*-Demethylation reactions

S-(−)-Nicotine cannot easily be *N*-demethylated to nornicotine, thus precluding a facile synthesis of S-(−)-nornicotine. Consequently, nornicotine enantiomers usually have to be prepared from either resolution of the racemate, or by enantioselective synthesis. Early literature has reported the formation of very low yields of S-(−)-nornicotine from the reaction of S-(−)-nicotine with either silver oxide-water or aqueous potassium

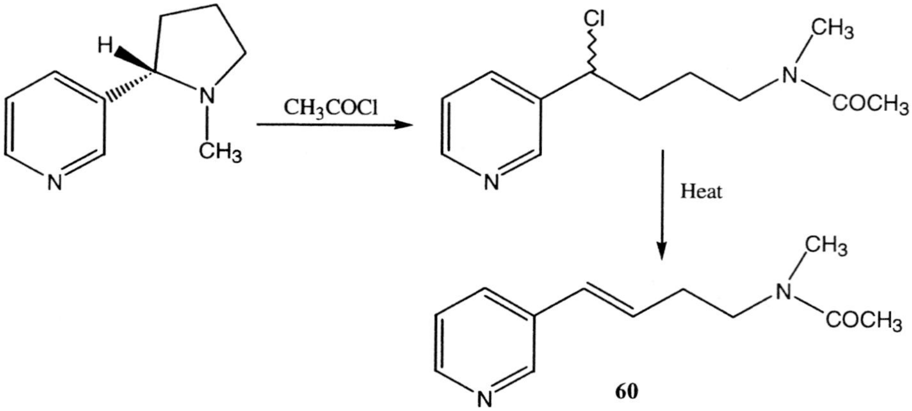

Scheme 2. Formation of metanicotine (**60**) from the reaction of nicotine with acetyl chloride.

permanganate [117]. Acylating agents such as benzoyl chloride, acetic anhydride, and ethyl chloroformate, all afford ring cleavage products. For example, when nicotine is heated with acetyl chloride, *N*-acetylmetanicotine [*N*-acetyl-*N*-methyl-4-(3-pyridinyl)-3-butene-1-amine] (**60**) is formed via the pathway shown in Scheme 2 [118]. A low yielding nicotine *N*-demethylation reaction has been reported [119]; in this reaction, S-(−)-nicotine is first oxidized to the *N*-1'-oxide with *m*-chloroperoxybenzoic acid, followed by treatment with ferrous sulfate. S-(−)-nornicotine is obtained in about 10% yield from this procedure. A 10% yield of nornicotine is also obtained from treatment of nicotine with $NaIO_4/RuO_4$ in $CCl_4$-water for 6 h at ambient temperature [63].

## D. *N*-alkylation and *N*-substitution reactions

*N*-Alkylation reactions of nicotine have been known for many years. Some of the very early studies were performed before the turn of the century by Kekule [120], and Stahlschmidt [121], who isolated nicotine dimethiodide (**61**) and nicotine diethiodide (**62**), respectively, and by Pictet and Genequand [122], who prepared two monomethiodides of nicotine, viz. *N*-1-methylnicotinium iodide (**63**) and *N*-1'-methylnicotinium iodide (**64**). *N*-Methylation of nicotine is problematic, because of the presence of two basic nitrogens in the molecule, both of which can be quaternized. The pKa values of these two nitrogens are very different ($pKa_N = 3.04$, $pKa_{N'} = 7.84$), which would lead one to believe that reacting nicotine with methyl iodide should afford predominantly the mono-*N*-1'-methylated product. However, one must not confuse *basicity* with *nucleophilicity*. Previous workers have experienced difficulty in preparing the two methiodides in a pure form, and most earlier samples of *N*-1'-methylnicotinium iodide were probably mixtures of the *N*-1- and *N*-1'-methiodides, perhaps also contaminated with some of the dimethiodide product. Seeman et al. [123] have shown that when nicotine is reacted with one equivalent of methyl iodide in either methanol or acetonitrile, a 2.5:1 mixture of *N*-1-methiodide: *N*-1′-methiodide is obtained. Subsequent studies [124] have determined that in the *N*-methylation of nicotine, the presence of the pyridine ring acts to destabilize the *N*-1′-iodomethylation transition state, thus depressing *N*-1′-methylation and allowing a significant amount of *N*-1-methylation to occur. These studies have also provided convenient methodologies for the preparation of pure samples of the two mono-methiodides of nicotine. S-(−)-*N*-1-Methylnicotinium iodide can be prepared free from the *N*-1′-isomer by reaction of S-(−)-nicotine with two equivalents of methyl iodide in acetic acid at room temperature for 3 days. S-(−)-*N*-1'-Methylnicotinium iodide can be isolated uncontaminated with the *N*-1-isomer by continuous extraction of a 2.5:1 mixture of the *N*-1'- and *N*-1-methiodides with chloroform [123].

Shibagaki et al. [125] have examined the effect of increasing steric bulk of the alkyl halide on regioselectivity of *N*-alkylation of nicotine. They found that alkylations of nicotine with alkyl iodides greater in steric volume than methyl iodide, i.e. ethyl iodide, propyl iodide, isopropyl iodide, benzyl iodide, afforded exclusively the *N*-1-alkylated product, suggesting that N'-alkylation is more sensitive to increasing steric effects than *N*-alkylation. However, as might be predicted, the reactivity of the alkyl halide decreased significantly with increasing steric bulk. A number of *N*-1-alkyl derivatives of S-(−)-nicotine have recently been prepared which are selective inhibitors at neuronal nicotinic acetylcholine receptor subtypes [126].

Direct *N*-methylation of nornicotine with methyl iodide forms mixtures of methylated products, affording only low yields of nicotine. The most efficient way to *N*-1'-methylate nornicotine and other noralkaloids is via reaction with formic acid or formic acid /formaldehyde [103,127–129]. Another mild and very high yielding method is to treat nornicotine with formaldehyde and sodium cyanoborohydride in aqueous acetonitrile or methanol [130,131]. In a similar manner, *N*-1'-isopropylnornicotine can be prepared from the reaction of nornicotine with acetone and sodium cyanoborohydride in methanol [132]. These reactions retain the configurational integrity at the 2'-carbon. Other alkyl groups can be introduced at *N*-1' by treatment with base, e.g. $Na_2CO_3$, and alkyl halide [133–135]. Note, strong bases such as butyl lithium should be avoided in these reactions, because concomitant alkylation of the pyridine ring can occur. Nornicotine and anabasine react with ethylene oxide in methanol to afford good yields of the corresponding *N*-1'-(2-hydroxyethyl) derivatives **65** and **66a**, respectively [136–138], and treatment of anabasine with 1,2-epoxypropane affords *N*-1'-(2-hydroxypropyl)anabasine (**66b**) [138]. Myosmine can be converted into racemic nicotine by treatment with acetic/formic anhydride [139].

There are many examples of *N*-acylation reactions of nornicotine and the related alkaloids, anabasine and anatabine in the literature. These reactions can be carried out with a free carboxylic acid in the presence of coupling agents such as *N*,*N*'-carbonyldiimidazole [140]; with acyl chlorides [141] or acid anhydrides [112]; or via reactions with carboxylic esters [142]. Nornicotine also reacts with phenylisothiocyanate to afford the *N*-1'-carbothioic acid anilide derivative **67** [133,143], and anabasine has been reported to react with ketene to afford *N*-1'-acetylanabasine (**68**) [144]. The phosphoramidate analogue **69** of anabasine has been prepared via reaction with phosphonic acid diisopropyl ester [145].

Other significant reactions of noralkaloids are: the reaction of nornicotine and anabasine with sodium nitrite to afford the corresponding *N*-1'-nitroso derivatives **70** and **71**, respectively [76,133], which are constituents of tobacco and suspected carcinogens; and the formation of *N*-1'-formylnornicotine (**72**) from the reaction of nornicotine with N,*N*-dimethylformamide in the presence of hydrous zirconium oxide [146]. *N*-1'-Formylnornicotine is also a minor tobacco plant constituent [1].

## E. Ring-opening reactions

The reactions of nicotine with acylating agents such as acetic acid anhydride [147], trichloroacetic acid anhydride [118], acetyl chloride [118], benzoyl chloride [148], ethyl orthochloroformate [118] and phenyl orthochloroformate [118], all afford ring opened products, which have been identified as the appropriate *N*-acylmetanicotine (**73**), or *N*-acyl-*N*-methyl-4-chloro-4-(3-pyridinyl)butane-1-amine (**74**) if the acylating reagent is an acyl chloride. The reaction of nicotine with methyl iodide in methanol at 100°C followed by hydrogenation of the reaction product over platinum black, affords the ring-opened product, *N*,*N*-dimethyl-4-(*N*-methyl-3-piperidinyl)-butylamine (**75**) [149]. Kisaki et al. have reported the ring opening of nicotine-*N*-1'-oxide with acetic anhydride to form *N*-methyl-*N*-[4-oxo-4-(3-pyridylbutyl)]acetamide (**76**) [45,52], and the *N*-1'-oxide also undergoes molecular rearrangement to 6S-2-methyl-6-pyridine-2-yl-1,2-oxazinane (**77**) when heated at 150–200°C [52,150–152]. A 91% yield of 4-(6-hydroxy-3-pyridyl)-4-oxobutyric acid (**78**) has been reported from incubation of nicotine with a *Pseudomonas* species at 30°C for 15 h [84].

## F. Ring substitution reactions

The pyridine moiety is a common structural feature in the nicotine alkaloids, and thus requires some discussion with respect to its chemical properties. The π-electron density at C-5 of the pyridine ring is higher than that at C-2, C-4 and C-6, as a consequence of nitrogen being more electronegative than carbon. Thus, there is a drift of the π-electrons towards the nitrogen atom. This drift denudes the C-2, C-4 and C-6 positions of electrons by a very strong resonance interaction, which can be compared with the similar delocalization of electrons from the C-2 and C-4 positions of the ring in nitrobenzene.The *inductive* effect of the nitrogen atom in the pyridine ring tends to reduce the π-electron density at C-5; however, this is very small in comparison to the reduction at C-2, C-4 and C-6. Thus, nucleophilic substitution in the pyridine ring should favor positions C-2, C-4, and C-6.

Electrophilic substitutions in the pyridine ring of nicotine are usually difficult to carry out compared to similar reactions with benzenoid compounds. This is due to the presence of the basic nitrogen in the heteroaromatic ring, which on protonation, greatly accentuates its electronegative properties and its ability to withdraw electrons from the ring. Thus,

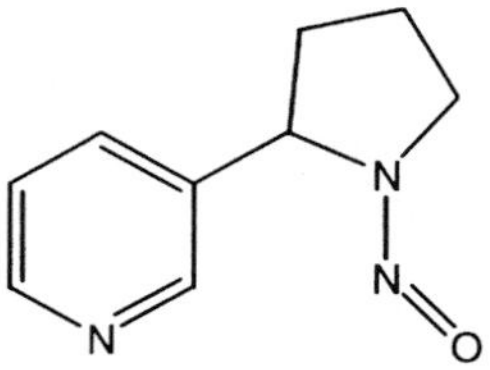

70

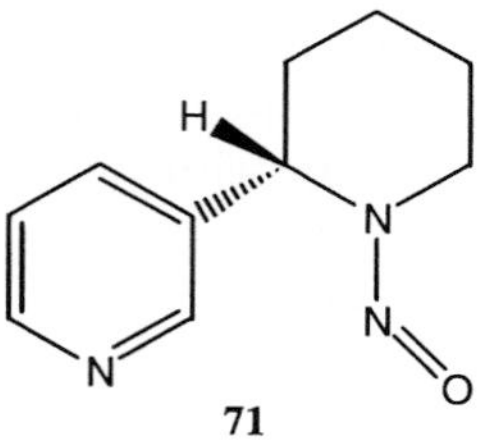

71

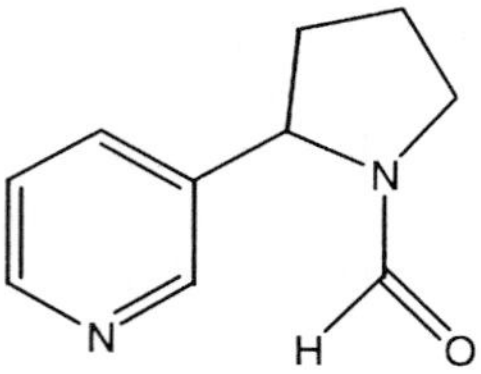

72

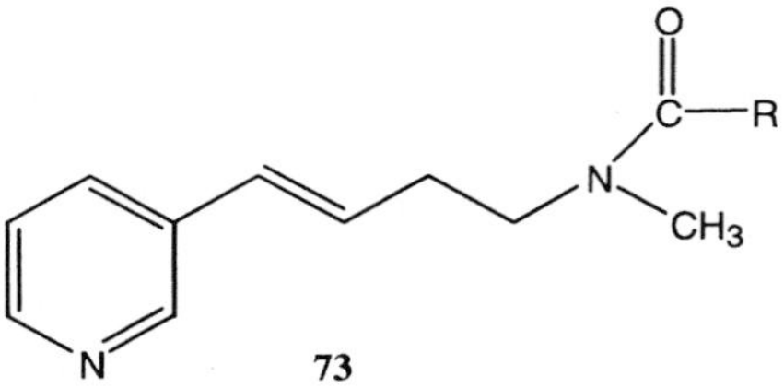

73

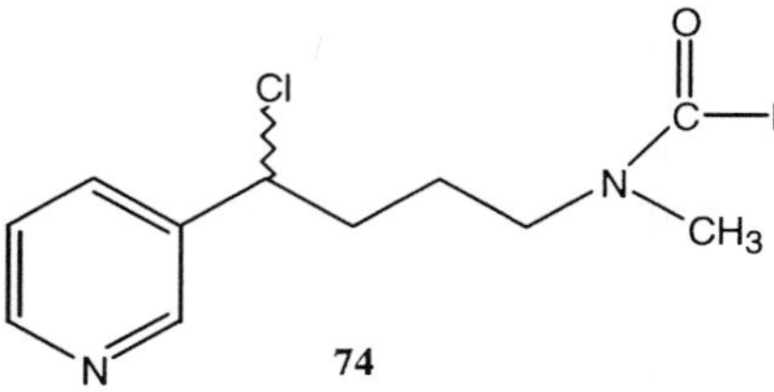

74

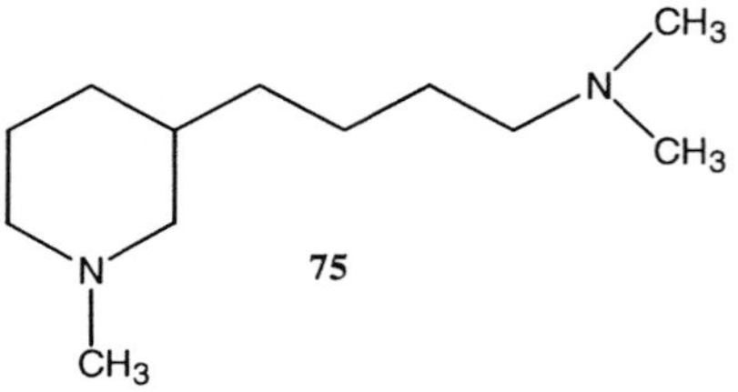

75

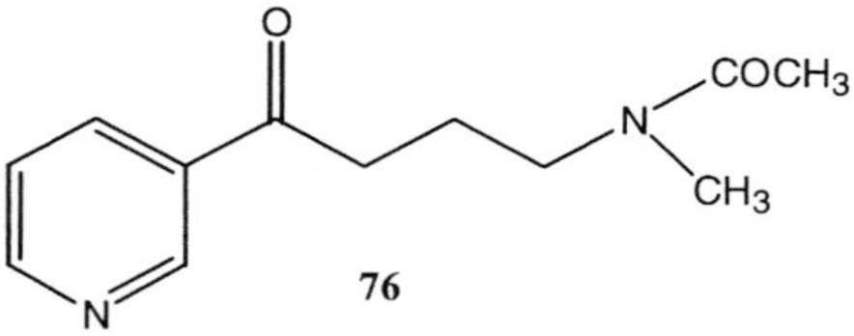

76

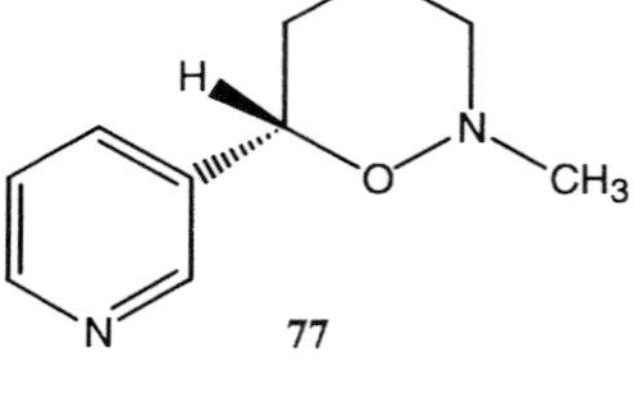

77

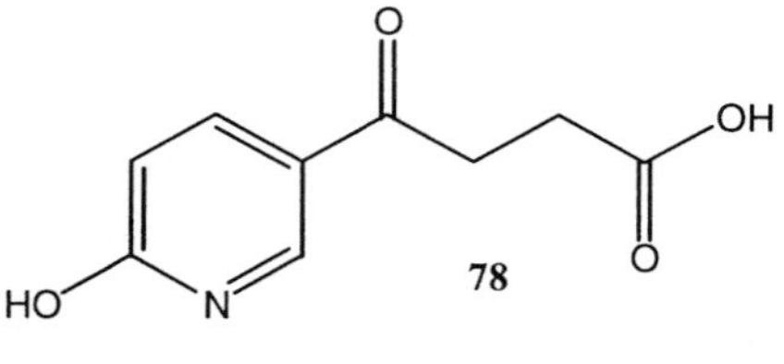

78

since the most electronegative position in the pyridino ring is the nitrogen atom, attack by acids or electrophilic reagents would initially form a pyridinium cation. Subsequent reaction of this cation with the electrophilic reagent would be predicted to be very slow, and to occur exclusively at C-5 of the pyridine ring, based upon $\pi$-electron densities.

Chemical reactions which introduce substituents into the pyridine ring of nicotine and nicotine alkaloids may retain the configurational integrity of the parent molecule, and thus, may constitute useful and expedient ways of generating stereochemically defined derivatives for more meaningful pharmacological and toxicological structure-activity studies, or for generating

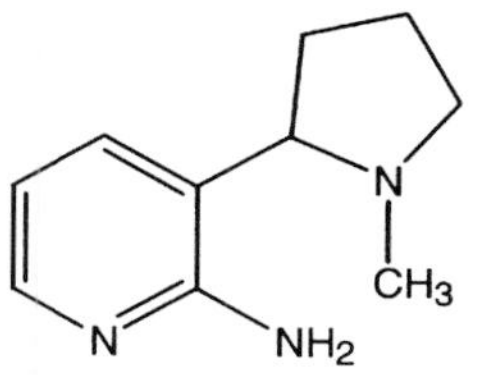

79a

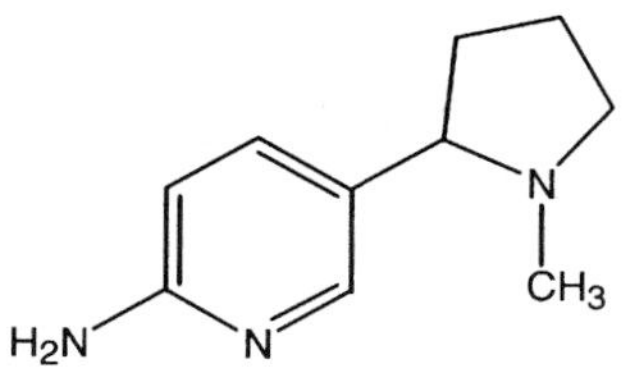

79b

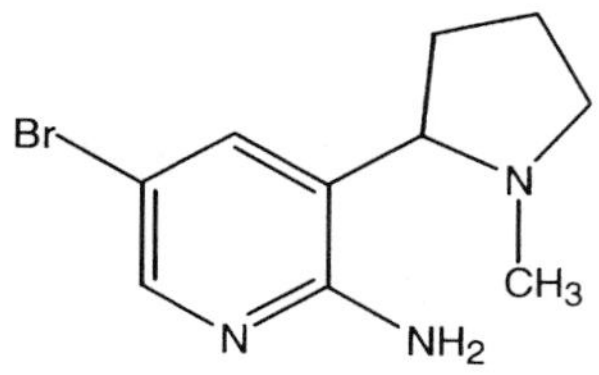

80

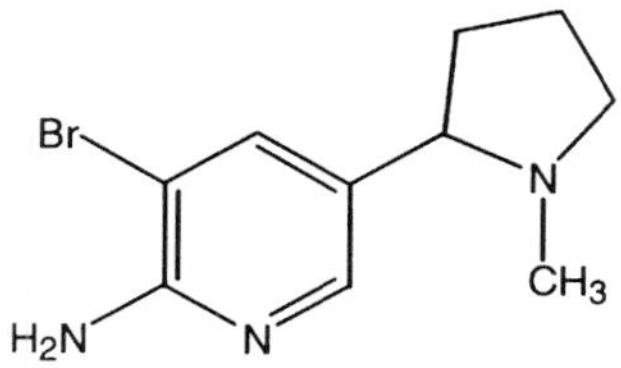

81

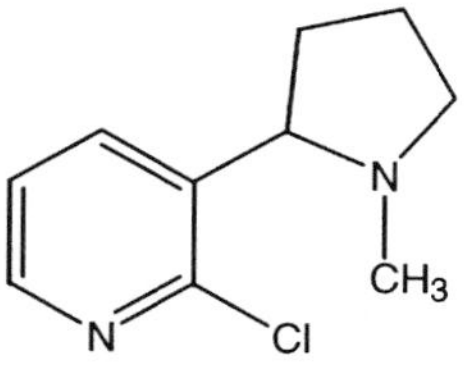

82

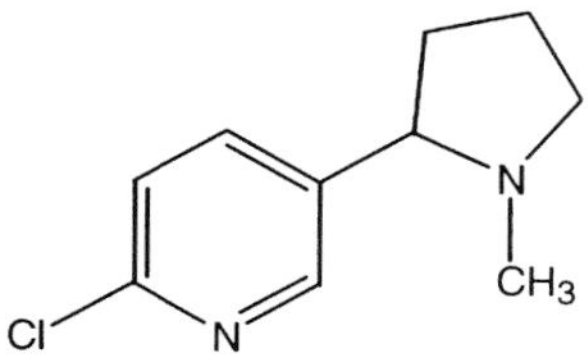

83

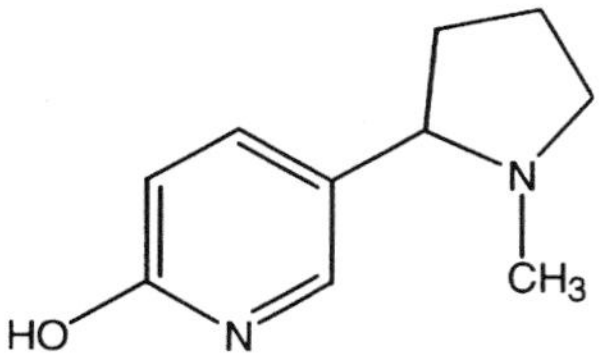

84

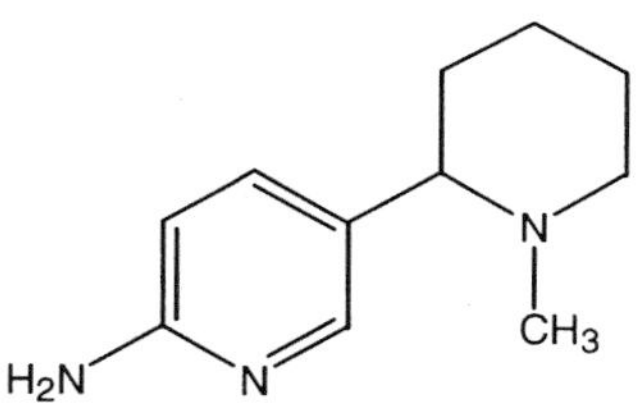

85

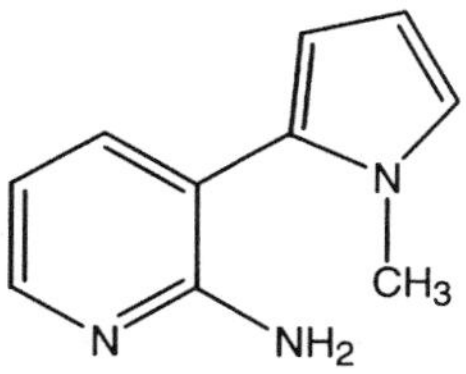

85a

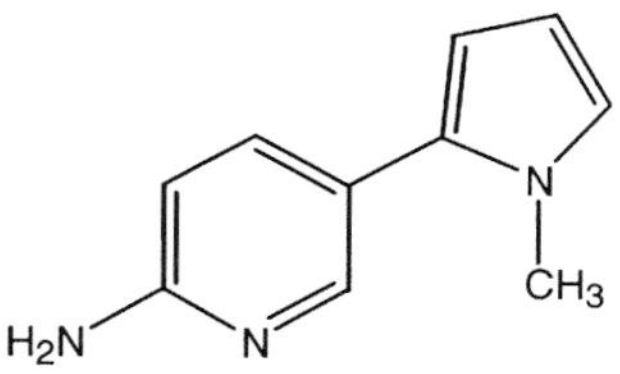

85b

useful intermediates that can be subsequently transformed into configura-
tionally pure products, without the need for the development of more
time-consuming enantioselective syntheses. One of the earliest nicotine
substitution reactions was described by Tshitschibabin, who prepared the
2-amino and 6-amino analogues of S-(−)-nicotine (**79a** and **79b**, repectively)
via reaction with sodamide in toluene at reflux [153–156]. These amino
analogues can be brominated at the 5-position to afford 2-amino-5-bromo
(**80**), and 6-amino-5-bromonicotines (**81**) [157], both of which can also be
converted to 2-chloro- (**82**) and 6-chloronicotines (**83**) via diazotization
followed by addition of HCl. Also, 6-hydroxynicotine (**84**) has been prepared
from diazotization of 6-aminonicotine followed by hydrolysis with water
[157]. It is not known whether these substituted nicotines retain the 2′S
configuration of the original nicotine molecule in their structures. Anabasine

has also been aminated at the 6-position via Tschischibabin chemistry [158, 159], and β-nicotyrine affords a mixture of 2-amino and 6-amino products (**85a** and **85b**, respectively [160].

One of the most studied pyridine substitution reactions of nicotine is the reaction with lithium alkyls in a variety of solvents. Haglid [161] treated nicotine with an equimolar amount of methyllithium and obtained 6-methyl-nicotine (**86**) as the major product, together with 4-methylnicotine (**87**), and *trace amounts* of 2-methylnicotine (**88**). These results were essentially

|   **86**   |   **87**   |   **88**   |

'confirmed' by other workers [162,163] some years later. However, sub-sequent studies by Seeman et al. [164] showed that the major isomers formed in this reaction are the 2-methyl and 6-methyl isomers together with only trace amounts of 4-methylnicotine. 4-Methylnicotine is only observed as a significant product in radical methylation reactions. These results are consistent with the well-documented ability of alkyl lithium reagents to attack in a regioselective manner the 2- (or 6-) position of a 3-substituted pyridine [165]. Seeman et al. also showed that the product composition in the reaction is strongly dependent on the experimental conditions, i.e. the effect of temperature, solvent, and the relative reagent concentration. Inter-estingly, these methyllithium reactions lead to partially racemized methylnicotines, even though the recovered nicotine has almost complete retention of optical activity. Also, when optically pure S-(−)-6-methylnico-tine is treated with methyllithium, it is recovered with *complete* retention of optical activity. These observations were used to propose a novel reversible ring-opening mechanism for methyllithium reactions with nicotine (Scheme 3). Subsequent work has shown that a variety of *n*-alkyl, branched alkyl, hydroxyalkyl and cyclooalkyl substituents can be introduced into the pyridine ring of nicotine via this methodology [138,166–168]. However, reaction of nicotine with *tert*-butyllithium has been shown to afford pyrrolidine ring-opened alkylated products in addition to 6-*t*-butylnicotine and 6-ethylnicotine [166] (Scheme 4).

Reaction of S-(−)-nicotine with *t*-butyl hydroperoxide in the presence of FeSO$_4$.7H$_2$O/H$_2$SO$_4$ afforded low yields of a mixture of 2-methyl, 4-methyl, 6-methyl- and 4,6-dimethylnicotines, which could be separated by chroma-tography [167], and treatment of nicotine with methanol and ammonium

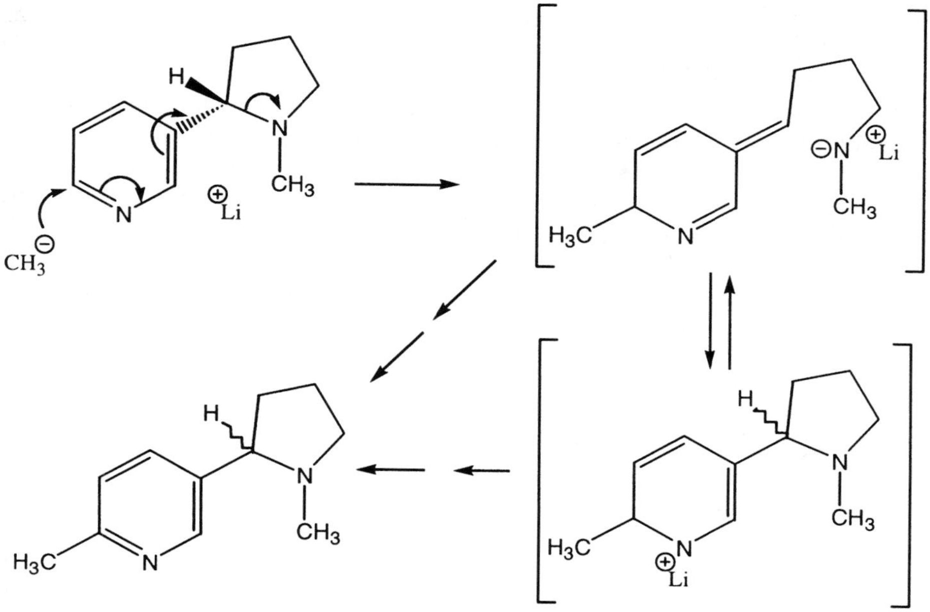

Scheme 3. Mechanism of 6-methylation of nicotine with methyllitium.

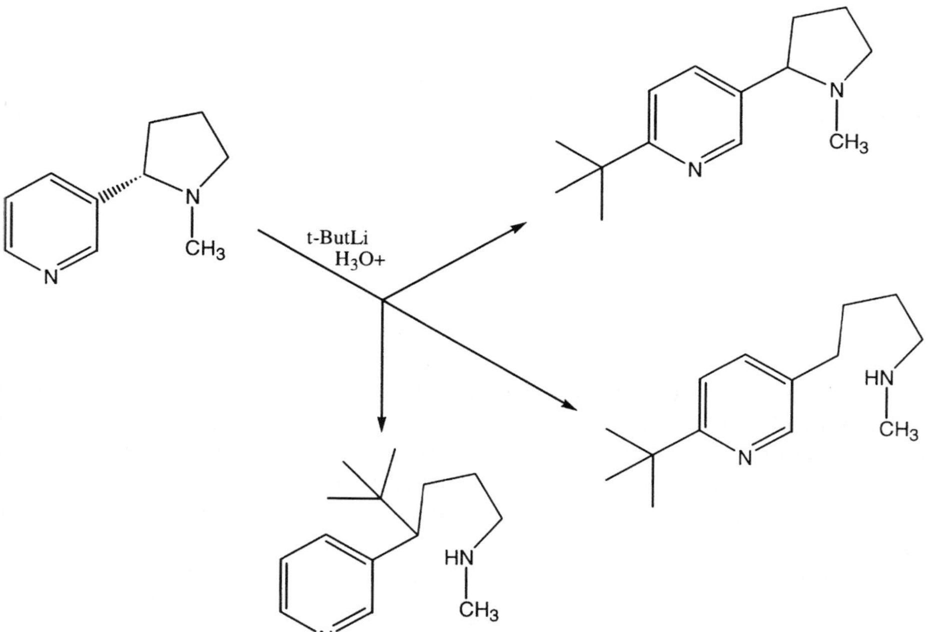

Scheme 4. C-alkylated products from the reaction of nicotine with *t*-butyl lithium.

peroxydisulfate/conc. $H_2SO_4$ affords a low yielding mixture of 4-methoxy (**89**) and 6-methoxynicotines (**90**) [138]. This latter reagent has been utilized for the preparation of alkylated nicotines in the reaction of nicotine with alkanoic acids. Thus, treatment of S-(−)-nicotine with propanoic acid in the presence of $(NH_4)(NH_4)_2S_2O_8$/$AgNO_3$ affords a mixture of 2-ethyl, 4-ethyl, 6-ethyl- and 2,4-diethylnicotines [168], and reaction with butanoic acid under similar conditions gives a mixture of 2-propyl, 6-propyl-, 2,6-dipropyl and 4,6-dipropylnicotines [168]. The use of branched alkanoic or cycloalkanoic acids generally leads to reasonable yields of the appropriate 6-alkylated nicotine; thus isobutanoic acid and 2,2-dimethylpropanoic acid afford yields of 41% and 42%, respectively, of the corresponding 6-alkylnicotine [168]. Reaction of S-(−)-cotinine with benzoyl peroxide/TFA and methanol is reported to afford S-(−)-6-methoxycotinine (**91**) [167].

Abramovitch has reported the reaction of S-(−)-nicotine with phenyl lithium, which affords a mixture of 2-phenyl and 6-phenylnicotines (**92** and **93**, repectively) [169,170], however, no subsequent studies on this reaction have been reported.

Pyrrolidine ring substitutions in the nicotine molecule are relatively rare. S-(−)-nicotine can be brominated with bromine in acetic acid to afford 2'S-3',3'-dibromocotinine (**31**) [55–57]; this compound is an intermediate in the synthesis of S-(−)-cotinine from S-(−)-nicotine. Anabaseine reacts with N-chlorosuccinimide in $CCl_4$ to afford 3',3'-dichloroanabaseine (**94**) [171], and also reacts with benzaldehyde and aromatic ring-substituted benzaldehydes in the presence of mineral acid to afford the 3'-benzylidene derivative (**95**) [172]. Zoltecwicz et al. have prepared a whole series of phenyl substituted analogs (see structures **96–111**) [172]. Interest has been focused on these benzylidene derivatives because of their nicotinic receptor binding properties

[173]. Acheson has reported the preparation of 'thiocotinine' (nicotine-5-thione) (**112**) in 77% yield, from the reaction of cotinine with phosphorus pentasulfide at 130°C [118].

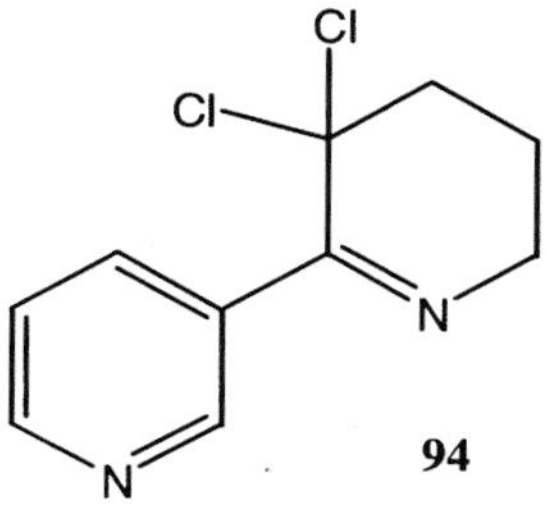

**94**

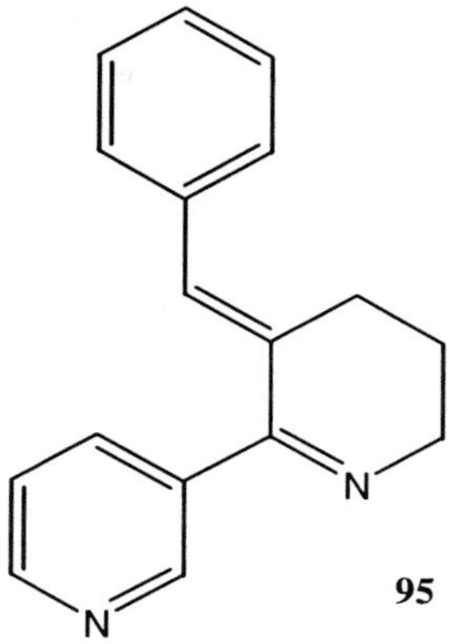

**95**

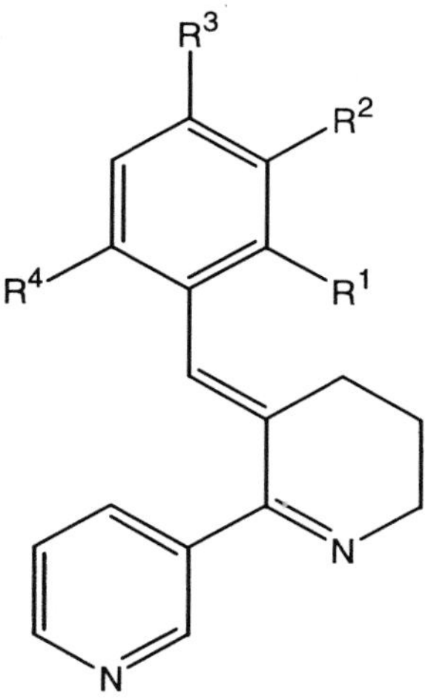

**96**: $R_1$=$OCH_3$, $R_2$=$R_3$=$R_4$=H
**97**: $R_1$=$R_3$=$R_4$=H, $R_2$=$OCH_3$
**98**: $R_1$=$R_2$=$R_4$=H, $R_3$=$NH_2$
**99**: $R_1$=$R_2$=$R_4$=H, $R_3$=$N(CH_3)_2$
**100**: $R_1$=$R_2$=$R_4$=H, $R_3$=$N(C_2H_5)_2$
**101**: $R_1$=$R_2$=$R_4$=H, $R_3$=Cl
**102**: $R_1$=$R_2$=$R_4$=H, $R_3$=CN
**103**: $R_1$=$R_2$=$R_4$=H, $R_3$=$CH_3$

**104**: $R_1$=$R_2$=$R_4$=H, $R_3$=$OCH_3$
**105**: $R_1$=$R_2$=$R_4$=H, $R_3$=$NO_2$
**106**: $R_1$=$R_3$=$CH_3$, $R_2$=$R_4$=H
**107**: $R_1$=$R_3$=$OCH_3$, $R_2$=$R_4$=H
**108**: $R_1$=$R_3$=Cl, $R_2$=$R_4$=H
**109**: $R_1$=$R_4$=H, $R_2$=$OCH_3$, $R_3$=OH
**110**: $R_1$=$R_3$=$R_4$=$CH_3$, $R_2$=H
**111**: $R_1$=$R_3$=$R_4$=$OCH_3$, $R_2$=H

β-Nicotyrine undergoes a number of pyrrole ring substitution reactions; treatment with acetic anhydride /acetic acid and AlCl$_3$ affords a mixture of 4'-acetyl- and 5'-acetyl-β-nicotyrine (**113** and **114**, respectively) [174]. The 5'-acetyl derivative can also be prepared from the reaction of β-nicotyrine with acetic anhydride and aqueous HI/I$_2$ [ 174]. Reaction of β-nicotyrine with iodine in dilute aqueous NaOH affords 3'-iodo-β-nicotyrine (**115**) [175]. An interesting dimeric product, **116**, is formed when β-nicotyrine is reacted with sulfur at 140°C [81], and Acheson has reported the formation of 3,3'-(1,1'-dimethyl-5,5'-sulfanediyl-dipyrrol-2-yl)*bis*-pyridine (**42**) from the reaction of β-nicotyrine with sulfur monochloride in dichloromethane [82].

The relative acidity of the 3'-hydrogens in the cotinine molecule provides for a number of methodologies for introducing structural moieties into the pyrrolidine ring of this molecule. 3'-Alkylation of cotinine can be carried out with sodamide/FeCl$_3$ in liquid ammonia and the appropriate alkyl halide; using this procedure, several 3'-alkyl and 3',3'-dialkyl analogues of cotinine have been prepared, including 3'-(2-propenyl)-, 3'-(N,N-dimethylethyl)-, and 3'-(N,N-dimethylpropyl)cotinine (see structures **117–119**) [118]. 3',3'-dialkyl products can also be obtained (**127a, 127b**). 3'-Alkylation of cotinine can also be carried out with LDA/THF and an appropriate alkyl halide at $-70°C$; 3'-methyl-, 3'-ethyl-,and 3'-benzylcotinine (**120, 121** and **122**, respectively) have been prepared via this procedure [62], as well as the *tert*-butyl ester of 3'-cotininylacetic acid (**123**) [138]. Reaction of cotinine with LDA/THF and formaldehyde affords 3'-hydroxymethylcotinine (**124**) [62]. Chavdarian has reported the synthesis of 3'-phenylselenenylcotinine (**125**) from the reaction of cotinine with benzeneselenenyl chloride (**126**) and LDA/butyl lithium at $-70°C$ [176].

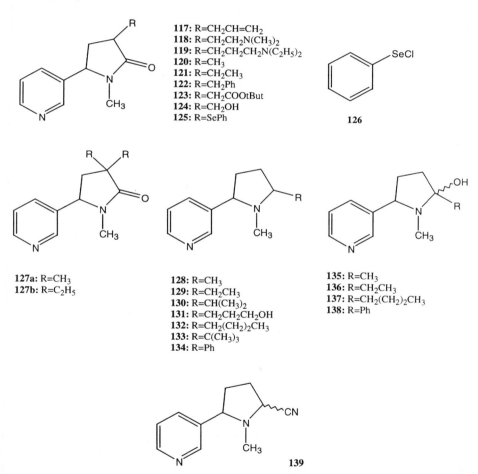

117: R=CH$_2$CH=CH$_2$
118: R=CH$_2$CH$_2$N(CH$_3$)$_2$
119: R=CH$_2$CH$_2$CH$_2$N(C$_2$H$_5$)$_2$
120: R=CH$_3$
121: R=CH$_2$CH$_3$
122: R=CH$_2$Ph
123: R=CH$_2$COOtBut
124: R=CH$_2$OH
125: R=SePh

126

127a: R=CH$_3$
127b: R=C$_2$H$_5$

128: R=CH$_3$
129: R=CH$_2$CH$_3$
130: R=CH(CH$_3$)$_2$
131: R=CH$_2$CH$_2$CH$_2$OH
132: R=CH$_2$(CH$_2$)$_2$CH$_3$
133: R=C(CH$_3$)$_3$
134: R=Ph

135: R=CH$_3$
136: R=CH$_2$CH$_3$
137: R=CH$_2$(CH$_2$)$_2$CH$_3$
138: R=Ph

139

Shibagaki et al. have developed a procedure for the synthesis of 5'-alkylnicotines via reaction of cotinine with an alkyl halide in the presence of bromocresol green/aqueous HCl/sodium cyanoborohydride [177]. 5'-Methyl-, 5'-ethyl-, 5'-isopropyl-, 5'-(3-hydroxypropyl)-, 5'-*n*-butyl-, 5'-*t*-butyl-, and 5'-phenylnicotines (see structures **128–134**) have been prepared in this way. The products obtained are diastereomeric mixtures of the 5'R- and 5'S-isomers. Lin et al. [62] have shown that treatment of cotinine with alkyl lithium reagents in diethyl ether for 1 h, affords the corresponding 5'-alkyl-5'-hydroxynicotine. 5'-Methyl-5'-hydroxy-, 5'-butyl-5'-hydroxy-, and 5'-phenyl-5'-hydroxynicotines (see structures **135–138**) have been prepared in this manner. Also, 5'-cyanonicotine (**139**) is formed when cotinine is reacted with sodium bis-(2-methoxyethoxy)aluminum hydride in toluene (Red-Al) followed by treatment with KCN [116].

Treatment of cotinine with phosphorous oxychloride and aromatic amino compounds has been shown to afford a variety of 5'-imino analogs of nicotine of general structure (**140**) [118]. Similar condensation products can be formed with *N,N*-dimethyl- and *N,N*-diphenylhydrazines [118] (see structures **141** and **142**).

**140a:** R=H
**140b:** R=Cl
**140c:** R=NO$_2$

**141:** R=CH$_3$
**142:** R=Ph

## G. Racemization and resolution

Since nicotine occurs in nature as the S-(−)-isomer only, the desire to obtain samples of the R-(+)-isomer has led to the development of procedures that convert S-(−)-nicotine into its racemate, from which the unnatural isomer can be resolved. This process precludes the necessity to develop lengthy enantioselective synthetic routes starting from achiral precursors. The availability of the R-(+)-isomer is of value to pharmacologists and drug biotransformation scientists, who have an interest in stereoselective processes (i.e. receptor interactions, enantioselective metabolic pathways, behavioral pharmacology, etc.). In addition it has been shown that S-(−)-nicotine in tobacco is converted into small amounts of R-(+)-nicotine during the smoking of a cigarette [2].

An efficient procedure for the racemization of S-(−)-nicotine has been reported by Bowman et al. via treatment with sodium hydride in refluxing xylene for 6 h [178]. The resulting racemic product is then resolved into its pure optical isomers by formation of the salts with either D- or L-O,O'-di-*para*-toluoyl tartaric acid. This procedure is suitable for scale-up, and has now become a method of choice for the preparation of R-(+)-nicotine. It should be noted that when *N*-methylanabasine is treated with sodium in xylene at reflux for 30 h, two dimeric products identified as 2,2-*bis*-*N*-

methylanabasine (**143**) and 4,4-*bis*-*N*-methylanabasine (**144**), are formed [179]. Similar treatment of anabasine afforded a mixture of the 4,4-*bis*- and 4,6-*bis*-dimers of anabasine (**145** and **146**, respectively) [179]. Racemization of S-(−)-anabasine has been reported to occur by treatment with aqueous $H_2SO_4$ at 200°C [180]. Other resolution procedures described in the literature involve the use of microorganisms; for example, incubating either (±)- or S-(−)-nicotine with the microorganism *Pseudomonas putida* for 3 days at 30°C affords R-(+)-nicotine in 78.5% yield [181].

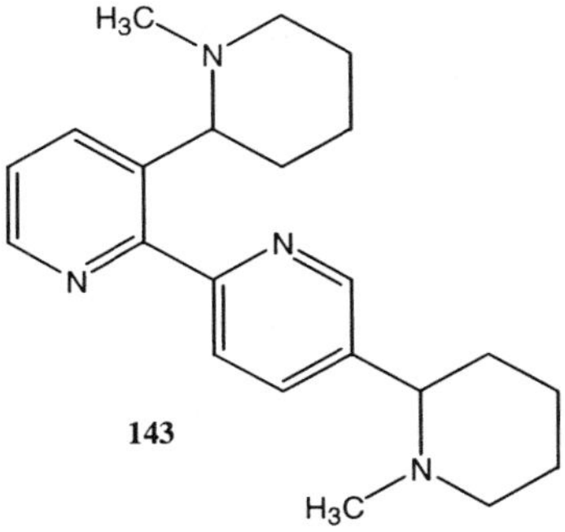

143

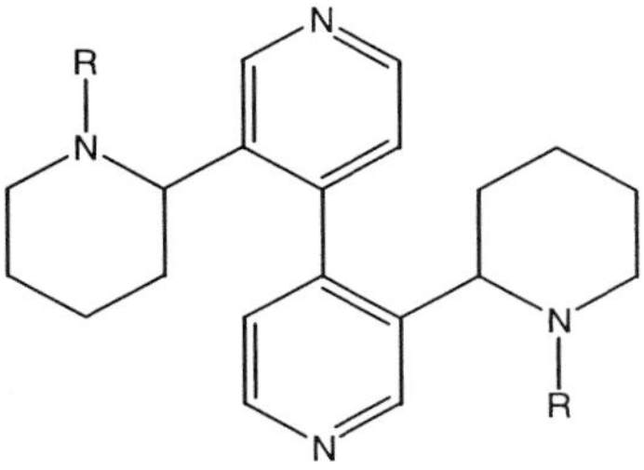

144: R=CH$_3$
145: R=H

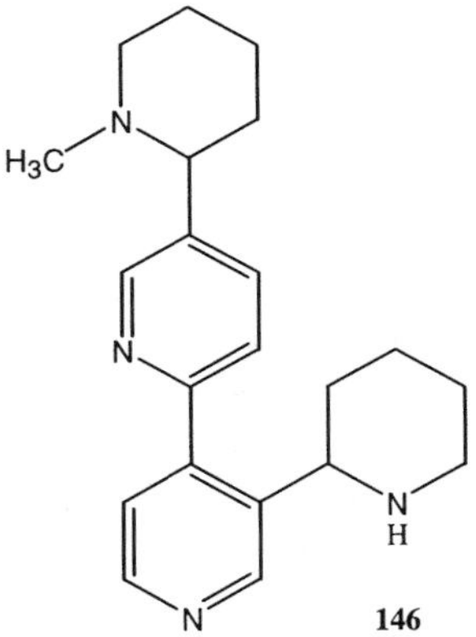

146

Unlike nicotine, nornicotine exists as a mixture of both enantiomers in tobacco. Pure optical isomers have been obtained by either enantioselective synthesis [182–184], from *N*-demethylation of S-(−)-nicotine [119], by preparation of diastereomeric derivatives followed by chromatographic separation [185], or by chromatographic separation of optical isomers on a chiral matrix [186]. Other reported procedures involve resolution of racemic 5-bromonornicotine followed by removal of the 5-bromo group by hydrogenolysis [184], and resolution of norcotinine followed by reduction with 1.0 M BH$_3$/THF [115]. Resolution of racemic norcotinine can be achieved *via* fractional crystallization of the d-10-camphorsulfonate salt [57].

# V. CHEMICAL SYNTHESIS OF NICOTINE AND TOBACCO RELATED ALKALOIDS

There is a wealth of information on the synthesis of nicotine and other tobacco alkaloids, and it is not the purpose of this work to cover all the known literature in this area. The following examples of synthetic pathways to the tobacco alkaloids focus on representative enantioselective routes, which may be of particular value to pharmacologists and medicinal chemists, as a means for obtaining sufficient quantities of optically pure minor alkaloids for pharmacokinetics, and pharmacological and toxicological evaluation.

## A. Nicotine and nornicotine

Since S-( − )-nicotine is a relatively simple molecule, and is readily available and inexpensive, there are few examples of its enantioselective synthesis. Also, since S-( − )-nicotine can be racemized [178,187] and the R-( − )- and S-( − )-isomers easily resolved via their D- and L-di-*para*-toluoyltartaric acid salts [178], a ready supply of both enantiomers is available. One example of an enantioselective synthesis of S-( − )-nicotine described by Chavdarian et al. [188], which utilizes the amino acid L-proline as a precursor, is illustrated in Scheme 5. The strategy in this approach is to utilize a precursor molecule that represents the pyrrolidine fragment of the nicotine structure and contains the desired configuration at the 2-position. The synthesis is then designed to fabricate the pyridine ring around this moiety, without disturbing the configuration at C-2. L-Proline serves this purpose well, and is readily available in both enantiomeric forms. Thus, L-proline is converted into L-*N*-methylprolinol either by *N*-formylation with formic acid and acetic anhydride, followed by LAH reduction, or by LAH reduction of L-proline to L-prolinol, and *N*-methylation with lithium butyl/methyl iodide. Conversion of L-*N*-methyl prolinol to the chloro analogue **147** with thionyl chloride, followed by reaction with sodium cyanide affords the key intermediate **148**, which now contains an appropriate functionality for building on the 3-pyridyl moiety. Reaction of **148** with the unsaturated aldehyde **149** in the presence of strong base gives the 1,2-addition product **150** in modest yield. Treatment of **150** with HBr-acetic acid results in cyclization and aromatization, to give 2-bromonicotine. Removal of the bromo group by hydrogenation over palladium chloride affords S-( − )-nicotine.

In terms of usefulness, an enantioselective synthesis of nornicotine is desirable, since S-( − )-nornicotine can be *N*-alkylated with butyllithium/ alkyl halide to afford a variety of *N*-substituted nornicotines in which the integrity of the 2′-configuration can be maintained. Thus, *ipso facto*, an enantioselective preparation for nornicotine can also be applied to the synthesis of the optical isomers of nicotine. In addition, pure enantiomers of nornicotine are not commercially available, since the natural alkaloid is a mixture of the two optical isomers. Several procedures are described in the literature for the preparation of S-(−)-nornicotine. Jacob [184] has reported a total synthesis of nornicotine enantiomers from ethyl 5-bromonicotinate (**151**) (see Scheme 6), which is initially reacted with *N*-vinylpyrrolidin-2-one (**152**) to afford 5-bromomyosmine (**153**). Sodium borohydride reduction of 5-bromomyosmine affords (±)-5-bromonornicotine (**154**), which can be efficiently resolved into its enantiomers via the (+)- and (−)-α-methoxy-α-trifluoromethylphenylacetic acid (MTPA) salts. Hydrogenolysis of the each of the 5-bromo-enantiomers over palladium-on–charcoal affords S-(−)- and R-

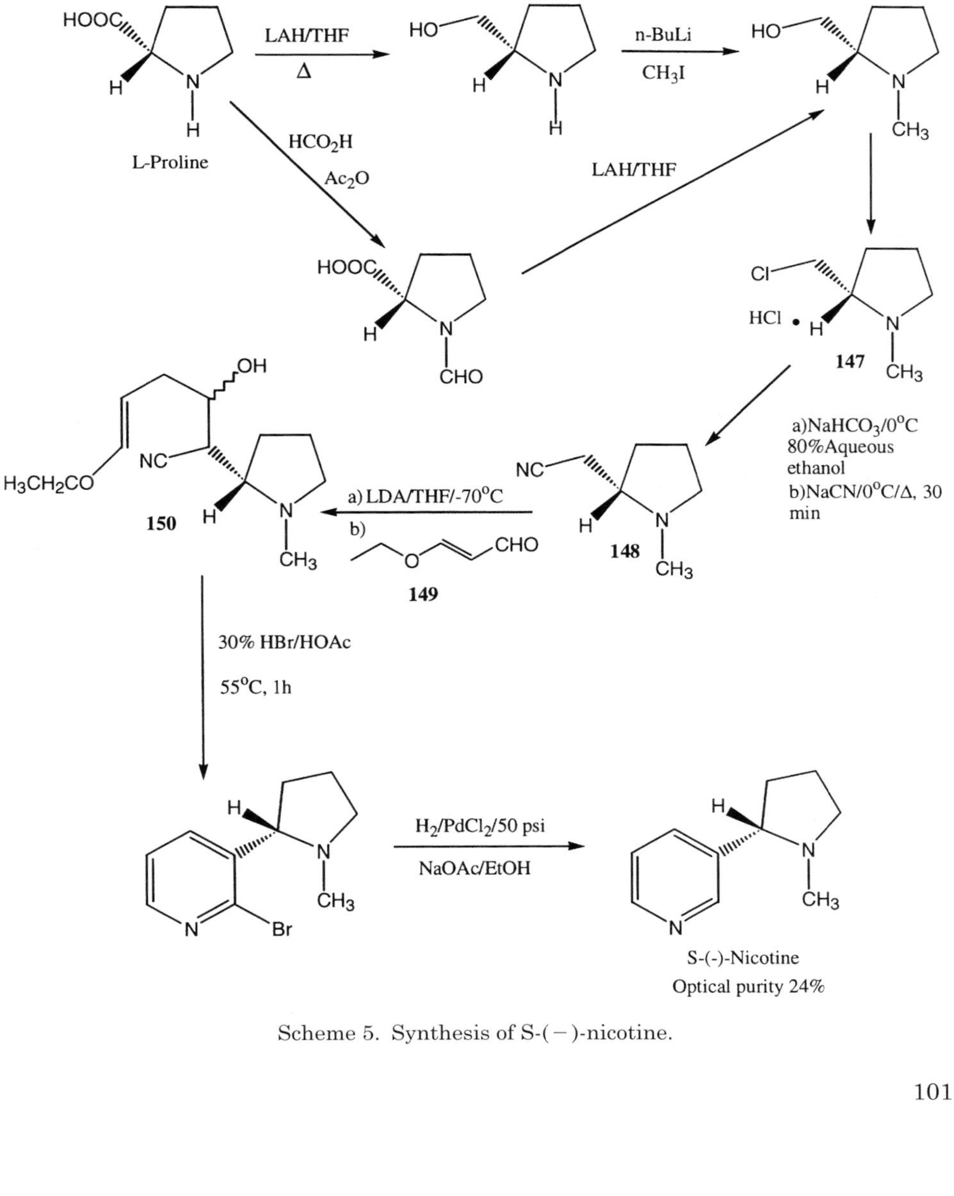

Scheme 5.  Synthesis of S-(−)-nicotine.

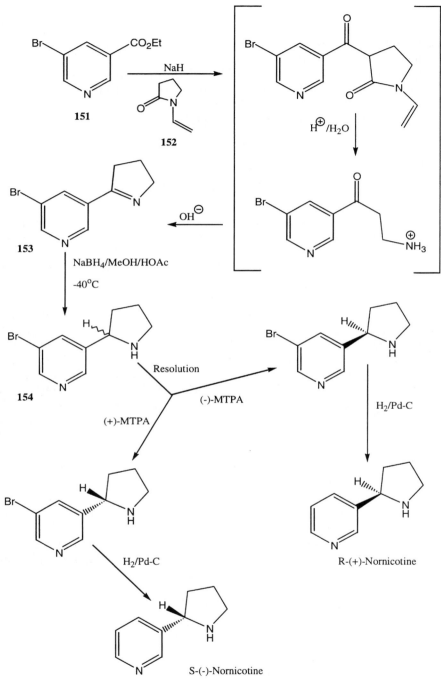

Scheme 6. Synthesis of S-(−)- and R-(+)-nornicotine.

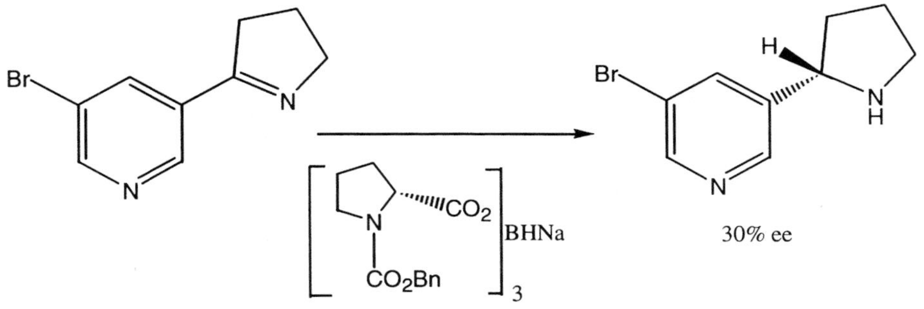

**155**

Scheme 7. Synthesis of 5-bromonornicotine.

(+)-nornicotine in high enantiomeric purity. Recently, Bleicher et al. [189] have reported the conversion of 5-bromomyosmine to S-(−)-nornicotine via reaction with sodium borohydride complex **155** (Scheme 7); the product had an ee of 30%.

Joyce and Leete [119] have described a low yielding but potentially useful preparation of S-(−)-nornicotine from S-(−)-nicotine (Scheme 8), which involves initial conversion of S-(−)-nicotine into its N-1'-oxide with *meta*-chloroperoxybenzoic acid (mCPBA). The N-1'-oxide is then N-demethylated by treatment with ferric nitrate and L-(+)-tartaric acid at pH 6.8. While the yield of this reaction is on the low side, the availability and inexpensiveness of S-(−)-nicotine, and the fact that the procedure involves only two steps, makes this a viable synthesis. Another more recent report [115] describes the synthesis of S-(−)- and R-(+)-nornicotines in high enantiomeric purity from the S-(−)- and R-(+)-norcotinines (see below), via reduction with $BH_3$/ THF, and Seeman et al. [190] have prepared nornicotine enantiomers via HPLC separation of their diastereomeric urethane derivatives (**156a** and **156b**). Garvey et al. [191] have separated nornicotine enantiomers as their N-(t-butyloxycarbonyl) analogues on a Chiralcel-OD (Diacel) column utilizing hexane/isopropanol (3:1) as the mobile phase (see Scheme 9). Recently, Swango et al. [182,183] have reported an efficient and versatile total synthesis of nornicotine enantiomers from readily available 3-aminomethylpyridine (**157**) (Scheme 10), based upon an earlier achiral synthesis [192]. For the synthesis of S-(−)-nornicotine, 3-aminopyridine was condensed with the chiral auxiliary agent, 1R,2R,5R-(+)-2-hydroxy-3-pinanone

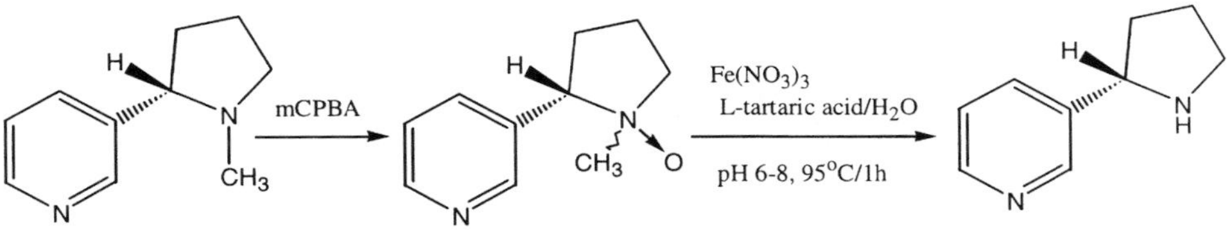

Scheme 8. N-demethylation of S-(−)-nicotine to S-(−)-nornicotine.

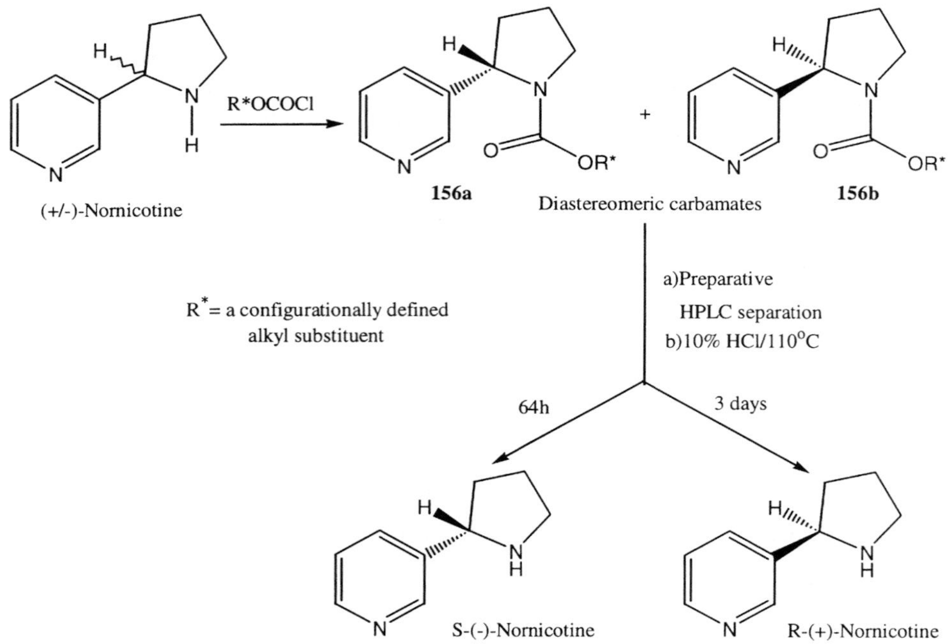

Scheme 9. Preparation of nornicotine enantiomers from racemic nornicotine.

(158) to form the ketimine intermediate 159. Treatment of 159 with LDA at −78°C followed by C-alkylation with 3-bromopropan-1-ol and subsequent hydrolysis of the ketimino group with hydroxylamine hydrochloride, afforded the primary amine 160. The intermediate 160 could then be converted into S-(−)-nornicotine by treatment with HBr in a pressure vessel followed by base-catalyzed intramolecular cyclization. R-(+)-Nornicotine was prepared in a similar manner from 1S,2S,5S-(−)-2-hydroxy-3-pinanone. The products obtained from these procedures had ee's in the range 81–91%. This procedure can also be adopted for the enantioselective synthesis of other minor tobacco alkaloids and related compounds, and a modified version has recently been developed for the enantioselective synthesis of anabasine [193] (see section IVc).

## B. Myosmine and anabaseine

Myosmine has been prepared via a number of synthetic routes. Some of the more recent syntheses are discussed below. The majority of these approaches utilize readily available starting materials. Seeman [194] prepared myosmine in 17% yield from the reaction of N-vinylpyrrolidone (161) with 3-pyridyllithium followed by N-deprotection with perchloric acid (Scheme 11). A similar approach has been reported [195] utilizing N-BOC-protected pyrrolidin-2-one (162) and afforded an overall yield of 40% (see Scheme 12).

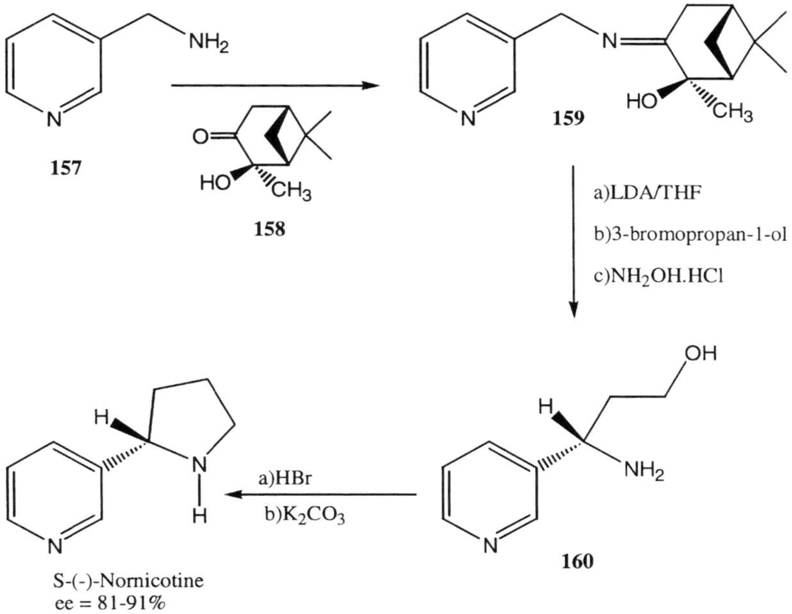

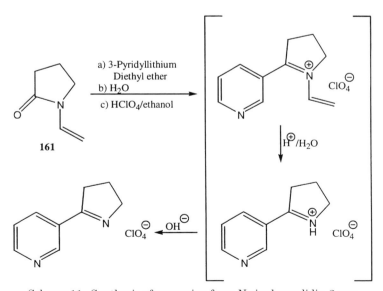

Scheme 10. Enantiomeric synthesis of nornicotine via a 3-hydroxypinanaone chiral auxiliary agent.

Scheme 11. Synthesis of myosmine from *N*-vinylpyrrolidin-2-one.

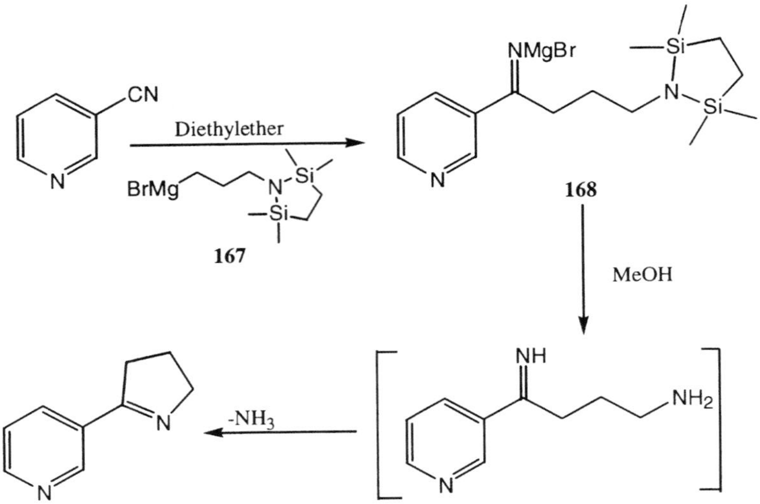

Scheme 15. Synthesis of myosmine from 3-cyanopyridine.

with *N*-BOC-protected 5-ethoxypyrrolidin-2-one (**174**) and LDA, to afford 1,4-dioxo-1-(3-pyridyl)butane (**175**), which could be cyclized to β-nicotyrine in 57% yield by treatment with methylamine in aqueous methanol (Scheme 18). A more direct and higher yielding synthesis of β-nicotyrine has been reported by Minoto et al. [201], involving treatment of 3-bromopyridine (**173**) with 1-methyl-2-pyrrolyl-zinc chloride (**176**) in the presence of PdCl₂(dppb) as catalyst (Scheme 18). β-Nicotyrine was obtained in one step, in 66% yield. Heating S-(−)-nicotine at 500°C in the gas phase over hydrous zirconium oxide, affords β-nicotyrine in 73% yield [202].

## D. Norcotinine and cotinine

The established method of synthesis of norcotinine, which is still currently being utilized, was developed in the 1960s [57,203]. The method involves the conversion of ethyl nicotinate (**177**) into the γ-ketoacid **178**, by condensation with diethyl succinate followed by decarboxylation to **179**, and reaction of **179** with hydroxylamine hydrochloride to form the oxime, **180**. Reduction of

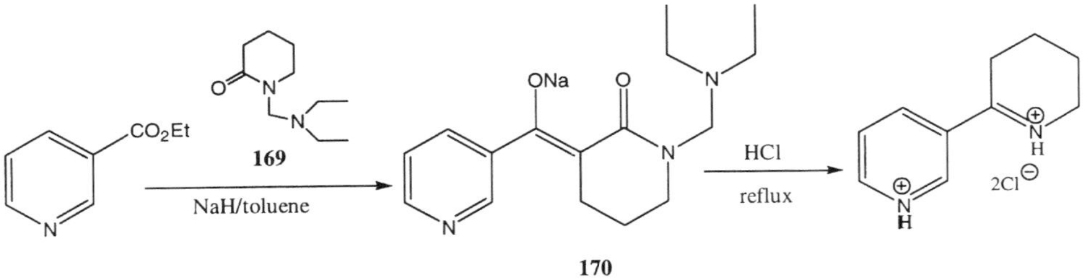

Scheme 16. Synthesis of anabaseine from ethyl nicotinate.

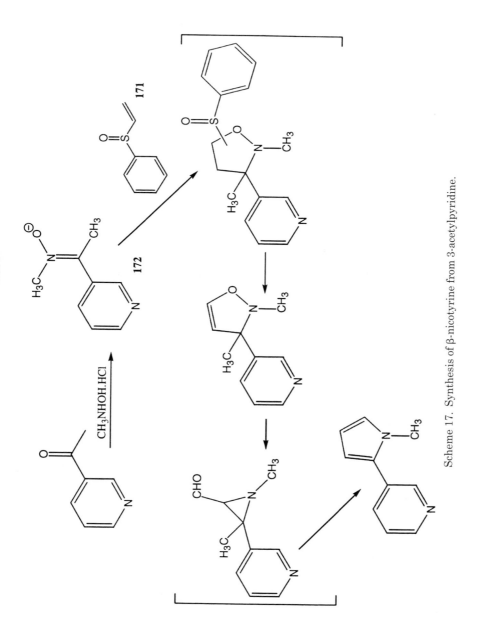

Scheme 17. Synthesis of β-nicotyrine from 3-acetylpyridine.

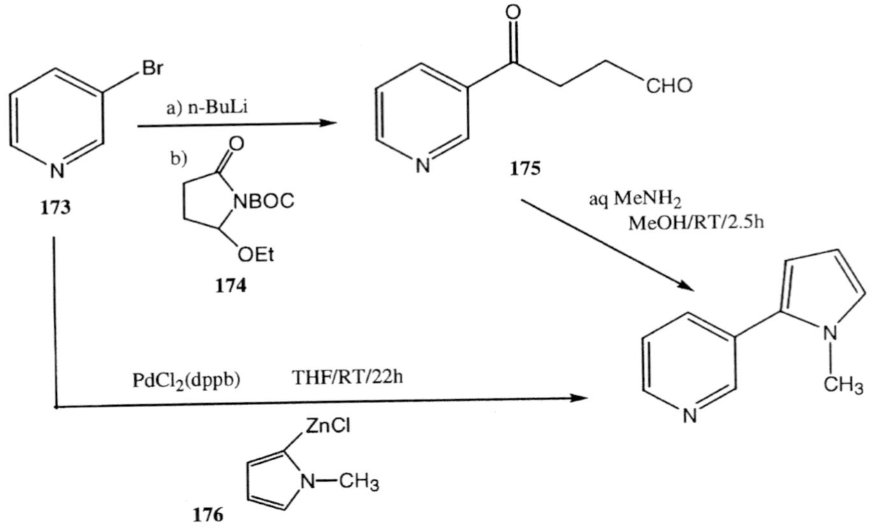

Scheme 18. Synthesis of β-nicotyrine from 3-bromopyridine.

oxime **180** with zinc/acetic acid, followed by thermal cyclization of the resulting aminoacid **181**, affords (±)-nornicotine. Alternatively, the γ-ketoacid can be converted directly to norcotinine by hydrogenation with a mixture of hydrogen and ammonia over Raney nickel at 80 atmospheres pressure and 130°C for 7 h (see Scheme 19). Resolution of the racemic product into the R-(+)- and S-(−)-enantiomers can be achieved with d-10-camphorsulfonic acid.

The standard method of preparation of cotinine is from 3′,3′-dibromocotinine, **182**, which is obtained from treatment of nicotine with bromine in acetic acid. Reductive debromination of **182** with zinc-HCl or hydrogenation over palladium-calcium carbonate catalyst, affords cotinine in excellent yields, and since the 2′-configuration of the starting nicotine is not changed during this procedure, the method can be used for the preparation of both S-(−)- and R-(+)-cotinine, depending upon which enantiomer of nicotine is used as starting material [58] (see Scheme 20).

## E. Nicotine-*N*-1′-oxide

Early preparations of nicotine-*N*-1′oxide via oxidation of nicotine with hydrogen peroxide or peracids, afforded a product that was a mixture of the 1′R,2′S- and 1′S,2′S-diastereomers [41–46], from which the 1′-epimers had to be separated by chromatographic means. Brandange et al. [47] have developed a procedure in which small amounts of alkali molybdates(VI) or alkali tungstates(VI) are added to hydrogen peroxide-catalyzed oxidations of S-(−)-nicotine, forming peroxomolybdates or peroxotungstates in situ. Nicotine has been shown by NMR spectroscopy to exist predominantly as the *trans*-invertomer form in solution [124]. However, oxidizing agents such as hydrogen peroxide or m-chloroperoxybenzoic acid afford a value of ~0.5 for the ratio of 1′R,2′S-epimer : 1′S,2′S-epimer in the oxidized product. Thus, these oxidizing reagents are believed to react almost five times faster with the *cis*-invertomer than with the predominant *trans*-invertomer. The strategy of Brandange et al. was to utilize a more sterically demanding oxidation reagent. Thus, in the presence of 0.05 equivalents of $Na_2WO_4$, hydrogen peroxide oxidation of S-(−)-nicotine afforded 1′R,2′S- nicotine-N-1′-oxide contaminated with only 2% of the N-1′S,2′S-epimer (see Scheme 21). 1′S,2′S-Nicotine-N-1′-oxide can be obtained by treatment of S-(−)-nicotine with m-chloroperoxybenzoic acid in chloroform at 0°C; the product is contaminated with less than 2% of the N-1′R,2′S-epimer after Dowex chromatography [47].

## F. Anabasine

The simplest means for preparing S-(−)-anabasine is by catalytic hydrogenation of S-(−)-anatabine over palladium catalyst [75]. However, the

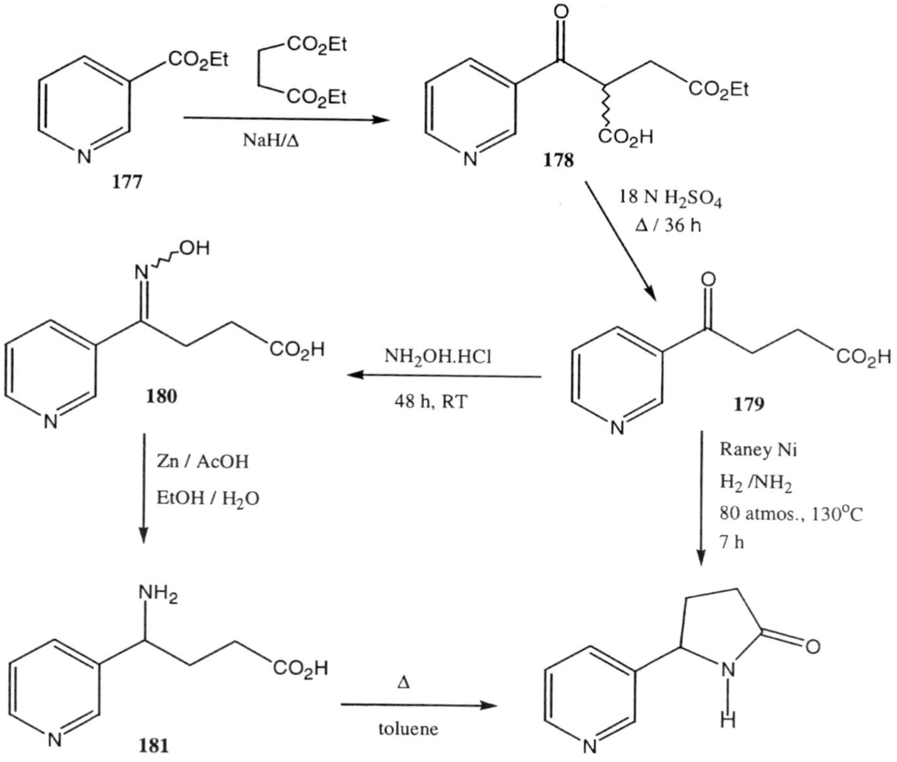

Scheme 19. Synthesis of norcotinine.

Br$_2$-HBr Complex

**182**

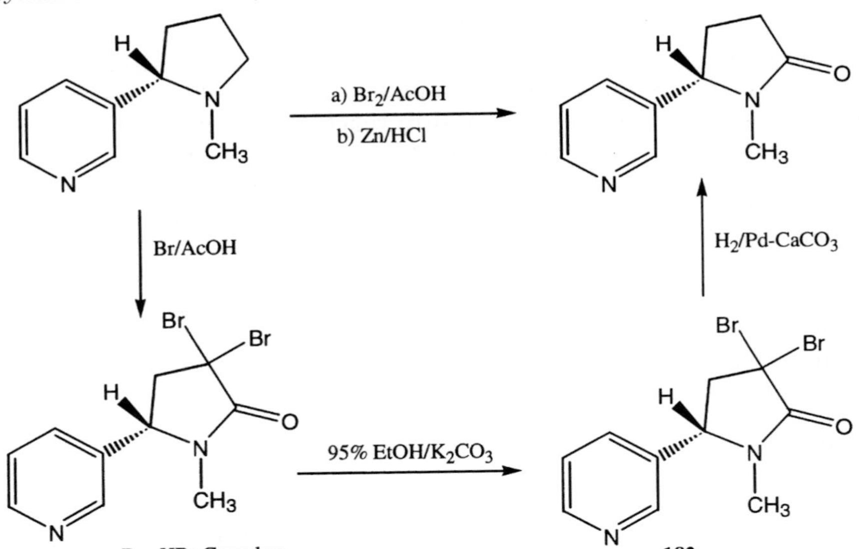

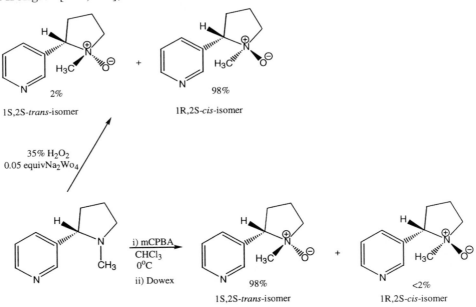

Scheme 20. Synthesis of S-(−)-cotinine from S-(−)-nicotine.

unavailability of S-(−)-anatabine makes this procedure of little utility. A rather elegant synthesis of S-(−)-anabasine has been reported by Kunze and Pfrengle [204,205], who utilized a carbohydrate chiral template in a

Scheme 21. Synthesis of nicotine-1′-*N*-oxide isomers.

stereoselective tandem Mannich-Michael reaction (Scheme 22). Thus, the tetra *O*-pivaloyl derivative of an amino sugar, **183**, was reacted with pyridine-3-carboxaldehyde (**184**) to afford the imine **185**, which was then treated with the trimethylsilyl-protected *cis*-diene, **186**, to give predominantly epimer **187**, contaminated with only 4% of the alternative epimer, **188**. Acid-catalyzed intramolecular cyclization of **187** to **189**, followed by reduction with L-Selectride (lithium tri-*sec*-butylborohydride in THF), afforded **190**, which could then be *N*-deprotected with HCl/methanol to

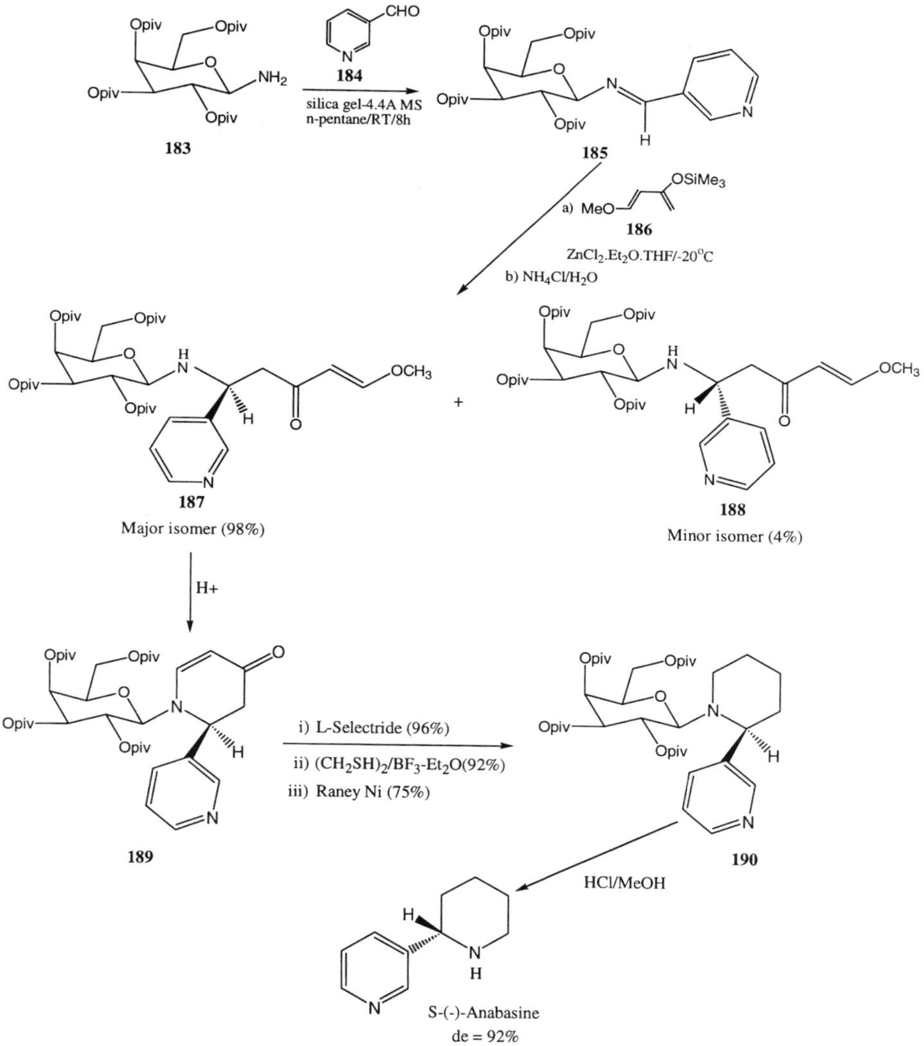

Scheme 22. Synthesis of S-(−)-anabasine.

afford S-(−)-anabasine. The S:R ratio in the product was 20:1. Hattori and Yamamoto [206,207] have described a synthesis of anabasine involving an enantioselective aza-Diels-Alder reaction in the presence of a chiral boron reagent (Scheme 23). This synthesis utilizes *N*-benzyl-3-pyridylimine (**191**) as starting material in a reaction with the cis-diene **192**, in the presence of the chiral boron reagent, **193**, to give **194**, which is then reduced with L-Selectride to afford S-(−)-anabasine in high enantiomeric purity. In a more recent report [193], a variant of the nornicotine synthesis described by Swango [181,182] affords anabasine in low yield, but excellent enantiomeric purity (ee>95%) (see Scheme 24).

## G. Anatabine

Anatabine has been synthesized via a stereoselective route by Mehmandoust et al. [208]. The method involves reacting R-(−)-phenylglycinol (**195**) with Zincke's salt (**196**) in propanol overnight, to afford a good yield of the pyridinium salt **197**, which could then be reduced with NaBH$_4$ in ether/5M

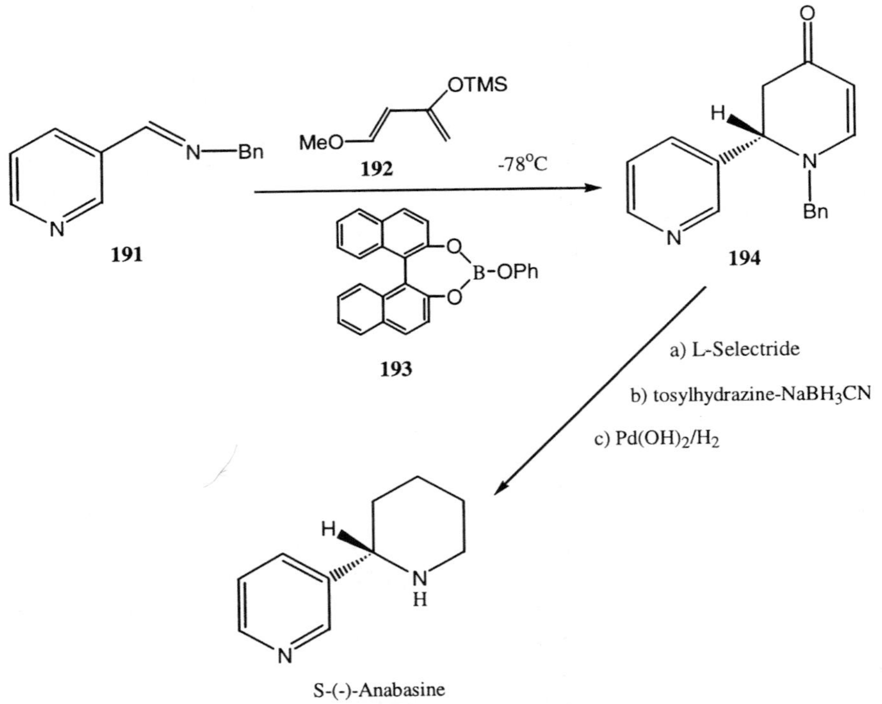

Scheme 23. Enantioselective synthesis of S-(−)-anabasine.

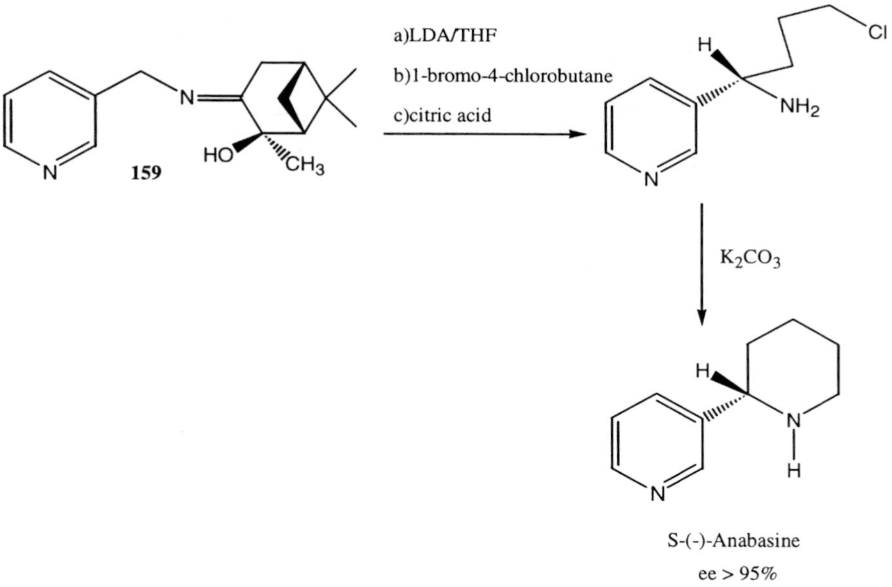

Scheme 24. Enantiomeric synthesis of anabasine.

NaOH at 0°C to afford the bicyclic intermediate, **198**. Reaction of **198** with 3-pyridylmagnesiumbromide (**199**) generated a product which was a mixture of the two epimers **200** and **201**, in the ratio 4:26. Swern oxidation of epimer **201** followed by treatment with 2,4-dinitrophenylhydrazine in propanolic acetic acid, afforded R-(+)-anatabine in 50% yield (see Scheme 25). The use of S-(+)-phenylglycinol instead of the R-(−)-isomer will, of course, produce natural S-(−)-anatabine as the major product.

## H. 2,3′-Bipyridyl and nicotelline

There are a number of syntheses described in the literature for preparation of 2,3′-bipyridyl. Several of these utilize anabasine [76,77] or anatabine [78,75] in dehydrogenation reactions. Kempe [209] has described the preparation of 2,3′-bipyridyl from treatment of 3′,3′-dichloroanabaseine (**94**) with NaOMe/MeOH in 91–95% yield. A number of convenient methods of synthesis utilizes pyridyl organometallic compounds in reactions with halogeno-pyridines [210–216]. For example, reaction of 3-pyridyllithium with 2-bromopyridine in the presence of tetrakis[iodo-(tributylphosphine-)copper(I) and oxygen affords an 85% yield of 2,3′-bipyridyl. Examples of other procedures are provided in Scheme 26.

Utilizing a similar methodology to the synthesis of 2,3′-bipyridyl, the preparation of nicotelline has been reported by Yamamoto et al. [213] in 48%

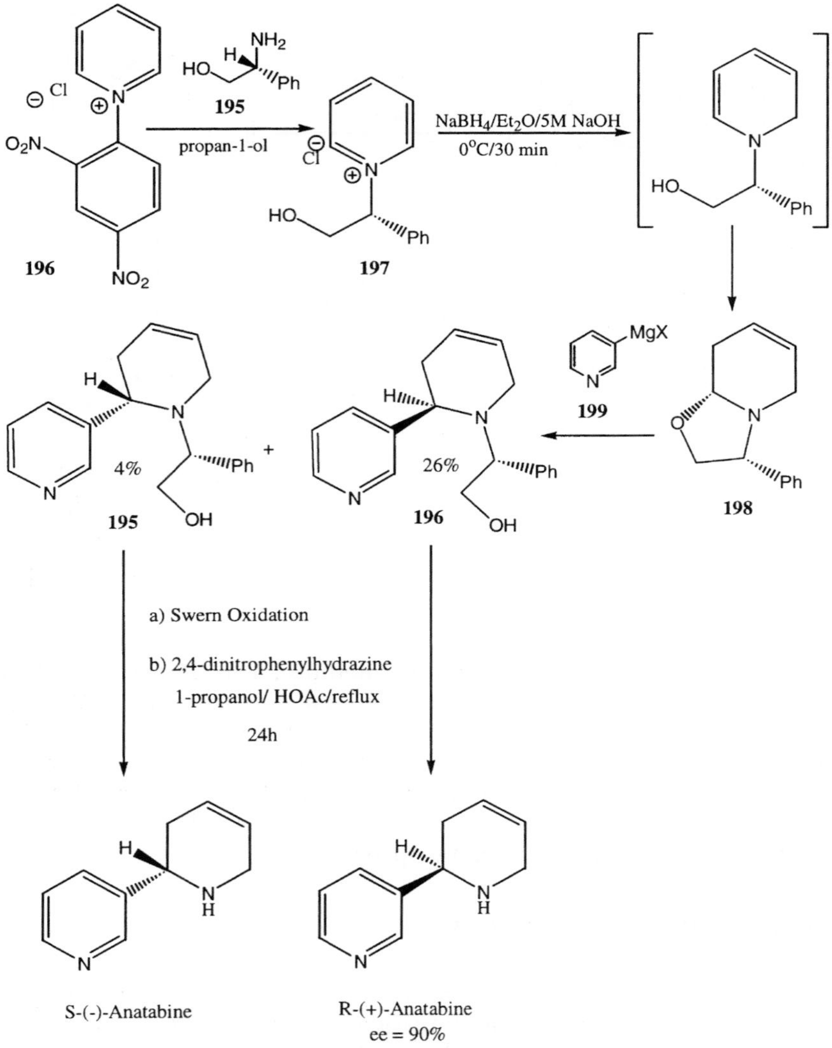

Scheme 25. Synthesis of anatabine.

yield, via the reaction of 3-bromopyridine (**173**) with 2,4-di-(trimethyl-stannyl)pyridine (**202**) in xylene, using Pd(Ph$_3$P)$_4$ as catalyst (Scheme 27).

## VI. SYNTHESIS OF ISOTOPICALLY LABELLED FORMS OF TOBACCO ALKALOIDS

The following syntheses of isotopically labelled tobacco alkaloids are provided for those readers who have an interest in the use of these

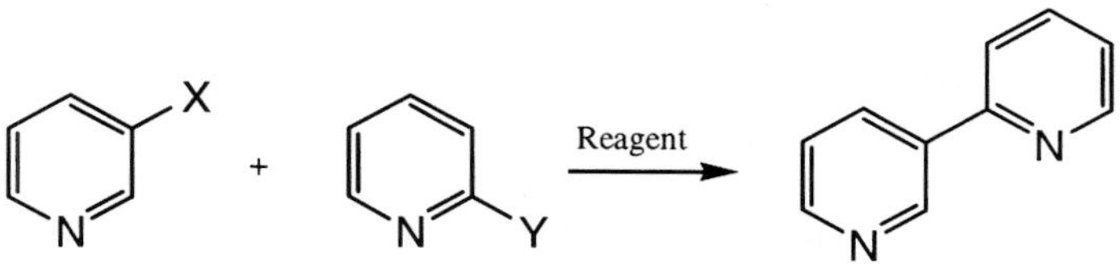

| X | Y | Reagent | Yield |
|---|---|---|---|
| Li | Br | Cu(t-ButP)I, O$_2$ | 80% |
| I | Br | active Zn/Pd(Ph$_3$P)$_4$ | ? |
| B(Et)$_2$ | Cl | KOH/(n-But)$_4$N$^+$Br$^-$/Pd(Ph$_3$P)$_4$ | 84% |
| Sn(Me)$_3$ | Br | Pd(Ph$_3$P)$_4$/xylene | 57% |
| I/Br | Sn(Me)$_3$ | Pd(Ph$_3$P)$_4$/xylene | 63% |
| MgBr | OSPh | THF/3h | 63% |
| Br | O$_2$OSCF$_3$ | Pd(Ph$_3$P)$_4$/LiCl/hexamethylditin | 53% |

Scheme 26. Synthesis of 2,3'-bipyridine.

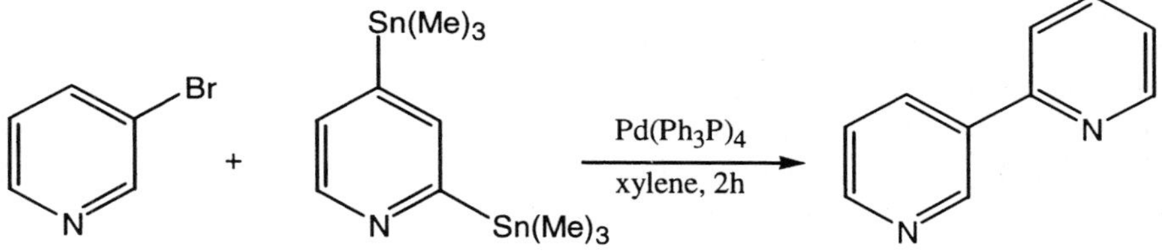

Scheme 27. Synthesis of nicotelline.

compounds for studying their biotransformation, distribution and pharmaco-kinetics of tobacco alkaloids. It is not meant to be a comprehensive review of synthetic work in this area, but rather a compilation of some of the more representative types of labelling procedures developed for tobacco alkaloids. As might be expected, the bulk of the literature in this area deals with isotopic labelling of the nicotine molecule.

## A. Stable isotope-labelled alkaloids

Stable isotopes of nicotine and other tobacco alkaloids have found good use in drug metabolism and bioavailability studies in smokers, and in determining the pharmacokinetics of specific tobacco alkaloids via various routes of adminstration. These studies often involve intravenous administration of the isotopically labelled alkaloid, and the use of mass spectrometric techniques to distinguish the labelled form from the unlabelled, natural form. In this way, the clinical pharmacokinetic parameters can be determined in the presence of the natural alkaloid, and absolute bioavailability can be determined [217]. In these types of studies, it is important to demonstrate bioequivalence of the natural and isotopically labelled compounds. Also, the labelled form of the alkaloid must have a sufficient mass difference from the natural form to overcome interference by the small amount of the naturally occurring carbon-13 isotope found in all organic compounds. Other factors to consider are: (i) the specific metabolic pathway(s) to be studied, since the isotopic label(s) need to be placed in the alkaloid molecule in a position that is not compromised by the metabolic chemistry, i.e. $N$-CD$_3$ labelled nicotine cannot be used for studying the metabolism of nicotine to nornicotine; (ii) the kinetic isotope effect, because labelling the molecule at sites where metabolic transformations occur may slow the rate of metabolism compared to the natural compound, and thus compromise its pharmacokinetics; (iii) the configurational purity of the isotopically labelled alkaloid, since most nicotine alkaloids exist as pure enantiomers in tobacco, and it is well documented that nicotine exhibits stereoselectivity in drug biotransformation pathways, thus precluding the use of racemic compounds in these types of studies; (iv) the fragmentation characteristics of the alkaloid if mass spectroscopic methodology is to be employed, especially if a fragment ion rather than the molecular ion is to be monitored; and (v) the ease of synthesis of the isotopically labelled alkaloid, if it is not commercially available.

Deuterated analogs of tobacco alkaloids have also been utilized in NMR studies designed to determine the spectroscopic properties of these molecules, and their preferred conformations in solution [13].

*(i) Deuterated alkaloids structurally related to nicotine*
*Pyrrolidine ring deuterated compounds.* One of the simplest ways of labeling the nicotine molecule in the pyrrolidine ring is through $N$-deuteromethylation of nornicotine. This procedure can be carried out on enantiomerically pure nornicotines, usually without loss of optical purity in the final product. Several reagents have been utilized. The use of CD$_3$I in solvents such as methanol or diethyl ether, usually affords low yields of nicotine, due to formation of other products from competing $N$-methylation reactions [114]. [$N'$-CD$_3$]-Nicotine has been prepared by treating nornicotine with ethyl chloroformate followed by reduction with LiAlD$_4$ in THF [218], by reacting nornicotine with DCO$_2$D and aqueous D$_2$CO under reflux for 16 h [13] or by $N'$-alkylation with butyl lithium/CD$_3$I [219]. Another reported method involves initial alkylation of myosmine with CD$_3$I, followed by sodium cyanoborohydride reduction of the quaternary ammonium myosmine intermediate, (**203**) [219]. [$N'$-CHD$_2$]-Nicotine has been prepared from the reaction of nornicotine with HCOOH/D$_2$CO [129]. These reactions are summarized in Scheme 28.

Another source of deuterated nicotines is from readily available S-(−)-cotinine. [5′,5′-d$_2$]-S-(−)-Nicotine (**204**) can be obtained from treatment of cotinine with LiAlD$_4$ in THF [218], and heating a mixture of cotinine and K$_2$CO$_3$ in D$_2$O at 100°C for 10 days affords [3′,3′-d$_2$]-cotinine (**205**), which can then be reduced with either LiAlH$_4$ or LiAlD$_4$ to afford [4′,4′-d$_2$]- and [4′,4′,5′,5′-d$_4$]-S-(−)-nicotine (Scheme 29, **206** and **207**, respectively) [55,114,220]. These deuterated forms of S-(−)-nicotine have also been prepared from 3′,3′-dibromocotinine (**182**) via reduction with zinc in CH$_3$CO$_2$D/D$_2$O followed by reaction with either LiAlH$_4$ or LiAlD$_4$ [55]. Nguyen and Castagnoli have reported the synthesis of [2′,5′,5′-d$_3$]-nicotine from [5′-d]-cotinine via reduction with LiAlD$_4$ [218], and have also shown that the use of LiAlD$_4$ in reduction reactions of pyridino compounds results in partial incorporation of deuterium into the pyridine ring, which can amount to as much as 20%. Peterson et al. [116] have reported a synthesis of the [5′-d]-labelled epimers of S-(−)-nicotine (see structures **208** and **209**)

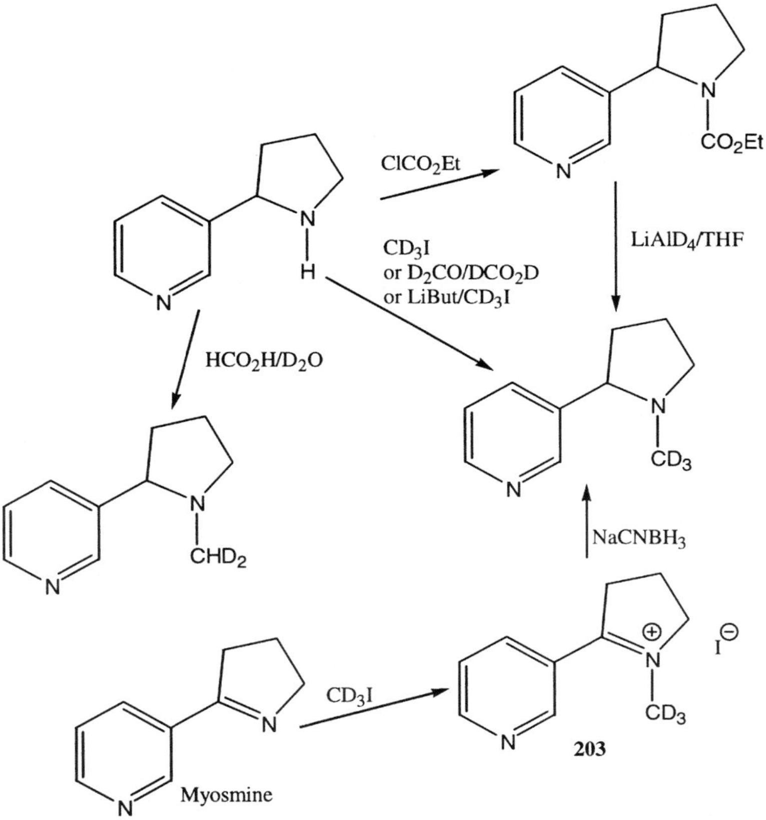

Scheme 28. Synthesis of di- and trideuteromethylnicotines.

from S-( − )-cotinine via the route illustrated in Scheme 29. S-( − )-Cotinine is initally reduced with sodium bis-(2-methoxyethoxy)aluminum hydride (Red-Al)/KCN to afford 5′-cyanonicotine (**135**), which can then be converted into the iminium salt **210**, via treatment with perchloric acid. Catalytic hydrogenation of **210** with $D_2$ in $D_2O$ over platinum dioxide affords stereoselective reduction to epimer **208**. Epimer **209** is prepared by reduction

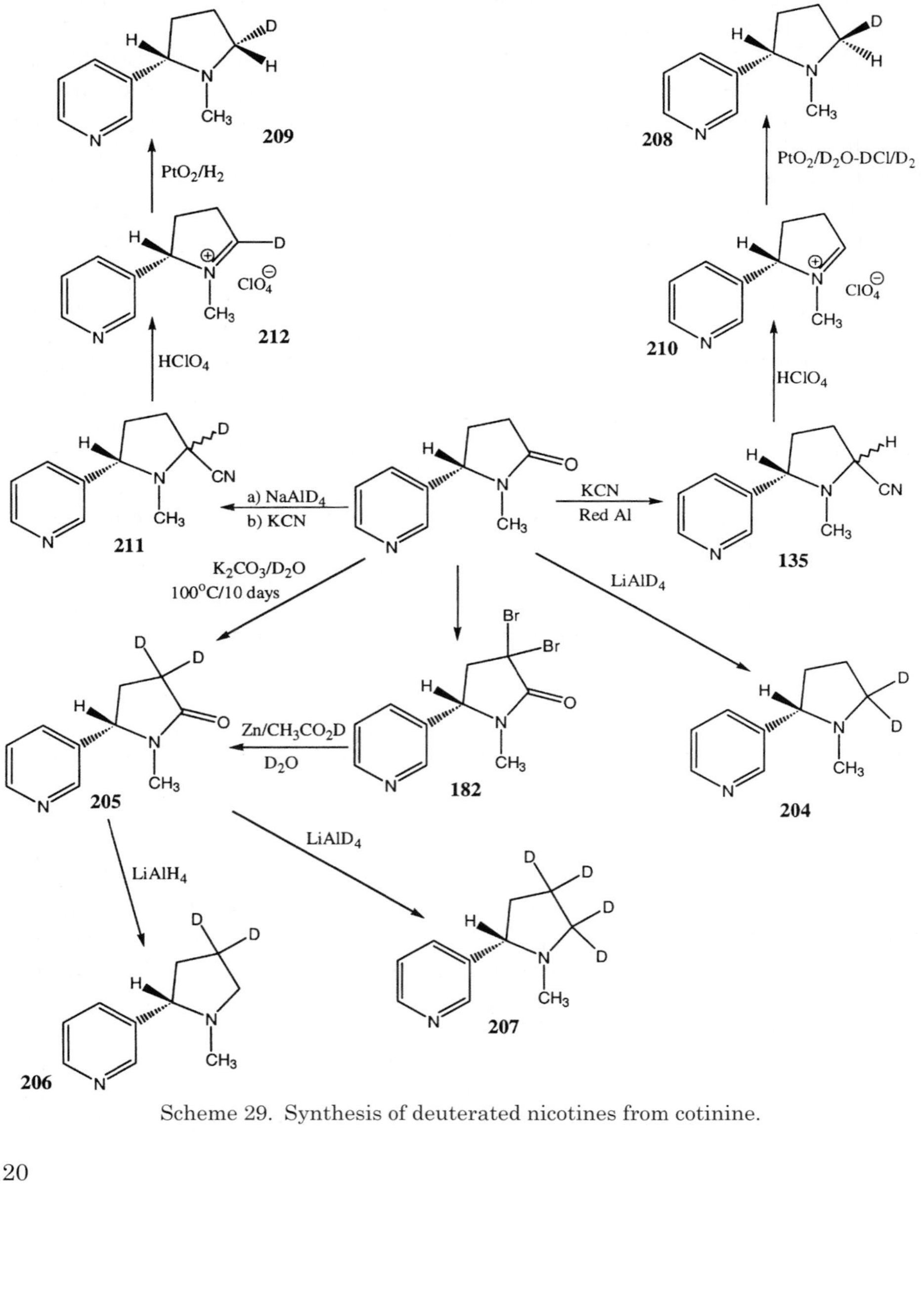

Scheme 29. Synthesis of deuterated nicotines from cotinine.

of S-(−)-cotinine with NaAlD$_4$/KCN to afford [5′-d]-2′S-5′-cyanonicotine (**211**), which is then converted to the corresponding [5′-d]-iminium salt **212**, followed by catalytic hydrogenation with H$_2$ in H$_2$O over platinum dioxide. These syntheses are summarized in Scheme 29.

Since nornicotine is obviously a useful starting material for the synthesis of nicotines containing an N-CD$_3$ moiety it would be advantageous to know which deuterated forms of nornicotine can be prepared, so that specific patterns of deuteration in the nicotine molecule can be constructed to suit the design of metabolic and pharmacokinetic experiments. In this respect, the synthesis of a variety of deuterated nornicotines is reviewed below.

Facile reduction of myosmine either with NaBD$_4$, or by hydrogenation with D$_2$ over palladium-on-charcoal in ethyl acetate, affords (±)-[2′-d]-nornicotine (**213**) [114,129,218]. Treatment of myosmine with K$_2$CO$_3$/D$_2$O, or CF$_3$CO$_2$D leads to deuterium exchange at C-3′ [13,218]; the latter reaction affords almost quantitative yields of [3′,3′-d$_2$]-myosmine (**214**). This product can then be reduced with either NaBH$_4$ or NaBD$_4$ to afford [3′3′-d$_2$]- or [2′,3′,3′-d$_3$]-nornicotine (**215** and **216**, respectively) (see Scheme 30). Deuterated nornicotines can also be prepared from norcotinine, which can be reduced with LiAlD$_4$ to give [5′,5′-d$_2$]-nornicotine (**217**); only low yields have been reported (∼20%) for this procedure [13]. Norcotinine can undergo base-catalyzed exchange in D$_2$O to afford [3′,3′-d$_2$]-norcotinine (**218**), which can then be converted in low yield to either [4′,4′-d$_2$]-nornicotine (**219**) or [3′,3′,4′,4′-d$_4$]-nornicotine (**220**), via reduction with LiAlH$_4$ and LiAlD$_4$, respectively [13] (see Scheme 31). The availability of the above deuter-onornicotines makes feasible the preparation of deuteronicotines labelled at C-2′ and C-3′ (i.e. structures **221** and **222**) from treatment of the appropriately labelled deuteronornicotine with formaldehyde/formic acid. The synthesis of [3′,3′,4′,4′-d$_4$]-nicotine (**223**) has not been reported.

*Pyridine ring deuterated compounds.* Syntheses of pyridine ring deuterated tobacco alkaloids have been described, and these compounds can be prepared directly from S-(−)- or R-(+)-nicotine by deuterium exchange reactions at the 2- and 6-positions of the pyridine ring. However, these forms are generally considered to be less stable than pyrrolidine ring-labelled compounds. Thus, [2-d]-S-(−)-nicotine (**224**) can be prepared by treating S-(−)-nicotine with zinc in D$_2$SO$_4$ and [2,6-d$_2$]-S-(−)-nicotine (**225**) is formed when S-(−)-nicotine is reacted with D$_2$O at 220°C [221] (Scheme 32). Munson and Hodgkins have reported a synthesis of [2,4,5,6-d$_4$]-nornicotine (**226**) and its N′-nitroso derivative (**227**) [222] (Scheme 33). The nor-compound was prepared from pyridine-d$_5$, which was reacted with bromine/SO$_3$-D$_2$SO$_4$ to afford [2,4,5,6-d$_4$]-3-bromopyridine (**228**). This compound could be converted, without apparent loss of deuterium, to [2,4,5,6-d$_4$]-nicotinic acid (**229**), followed by esterification with SOCl$_2$/ethanol and reaction of the resulting ester with LDA/N-trimethylsilylpyrrolidin-

2-one **(230)** to afford [2,4,5,6-d$_4$]-myosmine **(231)**. Reduction of this product with NaBH$_4$ in acetic acid afforded (±)-[2,4,5,6-d$_4$]-nornicotine **(226)**. Jacob and Benowitz [217] have utilized a similar procedure for the synthesis of a number of pyirdine ring deuterated alkaloids (see Scheme 34). In their approach, pyridine-d$_5$ was treated with bromine to afford [2,4,6-d$_3$]-3,5-dibromopyridine **(232)**, which was then reacted with LDA and *N*-vinylpyrrolidone to afford [2,4,6-d$_3$]-5-bromomyosmine **(233)**. Sodium borohydride reduction of this product afforded (±)-[2,4,6,-d$_3$]-5-bromonornicotine **(234)**. The pure enantiomers of this compound could be obtained by resolution of their MTPA salts, and the S-(−)-enantiomer was then debrominated with zinc in CH$_3$OD to afford [2,4,5,6-d$_4$]-S-(−)-nornicotine **(235)**. Subsequent *N*-methylation with formaldehyde/NaBH$_4$ gave [2,4,5,6-d$_4$]-S-(−)-nicotine **(236)**. [2,4,5,6-d$_4$]-S-(−)-Cotinine **(237)** was also prepared from this product via treatment with bromine/acetic acid/zinc.

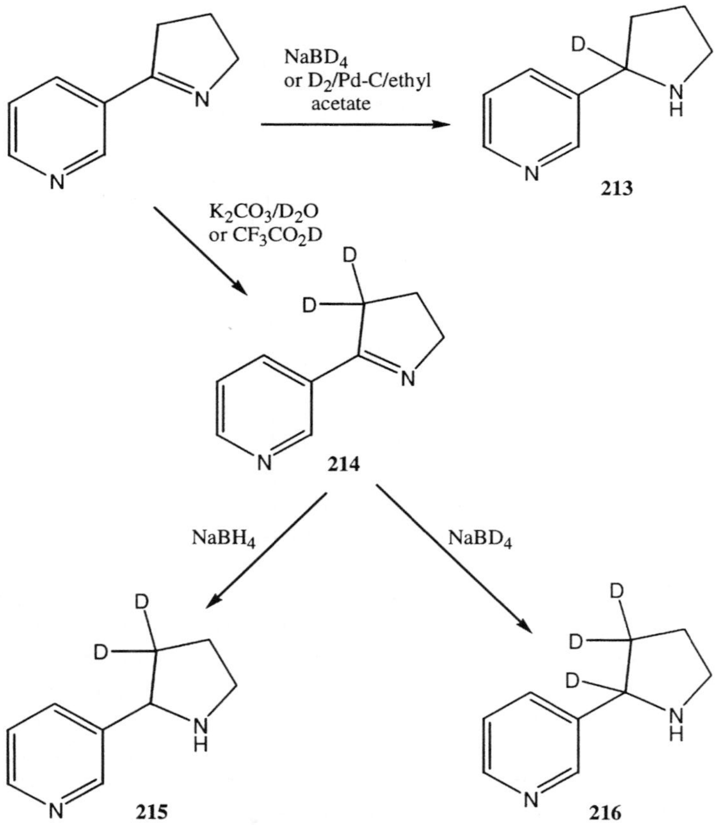

Scheme 30. Synthesis of deuterated nornicotines from myosmine.

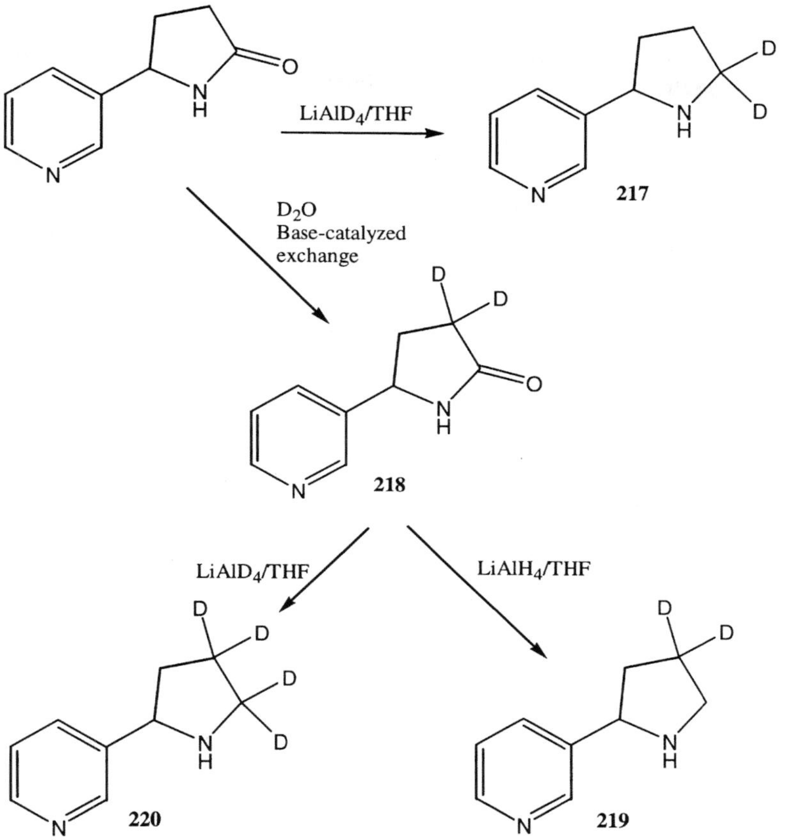

Scheme 31. Synthesis of deuterated nornicotines from norcotinine.

## (ii) Deuterated alkaloids structurally related to anabasine

Relatively few studies have been reported on the synthesis of deuterium-labelled forms of the piperidine alkaloids anabasine, *N*-methylanabasine, anabaseine and anatabine. Both anatabine and anabaseine can serve as precursors for the synthesis of deuterium-labelled forms of anabasine, which in turn can be converted, by facile *N*-methylation or trideuteromethylation, to the corresponding *N*-methylanabasine. Thus, catalytic hydrogenation of anabaseine with $D_2$ over palladium-on-charcoal in ethyl acetate, affords (±)-[2′-d]-anabasine in good yield [114]. Leete and Mueller [223] have also reported the synthesis of [2,2′-$d_2$]-anatabine (**238**) via a biomimetic dimerization of (±)-[2-d ]-baikiain (**239**) (see Scheme 35), and Yang and Tanner [102] have recently described a facile route to the [3′-d]- and [6,6′-$d_2$ ]-labelled forms of anatabine and anabasine (see structures **240**, **241**, and **242**, **243**, respectively) from the reaction of pyridine with either $LiAlH_4$ followed by $D_2O$, or with $LiAlH_4$ followed by water, respectively (Scheme 36). This reaction initially involves the formation of the intermediate $N-1,2/N-1,4$-dihydropyridyl aluminate complex (**244**), which affords anatabine on decomposition with water.

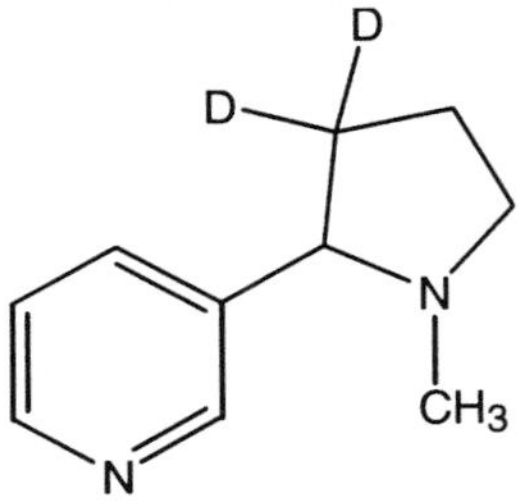

**221**

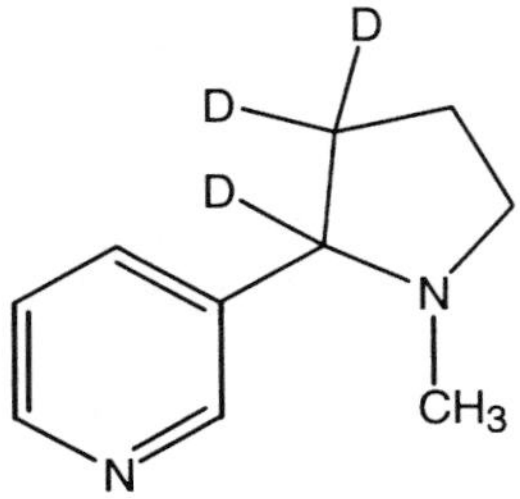

**222**

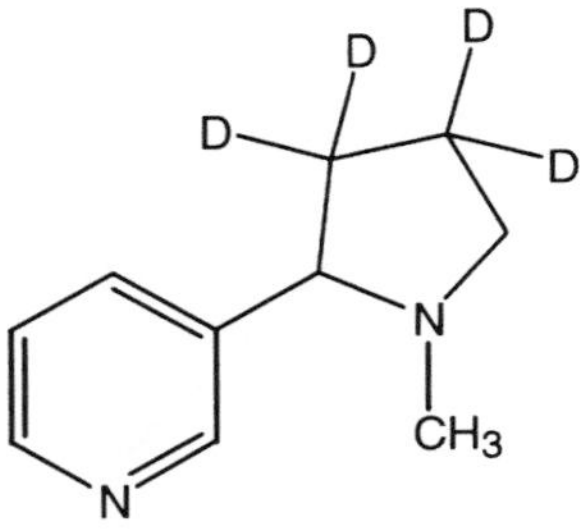

**223**

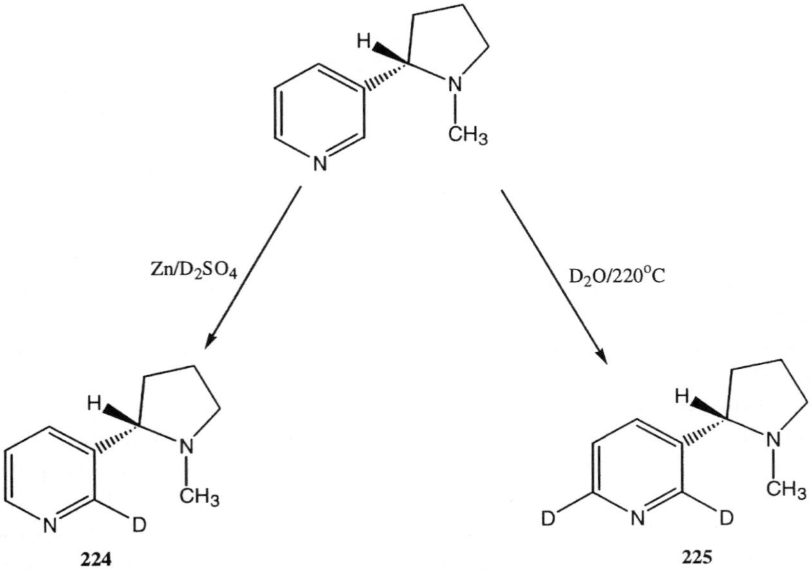

Zn/D₂SO₄

D₂O/220°C

**224**

**225**

Scheme 32. Synthesis of [2-d]- and [2,6-d₂]-S-(−)-nicotines.

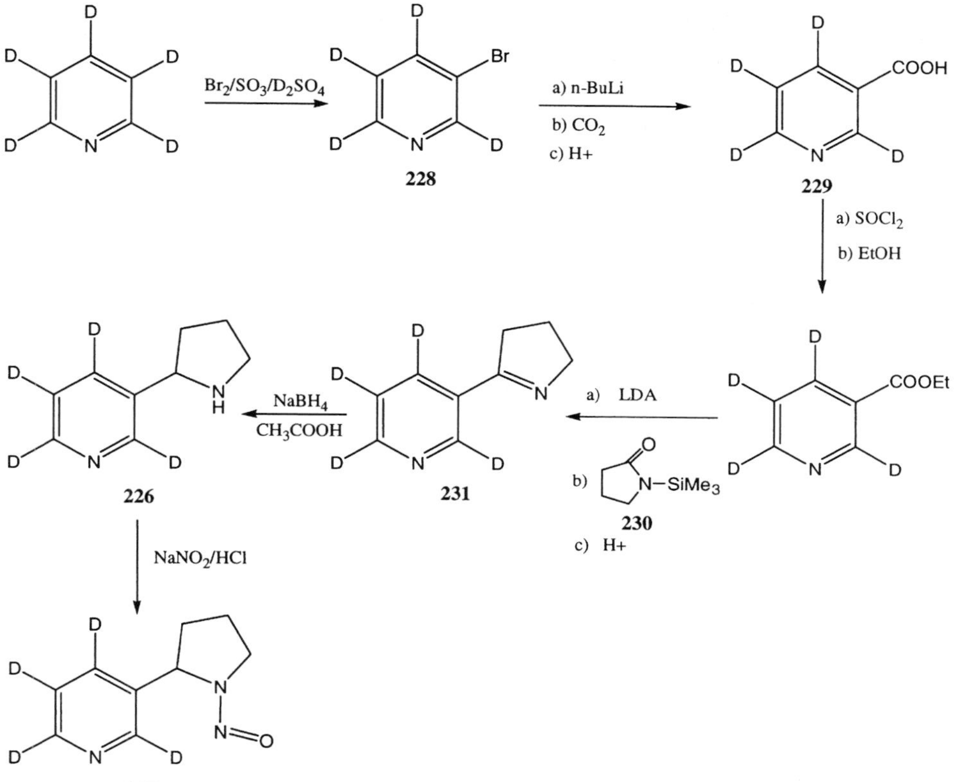

Scheme 33. Synthesis of 2,4,5,6-tetradeuteronornicotine and its *N*-nitroso derivative.

*(iii) Carbon-13 labelled compounds*

Syntheses of $^{13}$C-labelled tobacco alkaloids are not as popular as syntheses of deuterated forms, since $^{13}$C synthesis suffers from the lack of availability or high cost of $^{13}$C precursor molecules. Also, they are not often utilized in drug metabolism studies due to the difficulty in preparing such compounds, and the fact that introduction of a $^{13}$C label into the alkaloid structure adds only one mass unit to the molecular weight of the compound. However, such compounds have found utility in spectroscopic work, and in studies of secondary plant metabolism.

Preparation of nicotine labelled with $^{13}$C in the *N'*-methyl group does not present a synthetic challenge, since several $^{13}$C-one-carbon precursors are commercially available, and procedures established for the preparation of *N'*-trideuteromethyl analogs of tobacco alkaloids can be utilized. Nakane and Hutchinson [224] have reported a ten-step synthesis of (±)-nicotine labelled with $^{13}$C at three sites in the molecule, i.e. the *N'*-methyl group, C-2' and C-3' (Scheme 37). The synthesis starts from [1,2-$^{13}$C$_2$]-acetate (245),

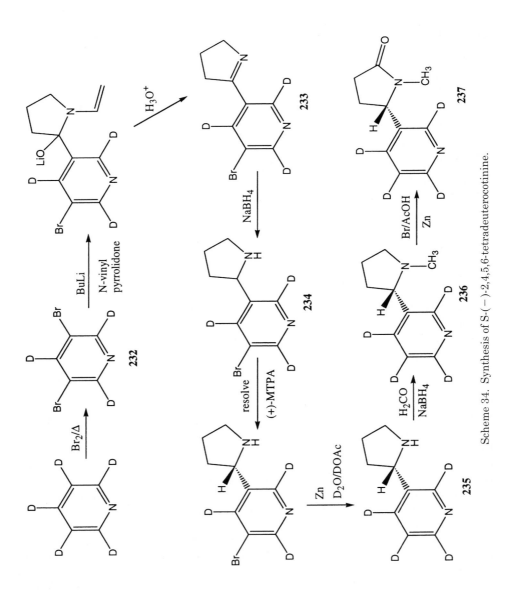

Scheme 34. Synthesis of S-(−)-2,4,5,6-tetradeuterocotinine.

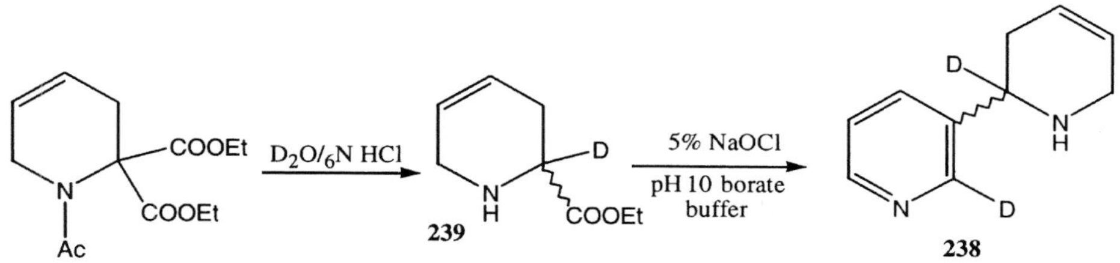

Scheme 35. Synthesis of dideuteroanatabine.

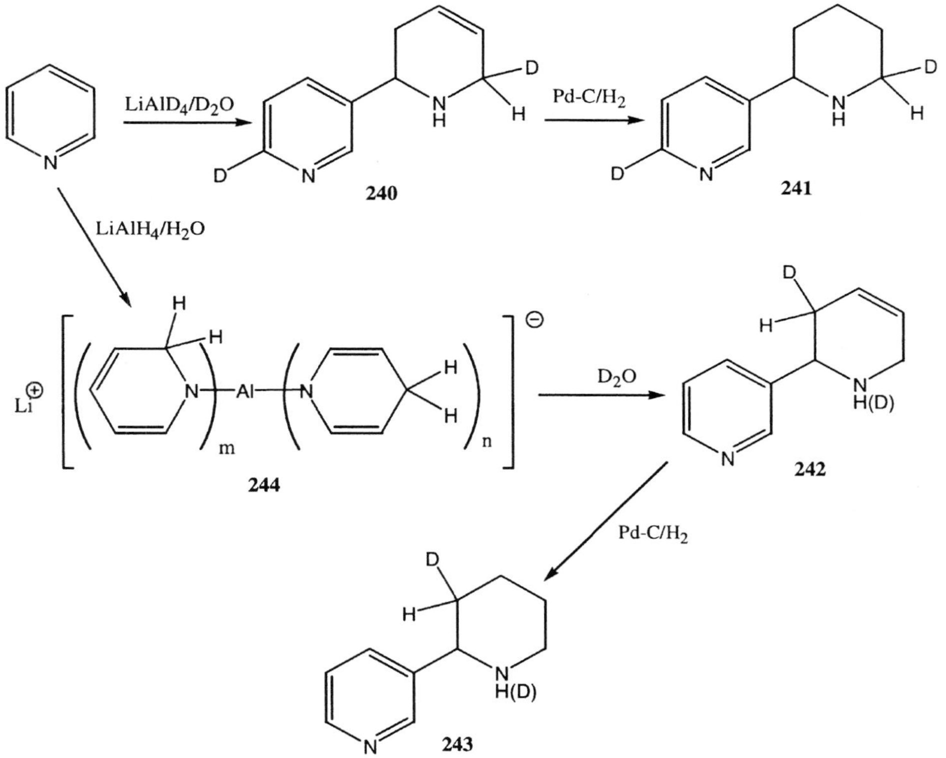

Scheme 36. Synthesis of deuteroanatabine and anabasine.

which is converted to $[1,2-^{13}C_2]$-ethyl iodide (**246**) by standard procedures. This product is then reacted with 3-pyridyllithium to afford $[1',2'-^{13}C_2]$-3-ethylpyridine (**247**). Oxidation of this product to the corresponding 3-acetylpyridine (**248**) and reaction with methoxylamine affords the oxime **249** in 23% yield. Reaction of **249** with sulfoxide **250** affords the Michael product **251**, which can be reduced to the primary amine **252** with diborane/THF. Deprotection of the unmasked thioacetal in formic acid affords the

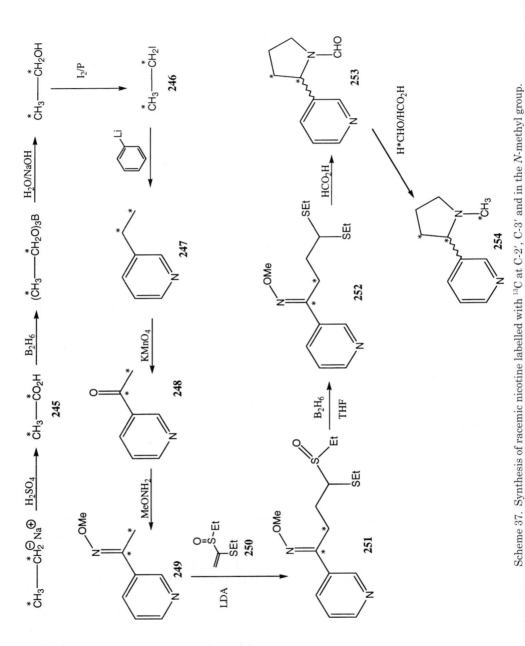

Scheme 37. Synthesis of racemic nicotine labelled with $^{13}C$ at C-2', C-3' and in the $N$-methyl group.

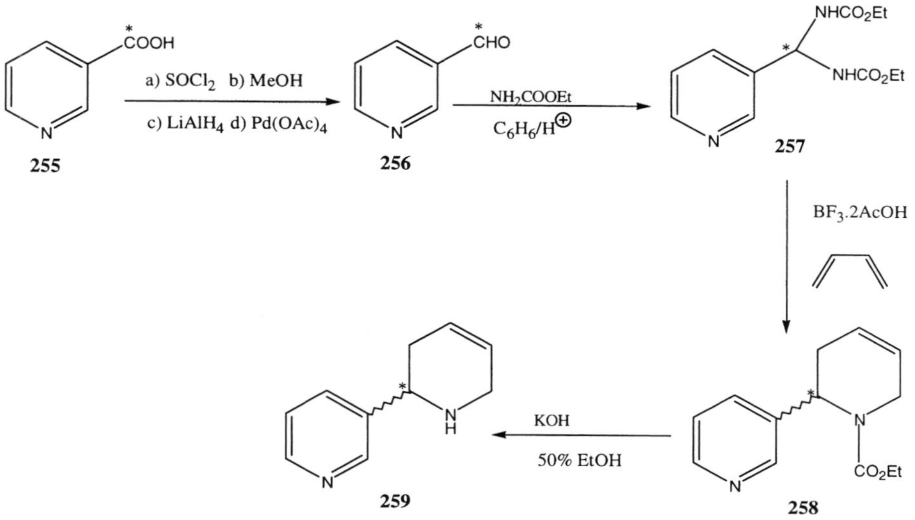

Scheme 38. Synthesis of 2'-$^{13}$C racemic anatabine from $^{14}$C-nicotinic acid.

$N'$-formyl derivative of nicotine (253), which can then be reduced with [$^{13}$C]-formaldehyde and formic acid to give [2',3'-$^{13}$C$_2$, $N'$-$^{13}$CH$_3$]-nicotine (254) in an overall yield of less than 5% from ethyl iodide.

Anatabine labelled at C-2' has been reported by Leete et al. [225] using [$^{13}$C-carboxyl]-nicotinic acid (255) as a precusor (Scheme 38). The $^{13}$C-nicotinic acid is initially reduced to 3-pyridincarboxaldehyde (256) with SOCl$_2$/LiAlH$_4$/Pd(OAc)$_4$ followed by reaction with ethyl carbamate in benzene to afford 257. Treatment of 257 with 1,3-butadiene and BF$_3$/acetic acid affords the $N$-carboxyethyl derivative of anatabine 258, which is then $N$-deprotected with KOH/ethanol to give [2'-$^{13}$C]-anatabine (259). This procedure has also been used for the synthesis of [2'-$^{14}$C]-anatabine (see later). Leete and Mueller [223] have also described a biomimetic synthesis of anatabine labelled with $^{13}$C at the C-2 and C-2' positions of the molecule (Scheme 39). The precursor, [2-$^{13}$C]-diethyl 2-acetamidomalonic acid (260) is initially reacted with *cis*-dichlorobut-2-ene (261), to afford the addition product 262, in 70% yield. Treatment of 262 with LHDMS [lithium *bis*-(trimethylsilyl)amide] to form 263, followed by $N$-deprotection and decarboxylation with 6 N HCl gives (±)-[2-$^{13}$C]-baikiain (264), which can then be dimerized to [2,2'-$^{13}$C$_2$]-anatabine (265).

Nicotine labelled with $^{15}$N-has been reported [226]. Edwards et al. have prepared racemic [$^{15}N'$]-nicotine and [$^{15}N'$]-nornicotine from cyclopropyl 3-pyridyl ketone (266) by reaction with [$^{15}$N]-formamide and MgCl$_2$ in ethoxyethyl ether to afford [$^{15}N'$]-$N'$-formylnornicotine (267), which is then $N'$-deformylated in aqueous mineral acid to give [$^{15}N'$]-nornicotine (268). $N'$-Methylation with HCO$_2$H/H$_2$CO afforded [$^{15}N'$]-nicotine (269) (Scheme 40).

129

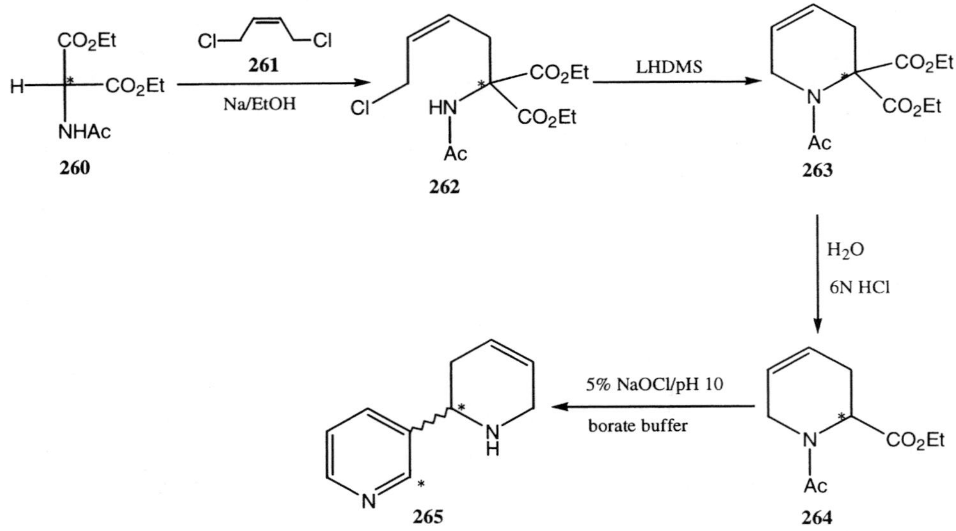

Scheme 39. Synthesis of racemic anatabine labelled at C-2 and C-2' with $^{13}$C.

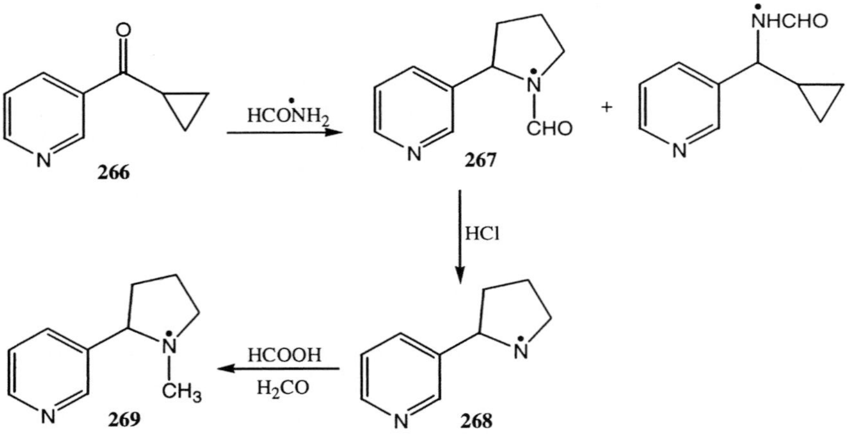

Scheme 40. Synthesis of [$^{15}N'$]-nicotine and [$^{15}N'$]-nornicotine.

## B. Radiolabelled tobacco akaloids

Radiolabelled forms of nicotine have utility in metabolic studies, and in other studies where an accounting of the total mass of administered compound is required. Such mass balance studies are particularly important in bio-transformation studies where a determination of the *total metabolic fate* of the tobacco alkaloid is required. Earlier studies on nicotine metabolism in both animals and humans failed to identify significant amounts of the radiolabel in urine and plasma after administration of both tritiated and $^{14}$C-labelled nicotine. This observation ultimately lead to the discovery of the glucuronic acid conjugates, which are now established as major Phase II metabolites of nicotine. The advent of HPLC, and the availability of flow-through radioactivity detectors capable of monitoring the radioactive effluent stream from an HPLC column, has made the use of radiolabeled compounds in metabolic studies much more routine, and often the method of choice when analytical sensitivity and mass balance are major concerns (see Chapter 7).

The preparation of tritiated tobacco alkaloids presents the same chemical challenges as those seen in the synthesis of deutero compounds, except that reactions are usually carried out on the microscale level, and products are rarely isolated and characterized. Thus, determination of homogeneity on several chromatographic systems is required, and both radiochemical and chemical purity should be established. [³H-Methyl]-nicotine, [³H-2']-nicotine and [³H₂-4',4']-nicotine have been prepared via routes previously described for the corresponding deuterated compounds (see above) [4,56,227].

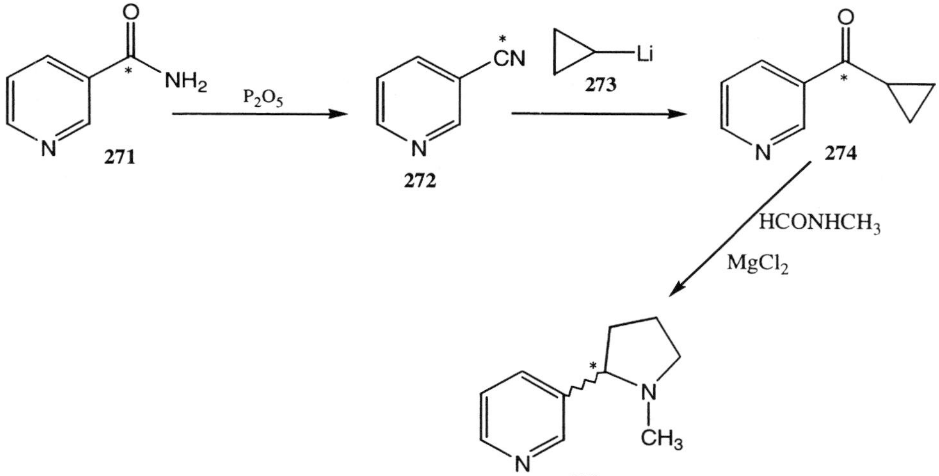

Scheme 41. Synthesis of C-2'-¹⁴C-labelled racemic nicotine from ¹⁴C-nicotinamide.

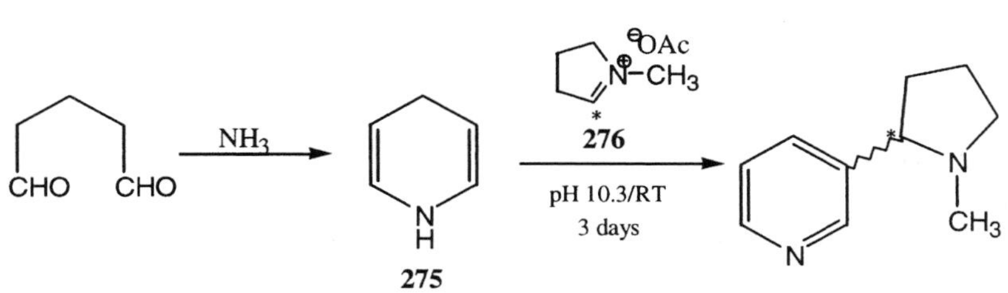

Scheme 42. Biomimetic synthesis of 2'-¹⁴C-racemic nicotine.

131

The synthesis of [14]C-labelled compounds is usually more challenging than the preparation of similar [13]C-labelled forms, because of the lack of appropriate precursor molecules, and the isolation difficulties resulting from the scale-down of such reactions. It is often prudent to develop a synthetic strategy where the [14]C-label is introduced into the precursor molecule at the latest point in the synthesis. ($\pm$)-[$N'$-[14]C-Methyl]-nicotine has been prepared by Leete and Bell [228] by $N$-methylating nornicotine with [[14]C]-methyl iodide in diethylether; a 23% radiochemical yield was obtained. Berger et al. [229] have synthesized ($\pm$)-[$N'$-[11]C-methyl]-nicotine via reaction of ($\pm$)-nornicotine with [[11]C]-formaldehyde, followed by reduction with sodium cyanoborohydride, and Saji et al. [230] have recently reported the synthesis of S-($-$)- and R-($+$)-[$N'$-[11]C]-nicotine via reaction of the appropriate nornicotine enantiomer with [[11]C]-methyl iodide for 5 minutes in DMF/DMSO at 115°C. Carbon-11 is a short-lived radionuclide used in positron emission tomography; the above [11]C-labelled nicotine enantiomers were utilized in studies designed to determine the regional distribution of brain nicotinic receptors in the mouse (see also Langstrom et al. [231] and Sinver et al. [232].

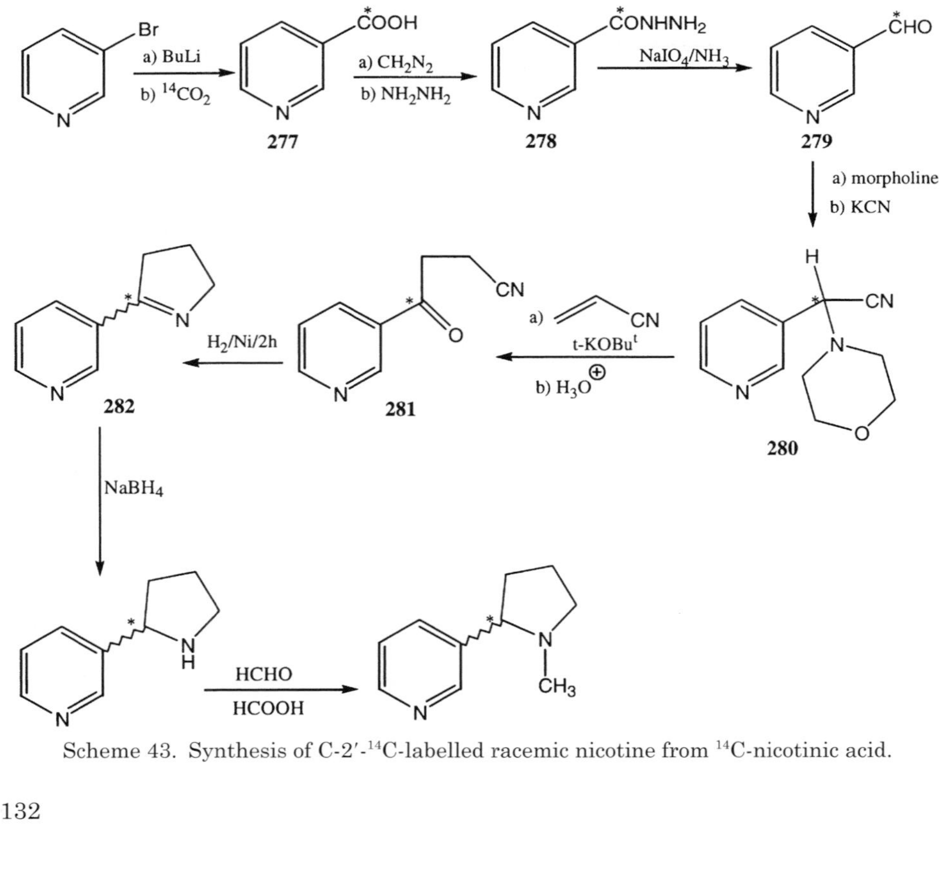

Scheme 43. Synthesis of C-2'-[14]C-labelled racemic nicotine from [14]C-nicotinic acid.

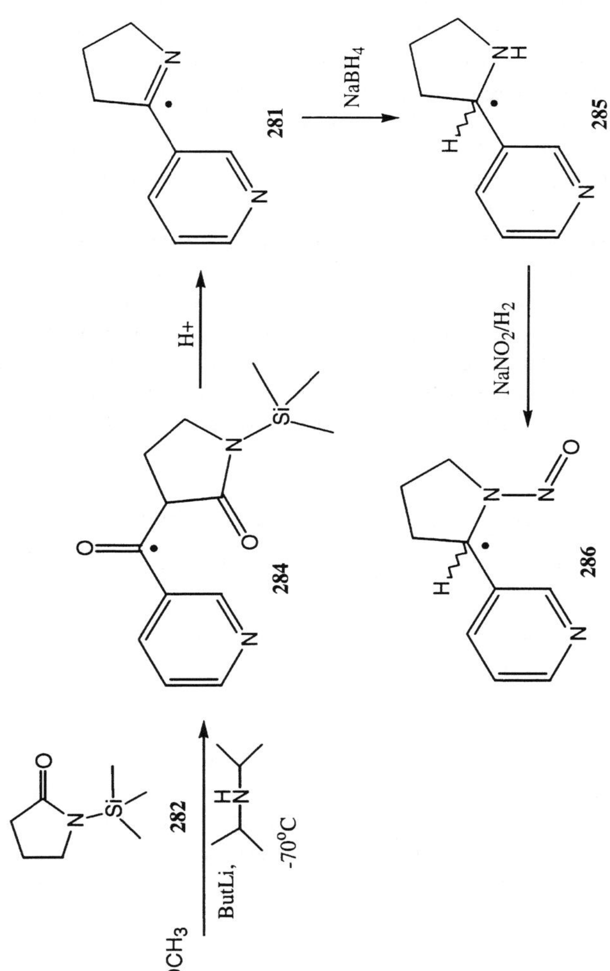

Scheme 44. Synthesis of [2′-¹⁴C]-myosmine, [2′-¹⁴C]-nornicotine, and [2′-¹⁴C]-*N*-nitrosonornicotine from methyl [¹⁴C-carboxyl]-nicotinate.

Several methods for the preparation of (±)-[2′-$^{14}$C]-nicotine (270) are described in the literature. Comes et al. [233] reacted [$^{14}$C-carboxyl]-nicotinamide (271) with $P_2O_5$ to obtain [$^{14}$C-cyano]-3-cyanopyridine (272), which was then reacted with cyclopropyllithium (273) to afford [$^{14}$C-carboxyl]-cyclopropyl 3-pyridyl ketone (274). Reaction of this keto-intermediate with *N*-methylformamide/magnesium chloride afforded (±)-[2′-$^{14}$C]-nicotine) in 11% overall yield (Scheme 41). A less efficient biomimetic synthesis affords (±)-[2′-$^{14}$C]-nicotine in 7% yield, from the reaction of 1,4-dihydropyridine (275) with [5-$^{14}$C]-$^1\Delta$-methylpyrrolinium acetate (276) at pH 10.3 and ambient temperature for 3 days [234] (Scheme 42). An eight-step synthesis from 3-bromopyridine has also been reported [4] (see Scheme 43); in this synthesis, 3-bromopyridine is converted into [$^{14}$C-carboxyl]-nicotinic acid (277) by reaction with butyl lithium and $^{14}CO_2$. The nicotinic acid is then converted, in a two step procedure, to the 3-carboxyhydrazido derivative 278, and then to [$^{14}$C-carboxyl]-2-pyridine carboxaldehyde (279). Reaction with morpholine and KCN to give 280, followed by Michael reaction with acrylonitrile, and acid treatment, afforded the ketocyano intermediate, 281, which could be reductively cyclized by hydrogenation over nickel catalyst to give [2′-$^{14}$C ]-myosmine (282). (±)-[2′-$^{14}$C]-Nornicotine and (±)-[2′-$^{14}$C]-nicotine were then obtained via standard procedures. Although this synthesis has a very low overall yield, due to the number of steps involved, it starts from an inexpensive $^{14}$C precursor, and generates three [2′-$^{14}$C]-labelled tobacco alkaloids. Hu et al. [235] has reported a synthesis of [2′-$^{14}$C]-myosmine, [2′-$^{14}$C]-nornicotine, and [2′-$^{14}$C]-*N*-nitrosonornicotine from methyl [$^{14}$C-carboxyl]-nicotinate (see Scheme 44) utilizing a procedure developed by Korte and Schulze-Steiner [236]. *N*-Trimethylsilyl-2-pyrrolidone (283) is reacted with methyl [$^{14}$C-carboxyl]-nicotinate (284) to form the intermediate [$^{14}$C-carbonyl]-3-nicotinoyl-*N*-trimethylsilyl-2-pyrrolidone (285), which is converted to [2′-$^{14}$C]-myosmine (282) on treatment with acid. Subsequent reduction with $NaBH_4$ afforded (±)-[2′-$^{14}$C]-nornicotine (286), which could be converted into [2′-$^{14}$C]-*N*-nitrosonornicotine (287) by reaction with $NaNO_2$/HCl.

The synthesis of (±)-[5-$^{125}$I]-5-iodonicotine has been described by Chan et al. [237] via reaction of 5-bromonicotine with Na$^{125}$I in ethanol/acetic acid containing copper sulfate as catalyst, at 121°C for 35 min. The radiochemical yield from this reaction was 25%. $^{125}$I is a single photon gamma emitter; thus nicotine labeled with $^{125}$I has utility in imaging and brain uptake studies.

# REFERENCES

1 Bush LP, Fannin FF, Chelvarajan RL Burton HR. Biosynthesis and metabolism of nicotine and related alkaloids. In: *Nicotine and Related Alkaloids-Absorption, Distribution, Metabolism, Excretion*, Gorrod JW, Wahren J (Eds.), Chapman and Hall, London, pp. 1–30, 1993.

2 Crooks PA, Godin CS, Pool WF. Enantiomeric purity of nicotine in tobacco smoke. *Med. Sci. Res.*, 1992: **20**; 879–880.

3 Kisaki T, Tamaki E. Phytochemical studies on the tobacco alkaloids: I. Optical rotatory power of nornicotine. *Arch. Biochem. Biophys.*, 1966: **92**; 351–355.

4 Leete E, Chedekel MR. Metabolism of nicotine in *Nicotiana glauca*. *Phytochem.*, 1974: **13**; 1853–1859.

5 Dwoskin LP, Teng LH, Buxton ST, Ravard A, Deo N, Guo YS, Crooks PA. Minor alkaloids of tobacco release [³H]-dopamine from superfused rat striatal slices. *Eur. J. Pharmacol.*, 1995: **276**; 195–199.

6 Kem WR. *Handbook of Natural Toxins, Marine Toxins and Venoms*, Tu AT (Ed.), Marcell-Dekker, New York, Vol. 3, Chapter 15, 1988.

7 Kim, HS, Jeffrey GA. The crystal structure of a 1:1 nicotine-salicylic acid complex (nicotinyl salicylate). *Acta Cryst.*, 1971: **B27**; 1123–1131.

8 Pullman B, Courriere P, Coubiels JL. Quantum mechanical study of the conformational and electronic properties of acetylcholine and its agonists muscarine and nicotine. *Mol. Pharmacol.*, 1971: **7**; 397–405.

9 Kier LB. A molecular orbital calculation of the preferred conformation of nicotine. *Mol. Pharmacol.*, 1968: **4**; 70–76.

10 Testa B, Jenner P. Circular dichroic determination of the preferred conformation of nicotine and related chiral alkaloids in aqueous solution. *Mol. Pharmacol.*, 1973: **9**; 10-16.

11 Chynoweth KR, Ternai B, Simeral LS, Maciel GE. Nuclear magnetic studies of the conformation and electronic distributions in nicotine and in acetylcholine. *Mol. Pharmacol.*, 1973: **9**; 144–151.

12 Ohashi M, Morishima I, Yonezawa T. Application of proton NMR shift reagents to the stereochemical analysis of nicotine. *Bull. Chem. Soc. Jpn.*, 1971: **44**; 576–577.

13 Pitner TP, Edwards III WB, Bassfield RL, Whidby, JF. The solution conformation of nicotine. A ¹H and ²H nuclear magnetic resonance investigation. *J. Amer. Chem. Soc.*, 1978: **100**; 246–251.

14 Seeman JI. Recent studies in nicotine chemistry, conformational analysis, chemical reactivity studies and theoretical modeling. *Heterocycles*, 1984: **22**; 165–193.

15 Zoltewicz JA, Bloom LB, Kem WR. Quantitative determination of the ring-chain hydrolysis equilibrium constant for anabaseine and related tobacco alkaloids. *J. Org. Chem.*, 1989: **54**; 4462–4468.

16 Badgett CO, Eisner A, Walens HA. Distribution of pyridine alkaloids in the system buffer-*t*-amyl alcohol. *J. Amer. Chem. Soc.*, 1952: **74**; 4096–4098.

17 Ghose AK, Crippen GN. Atomic physicochemical parameters for three-dimensional-structure-directed quantitative structure-activity relationships. 2. Modeling dispersive and hydrophobic interactions. *J. Chem. Inf. Comput. Sci.*, 1987: **27**; 21–35.

18 Viswanadhan VN, Ghose AK, Revankar GR, Robins RK. Atomic physicochemical parameters for three-dimensional structure directed quantitative structure-activity relationships. 4. Additional parameters for hydrophobic and dispersive interactions and their application for an automated superposition of certain naturally occurring nuleoside antibiotics. *J. Chem. Inf. Comput. Sci.*, 1989: **29**; 163–1172.

19  Broto P. Molecular structures: perception, autocorrelation descriptor and SAR studies. System of atomic contributions for the calculation of the *n*-octanol/water partition coefficients. *Eur. J. Med. Chem., Chim. Ther.*, 1984: **19**; 71–78.

20  Perfetti TA. Structural study of nicotine salts. *Beitrage Tabak. Int.*, 1983: **12**; 43–54.

21  Teng LH, Buxton ST, Crooks PA Dwoskin LP. Lobeline and nicotine evoke [³H]-overflow from rat striatal slices preloaded with [³H]dopamine: differential inhibition of synaptosomal and vesicular [³H]dopamine uptake. *J. Pharmacol. Exp. Therap.*, 1997: **80**; 1432–1444.

22  Dwoskin LP, Buxton ST, Jewell AL, Crooks PA. S-(−)-Nicotine increases dopamine release in a calcium-dependent manner from superfused rat striatal slices. *J. Neurochem.*, 1993: **60**; 2167–2174.

23  Kem WR. Structure and action of nemertine toxins. *Am. Zool.*, 1985: **25**; 99–111.

24.  Dwoskin LP, Teng LH, Buxton ST, Crooks PA. S-(−)-Cotinine, the major brain metabolite of nicotine, stimulates nicotinic receptors to evoke [³H]dopamine release from rat striatal slices in a calcium-dependent manner. *J. Pharmacol. Exp. Therap.*, 1999: **228**; 905–911.

25  Ciamician G, Silber P. Chemische lichtwirkungen, XXXI. Autooxydationen, VIII. *Chem. Ber.*, 1915: **48**; 181–187.

26  Wada E, Kisaki T, Saito K. Autooxidation of nicotine. *Arch. Biochem. Biophys.*, 1959: **79**; 124–130.

27  Caldwell WS, personal communication.

28.  Weidel H. Zur kenntnifs des nicotins, *Justus Liebigs Ann. Chem.*, 1873: **165**; 328–349.

29  Huber C. Vorläufige notiz über einige derivate des nicotins. *Justus Liebigs Ann. Chem.*, 1867: **141**; 271.

30  Huber C, Vorläufige Mittheilung. *Chem. Ber.*, 1870: **3**; 849.

31  Laiblin R von. Zur kenntniss des nicotines. *Chem. Ber.*, 1877: **10**; 2136–2140.

32  Laiblin R von. Ueber nicotin und nicotinsäure. *Justus Liebigs Ann. Chem.*, 1879: **196**; 129–182.

33  Pictet A, Sussdorf, *Org. Synth. Coll. Vol.* I, New York, S378, 1932.

34  Weidel H. Zur kenntnifs des nicotins. *Justus Liebigs Ann. Chem.*, 1873: **165**; 328–349.

35  McElvan SM, Adams R. Synthesis of a new bicyclic nitrogen ring. Isogranatanine derivatives. Preparation of an isomer of homococaine. *J. Amer. Chem Soc.*, 1923: **45**; 2738–2750.

36.  Woodward CF, Badgett CO, Kaufman JG. Chem-catalytic liquid-phase oxidation of nicotine, β-picoline and quinoline to nicotinic acid. *Ind. Eng. Chem.*, 1944: **36**; 544–546.

37  Matveev BV. Catalytic oxidation of anabasine. *J. Gen. Chem. USSR*, **17**; 482–484: 1947. *Chem. Abs.*, 1948: **42**; 571h.

38  Leete E. Tobacco alkaloids and related compounds. Part 49. Spermidine: an indirect precursor of the pyrrolidine rings of nicotine and nornicotine in *Nicotiana glutinosa. Phytochem.*, 1985: **24**; 957–960.

39  Lund H. Pyridylnitropyrazole, an oxidation product of nicotine. The position of the nitro group. *J. Chem. Soc.*, 1933: 686–687.

40  Gough GAC, King H. 4-Nitro–5-(3-pyridyl)pyrazole, a new oxidation product of nicotine. Part I. *J. Chem. Soc.*, 1931: 2968–2971.

41 Beckett AH, Jenner P, Gorrod JW. Characterization of the diastereomers of nicotine-1′-*N*-oxide, a metabolite of nicotine, and other possible oxidation products by nuclear magnetic resonance spectroscopy. *Xenobiotica*, 1973: **3**; 557–562.

42 Johnson AW, King TJ, Turner JR. New transformation products of nicotine. *J. Chem. Soc.*, 1958: 3230–3231.

43 Taylor EC, Boyer NE. Pyridine-1-oxides. IV. Nicotine-1-oxide, nicotine-1′-oxide, and nicotine-1,1′-dioxide *J. Org. Chem.*, 1959: **24**; 275–277.

44 Booth J, Boyland E. The metabolism of nicotine into two optically-active stereoisomers of nicotine-1′-oxide by animal tissues in vitro and by cigarette smokers. *Biochem. Pharmacol.*, 1970: **19**; 733–742.

45 Kisaki T, Ihida M, Wada E. A new synthesis of *N*-methylmyosmine. *Bull. Agric. Chem. Soc. Japn.*, 1959: **23**; 454–455.

46 Phillipson JD, Handa SS. Nicotine *N*–1-oxides. *Phytochem.*, 1975: **14**; 2683–2690.

47 Brandage S, Lindblom L, Samuelsson D. Stereochemistry of oxidation of nicotine to its 1′-*N*-oxides. The action of tungstate (VI) and molybdate (VI). *Acta Chemica. Scand.*, 1977: **B31**; 907–922.

48 Seaton QF, Lawley CW, Akers HA. The reduction of aliphatic and aromatic *N*-oxides to the corresponding amines with titanium(III) chloride. *Anal. Biochem.*, 1984: **138**; 230–241.

49 Schmidt B, Neitmeyer V. 6-Pyridylnicotine and *bis*-6,6′-nicotine-new chiral 2,2′bipyridines. *Synthesis*, 1998: **1**; 42–44.

50 Takeshita M, Yoshida S. Reduction of *N*-oxides with bakers' yeast. *Heterocycles*, 1990: **30**; 871–874.

51 Shibagaki M, Matsushito H, Kaneko H. The synthesis of 4-aminonicotine and 4-aminocotinine. *Heterocycles*, 1985: **23**; 1681–1684.

52 Kisaki T, Ihida M, Tamaki E. Chemistry of *N*-oxides of nicotine and myosmine. *Bull. Agric. Chem. Soc. Japn.*, 1960: **24**; 719–728.

53 Moran DB, Morton GO, Albright JD. Synthesis of (pyridinyl)–1,2,4-triazolo[4,3a]-pyridines. *J. Hetero. Chem.*, 1986: **23**; 1071–1077.

54 Plaquevant J-C, Chichaoui I. Reduction régiospécifique des bipyridines. *Tet. Lett.*, 1993: **34**; 5287–5288.

55 Botte M, Mabon F, Mouillour ML, Robins RJ. Biosynthesis of nornicotine in root cultures of *Nicotiana alata* does not involve oxidation at C-5′ of nicotine. *Phytochem.*, 1997: **36**; 117–122.

56 Vincek WC, Martin BR, Aceto MD, Tripathi HL, May EL, Harris, LS. Synthesis of 4′,4′-ditritio-(+)-nicotine: comparative binding and distribution studies with natural enantiomer. *J. Pharm. Sci.*, 1981: **70**; 1292–1293.

57 McKennis H, Turnbull LB, Bowman ER. Demethylation of cotinine in vivo. *J. Amer. Chem. Soc.*, 1959: **81**; 3951–3954.

58 Bowman ER, McKennis H. (−)-Cotinine. *Biochem. Prep.*, 1963: **10**; 36–39.

59 Moriarty RM, Vaid RK, Duncan MP, Ochiai M, Inenaga M, Nagao Y. Hypervalent iodine oxidation of amines using iodosobenzene: synthesis of nitriles, ketones and lactams. *Tet. Lett.*, 1988: **29**; 6913–6916.

60 Chauncey MA, Ninomiya S-I., Metabolic studies with model cytochrome P 450 systems. *Tet. Lett.*, 1990: **31**; 5901–5904.

61 Jacob III P, Shulgin AT, Benowitz NL. Synthesis of (3′R, 5′S)-*trans*-3-hydroxy-cotinine, a major metabolite of nicotine. Metabolic formation of 3′-hydroxycotinine in humans is highly stereoselective. *J. Med. Chem.*, 1990: **33**; 1888–1891.

62  Lin N-H, Carrera GM, Anderson DJ. Synthesis and evaluation of nicotine analogs as neuronal nicotinic acetylcholine receptor ligands. *J. Med. Chem.*, 1994: **37**; 3542–3553.

63  Perrone R, Carbonara G, Tortorella V. Chemical studies on drug metabolism oxidation with ruthenium tetroxide of some medicinal alicyclic *N*-methylamines. *Arch. Pharm.*, 1984: **317**; 21–27.

64  Barton DHR, Boivin J, Gaudin D, Jankowski K. On the GIF oxidation of alicyclic tertiary amines. *Tet. Lett.*, 1989: **30**; 1381–1382.

65  Wenkert E, Angell EC. Preparation of lactams via oxidation of cyclic tertiary and secondary amines with Hg(II)-EDTA complex in alkali medium. *Synth. Comm.*, 1988: **18**; 1331–1338.

66  Akabori S, Suzuki T. *Proc. Imp. Acad. Tokyo*, **5**; 255–256. Katalysche übertragung von wasserstoff zwischen organischen verbindungen. *Chem. Zentralblt.*, 1929: **II**; 2033.

67  Cislak FE, Wheeler WR. Catalytic oxidation of nicotine. Reilly Tar and Chemical Corp. *US Patent* 2,456,380, 1941. *Chem. Abs.*, 1949: **43**; 1811b.

68  Cislak FE, Wheeler WR. Catalytic oxidation of substituted pyridine bases. Reilly Tar and Chemical Corp. *US Patent* 2,437,998, 1941. *Chem. Abs.*, 1948: **42**; 4204f.

69  Woodward CF, Badgett CO, Willaman JJ. Nicotine to nicotinonitrile-catalytic vapor-phase oxidation. *Ind. Eng. Chem.*, 1944: **36**; 540–544.

70  Woodward CF, Badgett CO, Haines PG. Oxidation of nicotine to 1-methyl-2-(3-pyridyl)pyrrole. *US Patent* 2,432,642:1943. *Chem. Abs.*, 1948: **42**; 2624d.

71  Woodward CF, Eisner A, Haines PG. Pyrolysis of nicotine to myosmine. *J. Amer. Chem. Soc.*, 1944: **66**; 911–914.

72  Wibaut JP, Overhoff J. The catalytic hydrogenation of nicotine. A suitable method for the preparation of *N*-methyl-(3-pyridyl)-2-pyrrole (3,2′-nicotyrine). *Recl. Trav. Chem. Pays-Bas*, 1928: **47**; 935–939.

73  Shibugaki M, Takahashi K, Kuno H, Matsushita H. Preparation of nicotyrine via catalytic dehydrogenation of nicotine. *Agric. Biol. Chem.*, 1988: **10**; 2651–2652.

74  Pictet A. *N*-Methyl-pyrrolidin aus nicotin. *Chem. Ber.*, 1905: **38**; 1951–1952.

75  Späth E, Kesztler F. L-Anatabin, ein neues tabakalkaloid (XI. Mitteil. über tobakbasen). *Chem. Ber.*, **70**; 239–243:1937. Über das vorkommen von d,l nornicotin, d,l anatabin und l-anabasin im tabak (XII. Mitteil. über tabak-alkaloide). *Chem. Ber.*, 1937: **70**; 704–709.

76  Orechof A, Menschikoff G. Über die alkaloide von *Anabasis aphylla*. *Chem. Ber.*, 1931: **64**; 266–274.

77  Smith CR. Occurrence of anabasine in *Nicotiana glauca* R. Grah. (*Solanaceae*). *J. Amer. Chem. Soc.*, 1935: **57**; 959–960.

78  Wada E, Kisaki T, Ihida M. Tobacco alkaloids in the root and sap of some *Nicotiana* plants. *Arch. Biochem. Biophys.*, 1959: **80**; 258–263.

79  Auterhoff H, Huth W. Reactions of nicotyrine with 4-(dimethylamino)benzaldehyde. *Arch. Pharm.*, 1971: **304**; 288–296.

80  Morton AA, Horvitz D. The dehydrogenation of nicotine in toluene as solvent. *J. Amer. Chem. Soc.*, 1935: **57**; 1860–1861.

81  Wawzonek S, Hansen GR. The action of sulfur on cyclic amines. *J. Org. Chem.*, 1966 **31**; 3580–3582.

82  Acheson RM, Ferris MJ, Critchley SR, Watkins DJ. Identification of the product from nicotine and sulphur as *bis*-1-methyl-2-(3-pyridyl)pyrrol-3-yl disulphide. *J.*

*Chemical Properties of Nicotine and Other Tobacco-related Compounds*

*Chem. Soc., Perk. Trans.*, 1980: **2**; 326–329.

83  Uchida S, Maeda S, Kisaki T. Conversion of nicotine into nornicotine and *N*-methylmyosmine by fungi. *Agric. Biol. Chem.*, 1983: **47**; 1949–1954.

84  Rodiut J-P, Wellig A, Kiener A. Renewable functionalized pyridines derived from microbial metabolites of the alkaloid (S)-nicotine. *Heterocycles*, 1997: **45**; 1687–1702.

85  Hochstein LI, Rittenburg SC. Bacterial oxidation of nicotine (I) nicotine oxidation by cell free preparations (II) isolation of the first oxidative product and its identification as (l)-6-hydroxynicotine. *J. Biol. Chem.*, 1959: **234**; 151–156.

86  Gloger M, Decker K. Microbial preparation of the optical antipodes of 6-hydroxy-nicotine. *Z. Naturforsche*, 1969: **24b**; 140.

87  Decker K, Eberwein H, Gries FA, Bruehmueller M. Decomposition of nicotine by bacterial enzymes. IV. L–6-Hydroxynicotine as the first intermediate. *Biochem Z.*, 1961: **334**; 227–244.

88  Wada E, Yamasaki K. Degradation of nicotine by soil bacteria. *J. Amer. Chem. Soc.*, 1954: **76**; 155–157.

89  Frankenburg WG, Vaitekumas AA. Chemical studies on nicotine degradation by microorganisms derived from the surface of tobacco seeds. *Arch. Biochem. Biophys.*, 1954: **58**; 509–512.

90  Wada E. Microbial degradation of nornicotine. *Arch. Biochem. Biophys.*, 1956: **64**; 244–246.

91  Wada E. Microbial degradation of the tobacco alkaloids, and some related compounds. *Arch. Biochem. Biophys.*, 1957: **72**; 145–162.

92  Windus W, Marvel CS. The reduction of nicotine and some derivatives of hexa- and octahydronicotines. *J. Amer. Chem. Soc.*, 1930: **52**; 2543–2546.

93  Pinner A. Über nicotin (6. Mittheilung). *Chem. Ber.*, 1893: **26**; 765–771.

94  Liebrecht A. Reduction des nicotins. *Chem. Ber.*, 1886: **19**; 2587–2598.

95  Blau F. Zur constitution des nicotins. *Chem. Ber.*, 1893: **26**; 628–633.

96  Britton EC, Horsley LH. Dow Chem. Co. 1-Substituted 3-(1-methyl-2-pyrrolidine)-piperidine *US Patent*, 2,834,784, 1958.

97  Harlan WR, Hixon RM. Catalytic reduction of nicotine and metanicotine. *J. Amer. Chem. Soc.*, 1930: **52**; 3385–3388.

98  Orechoff A, Brodsky D. Über die alkaloide von *Anabasis aphylla*, VI. Mitteil.: über die hydrierung des anabasins. *Chem. Ber.*, 1933: **66**; 466–468.

99  Sadykov AS, Otroschtschenko OS. Syntheses on the basis of anabasine. XIII. Hydrogenation of anabasine. *J. Gen Chem. USSR (Engl.)*, 1959: **29**; 2400–2401.

100  Forostyan YN, Lazur'evskii GV. Syntheses based on hydrogenated anabasine. *Trudy Chim. Prirdn. Soeden, Kisinevsk. Univ.*, 1959: **2**; 53–54. *Chem. Abs.*, 1961: **55**; 27307b.

101  Späth E, Kesztler F. l-Anatabin, ein neues tabakalkaloid (XI. Mitteil. über tabak-basen). *Chem. Ber.*, 1937: **70**; 239–242.

102  Yang CM, Tanner DD. A simple synthesis of (±)-1,2,3,6-tetrahydro-2,3′-bipyridine (anatabine) and (±)-3-(2-piperidinyl)pyridine (anabasine) from lithium aluminum hydride and pyridine. *Can. J. Chem.*, 1997: **75**; 616–620.

103  Alberici GF, Andrieux J, Adam G, Plat MM. Synthesis of tobacco alkaloids via tertiary azides. *Tet. Lett.*, 1983: **24**; 1937–1940.

104  Murakoshi I. Synthesis of cyclic nitrogenous compounds from amino acids. V. Synthesis of 2-pyrroline derivatives from 4-aminobutyric acid and L-glutamic acid.

*Yakugaki Zosshi.*, **77**; 1062–1064: 1957. *Chem. Abs.*, 1958: **52**; 5376.

105 Schoepf C, Komzac A, Braun F, Jacobi E. Über die polymeren des Δ¹-piperideins *Justus Liebigs Ann. Chem.*., 1948: **559**; 1–42.

106 Smith CR. Neonicotine and isomeric pyridylpiperidines. *J. Amer. Chem. Soc.*, 1931: **53**; 277–281.

107 Lunn G. Preparation of piperidinylpyridines via selective reduction of bipyridines with nickel-aluminum alloy. *J. Org. Chem.*, 1992: **57**; 6317–6320.

108 Koning AJ De, Budzelaar PHM, Boersma J, Kerk GJM Van Der. Specific and selective reduction of aromatic nitrogen heterocycles with the *bis*-pyridine complexes of *bis*(1,4-dihydro-1-pyridyl)zinc and *bis*(1,4-dihydro-1-pyridyl)magnesium. *J. Organomet. Chem.*, 1980: **199**; 153–170.

109 Späth E, Kuffner F. Ein vereinfachung der Pictetschen nicotin-synthese (II. Mitteil. über tabak-basen). *Chem. Ber.*, 1935: **68**; 494–497.

110 Overhoff J, Wibaut JP. The catalytic hydrogenation of pyridine derivatives (methyl- and phenyl-pyridines, quinoline, pyridylpyrroles, nicotyrines, nicotine). *Recl. Trav. Chem. Pays-Bas.*, 1931: **50**; 957–980.

111 Wibaut JP, Hackman JT. Reduction of 3,2'-nicotyrine to dihydronicotyrine and inactive nicotine-disproportionation of dihydronicotyrine. *Recl. Trav. Chem. Pays-Bas.*, 1932: **51**; 1157–1165.

112 McKennis H Jr, Turnbull LB, Wingfield HN Jr, Dewey LJ. Metabolites of nicotine and a synthesis of nornicotine. *J. Amer. Chem. Soc.*, 1958: **80**; 1634–1636.

113 Zymalkowski F, Trenktrog B. A novel synthesis of nicotine. *Arch. Pharm.*, 1959: **292**; 9–14.

114 Duffield AM, Budzikiewicz H, Djerassi C. Mass spectrometry in structural and stereochemical problems. LXXII. A study of the fragmentation process of some tobacco alkaloids. *J. Amer. Chem. Soc.*, 1965: **87**; 2926–2932.

115 Ravard A, Crooks PA. Chiral purity determination of tobacco alkaloids and nicotine-like compounds by ¹H NMR spectroscopy in the presence of 1,1'-binaphthyl-2,2'-diylphosphoric acid. *Chirality.*, 1996: **8**; 295–299.

116 Peterson LA, Trevor A, Castagnoli N Jr. Stereochemical studies on the cytochrome P-450 catalyzed oxidation of (S)-nicotine to (S)-nicotine Δ1'(5')-iminium species. *J. Med. Chem.*, 1987: **30**; 249–254.

117 Späth E, Marion L, Zajic E. Synthese des l-nor-nicotine (IV. Mitteil. über tabakbasen). *Chem. Ber.*, 1936: **69**; 251–255.

118 Acheson RM, Ferris MJ, Sinclair NM. Transformations involving the pyrrolidine ring of nicotine. *J. Chem. Soc., Perk. Trans.*, 1980: **1**; 579–585.

119 Joyce NJ, Leete E. The formation of 1-methyl-3-nicotinylpyrrolidine from nicotine-1'-oxide. *Heterocycles.*, 1989: **29**; 1335–1342.

120 Planta A von, Kekule A. Beiträge zur kenntnifs einiger flüchtigen basen. *Ann. Pharm. Chem.*, 1853: **87**; 1–11.

121 Stahlschmidt. Untersuchung über einige methylirte organische basen. *Chem. Pharm.* 1854: **90**; 218–228.

122 Pictet A, Genequand P. Ueber die jodomethylate des nicotins. *Chem. Ber.*, 1897: **30**; 2117–2125.

123 Seeman JI, Whidby JF. The iodomethylation of nicotine. An unusal example of competitive nitrogen alkylation. *J. Org. Chem.*, 1976: **41**; 3824–3826.

124 Seeman JI. Recent studies in nicotine chemistry. Conformation analysis, chemical reactivity studies, and theoretical modeling. *Heterocycles.*, 1984: **22**; 165–193.

125 Shibagaki M, Matsushita H, Shibata S, Saito A, Tsujino Y, Kaneko H. The selectivity in *N*-alkylation of nicotine. *Heterocycles.*, 1982: **19**; 1641–1645.

126 Crooks PA, Ravard A, Teng LH, Dwoskin LP. Inhibition of nicotine-evoked dopamine release by pyridino *N*-substituted analogues: a new class of nicotinic antagonists. *Drug Dev. Res.*, 1995: **36**; 71–82.

127 Späth E, Zajic E. Über das l-nor-nicotin (III. Mitteil. über tabak-basen.). *Chem. Ber.*, 1935: **68**; 1667–1670.

128 Späth E, Kainrath E. Uber die Pictetsche nicotin-synthese (XV. Mitteil. über tabak-alkaloide). *Chem. Ber.*, 1938: **71**; 1276–1281.

129 Jones JP, Trager WF, Carlson TJ. The binding and regioselectivity of reaction of (R)- and (S)-nicotine with cytochrome P-450cam: parallel experimental and theoretical studies. *J. Amer. Chem. Soc.*, 1993: **115**; 381–387.

130 Murahashi S-I,, Mitsui H, Shiota T, Tsuda T, Watanabe S, Tungstate-catalyzed oxidation of secondary amines to nitrones. α-Substitution of secondary amines via nitrones. *J. Org. Chem.*, 1990: **55**; 1736–1744.

131 Jacob P III, Benowitz NL, Copeland JR, Risner ME, Cone EJ. Disposition kinetics of nicotine and cotinine enantiomers in rabbits and beagle dogs. *J. Pharm. Sci.*, 1988: **77**; 396–400.

132 Leete E. *N'*-Isopropylnornicotine: its formation from nicotine in aged leaves of *Nicotiana tabacum*. *Phytochem.*, 1981: **20**; 1037–1040.

133 von Braun J, Weissbach K. Entalkylierung tertiärer amine durch organische säuren, II. Mitteilung: nicotin. *Chem. Ber.*, 1930: **63**; 2018–2026.

134 Seitz G, Tegethoff R. Heterocyclically annulated indolizines by intramolecular [3+2]-cycloaddition with nicotinium dicyanomethylides. *Arch. Pharm.*, 1993: **326**; 443–446.

135 Klimek C, Seitz G. Synthesis of conformationally restricted anabasine analogs. Heterocyclically annulated indolizidines by intramolecular [3+2]cycloaddition with anabasinium-dicyanomethylides. *Sci. Pharm.*, 1996: **64**; 497–502.

136 Sadykov AS, Ashrapova N. Syntheses from anabasine. III. Condensation of aliphatic oxides with anabasine. *Zh. Obshsch. Khim.*, **17**; 1212–1215: 1948. *Chem. Abs.*, 1948: **42**; 1601d.

137 Sadykov AS. Ostroshchenko OS. Syntheses based on anabasine. V. Products of transformation of *N*-(2-hydroxyethyl)anabasine. *Zh. Obshsch. Khim.*, 1953: **23**; 968–971.

138 Seeman JI, Chavdarian CG, Secor HV, Osdene TS. Preparation of hydroxyalkyl-substituted nicotinoids. *J. Org. Chem.*, 1986: **51**; 1548–1551.

139 Brunner H, Kuerzinger A, Mahboobi S, Wiegrebe W. Enantioselective hydrosilation and hydrogenation of alkaloid precursors. *Arch. Pharm.*, 1988: **321**; 73–76.

140 Miyano M, Yasumatsu N, Matsushita H, Nashida K. 1'-(6-Hydroxyoctanoyl)-nornicotine and 1'-(7-hydroxyoctanoyl)nornicotine, two new alkaloids from Japanese domestic tobacco. *Agric. Biol. Chem.*, 1981: **45**; 1029–1033.

141 Weymann M, Pfrengle W, Schollmeyer D, Kunz H. Enantioselective synthesis of 2-alkyl-, 2,6-dialkylpiperidines and indolizidine alkaloids through diastereoselective Mannich-Michael reactions. *Synthesis*, 1997: **10**; 1151–1160.

142 Cox RH, Kao J, Secor HV, Seeman JI. Assesment of isolated electronic effects on conformation. NMR analysis of nicotine and related compounds and ab initio studies on model compounds. *J. Mol. Struct.*, 1986: **140**; 93–106.

143 Ehrenstein M. Alkaloids of tobacco. *Arch. Pharm.*, 1931: **269**; 627–659.

144 Svetkin YV. Reaction of ketene with nitrogenous bases. IV. Acetylation of *Anabasis* alkaloids with ketene. *Zh. Obshsch. Khim.*, 1957: **27**; 2570–2572. *Chem. Abs.*, 1958: **52**; 7335d.

145 Gazaliev AM, Zhurinov MZh, Tilyabaev Z, Dalemov DN, Mukanova DK, Dyusembaev SA. Synthesis of organophosphorous derivatives of anabasine and their cholinergic activity. *Khim. Prir. Soedin.*, 1989: **4**; 584–585. *Chem. Abs.*, 1990: **112**; 50550f.

146 Takahashi K, Shibagaki M, Matoushita H. Formylation of amines by dimethylformamide in the presence of hydrous zirconium oxide. *Agric. Biol. Chem.*, 1988: **52**; 853–854.

147 Etard A. *Bull. Soc. Chim. France*, 1894: **3**; 109-112.

148 Pinner A. Sue la saturation des azoles de la nicotine et sur l'acetylnicotine. *Arch. Pharm.*, 1895: **233**; 586-589.

149 Hromarka O. Über verbindungen hydrierten nicotine. *Chem. Ber.*, 1942: **75**; 522–530.

150 Rayburn CH, Harlan WR, Hanmew HR. Rearrangement of nicotine oxide. *J. Amer. Chem. Soc.*, 1950: **72**; 1721–1723.

151 Carruthers W, Johnstone RAW. Thermal rearrangement of nicotine 1'-oxide and related compounds. *J. Chem. Soc.*, 1965: 1653–1658.

152 Jacob P III, Benowitz NL, Yu L, Shulgin AT. Determination of nicotine-*N*-oxide by gas chromatography following thermal conversion to 2-methyl-6-(3-pyridyl)-tetrahydro-1,2-oxazine. *Anal. Chem.*, 1986: **58**; 2218–2221.

153 Tschischibabin AE, Kirssanow AW. Aminierung des nicotins mit natrium- und kaliumamid. *Chem. Ber.*, 1924: **57**; 1163–1168.

154 Tschischibabin AE, Bucholz LA. 2(or 6)-Aminonicotine. *Zh. Russ. Fiz.-Khim. O.-va.*, **50**; 548–552: 1949. *Chem. Abs.*, 1924: **18**; 1502.

155 Sadykov AS, Otroshchenko OS, Yusupov MK. Syntheses based upon anabasine. VI. Amination of anabasine, *N*-methylanabasine and nicotine. *Zh. Obshsch. Khim.*, **23**; 980–982: 1953. *Chem. Abs.*, 1954: **48**; 7616f.

156 Ohnishi A, Kodama H, Kato K. *N*-Glycoside of nicotine derivative *N*-[5-(*N'*-methyl–2'-pyrrolidinyl)–2-pyridine]-β-D-glucopyranoside. *Agric. Biol. Chem.*, 1982: **46**; 831–832.

157 Rhondahl L. Synthetic analogues of nicotine. *Doctoral Thesis*, Royal Institute of Technology, Stockholm, Sweden, pp. 8–9, 1980.

158 Katznel'son MM, Kabachnik MI. Introduction of the amino group into compounds of the alkaloid series by the action of sodium and potassium amides. I. α-Aminoanabasine. *Bull. Soc. Chim. France*, **2**; 521–523: **2**; 576–579:1935. *Chem. Abs.*, 1934: **28**; 4059.

159 Khromov-Borisov NV, Vatkina EG. A method of isolation of 6-aminoanabasine and lupinine from a technical mixture of anabasine and lupinine. *Zh. Obshsch. Khim.*, **25**; 1161–1162: 1955. *Chem. Abs.*, 1956: **50**; 3442b.

160 Clemo GR, Swann GA. A contribution to the study of nicotine and the synthesis of 7-azaindole derivatives. *J. Chem. Soc.*, 1945: 603–607.

161 Haglid F. The methylation of nicotine with methyl-lithium. *Acta Chem. Scand.*, 1967: **21**; 329–334.

162 Leete E, Leete SAS. Synthesis of 4-methylnicotine and an examination of its possible biosynthesis from 4-methylnicotinic acid in *Nicotiana tabacum*. *J. Org.*

*Chem.*, 1978: **43**; 2122–2125.

163 Itokawa H, Inaba T, Haruta R, Kameyama S. Radical methylation and radical hydroxymethylation of nicotine and quinine. *Chem. Pharm. Bull.*, 1978: **26**; 1295–1297.

164 Seeman JI, Secor HV, Howe CR, Chavdarian CG, Morgan LW. Organometallic methylation of nicotine and nicotine *N*-oxide. Reaction pathways and racemization mechanisms. *J. Org. Chem.*, 1983: **48**; 4899–4904.

165 Hauck AE, Giam C-S. Regioselective nucleophilic addition of organolithium compounds to 3-(4,4-dimethoxyoxazolin-2-yl)pyridine. *J. Chem. Soc., Perk. Trans.*, 1980: **1**; 2070–2076.

166 Seeman JI, Chavdarian CG, Kornfeld RA, Naworal JD. Nicotine chemistry. The addition of organolithium reagents to (−)-nicotine. *Tetrahedron*, 1985: **41**; 595–602.

167 Chavdarian CG, Seeman JI. The *tert*-butylation of nicotine: novel reaction pathways and racemization studies. *Tet. Lett.*, 1982: **23**; 2519–2522.

168 Secor HV, Chavdarian CG, Seeman JI. The radical and organometallic methylation of nicotine and nicotine *N*-oxide. *Tet. Lett.*, 1981: **22**; 3151–3154.

169 Abramovitch RA, Notation AD, Seng GC. Orientation in the reaction of phenyllithium with 3-substituted pyridines *Tet. Lett.*, 1959: **8**; 1–3.

170 Abramovitch RA, Seng GC, Notation AD. Arylpyridines. Part I. Orientation in the reaction of phenyllithium with some 3-substituted pyridines. *Can. J. Chem.*, 1960: **38**; 761–700.

171 Kempe NG, Keppens MA, Fonke G. A new efficient synthesis of pyridines. *J. Chem. Soc., Chem. Commun.*, 1996: **5**; 635–636.

172 Zoltecwicz JA, Prokai-Tetrai K, Bloom LB. Long range transmission of polar effects in cholinergic 3-arylidene anabasines. Conformations calculated by molecular modeling. *Heterocycles*, 1993: **35**; 171–180.

173 Kem WR. Occurrence of anabaseine in paranemertes and other nemertineans. *Toxicon*, 1971: **9**; 23–32.

174 Haines PG, Eisner A. Acetylation of 3,2′-nicotyrine. *J. Amer. Chem. Soc.*, 1950: **72**; 4618–4619.

175 Pictet A, Crépieux P. Ueber die hydrirung des nicotyrins. *Chem. Ber.*, 1898: **31**; 2018–2022.

176 Chavdarian CG. Optically active nicotine analogues. Synthesis of (S)-2,5-dihydro-1-methyl-2-(3-pyridyl)pyrrole ((S)-(−)-3′,4′-dehydronicotine). *J. Org. Chem.*, 1983: **48**; 1529–1531.

177 Shibagaki M, Matsushita H, Kaneko H. The synthesis of 5′-alkylnicotines. *Heterocycles*, 1986: **24**; 423–428.

178 Bowman ER, McKennis H Jr, Martin BR. A convenient method for the preparation of racemic nicotine. *Synth. Comm.*, 1982: **12**; 871–876.

179 Ziyaev AA, Sadykov AS. Dimerization of some β-substituted pyridine bases in the presence of metallic sodium and Raney nickel catalysts. *Isz. Akad. Nauk. SSSR Ser. Khim.*, 1983: **11**; 2607–2611. *Chem. Abs.*, 1984: **100**; 138528.

180 Orechov AP, Norkina SS. Alkaloids of *Anabasis aphylla* L. V. *N*-Aminoanabasine and racemic anabasine. *Chem. Ber.*, 1932: **65**; 1126–1130.

181 Edwards WB III, McCuen R. Preparation of optically pure (R)-(+)-nicotine. Studies on the microbial degradation of nicotinoids. *J. Org. Chem.*, 1983: **48**; 2484–2487.

182 Swango JH, Bhatti BS, Qureshi MM, Crooks PA. A novel enantioselective synthesis of (S)-(−)- and (R)-(+)-nornicotine via alkylation of a chiral 2-hydroxy-3-pinanone ketimine template. *Chirality*, 1999: **11**; 316–318.

183 Swango JH, Qureshi MM, Crooks PA. A novel enantioselective synthesis of (S)-(−)- and (R)-(+)-nornicotine via alkylation of a chiral pinanone ketimine template. *Pharm. Research*, 1997: **14**; S-695.

184 Jacob P III. Resolution of (±)-5–bromonornicotine, synthesis of (R)- and (S)-nornicotine of high enantiomeric purity. *J. Org. Chem.*, 1982: **47**; 4165–4167.

185 Seeman JI, Chavdarian CG, Secor HV. Synthesis of the enantiomers of nornicotine. *J. Org. Chem.*, 1985: **50**; 5419.

186 Armstrong DW, Spino LA Han SM, Seeman JI, Secor HV. Enantiomeric resolution of racemic nicotine and nicotine analogues by microcolumn liquid chromatography with β-cyclodextrin inclusion complexes. *J. Chromatogr.*, 1987.: **411**; 490–493.

187 Aceto MD, Martin BR, Uwaydah IM, May EI, Harris LS, Izazola-Conde C, Dewey WL, Bradshaw TJ, Vincek WC. Optically pure (+)-nicotine from (±)-nicotine and biological comparisons with (−)-nicotine. *J. Med. Chem.*, 1979: **22**; 174–177.

188 Chavdarian CG, Sanders EB, Bassfield RL. Synthesis of optically active nicotinoids. *J. Org. Chem.*, 1982: **47**; 1069–1073.

189 Bleicher LS, Cosford NDP, Herbaut A, McCallum JS, McDonald IA. A practical and efficient synthesis of the selective neuronal acetylcholine-gated ion channel agonist (S)-(−)-5–ethynyl-3-(1–methyl-2-pyrrolidinyl)pyridine maleate (SIB-1508Y). *J. Org. Chem.*, 1998: **63**; 1109–1118.

190 Seeman JI, Chadvarian CG, Secor HV. Synthesis of the enantiomers of nornicotine. *J. Org. Chem.*, 1985: **50**; 5419–5421.

191 Garvey DS, Wasicak JT, Decker MW, Brioni JD, Buckley MJ, Sullivan JP, Carrera GM, Holladay MW, Arneric SP, Williams M. Novel isoxazoles which interact with brain cholinergic channel receptors have intrinsic cognitive enhancing and anxiolytic activities. *J. Med. Chem.*, 1994: **37**; 1055–1059.

192 Deo MN, Crooks PA. Regioselective alkylation of *N*-(diphenylmethylidine)-3-(aminomethyl)pyridine: a simple route to minor tobacco alkaloids and related compounds. *Tet. Lett.*, 1996: **37**; 1137–1140.

193 Jones MD, Qureshi MM, Dwoskin LP, Crooks PA. A novel enantioselective synthesis of R-(+)- and S-(−)-anabasine via alkylation of a chiral hydroxypinanone ketimine template. *Pharm. Sci*, 1998: **1**; S-574.

194 Seeman JI. A new pyrroline synthesis: the use of an *N*-vinyl moiety as an NH-protecting group. *Synthesis*, 1977: 498–500.

195 Giovanni A, Savoia D, Umani-Ronchi A. Organometallic ring-opening reactions of N-acyl and N-alkoxycarbonyl lactams. Synthesis of cyclic imines. *J. Org. Chem.*, 1989: **54**; 228–234.

196 Kimpe NG, Keppens MA, Stevens CV. Synthesis of cyclic imines via ethylenetetramethyldisilyl-protected ω-aminoimines. Application to the synthesis of alkaloids. *Tet. Lett.*, 1993: **34**; 4693–4696.

197 Keppens MA, Kimpe ND, Fonke G. Synthesis of 2-aryl-1-pyrrolines from arylnitriles. *Synth. Commun.*, 1996: **26**; 3097–3102.

198 Zoltewicz JA, Cruskie MP Jr. A superior synthesis of cholinergic anabaseine. *Org. Prep. Proc. Int.*, 1995: **27**; 510–512.

199 Mullen GB, Georgiev V St. Synthesis of α- and β-nicotyrines. Use of phenyl vinyl sulfoxides as a marked equivalent of acetylene dipolarophile. *J. Org. Chem.*, 1989:

54; 2476–2478.

200 Savoia D, Concialini V, Roffia S, Tarsi L. Organometallic reactions of ω-acyl lactams. A new route to γ-ketoaldehydes from 5-ethoxy-2-pyrrolidione. *J. Org. Chem.*, 1991: **56**; 1822–1827.

201 Minoto A. Tamao K, Hayashi T, Suzuki K, Kumada M. Palladium-phosphine complex catalysed cross-coupling reaction of 1-methyl-2-pyrrolyl-magnesium bromide and zinc chloride with organic halides. *Tet. Lett.*, 1981: **221**; 5319–5322.

202 Shibagaki M, Takahashi K, Kaho H, Matsushita H. Preparation of nicotyrine via catalytic dehydrogenation of nicotine. *Agric. Biol. Chem.*, 1988: **52**; 2651–2652.

203 McKennis H, Schwartz SL, Turnbull LB, Tamaki E, Bowman ER. The metabolic formation of γ-(3-pyridyl)-γ-hydroxybutyric acid and its possible intermediary role in the mammalian metabolism of nicotine. *J. Biol.Chem.*, 1964: **239**; 3981–3989.

204 Kunz H, Pfrengle W. Carbohydrates as chiral templates: stereoselective tandem Mannich–Michael reactions for the synthesis of piperidine alkaloids. *Angew. Chem. Int. Ed.*, 1989: **28**; 1067–1068.

205 Pfrengle W, Kunz H. Hetero-Diels-Alder reactions on a carbohydrate template: stereoselective synthesis of (S)-anabasine. *J. Org. Chem.*, 1989: **54**; 4263–4265.

206 Hattori K, Yamamoto H. Asymmetric aza-Diels-Alder reaction: enantio- and diastereoselective reaction of imine mediated by chiral Lewis acid. *Tetrahedron*, 1993: **49**; 1749–1760.

207 Hattori K, Yamamoto H. Asymmetric aza-Diels-Alder reaction mediated by chiral boron reagent. *J. Org. Chem.*, 1992: **57**; 3264–3265.

208 Mehmandoust M, Marazano C, Das BC. A stereoselective route to enantiomeric 2-alkyl-1,2,3,6–tetrahydropyridines. *J. Chem. Soc. Chem. Commun..* 1989: 1185–1187.

209 Kempe N De, Keppens M, Fonck. G. A new efficient synthesis of pyridines. *J. Chem. Soc., Chem. Commun.*, 1996: 635–636.

210 Goshaev MG. Synthesis of 3-alkyl- and 3-arylpyridines. *Khim. Geterotsikl. Soedin.*, 1992: **8**; 1143–1144.

211 Sakamoto T, Kondo Y, Murato N, Yamanaka H. Pyridinylzinc halides by oxidative addition of active zinc with halopyridines. *Tet. Lett.*, 1992: **33**; 5373–5374.

212 Ishikura M, Kamada M, Terashima M. An efficient synthesis of 3-hetero-arylpyridines via diethyl-(3-pyridyl)borane. *Synthesis*, 1984: **11**; 936–938.

213 Yamamoto Y, Azuma Y, Mitoh H. General method for synthesis of bipyridines; palladium catalysed cross coupling reaction of trimethylstannyl-pyridines with bromopyridines. *Synthesis*, 1986: **7**; 564–565.

214 Gronowitz S, Bjoerk P, Malm J, Höernfeldt A-B. The effect of some additives on the Stille Pd°-catalyzed cross coupling reaction. *J. Organomet. Chem.*, 1993: **460**; 127–130.

215 Furukawa N. Shibutani T. Fujihara H. Preparation of pyridyl Grignard reagents and cross coupling reactions with sulfoxides bearing azaheterocycles. *Tet. Lett.*, 1987: **28**; 5845–5848.

216 Hitchcock SA, Mayhugh DR, Gregory GS. Selectivity in palladium (0)-catayzed cross-coupling reactions: application to a tandem Stille reaction. *Tet. Lett.*, 1995: **36**; 9085–9088.

217 Jacob P III, Benowitz NL. Pharmacokinetics of (S)-nicotine and metabolites in humans. In: *Nicotine and Related Alkaloids: Absorption, Distribution, Metabolism, Excretion.* Gorrod JW, Wahren J (Eds.), Chapman and Hall, London, Chap. 10; pp.

197–218, 1993.

218 Nguyen TL, Castagnoli N. Jr. The synthesis of deuterium labelled tobacco alkaloids: nicotine, nornicotine and cotinine. *J. Labelled Comp. Radiopharm.*, 1978: **14**; 919–934.

219 Seeman JI, Secor HV, Forrest G. Convenient synthesis of N-CD$_3$ labelled nicotine and nicotine analogues. *J. Labelled Comp. Radiopharm.*, 1979: **16**; 387–395.

220 Dagne E. Gruenke L. Castagnoli N. Jr. Deuterium isotope effects in the in vivo metabolism of cotinine. *J. Med. Chem.*, 1974: **17**; 1330–1333.

221 Liehr JG, Schulze P, Richter WJ. Ringschlussreaktionen in der elektronenstoss-induzierten fragmentierung des nicotins. *Org. Mass Spectrom.*, 1973: **7**; 45–51.

222 Munson JW, Hodgkins TG. Synthesis of nornicotine-2,4,5,6-d$_4$ and its N′-nitroso derivative. *J. Labelled Comp. Radiopharm.*, 1977: **13**; 461–469.

223 Leete E, Mueller ME. Biomimetic synthesis of anatabine from 2,5-dihydropyridine produced by the oxidative decarboxylation of baikiain. *J. Amer. Chem. Soc.*, 1982: **104**; 6440.

224 Nakane M. Hutchinson CR. Biosynthetic studies of secondary plant metabolites with $^{13}CO_2$ . *Nicotiana* alkaloids. 2 New synthesis of nornicotine and nicotine. Quantitative carbon-13 NMR spectroscopic analysis of [2′,3′,N-CH$_3$-$^{13}$C$_3$] nicotine. *J. Org. Chem.*, 1978: **43**; 3922–3931.

225 Leete E, Random KC, Riddle RM. The metabolism of anatabine to α, β-dipyridyl in *Nicotiana* species. *Phytochem.*, 1979: **18**; 75–78.

226 Edwards WB III, Glenn DF, Green F, Newman RH. The preparation of tobacco constituents incorporating stable isotopes. I. The synthesis of d,l-nornicotine-1′-$^{15}$N and d,l-nicotine-1′-$^{15}$N. *J. Labelled Comp. Radiopharm.*, 1977: **14**; 255–261.

227 Vincek WC, Martin BR, Aceto MD, Bowman, ER. Synthesis and preliminary binding studies of 4′,4′-ditritio-(−)-nicotine of high specific activity. *J. Med. Chem.*, 1980: **23**; 960–962.

228 Leete E, Bell VM. The biogenesis of the *Nicotiana* alkaloids. VIII. The metabolism of nicotine in *N. tobacam*. *J. Amer. Chem. Soc.*, 1959: **81**; 4358–4359.

229 Berger G, Maziere M, Knipper R, Prenant C, Comar D. Automated synthesis of $^{11}$C-labelled radiopharmaceuticals: imipramine, chlorpromazine, nicotine and methionine. *Int. J. Appl. Rad. Isotopes*, 1979: **30**; 393–399.

230 Saji H, Magata Y, Yamada Y, Tajima K, Yonikura Y, Konishi J, Ohmomo Y, Yokoyama A. Synthesis of (S)-N-[methyl-$^{11}$C] nicotine and its regional distribution in the mouse brain: a potential tracer for visualization of brain nicotinic receptors by positron emission tomography. *Chem. Pharm. Bull.*, 1992: **40**; 734–736.

231 Langström B, Antoni G, Halldin C, Svärd H, Bergson, G. The synthesis of some $^{11}$C-labelled alkaloids. *Chemica Scripta*, 1982: **20**; 46–48.

232 Sihver W, Fasth KJ, Ogren M, Nordberg A, Watanabe Y, Langstrom B. Development of "C-nicotinic receptor ligands to characterize nicotinic acetyl choline receptors in the central nervous system. *J. Labelled Comp. Radiopharm.*, 1997: **40**; 592–594.

233 Comes RA, Core MT, Edmonds MD, Edwards WB III, Jenkins RW Jr. The preparation of carbon-14 labelled tobacco constituents. II. The synthesis of DL-nicotine (2′-$^{14}$C). *J. Labelled Comp. Radiopharm.*, 1973: **9**; 253–259.

234 Leete E. Biomimetic synthesis of nicotine. *J. Chem. Soc., Chem. Commun.* 1972: 1091.

235 Hu MW, Bondinell WE, Hoffman D. Chemical studies on tobacco smoke XXIII. Synthesis of carbon-14 labelled myosmine, nornicotine and *N'*-nitrosonornicotine. *J. Labelled Comp. Radiopharm.*, 1974: **10**; 79–88.

236 Korte F, Schultze-Steiner HJ. Acyl-lacton-umlagerung, XXII. Umlagerung von α-aroyl-pyrrolidonen in konz. Salzsäuer zu pyrrolinderivaten. *Chem. Ber.*, 1962: **95**; 2444–2452.

237 Chan SM, Basmadjian GP, Marten DF, Sadek SA, Magarian RA, Grunder JR Ice RD. Synthesis and biodistribution of radioiodinated 5-iodonicotine. *J. Labelled Comp. Radiopharm.*, 1983: **20**; 1017–1025.

*Chapter 5*

# The physical chemistry of nicotine

Joseph L. Banyasz

*Philip Morris USA, Research Development and Engineering Center, P.O. Box 26583, Richmond, VA 23261-6583, USA*

## I. INTRODUCTION

The genus of tobacco, *Nicotiana*, and subsequently its principal alkaloid, nicotine, were named after Jean Nicot, the French ambassador to the court in Lisbon in 1559. Nicot sent both seeds and plants to the French court, in particular to the queen mother Catherine de Medici. Nicot promoted tobacco as an herbal panacea. Catherine de Medici was responsible for its cultivation in France. It is interesting to note that in 1572 there was a proposal to name tobacco Medicée after the queen mother [1,2]. The alkaloid might have a rather different name had that proposal carried the day.

Early investigators focused largely on potential medicinal applications of tobacco. By the end of the eighteenth century chemistry had progressed to the point that isolation of the active principles of medicinal plants could be contemplated. The first success came with the isolation of morphine from opium in 1803. Caffeine was isolated from coffee in 1820. Cerioli in Italy and Vauquelin in France produced oils of tobacco in 1807 and 1809 respectively. Posselt and Reimann in Germany finally isolated nicotine in 1828 [1,2].

The next milestone, the molecular structure of nicotine, required another 65 years to reach. This period saw explosive growth in the understanding of chemistry. The tetrahedral nature of the carbon atom was recognized by Le Bel and van't Hoff. Stereochemical studies led to a theory of optical isomerism by van't Hoff and Le Bel in 1875. Kekule deduced the structure of benzene and laid the foundations for the understanding of aromatic systems. In 1891 Emil Fischer of the University of Würzburg announced his proof of the configuration of (+)-glucose, one of the crowning achievements of nineteenth century organic chemistry, for which Fischer was awarded the Nobel Prize in 1901. A. Pinner presented the structure of nicotine to the German Chemical Society in 1893 [3].

The synthesis of nicotine was achieved relatively quickly after its structure was elucidated. Auerbach and Wolffenstein produced nicotine by reduction of nicotine-*N*-oxide in 1901. However, the first real synthesis was published in 1911 by Auzies who prepared nicotine by passing a mixture of pyrrole, ammonia and butadiene over a catalyst at 350°C [1].

One cannot speak of physical chemistry as such prior to the 1880s. During the nineteenth century it became increasingly obvious that the discoveries in physics had a direct bearing on and applicability to chemistry. The need became clear for a field that applies physical laws to chemical phenomena. This provided the impetus for Ostwald, van't Hoff and Arrhenius to systematize the discipline that is today called physical chemistry and, in 1881, to found Zeitschrift für physikalische Chemie. The birth of physical chemistry as a distinct discipline is generally dated to the first issue of that journal. The Arrhenius theory of electrolytic solutions, one of the early triumphs of the new discipline, was published in Zeitschrift für physikalische Chemie in 1877.

In the last two decades of the nineteenth century Berthelot and others carried out extensive calorimetric investigations of the heats of chemical reactions in hopes of developing a criterion of spontaneity. During this period the specific heat and the heats of solution, neutralization and combustion of nicotine were determined [4,5]. These might be called the first physical chemical measurements made with nicotine. Kharasch at the National Bureau of Standards (US) repeated the determination of the heat of combustion in 1929 [1]. These appear to be the only calorimetric data ever to have been published for nicotine. Unfortunately, they predate the development of what would now be considered reliable calorimeters.

During the 1890s phase studies by Pickering and others led to the discovery of a number of clathrate hydrates including those of amines. Hudson in 1904 determined the oval shaped, closed loop phase diagram of the nicotine-water system which was to become the textbook example of a system with both a lower and an upper critical solution temperature [1]. The phase diagram was interpreted in terms of a stoichiometric hydrate. This led to a long and, to date, unsuccessful quest for the isolation of a hydrate of nicotine.

In the early decades of this century nicotine disappears from the forefront of chemistry despite continuing interest by the biological sciences. Physical chemical studies become rare and are frequently driven by applications. The application of nicotine as a fumigant or pesticide spurred interest in its vapor pressure leading to several published studies. The partitioning of nicotine between water and organic solvents was investigated for the same reason. The ionization constants of nicotine have been determined by a number of workers using a variety of techniques over a span of several decades. Unfortunately none of these data can be considered completely reliable. More recently, there was considerable activity, by NMR and computation, that led to an understanding of the conformation of nicotine in aqueous solution.

Nicotine has received scant attention from physical chemists since the very early studies. The reasons for this are not difficult to fathom. Its toxicity aside, nicotine is a difficult compound to work with. It has defied all attempts to crystallize it over the past century by forming a glass. Pure nicotine is hygroscopic and precautions must be taken to keep out water. It oxidizes readily on exposure to light and oxygen. Consequently it is purified by vacuum distillation and is best stored free of oxygen in the dark at low temperature. Even under these conditions it has a finite lifetime and must be repurified as needed.

Many authors cite the measured value of a physical property as a criterion of the purity of their nicotine sample. Unfortunately precise and accurate values for the physical properties of high purity nicotine are not available. Furthermore, authors rarely state the precision of the measurement or the reference value to which the measurement is to be compared. The best single indicator that high purity nicotine was used is a statement that the sample was vacuum distilled to yield a clear, colorless liquid. Nicotine upon oxidation becomes increasingly yellow then brown in color.

The chemistry of nicotine was ably and thoroughly reviewed by Jackson in 1941 [1]. Most of the physical chemical data available for nicotine are summarized in that work.

This chapter will focus on two objectives. The first is to determine what reasonably reliable data are available for the properties of nicotine. This was achieved with some success for the vapor pressure, the specific rotation and the ionization constants. The second is to examine the behavior of nicotine in water in light of more recent developments in the understanding of aqueous solutions. A number of thermodynamic parameters were calculated from available data for aqueous nicotine.

It must be kept in mind that the majority of the experimental results cited in this chapter were determined 50 to 100 years ago. It hardly needs to be pointed out that the arsenal of experimental techniques, as well as accuracy and precision, have grown vastly in the intervening decades.

It is hoped that, despite its irksome nature, nicotine will be shown to be an interesting molecule more deserving of attention from the physical chemist than it has hitherto received.


## II. CONCENTRATION UNITS

Concentrations in this chapter are expressed in terms of the weight percent, molal and mole fraction scales. The $m$ or molal scale, generally used in dilute solution, is defined as:

$m$ = moles solute/kilograms solvent

The mole fraction of nicotine in a binary aqueous solution is given by:

$$X_2 = \frac{n_2}{n_1 + n_2}$$

where $X$ and $n$ stand for mole fraction and mole number, respectively. The subscripts 1 and 2 in all cases stand for water and nicotine, respectively. The mole fraction scale is particularly convenient in concentrated solutions where the distinction between solvent and solute blurs. Most thermodynamic laws of solutions or mixtures are stated in terms of mole fractions. Mole percent, which is mole fraction multiplied by 100, is used interchangeably with mole fraction.

These units are favored for physical chemical applications because they are strictly on a weight basis. Weights can ordinarily be measured with greater precision than volumes. Conversion from the molal to the mole fraction scale is simple:

$$X_2 = \frac{m}{55.5 + m} \tag{1}$$

The same conversion for the more extensively used M or molar scale (i.e. moles solute/liter solution) requires a knowledge of the density of the solution for which a precise value is often not available.

## III. THE VAPOR PRESSURE OF NICOTINE

Several studies of the vapor pressure of nicotine as a function of temperature are available in the literature. Young and Nelson's determinations cover the temperature range of 60–250°C [6]. Vapor pressures up to 100°C were determined by the air saturation method. Above that temperature static measurements were done with an isoteniscope. Gorbachev published an overlapping data set based on boiling points valid over the range 100–230°C [7]. His results are in good agreement with Young and Nelson.

There is a paucity of data near room temperature. Harlan and Hixon [8] published air saturation concentrations over the range of 25–40°C from which approximate vapor pressure values may be calculated. Young and Nelson used this approximate data set to extend their temperature range. More recently, Lewis presented an equation pertaining to the room temperature range based, in part, on an unpublished data set [9].

Fortunately, a study by Norton, Bigelow and Vincent appears to definitively establish the vapor pressure of nicotine at 25°C [10]. The nicotine was vacuum distilled and the purity of different samples prepared during the course of the study was maintained by monitoring specific rotation. Temperature control is stated to be within ±0.02°C. Vapor pressures were determined by the gas saturation method using nitrogen so as to avoid oxidation of nicotine. The vapor pressure of nicotine at 25°C was determined to be 0.0425±0.0003 mm Hg based on ten replicates involving five different nicotine samples.

The results of Young and Nelson, Gorbachev and Norton et al. comprise an internally consistent data set over the temperature range 25–250°C. This set was used here to determine expressions for the vapor pressure of nicotine as a function of temperature.

Since curvature is apparent in the data, the following commonly used equations were tested as to goodness of fit:

$$\log P = A + \frac{B}{T} + CT$$

and

$$\log P = A + \frac{B}{T} + C \log T$$

where $T$ is the absolute temperature and $P$ is in units of mm Hg. The second form proved to give a somewhat better fit yielding:

$$\log P = 43.745 - \frac{4929}{T} - 11.561 \times \log T \tag{2}$$

with a standard error of estimate of 0.025. The mean relative error in vapor pressure is 4.1%. The result is shown in Fig. 1.

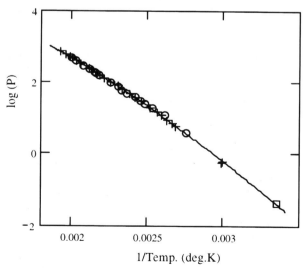

Fig. 1 The vapor pressure of nicotine. Circles: Young and Nelson [6]; +'s: Gorbachev [7]; squares: Norton et al. [10].

Given the distribution of the data, eq. 2 is weighted toward the higher temperatures. The low temperature range, 25–120°C, is well represented by the linear relationship:

$$\log P = 9.541 - \frac{3259}{T} \tag{3}$$

The mean relative error in vapor pressure is 5.5% for eq. (3). The low temperature results are shown in Fig. 2.

Equation (2) yields a value of 15.7 Kcal/mole for the heat of vaporization of nicotine at 25°C as compared to 14.9 Kcal/mole obtained from eq. (3).

Lewis, based on the low temperature measurements of Young and Nelson augmented with recent unpublished data over the approximate temperature range of 20 to 50°C, gives the following expression for the saturated vapor phase concentration of nicotine:

$$\log C(g/cc) = 4.47 - \frac{3260}{T} \tag{4}$$

Given the low vapor pressures involved, Lewis' concentration units can be converted to pressures in mm Hg using the ideal gas law:

$$P(\text{mmHg}) = C \cdot \frac{1000}{162.23} \cdot R \cdot T \cdot 760 \tag{5}$$

Figure 2 shows the data of Young and Nelson, Gorbachev and Norton et al. superimposed on the line calculated from eq. (3). The diamonds show the vapor pressures calculated from Lewis' equation at 20° interval starting at 20°C. Equation (3) and Lewis' eq. (4) give virtually identical results.

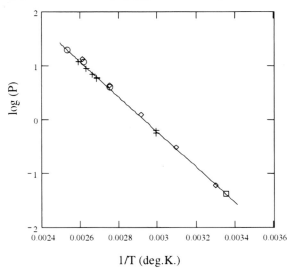

Fig. 2 The vapor pressure of nicotine from 25 to 120°C. Circles: Young and Nelson [6]; +'s: Gorbachev [7]; square: Norton et al. [10]; diamonds: calculated from Lewis' eq. (9).

There is close agreement between four independent data sets. The vapor pressure appears to be the most firmly established physical property of nicotine.

## IV. SPECIFIC ROTATION

Values published for the specific rotation, $[\alpha]_D^{20}$, of nicotine show considerable variability. This is illustrated by the rather different values obtained by two careful early researchers. Landolt in 1893 published a value of $-161.55$ [11] whereas Jephcott in 1918 obtained a value of $-168.5$ [12]. This discrepancy has persisted in the literature. Heller and Curmé, in a review of optical rotation in 1971, cited Landolt's value [13]. The most recent edition of the Merck Index quotes Jephcott's number [14]. Since a number of workers have used $[\alpha]_D^{20}$ to characterize their nicotine samples, resolution of the discrepancy is matter of some interest.

Jephcott in 1918 noted that the value obtained for the specific rotation was significantly affected by the method used to purify nicotine [12]. Most workers up to that time, Landolt among them, used distillation under hydrogen to purify their samples. The specific rotations in this case ranged from $-161$ to $-165°$. Ratz, like Jephcott, purified by vacuum distillation and obtained an $[\alpha]_D^{20}$ close to Jephcott's [12]. There is further corroboration. Norton et al., who took great pains to ensure the purity of their nicotine samples by vacuum distillation, quote a value of $-168.9 \pm 0.2°$ for $[\alpha]_D^{20}$ [10].

Nicotine can be reduced by hydrogen over a platinum catalyst at room temperature [1,15]. The possibility exists that, at the elevated temperatures required for distillation at atmospheric pressures, some reduction might take place in the absence of added catalysts. Furthermore, Harlan and Hixon noted in 1930 that there is not a sufficient difference between the boiling points of nicotine and its reduction products to allow for a sharp separation by distillation [15]. Distillation under hydrogen could, thus, yield nicotine contaminated by variable amounts of reduction products. This hypothesis would account for the variability of the $[\alpha]_D^{20}$ values determined for samples distilled under hydrogen. Given the analytical techniques available today, confirmation should not be difficult

Jephcott's value would, thus, appear to be the most trustworthy. Jephcott data, based on four separate and distinct preparations of pure nicotine, yielded an average value of $-168.5°$ for $[\alpha]_D^{20}$ with a standard deviation of $0.1°$.

## V. THE IONIZATION CONSTANTS OF NICOTINE

The nicotine molecule, shown below, contains two nitrogen atoms capable of accepting a proton.

The Structure of Nicotine

The pyrrolidine nitrogen is considerably more basic than the pyridine nitrogen which is part of an aromatic ring. $K_{a1}$ pertains to the ionization of the pyridine nitrogen while $K_{a2}$ refers to the pyrrolidine nitrogen.

The ionization constants, as presented in this chapter, are mixed acid dissociation constants defined as:

$$K_{a1} = \frac{[NicH^+] \times a_{H^+}}{[NicH_2^{2+}]}$$

$$K_{a1} = \frac{[Nic] \times a_{H^+}}{[NicH^+]} \tag{6}$$

The nicotine species are expressed as concentrations while the hydrogen ion term is an activity as calculated from:

$$a_{H^+} = 10^{-\text{pH}} \tag{7}$$

To convert the equilibrium constants to the pure concentration form, the ionic strength dependent value of the hydrogen ion activity coefficient, $\gamma_{H^+}$, must be taken into account:

$$[H^+] = \frac{a_{H^+}}{\gamma_{H^+}} \tag{8}$$

At 0.1 M ionic strength, the value of $\gamma_{H^+}$ is 0.83 as calculated from the Debye-Hückel equation corrected for finite ionic size.

A.E. Martell, in the introduction to his compilation of critical stability constants, lists the criteria for establishing the reliability of equilibrium constant data [16]. These include specification of the purity of the ligand, temperature, ionic strength and supporting electrolyte. Frequent deficiencies include lack of calibration of the potentiometric apparatus and failure to define the equilibrium quotient reported. For nicotine one should add the lack of providing a means of excluding carbon dioxide from the titration. All of the data reported for nicotine to date appear to have been derived from the titration of free base nicotine with acids. The pH of the starting solutions is ten or higher. Carbon dioxide absorption is, thus, a problem particularly at lower temperatures.

In terms of the above definition there appear to be no reliable equilibrium constant data for nicotine. Every report is deficient with regard to at least two of the criteria. Failure to control or specify ionic strength and providing no means for excluding carbon dioxide are almost universal.

The review by Jackson summarizes the work prior to 1940 [1]. A number of studies are cited which utilize indicator as well as potentiometric methods. No mention is made of carbon dioxide exclusion or ionic strength. The values reported at 15 and 20°C are 8.19 and 8.11, respectively, for $pK_{a2}$ and 3.49 and 3.22, respectively, for $pK_{a1}$.

Unfortunately there has been very little activity in the field since Jackson's review. Fowler, in 1954, published values for the ionization constants of nicotine at 0, 25 and 40°C based on the titration of 0.1 N nicotine solutions with acid [17]. A glass and calomel electrode pair were used in conjunction with a vacuum tube potentiometer to determine pH. The nicotine was vacuum distilled repeatedly to achieve high purity. A multi-point calibration was performed to standardize the electrodes. However, no provisions were made to exclude carbon dioxide and ionic strength was not controlled. The values reported at 0, 25 and 40°C were 8.20, 8.01 and 7.72, respectively for $pK_{a2}$ and 3.55, 3.41 and 3.22, respectively, for $pK_{a1}$.

The most systematic study to date was published in 1980 by Gonzalez, Monge and Whittembury [18]. The value of $pK_{a2}$ was determined as a function of temperature and ionic strength using sodium chloride as the supporting electrolyte. Carbon dioxide was excluded by bubbling carbon dioxide free nitrogen through the titration cell. Unfortunately, no mention is made of the purity of the nicotine or of the details of the titration apparatus nor is the equilibrium quotient defined beyond calling it $pK_a$. No description is given of the method of electrode standardization which should involve a three point calibration at each temperature. The values reported for $pK_a$, which is clearly $pK_{a2}$, are shown in Table 1.

Though no mention is made of it, it is safe to assume that the titration apparatus consisted of a glass electrode, in the form of a combination electrode or a glass and reference electrode pair, in conjunction with a pH meter. It is well known that glass electrodes respond to sodium ions at high pH thereby introducing an error into pH readings [19]. The pH at which this effect becomes significant varies widely with the type of glass and depends on temperature as well as sodium concentration. In the absence of any information as to the sodium correction for the electrode used, the results shown in Table I for 0.557 M NaCl must be considered suspect particularly at the higher temperatures. There is less cause for concern about the $pK_a$ values in 0.145 M NaCl.

TABLE 1

The ionization constants of nicotine

| Temperature °C | pKa1 (no supporting electrolyte) | pKa2 (no supporting electrolyte) | pKa2 (0.145 M NaCl) | pKa2 (0.557 M NaCl) | Reference |
|---|---|---|---|---|---|
| 0 | 3.55[a] | 8.20 | | | [17] |
| 10 | | 8.28 | 8.42 | 8.51 | [18] |
| 15 | | 8.20 | 8.32 | 8.44 | [18] |
| 20 | | 8.06 | 8.20 | 8.34 | [18] |
| 25 | 3.41[a] | 8.01 | (8.07)[b] | | [17] |
| 25 | | 7.98 | 8.10 | 8.29 | [18] |
| 30 | | 7.86 | 8.00 | 8.14 | [18] |
| 35 | | 7.80 | 7.92 | 8.05 | [18] |
| 40 | 3.22[a] | 7.72 | (7.79)[b] | | [17] |
| 40 | | 7.74 | 7.81 | 8.00 | [18] |

[a] ionic strength = 0.15 M
[b] corrected to 0.15 M ionic strength

The temperature dependence of $pK_{a,2}$ in 0.145 M NaCl is shown in Fig. 3. Regression of the data to the equation:

$$pK_{a2} = pK^0_{a2} + \frac{\Delta H_{a2}}{2.303RT} \tag{9}$$

yielded:

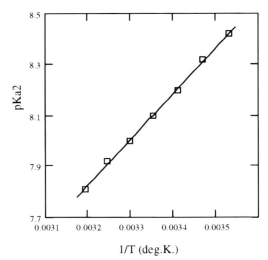

Fig. 3 The second ionization constant of nicotine at I=0.145 M (NaCl). Data points: Gonzalez et al. [18].

$$pK_{a2} = 2.088 + \frac{1793}{T} \tag{10}$$

where $R$ and $T$ are the gas constant and the absolute temperature, respectively. Equation (10) yields a value of 8.2 kcal/mole for $\Delta H_{a2}$ which is in good agreement with the calorimetrically obtained value of 8.05 [1] kcal/mole measured for the first equivalent of nicotine.

The values of $pK_{a2}$ determined by earlier workers at 15, 20 and 25°C, though measured at lower but undefined ionic strengths, agree within about 0.1 with the corresponding values in 0.145 M NaCl. Fowler's value at 0°C is in more serious disagreement possibly due to carbon dioxide absorption.

It has been shown that the $pK_a$ values of a variety of amines exhibit a linear dependence on the ionic strength, I [20–22]. From the data in reference 22 one can estimate that the values of $dpK_a/dI$ for pyrrolidine are 0.63 and 0.70 at 25 and 40°C, respectively. These factors may be used to correct Fowler's $pK_a$ values to the ionic strength of Gonzalez et al. Fowler titrated 0.05 M nicotine solutions with hydrochloric acid which means that, at the half neutralization point of the first equivalent, i.e. where pH=$pK_{a2}$, the ionic strength would have been 0.025 M. Gonzalez et al. titrated 0.01 M nicotine with hydrochloric acid which would have resulted in an ionic strength of 0.1475 at the first half neutralization point. The ionic strength differential is thus 0.1225 M. Using the correction factors above, one calculates 8.09 and 7.81, at 25 and 40°C, for Fowler's $pK_{a2}$ values at the ionic strength of Gonzalez et al. The corresponding values determined by Gonzalez et al. are 8.10 and 7.81. The agreement is now very good.

The relative lack of sensitivity of $pK_{a2}$ to ionic strength is attributable to the fact that the reaction involves the addition of charge to a neutral molecule. Since the second proton-transfer equilibrium adds a second charge to an already positively charged species, the value of $pK_{a1}$ for nicotine is expected to be considerably more sensitive to ionic strength. However, Fowler's titration would have resulted in an ionic strength of 0.1125 M at the second half equivalence point which is not too far from the ionic strength regime of Gonzalez et al.

The combined data sets yield $pK_{a1}$=3.41 and $pK_{a2}$=8.10 at 25°C and 0.15 M ionic strength. The value of $pK_{a2}$ appears to be established with some level of confidence. The agreement with Fowler, who took measures to ensure nicotine purity and carried out multipoint calibrations, lends credence to the Gonzalez results. The close agreement between the Gonzalez $\Delta H_{a2}$ value and calorimetric measurement, mentioned earlier, further support this conclusion. Given the close agreement with regard to $pK_{a2}$, there is no apparent reason to question Fowler's $pK_{a1}$ value. Corroboration is lacking in this case, however.

It is worth noting that the older literature frequently cites 'dissociation constants' for nicotine rather than $pK_a$ values. These are, in actuality, hydrolysis constants which are defined as:

$$K_1 = \frac{[NicH^+] \times a_{OH^-}}{[Nic]} = \frac{K_W}{K_{a2}}$$

$$K_2 = \frac{[NicH_2^{2+}] \times a_{OH^-}}{[NicH^+]} = \frac{K_W}{K_{a1}} \tag{11}$$

where $K_W$ is the ion product of water. This convention persists in some reference works [14]. At 25°C, $pK_W = 13.9965$ [23]. The values of the 'dissociation constants' are, thus, $pK_1 = 5.90$ and $pK_2 = 10.59$. Confusion of the two conventions leads to a serious overestimation of the base strength of nicotine. Unfortunately, equilibrium constants are frequently quoted without specifying the equilibria to which they pertain.

## VI. STERIC HINDRANCE AT THE PYRROLIDINE NITROGEN

The pyrrolidine nitrogen is nearly five orders of magnitude more basic than the pyridine nitrogen. It should follow that the pyrrolidine nitrogen is the preferred site for both hydrogen bonding and metal ion coordination. However, this is not the case. IR measurements in carbon tetrachloride have shown that phenols hydrogen bond to nicotine exclusively at the pyridine nitrogen [24]. NMR and IR results for the Zn(II), Cd(II) and Hg(II) complexes of nicotine in aqueous solution indicate that it is a monodentate ligand with coordination taking place at the pyridine nitrogen [25]. The shifting of the expected hydrogen bonding site to the weaker base results from steric hindrance at the pyrrolidine nitrogen.

Steric hindrance manifests itself in the thermodynamics associated with the second ionization constant of nicotine. The $pK_a$ value of N-methylpyrrolidine is 10.46 at 25°C and zero ionic strength [26]. Extrapolation of the data for nicotine at 25°C to zero ionic strength yields a value of 7.97 for $pK_{a2}$. The presence of the pyridine ring thus reduces the base strength of the pyrrolidine nitrogen by a factor of about 300. The heats of neutralization are comparable with values of $-9.05$ [26] and $-8.2$ kcal/mole for N-methylpyrrolidine and nicotine, respectively. The large difference in base strength is consequently due to the entropies of protonation. The value of the entropy of protonation for N-methylpyrrolidine is 17.5 cal/deg mole as compared to 8.9 cal/deg mole for nicotine. The large negative contribution by the pyridine ring to the entropy of protonation could be the result either of steric hindrance to the proton transfer reaction at the pyrrolidine nitrogen or the loss of conformational freedom by the nicotinium ion once it is formed. The two possibilities are further discussed below.

A study of the ionization potentials of nicotine and related compounds, including pyrrolidine and N-methylpyrrolidine, by Kuhn, Lilly-Leister, Kao and Lilly sheds further light on the basicity of the pyrrolidine nitrogen in nicotine [27]. The ionization potential (IP) is defined as the energy required to remove an electron from a molecule in the ground state to form a molecule ion in the ground state. The first ionization potential refers to the first electron removed which is, of necessity, the most loosely held. In nicotine this is one of the lone pair electrons of the pyrrolidine ring. A base may be defined in the broadest sense as an electron pair donor. It follows that the IP, which is a measure of how easily an electron is completely removed, should also be an indicator of how readily it is shared with an electrophile. Consequently, there is an inverse correlation between the IP and base strength. Furthermore, the IP value is not affected by the steric factors which can play a major role in proton transfer processes in solution. The IP is, thus, a measure of what may termed 'intrinsic' base strength. It has been correlated to gas-phase basicity.

The results of Kuhn et al. show that $N$-methylpyrrolidine has a lower IP than pyrrolidine. This is expected given the well known electron donating properties of the methyl group. The result should be an increase in the base strength and consequently an increase in the $pK_a$ value of the pyrrolidine nitrogen. However, the opposite is observed. The addition of the methyl group has the secondary effect of increasing the crowding at the pyrrolidine nitrogen and thereby hindering the proton transfer process. The steric factor overwhelms the inductive effect of the methyl group leading to a lower $pK_a$ value.

The IP values of $N$-methylpyrrolidine and nicotine are, within experimental error, identical. This argues that the pyridine ring has no inductive effect on the pyrrolidine nitrogen. All else being equal, the $pK_{a2}$ value of nicotine should be identical to the $pK_a$ value of $N$-methylpyrrolidine. The inescapable conclusion is that the 300 fold difference is the result of a steric effect.

Closer examination of possible steric effects requires an understanding of the solution configuration of aqueous nicotine.

This has been well established by number of experimental and theoretical investigations [28–33].

The solution configuration of nicotine as calculated by R.W. Dwyer via MOPAC geometry optimization using the AM1 potential function is shown below [34]. The calculated configuration is in close agreement with the experimental results as well as the lowest energy configuration calculated by Kuhn et al. using the molecular-orbital-molecular-mechanics approach.

The pyridine and pyrrolidine rings are perpendicular to each other with relatively free rotation about the C(2')–C(3) bond. The most highly favored conformation has the C(2')–H(2') bond in the plane of the pyridine ring. The pyrrolidine ring, which exhibits considerable conformational flexibility, is preferentially in an envelope conformation. Rapid inversion takes place about the pyrrolidine nitrogen. The methyl group can point either at the pyridine ring in the *cis* configuration or away from it in the *trans* configuration. The *trans* configuration is favored over the *cis* by a factor of about 10 : 1. Koo and Kim [35] published the x-ray analysis of the structure of a crystal of nicotine dihydroiodide. The general features of the crystal structure are in good agreement with the solution results. The planes of the rings form an angle of 94° with each other and the *N*-methyl group is *trans* to the pyridine ring.

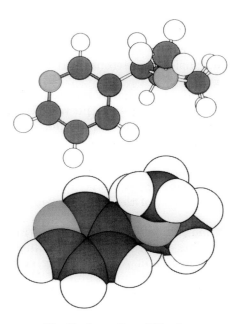

The Conformation of Nicotine

In the *trans* configuration the lone pair of the pyrrolidine nitrogen points at the pyridine ring which hinders the approach of an electrophile. The lone pair is much more accessible in the less favored *cis* conformer. However, the nicotine molecule spends only about 10% of its time in this conformation. The resulting reduction of time average lone pair density makes attack on the *cis* conformer less attractive. The size of the attacking group is a critical consideration. Iodomethylation experiments [36] have shown the ratio of *cis* methylation (of the *trans* conformer) to *trans* methylation is 3 : 2. Statistically one would expect a ratio of 10 : 1. The influence of the steric effect is clear, though methyl iodide appears to be small enough to leave methylation of the more stable *trans* conformer the preferred route. In the case of the much larger phenols, the pyridine ring blocks access to the *trans* conformer while the reduced time-average lone pair density of the *cis* conformer prevents hydrogen bond formation. Consequently phenol hydrogen bonds to the pyridine nitrogen.

The proton in aqueous solution is commonly written as the hydronium ion, $H_3O^+$, to denote that it is hydrated. However, this convention understates the extent of the hydration. Several water molecules appear to participate in the hydration of the proton with the tetrahydrate $H_9O_4^+$ being particularly stable. According to some estimates as many as eight water molecules may be involved in the hydrate [37]. The hydrogen ion is quite mobile within this framework moving about with a bond skipping mechanism which leads to its very high diffusion coefficient in water. The charge is delocalized over the hydrogen bonded network giving the hydrated proton a very large apparent size. Noyes [38] has suggested an open structure for the hydrated proton with three waters hydrogen bonded to a central hydronium ion to form $H_9O_4^+$ with one or two additional associated but not hydrogen bonded water molecules. The size of the hydrated proton was estimated to be about 3 Å which is larger than one would expect for methyl iodide.

The protonation of the pyrrolidine nitrogen, thus, involves the approach of a sizable structure to the lone pair. The pyridine ring should hinder the process. The effect might, however, be ameliorated somewhat by the fact that the structure, though large, should deform readily and be able to shed some of its water molecules as it approaches the lone pair.

A second potentially significant factor is loss of conformational freedom in the nicotinium ion relative to the nicotine molecule. In the case of the alkylamines it has been argued that the charge resulting from protonation of the nitrogen annuls the hydrophobicity of the portions of the side chains in the immediate vicinity of the charge while repelling the nonpolar portions farther away [20,21]. The side chains assume a more rigid, extended conformation. The net result is a loss of conformational freedom leading to a negative contribution to the entropy of protonation. The effect in nicotine would be a reduction of the conformational flexibility of the pyrrolidine ring. However, this cannot explain the difference between nicotine and *N*-methylpyrrolidine since the latter would be affected in the same way as nicotine. The explanation must be sought in the pyridine ring. Carbon-13 NMR results show that protonation of the pyrrolidine nitrogen leads to a large upfield shift of the pyridine C(3) resonance indicating an increase in electron density at that position [39]. If the nicotinium ion retains the *trans* conformation, the proton attached to the pyrrolidine nitrogen will be in the vicinity of the pyridine C(3) carbon. Calculations by R.W. Dwyer show that this is, indeed, the case. The increased electron density at C(3) is, thus, induced by the positive charge on the pyrrolidine nitrogen. The interaction between the charge and the induced dipole should raise the barrier to rotation about the C(3)–C(2′) bond thereby reducing conformational freedom. This would manifest itself as a negative contribution to the entropy of protonation. The polarization effect should also reduce the lone pair density on the pyridine nitrogen resulting in a reduction of its base strength.

## VII. THE BEHAVIOR OF NICOTINE IN AQUEOUS SOLUTION

Nicotine is an amphiphilic molecule, i.e. part hydrophilic and part hydrophobic. It contains two nitrogen atoms which can hydrogen bond with water. However, the rest of the molecule is hydrocarbon like in nature and therefore hydrophobic. The tension between these opposing tendencies leads to the complex solution behavior of nicotine.

The behavior of amphiphilic solutes has been extensively investigated by many workers [40–45]. The body of knowledge derived from these investigations is essential to an understanding of the solution behavior of nicotine. The pertinent data for nicotine solutions is limited to partial vapor pressure and density measurements and a few very old calorimetric determinations. Fortunately these suffice to shed some light on the behavior of nicotine at least in a qualitative sense.

The solution behavior of apolar or hydrophobic molecules can only be understood in context of the structure of water. The structure of water is the subject of a vast body of literature spanning several decades and is beyond the scope of this review. Fortunately, the seminal concept of the 'flickering cluster' suffices for our purposes.

The formation of a hydrogen bond increases the polarization of a water molecule which makes the formation of further hydrogen bonds more likely. Frank and Evans first recognized the cooperative nature of hydrogen bond formation in water [46]. Under the influence of a negative energy fluctuation, a cluster of extensively hydrogen-bonded water molecules rapidly forms only to dissolve with the next positive energy fluctuation. Based on the structural relaxation time of water, the average lifetime of a cluster is on the order of $10^{-12}$ s. The lifetime, though short, is long compared to the period of a molecular vibration. Liquid water is thus characterized by ice-like clusters which are continuously flickering into and out of existence.

The presence of an apolar solute increases the propensity of water molecules to hydrogen bond. At the same time, the proximity of an apolar solute insulates an existing cluster from the positive energy fluctuations that melt it. The net result is an increase in ice-like structure. Frank and Evans coined the term 'iceberg formation' to describe the effect. It should be pointed out that they also cautioned that the term 'iceberg', while implying an increase in water structure, did not necessarily mean ice structure [46].

Given the open nature of water structure, apolar solutes reside in the available cavities, the walls of which are reinforced by extensive hydrogen bonding. The clathrate hydrates formed by hydrophobic or amphiphilic solutes, e.g. aliphatic amines, provide a model for the cage structures. X-ray crystallography has shown that the clathrates consist of fused polyhedra of hydrogen bonded water molecules with the apolar guest molecules residing in the cavities [47]. The clathrate hydrates are, thus, exclusion compounds formed by the increased association of water molecules induced by the solute. Jeffrey and McMullan have shown that the basic structural unit is the $H_{40}O_{20}$ pentagonal dodecahedron which can be arranged in a variety of ways to form the host lattice [48-50]. In the case of a large guest molecules it may expand to a larger polyhedron. The hydrophilic portion of an amphipilic guest may interact with the lattice. With primary amines, the amine functional group may be incorporated into the cage structure while the nonpolar moiety resides in the cavity. The more hindered *t*-butylamine resides entirely within the cavity [47]. The clathrate hydrates span a wide range of stoichiometry and stability with hydration numbers ranging from 3.5 to as high as 40 and melting points ranging from $-35$ to 25°C [47,50].

Just as water is characterized by a lingering ice structure so solutions of hydrophobic or amphiphilic solutes are characterized by a lingering clathrate structure. However, despite the extensive effort invested, the search for rigid clathrate structures in solution has generally not been successful. Nevertheless, the evidence supports that apolar solutes reside in hydration shells distinct from bulk water. The hydration shells may be thought of as clathrate structures in a time average sense. The water molecules in the shell are characterized by increased hydrogen bonding and decreased mobility. Dielectric relaxation and NMR experiments have shown that rotational correlation times are increased with the magnitude of the effect dependent on the size of the apolar moiety and its concentration [51,52]. Translational mobility is also reduced. Some calculations have indicated an increase in five membered rings of water molecules, characteristic of clathrate hydrates [53].

At low concentration the amphiphilic solute may be thought of as enclosed in a distinct and separate hydration shell. However, as solute concentration increases the system seeks to reduce the unfavorable entropy by aggregation which generally appears to involve the sharing of water molecules by fusing cages. Eventually the required excess of water is depleted. Beyond this point the solution becomes increasingly micellar in nature and is characterized by heterogeneity at the microscopic level. A tendency toward phase separation develops.

The next several sections examine the available thermodynamic data for nicotine in context of the solution properties of amphiphiles.

## VIII. THE ACTIVITY COEFFICIENT OF NICOTINE IN AQUEOUS SOLUTION

Solutions of nicotine in water are far from ideal in behavior. While nicotine-water interactions are negligible, the presence of nicotine results in increased association between water molecules. As concentration increases nicotine aggregation becomes significant. These phenomena have a marked influence on solution behavior. The deviation from ideality is most directly described by activity coefficients. For solutions with volatile components activity coefficients are conveniently derived from vapor pressure data.

The laws governing the behavior of ideal systems are generally straight-forward. The vapor pressure of a component above an ideal solution is given by Raoult's Law:

$$P_i = X_i P_i^0 \tag{12}$$

where $X_i$ is the mole fraction of the $i$-th component in the solution, $P_i$ is the vapor pressure of the $i$-th component above the solution and $P_i^0$ is the vapor pressure of pure component $i$. Thus, if aqueous nicotine were an ideal system, we could calculate the nicotine vapor pressure at any concentration given the vapor pressure of pure nicotine. However, nicotine is not an ideal solute and the calculation would be seriously in error. A correction factor must be applied to eq. 12. In order to maintain the simple form of the ideal law, the established convention is to apply the correction factor in the form of an activity coefficient $\gamma$. The corrected vapor pressure law for nicotine is, thus, given by:

$$P_i = \gamma_i X_i P_i^0 \tag{13}$$

The deviation of $\gamma$ from unity is a measure of the departure from ideality. For hydrophobic solutes is generally greater than unity indicative of the desire of the solute to escape from the aqueous phase.

Activity coefficients are generally concentration dependent and must be determined from experimental data. For volatile compounds they are most conveniently determined from vapor pressure data as a function of concentration. In the case of aqueous nicotine at 25°C we may write:

$$\gamma_2 = \frac{P_2}{X_2 P_2^0} = \frac{P_2}{0.0425 X_2} \tag{14}$$

Norton et al. determined the vapor pressures of nicotine and water at 25°C over solutions spanning the full concentration range of 0–100 mole percent nicotine [10]. They give the concentration dependences of the vapor pressures in mm Hg as:

$$\log P_1 = 1.3714 + 1.50 \log X_1 + 0.643 X_2 - 0.709 X_2^2 + 1.23 X_2^3 \tag{15}$$

$$\log P_2 = -1.3716 + 0.02 \log X_2 - 0.643 X_1 + 1.14 X_1^2 - 1.23 X_1^3 \tag{16}$$

Equation (15) represents the water vapor pressure data well for solutions ranging from 2–90 mole percent nicotine. Equation (16) works well for the nicotine vapor pressure for solutions above two mole percent nicotine. The nicotine vapor pressure is shown in Fig. 4 where the solid line was calculated using eq. 16.

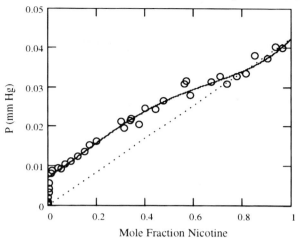

Fig. 4 The vapor pressure of aqueous nicotine. After Norton et al. [10]. The dotted line represents ideal behavior.

In the dilute region, i.e. less than two mole percent nicotine, the vapor pressure of nicotine rises dramatically with concentration. Norton et al. suggest that Henry's Law is obeyed in this region. However, inspection of the data does not support that conclusion. Henry's Law, usually written in terms of molality, requires a linear increase in vapor pressure with concentration:

$$P_2 = k'm \tag{17}$$

where $k'$ is the Henry's Law constant. Instead, for solutions up to 0.9 molal, i.e. 1.6 mole percent nicotine, the vapor pressure data are well represented by:

$$P_2 = 2.600 \times 10^{-3} \, m^{0.5} + 2.115 \times 10^{-2} \, m - 1.489 \times 10^{-2} \, m^{1.5} \tag{18}$$

Figure 5 shows the data of Norton et al. superimposed on a line calculated using eq. (18). Table 2 lists the calculated and experimental values of the nicotine vapor pressure. Equation (18) represents the data well up to a concentration of 0.9 molal but begins to deviate sharply above that concentration.

Equations (14) and (18) may be used to calculate the activity coefficient of nicotine in solutions up to 0.9 m in concentration. A tabulation of values at various concentrations is shown in Table 3. The activity coefficient ranges from a value of 134 at a concentration of $10^{-3}$ molal to 13 at 0.9 molal (12.7% nicotine). Clearly, Raoult's Law is not obeyed anywhere in the water rich region. Nicotine approaches Raoult's Law behavior at a concentration of about sixty mole percent, i.e. 93% nicotine.

Ideal behavior is defined by the straight line in Fig. 4 going from the origin to the limiting value of 0.0425 mm in pure nicotine. The strong positive deviation from ideality exhibited by nicotine is characteristic for amphiphilic

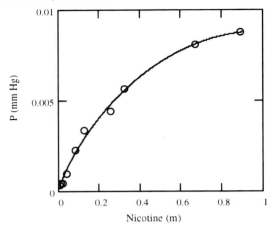

Fig. 5 Vapor pressure of aqueous nicotine at low concentration. Points calculated from data of Norton et al. [10]. Line obtained by regression to eq. (18).

solutes. It may be viewed as resulting from the tendency of the solute to escape from the aqueous phase. The sharp drop in activity coefficient with increasing concentration in the dilute region is a further characteristic of amphiphilic solutes and is generally interpreted to be the result of solute aggregation.

## IX. AZEOTROPISM IN THE NICOTINE-WATER SYSTEM

Kelly, MacCarthaig and Reilly et al. first noted a nicotine-water azeotrope [54]. Kelly, O'Connor and Reilly subsequently confirmed the azeotrope at

TABLE 2

Calculated versus experimental vapor pressures of nicotine in aqueous solutions

| Nicotine concentration | | Vapor Pressure[a] | Calculated vapor |
|---|---|---|---|
| Percent | molal | (mm $\times 1000$) | Pressure[b] (mm $\times 1000$) |
| 0.108 | 0.0067 | 0.35 | 0.35 |
| 0.186 | 0.0115 | 0.45 | 0.50 |
| 0.281 | 0.0174 | 0.49 | 0.68 |
| 0.648 | 0.0402 | 1.0 | 1.3 |
| 1.318 | 0.0823 | 2.27 | 2.1 |
| 2.0 | 0.126 | 3.41 | 2.9 |
| 4.0 | 0.257 | 4.42 | 4.8 |
| 5.0 | 0.324 | 5.7 | 5.6 |
| 9.8 | 0.670 | 8.1 | 8.1 |
| 12.6 | 0.889 | 8.8 | 8.8 |

[a] Reference 10
[b] Calculated from eq 16

TABLE 3

The activity coefficient of aqueous nicotine

| Weight percent | Molality | Molarity | Mole percent | Activity coefficient |
|---|---|---|---|---|
| 0.016 | 0.001 | 0.001 | 0.0018 | 134 |
| 0.081 | 0.005 | 0.005 | 0.009 | 74 |
| 0.162 | 0.01 | 0.01 | 0.018 | 60 |
| 0.805 | 0.05 | 0.05 | 0.09 | 38 |
| 1.596 | 0.1 | 0.098 | 0.18 | 32 |
| 7.503 | 0.5 | 0.468 | 0.89 | 19 |
| 12.74 | 0.9 | 0.793 | 1.6 | 13 |

2.45% nicotine with a boiling point of 99.6°C at 760 mm [55]. Smith used the nicotine azeotrope, which he determined to be 2.5% nicotine boiling at 99.988°C, to separate nicotine from nornicotine and anabasine which do not form azeotropes with water [56]. Fowler determined the azeotrope to be at 2.52% nicotine with a boiling point of 99.85°C at 760 mm and showed that its nicotine content decreases as the pressure is reduced with no azeotrope being detectable at pressures below 400 mm [57].

The nicotine-water azeotrope is of the minimum boiling type. It is the result of the strong positive deviation of dilute nicotine solutions from Raoult's Law which causes a maximum in the vapor pressure versus concentration curve with a corresponding minimum in the boiling point curve. Such systems have also been termed 'pseudo-azeotropes' to distinguish them from maximum boiling azeotropes caused by a minimum in the vapor pressure curve resulting from interactions, e.g. hydrogen bonding, between components.

## X. THERMODYNAMICS OF MIXING

The thermodynamic parameters associated with the mixing of nicotine with water reflect the increased association of water molecules due to the influence of nicotine. As suggested above the effects have similarities to the freezing of water. The increased hydrogen bonding between water molecules results in an exothermic heat of mixing. The increased structuring of water is reflected in a sizable negative entropy.

To obtain a perspective of the magnitude of the deviation from ideality it is necessary to first define the thermodynamics of mixing for ideal solutions. In an ideal solution the interactions between components is the same as those between molecules in the pure components. The heat of mixing is, thus, zero. However, the mixing, even of ideal components, does lead to an increase in disorder which results in a positive entropy of mixing. For a binary solution the associated entropy change is given by:

$$\Delta S_m^0 = -R(X_1 \ln X_1 + X_2 \ln X_2) \tag{19}$$

It should be noted that the entropy change is positive despite the negative sign because the mole fractions are less than one. The free energy of mixing for an ideal solution at constant temperature can now be defined:

$$\Delta G_m^0 = \Delta H_m^0 - T\Delta S_m^0 = RT(X_1 \ln X_1 + X_2 \ln X_2) \tag{20}$$

For non-ideal solutions the thermodynamic parameters are generally written as the sum of the ideal contribution and an excess term which represents the deviation from ideality. Thus the free energy may be written:

$$\Delta G_m = \Delta G_m^0 + \Delta G^E \tag{21}$$

The excess free energy of mixing can be split into the contributions from each of the components of a solution. For a binary solution:

$$\Delta G^E = X_1 \mu_1^E + X_2 \mu_2^E \tag{22}$$

The $\mu^E$ terms on the right hand side of eq. (22) are the excess chemical potentials of the components which may be further defined as:

$$\mu_i^E = RT \ln \gamma_i \tag{23}$$

It follows from eqs (22) and (23) that the excess free energy of mixing may be calculated if all activity coefficients are known.

The values of $\Delta G^E$ for aqueous nicotine at 25°C were calculated as a function of concentration from the vapor pressure data of Norton et al. [10]. The result is shown in Fig. 6. The value of $\Delta G^E$ is small and positive over the entire concentration range peaking near the equimolar point. The positive sign indicates that the excess contribution to the mixing process is thermodynamically unfavorable.

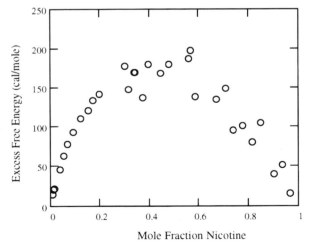

Fig. 6 Excess free energy of aqueous nicotine. Calculated from data of Norton et al. [10].

The measurements of the heat of mixing nicotine with water are very old and very few. However, the facts that the measurements are straightforward and the magnitudes of the measured heats are sizable lend credibility to the results. In 1889 Colson determined that 6.6 Kcal of heat is evolved at 15°C when one mole of nicotine is mixed with two liters of water [4]. Berthelot and Andre, in 1899, determined a value of $-7.05$ Kcal for the heat of mixing one part of nicotine with 25 parts water at 11°C [5]. Since the ideal heat of mixing is zero, the measured heats represent the excess heat.

With the free energy and enthalpy in hand, the entropy can be calculated. Unfortunately the free energy and the heats are at different temperatures. However, it can be shown that the free energy is negligible compared to the enthalpy. The final nicotine concentration in Colson's determination is 0.9 mole percent. The data of Norton et al. yield a value of 10 cal/mole for the excess free energy of mixing at that concentration. Even if the excess free energy were to increase by an order of magnitude upon cooling from 25 to 15°C, it would remain negligible compared to Colson's value for the heat of mixing. Consequenly one can safely make the approximation $\Delta S^E = \Delta H^E / T = -23$ cal/mole-degree at 15°C.

The thermodynamic parameters for mixing nicotine with water clearly meet the expectations for an amphiphilic solute as outlined at the beginning of this section. The sizeable exothermic heat of mixing is consistent with increased hydrogen bonding between water molecules. The large negative entropy is indicative of the increased structure induced in water by the nicotine molecule.

## XI. PARTIAL MOLAR VOLUME

The volumetric behavior of amphiphilic solutes has been thoroughly studied and is well understood [40,58–61]. The ideal case is again taken as the point of departure. The volume of an ideal, binary solution is given by:

$$V = n_1 V_1^0 + n_2 V_2^0 \tag{24}$$

The $V_0$ terms in eq. (24) are molar volumes obtained from the molecular weights and densities of the components.

Densities are frequently reported in terms of a specific gravity, e.g. $D_4^{20}$ the density of the substance at 20°C relative to the density of water at 4°C. Jackson, in his review [1], tabulates 22 separate determinations of the $D_4^{20}$ value of nicotine. These yield an average value of 1.01006 with a standard deviation of 0.00119. The variability is rather large. For reasons identical to those outlined in the specific rotation section, the $D_4^{20}$ value of 1.00925 obtained by Jephcott and Ratz is used here [12]. Given the value of the absolute density of water at 4°C, one obtains a value of 1.00922 g/cc for the absolute density of nicotine at 20°C. The molar volume of nicotine is, thus, 160.7 cc/mole.

For non-ideal solutions the constant molar volumes in eq. (24) are replaced by the concentration dependent partial molar volumes:

$$V = n_1 \bar{V}_1 + n_2 \bar{V}_2 \qquad (25)$$

Partial molar volumes are defined as:

$$\bar{V}_i = \left( \frac{\partial V}{\partial n_i} \right)_{n_j, T, P} \qquad (26)$$

Thus, the partial molar volume may be viewed as the volume change that takes place when one mole of a component is added to an infinite amount of solution.

Density data is most conveniently analyzed in terms of the apparent partial molar volume. For aqueous nicotine this is given by:

$$\Phi_2 = \frac{1}{m} \left( \frac{1000 + 162.23m}{d} - \frac{1000}{d_0} \right) \qquad (27)$$

where $m$ is the molality of nicotine, $d$ the density of the solution, $d_0$ the density of water and 162.23 is the molecular weight of nicotine. Equation (27) assumes a constant volume for the solvent and attributes all volume changes to the solute. The partial molar volume may then be obtained from the apparent partial molar volume via:

$$\bar{V}_2 = \Phi_2 + m \left( \frac{d\Phi_2}{dm} \right) \qquad (28)$$

The value of the partial molar volume of a solute at infinite dilution has particular significance. It is usually expressed as an excess value, i.e. the partial molar volume at infinite dilution less the molar volume:

$$V_i^{0,E} = \bar{V}_i^0 - V_i^0 \qquad (29)$$

The value of $V_2^{0,E}$ in aqueous solutions appears to be invariably negative reflecting the openness of the structure of water. Its magnitude may be interpreted as a measure of the compatibility of the solute with the structure of water [62].

The partial molar volume curve of amphiphilic solutes goes through a minimum in the water rich regime. The more hydrophobic the solute, the steeper the initial drop in partial molar volume and the lower the concentration at which the curve goes through the minimum. The existence of a minimum in the partial molar volume curve is generally considered diagnostic for the building of a hydration shell or water cage around the solute. The initial slope of the partial molar volume curve, i.e. $-(\partial \bar{V}_2 / \partial X_2)$, is

172

a frequently cited parameter for amphiphilic solutes. It is taken to be a measure of solute-solute interactions. The magnitude of the initial slope varies with the number and size of apolar side chains. Franks has suggested that the hydration shells of the solute molecules initially reinforce [40]. The requisite excess of water is depleted at the minimum and the shells begin to interfere as the concentration increases.

The values of $\Phi_2$ for nicotine were calculated as a function of concentration using the density data of Jephcott [12] which cover a wide concentration range. With amines a correction is frequently made to take into account the reduction of free base concentration by hydrolysis. Since the base strength of nicotine is low, the hydrolysis correction is insignificant and was omitted. The results were fitted by regression to the following polynomial:

$$\Phi_2 = a_0 + a_1 m^{0.5} + a_2 m + a_3 m^{1.5} + a_4 m^2 \tag{30}$$

This function has no theoretical significance but can readily be differentiated to yield the partial molar volume via eq. (28).

Figure 7 shows the values of $\Phi_2$ superimposed on the regression line with the calculated partial molal volume shown by the dotted line. Despite the scatter, the data clearly demonstrate the existence of a minimum.

The partial molal volume of nicotine as a function of mole fraction is shown in Fig. 8. The solid line, obtained from the density data of Jephcott at 20°C, goes through a minimum at 1.7 mole percent. Given the scatter in the data and the fact that the position of the minimum is influenced by the function chosen for curve fitting, an estimate of 2.0±0.5 mole percent may be more realistic. The dashed line was calculated from the more precise data of Campbell, Kartzmark and Falconer [63] at 25°C which, unfortunately, does

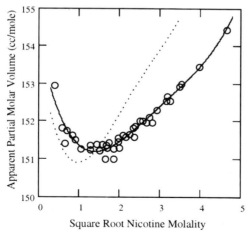

Fig. 7 Apparent partial molar volume of nicotine at concentrations up to 25 molal. Points calculated from data of Jephcott [12]. Solid line obtained by regression to eq. (30). Dotted line is partial molar volume as calculated by eq. (28).

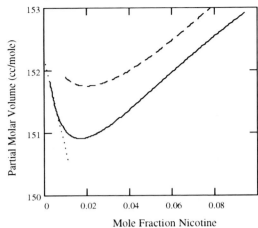

Fig. 8 Partial molar volume of nicotine. Solid line: calculated from data of Jephcott [12]. Dotted line: initial slope calculated from Jephcott's data [12]; dashed line: calculated from data of Campbell et al. [63].

not extend below 10% nicotine, i.e. 1.2 mole percent. This yields a minimum at 1.9 mole percent nicotine. Taken together the results indicate that the partial molar volume of nicotine goes through a minimum at about two mole percent nicotine. The hydration shell of nicotine should, thus, contain about fifty water molecules.

The dashed line in Fig. 8 lies about 0.9 cc higher at its minimum than the solid line. This reflects the temperature difference of 5°C between the conditions at which the two data sets were measured.

The dotted line in Fig. 8 shows the initial slope of the partial molar volume curve obtained from Jephcott's data. The estimation of the initial slope is not a straightforward matter. Franks has shown that the partial molar volume of alcohols does not decrease in linear fashion from the intercept [40]. However, since data at extreme dilution are rarely available, the usual procedure is a linear extrapolation of the partial molar volume data in the dilute region. That practice was followed here.

The initial slope was determined directly from Jephcott's data without resort to eq. (30). The values of the differential $\Delta V/\Delta X_2$, evaluated per concentration interval, were graphed against the average value of $X_2$ over the interval. Regression to a straight line then yielded a slope of $-150\pm60$ with an intercept of $152.2\pm0.3$. The stated error bars are the standard errors of the slope and intercept. The graph is shown in Fig. 9. The result is in good agreement with the calculated partial molar volume curve as shown in Fig. 8 where the dotted line shows the initial slope.

Given the intercept, the excess partial molar volume of nicotine at infinite dilution is about $-9$ cc/mole. This result, as well as the initial slope, are consistent with the results obtained for a number of amines [58,59].

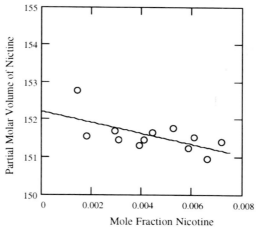

Fig. 9 Initial slope of partial molar volume of nicotine with concentration calculated from the density data of Jephcott [12].

## XII. DOES NICOTINE FORM A CLATHRATE HYDRATE?

Over the years there has been intermittent interest in potential hydrates of nicotine. Attention has been focused with no success on the nicotine rich region. This leaves open the question of the existence of a clathrate hydrate in the water rich region.

Many amines form clathrate hydrates. Based on the strength of the hydrophobic effect, nicotine appears to be comparable to a number of these. The values of $V_2^{0,E}$ and the initial slope of the partial molar volume curve for nicotine are compared to those published for a number of amines in Table 4 [59,60]. The nicotine values clearly fall in the same range as the listed amines. Ethylamine, dimethylamine, $n$-propylamine, the butylamines and diethylamine all form clathrate hydrates.

TABLE 4

Volumetric properties of some amines

| Solute | $\bar{V}_2^{0,E}$ (cc/mole) | $d\bar{V}_2/dX_2$ | Reference |
|---|---|---|---|
| Nicotine | $-8.5\pm0.3$ | $-150\pm60$ | |
| EtNH$_2$ | $-8.0$ | $-69$ | [58] |
| (Me)$_2$NH | $-8.0$ | $-102$ | [58] |
| $n$-PrNH$_2$ | $-8.5$ | $-159$ | [58] |
| $n$-BuNH$_2$ | $-10.1$ | $-165$ | [58] |
| $t$-BuNH$_2$ | $-12.2$ | $-362$ | [59] |
| (Et)$_2$NH | $-5.0$ | $-153$ | [58] |
| Pyrrolidine | $-5.8$ | $-112$ | [59] |
| Piperidine | $-6.8$ | $-154$ | [59] |

175

The composition and stability of the clathrate hydrates vary over a wide range. The concentration at which the partial molar volume curve goes through the minimum does not necessarily correspond to the composition of the hydrate. The size of the guest molecule may, however, give an indication of the composition of the hydrate. For the amines listed in Table 4, the hydration numbers range from 5.5 to 11.5 and the melting points from $-35$ to $-2°C$ [46]. The much larger tetra $n$-butyl ammonium salts form high hydrates with hydration numbers of about 40 and melting points ranging from 12 to 31°C [64–66].

Nicotine is a larger molecule than the simple amines listed in Table 4. Space filling models indicate that $t$-butylamine and $n$-butylamine in the coiled conformation are comparable in size to the $N$-methylpyrrolidine ring. The longest dimension of the nicotine molecule is comparable to that of the tetra n-butyl ammonium ion. This suggests that the clathrate hydrate of nicotine, if it exists, should have a large hydration number, perhaps approaching 40.

A hydration number of 40 corresponds to a nicotine concentration of 18%. The phase diagram of nicotine was investigated by a number of early workers [1] and, more recently, was re-examined by Campbell et al. [63]. No mention is made of a hydrate. It is unlikely that all of these studies would have missed a hydrate at 18% nicotine; one can only conclude that nicotine does not form a clathrate hydrate.

In the absence of phase data, one can only speculate about nicotinium salts; these cannot be precluded from forming clathrate hydrates particularly if the counterions can stabilize or form part of the cage structure. They should be comparable in size to the tetra $n$-butyl ammonium salts but are probably less hydrophobic.

## XIII. PARTITION COEFFICIENTS

The importance of the hydrophobic effect in biological systems has long been recognized. Around the turn of the century, Meyer and Overton used the oil-water partition coefficient as a means of ordering the narcotic action of simple chemicals [67]. However, several decades were to pass before this concept was quantified. Early work with phenoxyacetic acid derivatives showed that substituents which lower electron density at the ortho positions increase plant growth promoting activity. However, attempts at correlating activity to the constants of the Hammett equation failed. There appeared to be a missing factor. In 1962 Corwin Hansch and coworkers proposed that the missing factor is the rate of penetration into the plant cell. In order to penetrate the cell a molecule must first pass through the cell wall which is largely lipid in nature. This requires a balance of hydrophilic to lipophilic properties. The octanol-water partition coefficient was used as a measure of that balance. Hansch's model, which incorporated the partition coefficient as well as the Hammett constants, proved successful [68]. The model provided an approach to separating in principle three of the most important parameters governing biological activity: electronic effects, steric effects and penetration rate [69]. In 1965 Hansch et al. showed that, given constant electronic and steric effects, the biological activities of a series of drugs varied as the logarithm of the partition coefficient [70]. Since then partition coefficients have been determined for a large number of compounds [67,71]. Hansch's work provided the basis for the Quantitative Structure Activity Relation (QSAR) paradigm widely used in pharmacology, toxicology and pharmaceutical design [67,72].

Values have been determined for the partition coefficients pertaining to the distribution of nicotine between various organic solvents and water. However, before considering these, the factors which influence the value of the partition coefficient of nicotine will be briefly examined. The partition coefficient is written here as concentration in the organic phase over concentration in the aqueous phase.

The partitioning of nicotine is strongly affected by pH. The neutral form of nicotine distributes between water and organic solvents and water while the protonated nicotinium ions remain in the aqueous phase. One can write a pH dependent apparent partition coefficient, $K_{pH}$:

$$K_{pH} = \frac{K_0}{1 + \dfrac{a_{H^+}}{K_{a2}} + \dfrac{a_{H^+}^2}{K_{a1}K_{a2}}} \tag{31}$$

where $K_0$ is the pH independent partition coefficient of the neutral molecule. Given the values of $pK_{a1} = 3.41$ and $pK_{a2} = 8.10$, established in a previous section of this chapter, the partition coefficient of nicotine at the plasma pH level of 7.4 is reduced to about 17% of $K_0$. The value of the ionization constant is an important factor in biological activity. Consequently, the ionization constants of a compound are frequently determined along with the partition coefficient.

The effect of temperature on the kerosene-water partition coefficient of nicotine was examined by Claffey et al. over the range of 5–98°C [73]. A van't Hoff plot of their data is shown in Fig. 10. The partition coefficient increases with temperature approaching a limiting value at about the lower critical solution temperature shown by the vertical dotted line. The lower critical solution temperature is the lowest temperature at which nicotine is no longer miscible in all proportions with water. The linear portion of the plot yields an enthalpy of 7.6 kcal/mol. The results appear consistent with the effect of temperature on water structure. As the temperature increases, the extent of hydrogen bonding between water molecules decreases reducing the hydration of nicotine and forcing it into the kerosene. As the lower critical

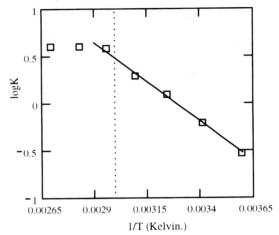

Fig. 10 The effect of temperature on the kerosene-water partition coefficient of nicotine. Data points from reference [73]. Dotted vertical line shows the lower critical temperature of the nicotine-water system.

solution temperature is approached hydration collapses. The high temperature limit should, thus, represent the partition coefficient between kerosene and water in the absence of hydrophobic hydration.

Oakley and Swarbrick [74] examined the thermodynamics of the partitioning of nicotine between a number of organic solvents and water. Their results are shown in Table 5 along with the values calculated for kerosene-water from the data in reference [75].

It was shown in a previous section of this chapter that the enthalpy of hydration for nicotine is $-6.6$ kcal/mol and the entropy of hydration was estimated to be about $-23$ cal/deg-mol. Comparison of these values with the results in Table 5 suggests that nicotine is completely stripped of its hydration in miglyol 812, isopropyl myristate and kerosene. Apparently it suffers only a small reduction of its hydration shell in *n*-butanol. This is

TABLE 5

Thermodynamics of the partitioning of nicotine

| Organic phase | $\Delta H$ (kcal/mol) | $\Delta S$ (cal/deg-mol) | Reference |
|---|---|---|---|
| *n*-butanol | 0 | 7.2 | [74] |
| *n*-octanol | 3.38 | 17.8 | [74] |
| mioglyol 812 | 6.22 | 23.2 | [74] |
| isopropyl myristate | 6.85 | 24.2 | [74] |
| kerosene | 7.6 | 24.8 | calculated from data in reference [75] |

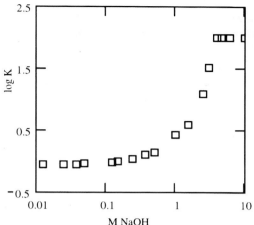

Fig. 11 The effect of excess sodium hydroxide on the log of the partition coefficient of nicotine between kerosene and water. Data points from reference [75].

consistent with the high solubility of water in n-butanol (9.3 M) [74]. Octanol appears to be an intermediate case.

Salt concentration is a the third factor that can have a sizeable effect on the partition coefficient of nicotine. Norton studied the effect of excess sodium hydroxide on the partition coefficient of nicotine between lubricating oil and water [75]. His results are shown in Fig. 11. The value of $\log K_o$ in dilute solution is $-0.045$, then begins to rise as the sodium hydroxide concentration exceeds 0.25 M and finally approaches a limiting value of 2.0 in 4 M sodium hydroxide. Since nicotine is present as the neutral molecule over the whole experimental range, this is clearly not a pH effect but is, rather, analogous to the salt effect. The effects of various salts on the solubility of nicotine in water have been tested. Cuvelier examined the effectiveness of a series of ammonium salts for salting nicotine out of aqueous solution [76]. The relative effectiveness of the anions is: sulfate > oxalate > phosphate > carbonate ~ acetate > fluoride. The sulfate ion has also been shown to be more effective than hydroxide ion at forcing nicotine out of water [77].

Finally, the concentration of nicotine itself can affect the value of the partition coefficient. However, this occurs in more concentrated solutions and will be considered in the next section.

The partition coefficients of neutral nicotine in selected solvent systems are shown in Table 6 along with the $pK_a$ values where determined. There is good agreement with respect to the n-octanol-water partition coefficient which is the most important in QSAR applications. The study by Li, Li and Gorrod [79] was presumably carried out at room temperature which should not be very different from 20°C. The n-octanol partition coefficient is larger than the cyclohexane coefficient. Nicotine, a polar molecule, is expected to

TABLE 6

The partition coefficients of neutral nicotine for organic solvents and water

| Temperature (°C) | Organic phase | log $K_o$ | $pK_{a1}$ | $pK_{a2}$ | Reference |
|---|---|---|---|---|---|
| 20 | *n*-octanol | 1.17 | 3.11 | 8.05 | [78] |
| – | *n*-octanol | 1.17 | 3.42 | 8.21 | [79] |
| – | cyclohexane | 0.36 | 3.42 | 8.21 | [79] |
| 17 | trichloroethylene | 1.01 | – | – | [80] |
| 20 | kerosene | – 0.20 | – | – | [73] |
| 20 | *n*-butanol | 3.60 | – | – | calculated from thermodynamic data in reference [74] |
| 20 | *n*-octanol | 1.37 | – | – | calculated from thermodynamic data in reference [74] |
| 20 | isopropyl myristate | 0.18 | – | – | calculated from thermodynamic data in reference [74] |

prefer a more polar organic solvent. A second reason might be that *n*-octanol contains more water at equilibrium than cyclohexane. The trichloroethylene and kerosene partition coefficients are of interest primarily for industrial applications such as nicotine extraction. The *n*-butanol partition coefficient is very large due to the presence of a large amount of water in the organic phase. Oakley and Swarbrick found that the monoprotonated nicotinium ion also distributes into *n*-butanol [74].

## XIV. CONCENTRATED SOLUTIONS

A number of the properties of nicotine solutions go through maxima at concentrations above two mole percent. These have been taken to be evidence for the formation of hydrates of nicotine. However, to date, no nicotine hydrates have been isolated.

The partition coefficient of nicotine between water and organic solvents, defined as concentration in the organic phase over concentration in the aqueous phase, goes through a minimum at eleven mole percent [73,75]. The position of the minimum appears to be independent of the organic phase suggesting that a process in the aqueous phase is responsible. Figure 12 shows the curve calculated from Norton's data [75] for the partitioning of nicotine between kerosene and water. The effect is quite large. At the minimum, nicotine is about a factor of seven less available to the organic phase than in dilute solution. Norton suggested that the minimum is indicative of a hydrate which would have a water to nicotine ratio of 8 : 1.

It has long been known that the density of aqueous nicotine passes through a maximum. The density maximum falls between 19 and 25 mole percent, depending on which data set is used. Figure 13 shows the curve

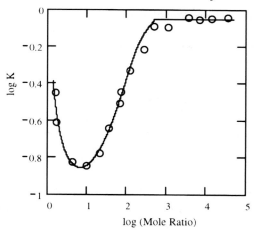

Fig. 12 The logarith of the partition coefficient of nicotine between water and kerosene. Data points from Norton [75].

calculated from Jephcott's measurements at 20°C [12]. The viscosity also passes through a maximum in the same concentration range. The effect is quite large with a relative viscosity of 20 at the peak [1]. Dezelic et al. have shown that the derivative of the refractive index goes through a maximum at 25 mole percent [81]. This finding is particularly significant since maxima in the derivative of a property are stronger indicators of structural effects than maxima in the property itself. Campbell et al. have shown that nicotine solutions stop freezing between 17 and 21 mole percent [63]. This suggests that the last vestiges of water structure have disappeared and would also appear to preclude the existence of a crystalline trihydrate of nicotine.

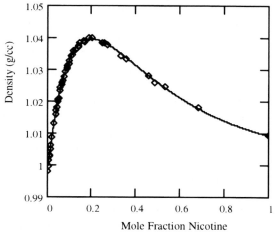

Fig. 13 The density of aqueous nicotine. Data points from Jephcott [12].

There is no question that significant structural changes are taking place in nicotine solutions as the water to nicotine ratio passes through $3:1$. Just what these changes are is less clear. Suggestions of specific complex formation of the sort:

$$m\text{Solute} + n\text{H}_2\text{O} \rightleftharpoons (\text{Solute})_m(\text{H}_2\text{O})_n$$

have been made for nicotine as well as other amphiphilic solutes. Such mechanisms have, in general, been found to be inadequate [41].

As solute concentration increases, solutions of amphiphiles tend to become micellar in nature exhibiting microscopic heterogeneity which readily leads to phase separation. The behavior of $t$-butyl alcohol (TBA) is illustrative. Based on a light scattering studies, Iwasaki and Fujiyama [82] proposed that, in dilute solution, TBA molecules are enclosed in polyhedra of 21 water molecules which are dispersed in bulk water. When the concentration rises above $X_2 = 0.045$ the solution changes drastically. It now consists of clusters of TBA molecules interspersed with hydrate regions comprised of five fused water polyhedra each containing a guest TBA molecule. A more recent X-ray diffraction study by Nishikawa and Iijima has confirmed the light scattering results [83]. Based on the X-ray results at a TBA mole fraction of 0.039, they proposed that TBA is enclosed in a polyhedron of 24–28 water molecules with the cage free to move about in the bulk phase. As the concentration increases the polyhedra fuse to form larger aggregates. At a TBA mole fraction of 0.17 the solution consists of the hydrate aggregates dispersed in bulk TBA.

The equilibrium state in such a system is best not represented as some fixed distribution but rather as a set of fluctuations about a mean. If the system is perturbed by changing temperature, pressure or concentration, it moves to a new equilibrium represented by a new set of concentration fluctuations. The transition is diffusion controlled [41].

The behavior of nicotine is, in all likelihood, explained by some analogous picture involving concentration fluctuations. The data presently available are insufficient to provide any detailed understanding.

## XV. PHASE DIAGRAM OF NICOTINE-WATER SYSTEM

Hudson in 1904 first determined the oval shaped, closed loop phase diagram for the nicotine-water system that became the textbook example of a system with both a lower critical solution temperature (LCST) and an upper critical solution temperature (UCST). Hudson found the LCST to be about 61°C and the UCST to be about 210°C. He found both critical compositions to be the same: 32.2 weight percent nicotine. Tsakalatos repeated the measurements of Hudson and found the critical temperatures to be 60.8 and 208°C and the critical composition, at both temperatures, to be 34 weight percent. Leone repeating the measurements again in 1926 was in close agreement with Hudson [1,63].

Campbell et al. repeated the nicotine water measurements in 1958 as part of a study of the nicotine-methylethyl ketone-water system [63]. The lower portion of their phase diagram was in reasonable agreement with the earlier measurements with an LCST of 61.5°C and a critical composition of 36 weight percent nicotine. The upper portion of the curve was, however, in serious disagreement with previous measurements. They found the UCST to be 233°C and the critical composition to be 40%. The reasons for the discrepancy are not clear. The authors offer no explanation but rather suggest that earlier measurements were seriously in error.

Figure 14 shows the phase diagram of the nicotine-water system. The two phase domain is enclosed within the oval shaped loop. The solid line shows the early measurements [84]. The dotted line through the solid loop at 90°C shows the phase inversion in the two phase region. Below the dotted line the nicotine rich phase is heavier. Above the dotted line the water rich phase is on the bottom. The measurements of Campbell et al. are shown by the dashed line [63].

The hydrophobic nature of nicotine was understood by early researchers though the term itself had not yet been introduced. It was proposed that, while nicotine itself is insoluble in water, it forms a hydrate which is miscible with water. The breaking of the hydrate at the LCST would then result in phase separation with true miscibility presumably occurring at the UCST. The search for the elusive hydrate of nicotine was in large part motivated by the desire to explain the phase diagram.

In terms of the contemporary view, the existence of a discrete hydrate is not necessary. Nicotine solutions at the appropriate concentrations are understood to be heterogeneous at the microscopic level with an intrinsic tendency to phase separation.

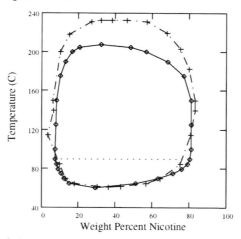

Fig. 14 Phase diagram of nicotine-water system. Dashed line: data of Campbell et al. [63]; Solid line: early measurements [84]; dotted line: phase inversion.

Kehiain [85–87] in a series of publications presented a thermodynamic analysis of the nicotine phase diagram. A summary of the derivation and the results are presented below.

The free energy for mixing the components of a binary solution may be written:

$$\Delta G_{mix} = RT(X_1 \ln X_1 + X_2 \ln X_2) + wX_1X_2 \tag{32}.$$

where the first term on the right hand side is the free energy of mixing for an ideal solution and the second term is the excess free energy of mixing, $\Delta G_{mix}^E$, which is a measure of the deviation from ideality. Kehian assumes the excess term to be of the form:.

$$\Delta G_{mix}^E = A(T)X_2(1 - X_2) \tag{33}$$

where $A(T)$ is a function of temperature only. (Note that $X_1 + X_2 = 1$) Kehiain states that for systems with two critical solution temperatures $A(T)$ is of the form:

$$A(T) = \alpha + \beta T + \gamma \cdot T \ln(T) \tag{34}$$

where the constants $\alpha$, $\beta$ and $\gamma$ are determined from the critical solution temperatures.

The critical solubility curve is given by all those points for which:

$$\left(\frac{\partial \Delta G_{mix}}{\partial X_2}\right) = 0$$

Given eqs (32) and (33), this leads to the following expression for the solubility curve:

$$A(T) = RT \frac{1}{2X_2(1 - X_2)} \tag{35}$$

which ultimately yields:

$$X_2 = \frac{-1 \pm \sqrt{1 + 2RT/A(T)}}{2} \tag{36}$$

The two roots represent the limbs of the solubility curve which is symmetric about $X_2 = 0.5$. That symmetry is built in by the assumption of eq. (33).

The phase diagram of the nicotine water system when graphed against mole fraction nicotine is rather asymmetric. Figure 6 shows that the excess free energy of mixing is not symmetric about the $X_2 = 0.5$ line. Thus, eq. (33) is not valid. However, Kehiain circumvents the symmetry problem by transforming coordinates to $Z$-fractions. The $Z$-fractions are defined in terms of mole fractions:.

$$Z_1 = \frac{QX_1}{QX_1 + X_2}$$

$$Z_2 = \frac{X_2}{QX_1 + X_2} \tag{37}$$

The term $Q$ is an asymmetry factor the value of which is determined from the upper and lower temperatures of the closed solubility loop at a given mole fraction. Since the value of $Q$ is nearly concentration independent, the average value may be used in eq (37). The phase diagram plotted against Z-fraction nicotine is a nearly symmetric figure about the $X_2 = 0.5$ line. Equations (32) to (36) are recast in terms of Z-fractions to predict the solubility loop.

The results are shown in Fig. 15. The dashed line shows the highly asymmetrical solubility curve in terms of mole fraction. The solid line depicts the nearly symmetrical figure that results from graphing against the Z-fraction. The dotted line shows the calculated solubility loop. The agreement between the calculated and the experimental curves is reasonably good.

## CONCLUSION

As stated in the introduction, the vast majority of the physical chemical data available for nicotine is very old. In virtually every instance, there is a serious need for updating using contemporary techniques. The vapor pressure of neat nicotine appears to be the only exception. Though the value of $pK_{a2}$ may be stated with some confidence, a systematic redetermination of the ionization constants with the proper controls (i.e. purity, ionic strength, carbon dioxide exclusion and proper electrode calibration) would be highly desirable. Calorimetric data with regard to heats and heat capacities of

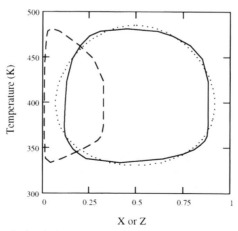

Fig. 15 Phase diagram of nicotine-water system. Dashed line: experimental curve vs. mole fraction; Solid line: experimental curve vs. Z-fraction; Dotted line: calculated curve vs. Z-fraction.

solution are essential to a thermodynamic understanding of aqueous nicotine. The precision of density data has increased vastly since Jephcott's day. Instrumental techniques, such as light scattering, ultrasound absorption, dielectric relaxation and NMR, which have proven highly informative with regard to other amphiphilic molecules, have yet to be applied to the nicotine-water system.

Given its importance at physiological pH, it is unfortunate that there are virtually no physical chemical data available for the nicotinium ion.

The amphiphilic nature of nicotine is of interest not only from a physical chemical point of view but potentially has implications for biological systems as well. It is well known that amphiphilic alcohols can have strong hydrophobic interactions with proteins [88].


## ACKNOWLEDGEMENTS

I am indebted to Bill Dwyer for the calculation of the optimum conformation of the nicotine molecule shown in this chapter. I am grateful to Cliff Lilly, Ken Shafer and Jerry Whidby for their helpful discussions and suggestions. Thanks are due to Richard Cox, Bill Dwyer, Cliff Lilly, Richard Jupe, Peter Lipowicz, and Jerry Whidby for reading the manuscript, to Kathy Mitchell for coordinating the internal review, and to Lynn Larson for obtaining copies of many rather inaccessible references.


## REFERENCES

1 Jackson KE. Alkaloids of tobacco. *Chem. Rev.*, 1941: **29**; 123–197.
2 Goodman J. *Tobacco in History: the Cultures of Dependence*. London and New York: Routledge, 1993.
3 Pinner A. Ueber Nicotin. Die Constitution des Alkaloids. *Chemische Berichte*, 1893: **26**; 292–305.
4 Colson A. Sur une application de la thermochimie. *Compt. Rend.*, 1889: **109**; 743–745.
5 Berthelot MM, André G. Nouvelles recherches sur les chaleurs de formation et de combustion de divers composés azotés et aultres. *Compt. Rend.*, 1899: **128**; 959–971.
6 Young HD, Nelson OA. Vapor pressures of fumigants IV-Vapor pressure of nicotine. *Ind. Eng. Chem.*, 1929: **21**; 321–322.
7 Gorbachev SV. Uprogost' para anabazina i nikotina. *Zh. Prikl. Khim.*, 1934: **7**; 388–391.
8 Harlan WR, Hixon RM. Volatility of nicotine. *Ind. Eng. Chem.*, 1928: **20**; 723–724.
9 Lewis DA. *The Evaporation and Diffusion of Nicotine from Mainstream Tobacco Smoke*. Ph. D. Thesis: University of Essex, 1994.

10 Norton LB, Bigelow CR, Vincent WB. Partial vapor pressures from nicotine solution at 25°. *J. Amer. Chem. Soc.*, 1940: **62**; 261–264.

11 Landolt A. Erste Abhandlung ueber die Ermittelung der specifischen Rotation activer Substanzen. *Ann.*, 1878: **189**; 241–337.

12 Jephcott H. The physical constants of nicotine. part i. specific rotatory power of nicotine in aqueous solution. *J. Chem. Soc.*, 1918: **115**; 104–108.

13 Heller W, Curmé HG. *Optical Rotation - Experimental Techniques and Physical Optics.* In: *Physical Methods of Chemistry Part IIIC Optical, Spectroscopic, and Radioactivity Methods.* Weissberger A and Bryant WR (Eds.), New York: Wiley-Interscience, 1972.

14 *The Merck Index - Twelfth Edition.*, Budavari, S (Ed.), Whitehouse Station, NJ: Merck & Co., Inc., 1996.

15 Harlan WR, Hixon RM. Catalytic reduction of nicotine and metanicotine. *J. Amer. Chem. Soc.*, 1930: **52**; 3385–3388.

16 Smith RM, Martell AE. *Critical Stability Constants Volume 2: Amines*, New York and London: Plenum Press, 1990.

17 Fowler RT. A redetermination of the ionization constants of nicotine. *J. Appl. Chem.*, 1954: **4**; 449–452.

18 Gonzalez E, Monge C, Whittembury J. Ionization constants of 5,5'-dimethyl–2,4-oxazolidinedione (dmo) and nicotine at temperatures and NaCl concentrations of biological interest. *Acta Cient. Venezolana*, 1980: **31**; 128–130.

19 Bates RG. *Determination of pH: Theory and Practice.* New York: John Wiley and Sons, Inc., 1964.

20 Cox MC, Everett DH, Landsman DA, Munn RJ. The thermodynamics of the acid dissociation of some alkylammonium ions in water. *J. Chem. Soc. (B)* 1968; 1373–1379.

21 Timimi BA, Everett DH. The thermodynamics of the acid dissociation of some amino-alcohols in water. *J. Chem. Soc. (B)* 1986; 1380–1386.

22 Hetzer HB, Bates RG, Robinson RA. Dissociation constant of pyrrolidinium ion and related thermodynamic quantities from 0 to 50°. *J. Phys. Chem.*, 1963: **67**; 1124–1127.

23 *CRC Handbook of Chemistry and Physics 51st Edition.*, RC Weast, ed., Cleveland OH: The Chemical Rubber Company, 1970.

24 De Taye J, Zeegers-Huyskens Th. Infrared study of the proton acceptor ability of nicotine. *Bull. Soc. Chim. Belg.*, 1987: **96**; 1–6.

25 Munir C, Zaidi MI, Yousaf SM. Zinc, cadmium and mercury as extractants of nicotine from tobacco leaves. *Main Group Met. Chem.*, 1994: **17**; 673–677.

26 Smith RM, Martell AE. *Critical Stability Constants Volume 2: Amines*, p 337, New York: Plenum Press, 1975.

27 Kuhn WF, Lilly-Leister D, Kao J, Lilly AC. Comparison of theoretical and experimental ionization potentials of nicotine and related molecules. *J. Mol. Struc.*, 1989: **212**; 37–44.

28 Pitner PT, Whidby JF, Edwards WB. Anisotropic reorientation of nicotine in solution. a carbon–13 spin-lattice relaxation time study of molecular conformation. *J. Am. Chem. Soc.*, 1980: **102**; 5149–5150.

29 Whidby JF, Edwards WB, Pitner TP. Isomeric nicotines. their solution conformation and proton, deuterium, Carbon–13, and Nitrogen–15 Nuclear Magnetic Resonance. *J. Org. Chem.*, 1979: **44**; 794–798.

30  Pitner TP, Edwards WB, Bassfield RL, Whidby JF. The solution conformation of nicotine. a $^1$H and $^2$H nuclear magnetic resonance investigation. *J. Am. Chem. Soc.*, 1978: **100**; 246–251.

31  Pitner TP, Seeman JI, Whidby JF. Assignment and solvent dependence of the carbon–13 nuclear magnetic resonance spectrum of nicotine. *J. Heterocycl. Chem.*, 1978: **15**; 585–587.

32  Whidby JF, Seeman JI. The configuration of nicotine. A nuclear magnetic resonance study. *J. Org. Chem.*, 1976: **41**; 1585–1590.

33  Pullman B, Courriére Ph., Coubeils JL. Quantum mechanical study of the conformational and electronic properties of acetylcholine and its agonists muscarine and nicotine. *Molec. Pharm.*, 1972: **7**; 397–405.

34  Dwyer, RW. Unpublished Results, 1998.

35  Koo CH, Kim HS. The crystal structure of nicotine dihydroiodide. *J. Korean Chem. Soc.*, 1965: **9**; 134–141.

36  Seeman JI, Secor HV, Chadvarian CG, Sanders EB, Bassfield RL, Whidby JF. Steric and conformational effects in nicotine chemistry. *J. Org. Chem.*, 1980: **46**; 3040–3048.

37  Hepler LG, Woolley EM. Hydration Effects and Acid-Base Equilibria.. In: *Water: A Comprehensive Treatise Volume 3: Aqueous Solutions of Simple Electrolytes.*, F. Franks (Ed.), New York: Plenum Press 1977.

38  Noyes RM. Assignment of individual ionic contributions to properties of aqueous ions. *J. Am. Chem. Soc.*, 1964: **86**; 971–979.

39  Slaven RM. The Carbon–13 and proton NMR spectra of nicotine in aqueous media. *J. Heterocyc. Chem.*, 1984: **21**; 1329–1332.

40  Franks F, Ives DJG. The structural properties of alcohol-water mixtures. *Quart. Rev. Chem. Soc.*, 1966: **20**; 1–44.

41  Blandamer MJ, Waddington D. Ultrasonic absorption properties of binary aqueous mixtures. *Advan. Mol. Relax. Process.*, 1970: **2**; 1–40.

42  Patterson D. Structure and thermodynamics of non-electrolyte mixtures. *J. Soln. Chem.*, 1994: **23**; 105–120.

43  Matubayasi N, Reed LH, Levy RM. Thermodynamics of the hydration shell. 1. Excess energy of a hydrophobic solute. *J. Phys. Chem.*, 1994: **98**; 10640–10649.

44  Lee SH, Rossky PJ. A comparison of the structure and dynamics of liquid water at hydrophobic and hydrophillic surfaces - a molecular dynamics simulation study. *J. Chem. Phys.*, 1994: **100**; 3334–3345.

45  Garcia-Tarres L, Guardia E. Hydration and dynamics of a tetramethylammonium ion in water: a computer simulation study. *J. Phys. Chem. B*, 1998: **102**; 7448–7454.

46  Frank HS and Evans MW. Free volume and entropy in condensed systems iii. entropy in binary liquid mixtures; partial molal entropy in dilute solutions; structure and thermodynamics in aqueous electrolytes. *J. Chem. Phys.*, 1945: **13**; 507–532.

47  McMullen RK, Jordan TH and Jeffrey GA. Polyhedral clathrate hydrates. xii. the crystallographic data on hydrates of ethylamine, dimethylamine, trimethylamine, *n*-propylamine (two forms), iso-propylamine, diethylamine (Two Forms), and *tert*-butylamine. *J. Chem. Phys.*, 1967: **47**; 1218–1234.

48  Jeffrey GA, Feil D and Mcmullen R. Polyhedral clathrate "ice" structures in high hydrates. *CA* 1963: **58**; 7454b.

49 Jeffrey GA. Structural factors in the formation of clathrate hydrates. *CA* 1963: **58**; 10962h.

50 McMullen R and Jeffrey GA. Hydrates of the tetra *n*-butyl and tetra *i*-amyl quaternary ammonium salts. *J. Chem. Phys.*, 1959: **31**; 1231–1234.

51 Hallenga K, Grigera JR and Berendsen HJC. Influence of hydrophobic solutes on the dynamic behavior of water. *J. Phys. Chem.*, 1980: **84**; 2381–2390.

52 Ludwig R. NMR relaxation studies in water-alcohol mixtures: the water rich region. *Chem. Phys.*, 1995: **195**; 329–337.

53 Head-Gordon T. Is water structure around hydrophobic groups clathrate-like? *Proc. Natl. Acad. Sci. USA* 1995: **92**; 8308–8312.

54 Kelly DF, MacCarthaig D and Reilly J. Distillation 'constants'. *Nature* 1941: **148**; 438–439.

55 Kelly DF, O'Connor MJ and Reilly. An azeotrope of nicotine and water. *J. Chem. Soc.*, 1942: **2**; 511–513.

56 Smith CR. Azeotropism in the system nicotine-water. *Ind. Eng. Chem.*, 1942: **34**; 251–252.

57 Fowler RT. Azeotropism in binary solutions at reduced pressures. I. The nicotine-water system. *J.S.C.I., Supplementary Issue, No. 2, 1950;* S65-S69.

58 Verrall RE and Conway BE. Partial molar volumes and adiabtic compressibilities of tetraalkylammonium and aminium salts in water. II. Volume and volume change relationships. *J. Phys. Chem.*, 1966: **70**; 3961–3969.

59 Kaulgud MV and Patil KJ. Volumetric and isentropic compressibility behavior of aqueous amine solutions. I. *J. Phys. Chem.*, 1974: **78**; 714–717.

60 ibid. II. ibid. 1976; **80**; 138–143.

61 Wen WY and Saito S. Apparent and partial molal volumes of five symmetrical tetraalkylammonium bromides in aqueous solution. *J. Phys. Chem.*, 1964: **68**; 2639–2644.

62 Franks F and Quickenden MJ. Partial molar volumes of cyclic ethers in dilute aqueous solution at 25°. *J. Chem. Soc. Chem. Com.*, 1968; 388–390.

63 Campbell AN, Kartzmark EM and Falconer WE. The system: nicotine – methylethyl ketone – water. *Can. J. Chem.*, 1958: **36**; 1475–1486.

64 Fowler DL, Loebenstein WV and Kraus CA. Some unusual hydrates of quaternary ammonium salts. *J. Am. Chem. Soc.*, 1940: **62**; 1140–1142.

65 Bonamico M, Jeffrey GA and McMullan RK. Polyhedral Clathrate Hydrates. III. The structure of the tetra-*n*-butyl ammonium benzoate hydrate. *J. Chem. Phys.*, 1962: **37**; 2219–2231.

66 McMullan RK, Bonamico M and Jeffrey GA. Polyhedral clathrate hydrates. V. The structure of tetra-*n*-butyl ammonium fluoride hydrate. *J. Chem. Phys.*, 1963: **39**; 3295–3310.

67 Hansch C, Kim D, Leo AJ Novellino E, Silipo C and Vittoria A. Toward a quantitative comparative toxicology of organic compounds. *Crit. Rev. Toxicol.*, 1989: **19(3)**; 185–226.

68 Hansch C, Maloney PP and Fujita T. Correlation of biological activity of phenoxyacetic acids with hammett substituent constants and partition coefficients. *Nature* 1962: **194**; 178–180.

69 Hansch C, Muir RM, Fujita T, Maloney PP, Geiger F and Streich M. The correlation of biological activity of plant growth regulators and chloromycetin derivatives with

hammett constants and partition coefficients. *J. Am. Chem. Soc.*, 1963: **85**; 2817–2824.

70 Hansch C, Steward AR, Junkichi I and Deutsch EW. The use of a hydrophobic bonding constant for structure-activity relationships. *Mol. Pharmacol.*, 1965: **1**; 205–213.

71 Hansch C, Leo A and Hoekman D. Exploring QSAR hydrophobic, electronic and steric constants, Washington DC: American Chemical Society 1995.

72 Quantitative Approaches to Drug Design, JC Dearden, ed., Amsterdam: Elsevier 1983.

73 Claffey JB, Badgett CO, Skalamera JJ and MacPherson Phillips GW. Nicotine extraction from water with kerosene. *Ind. Eng. Chem.*, 1950: **42**; 166–171.

74 Oakley DM and Swarbrick J. Effects of ionization on the percutaneous absorption of drugs: partitioning of nicotine into organic liquids and hydrated stratum corneum. *J. Pharm. Sci.*, 1987: **76**; 866–871.

75 Norton LB. Distribution of nicotine between water and petroleum oils. *Ind. Eng. Chem.*, 1940: **32**; 241–244.

76 Cuvelier BVJ. Einfluss der Ammoniumsalze auf die Loeslichkeit des Nicotins. *Z. anal. Chem.*, 1936: **105**; 325–328.

77 *International Critical Tables. Volume III.*, p 409, E.W. Washburn ed., New York and London: McGraw-Hill Book Company, Inc., 1928.

78 Fujita T, Nakajima M, Soeda Y and Yamamoto I. Physicochemical properties of biological interest and structure of nicotine and its related compounds. *Pestic. Biochem. Physiol.*, 1971: **1**; 151–162.

79 Li N-Y, Li N and Gorrod JW. Determination of partition coefficients and ionisation constants of (S)(−)-nicotine and certain metabolites. *Med. Sci. Res.* 1992: **20**; 901–902.

80 Reilly J, Kelly DF and O'Connor M. The distribution of nicotine between trichloroethylene and water. *J. Chem. Soc.*, 1941; 275–278.

81 Dezelic M, Stancic B and Grujic-Vasic J. Determination of the composition of nicotine hydrate with physicochemical methods. *Bull. Chem. Soc. Chem. Bosnia Hercegovina* 1953: **2**; 19–27.

82 Iwasaki K and Fujiyama T. Light-scattering study of clathrate hydrate formation in binary mixtures of *tert*-butyl alcohol and water. *J. Phys. Chem.*, 1977: **81**; 1908–1912.

83 Nishikawa K and Iijima T. Structural study of *tert*-butyl alcohol and water mixtures by X-ray diffraction. *J. Phys. Chem.*, 1990: **94**; 6227–6231.

84 *International Critical Tables. Volume III.* E.W. Washburn (Ed.), New York and London: McGraw-Hill Book Company, Inc., 1928, p. 392.

85 Kehiaian H. Studies on the Liquid - Liquid Equilibrium. I. Binary systems with upper and lower critical solubility temperatures. *Bull. Acad. Polon. Sci., Ser. Sci. Chim.*, 1962: **10**; 569–577.

86 ibid., II. The function of the solubility curve asymmetry for some binary systems with closed miscibility gaps. *ibid.* 579–580.

87 ibid., III. An equation for the coexistence curve of binary systems with upper and lower critical solubility temperatures. *ibid.* 585–589.

88 Eagland D. Nucleic Acids, Peptides and Proteins. In: *Water: A Comprehensive Treatise Volume 4. Aqueous Solutions of Amphiphiles and Macromolecules,* F. Franks (Ed.), New York and London: Plenum Press, 1975.

*Chapter 6*

# Use of gas chromatographic and mass spectrometric techniques for the determination of nicotine and its metabolites

Peyton Jacob, III[1] and Gary D. Byrd[2]

[1] *Division of Clinical Pharmacology, San Francisco General Hospital Medical Center, Department of Medicine and Drug Dependence Research Center, Langley Porter Psychiatric Institute, University of California, San Francisco, CA 94110, USA*
[2] *R. J. Reynolds Tobacco Company, Biological Chemistry Division, Winston-Salem, NC 27102-1236, USA*

## I. SCOPE

In this chapter, the application of gas chromatographic methods with emphasis on mass spectrometric detection for the determination of nicotine and its metabolites is examined. The coupling of a high resolution separation technique such as gas chromatography (GC) with mass spectrometry (MS) produces one of the most effective analytical tools used in chemistry today. Fundamental descriptions of GC and MS methodologies are presented here. Qualitative and quantitative strategies for GC–MS are discussed and applications are presented to illustrate these points.

## II. INTRODUCTION

### History of GC for determination of nicotine and related alkaloids

Gas chromatography is a highly suitable method for determination of nicotine, and has been employed for many years to measure nicotine concentrations in a variety of matrices. Nicotine is very stable thermally, relatively volatile, and has good chromatographic properties on a variety of

191

GC stationary phases. Furthermore, GC generally offers excellent resolution from other materials present in the sample, especially with modern high-resolution capillary columns, and a variety of detectors are suitable for nicotine determination.

The first report of GC for analysis of nicotine appears to be a paper published in 1959 by Quin, which describes the separation of nicotine and related alkaloids in tobacco smoke using columns packed with polyglycol stationary phases [1]. In 1966, Beckett and Triggs [2] reported a GC method for determination of nicotine and its metabolite cotinine in urine, and since that time numerous GC and GC–MS methods have been published.

Most of the early methods employed packed GC columns. During the 1980s, there were rapid advances in capillary GC column technology, in particular the advent of durable fused silica columns with highly stable cross-linked stationary phases. Not only do capillary columns provide greatly enhanced resolution compared to packed columns but, due to low column bleed and low carrier gas flow rates, they provide much more stable baselines during temperature programming than do packed columns. Consequently, most GC applications, including determinations of nicotine and its metabolites, are now performed using fused silica capillary columns.

## Development of GC–MS

Mass spectrometry was developed as an analytical tool for organic chemists during the middle part of this century [3]. At first, the technique was thought to be too burdensome for routine analysis due to size, cost, and operation. Technological advances, however, eventually overcame this perception and the advantages of mass spectrometry became obvious. The mass spectrometer can produce an identifiable fingerprint (mass spectrum) from a couple of nanograms of material. When monitoring only specific ions, detection of picogram quantities of a compound is possible. These features make mass spectrometry an attractive detection system, especially when identification and quantitation is required on trace components in a complex matrix. Mass spectrometry has been widely applied to the study of nicotine and its metabolites in biological samples. One of the earliest reports on the use of mass spectrometry to detect nicotine exposure was by Horning et al. [4] who demonstrated that the mass spectrometer could find signals unique to nicotine and its major metabolites, cotinine and *trans*-3'-hydroxycotinine, in extracts of smokers' urine.

The coupling of mass spectrometry to GC seemed possible since both techniques were restricted to volatile samples. There were, however, several technical challenges. First, the effluent at the end of a GC column is near atmospheric pressure, while the ion source of a mass spectrometer must be kept under high vacuum ($10^{-5}$ Torr). With packed GC columns, attempts to mate the two resulted in various separator interfaces that reduced the amount of carrier gas before the eluents arrived in the mass spectrometer. By the 1970s, successes with these systems demonstrated the power of the technique [5]. Later, the low flows of capillary GC columns and more efficient mass spectrometer pumping systems permitted direct placement of the end of the column in the ionization source with an improvement in ionization efficiency. The need to obtain mass spectra rapidly 'on the fly' as compounds eluted from the GC represented another challenge. Early mass spectrometers were primarily magnetic sector instruments. Though stable and capable of high mass resolution, magnets were resistant to rapid scanning. The development of the quadrupole mass analyzer, where scans are obtained by changing electrical voltages rather than magnetic fields, facilitated rapid scanning. In addition, advancements in computers enabled the entire operation to be computer-controlled and permitted storage of the vast amounts of data produced by full scanning GC–MS. Today, the abundance of well-designed commercial systems has allowed highly sensitive GC–MS applications to become ubiquitous in analytical laboratories. They are often part of definitive methods used in clinical research.

## III. INSTRUMENT COMPONENTS

### Overview of GC

A gas chromatograph is a device for separating substances that are volatile enough to be vaporized. It consists of an injection system for introducing the sample, a column in which the separation occurs, and a detector. An inert gas, called the carrier gas, which is usually helium but occasionally hydrogen or nitrogen, is passed through the system. The separation occurs by partitioning of the substances in a mixture between the vapor phase moving through the system and the material in the column. Those substances that are less volatile and/or have greater affinity for the column material move more slowly, and this effects the separation. The material in the column is called the stationary phase. In packed columns, the stationary phase is usually a non-volatile liquid or polymer adsorbed on a silica-based, finely divided powder. Packed columns are generally constructed of glass or stainless steel, and typically are about 2 to 8 millimeters in diameter and 1 to 4 meters in length. In capillary columns, the stationary phase is usually a silicone bonded to the walls of the column, which is made of fused silica. As the name implies, capillary columns are small in diameter, typically 0.05 to 0.5 mm inside diameter (I.D.), with lengths ranging from 5 to 50 meters. A variety of detectors are available. Some will respond to nearly all organic compounds, and others provide very high sensitivity and selectivity for certain classes of compounds.

## Injection systems

A variety of methods are available for introduction of samples into gas chromatographs [6,7]. In early instruments, this usually consisted of a metal tube that was heated electrically to vaporize the sample. Although this works well for many compounds, it is not suitable for trace analysis of chemically labile compounds or basic substances such as nicotine that might decompose or be strongly adsorbed on metal surfaces. Glass-lined injection ports, often deactivated by silanization or by other treatments, are generally preferable for such substances.

Direct or on-column injection onto the GC column is another technique that can be used for labile compounds [7]. This is easily accomplished with packed columns, but may be more difficult with narrow bore capillary columns, particularly by autosampler. Some autosamplers are capable of injection into columns with I.D. ≥ 0.25 mm; otherwise, manual injection using narrow gauge stainless steel or fused silica needles can be performed.

Headspace sampling for volatile substances, and sample enrichment by adsorption-desorption are other techniques used to introduce samples into a gas chromatograph, but these techniques do not appear to have been widely used for nicotine and related compounds.

Due to its basicity, nicotine can be strongly adsorbed to metal or glass surfaces, and this occurring in an injection port or GC column could lead to tailing peaks and loss of sensitivity. Deactivation of glass injection port liners and capillary columns has been recommended to reduce the adsorption of nicotine. For example, Stehlik et al. used Carbowax 1000 deactivated glass beads in the injection port to suppress adsorption [8]. Similarly, Jacob et al. deactivated glass injection port liners with polyethylene glycol 4000, and used a small plug of polyethylene glycol 4000 deactivated glass wool in the injection port liners to aid sample vaporization and prevent non-volatile impurities from entering the GC column [9].

## GC columns

Because nicotine and most of its metabolites are basic compounds, the column must be free of significant acidic sites to avoid peak tailing or losses. For this reason, most early methods were carried out with packed columns which utilized stationary phases that were base-deactivated. For example, a stationary phase of Carbowax and potassium hydroxide was frequently used in early methods [2,10], because nicotine and its major metabolite cotinine both have good peak symmetry on this phase. Base-deactivated silicone stationary phases of the OV-1 and OV-17 types have also been employed for determination of nicotine and metabolites [11].

Modern fused-silica capillary columns generally utilize cross-linked stationary phases that are frequently bonded to the column surface. This stabilizes the stationary phase and masks acidic silanol sites present on the column surface. For this reason, chromatographic separation of nicotine and some of its metabolites can be carried out with excellent peak symmetry and high sensitivity using fused silica capillary columns.

## Detectors

*Thermal conductivity detector (TCD).* The TCD was one of the first employed in gas chromatography. The principle of detection is that analytes will generally have lower thermal conductivity than the carrier gas. It is rugged, but is not very selective compared to other detectors [12]. An advantage of the TCD is that it is non-destructive, making it useful for preparative separations.

*Flame ionization detector (FID).* The FID was also developed in the early days of GC. Detection is based upon the increase in electrical conductivity caused by generation of ions when the analyte passes through a flame of burning hydrogen. It has been widely employed for determination of nicotine, related alkaloids, and nicotine metabolites [13]. The FID has good sensitivity and, with few exceptions, is a universal detector for organic compounds; but since all organic compounds give a response, it is not very selective and co-eluting components in a sample will interfere with the analysis. Consequently, for determination of nicotine and its metabolites in complex matrices, a more selective detector such as a nitrogen-phosphorus detector or a mass spectrometer is generally preferable.

*Nitrogen-phosphorus detector (NPD).* This detector, formerly called the alkali flame ionization detector (AFID), became available in the 1960s, and is still widely used for determination of drugs, metabolites, and pesticides. As the name implies, it is selective for compounds containing nitrogen or phosphorus. This detector consists of an electrically heated ceramic bead containing a rubidium salt, over which column effluent mixed with hydrogen and air is passed. A 'low temperature plasma' is generated, and rubidium ions combining with nitrogen- or phosphorous-containing pyrolysis products leads to increased electrical conductivity, which is monitored. The selectivity for organic compounds containing nitrogen compared to organic compounds not containing nitrogen or phosphorus is on the order of 50:1 or more. This selectivity is a major advantage for determination of nitrogen-containing compounds in complex mixtures, since co-eluting substances not containing nitrogen or phosphorus will not interfere unless their concentrations are much higher than those of the analytes. Since nicotine contains two nitrogen atoms, the NPD has been widely used for GC determination of nicotine and its metabolites [13].

*Electron capture detector (ECD).* This detector is very sensitive and selective for compounds containing electronegative groups, such as halogens and nitro groups. If the analyte does not have such a substituent, but does have a functional group that can be derivatized, the analyte can be converted to a derivative that can be determined with high sensitivity using an ECD. The principle of operation is absorption of electrons, generated from a beta-emitter such as [63]Ni, by the analyte. The ECD has occasionally been used for determination of nicotine. For example, Hartvig et al. reported a method for determination of nicotine in plasma based on the reaction with 2,2,2-trichloroethyl chloroformate which results in cleavage of the pyrrolidine ring of nicotine, producing an *N*-(2,2,2-trichloroethyl) carbamate derivative [14]. More recently, heptafluorobutyric anhydride was used to produce a hepta-fluorobutyryl derivative of nicotine for electron capture detection [15].

*Mass spectrometric detection (MSD).* Combined gas chromatography-mass spectrometry (GC–MS) has been used for analysis of organic compounds for over 30 years. The mass spectrometer can be used in the scan mode to provide spectra of compounds as they elute from the GC column, or it can be used to monitor ions characteristic of the analytes (selected ion monitoring or SIM mode) to provide ion chromatograms used for quantitation. Thus, the mass spectrometer is used as a very selective and sensitive detector for the GC. Although mass spectrometers are much more expensive than conventional GC detectors, they may be advantageous for high sensitivity assays of complex matrices. By tuning the mass spectrometer to monitor an ion characteristic of the analyte, interference from coeluting matrix impurities may be eliminated. In addition, mass spectrometry is essential for studies employing stable isotopes, such as drug bioavailability studies, in which the chromatographic system will not separate the different isotopomers of the analyte [16]. GC–MS is being used with increasing frequency for the determination of nicotine and metabolites due to the development of rugged and relatively inexpensive quadrupole instruments. Because of the versatility and sensitivity of mass spectrometry and its wide applicability to studies of nicotine, related alkaloids, and their metabolites, GC–MS is the major focus of this chapter.

## Mass spectrometry

*Overview*
A mass spectrometer measures the mass (*m*) to charge (*z*) ratio (written as *m*/*z*) of ions formed by an ionization process in a high vacuum. For samples introduced into a mass spectrometer (eluting compounds from a GC column in this case), these ions are charged molecules and fragments produced from the sample. A graph where *m*/*z* is plotted on the X-axis and abundance is plotted on the Y-axis forms the 'mass spectrum' to describe how the ions are distributed. This mass spectrum is a fingerprint of the molecule's structure.

A basic mass spectrometer system with a GC inlet is shown schematically in Fig. 1. Sample molecules M elute from the GC and are introduced to the

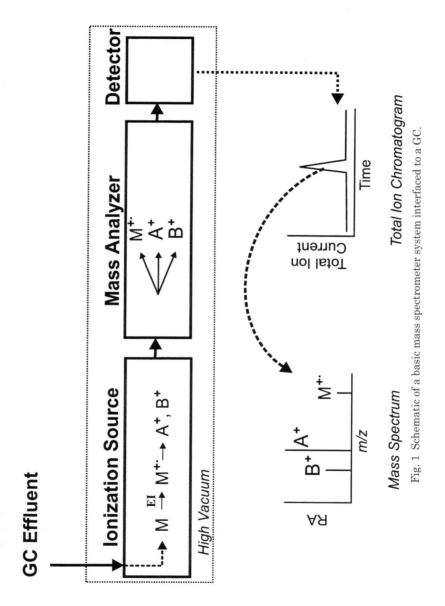

Fig. 1 Schematic of a basic mass spectrometer system interfaced to a GC.

high vacuum realm of the mass spectrometer where ions are produced, analyzed, and detected. The high vacuum is necessary to provide an adequate mean free path for the ions to move about without colliding with air molecules. The output signal is frequently plotted as a total ion chromatogram (TIC) which is a plot of retention time versus a summation of the abundances of all ions produced (the total ion current). Each point in the TIC represents a complete mass spectrum and can be extracted to show the distribution of ion masses and relative abundances for the compounds eluting at all retention times.

As shown in Fig. 1, the mass spectrometer has three distinct components: an ionization source, a mass analyzer, and a detector. This section will summarize the fundamentals of popular mass spectrometer hardware and techniques now in use with some mention of emerging fields. More detailed coverage of mass spectrometer instrumentation with regard to GC may be found in some recent texts [17,18].

## Ionization sources

The conventional and most widely applied ionization process is called 'electron impact' or EI. In a GC–MS with an EI source, sample molecules eluting from the GC pass into the high vacuum ($10^{-5}$ Torr) of the ion source where they are showered by high energy (usually 70 eV) electrons emitted from a heated filament. Through collision, an electron is displaced from the sample molecule to produce a positive ion. The removal of a single electron from the sample molecule produces an odd electron species called the 'molecular ion' and written $M^{+\cdot}$. The mass of this ion is the molecular weight of the compound and is perhaps the most analytically useful ion. The molecular ion receives excess energy in the ionization process which leads to fragmentation of a certain percentage of molecular ions. Figure 1 illustrates a simple case where two fragments, $A^+$ and $B^+$, are formed. For the most part in EI, each molecule and fragment has a charge of one so that $m/z$ is the same as the ion's mass. Neutral or radical species are also produced but, since they are uncharged, these are not detected by the mass spectrometer.

The advantages of EI are that it can produce both fragments and molecular ions, and the spectra are fairly reproducible so they can be matched by computerized searches on commercial databases for identification. A disadvantage of EI is the occasional lack of molecular ion for samples that undergo extensive fragmentation. Also, the EI mass spectra of isomers can be indistinguishable from each other.

An alternative ionization method called 'chemical ionization' (CI) can be used in cases where the molecular ion is weak or not present in the mass spectrum. The process derives its name from the use of gas phase chemical reactions to produce ions from the sample [19]. There are two types of chemical ionization, 'positive ion chemical ionization' (PICI or PCI) and 'negative ion chemical ionization' (NICI or NCI). In both cases, the ionization is softer, resulting in less fragmentation. PICI is most often accomplished through a gas phase proton transfer reaction. The ion source is flooded with a reagent gas, usually methane, at a relatively high pressure (1 Torr). During electron bombardment under these conditions, a series of gas phase reactions occur as depicted below for reagent gas methane and sample molecule M:

$$CH_4 \xrightarrow{EI} CH_4^{+\cdot}$$

$$CH_4^{+\cdot} + CH_4 \rightarrow CH_5^+ + CH_3^{\cdot}$$

$$CH_5^+ + M \rightarrow CH_4 + (M+H)^+$$

Methane molecular ions formed in the high pressure source collide with neutral reagent molecules ($CH_4$) and produce protonated methane. $CH_5^+$ acts as a strong Bronsted acid and transfers a proton to the sample molecule to produce the protonated molecular adduct $(M+H)^+$. PICI works best for samples containing a heteroatom; thus, nicotine and its metabolites readily form protonated adducts by this ionization method. PICI mass spectra are characterized by fewer fragment ions and more abundant molecular adduct ions. In addition, sensitivity is enhanced relative to EI. The higher source pressures used, however, do contaminate the instrument more rapidly and result in more frequent maintenance. Also, some modifications to the source are usually required when switching from EI to CI operation.

NICI sources operate under high pressure conditions similar to those for PICI with voltages switched to detect negative ions. The underlying ionization processes, however, are somewhat different. Several processes can occur to form negative ions. A common method is 'electron capture' which uses the reagent gas as a buffer to reduce the energy of electrons such that a molecule with a suitable electron affinity can capture an electron as shown below:

$$M + CH_4 + e^- \rightarrow M^{-\cdot} + CH_4$$

This process is rather selective, since not every molecule will readily form a stable $M^{-\cdot}$. Underivatized nicotine, for example, will not. Another process is adduct formation where a background anion such as $Cl^-$ will attach to a molecule to form $(M+Cl)^-$. A variety of ion-molecule reactions are possible and some can be rather complex. Proton abstraction is common and works well with samples with an acidic proton. Negative ion formation can be enhanced using derivatization to form an analog with a high electron affinity. The sensitivity and selectivity of such assays can be very high. For example, one group reports conversion of nicotine using heptafluorobutyric anhydride to a stable electron scavenger derivative that can be detected at the femtogram ($10^{-15}$ g) level on column [20].

## Mass analyzers

The ions formed in the source must be separated by their $m/z$ ratio and detected. Mass analyzers can operate in one of two modes. In the scanning mode, the analyzer continuously scans $m/z$ ratios over a set range in a specific time. For example, scanning $m/z$ 40 to 400 in 1 second is a good general scanning parameter for capillary column GC–MS. Few important ions occur below $m/z$ 40 and, although there are some high temperature GC applications for large molecules (>1,000 Daltons), most GC applications involve compounds with molecular weights <400 Daltons. For high resolution GC, the scan must be on the order of 1 s or less to define the peak which may be only a few seconds in width. Scanning mode is normally used to identify several compounds in a single GC run. In the selected ion monitoring mode (SIM), the mass analyzer jumps between only a few ions characteristic of a compound. Since more time is spent detecting ions rather than scanning areas where there are no useful ions, detection by SIM is 10 to 100 times more sensitive than the scanning mode. The cycle time can be much faster also, on the order of milliseconds, which better defines the chromatographic peak shape for quantitative purposes. SIM is usually done when the identity of the compound is known and determination at trace levels in a complex matrix is desired.

The earliest mass spectrometers used a magnetic sector to deflect ion currents into a detector system. Ions accelerated into the magnetic sector by a high voltage are dispersed by the magnetic field according to their $m/z$ to produce a mass spectrum. The first GC–MS systems were built around these instruments. Early magnetic sector instruments were large, heavy, expensive, and relatively slow in scanning a large $m/z$ range.

The most popular analyzer system used today is the quadrupole mass analyzer. These compact systems are relatively cheap, durable, and can scan very rapidly. The quadrupole assembly consists of four parallel rods arranged equidistant around a central axis. DC and RF voltages are applied to opposite pairs of rods to form a fluctuating electric field. Ions formed in the source are directed and focused into this field by electrostatic lenses. Only ions with a particular $m/z$ ratio as determined by the DC and RF field strength can pass through to the detector. Thus, the analyzer scans for particular ions by ramping these parameters. At the end of the quadrupole is a detector, usually an electron multiplier. Ions impinge on a surface to produce an electron cascade effect resulting in an electric current that is converted into data.

By using quadrupole analyzers in series, tandem mass spectrometry [21] with GC is possible. (Although sector instruments with magnetic and electrostatic fields can be used as stages, quadrupoles are more often used.) These systems operate by selecting a particular ion from the first stage analyzer and sending it into a second stage where the ion collides with an inert gas to produce further fragmentation. A third stage analyzes the resulting fragments, thus producing a mass spectrum from a precursor ion. This technique is referred to as 'MS/MS' or 'collision-induced dissociation (CID)' and is often used with soft ionization such as PICI to produce fragments from molecular adducts. An example is given below for a molecular adduct produced by PICI:

| Source | Analyzer 1 | Collision Cell | Analyzer 2 |
|---|---|---|---|
| $M + CH_5^+ \rightarrow$ | $(M+H)^+ \rightarrow$ | Argon, eV $\rightarrow$ | $A^+, B^+$ |
| | 'Precursor' | | 'Products' |

The selected ion is usually referred to as the 'precursor' or 'parent' and its fragments as 'products' or 'daughters'. Precursors and products can be linked to establish fragmentation pathways, thereby producing additional structural information. MS/MS systems can also be set to monitor specific fragmentation pathways, a process called 'multiple reaction monitoring' or MRM. This is analogous to SIM for a single quad system except it is even more selective for a particular analyte and can be used with very complex samples. More details and examples of MS/MS are given in the next chapter on LC–MS. While the cost of an MS/MS system is much more than for a single stage mass spectrometer system, the increased structural information and selectivity of these instruments is very attractive. They have become workhorse systems for many pharmaceutical laboratories, particularly when interfaced to LC.

A more recent development in mass analyzers that is increasingly found in analytical laboratories is the ion trap [22]. In an ion trap, samples are ionized and contained in a circular electromagnet until swept into a detector by an applied DC/RF frequency. When coupled directly to a GC, ions are formed by EI just as for a quadrupole system. Ion traps are easily switched between EI and CI modes, and are similar to quadrupoles in sensitivity, size, and cost. They are also capable of MS/MS analysis. MS/MS in the ion trap is carried out by colliding the ions while trapped either with themselves or a collision gas. The product ions are then detected. In addition, MS/MS on the product ions may be performed such that MS/MS/MS . . . ($MS^n$) is possible.

Somewhat related to the ion trap but older in use is Fourier transform mass spectrometry (FT–MS) [23]. Ions are produced by EI in a cubic cell consisting of opposing pairs of trapping plates, transmitting plates, and receiving plates. The ions formed are trapped by a high constant magnetic field and electrostatic trapping plates. Each ion undergoes cyclotron motion at a frequency determined by its $m/z$ ratio. All the ions are excited (resulting in increased radius of motion) simultaneously by applying a burst of RF energy over a range of frequencies corresponding to the cyclotron frequencies of the ions. The ions are detected simultaneously by measuring the image current induced on the detection plates of the cell. The frequencies form a beat pattern and, by Fourier transformation, the individual cyclotron frequencies of the ions are determined and the $m/z$ values produced. Many operational aspects of FT–MS are similar to FT–NMR. High mass resolution (discussed later) is possible with these systems as is $MS^n$. Applications of GC–FT–MS have been limited primarily due to the high cost of these instruments.

Time of flight (TOF) mass spectrometer systems have shown a comeback recently and have been interfaced to GCs [24]. TOF analyzers work on the simple principle that ion velocity is mass dependent. Ionization of compounds eluting from the GC usually occurs through a burst of high voltage electrons. Ionized sample ions are accelerated with a high voltage down a tube and the different $m/z$ ratios are separated in time. The arrival time at the detector is based on $m/z$ and, thus, a mass spectrum is produced. Naturally, scan speed is very fast and mass range is unlimited. GC–MS applications with TOF are presently uncommon but this may change.

High resolution mass analyzers can measure masses to within a thousandth of a mass unit. The utility of this is based on the fact that atomic masses, while close to integral values, are not truly integral. While $^{12}C$ is assigned 12.000000, $^{16}O$ is actually 15.994915, $^{1}H$ is 1.007825, $^{14}N$ is 14.003074 , and so on. Thus, high resolution mass measurements can lead to determination of elemental composition. For example, the exact mass of nicotine ($C_{10}H_{14}N_2$) is 162.115699. Using a high resolution mass spectrometer, its molecular ion would be distinguishable from that of another compound with a nominal $m/z$ 162 but different molecular formula such as *p*-nitrobenzyl cyanide ($C_8H_6N_2O_2 = 162.042928$). Like tandem mass spectrometry, high resolution mass spectrometers offer more selectivity than low resolution mass analyzers by monitoring very narrow mass ranges. Quadrupole analyzers are not capable of resolving power much in excess of 1 mass unit and cannot produce high resolution spectra. The most commonly used high resolution system is a double focusing instrument with an electrostatic sector to first select ions of a specific kinetic energy before analysis by magnetic sector. As mentioned earlier, FT–MS is also capable of high resolution mass spectra as are some TOF systems. The selectivity of high resolution mass spectrometry can provide assays with very high sensitivity by eliminating potential interferences, and this has been used to detect femtogram levels of nicotine [25].

## IV. SAMPLE PREPARATION

Some sort of sample cleanup is generally required prior to analysis by gas chromatography. Non-volatile, polar substances such as those present in biologic samples may build up in the injection port and column, causing degradation in performance. Sometimes extensive extraction/cleanup procedures are necessary to remove matrix impurities that would coelute with the analyte. Generally, liquid–liquid or solid phase extraction procedures are used for samples of biologic origin.

## Extraction procedures

A wide variety of procedures have been devised for the extraction of nicotine and its metabolites from sample matrices. Considerations include appropriate solvents for efficient recovery, the complexity of the matrix, what sort of clean-up steps are necessary to eliminate interfering substances, and suitability of the method for extraction of large numbers of samples, if that will be done. By using solvents that have a density less than water and which will not freeze at $-78°C$, phase separations can be carried out using the 'freeze and pour' technique, freezing the aqueous layers in a dry ice cooled bath and pouring organic layers (upper phase) into new tubes, thus greatly facilitating the extraction of large batches of samples [9].

Procedures that utilize back-extraction into aqueous acid are more time-consuming than one step extraction procedures, but they will produce cleaner extracts. This can be important for eliminating interfering substances, especially when high sensitivity is required, and trace impurities could interfere. Use of cleaner extracts may also prolong the life of the column and injection port liners. Another consideration is the choice of solvent for the final extraction, if evaporation is necessary to provide a concentrated sample for maximizing sensitivity. Nicotine is relatively easy to extract using a variety of solvents or solid-phase procedures, but its metabolites are more polar and therefore more difficult to extract from aqueous samples.

Early GC methods for nicotine determination used diethyl ether as the extraction solvent, with a back-extraction into aqueous acid to remove neutral and acidic impurities [11,26]. The metabolite cotinine is much more polar than is nicotine, and cotinine is not extracted efficiently with diethyl ether. Frequently, methylene chloride [10] has been employed for cotinine extraction, which resulted in an 88% extraction recovery from aqueous solution compared to 8% for diethyl ether [27]. Metabolites that are more polar than cotinine, such as *trans*-3'-hydroxycotinine, require more strenuous conditions for extraction. For example, 'salting out' with concentrated solutions of potassium carbonate and organic solvents such as butyl acetate, methylene chloride, acetonitrile or mixtures of isopropyl alcohol and methylene chloride have been employed [28–30].

Ether and methylene chloride are quite volatile (bp ~40°C), and this can be a problem if these solvents are not evaporated and replaced with a less volatile one. Solvents with a high vapor pressure, such as ether or methylene chloride, may not be picked up by syringes of autosamplers, unless the autosampler is capable of cooling the samples and syringe. To circumvent this problem, Jacob et al. utilized mixtures of toluene and 1-butanol (bp>100°C) to simultaneously extract nicotine and cotinine from biologic fluids [9]. If necessary to maximize sensitivity, the final extracts can be concentrated to a small volume in autosampler vials by placing them in a heated metal block. A word of caution is in order concerning evaporation of extracts containing nicotine. Nicotine is sufficiently volatile that it will be lost if extracts are taken to dryness, unless an acid is added to convert nicotine to a nonvolatile salt. Nicotine metabolites, with the possible exception of nornicotine, are sufficiently nonvolatile that extracts can be generally taken to dryness without significant losses.

In addition to a variety of liquid–liquid extraction methods, solid-phase extraction has been applied to nicotine and metabolites from plasma [8,31], urine [8], and meconium [32] prior to GC or GC–MS analysis. Solid-phase extraction may be especially advantageous for very polar metabolites, such as nicotine-*N'*-oxide [42]. Solid-phase extraction has been employed for the metabolite nornicotine and for the nicotine-related alkaloids anabasine and anatabine [33].

As discussed above, nicotine and its metabolites, being basic substances, may be adsorbed on glass surfaces, and this can be a problem during extraction. Some investigators routinely silanize tubes and autosampler vials [34]. Grubner et al. reported that addition of ammonia to reagents greatly reduced adsorption of nicotine to the glass surfaces of tubes used in the extraction process [35], and addition of ammonia to the aqueous bases used for extractions has been employed by other investigators [9]. Treating autosampler vials with aqueous tetrasodium EDTA has been reported to suppress adsorption of nicotine and cotinine [9], as well as other basic drugs [36], to glass surfaces of autosampler vials.

## Derivatization

Derivatization of an analyte may be carried out to improve the chromato-graphic properties or to enhance detector response [37,38]. Molecules with polar functional groups such as hydroxy, primary or secondary amino, or carboxylic acid may be strongly adsorbed to polar sites on injection ports and columns, leading to tailing peaks and poor sensitivity. Conversion of these groups to less polar derivatives, such as by replacing an active hydrogen atom with an acyl or alkyl group, can improve chromatography and sensitivity. Sometimes derivatization is used to improve intrinsic detector response, such as reaction with a halogenated reagent to produce a derivative that gives high sensitivity with an electron capture detector.

Nicotine and cotinine do not have active hydrogen atoms and they separate well on a variety of stationary phases, so derivatization of these substances is rarely done. Nicotine, however, can be converted to halogen-ated acyl derivatives by reactions involving ring-opening with an acid chloride or anhydride. Such reactions have been used to produce derivatives that give high sensitivity with an electron capture detector [14,15] or for GC–MS using negative ion chemical ionization (NICI) [20].

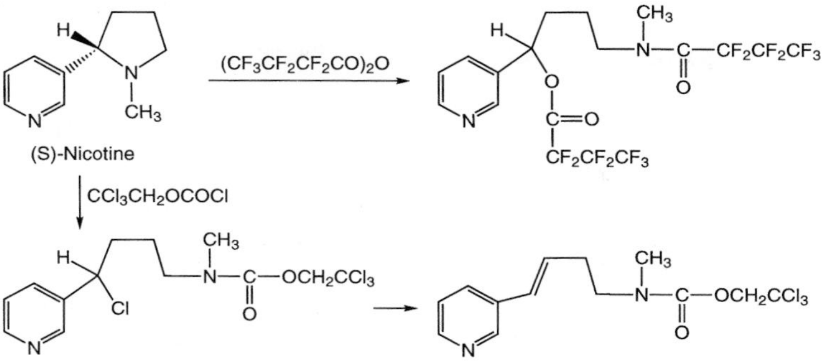

(S)-Nicotine

Most nicotine metabolites have polar functional groups and require derivatization for optimum chromatography and sensitivity. The major urinary metabolite, *trans*-3′-hydroxycotinine, is very polar and may not chromatograph well unless the chromatographic system is extremely inert. It has been determined by GC without derivatization [28,29], but conversion to a trialkylsilyl derivative prior to GC analysis can improve peak symmetry and sensitivity [30].

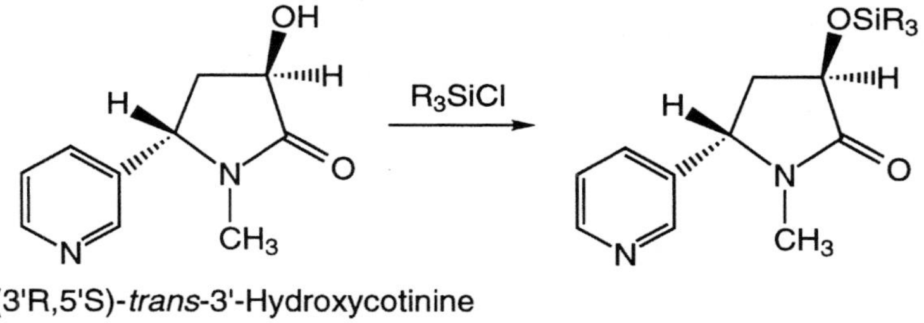

(3′R,5′S)-*trans*-3′-Hydroxycotinine

Nornicotine is both a tobacco alkaloid and a minor nicotine metabolite. It is a secondary amine and is readily converted to acyl or alkyl derivatives to improve chromatography. Neurath et al. converted nornicotine to a heptanoyl derivative for analysis by GC with NP detection [39].

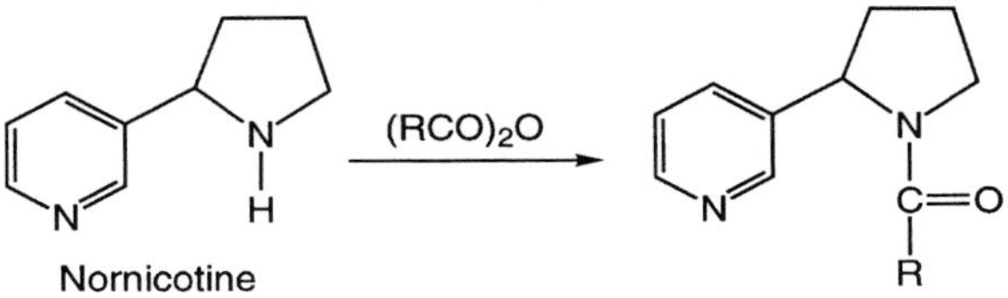

Nornicotine

Zhang et al. and Jacob et al. converted nornicotine and the chemically related tobacco alkaloids anabasine and anatabine to *N*-propyl derivatives by a reductive alkylation process for GC–NPD [40] and GC–MS [33] analysis.

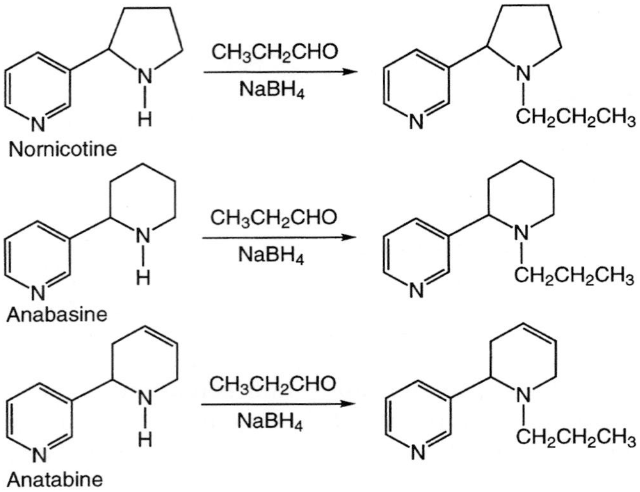

Nornicotine

Anabasine

Anatabine

Nicotine *N'*-oxide is very polar and thermally labile, and therefore cannot be analyzed directly by GC. Most investigators have measured this metabolite indirectly by reduction back to nicotine followed by GC analysis [41]. Jacob et al. made use of the thermal lability to convert nicotine-*N'*-oxide into a ring expansion product that has good GC properties [42].

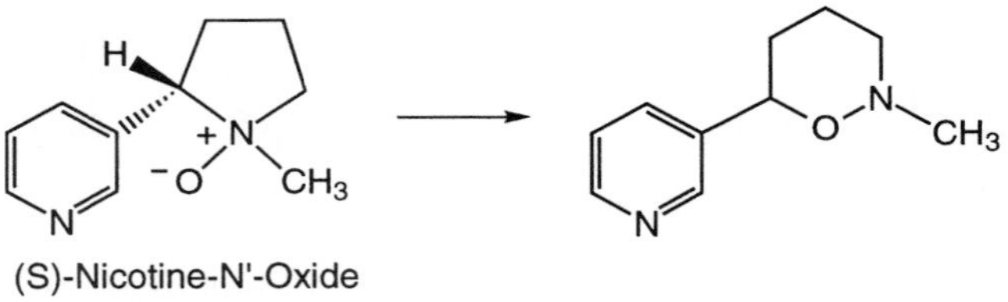

(S)-Nicotine-N'-Oxide

Metabolism of cotinine produces various carboxylic acid metabolites, including 4-(3-pyridyl)-4-oxobutyric acid and 3-pyridylacetic acid which are excreted in urine (Chapter 3). McKennis and co-workers converted these to methyl esters for GC analysis, but assays of biologic fluid samples were not reported [43]. Determination of these polar metabolites using HPLC has been recently reported [44].

## V. QUALITATIVE GC–MS

GC–MS is a two-dimensional technique that can identify a compound by both GC retention time and mass spectral features. GC alone can provide some qualitative information about a sample. For example, a narrow, single peak implies a single compound. Matching the retention time with a known standard can suggest identity. But retention time tells us little about the structure of a compound. By matching both retention time and mass spectral characteristics of an unknown with a standard, the identity of a compound can be made with high confidence. This section discusses the characterization of compounds by GC–MS, focusing on mass spectral features using nicotine and related compounds as examples.

## Interpreting mass spectra

The EI mass spectrum of nicotine is shown in Fig. 2. Nicotine has the molecular formula $C_{10}H_{14}N_2$ giving a nominal molecular weight of 162. Thus, we see $M^{+\cdot}$ at $m/z$ 162 in the mass spectrum. In general, the molecular ion is the highest mass ion in the spectrum and the most analytically important peak in the spectrum. The most intense peak, however, is referred to as the 'base peak'. The base peak in the mass spectrum of nicotine occurs at $m/z$ 84 and represents the pyrrolidine substructure unit of nicotine. The fragmentation pathway proposed for this ion is shown below where cleavage of the molecule occurs between the two rings:

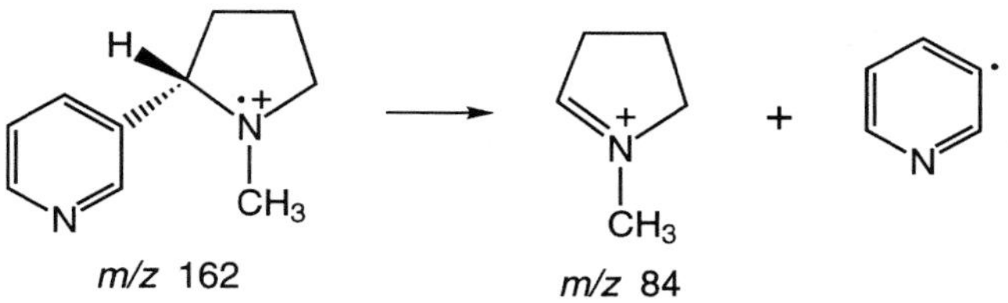

*m/z* 162                  *m/z* 84

The same type of fragmentation also occurs for cotinine to produce $m/z$ 98 as the base peak in its EI mass spectrum. Stable fragments show high abundance in the spectrum and from these we can gather information about molecular structure, solving the mass spectrum as a puzzle. The base peak is assigned a relative abundance of 100% and other peaks in the mass spectrum are normalized to this peak. For example, in the nicotine EI mass spectrum in Fig. 2, the molecular ion relative abundance is 12%. For interpretations of mass spectra, there are standard texts for organics [3,45] as well as for heterocyclics in particular [46]. Collections of nicotine-related compound spectra have been published [47–49].

Because EI mass spectra are somewhat reproducible, they can be searched against libraries of accumulated mass spectra. Most GC–MS data systems include spectral searching software for this purpose as well as one or more commercial libraries of thousands of compounds. The ability to create a user library is also included.

To illustrate the ability of mass spectra to differentiate between compounds, consider the EI mass spectrum of an isomer of nicotine known as metanicotine shown in Fig. 3. This compound has no molecular ion and a very different fragmentation pattern from that of nicotine. The base peak is now *m/z* 44 resulting in cleavage of the *N*-methyl group. Even though these compounds have the same molecular formula and similar properties, mass spectrometry can easily distinguish the two.

Chemical ionization can enhance the molecular weight information from a mass spectrum. Figure 4 shows the PICI mass spectra of metanicotine.

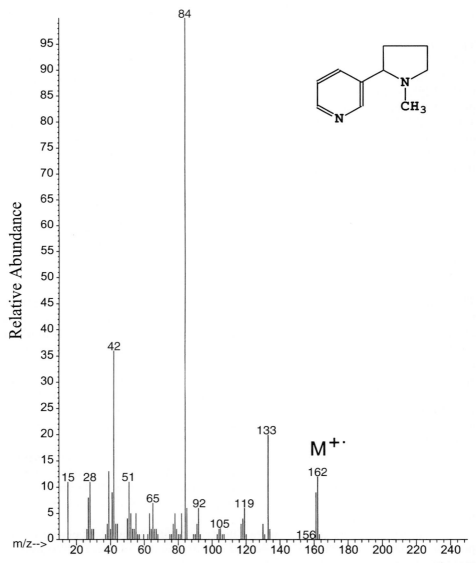

Fig. 2 Electron impact mass spectrum of nicotine. Data acquired on a quadrupole GC–MS system.

Compared to the EI mass spectrum of this compound (Fig. 2), the base peak is now (M+H)$^+$ at *m/z* 163.

## Ion chromatograms

When a full scan mass spectrum is acquired, it can be reprocessed to extract only responses for a particular ion. The resulting ion chromatogram is

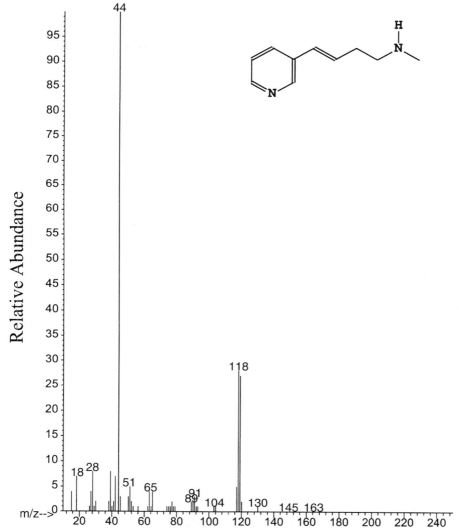

Fig. 3 Electron impact mass spectrum of metanicotine. Data acquired on a quadrupole GC–MS system.

sometimes called a 'reconstructed ion chromatogram' (RIC) or an 'extracted ion current profile' (EICP). By selecting one or more ions characteristic of a compound, the RIC can locate the response due to that compound in a complex background. For example, Fig. 5a is a complex TIC of a basic extract

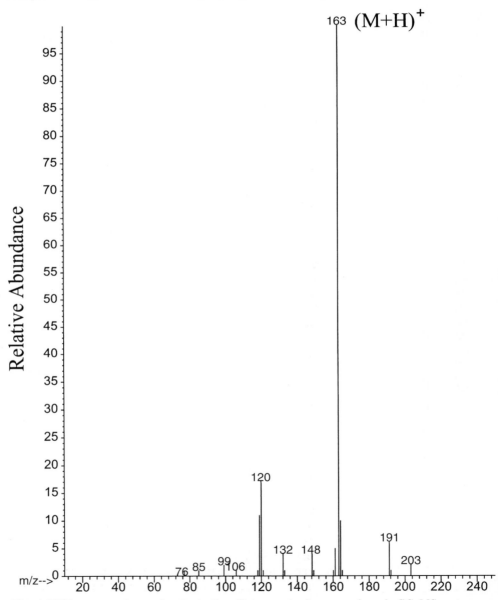

Fig. 4 PICI mass spectrum of metanicotine. Data acquired on a quadrupole GC–MS system.

from a sample of pooled smokers' urine. In Figs 5b and 5c, respectively, the RICs for $m/z$ 106 and 192 are plotted from that data. These are the responses for the base peak and $M^{+\cdot}$, respectively, for the nicotine metabolite

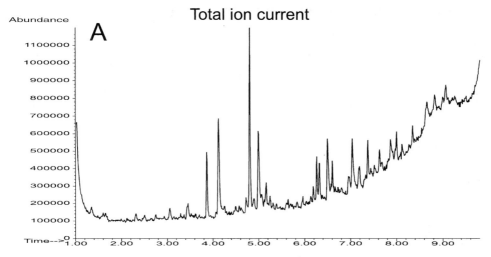

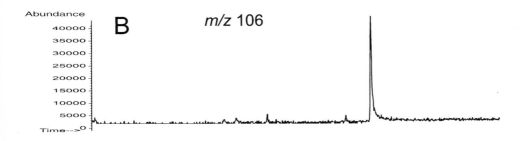

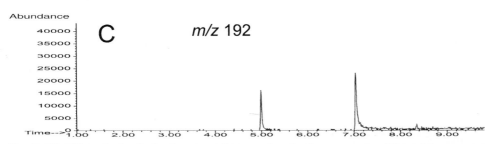

Fig. 5 GC–MS analysis of a basic extract from smokers' urine comparing (A) total ion current ($m/z$ 50–350) and extracted ion chromatograms for characteristic ions of 3HC: (B) $m/z$ 106 and (C) $m/z$ 192.

*trans*-3′- hydroxycotinine. If we obtain the mass spectrum at this time point (approximately 7 min here) and subtract background from it, we get a mass spectrum for that component, as shown in Fig. 6a. By comparing the retention time and mass spectrum with that of a synthetic standard, as shown in Fig. 6b, the presence of this metabolite is confirmed. Note that by operating the mass spectrometer in the SIM mode and monitoring only these two ions, the sensitivity for this metabolite would be greatly enhanced.

When looking for nicotine metabolites in biological fluids, the best approach is to compare extracts from smokers and nonsmokers. GC peaks that occur in the smokers and not in the nonsmokers are likely candidates;

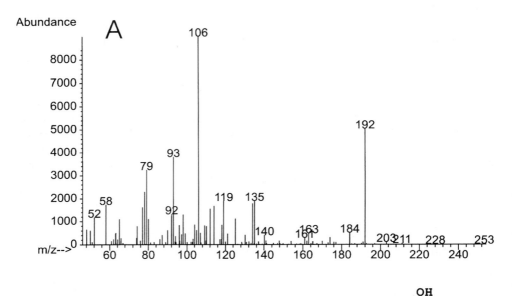

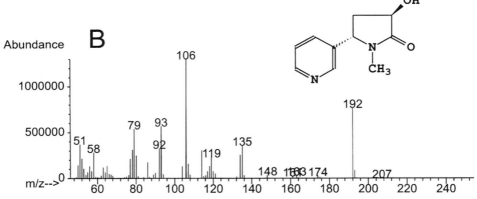

Fig. 6 Comparison of EI mass spectra of from (A) the TIC of an extract of smokers' urine (Fig. 5A) and (B) a synthetic standard of *trans*-3′-hydroxycotinine.

however, caution must be observed in that common medications or their metabolites may also appear in blood and urine.

## VI. QUANTITATION

### Standards

Pure standards that are easy to handle are important for any analytical method. Nicotine base is a liquid that darkens on standing, and is therefore not particularly convenient to use as a standard. The tartrate salt (also called the bitartrate or acid tartrate, which has the composition 1 mol nicotine : 2 mol (+)-tartaric acid : $2H_2O$), is a much more convenient standard, since it is a stable solid and can be easily weighed [27,50]. The major metabolite, cotinine, is a low melting solid, but it is hygroscopic and difficult to weigh. The perchlorate salt is a crystalline solid and, because it is very stable and non-hygroscopic, it is a convenient standard [50]. Nicotine 1'-*N*-oxide is usually synthesized as a mixture of diastereomers and is a viscous liquid, but it can be converted to a crystalline picrate [42]. The pure diastereomers of nicotine-1'-*N*-oxide can be separated by chromatographic methods [51–53] or prepared by stereoselective methods (Chapter 4). *Trans*-3'-hydroxycotinine is a crystalline solid [54], and can be purified by recrystallization from acetone [55] or converted to the perchlorate salt which appears to be very stable [56]. Nornicotine, like nicotine, darkens on standing, but it can be converted to crystalline picrate [57,58] or camsylate [59] salts. Most of the other nicotine metabolites are solids. See Chapter 4 for a discussion of methods for synthesizing nicotine metabolites.

### Peak measurement and data systems

In the early days of GC, detectors were generally interfaced with strip-chart recorders, and peak measurement (generally peak height) was often done manually using a ruler. Nicotine assays reported 20 or so years ago using manual peak measurement [26,27] gave quite good results, but of course this process was very time-consuming. In the early 1970s, computing integrators were introduced, and this greatly facilitated the data analysis process. In the 1980s, microcomputer-based data systems became available for chromatographic analysis, offering the advantage of peak integration, data analysis, and printing of chromatograms and reports post-run. Storage of data on disk also allows the chromatographer to reintegrate a chromatogram if the initial parameters do not provide a satisfactory baseline, or do not properly integrate closely eluting peaks.

### Peak height vs. peak area measurements

Based on chromatographic theory, under ideal conditions, quantitation using peak areas should give more accurate results than those based on peak heights, but practical considerations may be more important in the real world. Most chromatographers prefer to use peak areas, because if peak symmetry changes during a run due to column degradation or accumulation of active sites on injection port liners, peak areas will generally be less affected by changes in symmetry than will be peak heights. On the other hand, if an analyte is not completely resolved from another substance, the peak height may be less affected than the area. In most cases where peak resolution or symmetry changes are not a problem, peak heights and areas will give comparable results, as was shown for determination of some tobacco alkaloids in urine [33].

## Internal standard vs. external standard calibration

For quantitative gas chromatographic analyses, an internal standard is almost always employed. Added at the beginning of the analysis process, it can correct for losses occurring during extraction or in the chromatographic system, and makes it less critical to accurately measure the volumes of solvents and reagents used in the extraction and to precisely measure the volume of the aliquot injected into the chromatograph. Since quantitation is based on the ratio of analyte to internal standard, this ratio will remain constant even if there are variations in total recovery, provided an appropriate internal standard is used. Consequently, the vast majority of methods reported for GC determination of nicotine and its metabolites employ internal standards.

## Choice of internal standard

Variable losses of analyte can occur due to variation in the size of emulsion layers during extraction of samples leading to variable extract volume recovery, and losses occurring in the chromatographic system due to the presence of active sites. In order to correct for these losses, the internal standard must be chemically very similar to the analyte so that fraction of internal standard and analyte lost will be nearly identical.

The best internal standard will be a close structural analog of the analyte. If GC–MS is employed, a stable isotope-labeled analog of the analyte with nearly identical chemical properties can be employed, which is one of the advantages of GC–MS for quantitative analyses. These stable isotope-labeled analogs compensate for variabilities in sample preparation, injection, and instrumental parameters and greatly improve precision. Investigators have synthesized structural analogs and deuterium-labeled analogs of nicotine and its metabolites for use as internal standards [27,30,42,50,60–62], and some (deuterium-labeled nicotine and cotinine) are commercially available. The problems of using an inappropriate internal standard for nicotine and cotinine analyses have been documented [27,60,63].

## Calibration curves

Various methods have been used to obtain quantitative results from GC and GC–MS analyses [64,65]. The most frequently employed method, especially for analysis of biologic fluids, involves construction of a calibration curve by plotting instrument response versus concentration of standards. If an internal standard is used, instrument response will generally be the ratio of peak areas or peak heights of analyte/internal standard.

If the plot is linear, as is usually the case for most GC detectors, including mass spectrometers, it is often convenient to construct the plot by linear regression. Since simple linear regression analysis gives greater weight to higher values, it may be advantageous to utilize more than one standard curve (e.g. one for low and one for high levels) if a wide range of concentrations is to be measured. If working outside of the linear range of the detector, a linear calibration curve will not be appropriate. Most chromatography data software packages are capable of constructing linear, quadratic, hyperbolic, or point-to-point calibration curves using either the internal standard or external standard method.

## Sensitivity: limit of detection (LOD) and limit of quantitation (LOQ)

Sensitivity requirements are one of the first considerations in choosing a method. In general, the lower the concentrations and the more complex the matrix, the more demanding the assay will be. Sensitivity in practical terms will depend on both the intrinsic sensitivity of the detector, the quality of the chromatography (sharp peaks and absence of significant loss of analyte), and background resulting from matrix impurities. The LOD is the lowest concentration of analyte that can reliably be differentiated from background chromatographic noise. It is usually defined as a ratio of analyte signal (peak) to baseline noise of $2:1$ or $3:1$ [64,65]. The LOQ is the lowest concentration that can be measured with a stated level of confidence. For example, a conference of pharmaceutical scientists proposed that for analytical methods used in pharmacokinetic studies the precision (relative standard deviation or coefficient of variation, % CV) should not exceed 20%, and the mean value should be within ±20% of the actual value at the LOQ [66]. Another common method used to define LOD is based on the linear regression data from a calibration range [67]. The LOD is the analyte concentration giving a signal equal to the blank plus 3 standard deviation units. Similarly, LOQ is defined as the signal equal to the blank plus 10 standard deviation units.

## Validation

The way in which a method is validated will, of course, depend upon its application. If the results of the analyses are to be submitted to regulatory agencies, such as the US Food and Drug Administration or Environmental Protection Agency, specific validation procedures may be necessary, and Good Laboratory Practices (GLP) guidelines may need to be met.

For quantitative analyses, as a minimum one would generally want to determine precision, accuracy, and LOQ, and to get some idea of the specificity [66]. Within-run (intra-day) and between-run (inter-day) precision and accuracy are generally determined by carrying out replicate analyses of samples prepared from blank matrix spiked with the analyte. Three or four concentrations spanning the expected range are generally run. The LOQ will be the lowest level for which acceptable precision and accuracy are obtained.

The specificity of the assay is very important, but it is not always easy to determine. It requires anticipating potential sources of interference. For complex matrices, such as biological samples, blank matrices from several sources should be run to check for interferences. In addition, drugs, their metabolites, and other chemicals that might be used by, absorbed by, or formed in the person or other biological system that is the source of the sample should be tested to see if they lead to interference. Analysis of a set of samples by another method that is sufficiently different that it is unlikely to have the same interferences is another way to test the specificity of the method.

Stability of the analyte in the particular matrix should be determined. If samples are to be stored frozen, as is often done in biological studies, freeze-thaw stability may need to be determined in case samples need to be re-analyzed during the course of the study. Stability during storage is often done in the same fashion as for precision and accuracy, using blank matrix spiked with analyte spanning the expected range. Samples are analyzed periodically over a time period that is as long or greater than the time expected to complete analysis of the study samples.

## Chiral separations

Since enantiomers frequently have distinctly different biological properties, it is often important to carry out studies with compounds that have known stereochemical composition. Nicotine in tobacco is levorotatory and is virtually pure (S)-enantiomer [68], but a small amount of (R)-nicotine is found in tobacco smoke, presumably formed during combustion [69,70]. In a variety of animal and in vitro models, the (S)-isomer of nicotine is significantly more potent than the (R)-isomer [71]. Interestingly, nornicotine in tobacco is almost racemic [72]. Pharmacologic differences between the enantiomers of nornicotine have been reported [59].

In order to document the optical purity of synthetic enantiomers of nornicotine, Jacob used the chiral derivatizing agent *N*-(trifluoroacetyl-prolyl) chloride to prepare diastereomeric amides that were separated by packed column GC [58]. Subsequently, diastereomeric amides derived from nornicotine and (−)-camphanic acid chloride were separated on a capillary GC column. This latter method was extended to test the optical purity of nicotine, by demethylation followed by derivatization of the resulting nornicotine [73]. Separation of the enantiomers of nicotine by GC on a chiral column has also been reported [74].

Nornicotine

## VII. APPLICATIONS

Determination of nicotine, related alkaloids, and metabolites in specific matrices is the subject of other chapters in this monograph, and GC methods are frequently used. The reader is referred to those chapters for in-depth presentation of those applications. Here we will briefly discuss two GC–MS methods that illustrate the advantages of this technique for quantitation of nicotine and its metabolites.

### Simultaneous determination of nicotine and two metabolites in urine and blood by GC–MS

Voncken et al. [29] described a rapid and simple method using GC–MS to determine nicotine and its two major metabolites, cotinine and *trans*-3′-hydroxycotinine, in urine and blood. For urine, a 0.5 mL aliquot was mixed with saturated potassium carbonate to basify the sample. The salt also increases the polarity of the aqueous phase to assist the extraction of *trans*-3'-hydroxycotinine, which is quite polar and reluctant to leave an aqueous solution. The extracting solvent for urine was *n*-butyl acetate which also contained two internal standards, *N*-ethylnornicotine and *N*-ethylnorcotinine. For blood, 1 mL of serum or plasma was used and the extraction solvent was dichloromethane. Recovery for all three analytes in urine exceeded 80%. For blood, the recoveries ranged from 50% for *trans*-3'-hydroxycotinine to 77% for nicotine.

The GC–MS was operated first in the scanning mode to identify the analytes and then in the SIM mode for quantification. For a heavy smoker, these analytes occur at fairly high levels in the urine (μg/mL range) and dilution of the urine was necessary for determination of *trans*-3'-hydroxycotinine. Limits of detection were 10 ng/mL for *trans*-3'-hydroxycotinine and 2 ng/mL for nicotine and cotinine.

The simplicity of this method is attractive. Only one extraction step is involved and recovery is reasonable. The method as described in the paper would work well for smokers. With a couple of modifications, the limits of detection could be extended to below 1 ng/mL, so samples from nonsmokers exposed to low levels of tobacco smoke could be analyzed. For example, concentration of the extraction solvent is not mentioned and this could increase the amount of analyte injected on-column. Use of deuterated internal standards for each analyte would greatly improve the precision. This method would be a very good starting point for building more refined analyses.

Several rapid methods using GC–MS determination of nicotine and/or cotinine have been reported that are similar to the above method. Use of stable isotope labeled analogs as internal standards and extract concentration have extended the LOD to near 0.2 ng/mL for nicotine and cotinine in urine and plasma [75]. A solid phase extraction method for nicotine in plasma reports an LOD of 0.5 ng/mL [76]. Analysis of extracts of sweat patches for nicotine using GC–MS gave an LOD of 10 ng/patch [77].

## Determination of nicotine, cotinine, and deuterium-labeled analogs in biologic fluids by GC–MS

As discussed above, one of the major advantages of GC–MS compared with GC using less specific detectors is the ability of the mass spectrometry to distinguish substances that coelute on the GC column but produce ions with different masses. Mass spectrometry also allows the use of stable-isotope labeled internal standards to maximize precision and accuracy of assays. Mass spectrometry is required for most studies involving stable isotopes to distinguish labeled and unlabeled analytes. Stable isotope methods are often used in drug development for bioavailability studies. A tracer dose of the labeled drug can be administered intravenously to determine pharmacokinetic parameters, while at the same time the unlabeled drug can be administered in its usual dosage form, e.g. oral or transdermal. From the plasma clearance of labeled drug administered intravenously and plasma concentrations of the unlabeled drug administered in its usual dosage form, the bioavailabity, i.e. fraction reaching systemic circulation, can be determined.

Stable isotope methodology has been used to determine the bioavailability of nicotine administered transdermally [78] and to determine nicotine intake from cigarette smoking [79] and from other tobacco products [80]. To carry out these studies, a GC–MS method was developed for simultaneous determination of natural nicotine, nicotine–3',3'-d$_2$, and the metabolites cotinine and cotinine–4',4'-d$_2$ in human plasma [9]. Tetra-deutero analogs of nicotine and cotinine are used as internal standards. The method was designed to facilitate the analysis of the large numbers of samples generated in the pharmacokinetic studies. Extraction solvents (mixtures of toluene and 1-butanol) were chosen that remain liquid at $-78°C$ and are less dense than water, thus forming the upper layer, which allows easy separation of the extract using the 'freeze and pour' technique. Rather than using tedious pipetting to transfer extracts, whole racks of samples extracted using a multi-tube vortexer can be placed in a dry ice-acetone bath to freeze the aqueous layers, and organic extracts are transferred by pouring to a new tube or autosampler vial.

(S)-Nicotine-3',3'-d$_2$          (S)-Cotinine-4',4'-d$_2$

The GC–MS analysis is carried out using splitless injection onto a 0.2 mm ID, 12 m length fused silica capillary column with a (5% phenyl)-methylpolysiloxane stationary phase. The mass spectrometry is performed using electron ionization, and data is acquired in the SIM mode. For quantitation, the most abundant ions (base peaks) in the EI spectra of nicotine and nicotine-d$_2$, $m/z$ 84 and 86, respectively, and the molecular ions of cotinine and cotinine-d$_2$, $m/z$ 176 and 178, respectively, are selected. Calibration curves are constructed from plots of peak area ratios of analytes to internal standards, and are linear from 0–100 ng/mL for nicotine and nicotine-d$_2$, and from 0–1000 ng/mL for cotinine and cotinine-d$_2$. For nicotine and nicotine-d$_2$ over the concentration range 1–40 ng/mL, accuracy was 95–103% of the expected value, and precision (intra-day% CV) ranged from 0.8–3.9%. For cotinine and cotinine-d$_2$ in the concentration range of 10–400 ng/mL, accuracy was 91–99% and precision (% CV) was 0.8–5.5%. Limits of quantitation were 1 ng/mL for nicotine and nicotine-$d_2$, and 10 ng/mL for cotinine and cotinine-$d_2$.

Advantages of the method include: (1) simultaneous extraction and determination of nicotine, cotinine, and isotopomers with deuterium labels; (2) suitability of the method for analysis of large batches of samples; and (3) sensitivity adequate for characterization of nicotine and cotinine pharmacokinetics and for measuring concentrations typically found in plasma and urine of smokers and persons using nicotine-containing medications.

## ACKNOWLEDGMENTS

The authors are grateful to Alexander T. Shulgin for helpful discussions and to Kaye Welch for editorial assistance. Financial support from the National Institute on Drug Abuse, National Institutes of Health, USA (DA02277 and DA01696) and from the California Tobacco-Related Disease Research Program (4RT-0023) is gratefully acknowledged.

## REFERENCES

1 Quin LD. Alkaloids of tobacco smoke. I. Fractionation of some tobacco alkaloids and of the alkaloid extract of Burley cigarette smoke by gas chromatography. *J. Org. Chem.*, 1959: **24**; 911.

2 Beckett AH, Triggs EJ. Determination of nicotine and its metabolite, cotinine, in urine by gas chromatography. *Nature*, 1966: **211**; 1415–1417.

3 Biemann K. *Mass spectrometry: organic chemical applications*; McGraw-Hill: New York, 1962.

4 Horning EC, Horning MG, Carroll DI, Stillwell RN, Dzidic I. Nicotine in smokers, non-smokers and room air. *Life Sci.*, 1973: **13**; 1331–1346.

5 McFadden W. *Techniques of combined gas chromatography/mass spectrometry: applications in organic analysis*; Wiley: New York, 1973.

6 Gas chromatography: Biochemical, biomedical, and clinical applications; Clement RE (Ed.), John Wiley and Sons: New York, 1990.

7 Jennings WG. *Analytical gas chromatography*; Academic Press: Orlando, FL, 1987.

8 Stehlik G, Kainzbauer J, Tausch H, Richter O. Improved method for routine determination of nicotine and its main metabolites in biological fluids. *J. Chromatogr. B., Biomed. Sci. Appl.*, 1982: **232**; 295–303.

9 Jacob P, III, Yu L, Wilson M, Benowitz NL. Selected ion monitoring method for determination of nicotine, cotinine, and deuterium-labeled analogs. Absence of an isotope effect in the clearance of (S)-nicotine–3′–3′-$d_2$ in humans. *Biol. Mass Spectrom.*, 1991: **20**; 247–252.

10 Feyerabend C, Bryant AE, Jarvis MJ, Russell MAH. Determination of cotinine in biological fluids of non-smokers by packed column gas-liquid chromatography. *J. Pharm. Pharmacol.*, 1986: **38**; 917–919.

11 Hengen N, Hengen M. Gas-liquid chromatographic determination of nicotine and cotinine in plasma. *Clin. Chem.*, 1978: **24**; 50–53.

12 McNair HM, Bonelli EJ. *Basic gas chromatography*; Varian Aerograph: Palo Alto, CA, 1968.

13 Curvall M, Kazemi-Vala E. Nicotine and metabolites: Analysis and levels in body fluids. In: *Nicotine and Related Alkaloids: Absorption, distribution, metabolism and excretion*. Gorrod JW, Wahren J (Eds.), Chapman & Hall, London, 1993, 147–179.

14 Hartvig P, Ahnfelt N, Hammarlund M, Vessman J. Analysis of nicotine as a trichloroethyl carbamate by gas chromatography with electron-capture detection. *J. Chromatogr.*, 1979: **173**; 127–138.

15 Moore JM, Cooper DA, Kram TC, Klein RFX. Sensitive detection of nicotine after its novel perfluoroacylation and analysis using capillary gas chromatography-electron-capture detection. *J. Chromatogr.*, 1993: **645**; 273–281.

16 Baillie TA, Rettenmeier AW, Peterson LA, Castagnoli N, Jr. Stable isotopes in drug metabolism and disposition. In Annual Reports in Medicinal Chemistry, Vol. 19; Academic Press, New York, 1984, 273.

17 Johnstone RAW, Rose ME. *Mass spectrometry for chemists and biochemists*; 2nd Ed. Cambridge University: Cambridge, 1996.

18 McMaster M, McMaster C. *GC/MS: A practical user's guide*; Wiley-VCH: New York, 1998.

19 Munson MSB, Field FH. Chemical ionization mass spectrometry. I. General introduction. *J. Am. Chem. Soc.*, 1966: **88**; 2621–2630.

20 Cooper DA, Moore JM. Femtogram on-column detection of nicotine by isotope dilution gas chromatography/negative ion detection mass spectrometry. *Biol. Mass Spectrom.*, 1993: **22**; 590–594.

21 Busch KL, Glish GL, McLuckey SA. *Mass spectrometry/mass spectrometry: Techniques and applications of tandem mass spectrometry*; VCH Publishers: New York, 1988.

22 Charles MJ, Glish GL. Review of modern ion trap research. In: *Practical aspects of ion trap mass spectrometry, Vol. III* March RE, Todd JFJ (Eds), CRC Press, Boca Raton, FL, 1995.

23 Comisarow MB, Marshall AG. The early development of Fourier transformation cyclotron resonance (FT–ICR) spectroscopy. *J. Mass Spectrom.*, 1996: **6**; 581–585.

24 Haufler RE, Kerley EL. Miniaturized time-of-flight spectrometer for high-speed applications. 45th Annual Conference on Mass Spectrometry and Allied Topics. Palm Springs, CA, June 1–6, 1997.

25 Jones D, Curvall M, Abrahamsson L, Kazemi-Vala E, Enzell C. Quantitative analysis of plasma nicotine using selected ion monitoring at high resolution. *Biomed. Mass Spectrom.*, 1982: **9**; 539–545.

26 Feyerabend C, Levitt T, Russell MAH. A rapid gas-liquid chromatographic estimation of nicotine in biological fluids. *J. Pharm. Pharmacol.*, 1975: **27**; 434–436.

27 Jacob P, III, Wilson M, Benowitz NL. Improved gas chromatographic method for determination of nicotine and cotinine in biologic fluids. *J. Chromatogr.*, 1981: **222**; 61–70.

28 Neurath GB, Pein FG. Gas chromatographic determination of *trans*–3′-hydroxycotinine, a major metabolite of nicotine in smokers. *J. Chromatogr., Biomed. Appl.*, 1987: **415**; 400–406.

29  Voncken P, Schepers G, Schafer K-H. Capillary gas chromatographic determination of *trans*–3′–hydroxycotinine simultaneously with nicotine and cotinine in urine and blood samples. *J. Chromatogr.*, 1989: **479**; 410–418.

30  Jacob P, III, Shulgin A, Yu L, Benowitz NL. Determination of the nicotine metabolite *trans*–3′–hydroxycotinine in smokers using gas chromatography with nitrogen-selective detection or selected ion monitoring. *J. Chromatogr.*, 1992: **583**; 145–154.

31  Zins BJ, Sandborn WJ, Mays DC, Lawson GM, McKinney JA, Tremaine WJ et al. Pharmacokinetics of nicotine tartrate after single-dose liquid enema, oral, and intravenous administration. *J. Clin. Pharmacol.*, 1997: **37**; 426–436.

32  Ostrea EM, Knapp K, Romer A, Montes M, Ostrea AR. Meconium analysis to assess fetal exposure to nicotine by active and passive maternal smoking. *J. Pediatr.*, 1994: **124**; 471–476.

33  Jacob P, III, Yu L, Liang G, Shulgin AT, Benowitz NL. Gas chromatographic-mass spectrometric method for determination of anabasine, anatabine and other tobacco alkaloids in urine of smokers and smokeless tobacco users. *J. Chromatogr., Biomed. Applic.*, 1993: **619**; 49–61.

34  Verebey KG, DePace A, Mule SJ, Kanzler M, Jaffe JH. A rapid, quantitative GLC method for the simultaneous determination of nicotine and cotinine. *J. Analyt. Toxicol.*, 1982: **6**; 294–296.

35  Grubner O, First MW, Huber GL. Gas chromatographic determination of nicotine in gases and liquids with suppression of adsorption effects. *Anal. Chem.*, 1980: **52**; 1755–1758.

36  Jacob P, III, Elias-Baker BA, Jones RT, Benowitz NL. Determination of cocaine in plasma by automated gas chromatography. *J. Chromatogr.*, 1984: **306**; 173–181.

37  Blau K, Halket J. *Handbook of Derivatives for Chromatography, 2nd Edition*; John Wiley and Sons: Chichester, 1993.

38  Knapp DR. *Handbook of Analytical Derivatization Reactions*; John Wiley and Sons: New York, 1979.

39  Neurath GB, Orth D, Pein FG. Detection of nornicotine in human urine after infusion of nicotine. In: *Advances in pharmacological sciences, effects of nicotine on biological systems*, Adlkofer F, Thurau K (Eds.), Birkhauser Verlag, Basel, 1991, 45–49.

40  Zhang Y, Jacob P, III, Benowitz NL. Determination of nornicotine in smokers' urine by gas chromatography following reductive alkylation to $N'$-propylnornicotine. *J. Chromatogr.*, 1990: **525**; 349–357.

41  Beckett AH, Gorrod JW, Jenner P. The analysis of nicotine–1′-*N*-oxide in urine, in the presence of nicotine and cotinine, and its application to the study of in vivo nicotine metabolism in man. *J. Pharm. Pharmac.*, 1971: **23**; 55S–61S.

42  Jacob P, III, Benowitz NL, Yu L, Shulgin AT. Determination of nicotine–1′-*N*-oxide by gas chromatography following thermal conversion to 2-methyl-6-(3-pyridyl)-tetrahydro-1,2-oxazine. *Anal. Chem.*, 1986: **11**; 2218–2221.

43  McKennis H, Jr., Bowman ER, Dar MS. Studies on the separation of acidic metabolites of nicotine by gas chromatography. *Virginia J. Sci.*, 1967: **18**; 13–18.

44  Sai Y, Gorrod JW. Metabolic interconversion of acid metabolites of cotinine in vitro. *Drug Metabol. Drug Interact.*, 1995: **12**; 93–103.

45  McLafferty FW. *Interpretation of mass spectra*, 3rd Edn.; University Science Books: Mill Valley, CA, 1980.

46 Porter QN. *Mass Spectrometry of Heterocyclic Compounds*, 2nd Edn.; Wiley: New York, 1985.

47 Pilotti A, Enzell CR. Studies on the identification of tobacco alkaloids, their mammalian metabolites and related compounds by gas chromatography-mass spectrometry. *Beitr. Tabakforsch.*, 1976: **8**; 339–349.

48 Glenn DF, Edwards WB, III. Synthesis and mass spectrometry of some structurally related nicotinoids. *J. Org. Chem.*, 1978: **43**; 2860–2870.

49 Cowan DA, Damani LA, Gorrod JW. Metabolic *N*-oxidation of 3-substituted pyridines: Identification of products by mass spectrometry. *Biomed. Mass Spectrom.*, 1978: **5**; 551–556.

50 Jacob P, III, Benowitz NL, Shulgin AT. Synthesis of optically pure deuterium-labelled nicotine, nornicotine and cotinine. *J. Labelled Comp. Radiopharmaceut.*, 1988: **25**; 1117–1128.

51 Testa B, Jenner P, Beckett AH, Gorrod JW. A reappraisal of the stereoselective metabolism of nicotine to nicotine–1'-*N*-oxide. *Xenobiotica*, 1976: **6**; 553–556.

52 Sepkovic DW, Haley NJ, Axelrad CM. Short-term studies on the in vivo metabolism of *N*-oxides of nicotine in rats. *J. Toxicol. Environ. Hlth.*, 1986: **18**; 205–214.

53 Cashman JR, Park SB, Yang ZC, Wrighton SA, Jacob P, III, Benowitz NL. Metabolism of nicotine by human liver microsomes: Stereoselective formation of trans–nicotine–*N'*–oxide. *Chem. Res. Toxicol.*, 1992: **5**; 639–646.

54 McKennis H, Jr., Turnbull LB, Bowman ER, Tamaki E. The synthesis of hydroxycotinine and studies on its structure. *J. Org. Chem.*, 1963: **28**; 383–387.

55 Dagne E, Castagnoli N, Jr. Structure of hydroxycotinine, a nicotine metabolite. *J. Med. Chem.*, 1972: **15**; 356–360.

56 Jacob P, III, Shulgin AT, Benowitz NL. Synthesis of (3'R,5'S)–*trans*–3'–hydroxy-cotinine, a major metabolite of nicotine. Metabolic formation of 3'-hydroxycotinine in humans is highly stereoselective. *J. Med. Chem.*, 1990: **33**; 1888–1891.

57 Spath E, Kesztler F. Tobacco alkaloids. XII. Occurrence of d,l-nornicotine, d,l-anatabine, and l-anabasine in tobacco. *Chem. Ber.*, 1937: **70B**; 704–709.

58 Jacob P, III. Resolution of (±)-bromonornicotine. Synthesis of (R)- and (S)-nornicotine of high enantiomeric purity. *J. Org. Chem.*, 1982: **47**; 4165–4167.

59 Risner ME, Cone EJ, Benowitz NL, Jacob P, III. Effects of the stereoisomers of nicotine and nornicotine on schedule-controlled responding and physiological parameters of dogs. *J. Pharmacol. Exp. Ther.*, 1988: **244**; 807–813.

60 Curvall M, Kazemi-Vala E, Enzell CR. Simultaneous determination of nicotine and cotinine in plasma using capillary column gas chromatography with nitrogen–sensitive detection. *J. Chromatogr.*, 1982: **232**; 283–293.

61 Davis RA. The determination of nicotine and cotinine in plasma. *J. Chromatogr. Sci.*, 1986: **24**; 134–141.

62 Shulgin AT, Jacob P, III, Benowitz NL, Lau D. The identification and quantitative analysis of cotinine-*N*-oxide. *J. Chromatogr. Biomed. Applic.*, 1987: **423**; 365–372.

63 Thenot J-P, Hung A. Applications of the nitrogen detector to the analysis of steroid hormones and related compounds by gas chromatography. In: *National Bureau of Standards Special Publication 519, Trace Organic Analysis: A new Frontier in Analytical Chemistry*, Proceedings of the 9th Materials Research Symposium, April 10–13, 1978, NBS, Gaithersburg, MD, 1979.

64 Novak J. *Quantitative Analysis by Gas Chromatography*; Marcel Dekker, Inc.: New York, 1988.

65 Guiochon G, Guillemin CI (Eds.), *Quantitative Gas Chromatography: For Laboratory Analyses and On-Line Process Control*; Elsevier: Amsterdam, New York, 1988.

66 Shah VP, Midha KK, Dighe S, McGilveray IJ, Skelly JP, Yacobi A et al. Analytical methods validation: Bioavailability, bioequivalence and pharmacokinetic studies. *Pharmaceut. Res.*, 1992: **9**; 588–592.

67 Miller JC, Miller JN. *Statistics for Analytical Chemistry, 2nd Edn.*; Ellis Horwood Ltd., 115–117: Chichester, England, 1988.

68 Karrer P, Widmar R. Die Konfiguration des Nicotins. Optisch aktive Hygrinsaure. *Helv. Chim. Acta*, 1925: **8**; 364–368.

69 Klus H, Kuhn H. A study of the optical activity of smoke nicotines. *Fach. Mitt. Oesterr. Tabakregie.*, 1977: **17**; 331–336.

70 Pool WF, Godin CS, Crooks PA. Nicotine racemization during cigarette smoking. *The Toxicologist*, 1985: **5**; 232.

71 Martin BR, Tripathi HL, Aceto MD, May EL. Relationship of the biodisposition of the stereoisomers of nicotine in the central nervous system to their pharmacological actions. *J. Pharmacol. Exp. Ther.*, 1983: **226**; 157–163.

72 Kisaki T, Tamaki E. Phytochemical studies on the tobacco alkaloids. I. Optical rotary power of nornicotine. *Arch. Biochem. Biophys.*, 1961: **92**; 351–355.

73 Jacob P, III, Benowitz NL, Copeland JR, Risner ME, Cone EJ. Disposition kinetics of nicotine and cotinine enantiomers in rabbits and beagle dogs. *J. Pharmaceut. Sci.*, 1988: **77**; 396–400.

74 Perfetti TA, Coleman WM, III. Chiral-gas chromatography-selected ion monitoring-mass selective detection analysis of tobacco materials and tobacco smoke. *Beitr. Tabakforsch. Internat.*, 1998: **18**; 15–33.

75 James H, Tibazi Y, Taylor R. Rapid method for the simultaneous measurement of nicotine and cotinine in urine and serum by gas chromatography-mass spectrometry. *J. Chromatogr. B*, 1998: **708**; 87–93.

76 Davoli E, Stramere L, Fanelli R, Diomede L, Salmona M. Rapid solid-phase extraction method for automated gas chromatographic-mass spectrometric determination of nicotine in plasma. *J. Chromatogr. B Biomed. Sci. Appl.*, 1998: **708**; 312–316.

77 Kintz P, Henrich A, Cirimele V, Ludes B. Nicotine monitoring in sweat with a sweat patch. *J. Chromatogr. B Biomed. Sci. Appl.*, 1998: **708**; 357–361.

78 Benowitz NL, Chan K, Denaro CP, Jacob P, III. Stable isotope method for studying transdermal drug absorption: The nicotine patch. *Clin. Pharmacol. Ther.*, 1991: **50**; 286–293.

79 Benowitz NL, Jacob P, III, Denaro C, Jenkins R. Stable isotope studies of nicotine kinetics and bioavailability. *Clin. Pharmacol. Ther.*, 1991: **49**; 270–277.

80 Jacob P, III, Yu L, Shulgin AT, Benowitz NL. Minor tobacco alkaloids as biomarkers for tobacco use: Comparison of cigarette, smokeless tobacco, cigar and pipe users. *Am. J. Pub. Hlth*, 1999: **89**; 731–736.

*Chapter 7*

# Use of high-performance liquid chromatographic–mass spectrometric (LC–MS) techniques for the determination of nicotine and its metabolites

Peter A. Crooks[1] and Gary D. Byrd[2]

[1] *Division of Pharmaceutical Sciences, University of Kentucky, Lexington, KY 40536-0091, USA*
[2] *R. J. Reynolds Tobacco Company, Biological Chemistry Division, Winston-Salem, NC 27102-1236, USA*

## I. SCOPE

In this chapter, the application of HPLC methods for the determination of nicotine and its metabolites is examined with emphasis on mass spectrometric detection. Other important detectors such as radiometric and diode arrays detectors are covered in brief. A short history on the role of HPLC in studying nicotine metabolism is given and the operation of several LC–MS interfaces is described. Some information on the operation of mass spectrometers (mass analyzers and detectors) discussed in the previous chapter on GC–MS is relevant to LC–MS and will not be repeated in detail here. Likewise, quantitative aspects of LC–MS are analogous to those covered for GC–MS. Qualitative strategies for LC–MS are discussed here and applications that determine nicotine metabolites are presented to illustrate these points. Finally, practical considerations of LC–MS operation and cost are given.

## II. INTRODUCTION

### Why liquid chromatography?

As an analytical separation method, liquid chromatography (LC) can be applied to a greater variety of samples than gas chromatography (GC). This is due to the fact that GC requires the analyte to be volatilized while LC requires only dissolution of the analyte in some solvent prior to separation. For example, while nicotine and cotinine can be analyzed directly by GC methods, several important nicotine metabolites such as *N*-oxides and glucuronide conjugates are not thermally stable and cannot pass intact through a GC. Derivatization and conversion back to the aglycon can make these compounds suitable for GC; however, a direct method of analysis when applicable is preferable for reasons of simplicity.

## History of HPLC for determination of nicotine and its metabolites

High performance liquid chromatography (HPLC) emerged as a powerful analytical tool during the 1970s as progress was made in bonding durable phases to silica [1]. The first report of an HPLC method for nicotine and cotinine in smokers' urine was by Watson in 1977 [2]. He performed an extraction of basified urine and used a normal-phase isocratic method with UV detection. Compared with existing GC methods, this method was faster with both nicotine and cotinine eluting from the column in less than five minutes. Later, a rapid method utilizing XAD-2 resin for sample preparation followed by normal-phase HPLC and UV detection for final determination was reported and claimed detection limits for nicotine and cotinine of 2 ng/mL [3]. Reversed-phase HPLC analytical separations became more widespread and one was reported in 1981 for nicotine and other tobacco alkaloids [4]. HPLC coupled with radiometric detection played an important role in determining the extent of nicotine metabolism in studies that involved administration of radiolabeled nicotine to animals [5]. The metabolism of nicotine proved to be complex and HPLC methods with greater resolving power were developed. For example, a cation exchange method resolved numerous polar metabolites of nicotine in guinea pig urine [6]. Colorimetric assays were also applied to nicotine metabolites [7] though, like radiometry, identification of particular metabolites was often unclear. The coupling of HPLC to a more informative detector such as a mass spectrometer was clearly needed.

## Historical development of LC–MS

As shown in the previous chapter, GC–MS demonstrated the richness of information resulting from mass spectrometry coupled to chromatography. But the limitation of GC–MS to volatile and thermally stable compounds prohibited the direct analysis of entire classes of compounds. This included numerous compounds important in biological research such as peptides, proteins, and conjugated species. So, while GC–MS matured, research began on interfacing LC to mass spectrometry. Directing a liquid flow into the high vacuum source of a mass spectrometer was challenging. As chronicled in the

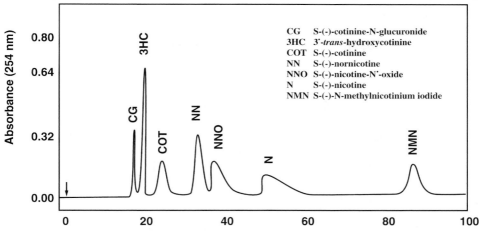

Fig. 2 Cation-exchange Partisil-10 SCX separation of nicotine and six of its metabolites. Mobile phase: primary – 0.2M sodium acetate/methanol (95/5, vol/vol) containing 0.02% vol/vol triethylamine, and adjusted to pH 5.5 with glacial acetic acid. The primary mobile phase was pumped through the column isocratically at 1.0 ml/min flow rate for 40 min, after which time the percentage of the secondary mobile phase [comprising 0.2 M sodium acetate/methanol (95/5, vol/vol) containing 1.0% vol/vol triethylamine, and adjusted to pH 5.5 with glacial acetic acid] was increased in a linear fashion from 0% to 100% over 10 min. Data are presented as UV absorption at 254 nm. Arrow indicates injection point.

bonded matrices exhibit weak retention characteristics, and are mainly used for the separation of carbohydrates; they are also less stable than other silica-bonded phases.

In addition to silica matrices, polymeric carbon-based resins are also used as HPLC column packings. The most popular non-silica matrix is a polystyrene-divinylbenzene co-polymer to which can be covalently attached various bonded phases containing a variety of functional moieties. Unlike the silica-based resins, which are stable only within the pH range 3–7, these packings are extremely resilient and can be utilized over the pH range 1–13. They also have a longer life than silica packings, are easier to clean and regenerate, and often afford superior peak shape for some separations. Polystyrene-divinylbenzene ion-exchange packings are less efficient than silica-based ion-exchange packings, but have the advantage of being more stable and more reproducible. Aluminum and zirconium columns are becoming more popular for their stability and broad useable pH range.

Chiral matrices in HPLC have proven to be very useful for the separation of optical isomers (enantiomers) of drug molecules. Such matrices are quite diverse in nature. Silica based chiral matrices consist of a chirally pure moiety attached to a short linker group, which is chemically bonded to the silicaceous hydroxyl groups of the matrix. The chiral moiety is designed in such a way that it presents to the racemic solute molecule a unique stereochemical 3-point interaction site, which will have greater affinity for one optical isomer vs. its structural antipode. Thus, the two enantiomers, which cannot be separated on achiral columns, will elute at different retention times, allowing separation and quantitation. A variety of chiral matrices are available; many are silica based, while other matrices consist of polysacharrides, proteins, and other natural matrices, or synthetic polymers. Chiral stationary phases can be divided three basic groups; (a) those that are hydrogen-bonding/$\pi$-complex forming phases, (b) cyclodextrin-bonded phases, and (c) natural and synthetic polymer phases. While the separation of nicotine isomers by chiral chromatography has been reported, no examples of the application of this method to the analysis of nicotine and nicotine metabolites in biological samples are known.

## Non-MS detectors (ultraviolet, photodiode array, electrochemical, radiometric, and fluorescence detectors)

*Ultraviolet detectors.* The most popular detector in HPLC analysis is the ultraviolet (UV) detector. This detector is utilized mainly for solutes that possess a chromophore absorbing in the practical region of the electromagnetic spectrum (i.e. 200–550 nm). Thus, these detectors, unlike refractive index detectors, are not universal. However, they have good sensitivity, and can be used with gradient elution.

Two basic types of UV detector are routinely used in HPLC; the first type is the fixed wavelength detector, which consists of a single wavelength source (e.g. the 254 nm mercury line), and is admirably suited for the analysis of aromatic compounds. Although operating at only one wavelength, these detectors are quite versatile, since UV absorption bands are usually quite broad, and it is not always necessary to monitor solute in the HPLC eluate at its maximum absorption wavelength, although this is recommended for maximum sensitivity. Thus, the limit of sensitivity of detection with UV detectors depends upon the extinction coefficient of the analyte at 254 nm, and also on the efficiency of the column. Fixed wavelengths other than the 254 nm mercury line can also be utilized by incorporating appropriate interference filters to isolate the desired mercury line (i.e., 280 and 365 nm), and zinc and cadmium gas discharge gas lamps have been used for wavelengths of 214 nm and 229 nm. The 214 nm detectors are particularly versatile, since a very large proportion of organic solutes absorb at this wavelength. However, care must be exercised in choosing a mobile phase that does not absorb in this region of the UV spectrum. UV detection is particularly suitable for the analysis of tobacco alkaloids and their analogs and metabolites, since the pyridine ring has a strong UV absorption band at 260–264 nm.

It should be mentioned that solutes without chromophores can be detected using indirect UV absorption. In this mode, the mobile phase contains chromophores but the solute does not and the change in absorption can be monitored when the solute passes through the detector.

The second widely used UV detector is the variable wavelength detector, which was developed to increase detection selectivity. These detectors incorporate a deuterium lamp and often a tungsten lamp to extend the wavelength range into the visible part of the electomagnetic spectrum. By setting the wavelength of the detector to the absorption maximum of the analyte under study, the limits of detection and quantitation can be extended. It may also be advantageous to set the detector at a wavelength remote from the λmax of the analyte, to minimize base line interference rather than maximize response, especially when similar UV-absorbing impurities are present.

While UV detectors are not useful for the analysis of non-chromophoric solutes, derivatives of non-UV-absorbing compounds can be prepared, which introduce a chromophoric moiety into the molecule, or shift the λmax of a compound to a more favorable wavelength for detection. Derivatization can take place as part of the sample preparation, or as a post-column event, occurring between the end of the column and the detector. Nicotine and nicotine metabolites have been converted to colored derivatives by pre-column reaction with barbituric acid and diethylthiobarbituric acid followed by reversed-phase HPLC and UV detection [7,17–19].

*Photodiode array detectors.* More recent UV detectors have been developed from the advances that have been made in fast photodiodes, low stray-light, and inexpensive gratings, coupled with the advent of microcomputers. The photodiode array detector allows the complete spectral scanning of column effluent every few seconds. Thus, a complete UV spectrum of each solute eluting from the column can be processed and stored electronically in a data station. The data can be recorded as three-dimensional surfaces, where the absorbance is a function of both wavelength and time. Thus, maximum information can be obtained on all the components in the eluate in a single analysis. This analytical methodology is particularly suited for drug metabolism studies, since metabolites often have absorption maxima at widely different wavelengths to those of the parent drug, and single wavelength detection may result in loss of sensitivity for some drug biotransformation products. Modern diode array detectors incorporate several hundred separate diodes, which span the whole range of the UV-visible range of the electromagnetic spectrum. Such detectors are now routinely used in HPLC analysis. They are especially useful for checking the known relative absorption of authentic compounds at different wavelengths, in order to confirm the identity of eluant components, and to determine the resolution or homogeneity of overlapping or superimposed peaks. A recently developed HPLC assay for the determination of nicotine metabolites in the plasma of smokers utilizes diode array detection for the quantitation of cotinine, 3-hydroxycotinine, and caffeine in a single analysis with good sensitivity [20].

*Electrochemical detectors.* Electrochemical (EC) detectors are utilized for the analysis of solutes that can be oxidized or reduced. The electron flow resulting from this chemical change constitutes the basis for the detector response. Thus the electrochemical detector is essentially a transducer which allows the direct conversion of chemical information into electrical current. The instantaneous current generated is measured as a function of time with a constant potential applied at a fixed electrode exposed to the flowing eluate, and is proportional to the concentration of solute in the cell. Since the electrochemical reaction occurs at the electrode surface, it is possible to reduce the cell volume to a thin layer of minimal thickness, and cell volumes ranging from 1 nL to 1 μL have been developed.

Modern EC detectors can be categorized into amperometric and coulometric types. With amperometric detectors, only about 5–10% of the total number of electroactive molecules is oxidized, whereas in coulometric detection, 100% conversion to the oxidization product is obtained. The oxidized form is usually unstable, and reacts further to afford a stable product, which passes out of the cell. It is important to note that coulometric detectors generally afford lower detection sensitivity than amperometric detectors, since an increase in electrode surface area is necessary, which is accompanied by an increase in background current and noise, and consequently lower signal-to-noise ratios.

Two types of electrode are commonly used in EC analysis, which are suitable for the analysis of both oxidizable and reducible substances. Carbon paste electrodes afford a lower background current, but cannot be used with mobile phases containing more than 20% organic modifier. Glassy carbon electrodes are more rugged, and are compatible with organic solvents. In reductive analyses, mercury and amalgamated gold can also be utilized as electrodes.

EC detection is gaining popularity as an analytical tool in HPLC. This stems from the fact that it is a very sensitive and selective technique with limits of detection in the picogram and femtogram range, rivaling those of fluorescence detectors. Because of this high sensitivity, EC detection has now become an established tool for the analysis of trace compounds in complex biological matrices and in environmental and pharmaceutical samples. The development of detectors with cell volumes of 1 nL and below has allowed their use in conjunction with capillary and microbore HPLC. However, some disadvantages are associated with EC detectors. Since they are very efficient antennas capable of picking up frequencies of 50–60 Hz, the detection apparatus has to be effectively shielded from extraneous noise by placing it inside a Faraday cage. Also, base-line stability can be adversely affected by pump oscillations and by temperature fluctuations. Thus, pump dampeners should be utilized, and strict temperature control should be exercised. Certain mobile phase contaminants, such as dissolved oxygen, and trace metal contaminants, etc. may also affect base line stability. It should be noted that EC detection is most successful and trouble-free if it is utilized continuously over a long period of time for the same analytical purpose, rather than used for a variety of applications over a short time period.

There are several reports of the use of HPLC–EC in the analysis of nicotine and nicotine metabolites [21–23]. This methodology has about the same sensitivity and limits of detection as GC, but may be useful for the determination of the more polar, non-volatile metabolites of nicotine, such as the N-methyl- and glucuronide conjugates. The electrochemical reaction involves a one-electron transfer with an irreversible secondary reaction involving solvent. It is postulated that nicotine initially loses a lone pair electron from the pyrrolidine ring nitrogen atom to afford a cation radical, which then rapidly extracts a hydrogen radical from the solvent to form N-1'-protonated nicotine. In the case of nicotine analysis, the presence of acetonitrile as a solvent component appears to be important, since it is likely that it is the active species in the solvent reaction, generating a resonance stabilized radical. The resulting N-1'-protonated nicotine is not reducible and is electroinactive.

*Radiometric detectors.* Before the advent of flow through liquid radioactivity scintillation cells, detection and measurement of radioactivity in eluents from liquid chromatography columns generally required the laborious collection and counting of successive fractions in static liquid scintillator counters. Problems of resolution, and analysis time, thus limited the usefulness of the HPLC analysis. Consequently, two basic types of flow through radioactivity detector were developed. Earlier detectors incorporated solid cells consisting of crystalline anthracene or anthracene-lithium glass. More modern solid cell detectors utilized a range of inorganic crystals. Solid cells have a disadvantage in that they are prone to contamination and have to be replaced constantly, washed, and repacked. Also, adsorption of radioactivity onto the scintillator can compromise the radiochromatogram obtained. Liquid flow through cells are now widely used, and appear to be the detector of choice for the analysis of unknown mixtures, such as metabolites in biological samples. This cell requires the mixing of the mobile phase from the HPLC column with scintillator fluid (usually xylene based) before entry into the liquid scintillator cell. This cell has the advantage of greater sensitivity, and continuous use, but suffers from the necessity for disposal of waste scintillator fluid (and the associated cost involved!).

The design of a modern liquid radioactivity detector is illustrated in Fig. 3. The HPLC effluent is usually passed through a primary detector (UV, RI, etc.) for the non-radioactive detection of solutes, which can often be added to the radioactive sample in sufficient quantities to provide markers on a second channel of the recorder unit. The effluent from the primary detector is then passed into a variable stream-splitter, which mixes part of the mobile phase with scintillation fluid pumped in via a mixing tee. The mixed scintillator fluid is passed through the liquid cell, and then voided to waste. The radioactivity in the sample can be recorded as counts per minute (cpm) or disintegrations per minute (dpm) with time on a second channel of the recorder, and compared in terms of retention time with standard UV or RI marker peaks generated on the first channel of the recorder. Usually, a small 'lag' time occurs between the two analyses, but this can be adjusted for by calculating the volume in the tubing connecting the two detectors.

Liquid flow through cell methodology is a continuous flow-monitoring device for radioactivity detection, which has obvious advantages. The cell responds to light emitted from scintillant that is mixed with column effluent.

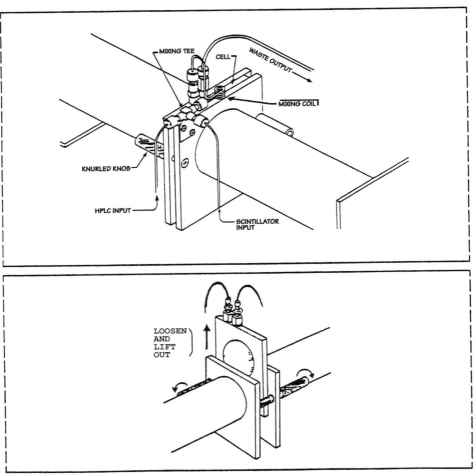

Fig. 3 Illustration of the design of a modern liquid flow-through radioactivity detector commonly used in high-performance liquid-radiochromatography.

The detector is able to integrate the dpm values of a large number of discrete independent measurements of samples equal in size to the volume of the flow through cell averaged over a period of time (for example 6 secs.). Flow cell size can be reduced to avoid band spreading, and can be increased and flow rate lowered for maximum sensitivity. Radioactivity detection can be monitored at specific time intervals, and radioactivity within each peak can be calculated and plotted to determine the fraction of the radioactivity in each peak, relative to the total amount of radioactivity in the sample. In addition, double-labeled compounds can be utilized in metabolic studies, and dual channel analysis of $^3H$ and $^{14}C$ can be obtained using this technique.

A number of studies have been reported on the use of HPLC with radioactivity flow through detection for the determination of nicotine biotransformation products in both in vivo and in vitro experiments [6,24–27]. Some of these studies are discussed later in more detail.

*Fluorescence detectors.* Most organic compounds do not fluoresce; however, when some molecules absorb high frequency electromagnetic radiation in the UV or near UV region, they become excited to higher electronic states. The excited molecules then decay from an excited singlet-state to ground level through either a triplet-state (whose lifetime is relatively long, allowing it to be deactivated via molecular collisions before a photon can be emitted) or through spontaneous emission of one or more photons. The frequency of the emitted photon is always lower than that of the absorbed photon. This latter process is called fluorescence, and affords a fluorescence spectrum that is a function of the excitation wavelength.

The basic design of a fluorescence detector consists of a UV light source, which is focused on the detector cell through which the column effluent containing the analyte flows. Light emitted in the perpendicular plane is collected by the photocell, and its intensity can be directly related to the concentration of analyte in the effluent. Basic fluorescence detectors generally utilize the emission line of a mercury lamp as the excitation wavelength, and the entire emission spectrum of the analyte is collected. In more modern programmable spectrofluorometers, both the excitation wavelength and the emission wavelength can be adjusted, affording more analytical sensitivity and selectivity, and the use of diode array technology allows the entire emission spectrum to be recorded as a function of time. More sophisticated instruments with greater sensitivity are now becoming available that are based upon laser-induced fluorescence technology.

Compared to most other detection methods, fluorescence detection offers greater detection sensitivity, lower limits of quantitation, and less interference from factors such as flow rate, temperature, and pressure. Limits of detection are generally in the mid to high pg/mL level, and in some cases can be in the fmol range. For compounds that are devoid of fluorescence properties, the formation of fluorescent derivatives of such compounds to improve assay sensitivity has received much attention. Several fluorescent derivatizing agents are available; these include dansylation reagents, phthaladehyde, and fluorescamine. Such derivatizations can be performed either pre-column or post-column. However, precolumn derivatization has the effect of reducing physicochemical differences in the analytes, and hence may reduce resolution in the column separation. Most derivatizations are performed post-column before passing through the detector. Post-column derivatizations need only increase the fluorescence of the species eluting from the column; thus, derivatizations resulting in more than one product, or even degradation products will not compromise the analysis. Fluorescence detection is particularly suitable for the detection and quantitation of drugs and their metabolites, since if the parent drug fluoresces, then its metabolites will most likely fluoresce.

## Mass spectrometry

*Overview*. Other than the interface to the chromatograph and larger vacuum pumping systems, there are few differences between mass spectrometry hardware for GC– and LC–MS. Thus, the mass analyzers and detectors described in the previous chapter on GC–MS are all applicable to LC–MS. Some commercial systems may be configured for either GC or LC interfaces with minor adjustments required when converting from one to the other.

*Interface/ionization sources*. Sample molecules are in solution as they emerge from the HPLC column. Before introduction into the high vacuum of the mass spectrometer, most of the solvent molecules must be removed and the sample molecules ionized. Heat, nebulization with gas, and differential pumping are techniques used to remove the solvent. Ionization of the sample molecules can occur by a variety of different methods as described in this section. Some methods, like particle beam, have distinct ionization steps while others, like electrospray, have ion formation inherent in the process. With the exception of particle beam, LC–MS ionization methods are "soft" like chemical ionization and produce mostly molecular adduct ions.

*Electrospray (ES)*. Perhaps one of the more intriguing methods of ionization, ES simply tranfers an existing ion in solution into the gas phase by using a high electric field [28]. Figure 4 is a schematic of an ES interface. The LC effluent passes through a capillary needle that is maintained at an high applied voltage (2–5 kV). At low flows (1–10 μl/minute), the high electrical field at the tip produces a mist of charged droplets at atmospheric pressure. For higher flow rates, a nebulizing gas delivered coaxially to the needle assists production of the mist. As evaporation decreases droplet size, ions are ejected as depicted below for sample molecules M and solvent molecules S:

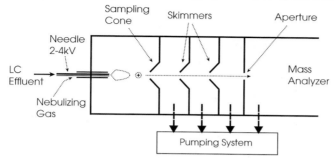

Fig. 4  Schematic diagram of ES ionization source interfaced to a mass spectrometer.

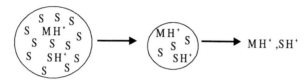

Of course, the analyte must already exist as an ion in solution but this is easily accomplished by adjusting the pH. Positive ions are shown above, but the process works equally well for negative ions. The ions are swept through a sampling cone and passed through a differential pumping system to remove solvent and nebulizing gas molecules before the ions enter the mass analyzer through a tiny aperture.

ES is a very gentle ionization method and can be used with either small or large molecules. Figure 5 shows an ES spectrum of the glucuronide conjugate of cotinine. This thermally labile compound produces $(M+H)^+$ as the base peak at $m/z$ 353. In comparison, a thermospray (discussed below) mass spectrum of this same compound yields mostly the protonated aglycon fragment [29–30]. An interesting feature of ES is the ability to produce ions with multiple charges. For macromolecules such as proteins, and the consequence of plotting $m/z$ in the mass spectrum, this permits the available mass range of the analyzer to extend to thousands of Daltons. (For example, a compound with $(M+H)^+$ at $m/z$ 5,000 would have $(M+5H)^{+5}$ at $m/z$ 1,000.) This feature is less useful for small molecules such as nicotine and its metabolites. ES works well with aqueous mobile phases and some organic solvents. To achieve the low flows necessary with ES (usually <100 µL/min) when using conventional HPLC methods with high flows (0.5–2 mL/min), the effluent is usually split prior to the interface.

*Atmospheric pressure chemical ionization (APCI).* APCI shares the same atmospheric pressure interface as ES but uses a different probe for producing ions [31]. Instead of the capillary needle in Fig. 4, APCI introduces the LC effluent to a heated tube (400–550°C) where the solvent and samples are volatilized at atmospheric pressure. A nebulizing gas (nitrogen) is used

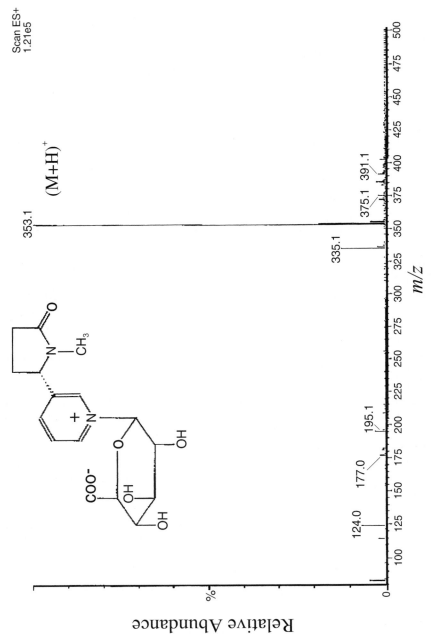

Fig. 5  ES mass spectrum of cotinine-*N*-glucuronide taken with a quadrupole mass analyzer.

to assist with volatilization. A discharge needle after the heated tube creates a plasma where reagent ions and electrons are produced. Chemical ionization occurs in the gas phase to produce protonated and other molecular adducts. APCI is rugged, reliable, and very sensitive due to efficient ionization. It works well for molecules with heteroatoms such as nicotine and its metabolites. The rapid heating does produce some thermal degradation of labile species. APCI can accommodate HPLC flows of 2 mL/min and works well with most types of solvents.

*Thermospray (TS).* One of the first widely applied interfaces developed for LC–MS was TS [32]. The LC effluent passes through a narrow metal tube whose tip is maintained at a high temperature (100–200°C) by resistive heating. The heated liquid volatizes and forms a spray in a reduced pressure chamber. Ions in solution are desorbed from the shrinking droplets in the spray; some gas phase chemical ionization reactions also occur. The LC effluent must contain an ionizing buffer such as ammonium acetate to produce ions and precise temperature control of the probe is crucial for TS operation. An optional filament placed near the probe tip can be used to increase ionization in some cases. The spray passes a conical aperture where ions are directed into the mass analyzer. TS works best with aqueous mobile phases such as those used in reversed-phase HPLC. Numerous applications of TS LC–MS have been published [8] including those for nicotine and its metabolites [9,33–34]. The strengths of TS are its simplicity and broad applicability to many types of samples. Like APCI, however, the high temperatures required for volatilization can degrade thermally labile samples. TS applications have decreased in the past few years due primarily to the success of APCI and ES techniques discussed previously.

*Particle Beam (PB).* The PB interface is not an ionization source as much as a solvent removal system prior to introducing the sample into a conventional EI source [35,36]. The LC effluent is first mixed with a gas (usually helium) to create an aerosol in a heated chamber. Selective evaporation and removal of solvent leaves a beam of sample molecules that pass through a series of apertures into the mass spectrometer source. There, ionization may occur by conventional EI or chemical ionization techniques. Unlike other LC–MS interfaces, PB produces EI spectra, which are reproducible and searchable by computer; this makes PB invaluable for identification. PB is limited to small molecular weight samples (<1,000 Daltons) and sensitivity can be low for certain applications. Sensitivity may be enhanced by using volatile organic buffers [37]. PB maintains a small market niche for certain industrial and environmental applications.

*Continuous flow fast atom bombardment (CFFAB).* This ionization technique is sometimes called dynamic FAB or liquid secondary ion mass spectrometry (SIMS). It allows the LC effluent to flow directly across the target area of a FAB probe [38]. The target is bombarded with fast atoms or ions and the sample is ionized by a desorption process. The LC effluent enters the probe at a very low flow rate (<10 μL/min). It is mixed with a matrix material such as glycerol to facilitate the ionization process. Like ES described above, it is a very gentle method of ionization. CFFAB is usually used on sector instruments with capillary LC columns for separation. CFFAB is not preferred as a general detector because ionization of the sample depends on the matrix. It is good for certain classes of compounds such as peptides. Because CFFAB tends to be sample specific, it has limited applications with no examples of analysis of nicotine or its metabolites.

*Mobile phase compatibility.* For all of these interfaces, an important consideration in coupling an HPLC method to a mass spectrometer is mobile phase compatibility. Popular LC–MS mobile phase solvents are water, methanol, and acetonitrile due to their wide range of solubilities and low molecular weights for reduced background ions in the mass spectrometer. Organic modifiers such as triethylamine, which are often used to improve chromatography, should be avoided since they can suppress ionization of sample molecules. While many of the interface/ionization sources listed here prefer a buffer at low concentrations (<20 mM), it must be a volatile organic buffer such as ammonium formate or ammonium acetate that will not leave deposits that block the apertures leading to the mass analyzer. Some recent work has been done on two stage orthogonal sampling that permits 'harsh' mobile phases like potassium phosphate buffers to be used (for example, the commercial 'Z-spray' interface offered by Micromass, Manchester, UK). This is extremely useful when adapting existing HPLC methods to a mass spectrometer.

*Mass analyzers.* LC–MS uses the same type of mass analyzers (quadrupoles, ion traps, sectors, FT–MS, TOF) that have been described in the previous chapter. Since LC–MS can ionize very large molecules, the upper scanning range of a mass analyzer is more important here than for GC–MS. Most sector and quadrupole instruments can achieve $m/z$ 2,000–4,000 as an upper limit. For applications on small molecules such as most nicotine metabolites, however, upper mass range is less critical since there are no singly charged analyte ions over $m/z$ 500. While there are some LC–MS systems built around sector instruments, most systems for LC–MS use quadrupole analyzers for the same reason as do GC–MS systems: they are relatively inexpensive, durable, and can scan rapidly. Recent reports on use of time-of-flight (TOF) analyzers for LC–MS have shown that by coupling them with either sectors [39], or quadrupoles [40], high resolution mass spectrometry on fragment ions is possible. In addition to exact mass measurements, TOF offers very rapid scanning and increased sensitivity. Such advantages have led to commercial quadrupole-TOF systems that, while expensive, hold much promise.

## III. SAMPLE PREPARATION

The quantitative measurement of nicotine and other tobacco alkaloids is often carried out on a wide range of biological matrices. These include plant material, liquid biological samples, such as urine, bile, sweat, tears, saliva, and cerebrospinal fluid, and solid/liquid mixtures, such as plasma, serum, blood, feces, and solid tissue samples.

### Plant material

Sample preparation of plant material initially involves drying the tissues. This can be carried out by drying in air, in a heated oven, vacuum oven, or by direct lyophilization. The latter method is preferred, since minimal chemical degradation (i.e. oxidation) occurs under freeze-drying conditions. The dried material is then powdered and subjected to an extraction procedure to isolate the alkaloidal components. This usually consists of treating the powder with aqueous mineral acid (at this stage, an internal standard can be added) homogenizing or sonicating the resulting mixture, centrifuging, and extracting the supernatant with an organic solvent such as chloroform or dichloromethane to remove non-alkaloidal components. The acidic aqueous layer is then adjusted to pH 11–12 with aqueous NaOH, and the free base form of the alkaloids extracted with organic solvent (usually chloroform, dichloromethane, or ethyl acetate). After drying and removal of solvent, the resulting alkaloidal residue is taken up in methanol and analyzed.

Some methodologies incorporate a Soxhlet extraction step by initially extracting the dried plant material contained in cellulose thimbles with hot organic solvent prior to analysis. Other more recent methodologies utilize small commercially available pre-packed columns to clean up the sample prior to analysis; this procedure shortens considerably the sample preparation time. Methods have also been developed for the rapid GC analysis of alkaloids in tobacco plants, which may have utility in high through-put screening studies [41].

### Animal material

Preparation of analytical samples from animal origin is dependent upon the nature of the particular matrix being investigated. Most biological samples are complex mixtures consisting of many different components (e.g. salts, proteins, lipids, nucleic acids, etc.) that may compromise the analytical procedure. The most common biological fluids analyzed are whole blood, plasma (or serum), and urine. Less common, are saliva, cerebrospinal fluid bile, sweat, and milk. Generally, the ease with which samples can be analyzed increases with their fluidity. Thus, cerebrospinal fluid is one of the least problematic fluids while whole blood and tissue are the most difficult.

*Blood.* Blood is a complex mixture of proteins, dissolved fats and solids, and suspended cells. Simple centrifugation can remove the red blood cells to afford a clear solution, the plasma. Care must be taken in ensuring that the red blood cells remain intact during separation, since rupturing the cells releases components that could make subsequent analysis more difficult. Thus, heating, freezing, mechanical manipulation and the use of buffers of high ionic strength, should be avoided. In this respect, if blood samples need to be diluted, isotonic saline is utilized. Blood coagulates on standing to form a clot and a clear liquid, the serum. Serum is, in most respects, similar to plasma but does not contain the soluble factors that lead to clotting; because of this, differences may occur in the analysis of drugs and metabolites in serum compared to plasma. Drugs and metabolites are generally not extracted directly from whole blood, but usually from plasma or serum.

*Plasma and serum.* The major problem with plasma and serum is the presence of large amounts of protein in the matrix. Protein macromolecules may bind to drugs and metabolites present in the sample and compromise their analysis. The presence of proteins may also affect the efficiency of the chromatography used in the assay procedure. Thus, methodologies to remove proteins from the sample and to release protein-bound drugs are required to allow the extraction of total drug for analysis.

While heating plasma samples to 90°C for 15 min or applying freeze-thawing cycles have been used to precipitate plasma proteins, these methods are not very efficient, may decompose the analyte, and are very time consuming. The most common methods for removal of plasma proteins prior to drug or metabolite extraction, is to denature and precipitate the protein and isolate the filtrate, or supernatant. The precipitated protein usually loses its drug binding properties on denaturation. Most popular protein precipitation agents are acidic reagents such as trichloroacetic acid, perchloric acid, and tungstic acid; however, these reagents are precluded in the analysis of acid-labile drugs and metabolites. Alternative reagents, such as ammonium sulfate or ammonium chloride, and organic solvents such as methanol, ethanol or acetonitrile, are more widely used. Usually, two volumes of organic solvent are sufficient to precipitate all plasma proteins. When added to plasma samples, an equal volume of acetonitrile saturated with sodium-bisulfate/sodium chloride affords an upper phase that can be analyzed directly [42]. Proteolytic enzymes can also be used for protein precipitation, which is useful for the analysis of chemically labile drugs and metabolites. For example, subtilisin, trypsin, and papain, have all been used for the preparation of plasma samples prior to analysis [43].

Lipid content in plasma and serum samples can also present a problem in analysis, if the content is high enough (fatty acid concentrations can vary with timing and nature of meals in clinical studies). Such samples will require a specific partition step in the work-up procedure to remove the lipids, and it is advisable to incorporate such a step as part of the general analytical protocol for all plasma samples.

While the binding of drugs and metabolites to protein represents a major analytical problem, such binding is reversible, and at an appropriate pH it should be possible to extract the drug with an organic solvent without the necessity for protein precipitation. Thus, if the partition of the unionized drug into the organic phase is sufficiently high, then the binding equilibrium can be shifted sufficiently to allow for efficient extraction of the drug from the aqueous environment. Such a procedure would entail adjusting the pH with a minimum amount of buffer, and then extracting the sample with a relatively large volume of organic solvent. This approach avoids a filtration or centrifugation step which makes possible its adoption as a routine procedure for the analysis of large numbers of samples, and a method of choice for laboratories supporting toxicology and clinical pharmacology studies.

The need for high throughput analysis of clinical samples has necessitated the search for simplifying the work-up of plasma samples prior to analysis. In many current methodologies, the initial precipitation and extraction step has been replaced by adsorption of the analyte in solution onto a solid matrix followed by elution with a much smaller volume of solvent. Several methods allow the raw plasma sample to be added to commercially available, short pre-packed cartridges of adsorbent, the nature of the matrix (i.e. charcoal, silica, reversed-phase, ion-exchange, etc.) being determined by the chemical composition of the drug(s) or metabolite(s) in the sample. Such a concentration and extraction procedure can been incorporated into a unified chromatographic analytical procedure to afford a completely automated elution sequence methodology, using appropriate column switching techniques. Examples of this technology are the Advanced Automation Sample Preparation (AASP) system of Waters Associates, and the Automatic Sample Preparation with Extraction Columns (ASPEC) from Gilson. The use of minicolumn cartridges combined with unified chromatographic elution schemes has been reported for a number of drugs and their metabolites in serum, plasma, saliva, and urine.

*Urine.* Compared to plasma, urine is relatively free from protein and lipids. Thus, it can usually be extracted directly with organic solvents, or in some cases, analyzed directly. Urine can be variable in appearance, being sometimes dark amber in color with overnight samples, or pale yellow to colorless for samples collected during the day. The composition is dependent largely on the diet. Since most endogenous components in urine are water soluble, drugs, which are usually lipid soluble, can be extracted relatively free from endogenous contaminants. Urine that is left to stand for long periods of time will slowly lose carbon dioxide, and will become more alkaline due to bacterial conversion of urea to ammonia, resulting in the precipitation of inorganic salts such as phosphates, and possible co-precipitation, or degradation of the analyte(s). Thus, stored urine may have a different composition compared to fresh urine, and the analytical methodology utilized and assessment of the stability of the analyte in these changed matrices must be taken into account. The use of mini-column cartridges in the analysis of drugs and metabolites in urine is currently the method of choice, and small centrifugation vials that incorporate a filtration/elution matrix are now available for the easy processing of urine samples for direct analysis.

*Saliva.* Saliva is relatively free from interfering substances. Although it is a somewhat viscid material with low viscosity, it can easily be extracted with organic solvents, and is amenable to direct analysis of analytes. For some drugs, levels of drug in saliva are thought to reflect the concentration of non-protein bound drug in the blood; for example, non ionized drugs such as steroids, should readily pass from plasma to saliva, and may represent a non-invasive method of monitoring. No significant problems have been reported in the analysis of saliva samples using existing methodologies developed for the analysis of plasma and serum. Direct analysis of nicotine and nicotine metabolites in saliva from smokers has been reported after a simple centrifugation step, followed by dilution with water [44]. Minor problems may occur with some drugs that bind to the mucoid proteins in saliva [45]. Normal saliva has a pH of 7.0–7.8; however, it should be noted that the pH of saliva has a dramatic effect on the distribution of drugs in saliva, and especially the amount of drug bound to protein. Thus the pH of the saliva sample should always be taken prior to analysis.

*Cerebrospinal fluid.* Analysis of drugs in cerebrospinal fluid is important for centrally acting drugs, since it provides concentrations of the drug and its metabolites in the vicinity of the receptors/enzyme active sites that are responsible for the drug's pharmacological properties. The concentrations of drug and metabolite(s), and the type of metabolite(s) detected may be quite different to those found in the plasma. While cerebrospinal fluid is more difficult to obtain, the techniques that have been developed for the analysis of drugs in plasma and serum appear to be readily applicable to this type of matrix.

*Bile.* Bile, like cerebrospinal fluid, is not frequently used for the analysis of drugs because of the difficulty in obtaining samples and the fact that the collection method itself may compromise the process under study. General methodologies for the analysis of plasma and serum samples can be utilized; however, the sample should be made alkaline before extraction with a mixture of dichloromethane:propanol. The residue can then be reconstituted in buffer prior to analysis.

*Milk.* Although not often reported, analysis of milk is important when one needs to determine if drug molecules can be transferred from mother to infant via this route. The major problem with this matrix is the presence of large amounts of fat in the sample (approximately 4.5% in mature human breast milk). Treatment of the sample with lipase enzymes prior to HPLC analysis has been reported to be effective for some drug analyses. However, it is recommended that a defatting step be included in any general work-up procedure; for example, removing the fats by washing 0.5 mL of breast milk, buffered at pH 3.0–4.0, with 10 mL of *n*-hexane and then following the usual procedure for plasma analysis appears to be an appropriate sample preparation methodology.

## IV. QUALITATIVE LC–MS

### Molecular weight information (molecular adducts)

Mass spectra produced by the different LC–MS interfaces all look basically the same (with the exception of PB with EI ionization) since they originate from soft ionization processes. The base peak is usually a molecular adduct like $(M+H)^+$ but $(M+NH_4)^+$ frequently occurs when an ammonium salt is used as a buffer. In addition, sodium and potassium present in biological samples or synthetic productions can produce $(M+Na)^+$ and $(M+K)^+$. Frequently, all of these adducts are produced with $(M+H)^+$ dominant. These adducts are very useful in determining the molecular weight but offer no information on structure.

### Structural information from fragmentation

Because of the limited information provided by molecular adducts, collision-induced dissociation (CID) of adduct ions is perhaps more critical in LC–MS than GC–MS. In the previous chapter, CID was explained using a tandem mass spectrometer or ion trap. But for LC–MS, a type of CID is also possible, prior to the analyzer for the atmospheric pressure interfaces (APCI and ES), using only a single stage mass analyzer. This in-source CID is accomplished by increasing the voltage on the sampling cone (see Fig. 4). The adduct ions produced in the relatively high pressure of the source are accelerated with sufficient energy so that collisions produce fragment ions before reaching the sampling cone. Figure 6 shows ES mass spectra of the metabolite nicotine-$N'$-oxide taken using a single stage quadrupole analyzer. With 5 V applied to the cone, the base peak is $(M+H)^+$ at $m/z$ 179 with a minor fragment at $m/z$ 161 (loss of $H_2O$). In addition, a proton-bound dimer appears at m/z 357, which is somewhat concentration-dependent. Increasing the voltage to 30 V produces fragments from opening of the pyrrolidine ring at $m/z$ 132 (probable loss of O and $CH_3NH_2$) and $m/z$ 130 (loss of an additional $H_2$). The intact pyrrolidine ring appears at $m/z$ 84 and little dimer is observed. With 45 V, the $m/z$ 132 fragment becomes the base peak. Thus, without the use of

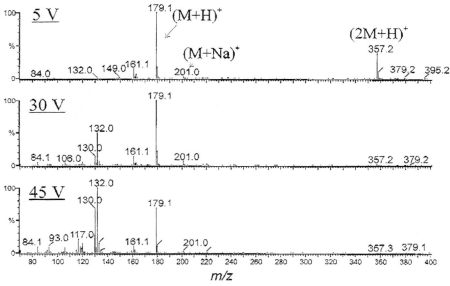

Fig. 6 APCI mass spectra of nicotine-$N'$-oxide taken on a quadrupole instrument at cone voltages of 5 V, 30 V, and 45 V.

tandem mass spectrometry, structural information from CID is possible. For specifically linking product ions with precursors, however, tandem mass spectrometry is required.

Tandem mass spectrometry, the linking of two mass analyzers with a collision cell, can operate in four different modes: product ion scan, precursor ion scan, constant neutral loss scan, and multiple reaction monitoring (MRM) mode. Either quadrupole or sectors can be used as stages, or ion trap technology can be applied as described in the previous chapter. The examples of each mode will be described here using a triple stage quadrupole mass analyzer.

Product ion scans are typically applied to molecular adduct ions as described in the previous chapter and result from CID. This is also known as 'MS/MS' since a mass spectrum is taken from an ion in a mass spectrum. The first quadrupole stage selects the ion of interest and transmits it to a collision cell containing an inert gas such as argon at a relatively higher pressure. Fragmentation of the ion is controlled by the voltage applied to the ion entering the cell. The fragment ions produced are sent to the third stage, a quadrupole mass analyzer that records the spectrum. As an example, the isobaric (same molecular weight) metabolites cotinine-$N$-oxide and *trans*-3'-hydroxycotinine give similar spectra by ES LC–MS with $(M+H)^+$ as the base peak at $m/z$ 193. Yet by selecting $m/z$ 193 and colliding it with argon, quite different CID spectra are obtained as shown in Fig. 7. For *trans*-

248

3′-hydroxycotinine, the major fragment is protonated pyridine at $m/z$ 80 while cotinine-$N$-oxide gives protonated pyridine-$N$-oxide at $m/z$ 96. It should be pointed out that when using in-source CID as described above to produce a fragment ion prior to the first analyzer in a tandem mass spectrometer, a type of MS/MS/MS is possible.

For precursor ion scans, the quadrupole after the collision cell is set to transmit a particular fragment ion and the first quadrupole scans for precursors. Thus, ions in the resulting mass spectrum are precursors of that fragment. Figure 8 is an example of a parent ion scan for $m/z$ 163 in the mass spectrum of nicotine-$N$-glucuronide. It is not surprising that the major parent ion is the protonated parent at $m/z$ 339; however, it is interesting

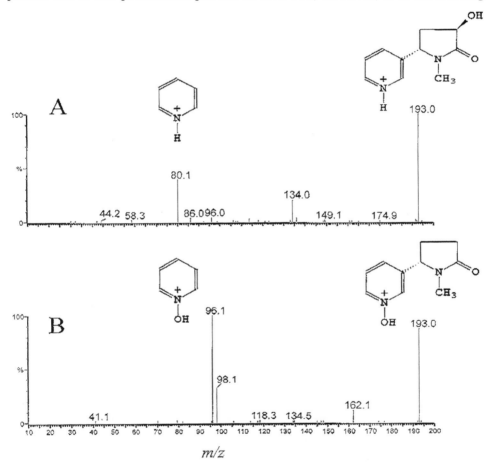

Fig. 7 CID mass spectra of (A) protonated *trans*-3′-hydroxycotinine and (B) protonated continine-$N$-oxide. Data were taken on a triple stage quadrupole mass spectrometer with ES source. Collision energy was 20 eV with argon as the collision gas at 0.001 millibar.

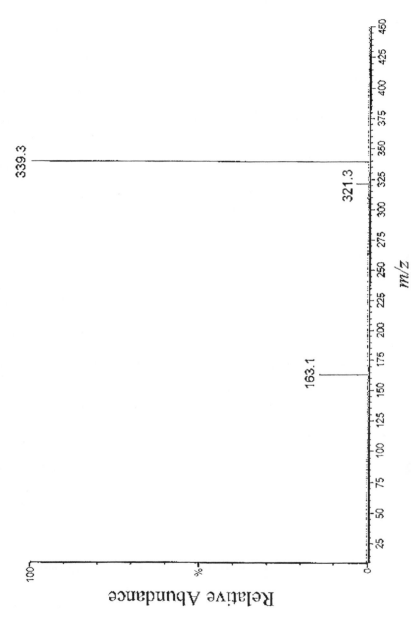

Fig. 8 Precursor ion scan for $m/z$ 163 on nicotine-*N*-glucuronide. Spectrum was taken on a triple stage quadrupole mass spectrometer with an ES source.

that some contribution to $m/z$ 163 comes from the protonated dehydrated compound at $m/z$ 321.

Similar to parent ion scanning is constant neutral loss scanning. As the name describes, a specific mass loss is monitored during the CID process. This is accomplished by scanning both the first and second analyzers with a constant mass offset between them. This could be applied, for example, to glucuronide conjugates which lose the glucuronic acid portion (176 Daltons) during CID.

The multiple reaction monitoring (MRM) mode is analogous to the SIM mode described previously but more specific. The first quadrupole is set to transmit a particular ion and the second quadrupole is set to transmit only a particular fragment. Like SIM, no time is wasted scanning over parts of the spectrum where no ions of interest occur. This mode is used when a particular analyte is targeted for detection. It is commonly used in quantitation schemes as shown in later examples and has very high sensitivity and selectivity.

## Chromatographic information

It is generally said that use of mass spectrometry for detection means that the chromatography does not have to be as good. But chromatography cannot be ignored since mass spectrometry cannot always eliminate chemical interferences with a sample. Also, chromatographic information from a good HPLC method is useful in establishing identity. The most obvious case is co-elution of unknowns and standards. An HPLC method with good resolution can also provide an indication of the complexity of a sample. Additional information on the polarity of analytes is available from using a reversed-phase method with a gradient that runs from mostly aqueous to mostly organic composition. In such a method, metabolites should elute before the parent compound with very polar materials like glucuronides or *N*-oxides eluting very early and slightly polar compounds such as demethylated molecules eluting just before the parent compound. HPLC method building is covered in several popular texts [46–48]. The general chromatography of alkaloids in particular is also described [49]. The initial stages of methods development can be done with a less specific detector such as UV absorbance.

## Qualitative strategies

The identification of a compound in a mixture usually begins with a full scan analysis after sufficient chromatographic resolution and retention has been achieved. The lower limit of the scan range is selected to avoid background ions from the mobile phase, usually $m/z$ 85 or higher (for example, the proton-bound dimer for acetonitrile occurs at $m/z$ 83). The upper range is determined by the compound of interest. Normally, more than twice the protonated molecular weight is useful in order to observe any dimers that might form. For small molecules such as nicotine, a scan range of $m/z$ 85-500 is usually sufficient.

As with GC–MS, reconstructed ion chromatograms (RIC's) can be used to locate specific components within a complex total ion chromatogram (TIC). To improve sensitivity for a particular component, selected ion monitoring of characteristic ions may be used. If tandem mass spectrometry is available, even more specificity is possible by using the MRM mode. For example, although $m/z$ 177 is specific for $(M+H)^+$ for cotinine, the CID transition of $m/z$ 177 → $m/z$ 80 is even more specific for that structure.

## Confirmation

Confirmation of the identity of a compound should involve matching both retention time and mass spectra with a known standard. This two dimensional approach provides high confidence in the identity of an unknown component in a sample. Confidence can be increased by using additional mass spectral techniques such as CID and also by using an alternative HPLC method.

## V. QUANTITATIVE LC–MS

Quantitative techniques for LC–MS are the same as for GC–MS described in the previous chapter. To emphasize two key points: (1) good chromatography is important for good quantitative data; (2) the use of labeled internal standards and operating the mass spectrometer in SIM or MRM modes greatly improves precision and takes advantage of mass selective detection.

## VI. APPLICATION EXAMPLES

### Serum cotinine by LC–MS/MS using APCI

The analysis of low concentrations of analytes in complex matrices may be approached in different ways. One approach seeks to perform extensive sample clean-up in order to present a relatively clean sample to the instrument. While this is labor intensive and requires meticulous attention to handling of sample, it permits the analytical system to operate longer and more reproducibly between servicing. A good example of this approach is the determination of cotinine in serum by the method of Bernert et al. [50]. They desired to measure exposure to nicotine for large numbers of samples from both smokers and non-smokers. A rapid, sensitive LC–MS/MS method with APCI was developed for serum cotinine that utilized sample extraction and concentration. Although sample preparation has many steps, the results of this method are impressive. The HPLC method had a retention time of less than one minute for the analyte and internal standard, which permitted samples to be injected every two minutes. Under routine operation, 100 samples a day were analyzed. A detection limit of 0.05 ng/mL was achieved which was sufficient to monitor serum cotinine in non-smokers exposed to second hand smoke.

Sample preparation began with addition of the internal standard, methyl-$d_3$-cotinine, to 1 mL serum. The sample was acidified and centrifuged to remove protein. The supernatant was basified and extracted with methylene chloride, dried, redissolved for transferal to a microvial, evaporated to dryness again, and finally reconstituted in 20 μL of toluene. For LC–MS/MS analysis, 10 μL were injected onto a 4.6 mm × 3 cm $C_{18}$ column for an isocratic separation with a flow rate of 1 mL/min of 80% methanol and 20% 2mM ammonium acetate. The effluent was introduced to an APCI source on a triple stage quadrupole mass spectrometer set to operate in the MRM mode. Ions monitored were $m/z$ 177 → $m/z$ 80 for cotinine and $m/z$ 180 → $m/z$ 80 for the internal standard. For a confirming transition for cotinine, $m/z$ 177 → $m/z$ 98 was also monitored.

The key features of this method's performance were sample extraction and concentration followed by very specific detection. Recoveries were 60–70% and most background materials were removed. Reducing the final volume to 0.02 mL resulted in a >10-fold concentration of the sample. Highly specific MRM analysis and focusing on only one analyte reduced chromatography requirements so that retention time could be minimal and an isocratic method could be used. The latter eliminated column equilibration time after the run, which is necessary for gradient methods. Thousands of serum samples were analyzed by this method but only for cotinine. The mass spectral parameters of this method could be modified to include nicotine since it could be extracted by this same procedure. Extension to other more polar metabolites of nicotine such as *trans*-3'-hydroxycotinine would be problematic as the method now exists since they are more difficult than cotinine to efficiently extract into an organic phase; a poor recovery would result from the use methylene chloride in the method described above.

## Determination of nicotine and several metabolites in smokers' urine, using LC–MS with TS

A different approach from the above is to perform minimal sample preparation and let the specificity and sensitivity of the instrument work to measure the analyte. Keeping the sample close to its original state minimizes chances of analyte loss prior to detection, especially when numerous analytes are involved. An example of such a method is that of McManus et al. [9]. Though neither as rapid or sensitive as the method above, it is ambitious in the scope of analytes it seeks to determine.

The objective of this method was to determine as many known urinary metabolites of nicotine as possible to assess nicotine exposure. The method was developed before the widespread availability of APCI and ES and relied on a TS interface to a single quadrupole instrument. Still, it permitted direct injection of urine from smokers and used a 10 min reversed-phase HPLC gradient method for separation. The $(M+H)^+$ for numerous known nicotine metabolites were detected by SIM and standards were used to establish retention times. Sample preparation was minimal with addition of methyl-$d_3$ cotinine as internal standard and filtration through a 0.2 μm mesh. Because no extractions were involved in this approach, extraction efficiency was not an issue and all identified metabolites could be reasonably determined. Only four of 17 known metabolites were detected in smokers' urine initially (*trans*-3'-hydroxycotinine, cotinine, nicotine-$N'$-oxide, norcotinine).

The strength of this method is the determination of nicotine and several metabolites in a single run. By determining multiple metabolites at the same time, a better indication of nicotine absorption is provided by this approach since some individuals may produce more of one metabolite over another in their urine. The limit of detection of approximately 20 ng/mL makes this method useful for smokers but not for non-smokers. Also, undiscovered minor nicotine metabolites would be difficult to find by this approach. The time and resources saved by this approach are offset partially by the introduction of many contaminants to the analytical system that damage columns and foul ionization sources so that performance is eventually compromised. A later application of the method to urine samples from 11 smokers [33] used enzymatic hydrolysis to indirectly account for glucuronide conjugates of nicotine, cotinine, and *trans*-3'-hydroxycotinine. With these additional metabolites accounted for, the amount of total nicotine absorption estimated by this approach was found to approximate that predicted by the FTC method. Further modification of this method has been reported using slightly faster chromatography and APCI tandem mass spectrometry [51]. MRM detection of six metabolites was applied to saliva and serum samples from smokers in addition to urine.

## Radiometric methods for metabolite detection

Radiometric analysis has been utilized in several studies of nicotine metabolism, both in vitro and in vivo [6,24–27,52–58]. This method of analysis has provided evidence that nicotine is enantioselectively bio-transformed by guinea pig tissue homogenates into its R-(+)-$N$-methylated metabolite, by utilizing radiolabeled forms of both S-(−)- and R-(+)-nicotine isomers in tissue homogenate incubations, and analyzing the metabolic products by flow through radiochromatography [6,24–25]. Similar methodology has also been used to completely elucidate the in vivo $N$-methylation pathway for nicotine in this animal species, in both single and double-radiolabeled studies [26–27,52–54]. Such metabolites are quaternary ammonium salts, which are very water soluble and not extractable into organic solvents; they are often difficult to determine by routine analytical methodologies.

Radiometric techniques have also been used to determine the total urinary and plasma metabolic profile in animals, after nicotine exposure [55–59]. The advantages of using this technique are that sensitivity is high, mass balance can easily be determined, analysis of very water-soluble bio-transformation products can be achieved (e.g. Phase II conjugation products), new metabolites can be easily be detected, low sample size can be utilized, gradient elution can be used, and analysis is free from interfering endogenous substances. In addition, double-labeled molecules can be utilized in dual channel analysis, which can provide more information on the characterization of unknown metabolites. Unfortunately, the administration of radiolabeled substances is often difficult to carry out in human clinical studies.

One major problem in developing radiometric HPLC analytical procedures for determination of metabolites of enantiomerically pure radiolabeled drugs, which is not widely recognized, is the chirodiastaltic problem [60–61]. When $^{14}$C-(±)-nicotine is coinjected onto a number of different *achiral* HPLC columns along with a non-radiolabeled UV-absorbing standard of S-(−)-nicotine, the radiolabel eluting from the column is divided into two distinct peaks. One of these radiolabeled peaks coelutes with the UV-absorbing S-(−)-nicotine standard, the other one elutes considerably later. This phenomenon is illustrated in Fig. 9 and appears to be due to a differential enantiomeric association between like and unlike optical isomers of nicotine in an enantiomeric mixture where one isomer is greatly in excess over the other. It seems likely that more than one physico-chemical property (i.e. solubility in the mobile phase, adsorption onto the chromatographic matrix, solute self-association characteristics, etc.) may be involved. The net result is that the less predominant isomer appears to fit less favorably into the mobile phase in which the order is built around its structural antipode; thus it has a longer retention time on the HPLC matrix, and is effectively resolved. The possibility of association of the nicotine molecules into dimers, trimers, or oligomers, etc. may also be considered to explain this phenomenon. A form of differential enantiomeric association between like and unlike pairs of enantiomers would require that 'homodimers' formed only from R-(+)-enantiomers would have different retention characteristics than 'heterodimers' formed from an S-(−)- and an R-(+)-isomer. While this is possible, no evidence exists for small scale association of nicotine molecules, either in solution or as a pure liquid.

Although chirodiastaltic effects can complicate the analysis of metabolites by radiometric detection, since under appropriate conditions, metabolites may occur as 'double peaks', or may not coelute with a UV-absorbing

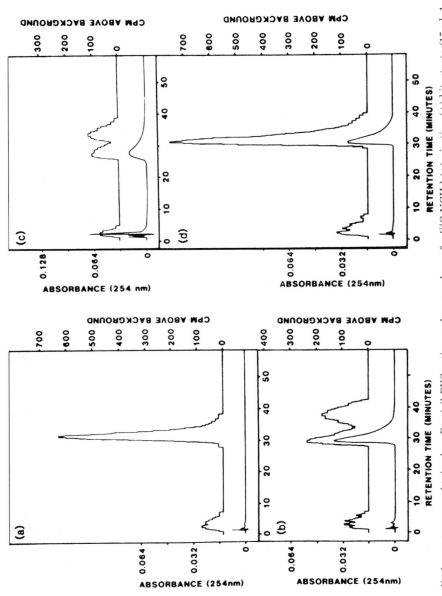

Fig. 9 Radiochromatograms obtained on Partisil-PXS cation-exchange column for [$^{14}$C-NCH$_3$]-($\pm$)-nicotine-($+$)-bitartrate (15 $\mu$l, 1 $\mu$Ci/ml): (a) alone; (b) with unlabelled ($-$)-nicotine (5 $\mu$l, 0.4 mg/ml); (c) with unlabelled ($+$)-nicotine (5 $\mu$l, 0.4 mg/ml); (d) with unlabelled racemic nitocine (5 $\mu$l, 0.4 mg/ml). Eluent: 0.06 m sodium acetate-methanol (70:30), Ph 6.8, 2.0 ml/min. Detection: 14C by Radiomatic Flo-1 HS; UV at 254 nm.

standard, this phenomenon has been successfully utilized in the determination of enantioselectivity in nicotine biotransformation pathways [26], and in the chromatographic resolution of radiolabeled forms of racemic nicotine [61].

## Pharmacokinetic studies

An extensive amount of work has been carried out over the last decade in studying the pharmacokinetics of nicotine in humans, and the bioavailability of nicotine via various routes of administration. Since the rate of absorption and distribution of nicotine profoundly affects the magnitude of nicotine's pharmacological response, distribution and metabolism studies are necessary in order to estimate nicotine intake from various sources of this alkaloid. Many methods have been developed for the analysis of human plasma and urine samples, and the advent of more sophisticated instrumentation and methodology has enabled the automated batch analysis of large numbers of samples from clinical studies. However, while these methods are both sensitive and selective, they still do not provide a means for the *simultaneous* determination of *all the major metabolites of nicotine* at levels found in both smokers and non-smokers.

Routine analytical procedures have been developed for plasma and urine nicotine, and for the analysis of primary metabolites such as cotinine, nicotine-*N'*-oxide, nornicotine, norcotinine, and for 3'-hydroxycotinine. These methods include GC, GC–MS, HPLC–UV, and HPLC–EC analysis (for reviews see [62–65]). In particular, the development of stable isotope methodology for determining the bioavailability of nicotine and nicotine metabolites by GC–MS has provided a wealth of data on the pharmacokinetic parameters and properties of the tobacco alkaloids [66,67].

In recent years, it has been shown that the nicotine metabolites mentioned above do not constitute all of the in vivo biotransformation products of nicotine. In fact, from early in vivo studies of mass balance using HPLC with radiometric detection, it was suggested that a glucuronide conjugate of 3'-hydroxycotinine might be a significant in vivo metabolite of nicotine [55]. Subsequent studies have shown that glucuronide metabolites of nicotine constitute almost 30% of total nicotine metabolites in smokers' urine, and include nicotine glucuronide, cotinine glucuronide and 3'-hydroxycotinine glucuronide [33]. Thus, these metabolites represent a major class of nicotine metabolite. The polarity and water solubility of such metabolites precludes their analysis by GC–MS. However, new developments in HPLC–MS technology [68], and the availability of deuterated forms of the glucuronide conjugates [30,69], have enabled these metabolites to be quantitated in smokers urine by direct determination.

HPLC with radiometric analysis has recently been used for the determination of brain nicotine and nicotine metabolites following acute and chronic peripheral radiolabeled nicotine administration in rats [70,71]. These studies have provided useful information on the pharmocokinetics of nicotine and nicotine metabolite transport into the central nervous system (CNS) from the periphery, as well as affording data on the metabolic fate of nicotine in brain. More recent results indicate that nicotine and at least five nicotine metabolites can be detected in brain after of a single bolus subcutaneous injection of $^{14}$C-nicotine [72]; the CNS pharmacokinetics for all five metabolites are shown in Fig. 10. Three of these metabolites have been identified as cotinine, norcotinine and nornicotine, and all have longer half-

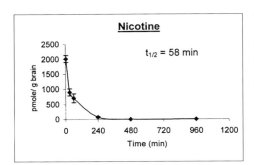

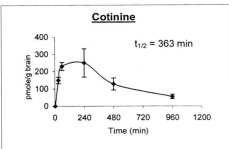

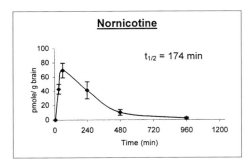

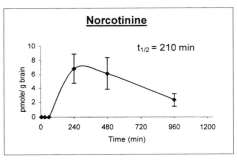

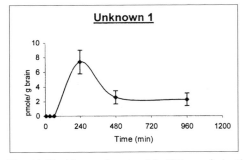

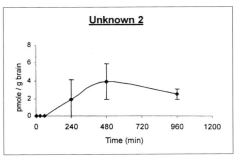

Fig. 10 Residence times and half-lives of nicotine and its metabolites in rat brain after a single subcutaneous injection of [2′-$^{14}$C]-nicotine (0.8 mg/kg, 50 μCi). Amounts were determined by HPLRC of brain supernatants on a Partisil-10 C$_8$ column.

lives than nicotine in the brain. Nornicotine, a pharmacologically active nicotine analog [73], was a major brain metabolite of nicotine in these studies. Thus, although it is only a very minor peripheral metabolite, nornicotine may contribute significantly to the neuropharmacological effects resulting from peripheral nicotine exposure [74].

## VII. PRACTICAL CONSIDERATIONS OF LC–MS

Improvements in commercial instruments in the past few years have made LC–MS as easy as GC–MS in operation. In addition, the cost of simple bench top systems are similar (approximately $100,000). There are two important considerations in order to get the most out of a LC–MS system. First of all, familiarity with HPLC is important, particularly for quantitative applications. Regardless of the detector, it is ultimately a LC peak that is integrated; sharp, well-resolved peaks provide the best data. Secondly, most of the interfaces produce molecular adducts that provide superb molecular weight information but little structural information. Thus, tandem mass spectrometry, as a triple stage system or an ion trap, may be worth the extra expenditure.

## VIII. RESOURCES

In addition to the periodicals and sites listed in the previous chapter, a very good web site for LC–MS is: http://www.lcms.com/

A book has been published recently that is very useful for people assessing LC–MS for their own needs [75].

## REFERENCES

1  Sander LC, Wise SA. Recent advances in bonded phases for liquid chromatography. *CRC Crit. Rev. Anal. Chem.*, 1987: **18**; 299–413.
2  Watson ID. Rapid analysis of nicotine and cotinine in the urine of smokers by high performance liquid chromatography. *J. Chromatogr.*, 1977: **143**; 203–6.
3  Maskarinec MO, Harvey RW, Caton JE. A novel method for the isolation and quantitative analysis of nicotine and cotinine in biological fluids. *J. Anal. Toxicol.*, 1978: **2**; 124–6.
4  Saunders JA, Blume DE. Quantitation of major tobacco alkaloids by high-performance liquid chromatography. *J. Chromatogr.*, 1981: **205**; 147–54.
5  Kyerematen GA, Damiano MD, Dvorchik BH, Vessell ES. Smoking-induced changes in nicotine disposition: application of a new HPLC assay for nicotine and its metabolites. *Clin. Pharmacol. Ther.*, 1982: **32**; 769–80.
6  Cundy KC, Crooks PA. High-performance liquid chromatographic method for the determination of *N*-methylated metabolites of nicotine. *J. Chromatogr. Biomed. Appl.*, 1984: **306**; 291–301.

7 Barlow RD, Thompson PA, Stone RB. Simultaneous determination of nicotine, cotinine and five additional nicotine metabolites in the urine of smokers using pre-column derivatisation and high-performance liquid chromatography. *J. Chromatogr.*, 1987: **419**; 375–80.

8 Yergey AL, Edmonds CG, Lewis IAS, Vestal ML. *Liquid chromatography/mass spectrometry*. New York:Plenum, 1990.

9 McManus KT, deBethizy JD, Garteiz DA, Kyerematen GA, Vesell ES. A new quantitative thermospray-LC/MS method for Nicotine and its metabolites in biological fluids. *J. Chrom. Sci.*, 1990: **28**; 510–516.

10 Ewing AE, Wallingford RA, Olefirowicz TM. Capillary electrophoresis. *Anal. Chem.*, 1989: **64**; 292A–303A.

11 Yang SS, Smetena I. Evaluation of capillary electrophoresis for the analysis of nicotine and selected minor alkaloids from tobacco. *Chromatographia*, 1995: **40**; 375–8.

12 Yang SS, Smetena I, Goldsmith AI. Evaluation of micellar ellectrokinetic capillary chromatography for the analysis of selected tobacco alkaloids. *J. Chromatogr. A*, 1996: **746**; 131–6.

13 Ralapati S. Capillary electrophoresis as an analytical tool for monitoring nicotine in AFT regulated tobacco products. *J. Chromatogr. B*, 1997: **695**; 117–29.

14 Lu GH, Ralapati S. Application of high-performance capillary electrophoresis to the quantitative analysis of nicotine and profiling of other alkaloids in ATF-regulated tobacco products. *Electrophoresis*, 1998: **19**; 19–26.

15 Nwosu CG, Crooks PA. Species variation and stereoselectivity in the metabolism of nicotine enantiomers. *Xenobiotica*, 1988: **18**; 1361–72.

16 Cundy KC, Godin CS, Crooks PA. Evidence of stereospecificity in the in vivo methylation of [$^{14}$C]-(+)-nicotine in the guinea pig. *Drug Metab. Disposit.*, 1984: **12**; 755–9.

17 Moore J, Greenwood M, Sinclair, N. Automation of a high-performance liquid chromatographic assay for the determination of nicotine, cotinine, and 3-hydroxy-cotinine in human urine. *J. Pharm. Biomed. Anal.*, 1990: **8**; 1051–4.

18 Kolonen SA, Puhakainen VJ. Assessment of the automated colorimetric and the high-performance liquid chromatographic methods for nicotine intake by urine samples of smokers, smoking low- and medium-yield cigarettes. *Clin. Chem. Acta*, 1991: **196**; 159–66.

19 Parviainen MT, Barlow RD. Assessment of exposure to environmental tobacco smoke using a high-performance liquid chromatographic method for the simultaneous determination of nicotine and two of its metabolites in urine. *J. Chromatogr.*, 1988: **431**; 216–221.

20 Ghosheh OH, de Leon J, Rogers T, Dwoskin LP, Crooks PA. An HPLC method for the quantitation of caffeine and major nicotine metabolites in human plasma. *Pharm. Sci.*, 1998: **1**; S574.

21 Mousa S, Houdi AA, Van Loon GR, Crooks PA. High-performance liquid chromatography with electrochemical detection for the determination of nicotine and *N*-methylnicotinium ion. *J. Chromatogr.*, 1985: **347**; 405–410.

22 Chien C-Y, Diana JN, Crooks PA. High performance liquid chromatography with electrochemical detection for the determination of nicotine in plasma. *J. Pharm. Sci.*, 1988: **77**; 277–279.

23 Chien C-Y, Diana JN, Crooks PA. Determination of plasma nicotine by high performance liquid chromatography. *LC-GC Mag.*, 1988: **6**; 53–55.

24 Cundy KC, Godin CS, Crooks, PA. Stereospecific in vitro *N*-methylation of nicotine in guinea pig tissues by an S-adenosylmethionine-dependent *N*-methyltransferase. *Biochem. Pharmacol.*, 1985: **34**; 281–4.

25 Cundy KC, Sato M, Crooks, PA. Stereospecific in vivo metabolism of nicotine in the guinea pig. *Drug Metab. Disposit.*, 1985: **13**; 175–85.

26 Kyerematen GA, Taylor LH, deBethizy JD et al. Radiometric-high-performance liquid chromatographic assay for nicotine and twelve of its metabolites. *J. Chromatogr.*, 1987: **419**; 191–203.

27 Kyerematen GA, Taylor LH, deBethizy JD et al. Pharmacokinetics of nicotine and 12 metabolites in the rat. *Drug Metab. Disposit.*, 1988: **16**; 125–9.

28 Kebarle P, Liang T. From ions in solution to ions in the gas phase. *Anal. Chem.*, 1993: **65**; 972A–86A.

29 Byrd GD, M. S. Uhrig MS, deBethizy JD, Caldwell WS, Crooks PA, Ravard A, Riggs RM. Direct determination of cotinine-*N*-glucuronide in urine using thermospray liquid chromatography/mass spectrometry. *Biol. Mass Spectrom.*, 1994: **23**; 103–7.

30 Caldwell WS, Greene JM, Byrd GD, Chang KM, Uhrig MS, deBethizy JD, Crooks PA, Bhatti BS, Riggs RM. Characterization of the glucuronide conjugate of cotinine: a previously unidentified major metabolite of nicotine in smokers' urine. *Chem. Res. Toxicol.*, 1992: **5**; 280–5.

31 Carroll DI, Dzidic I, Horning EC, Stillwell RN. Atmospheric pressure ionization mass spectrometry. *Appl. Spectrosc. Rev.*, 1981: **17**; 337–406.

32 Blakely CR, Carmody JJ, Vestal ML. A new liquid chromatograph/mass spectrometer interface using crossed-beam techniques. *Adv. Mass Spectrom.*, 1980: **8B**; 1616–23.

33 Byrd GD, Chang K-M, Greene JM, deBethizy JD. Evidence for urinary excretion of glucuronide conjugates of nicotine, cotinine, and *trans*-3'-hydroxycotinine in smokers. *Drug Metabol. Disp.*, 1992: **20**; 192–7.

34 Byrd GD, Davis RA, Caldwell WS, Robinson JH, deBethizy JD. A further study of FTC yield and nicotine absorption in smokers. *Psychopharmacology*, 1998: **139**; 291–9.

35 Willoughby RC, Browner RF. Monodispersed aerosol generation interface for coupling liquid chromatography with mass spectrometry. *Anal. Chem.*, 1984: **56**; 2626–31.

36 Cappiello A. Is particle beam an up-to-the-date LC–MS interface? *Mass Spectrom. Rev.*, 1996: **15**; 283–96.

37 Bellar TA, Behymer TD, Budde WL. Investigation of enhanced ion abundances from a carrier process in high-performance liquid chromatography particle beam mass spectrometry. *J. Am. Soc. Mass Spectrom.*, 1990: **1**; 92–8.

38 Caprioli RM. Continuous-flow fast atom bombardment mass spectrometry. *Anal. Chem.*, 1990: **62**; 477A–85A.

39 Keough T, Lacey MP, Ketcha MM, Batemen RH, Green MR. Orthogonal acceleration single-pass time-of-flight mass spectrometry for determination of the exact mass of product ions formed in tandem mass spectrometry experiments. *Rap. Comm. Mass Spectrom.*, 1997: **11**; 1702–8.

40 Morris HR, Paxton T, Panico M, McDowell R, Dell A. A novel geometry mass spectrometer, the Q-TOF, for low-femtomole/attomole-range biopolymer sequencing. *J. Protein Chem.*, 1997: **16**; 469–79.

41 Severson, F, McDuffie KL, Chaplin RF, Johnson AW. Rapid method for the analysis of tobacco nicotine alkaloids. *J. Chromatogr.*, 1981: **212**; 111–121.

42 Mathies, JC, Austin MA. Modified acetonitrile-protein precipitation method of sample preparation for drug assay by liquid chromatography. *Clin. Chem.*, 1980: **26**; 1760–5.

43 Holzbecher M, Ellenberger HA. Simultaneous determination of diphenhydramine, methaqualone, diazepam, and chlorpromazine in liver by use of enzyme digestion: comparison of digestion procedures. *J. Analyt. Toxicol.*, 1981: **5**; 62–66.

44 Byrd GD. Unpublished work.

45 Anavekar SN, Saunders, RH, Wardell, WM, Shoulson, I, Emmings, FG, Cook CE,Gringeri AJ. Parotid and whole saliva in the prediction of serum total and free phenytoin concentrations. *Clin. Pharmacol.*, 1978: **24**; 629–634.

46 Bildingmeyer BA. *Practical HPLC Methodology and Applications.* New York:Wiley 1992.

47 Snyder LR, Kirkland JJ, Glajch JL. *Practical HPLC Method Development.* New York:Wiley, 1997.

48 Neue UD. *HPLC Columns: Theory, Technology, and Practice.* New York:Wiley-VCH, 1997.

49 Popl M, Fahnrich J, Tatar V. *Chromatographic Analysis of Alkaloids.* New York: Marcel Dekker, 1990.

50 Bernert, Jr. JT, Turner WE, Pirkle JL, Sosnoff CS, Adkins JR, Waldrep MK, Ann Q, Covey TR, Whitfield WE, Gunter EW, Miller BB, Patterson, Jr. DG, Needham LL, Hannon WH, Sampson EJ. Development and validation of sensitive method for determination of serum cotinine in smokers and non-smokers by liquid chromatography/atmospheric pressure ionization tandem mass spectrometry. *Clin. Chem.*, 1997: **43**; 2281–91.

51 Byrd GD. LC–MS/MS method for profiling nicotine and its metabolites in biological fluids. *44th Annual Conference on Mass Spectrometry and Allied Topics*, Portland, OR, May 12–16, 1996.

52 Sato M, Crooks PA. *N*-Methylnornicotinium Ion, a new in vivo metabolite of R-(+)-nicotine. *Drug Metab. Disposit.*, 1985: **13**; 348–53.

53 Pool WF, Crooks PA. In vivo metabolism of R-(+)-[$^{14}$C-NCH3]-*N*-methylnicotinium ion in the guinea pig. *Drug Metab. Disposit.*, 1985: **13**; 578–81.

54 Pool WF, Crooks PA. Biotransformation of primary nicotine metabolites, III. Metabolism of R-(+)-[$^{3}$H-*N'*-CH$_3$; $^{14}$C-*N*-CH$_3$]-*N*-methylnicotinium acetate-use of double isotope studies to determine the in vivo stability of the *N*-methyl groups of *N*-methylnicotinium ion. *J. Pharm. Pharmacol.*, 1988: **40**; 758–62.

55 Pool WF, Houdi AA, Damani LA, Layton, JW, Crooks PA. Isolation and characterization of *N'*-methyl-*N'*-oxonicotinium ion, a new urinary metabolite of R-(+)-nicotine in the guinea pig. *Drug Metab. Disposit.*, 1986: **14**; 574–9.

56 Cundy KC, Crooks PA. Biotransformation of primary nicotine metabolites II. Metabolism of $^{3}$H-S-(−)-cotinine in the guinea pig: determination of in vivo urinary metabolites by high-performance liquid radiochromatography. *Xenobiotica*, 1987: **17**; 785–92.

57  Nwosu CG, Godin CS, Houdi AA, Damani LA, Crooks PA. Enantioselective metabolism during continuous administration of S-( − )- and R-(+)-nicotine isomers in guinea pigs. *J. Pharm. Pharmacol.*, 1988: **40**; 862–9.

58  Crooks PA, Sherrat AJ, Godin CS. Oxidative metabolism of nicotine enantiomers in cultured guinea pig hepatocytes. *Med. Sci. Res.*, 1992: **20**; 909–11.

59  Schepers G, Rustemeier K, Walk RA, Hackenberg. Metabolism of S-nicotine in non-induced and Aroclor-induced rats. *Eur. J. Drug Metabol. Pharmacokin.*, 1993: **18**; 187–97.

60  Crooks PA, Cundy KC. High performance liquid chromatography of radiolabeled nicotine enantiomers and their metabolism in the Guinea pig to *N*-methylated products. In: *Flow Through Radioactive Detectors in HPLC. Progress in HPLC*, Vol. 4, Parvez H, Kessler M, Parvez S (Eds.), Utrecht, Holland:VNU International Science Press, 1988, pp. 129–47.

61  Cundy KC, Crooks PA. Unexpected phenomenon in the high-performance liquid chromatographic analysis of racemic $^{14}$C-labeled nicotine: separation of enantiomers in a totally achiral system. *J. Chromatogr.*, 1983: **281**; 17–23.

62  Curvall M, Kazemi Vala E. Nicotine and metabolites: analysis and levels in body fluids. In: *Nicotine and Related Alkaloids: Absorption, Distribution, Metabolism, and Excretion.* Gorrod JW, Wahren J (Eds.), London: Chapman and Hall; 1993, pp. 147–79.

63  Jacob III P, Benowitz NL. Pharmacokinetics of (S)-nicotine and metabolites in humans. In: *Nicotine and Related Alkaloids: Absorption, Distribution, Metabolism, and Excretion.* Gorrod JW, Wahren J (Eds.), London: Chapman and Hall; 1993, pp. 197–218.

64  Benowitz NL, Jacob III P. Nicotine and cotinine elimination kinetics in smokers and non-smokers. *Clin. Pharmacol. Ther.*, 1993: **53**; 316–23.

65  Crooks PA. *N*-Oxidation, *N*-methylation and *N*-conjugation reactions of nicotine. In: *Nicotine and Related Alkaloids: Absorption, Distribution, Metabolism, and Excretion.* Gorrod JW, Wahren J (Eds.), London: Chapman and Hall; 1993, pp. 81–109.

66  Benowitz NL, Jacob III P, Denaro C. et al. Stable isotope studies of nicotine kinetics and bioavailability. *Clin. Pharmacol. Ther.*, 1991: **49**; 270–7.

67  Jacob III P, Yu L, Wilson, M. et al. Selected ion-monitoring method for determination of nicotine, cotinine, and deuterium-labeled analogs. Absence of an isotope effect in the clearance of (S)-nicotine–3′–3′-d$_2$ in humans. *Biol. Mass Spectrom.*, 1991: **20**; 247–52.

68  Byrd GD, Chang KM, Greene KM, de Bethizy JD. Determination of nicotine and its metabolites in urine by thermospray LC/MS. In: *The Biology of Nicotine.* Lippiello P, Collins AC, Gray JA, Robinson JH (Eds.), New York: Raven Press, 1992, pp. 71–83.

69  Byrd GD, Caldwell WS, Crooks PA, Ravard, A, Bhatti BS. Synthesis of (±)-*N*-β-D-glucopyranosyl-[methyl-d$_3$]-nicotinium inner salt and its use in the direct determination of nicotine-*N*-glucuronide in human biological samples. *J. Radiolab. Comp. Radiopharm.*, 1999; in review.

70  Crooks PA, Li M, Dwoskin LP. Determination of nicotine metabolites in rat brain after peripheral radiolabeled nicotine administration: detection of nornicotine. *Drug Metab. Disposit.*, 1995: **23**; 1175–7.

71 Crooks PA, Li M, Dwoskin LP. Metabolites of nicotine in rat brain after peripheral nicotine administration: cotinine, nornicotine and norcotinine. *Drug. Metab. Disposit.*, 1997: **25**; 47–54.

72 Ghosheh OH, Dwoskin LP, Crooks PA. Determination of the time course and concentrations of nicotine and its major brain metabolites following acute peripheral administration of nicotine. *Drug Metab. Disposit.*, 1999; in press.

73 Dwoskin LP, Buxton ST, Jewell AL, Crooks PA. S-(−)-Nornicotine increases dopamine release in a calcium-dependent manner from superfused rat striatal slices. *J. Neurochem.*, 1993: **60**; 2167–74.

74 Crooks PA, Dwoskin LP. CNS nicotine metabolites contribute to the neuropharmacological effects of nicotine and tobacco smoking. *Biochem. Pharmacol.*, 1997: **54**; 743–53.

75 Willoughby R, Sheehan E, Mitrovich S. *A Global View of LC-MS*. Pittsburgh:Global View, 1998.

*Chapter 8*

# Use of immunoassay techniques for the determination of nicotine and its metabolites

John J. Langone[1], Hilda B. Gjika[2] and Helen Van Vunakis[2]

[1] *Center for Devices and Radiological Health, Food and Drug Administration, Rockville, MD 20852, USA*
[2] *Department of Biochemistry, Brandeis University, Waltham, MA 02454, USA*

## I. BACKGROUND

Nicotine is present in tobacco and is also administered in non-tobacco forms via gum, skin patches and nasal sprays. It is metabolized by different enzymatic routes to produce several biotransformation products [1]. In several species, including humans, cotinine is a major metabolite (Fig. 1).

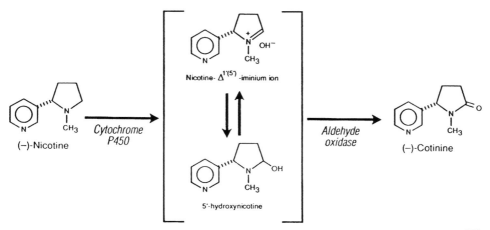

Fig. 1 Metabolism of (−)-nicotine to (−)-cotinine through the intermediate, nicotine-$\Delta^{1'(5')}$-iminium ion.

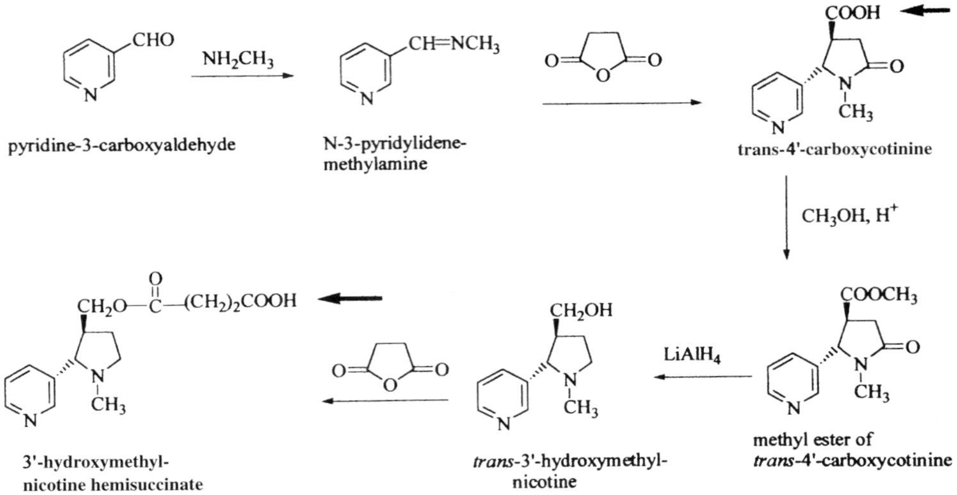

Fig. 2 Preparation of hapten derivatives: *trans*-4'-carboxycotinine and 3'-hydroxymethylnico-tine hemisuccinate are conjugated through their carboxyl groups to (a) macromolecules to prepare the immunogens or (b) to poly-L-lysine to coat microtiter plates for the ELISA and FIA assays.

the solution through two layers of Whatman #2 filter paper and evaporate under reduced pressure (rotary film evaporator) to obtain the imine as a yellow oil (30–35 g).

## Trans-4'-Carboxycotinine

Reflux *N*-3-pyridylidinemethylamine (25 g), prepared as above, and succinic anhydride (30 g) for 24 h in 100 ml xylene. After the mixture cools, decant the top layer and discard. Dissolve the residual brown oil in 5% sodium bicarbonate solution (approximately 300 ml), wash with two 250 ml portions of chloroform, and decolorize by swirling with 2 g activated charcoal. Filter the suspension through several layers of filter paper and remove traces of chloroform by heating the yellow filtrate on a steam bath. Carefully adjust the pH of the mixture to 4.7 with phosphoric acid to precipitate the product. Collect the crude carboxylic acid (15 g) by filtration and recrystallize from a minimum volume of boiling ethanol (yield 10–12 g of white or cream colored crystals, mp 193–195°).

Modification of the carboxyl group of *trans*-4'-carboxycotinine in three synthesis steps gives succinylated 3'-hydroxymethylnicotine, the derivative used to prepare the antigen for production of antibodies to nicotine [6].

## Methyl ester of trans-4'-Carboxycotinine

Add *trans*-4'-carboxycotinine (10 g) to methanolic sulfuric acid (2 N, 100 ml) and molecular sieve (5 g). Allow the mixture to stir at 20° overnight, then filter through two layers of filter paper. Add the filtrate carefully and in portions to an 8% solution of sodium bicarbonate (150 ml). Extract the solution with four 100-ml portions of chloroform. Dry the combined chloroform washes over anhydrous magnesium sulfate and concentrate on a rotary film evaporator to obtain the ester as a yellow oil. Crystallization from ether-acetone (3 : 1 v/v) gives 8.0 g of needles (mp 82–84°).

*Trans-3′-Hydroxymethylnicotine*
Add LiAlH$_4$ (3.6 g) *slowly* to 200 ml *dry* diethyl ether in a dried flask protected from moisture by a drying tube containing anhydrous sodium sulfate. Add a solution of the ester (6.0 g) in 25 ml ether, and allow the solution to stir at 20° overnight. Cool the solution in an ice bath and, while stirring, destroy excess LiAlH$_4$ by addition of cold water (3.0 ml added dropwise), followed by 15% sodium hydroxide (3 ml), and then water (10 ml). Filter the suspension through two layers of filter paper, dry the filtrate over anhydrous magnesium sulfate, and remove the ether by evaporation in a rotary film evaporator. The crude hydroxymethylnicotine is a pale yellow oil which can be used directly in the subsequent reaction. Alternatively, it can be purified by distillation through a short-path apparatus, bp 85–87° (10 μm).

*Succinylated trans-3′-Hydroxymethylnicotine*
Stir a solution of hydroxymethylnicotine (300 mg) and succinic anhydride (157 mg) in 30 ml dry benzene at 60° overnight and concentrate using a rotary film evaporator. The crude succinylated product is a semisolid that can be used without purification to prepare the antigen for production of antibodies to nicotine.

## Preparation of nicotine and cotinine macromolecular conjugates (antigens) for immunization

To prepare the antigens, a carbodiimide reaction is used to link the respective hapten derivatives via their free carboxyl groups to the free amino groups of macromolecules by an amide bond.

1-Ethyl-3-(3-dimethylaminopropyl)carbodiimide hydrochloride (EDCA) (10 mg) is added to an aqueous solution of succinylated hydroxymethylnico-tine or *trans*-4′-carboxycotinine (10 mg) and protein (KLH or human serum albumin) (10 mg) in a total volume of 1.0 ml water, and the pH adjusted to 6.8–7.2 with 0.01 N sodium hydroxide. After incubation at room temperature overnight, the antigen is separated from low-molecular-weight components by chromatography on a column (1.9 × 35 cm) of Sephadex G-50 wet-packed and eluted with saline-0.05 M phosphate buffer, pH 7.2. Fractions (1.5–2.5 ml) are collected and the tubes corresponding to blue dextran (Sigma Chemical Co., St. Louis, MO, USA) are pooled. Based on the difference in the absorption spectrum at 260 nm between the protein and the hapten-protein conjugates, the incorporation is 6–18 moles of hapten per mole of protein. The same hapten-protein conjugates are used for the production of polyclonal and monoclonal antibodies.

## Preparation of nicotine and cotinine poly-L-lysine conjugates for enzyme-linked immunosorbent assay (ELISA) and fluorescence immunoassay (FIA) [7]

A solution of succinylated hydroxymethylnicotine or *trans*-4'-carboxycotinine (1.5 mg) dissolved in 0.3 ml of 0.003 M phosphate, pH 6.35 (PB) is added to 1.0 ml of a 5 mg/ml solution of poly-L-lysine (PL; Mr 40 000) in PB at ambient temperature. EDCA (1.0 ml of 5 mg/ml solution in PB) is added dropwise with stirring and the solution allowed to stir at 4° for 16 h. The nicotine-PL and cotinine-PL conjugates are ready to use without purification. They may be stored for at least five months at 4° in the presence of 0.02% sodium azide, or frozen without azide in aliquots (e.g. 0.03 ml) at −80° for several months without loss of antibody binding activity.

## Production of antisera and monoclonal antibodies

Serum of an immunized animal contains a heterogeneous population of antibodies that vary in specificity and affinity, since they are the products of many different antibody-producing cells; i.e. the antibodies are polyclonal in origin. In contrast, monoclonal antibodies produced by hybidoma technology are chemically and immunologically identical (i.e. in terms of binding affinity and specificity) because they are produced by clones derived from single antibody-producing cells. Considerably less effort, skill and expense are required to obtain polyclonal antibodies, and they often prove to be excellent reagents for immunoassay.

Hybridoma technology is more complicated, time consuming and expensive than conventional immunization techniques. The procedure involves fusion of antibody-producing spleen cells with non-immunoglobulin (Ig)-secreting myeloma tumor cells, selection of individual hybrid cells (hybridomas) that secrete antibodies with desired specificity, and growing the hybridoma cells in tissue culture or in vivo (e.g. in ascites fluid). This technology has the advantage of yielding an unlimited supply of homogeneous antibody suitable for immunoassay [8].

*Immunization for production of polyclonal antibodies (antisera)*
New Zealand albino rabbits were injected in the foot pads and intramuscularly with 1 mg conjugate (1 ml in buffer (0.005 M sodium phosphate, pH 7.5, 0.15 M NaCl) emulsified in 1 ml complete Freund's adjuvant). Three weeks later they were bled from the ear vein. Every six weeks, they were given booster injections and bled one week later. Typically, one ml of

antiserum at the dilution used in the immunoassay (e.g. anti-cotinine, Table 2) is sufficient for $2 \times 10^4$ assays.

*Procedure for production of monoclonal antibodies [7]*
Monoclonal antibodies were prepared by immunizing female BALB/c mice with nicotine or cotinine bound to KLH prepared as described for preparation of rabbit antisera. Mice in each group received 100 µg of immunizing conjugate in 0.5 ml phosphate buffer emulsified with an equal volume of complete Freund's adjuvant by intraperitoneal and subcutaneous injections. Two additional booster injections were given 21 days apart. The mouse in each group with the highest titer of serum anti-nicotine or anti-cotinine, determined by double antibody radioimmunoprecipitation using [³H]nicotine or [³H]cotinine (see below; *Double-antibody RIA Procedure*) was given 50 µg of immunogen intravenously in 0.15 M saline three days before spleen cells were collected. Splenocytes were fused with non-Ig secreting murine myeloma cells (P3X63–Ag.8.653) according to conventional procedures using polyethylene glycol [8].

Hybridoma supernatants were also screened for anti-nicotine or anti-cotinine antibodies by the RIAs described below *(Double-antibody RIA; RIA procedure)* using [³H]nicotine or [³H]cotinine. Antibody-positive hybridomas

TABLE 2

Inhibition of [³H]Cotinine-anticotinine binding by cotinine

|  | 1 | 2 | 3 | 4 | 5 | 6 | 7 | 8 |
|---|---|---|---|---|---|---|---|---|
| Isogel Tris buffer (ml) | 0.6 | 0.6 | 0.5 | 0.5 | 0.5 | 0.5 | 0.5 | 0.5 |
| Inhibitor, cotinine (ml) | — | — | 0.1 | 0.1 | 0.1 | 0.1 | 0.1 | 0.1 |
| containing pg cotinine/0.1ml | — | — | 1,000 | 500 | 200 | 100 | 50 | 20 |
| [³H]Cotinine, (11,000 cpm/0.1 ml) | 0.1 | 0.1 | 0.1 | 0.1 | 0.1 | 0.1 | 0.1 | 0.1 |
| Normal rabbit serum (1:2000 in buffer) (ml) | 0.1 | — | — | — | — | — | — | — |
| Rabbit anticotinine (1:2000 in buffer) (ml) | — | 0.1 | 0.1 | 0.1 | 0.1 | 0.1 | 0.1 | 0.1 |
| Incubate at 37° for 60 min | | | | | | | | |
| Normal rabbit serum (1:25 in buffer) (ml) | 0.1 | 0.1 | 0.1 | 0.1 | 0.1 | 0.1 | 0.1 | 0.1 |
| Goat anti-rabbit γ-globulin (1:4) in buffer) (ml) | 0.1 | 0.1 | 0.1 | 0.1 | 0.1 | 0.1 | 0.1 | 0.1 |
| Incubate at 4° overnight | | | | | | | | |
| cpm in precipitate | 63 | 4430 | 929 | 1467 | 2374 | 2945 | 3635 | 3992 |
|  | 75 | 4370 | 960 | 1435 | 2311 | 2970 | 3585 | 4036 |
| Percentage inhibition | — | — | 79.8 | 68.1 | 47.5 | 33.3 | 18.2 | 8.9 |

Under the assay conditions described here, the repeatability (inter-assay coefficient of variation) for the cotinine range shown above was 6–10%.

(defined as producing supernatants in the screening procedure that precipitated >8 times background cpm using supernatants from myeloma cells as control) were cloned by the limiting dilution technique [8]. Each of nine nicotine and four cotinine hybridomas were re-selected and grown in ascites fluid of pristane-primed BALB/c mice as the source of antibodies used in the immunoassays. The anti-nicotine ascites fluids contained between 1.5–5.5 mg antibody/ml with binding affinities in the region of $10^8$/M. Anti-cotinine ascites contained between 0.6–1.0 mg antibody/ml with affinities in the range of $2–7.5 \times 10^8$/M. In the nicotine and cotinine microtiter plate ELISA and IFA, 1 ml of ascites fluid was sufficient for over $10^6$ assays.

## Immunoassay procedures

Since 1973, immunoassays for nicotine, cotinine and several other metabolites (Table 1) have been developed for determining levels of these compounds in physiological fluids and other test samples. The RIAs, ELISAs, and FIAs for nicotine and cotinine developed and used in our laboratories are described in detail.

### Double-antibody RIA

Originally applied to the measurement of insulin in plasma by Yalow and Berson [9], RIA today provides one of the most important techniques for the determination of biological molecules and their metabolites in physiological fluids. Among the desirable features of RIA are specificity, sensitivity, and the rapidity with which the analysis can be carried out.

The principle of RIA can be expressed as follows:

$$Ag^* + Ab \rightleftharpoons Ag^* - Ab$$
$$+$$
$$Ag$$
$$\updownarrow$$
$$Ag\text{-}Ab$$

Ag*    = free labeled antigen (analyte)
Ab     = antibody
Ag     = free unlabeled analyte (in standard solutions or in test sample to be analyzed).
Ag*-Ab = labeled antibody complex (bound analyte).
Ag-Ab  = unlabeled antibody complex.

Labeled and unlabeled analyte compete for a limited number of antibody combining sites. The extent of the competition serves as the basis for quantitative assay of the analyte in test samples. Since it is necessary to determine how much of the radioactive analyte is bound to antibody or how much is free, a variety of procedures have been developed to separate antigen-antibody complexes from unbound (labeled and unlabeled) antigen. These methods depend on differences in properties between the analyte that is free and analyte that is bound to its specific antibody.

In our RIAs, antigen-antibody complexes were separated from unbound ([³H]labeled and unlabeled) nicotine or cotinine by precipitation with a second species of antibody (e.g. goat anti-rabbit immunoglobulin) that formed insoluble complexes with antibodies present in the (e.g. rabbit) antisera.

### [³H]Nicotine and [³H]Cotinine

L-(−)-[*N*-Methyl-³H]Nicotine (60–87 Ci/mmol) prepared by methylation of nornicotine is available commercially, e.g. from DuPont NEN, Boston, MA, USA. L(−)[³H]Cotinine was prepared from [³H]nicotine as follows.
To 0.9 mCi/0.9 ml ethanol of L-(−)-[*N*-Methyl-³H]nicotine (81.5 Ci/mmol) add 20 μl of M HCl and evaporate the solvent under nitrogen. Add 0.5 ml of 0.5 M potassium phosphate buffer, pH 7.5, followed by 1.0 ml of rabbit liver preparation (12,000xg supernatant; 25 mg protein/ml [10]) and 4 mg of NADPH. Incubate the mixture at 37° with shaking. After 1.5 h add 0.5 ml of the rabbit liver preparation and another portion of NADPH, and incubate for an additional 1.5 h. Extract with $CH_2Cl_2$ (3×3 ml), centrifuge and combine the methylene chloride fractions. Purify the labeled cotinine by thin layer chromatography (TLC) [silica gel plates, solvent: ethyl acetate, methanol, glacial acetic acid (17:2:1); Rf for nicotine *N'*-oxide=0.01, nicotine=0.17, cotinine=0.5]. The [³H]cotinine is extracted from the TLC plate with methanol and stored at −20° [10].

### RIA procedure

The double-antibody technique was used to separate free labeled antigen from antibody-bound antigen [11]. The total reaction volume can be varied with larger volumes used to reduce non-specific interference (i.e. non-specific inhibition) that may occur with concentrated test samples. In a typical small volume reaction (0.4 ml), 0.1 ml portions of [³H]nicotine or [³H]cotinine (8,000–20,000 cpm), appropriately diluted antiserum and buffer (0.01 M Tris–0.15 M sodium chloride pH 7.4 containing 0.1% gelatin) are incubated at 37° for 1 h. For inhibition experiments, 0.1 ml-aliquots containing known amounts of standard or dilution of test sample are added in place of buffer. Goat anti-rabbit immunoglobulin (0.1 ml previously calibrated to be in antibody excess with respect to rabbit immunoglobulin) is added and the mixture incubated at 4° overnight. To control for non-specific binding, normal rabbit serum is used in place of immune serum. The precipitate is collected by centrifugation at 1000xg for 30 min at 5°, the supernatant decanted, and the walls of the tubes wiped dry. To count [³H]nicotine or [³H]cotinine, the precipitate is dissolved in 0.1 ml 0.1 N sodium hydroxide

273

before adding 2.5 ml scintillation fluid (e.g. *Bioflour* from Packard Instrument Company, Inc., Meriden, CT 06450, USA).

Table 2 shows part of a typical large volume protocol (1.0 ml) that may be used when assaying physiological fluids (serum, plasma, saliva, cerebrospinal fluid, urine, tissue extracts, enzymic reaction mixtures, etc.). As an example in the Table, purified cotinine is shown as the inhibitor, although the protocol would be the same with (appropriately diluted) test samples added as inhibitors in place of cotinine. As mentioned above the larger volume in this protocol minimizes nonspecific effects that may be encountered with samples such as urine that may differ in pH and ionic strength. For example, the maximum volume of adult urine and saliva usually analyzed without encountering non-specific interference is 10 µl. Depending upon the amount of analyte present in the test sample, higher dilutions (1 : 10–1 : 1,000) may be required to obtain inhibition values on the linear region of the standard curve. However, when the analyte is present in low concentrations, with the large volume protocol it is possible to assay up to 0.2 ml of serum (or plasma if EDTA is incorporated into the buffer) without prior extraction of the sample. When analyzing plasma, serum or other test samples, equivalent volumes of nicotine- or cotinine-free fluids should be added instead of buffer to construct the standard curve. All samples should be free of debris. Saliva specimens can be collected using *Salivette* (without citric acid, neutral No. 51.1534 from Sarsted, Inc.). Citric acid or lemon juice should not be used to stimulate saliva flow since it can alter the pH of the reaction mixture and give erroneous results.

Quantification in these assays is based on a standard curve obtained by graphing the percent inhibition of [$^3$H]nicotine or [$^3$H]cotinine binding to the respective antibody (i.e. anti-nicotine or anti-cotinine) vs. the known amounts of unlabeled analyte added. The protocol for the cotinine RIA is shown in Table 2. The amount of analyte in an experimental sample (added to the assay in place of the standard nicotine or cotinine) is determined by comparing its percent inhibition of immune binding with that obtained in the standard inhibition curve. The data may analyzed using one of several computer programs (e.g. *Assay Zap* Apple Macintosh assay analyzer from BIOSOFT, Cambridge, CB2 1LR, U.K.)

*ELISA and FIA* [7,12]
These non-isotopic solid phase assays using monoclonal antibodies provide a useful alternative to the fluid phase RIAs described above. Serum, plasma, saliva, and urine have been assayed routinely in the competitive inhibition format analogous to the RIA procedures, with quantification based on a standard inhibition curve. A comparative study [12] showed a strong correlation between cotinine levels in serum, plasma and saliva obtained with the monoclonal antibody non-isotopic assays, RIA with rabbit antisera, and gas chromatography.

*Procedure*

The ELISA and FIA are carried out in 96-well polyvinyl chloride or polystyrene (e.g. Immulon II from Dynatech) microtiter plates, respectively. Wells are coated by overnight incubation at 4° with between 0.004–0.12 μg conjugate based on PL added. These are optimal amounts that were determined by checkerboard titration of the monoclonal antibodies and the nicotine or cotinine PL conjugate [7,12]. The plates are washed once, then blocked by incubation at room temperature for 1 h with 0.2 ml casein buffer (0.01 M Tris-buffered, 0.15 M saline, pH 7.6 containing 0.01 M EDTA and 0.02% vitamin-free casein (e.g. Nutritional Biochemicals, Cleveland, Ohio, USA). We have shown that, compared with other proteins for blocking (gelatin, bovine serum albumin, ovalbumin, and chicken IgG), casein gives the most consistent and lowest background values (e.g. <0.05 absorbance units). After the plates are washed again, they are ready for use.

These assays are carried out in a total volume of 0.1 ml casein buffer. Maximum antibody binding is determined by incubating optimally diluted anti-nicotine or anti-cotinine monoclonal antibody (0.05 ml) plus 0.05 ml casein buffer in the nicotine-PL or cotinine-PL coated plates, respectively, overnight at 4°. The standard inhibition curve is obtained by adding serial dilutions of nicotine or cotinine in casein buffer (0.01 ml), or in the test article (e.g. saliva or serum) free of nicotine or cotinine. Test samples are added in place of standard nicotine or cotinine. Control wells should contain all reagents except anti-nicotine or anti-cotinine. Care should be taken to maintain the total reaction volume at 0.1 ml by adding buffer if necessary.

*Detection systems*

ELISAs for nicotine and cotinine were developed with detection systems using horseradish peroxidase-labeled staphylococcal protein A (HRP-SpA) and HRP-streptavidin-biotin (HRP-SAB) [7]. The FIA for cotinine used the combination of biotinylated sheep anti-mouse IgG-streptavidin-biotinylated alkaline phosphatase (AP–FIA) [12].

In the HRP-SpA ELISA, 0.1 ml optimally diluted rabbit anti-mouse IgG (optimal means determined by titration or from the manufacturer's instructions to be in excess with regard to the bound monoclonal antibody) in casein buffer is added to each well and the plates incubated at ambient temperature for 1.5 h. The plates are washed three times, incubated with 0.1 ml excess HRP-SpA (e.g. from Zymed Laboratories, South San Francisco, CA, USA; stock solution diluted 1/3000) for 1 h, then washed six times with 0.2 ml PBS (without casein)-0.05% Tween 20, pH 7.4. The substrate (0.04% o-phenyl-enediamine in 0.0 5 M phosphate-0.025 M citrate buffer, pH 5.0, containing 0.015% hydrogen peroxide) is added and incubated for 20–25 min. Sulfuric acid (4 $N$; 0.05 ml) is added to stop the reaction and $A_{490}$ values read in an ELISA reader (e.g. Bio-Tek Instruments, Burlington, VT, USA). Maximum antibody binding (i.e. in the absence of inhibitor) should gives $A_{490}$ between

1.0–1.2 with background (i.e. no anti-nicotine or anti-cotinine) values not to exceed 0.05 absorbance units.

In the HRP-SAB ELISA, 0.1 ml biotinylated sheep anti-mouse IgG (optimally diluted according to the manufacturer's instructions, or by titration) in casein buffer is added and the plates incubated at room temperature for 4 h. The washed plates are then incubated with 0.1 ml HRP-streptavidin (e.g. Zymed Laboratories; stock diluted 1/500) for 1.5 h, washed with PBS-Tween as above, and developed with substrate as in the HRP-SpA assay.

In the AP-FIA for cotinine, the plates are incubated with biotinylated sheep anti-mouse IgG, then sequentially for 1 h with streptavidin (0.5 µg in 0.1 ml) and biotinylated alkaline phosphatase (0.1 ml; Zymed Laboratories; 1/200 diluted stock). The substrate is 0.1 ml of 2.5 mM 4-methylumbelliferyl phosphate in 20% (v/v) aqueous diethanolamine, pH 10, containing 1 mM $MgCl_2$ and 9 mM levamisol. After 30 min at ambient temperature, fluorescence is measured at an excitation wavelength of 360 nm and an emission wavelength of 450 nm (Dynatech Microfluor instrument).

## Immunoassay sensitivity and specificity

Results in Table 2 show the extent to which cotinine competes with [³H]cotinine in a typical RIA. With anti-cotinine serum, 210 pg cotinine gives 50% inhibition (the lower limit of detection is <20 pg/tube). In contrast, nicotine is 2000 times less effective as an inhibitor in this system. Similarly, nicotine can be detected at low levels in the nicotine-anti-nicotine system, while cotinine is an ineffective inhibitor. About 100 compounds (nicotine metabolites and structurally-related molecules) have been tested in RIA and ELISA and generally found to be ineffective inhibitors at levels up to orders of magnitude higher than nicotine or cotinine [4,6,7,10,13].

In one exception, with the polyclonal anti-cotinine sera, *trans*-3′-hydroxy-cotinine was found to inhibit the reaction 2–30% depending on the antiserum tested [14–16]. This metabolite can be present in appreciable amounts in urine [17], even in amounts greater than cotinine itself. The urine values for cotinine assayed with most polyclonal antibodies tend to be overestimates. This may account for the fact that a comparison of RIA and chromatographic methods for cotinine usually agree when blood or saliva samples are analyzed but differ with urine samples [15]. Polyclonal antibodies may be useful in epidemiological studies since detection of the two closely related metabolites unique to nicotine (i.e cotinine and *trans*-3′-hydroxycotinine) makes the assay more sensitive. In contrast, the monoclonal antibodies, being homogeneous, were selected to bind only cotinine. The cross reaction with *trans*-3′-hydroxycotinine is uniformly less than 2% and gives more accurate values for cotinine levels [18].

In addition to testing related compounds of known structure as inhibitors, the specificity of antisera and monoclonal antibodies is apparent in other ways. In immunoassays, the analyte is quantified among thousands of naturally occurring compounds (e.g. in physiological fluids) and many synthetic compounds (e.g. from drug intake). The fact that physiological fluids from non-smokers are negative, speaks well for the specificities of the antibodies. However, the possibility that foodstuffs common to other cultures and that drugs yet to be synthesized may contain cross-reactive substances cannot be excluded. Unusual findings should be checked by an independent method, e.g. chromatography.

Even with reagents as specific as the anti-nicotine and anti-cotinine sera and monoclonal antibodies, occasionally a non-related compound may cross-react for reasons that can not be ascertained by looking at its chemical structure. For example, metyrapone, a drug whose use is limited to testing the ability of the pituitary to respond to a decreased concentration of plasma cortisol, shows approximately 15% cross-reactivity with the polyclonal anti-cotinine sera [13]. With the monoclonal antibodies the cross-reactivity is 100% [19].

## III. APPLICATIONS

Nicotine and/or cotinine levels have been determined by immunoassay in many different types of biological samples including serum and plasma [4,6,20–24], urine [6,25–28], saliva [24,26,29,30], amniotic fluid [31], cerebrospinal fluid [4], cord blood [4], meconium [32], semen [28,33], cervical mucus [34], follicular fluid [35,36], and hair [37,38]. This is not an exhaustive list of references, but they do give the reader a good idea of the types of samples that can be analyzed by this technique. In general, liquid samples free of debris can be assayed directly, while solid samples (e.g. hair) must be extracted prior to assay.

Immunoassays for nicotine, cotinine and other metabolites have also proven useful in metabolic, enzymatic and chemical studies, examples of which are summarized below.

### NICOTINE → COTININE

The RIAs for nicotine and cotinine have been used to obtain kinetic values for the conversion of nicotine to cotinine in in vitro systems containing microsomal cytochrome (CY) P-450, aldehyde oxidase, and necessary cofactors [6]. Because of the high specificity of the anti-cotinine antibodies, it is possible to detect the formation of less than 1% cotinine in incubation mixtures containing large amounts of nicotine. Suitably diluted aliquots of the incubation mixtures are assayed directly; there is no need to extract or process the samples before analysis.

The ELISAs for nicotine and cotinine were also used to study the role of rat liver cytochrome P-450 isoforms in the metabolism of nicotine to cotinine [39]. Rat liver CYP-450IIB1 efficiently converted nicotine to cotinine in the presence of cytosol, whereas purified CYP-450IA1 and CYP-450IIC6 were inactive. Because rabbit liver CYP-450IA1 was active, the conversion of nicotine to cotinine by CYP-450 enzymes shows some degree of species specificity. This evidence also suggests caution in extrapolating results with specific forms of P-450 enzymes in animal studies to nicotine metabolism in humans. Other studies with these assays have shown that cotinine binding to cytochrome P-450 from phenobarbital-induced rats and inhibition of functional activity in vitro are qualitatively like the effects of nicotine (although the apparent spectral dissociation constants and enzymatic inhibition constants for cotinine are higher).

$$\text{NICOTINE} \rightarrow \left[ \begin{array}{c} \textbf{NICOTINE-}\Delta^{1'(5')}\textbf{-IMINIUM ION} \\ \updownarrow \\ \textbf{5'-HYDROXYNICOTINE} \end{array} \right] \rightarrow \textbf{COTININE}$$

An active intermediate, nicotine-$\Delta^{1'(5')}$-iminium ion, formed during the conversion of nicotine to cotinine (Fig. 1), can be quantified at the pmole level [40,41]. The immunoassay specific for this intermediate permits the cytochrome P-450 nicotine 5'-hydroxylase activity to be singled out (rather than depending on measurements of nicotine loss or cotinine formation which can result from the actions of several different enzymes). The stability and reactions of the iminium ion can be monitored under conditions that resemble those found in vivo; an important consideration since polymerization occurs at higher concentrations. The iminium ion can covalently bind to macromolecules (e.g., microsomes, the sulfhydryl groups of polycysteine, and with as yet unidentified nucleophilic groups on nucleic acids), but can be quantified by RIA after it is released by treatment of the macromolecular adducts with perchloric acid to break covalent bonds.

A microsomal dehydrogenase distinguishable from the well-characterized cytosolic aldehyde oxidase by subcellular localization, requirement of NAD, and sensitivity to various enzyme inhibitors has been detected, but the relative importance of this alternative route for the generation of cotinine from nicotine-$\Delta^{1'(5')}$-iminium ion remains to be established [42].

## COTININE → γ-(3-PYRIDYL)-γ-OXO-*N*-METHYLBUTYRAMIDE

With specific antibodies to the product, it is possible to follow the oxidation of cotinine to the oxoamide and to characterize partially the enzyme system (i.e. $K_m$ value, pH optimum, and salt and cofactor dependence) in rabbit liver preparation responsible for this conversion [43].

## NORNICOTINE → *N'*-NITROSO NORNICOTINE

Antibodies to the product were used in RIA to monitor nitrosation of nornicotine and anabasine under chemical conditions and to determine the rate constants for the reactions. The in vitro trans-nitrosation of nornicotine by *N*-nitrosamines, C-nitro compounds, and *N*-nitrosamide and the inhibitory effects of some antioxidants on this reaction have also been studied [44,45].

## Nicotine + NADP or NAD → Corresponding Nicotine Dinucleotide Analog and Cotinine + NADP or NAD → Corresponding Cotinine Dinucleotide Analog

In the presence of NADase (NAD glycohydrolase, E.C.3.2.2.5), nicotine and cotinine can displace the nicotinamide moiety in NAD or NADP to form the corresponding dinucleotide analogs. Formation of the analogs can be followed by RIA both in vitro and in vivo. The high specificity of the respective antibodies makes it possible to detect the individual nucleotide analogs in the presence of structurally similar nucleotides, including the naturally abundant NAD, NADPH, and NMN [46,47].

## CONCLUSION

Details have been presented for three different types of quantitative immunoassays for nicotine and cotinine based on radioactivity, spectro-photometric and fluorescence measurements. Applications of immunoassay to other nicotine metabolites have also been described.

Cotinine is the compound commonly assayed to determine nicotine exposure from various sources (e.g. tobacco use, air contaminated with tobacco smoke, nicotine replacement products, etc.). As a biomarker, cotinine possesses several desirable properties. It has a considerably longer half-life than nicotine and is usually found in higher concentrations in the physiological fluids of the average smoker [6,48]. It is a relatively stable metabolite. While test samples should be stored frozen after collection to ensure accurate quantification, cotinine can still be detected, for example, in saliva samples that were not frozen and sent through the mail [49]. Cotinine is absent (or present in only minute amounts) in tobacco and in the nicotine replacement devices designed to deliver nicotine *per se*, i.e. chewing gum, skin patches, nasal sprays and inhalers. Its presence in physiological fluids clearly indicates nicotine absorption *and* metabolism have occurred. In the environment, nicotine can be absorbed in unexpected ways. For example, in occupations closely related to the harvesting and processing of tobacco, cotinine was found in significantly greater quantities in the urine of croppers [50]. These are the workers who harvest tobacco from the field and have dermal exposure to nicotine contained in the dew. Among the various tobacco workers, they have a significantly greater risk of suffering from Green Tobacco Sickness, a self-limiting disease with symptoms resembling those seen in individuals smoking their first cigarette, i.e. nausea, vomiting, dizziness, prostration, etc.

Immunoassays are continually being modified to make them more sensitive, rapid and easier to perform. For example, the RIAs for nicotine and cotinine are much more sensitive today with the availability of labeled nicotine of higher specific activity; i.e. $(-)$-[$N$-methyl-$^3$H]nicotine currently in use has a specific activity of 60–87 Ci/mmol compared with $(-)$-[$^3$H]nicotine, specific activity 2.4 Ci/mmol, used earlier [10,13]. ELISAs also are more sensitive when substrates with increased color yields are used.

Major modifications are in store for the future. For example, like enzymes and receptors, antibodies can be used as the biological component in biosensors. These devices (immunosensors) have the potential for increased sensitivity and rapid, simultaneous analysis of large numbers of test samples [51].

## REFERENCES

1 Gorrod JW, Schepers G. Biotransformations of nicotine and related compounds. *Analytical Determination of Nicotine and Related Compounds and their Metabolites.* JW Gorrod and P Jacob III (Eds), Elsevier Science, The Netherlands, 1999, pp. 45–67.

2 Jacob P, Byrd GD. Use of gas chromatographic and Mass Spectrometric techniques for the determination of nicotine and its metabolites. *Analytical Determination of Nicotine and Related Compounds and their Metabolites.* JW Gorrod and P Jacob III (Eds), Elsevier Science, The Netherlands, 1999, pp. 191–224.

3 Crooks P, Byrd GD. Use of high-performance liquid chromatographic and mass spectrometric techniques for the determination of nicotine and its metabolites. *Analytical Determination of Nicotine and Related Compounds and their Metabolites.* JW Gorrod and P Jacob III (Eds), Elsevier Science, The Netherlands, 1999, pp. 225–264.

4 Langone JJ, Van Vunakis H. Radioimmunoassay of pharmacologically active compounds: applications to nicotine and its metabolites. *Radioimmunoassay of Drugs and Hormones in Cardiovascular Medicine.* A. Albertini, M. Da Prada and B.A. Peskar (Eds), Elsevier/North-Holland Biomedical Press, 1979, pp. 55–70.

5 Cushman M, Castagnoli N, Jr. The synthesis of *trans*–3'-methylnicotine. *J. Org. Chem.,* 1972: **37**; 1268–1271.

6 Langone JJ, Gjika HB, Van Vunakis H. Nicotine and its metabolites. Radioimmunoassays for nicotine and cotinine. *Biochemistry,* 1973: **12**; 5025–5030.

7 Bjercke RJ, Cook G. Rychlik N, Gjika HB, Van Vunakis H, Langone JJ. Stereospecific monoclonal antibodies to nicotine and cotinine and their use in enzyme-linked immunosorbent assays. *J. Immunol. Meth.,* 1986: **90**; 203–213.

8 Galfre G, Milstein C. Preparation of monoclonal antibodies: strategies and procedures. *Meth. Enzymol.,* 1981: **73**; 1–46.

9 Yalow RS, Berson SA. Immunoassay of endogenous plasma insulin in man. *J. Clin. Invest.,* 1960: **39**; 1157–1175.

10 Langone JJ, Van Vunakis H. Radioimmunoassay of nicotine, cotinine, and γ-(3-pyridyl)-γ-oxo-*N*-methylbutyramide. *Meth. Enzymol.*, 1982: **84**; 628–640.

11 Midgley AR Jr., Hepburn MR. Use of the double-antibody method to separate antibody bound from free ligand in radioimmunoassay. *Meth. Enzymol.*, 1980: **70**; 266–274.

12 Bjercke RJ, Cook G, Langone JJ. Comparison of monoclonal and polyclonal antibodies to cotinine in nonisotopic and isotopic immunoassays. *J. Immunol. Meth.*, 1987: **96**; 239–246.

13 Van Vunakis H, Gjika HB, Langone JJ. Radioimmunoassay for nicotine and cotinine. World Health Organization International Agency for Research on Cancer. Lyon Environmental Carcinogens Methods of Analysis and Exposure Measurement. B Seifert, HJ van de Wiel, B Dodet, IK O'Neill (Eds.), Volume 12 (Indoor Air). IARC Scientific Publications 1993; No. 109: 293–299.

14 Schepers G, Walk RA. Cotinine determination by immunoassays may be influenced by other nicotine metabolites. *Arch. Toxicol.*, 1988: **62**; 395–397.

15 Watts RR, Langone JJ, Knight GJ, Lewtas J. Cotinine analytical workshop report: consideration of analytical methods for determining cotinine in human body fluids as a measure of passive exposure to tobacco smoke. *Environ. Hlth Perspect.*, 1990: **84**; 173–182.

16 Zuccaro P, Pichini S, Altieri I, Rosa M, Pellegrini M, Pacifici R. Interference of nicotine metabolites in cotinine determination by RIA. *Clin. Chem.*, 1997: **43(1)**; 180–181.

17 Neurath GB, Dünger M, Orth D, Pein FG. *Trans*–3′-hydroxycotinine as a main metabolite in urine of smokers. *Int. Arch. Occup. Environ. Hlth*, 1987: **59**; 199–201.

18 Langone JJ, Bjercke RJ. Idiotype-anti-idiotype hapten immunoassays: assay for cotinine. *Analyt. Biochem.*, 1989: **182**; 187–192.

19 Bjercke RJ, Hammond DK, Strobel HW, Langone JJ. Interaction of cotinine with rat liver hepatic microsomal P-450. Comparison with metyrapone and immunomodulation of cotinine and metyrapone binding by monoclonal anti-cotinine antibodies. *Drug Metab. Dispos.*, 1990: **18**; 759–764.

20 Haines CF Jr, Mahajan DK, Milijkovic D, Milijkovic M, Vesell ES. Radioimmunoassay of plasma nicotine in habituated and naive smokers. *Clin. Pharmacol. Ther.*, 1974: **16**(6); 1083–1089.

21 Castro A, Monji N, Malkus H, Eisenhart W, McKennis H Jr, Bowman ER. Automated radioimmunoassay of nicotine. *Clin. Chim. Acta*, 1979: **95**(3); 473–481.

22 Langone JJ, Van Vunakis H, Hill P. Quantitation of cotinine in sera of smokers. *Res. Commun. Chem. Pathol. Pharmacol.*, 1975: **10**; 21–28.

23 Zeidenberg P, Jaffe JH, Kanzler M, Levitt MD, Langone JJ, Van Vunakis H. Nicotine:cotinine levels in blood during cessation of smoking. *Comp. Psychiat.*, 1977: **18**; 93–101.

24 Van Vunakis H, Tashkin DP, Rigas B, Simmons M, et al. Relative sensitivity and specificity of saliva and serum cotinine in identifying tobacco smoking status of self-reported non-smokers and smokers of tobacco and/or marijuana. *Arch. Environ Hlth*, 1989: **44**; 53–58.

25 Matsukura S, Sakamoto N, Seino Y, Tamada T, Matsuyama H, Muranaka H. Cotinine excretion and daily cigarette smoking in habituated smokers. *Clin. Pharmacol. Ther.*, 1979: **25**(5 Pt 1); 555–561.

26 Langone JJ, Cook G, Bjercke RJ, Lifschitz MH. Monoclonal antibody ELISA for cotinine in saliva and urine of active and passive smokers. *J. Immunol. Meth.*, 1988: **114**; 73–78.

27 Wald NJ, Boreham J, Bailey A, Ritchie C, Haddow JE, Knight G. Urinary cotinine as marker of breathing other people's tobacco smoke. *Lancet*, 1984: I; 230–231.

28 Vine MF, Hulka BS, Margolin BH, Truong YK, Hu P, Schramm MM, et al. Cotinine concentrations in semen, urine and blood of smokers and non-smokers. *Am. J. Publ. Hlth*, 1993: **83**(9); 1335–1338.

29 Etzel RA. A review of the use of saliva cotinine as a marker of tobacco smoke exposure. *Preventive Med.*, 1990: **19**; 190–197.

30 Coultas DB, Howard CA, Peake GT, Skipper BJ, et al. Salivary cotinine levels and involuntary tobacco smoke exposure in children and adults in New Mexico. *Am. Rev. Respir. Dis.*, 1987: **136**; 305–309.

31 Van Vunakis H, Langone JJ, Milunsky A. Nicotine and cotinine in the amniotic fluid of smokers in the second trimester of pregnancy. *Am. J. Obstet. Gynecol.*, 1974: **120**; 64–66.

32 Ostrea EM, Jr., Knapp K, Romero A, Montes M, Ostrea AR. Meconium analysis to assess fetal exposure to nicotine by active and passive maternal smoking. *J. Pediat.*, 1994: **124**(3); 471–476.

33 Pacifici R, Altieri I, Gandini L, Lenzi A, Passa AR, Pichini S, et al. Environmental tobacco smoke: nicotine and cotinine concentration in semen. *Environ. Res.*, 1995: **68**; 69–72.

34 Sasson IM, Haley NJ, Hoffmann D, Wynder EL, Hellberg D, Nilson S. Cigarette smoking and neoplasia of the uterine cervix: smoke constituents in cervical mucus. *N. Engl. J. Med.*, 1985: **312**; 315–316.

35 Weiss T, Eckert A. Cotinine levels in follicular fluid and serum of IVF patients: Effect on granulosa-luteal cell function in vitro. *Human Reprod.*, 1989: **4**; 482–485.

36 Zenzes MT, Reed TE, Wang P, Klein H. Cotinine, a major metabolite of nicotine, is detectable in follicular fluids of passive smokers in in vitro fertilization therapy. *Fertil. Steril.*, 1996: **66**; 614–619.

37 Haley NJ, Hoffmann D. Analysis for nicotine and cotinine in hair to determine cigarette smoker status. *Clin. Chem.*, 1985: **31**; 1598–1600.

38 Eliopoulos C, Klein J, Phan MK, Knie B, Greenwald M, Chitayat D, Koren G. Hair concentrations of nicotine and cotinine in women and their newborn infants. *J. Am. Med. Assoc.*, 1994: **271**(8); 621–623.

39 Hammond DK, Bjercke RJ, Langone JJ, Strobel HW. Metabolism of nicotine by rat liver cytochromes P-450: Assessment utilizing monoclonal antibodies to nicotine and cotinine. *Drug Metab. Dispos.*, 1991: **19**; 804–808.

40 Obach RS, Van Vunakis H. Radioimmunoassay of nicotine-$\Delta^{1'(5')}$-iminium ion, an intermediate formed during the metabolism of nicotine to cotinine. *Drug Metab. Dispos.*, 1990: **18**(4); 508–513.

41 Obach RS, Van Vunakis H. Non-metabolic covalent binding of nicotine-$\Delta^{1'(5')}$-iminium ion to liver microsomes and sulfhydryl-containing polyamino acids. *Biochem. Pharmacol.*, 1990: **37**; 4601–4604.

42 Obach RS, Van Vunakis H. Nicotinamide adenine dinucleotide (NAD)-dependent oxidation of nicotine-$\Delta^{1'(5')}$-iminium ion to cotinine by rabbit liver microsomes. *Biochem. Pharmacol.*, 1990: **39**; R1–R4.

43  Langone JJ, Franke J, Van Vunakis H. Nicotine and its metabolites. Radio-immunoassay for γ-(3-Pyridyl)-γ-oxo-*N*-methylbutyramide. *Arch. Biochem. Biophys.*, 1974: **164**; 536–543.

44  Castonguay A, Van Vunakis H. Radioimmunoassays for *N*'-nitrosonornicotine and *N*'-acylnornicotine analogs. *Meth. Enzymol.*, 1982: **84**; 641–650.

45  Castonguay A, Van Vunakis H. Radioimmunoassay of *N*'-nitrosonornicotine, *Analyt. Biochem.*, 1979: **95**; 387–396.

46  Shen WC. Radioimmunoassay of nicotinamide nucleotide analogs of nicotine and cotinine. *Meth. Enzymol.*, 1982: **84**; 650–654.

47  Shen WC, Greene KM, Van Vunakis H. Detection by radioimmunoassay of nicotinamide nucleotide analogues in tissues of rabbits injected with nicotine and cotinine. *Biochem. Pharmacol.*, 1977: **26**(20); 1841–1846.

48  Benowitz NL. Pharmacology of nicotine: addiction and therapeutics. *Ann. Rev. Pharmacol. Toxicol.*, 1996: **36**; 597–613.

49  Cuckle HS, Van Vunakis H. The effectiveness of a postal smoking cessation 'kit'. *Commun. Med.*, 1984: **6**(3); 210–215.

50  Gehlbach SH, Perry LD, Williams WA, Freeman JI, Langone JJ, Peta LV, Van Vunakis H. Nicotine absorption by workers harvesting green tobacco. *Lancet*, 1975: **1**; 478–484.

51  Rogers KR, Mulchandani A (Eds). *Affinity Biosensors: Techniques and Protocols*. Totowa, NJ: Humana Press, 1998.

52  Matsushita H, Noguchi M, Tamaki E. Conjugate of bovine serum albumin with nicotine. *Biochem. Biophys. Res. Commun.*, 1974: **57**(4); 1006–1010.

53  Matsuyama H, Tamada T, Ishiguro T, Muranaka H. Radioimmunoassay of nicotine. *Biochem. Biophys. Res. Commun.*, 1975: **64**(2); 574–580.

54  Castro A, Prieto I. Nicotine antibody production: comparison of two nicotine conjugates in different animal species. *Biochem. Biophys. Res. Commun.*, 1975: **67**(2); 583–589.

55  Knight GJ, Wylie P, Holman MS, Haddow JE. Improved [$^{125}$I] radioimmunoassay for cotinine by selective removal of bridge antibodies. *Clin. Chem.*, 1985: **31**; 118–121.

*Chapter 9*

# Determination of nicotine in tobacco, tobacco processing environments and tobacco products

Michael F. Borgerding[1], Thomas A. Perfetti[1] and Suresh Ralapati[2]

[1] *R.J. Reynolds Tobacco Company, R&D Department, Winston-Salem, NC 27102, USA*
[2] *ATF/National Laboratory Center, Rockville, MD 20850-3188, USA*

## I. INTRODUCTION

From the viewpoint of the analyst, nicotine is a wonderful molecule. Its physical and molecular properties provide the analyst with a host of opportunities for qualitative and quantitative analysis. For example, the vapor pressure characteristics of nicotine are appropriate for analysis by gas chromatography and the presence of heteroatoms in the molecular structure permits selective detection with element specific detectors. Both reverse phase and normal phase liquid chromatography of nicotine are possible because of the solubility characteristics of nicotine and the fact that nicotine can be protonated. Nicotine's electronic and molecular structures make spectroscopic techniques such as ultraviolet spectroscopy and near infrared spectroscopy viable analytical techniques. While chromatographic and spectroscopic analyses are widely used in analytical chemistry, these techniques are just a few examples of the many approaches to nicotine analysis that are possible. In addition to unique molecular and physical properties that provide many different qualitative analysis opportunities, nicotine is an ideal molecule for quantitative analysis. It is a significant component of the tobacco leaf, often representing more than 98% of the total alkaloid content in tobacco and 1–3% of the total leaf mass.

Analysis of nicotine in tobacco has been conducted for many purposes. The chemical composition of tobacco crops varies from growing season to growing season and from one geographical region to the next. Tobacco nicotine

285

analysis provides one means of characterizing a new tobacco crop and understanding the raw material that will be included in the final products. The research and development of new tobacco breeding lines also requires tobacco nicotine analysis. The smoking quality of different tobacco varieties has been related to the alkaloid distribution found in the leaf. Nicotine and other (secondary) alkaloids have been analyzed for many years as new tobacco breeding lines are developed in order to provide tobacco varieties that are palatable to smokers. Another area in which the determination of tobacco nicotine is of critical importance is the classification of tobacco products for subsequent taxation. The presence and location of nicotine within the components of a smoking article are critical to whether the smoking article is classified as a cigarette, cigar or non-tobacco product, and to the amount of tax that is levied by the federal government.

Since the determination of nicotine in tobacco can be undertaken for many different reasons, there are several fundamental questions the analyst must consider before implementing a tobacco nicotine analysis. For example, is a qualitative analysis that determines the presence or absence of nicotine in a sample necessary, or is a quantitative analysis required to accurately determine the amount of nicotine in a sample? If a quantitative analysis is the goal, is it important to measure nicotine specifically, or is it desirable to measure the total alkaloid content of tobacco with the summary response expressed as nicotine? Will nicotine concentration in tobacco be expressed on a dry weight basis or on an 'as is' basis? While the purpose for conducting a particular test will largely determine the answers to the types of questions just posed, practical considerations will also shape the implementation of a tobacco nicotine analysis. The current availability of analytical instruments in the laboratory, the cost of any new instruments needed, sample throughput requirements, the level of analyst training required, sample preparation constraints, instrument sensitivity, the limit of detection required and safety considerations, among others, are practical considerations that will help determine what type of tobacco nicotine determination is conducted to address a particular need.

The legislative and regulatory history of tobacco control efforts is a substantial and varied topic that has been considered by all levels of government, e.g., at the federal, state and local levels in the United States [1]. Historically, laws have been enacted to regulate and control the potential fire hazard created by smoking, to increase revenues through taxation and to address the estimated health costs and risks of tobacco use. More recently, tobacco control efforts have included limitations on tobacco use in public places and strengthened restrictions on youth access to tobacco products. Thus, governmental based tobacco control efforts have generally affected consumers in a much more direct way than analytical chemists. Since 1996, however, the determination of nicotine in tobacco has become a specific requirement in some state regulations. With this new requirement, the scope and limitations of analytical methods that have been used for tobacco nicotine analysis must be carefully considered.

This chapter will deal with three topics, the determinations of nicotine in tobacco, nicotine in tobacco processing environments (concentrating primarily with processes in the cigarette industry), and regulatory aspects dealing with nicotine in tobacco products. The first topic will be presented as an historical perspective of the progress of researchers in tobacco science and achievements in analytical chemistry made in the determination of alkaloids in tobacco (particularly nicotine) from 1826 to the present. The second part of this chapter, on nicotine in tobacco processing environments, will describe tobacco processing that occurs within the tobacco industry, mainly the cigarette industry, and the determinations of nicotine that have been made in the individual process streams. This section will describe a typical tobacco manufacturing operation and present reasons why precise and accurate measurements of tobacco nicotine are necessary for quality assurance/quality control determinations. The third part of this chapter will highlight the current legislative and regulatory issues concerning the determinations of nicotine in tobacco and tobacco products.

## II. HISTORICAL ASPECTS

Tobacco is one of the most widely used commodities in the world. It has been studied extensively because of its scientific uniqueness, its economic importance in society, the health consequences of tobacco use, the economic and political importance of the industry it produced, because of its ability to generate massive revenues and due to governmental regulation [2,3]. Although, thousands of chemicals have been isolated from tobacco, nicotine has received a great deal of attention from the time it was first identified and determined. Nicotine is the principal alkaloid in all tobaccos grown and used commercially today [4].

The history associated with the discovery and use of 'oil' derived from tobacco dates back to about 1571 [3]. This 'oil' prepared by the French chemist Gohory undoubtedly contained some level of nicotine and was used as a remedy for diseases of the skin. In 1660 another French chemist, LeFevre, described means to steam distill tobacco to obtain oil that had medicinal uses [4]. In 1807 Cerioli discovered what he called the '*olio essenziale*' of tobacco [5]. In 1809 Vauquelin apparently made the same discovery. Both researchers described the oil as a volatile and colorless substance. Vauquelin recognized the basic nature of the material but failed to recognize its alkaloidal properties. He attributed the basicity of the material to the presence of ammonia.

In 1822, Hermbstadt confirmed Vauquelin's results of the presence of the oil described in sixteen different species of what is now known as *Nicotiana*

[5]. It was not until 1828 that Posselt and Reimann succeeded in isolating a pure sample of the oil and recognized it as an alkaloid. They characterized it as a water-clear liquid, boiling at 246°C, under atmospheric pressure and miscible with water, alcohol, and ether. They named the pure compound nicotine after Jean Nicot who introduced tobacco into the French court in about 1560 [6]. In 1826 Unverdorben isolated a water-soluble base from a dry distillation of tobacco [7]; the base contained nicotine. Melsens, in 1843, succeeded in isolating nicotine from the smoke of pipe tobacco and assigning the empirical formula, $C_{12}H_{14}N_2$ [8]. In 1893 Pinner reported on the final clarification of the constitution of nicotine, determined via degradation studies. Pinner's structural formula for nicotine was confirmed by Pictet's classical synthesis of nicotine in 1895. For many years nicotine was believed to be the only alkaloid in tobacco [9]. It was not until 1928 that Ehrenstein reported finding nornicotine in tobacco [10]. Anabasine was isolated from tobacco by Smith in 1935. Spath and Kesztler isolated anatabine in tobacco in 1937 [10]. A. Gautier and G. LeBon in 1892 were the first chemists to clearly recognize that additional alkaloids accompanied nicotine in tobacco smoke. These additional alkaloids were the secondary alkaloids and decomposition products of alkaloids. Unfortunately, Gautier and LeBon did not report further on their observations [11]. Wensuch and Scholler, in the early 1930s, worked diligently to separate the secondary alkaloids in tobacco smoke. Though they were not wholly successful they did discover and determine the formula for myosmine in cigar smoke in 1936 with the collaboration of Spath [11]. By 1936 Wensuch and Scholler had distinguished a large number of tobacco smoke bases in cigar and cigarette smoke by their behavior during steam distillation and extraction procedures [10,11]. By preparation of derivatives (picrates and picrolonates) and by qualitative organic reactions many assertions concerning the structure of the smoke bases were made. Until the study by Kuffner, Schick and Buhn in 1959, there was much confusion concerning the correct identity of the secondary alkaloids of tobacco smoke [11]. By the use of paper chromatography Kuffner and co-workers unraveled the confusion. Kuffner and co-workers analytically separated, by two-dimensional chromatograms, and determined, through derivatization, the presence of nineteen tobacco alkaloids in cigar tobaccos and cigar smoke [11].

## III.  DETERMINATION OF NICOTINE IN TOBACCO

The initial discoveries of the major alkaloids of tobacco (nicotine, nornicotine, anabasine and anatabine) illustrated the creativity and high degree of chemical knowledge of such notable chemists as Posselt, Reimann, Pictet, Pinner, Spath and Kuffner, to mention only a few. It is amazing that these chemists had the perseverance to diligently work long hours and complete the difficult tasks of separation, isolation, determination and quantitation without the aid of modern analytical techniques. After the initial identification and determination of nicotine, relatively few techniques or improvements to the chemistry of nicotine alkaloids were made, with the exceptions of the discoveries of the secondary alkaloids in 1928,1935 and 1937. Technology had not advanced at this point in time such that new methods and analytical techniques were available to chemists.

The pyrolysis studies of nicotine by Woodward, Eisner and Haines in 1944 began a new phase in the development of our knowledge of the chemistry of nicotine alkaloids [12]. The pyrolysis studies on nicotine afforded the opportunity to study the chemical reactivity of nicotine. Several new compounds were identified, and several new analytical techniques that had been recently developed were applied to the identification of nicotine and related alkaloids. Ultraviolet spectroscopy, paper chromatography, counter-current distribution techniques and new chemical procedures in the investigation of the chemistry of nicotine and its analogues opened the way for investigation of these materials in tobacco, processed tobaccos and tobacco smoke [12]. As additional compounds were identified in tobacco, processed tobaccos, tobacco fermentation processes, tobacco smoke and metabolites of tobacco and smoke, new methods of qualitative and quantitative detection of tobacco alkaloids were developed to meet the ever-exacting desires of chemists. The following paragraphs will describe the development of qualitative and quantitative methods developed and employed in the determination of nicotine and related alkaloids.

There are numerous analytical determinations described in the literature for the determination of nicotine. The first official method for the determination of nicotine in tobacco and tobacco products was the Kissling method. The Kissling method involves two steps. First, there is an extraction with ether and, second, this is followed by distillation with steam and titration. In 1909 Bertrand and Javillier were the first to publish a method for the determination of nicotine by a precipitation technique employing silicotungstic acid. In 1911 Chapin evaluated a number of published methods and concluded that Bertrand's method gave precise and accurate data on the quantity of nicotine in tobacco and tobacco products. This led to the adoption of a new official method for nicotine in tobacco and tobacco products by the Association of Official Agricultural Chemists (AOAC). The Kissling and silicotungtic acid methods for the determination of nicotine were accurate but time consuming and were subject to interference by ammonia and ammonium salts as shown by Ogg, Willits and Ricciuti in 1950.

In 1950, a significant departure from gravimetry was reported. Willits, Swaim, Connelly and Brice developed a spectrophotometric method for nicotine that was compared to the official AOAC method. It was shown that the spectrophotometric method was relatively rapid, equally reliable, showed no interference from ammonia and ammonium salts and eliminated the solubility errors inherent with the official AOAC method. Since that time many refinements have been made to the official AOAC method. Table 1 presents a representative sampling of research conducted on methods for the determination of nicotine and tobacco alkaloids in a variety of matrices from ~1900 to 1998. This table shows a progression of proposed refinements that parallels technological improvements in analytical chemistry in this century. Table 2 lists the current standard and official methods for the determination of nicotine in various matrices.

Somewhat arbitrarily, analytical method research conducted since 1950 can be considered as the basis for current laboratory methods as they are practiced today. Beginning with the spectrophotometric method developed by Willits, et al. in 1950, there have been many analytical methodologies applied to the determination of nicotine, or nicotine and other specific alkaloids, in tobacco. Examples of some of the many different approaches that have been applied are presented chronologically in Fig. 1 [13–26].

During the last fifty years, analytical methods reported for nicotine determination have employed the newest instrumental analysis techniques of the day. For example, a technical progression from colorimetry to gas chromatography, liquid chromatography, and finally capillary electrophoresis has occurred between 1950 and 1997. Regardless of the type of end-determination specified for a particular nicotine method, analytical chemists have faced similar challenges regarding accuracy, precision and practical application when developing new methods of analysis. First, quantitative measurement of nicotine in tobacco requires both the complete quantitative extraction of unprotonated nicotine and nicotine bound up as salts with organic acids from the tobacco matrix. Second, interfering substances must be removed from the tobacco extract or a selective detection mode (chemical or instrumental) must be applied before accurate and precise results can be obtained. Third, practical considerations such as sample throughput, simplicity, recovery and ruggedness must be optimized. The numerous avenues that tobacco chemists have successfully explored to address these challenges demonstrate their ingenuity and the versatility of the nicotine molecule.

The remainder of this section will focus on three examples of current analytical practices for the determination of nicotine. The first method is a continuous flow analysis method that is probably the most widely practiced method for the determination of nicotine today. The second is a gas chromatography method with which nicotine and other alkaloids are individually quantitated. The third method is intended for the determination of nicotine in smokeless tobacco products.

## Automated analysis by continuous flow analyzer

Automated analysis using a continuous flow approach was first developed and introduced in response to the needs of the clinical chemistry laboratory

TABLE 1

Representative sample of research conducted on methods for determination of nicotine and tobacco alkaloids

| Date | Authors | Description | Reference |
|------|---------|-------------|-----------|
| ca 1900 | AOAC | Kissling Method for determination of nicotine in tobacco and tobacco extracts is 1st official method for determination of nicotine. Ether extract of NaOH suspension of tobacco/tobacco extract. Remove ether. Steam-distill residue and titrate with standard solution of sulfuric acid to phenacetolin end point. | 1 |
| 1901 | Toth | Petroleum ether extraction via Soxhlet apparatus of dehydrated tobacco/tobacco extract, evaporate ether, dissolve residue with quantity of std. acid, back titrated with std. alkali. | 2 |
| 1904 | Emery | Determination nicotine in the presence of pyridine bases by a polariscope. | 3 |
| 1909 | Bertrand et al. | Extract nicotine from tobacco in boiling water containing HCl, precipitate nicotine with silicotungstic acid. Nicotine is released from the washed nicotine salt with calcined magnesia, nicotine is then steam-distilled and titrated with std. acid. | 4 |
| 1910 | USDA | USDA adopts the method of Bertrand and Javillier for determination of nicotine as it gives more accurate results vs the Kissling method. It is also deemed 'rapid, simple, more economical, and convenient'. | 5 |
| 1911 | USDA | AOAC adoption of Silicotungstic Acid Method (SAM) for the official determination of nicotine. Tobacco sample is steam-distilled from alkaline solution, and nicotine is precipitated with silicotungstic acid. The precipitated nicotine salt is filtered, washed, ignited, cooled and quantified by gravimetry. | 6 |
| 1928 | Pfyl et al. | One of earliest gravimetric determination of nicotine in tobacco and smoke via preparation of nicotine picrate. | 7 |
| 1939 | Avens et al. | Improvement to steam-distillation procedure of the AOAC silicotungstic acid determination for nicotine. | 8 |
| 1948 | Griffith et al. | Steam-distillation (Griffith Still Method) followed by gravimetric determination of nicotine in green and dry tobacco. | 9 |
| 1950 | Willits et al. | Nicotine in tobacco, tobacco extracts and distillates determined by ultraviolet spectrophotometry (UV). | 10 |
| 1950 | AOAC | Slight changes to the SAM adopted by AOAC. | 11 |

TABLE 1  Continued

| Date | Authors | Description | Reference |
|------|---------|-------------|-----------|
| 1955 | Cundiff et al. | Alternative to Willis method involving titration with perchloric acid in acetic acid to permit determination of nicotine and nornicotine. Steam-distillation is not required. | 12 |
| 1955 | Waltz et al. | Photometric determination of nicotine in tobacco and tobacco smoke. | 13 |
| 1955 | Leiserson et al. | Paper chromatography of nicotine and related compounds. | 14 |
| 1957 | Griffith | Modification to the Griffith Still Method to use a bank of stills to improve accuracy and reproducibility. | 15 |
| 1959 | Hasselbach | Polarographic determination of nicotine in tobacco. | 16 |
| 1960 | Sadler et al. | Automated determination of nicotine in steam-distilates by reaction of nicotine with cyanogen bromide and aniline to produce a color proportional to the concentration of nicotine. Method allows for measurement of water and nicotine analyses from same sample, permits absorbances to be obtained directly from a recorder and eliminates the need to compensate for background interferences. | 17 |
| 1961 | Koelle | Determination of nicotine in green tobacco employing the Kraft Spot Test. | 18 |
| 1966 | Kaburaki et al. | Gas chromatographic (GC) determination of nicotine and other middle and high boiling bases in mainstream smoke of cigarettes | 19 |
| 1966 | Harvey et al. | Determination of nicotine alkaloids in tobacco leaf, cigarette filler and particulate matter of smoke by acid-methanol extraction. | 20 |
| 1968 | Collins et al. | Determination of nicotine alkaloids in tobacco using an AutoAnalyzer. | 21 |
| 1969 | Harvey et al. | Automated determination of reducing sugars and nicotine alkaloids on a single tobacco leaf extract. | 22 |
| 1969 | Weeks et al. | GC separation and determination of tobacco alkaloids. | 23 |
| 1970 | Cano et al. | Determination of nicotine by GC. | 24 |
| 1970 | Cano et al. | GC assay for determination of nicotine. | 25 |
| 1970 | Harvey et al. | Field test estimation of reducing sugars and nicotine alkaloids in tobacco leaf. | 26 |
| 1970 | Derrick et al. | Routine determinations of total sugars and nicotine in tobacco using dual channel AutoAnalyzers and electronic data handling systems. | 27 |
| 1972 | Bush | Quantitative analysis of tobacco alkaloids by gas chromatography. | 28 |

TABLE 1 Continued

| Date | Authors | Description | Reference |
|------|---------|-------------|-----------|
| 1975 | Jurzysta | Densitometric method for the determination of nicotine, nornicotine and anabasine in tobacco | 29 |
| 1975 | Lyerly et al. | GC assay for determination of low levels of nicotine in tobacco. | 30 |
| 1976 | Davis | Automated procedure for determination of reducing sugars and nicotine alkaloids in tobacco and tobacco products using a cyanogen chloride colorimetric procedure and UV detection. | 31 |
| 1976 | Holzer et al. | Determination of organic compound (including nicotine) in exhaled tobacco smoke by gas chromatography - mass spectrometry (GC-MS) employing glass capillary columns. | 32 |
| 1976 | Pilotti et al. | Determination of nicotine and nicotine metabolites in blood by GC-MS, using multiple ion detection. | 33 |
| 1978 | Arakawa et al. | Rapid analysis of nicotine and nornicotine by gas chromatography, stainless steel columns. | 34 |
| 1978 | Browne et al. | Review of current methods and development of a new method for collection of mainstream and sidestream smoke constituents, including nicotine. | 35 |
| 1979 | Wagner et al. | Gas-liquid chromatographic determination of nicotine on Cambridge pads. Collaborative Study results, published in JAOAC. | 36 |
| 1979 | Rosa | Pyrolysis gas chromatographic estimate of tobacco alkaloids and neophytadiene. | 37 |
| 1980 | Green et al. | Review of current methods for collection and analysis tobacco and mainstream and sidestream smoke constituents, including nicotine. GC, GC-MS, gas liquid chromatography, liquid chromatography, GC-infrared spectrometry methods reviewed. | 38 |
| 1980 | Downs et al. | Application of NIR spectroscopy to determination of moisture, nicotine, sugars and stems in tobacco blends. | 39 |
| 1980 | Djordjevic | Spectrophotometric determination of nicotine in distillates obtained from alkaline treated tobacco. | 40 |
| 1980 | Shoffner et al. | Rapid method for determination of puff-by-puff smoke nicotine and smoke menthol. | 41 |
| 1981 | Gordon et al. | Results of a comparative study comparing the Cundiff-Markunas AOAC titration method and a newly developed GC method for the determination of nicotine and total secondary alkaloids (as nornicotine). | 42 |

TABLE 1 Continued

| Date | Authors | Description | Reference |
|------|---------|-------------|-----------|
| 1981 | Harvey et al. | On-line generation of cyanogen chloride as a replacement for cyanogen bromide in the determination of total alkaloids in tobacco. | 43 |
| 1981 | Muranaka et al. | Micro-determination of nicotine and its metabolites in biological fluids. | 44 |
| 1981 | Severson et al. | Rapid method for determination of tobacco alkaloids by glass capillary gas chromatography coupled with nitrogen-specific detection. | 45 |
| 1981 | Kogan et al. | Simultaneous determination of nicotine and cotinine in human plasma by nitrogen detection gas liquid chromatograghy (GLC). | 46 |
| 1981 | Hollweg et al. | Simultaneous determination of nicotine and water in tobacco smoke condensate by automatic gas chromatograghy. | 47 |
| 1982 | Verebey et al. | Rapid, quantitative GLC method for determination of nicotine and cotinine. | 48 |
| 1982 | Thompson et al. | Analyses of nicotine and cotinine in tissue by capillary gas chromatography and GC-MS. | 49 |
| 1982 | McClure et al. | Review of uses of Fourier analysis NIR diffuse reflectance spectroscopy in the tobacco industry, applications to nicotine discussed. | 50 |
| 1982 | Mizunuma et al. | Microanalysis of nicotine in urine and plasma by GC-MS. | 51 |
| 1982 | Hoffmann et al. | Use and detection techniques for $^{14}$C-nicotine in studies on the fate of nicotine during leaf curing and smoking. | 52 |
| 1982 | Higgins et al. | Sampling and analysis of cigarette smoke components using solid adsorbents (e.g., Tenax), nicotine mentioned. | 53 |
| 1982 | Kinnersley et al. | Determinations of nicotine in tobacco callus. | 54 |
| 1983 | Heinzer et al. | Capillary GC head space analysis of tobacco aroma, nicotine mentioned. | 55 |
| 1983 | Trundle et al. | GC determination of nicotine in human breast milk. | 56 |
| 1983 | Blume | Use of radioimmunoassay (RIA) for measuring low levels of nicotine in tobacco root suspensions and tissues. | 57 |
| 1984 | Williams et al. | Simple method for the collection and determination of nicotine in air by glass capillary GC with nitrogen selective detector. | 58 |
| 1984 | Sakaki et al. | Analysis of head space volatiles of tobacco using an ether trap. Ether solutions analyzed by GC and GC-MS. | 59 |

TABLE 1 Continued

| Date | Authors | Description | Reference |
|------|---------|-------------|-----------|
| 1984 | Sakaki et al. | Analysis of tobacco head space volatiles using Tenax or acivated carbon. Determination of volatiles, including nicotine, accomplished by thermal desorption GC. | 60 |
| 1984 | Dias et al. | Determination of nicotine in smokers' urine by thin layer chromatography (TLC) and Dragendo reagent for coloration. | 61 |
| 1984 | Sudan et al. | Determination of nicotine in allergenic extracts of tobacco by high performance liquid chromatography (HPLC). | 62 |
| 1984 | Atkinson et al. | Determination of nicotine in tobacco by circular dichroism employing a spectropolarimeter. | 63 |
| 1985 | Ayad et al. | Atomic absorption determination of nicotine in tobacco. | 64 |
| 1985 | Yaegashi et al. | Determination of nicotine in mulberry leaves by flame ionization GC. | 65 |
| 1985 | McClure | Application of NIR spectroscopy to determination of nicotine in tobacco. | 66 |
| 1986 | Murthy et al. | Determination of nicotine in Indian tobacco by HPLC. | 67 |
| 1986 | Xiao et al. | Determination of nicotine in cigarettes by TLC. | 68 |
| 1986 | Li et al. | Rapid extract and spectrophotometric method for the determination of nicotine in tobacco. | 69 |
| 1986 | Court | Review of HPLC techniques used in the determination of constituents in tobacco and tobacco smoke; nicotine included in review. | 70 |
| 1986 | Long | Review of robotic processing of tobacco and tobacco smoke analyses. An IBM 7565 robot prepares the samples for nicotine and sugar analysis by two Technicon RA 1000 Discrete Analysers. The cyanogen bromide - aniline colorimetric analysis for the determination of nicotine is employed. This is the first report of a robotic system for the analysis of nicotine. | 71 |
| 1987 | Tiburcio et al. | Review of analyses of alkaloids in tobacco callus by HPLC. | 72 |
| 1987 | Nyiredy et al. | High resolution HPLC separation of tobacco alkaloids employing gradient elution. | 73 |
| 1987 | Hassan | Liquid membrane electrodes for the selective determination of nicotine in tobacco plants. | 74 |

TABLE 1 Continued

| Date | Authors | Description | Reference |
|------|---------|-------------|-----------|
| 1987 | Gottscho et al. | Two standard (official) AOAC methods (Cundiff - Markunas and Griffith Still) for the determination of nicotine are compared on smokeless tobacco samples. Results indicate that neither AOAC method is suitable for analysis of nicotine in smokeless tobacco samples. Several other non-AOAC methods for determination of nicotine are discussed and are shown to be more suitable for the determination of nicotine in smokeless tobacco samples. Recommendations to revise the AOAC methods were presented. | 75 |
| 1987 | Dai et al. | Analytical applications for the determination of nicotine in tobacco products employing tetraphenylborate-nicotine-graphite-coated membrane and PVC membrane selective electrodes are presented. | 76 |
| 1988 | Nel | Comparison of high performance TLC and GC method for the determination of nicotine in dried tobacco. | 77 |
| 1988 | Finster et al. | Flow injection analysis (FIA) is employed for the determination of nicotine in tobacco extracts. | 78 |
| 1989 | Chung et al. | Liquid chromatography - Mass Spectrometry (LC-MS) is used to identify two nicotine biosynthetic intermediates. | 79 |
| 1990 | Kang et al. | Analytical results obtained from a new electronic sensor for rapid determination of tobacco nicotine are comparable to results obtained from silicotungstate gravimetric and UV spectrometry methods for nicotine. | 80 |
| 1991 | Mahanwal et al. | Determination and comparison of nicotine in tobacco products by derivative UV spectroscopy. | 81 |
| 1992 | Manceau et al. | HPLC procedure for the analysis of tobacco alkaloids in tobacco and cell suspensions. The yield of biotransformation which was established at 53.2% by GC was increased to nearly 100% by the new HPLC analysis. | 82 |
| 1992 | Li et al. | Determination of nicotine and benzene in tobacco extracts by derivative UV spectrophotometric method. | 83 |
| 1992 | Coleman | A GC- vapor phase IR-MS hyphenated technique for the separation and determination of volatile and semi-volatile components of supercritical fluid and methylene chloride extracts of tobacco. Nicotine is determined in this technique. | 84 |
| 1993 | McCalley | Reversed phase high performance liquid chromatographic determination of tobacco alkaloids. | 85 |

TABLE 1 Continued

| Date | Authors | Description | Reference |
|------|---------|-------------|-----------|
| 1994 | Lang et al. | An indirect method for the determination of nicotine in tobacco by the formation of specific nicotine metal salt precipitates. The metal content of nicotine-metal salts is determined by atomic absorption spectrometry, and the concentration of nicotine is indirectly calculated. | 86 |
| 1994 | Barber et al. | Determination of nicotine in sidestream cigarette smoke by surface-enhanced Raman scattering spectroscopy (SERS). | 87 |
| 1994 | Rai et al. | An UV spectrophotometric determination of nicotine based on the bromination reaction of nicotine to form dibromonicotine which reacts with potassium iodide in the presence of starch to form a water soluble blue complex. Max UV asborbance at 580 nm. | 88 |
| 1994 | Munir et al. | Nicotine-metal complexes are formed from tobacco extracts, nicotine is recovered by removal of metal as its sulfide, nicotine is determined by several known analytical methods. This method is based on stereoselective reactions of metal ions with nicotine, a natural ligand. | 89 |
| 1995 | Wooten | Nicotine in the leaf and stems of flue-cured and burley tobacco was determined by solid phase $^{13}$C CPMAS NMR. | 90 |
| 1995 | Zhang et al. | Nicotine in cigarettes is determined by a multiplex gas chromatographic technique. | 91 |
| 1995 | Hana et al. | Application of artificial neural networks in the estimation of nicotine in tobacco from NIR data. | 92 |
| 1995 | Yang et al. | Analysis of nicotine and secondary alkaloids of tobacco by capillary electrophoresis. | 93 |
| 1995 | Ye et al. | Indirect determination of nicotine by atomic emmission spectrometry. | 94 |
| 1996 | Ravard et al. | Determination of the chiral purity of nicotine and related alkaloids by the use of $^{1}$H NMR spectroscopy in the presence of 1, 1′-binaphthyl-2, 2′-diylphosphoric acid. | 95 |
| 1996 | Yang et al. | Analysis of selective tobacco alkaloids by micellar electrokinetic capillary chromatography. | 96 |
| 1996 | Guo et al. | A rapid, direct determination of nicotine in tobacco and cigar smoke by linear-sweep polarography. | 97 |
| 1997 | Ralapati | Application for capillary electrophoresis in the determination of nicotine in ATF (Bureau of Alcohol, Tobacco and Firearms) regulated tobacco products. | 98 |

TABLE 1 Continued

37. Rosa N. Pyrolysis gas chromatographic estimation of tobacco alkaloids and neophytadiene. *J. Chromatogr.*, 1979; **171**: 419–23.

38. Green CR, Colby DA, Cooper PJ, Heckman RA, Lyerly LA, Thorne FA. Advances in analytical methodology of leaf and smoke. *Tobacco Chem. Res. Conf. (TCRC) 34th*, 1980; Richmond (VA).

39. Downs DR, Long TM. Application of NIR spectroscopy to moisture nicotine sugars and stem determination in tobacco blends. *Int. Tobacco Sci. Cong.*, 1980; Manila.

40 Djordjevic M. Parallel investigations of nicotine content in tobacco and smoke condensate of cigarettes of different quality. *Arhiv. Za Poljoprivredne Nauke*, 1980; **41(143)**: 409–12.

41. Shoffner RA, Ireland MS. Rapid analysis of menthol and nicotine in smoke and the effects of air dilution on delivery. *Tobacco Chem. Res. Conf. (TCRC) 34th*, 1980; Richmond (VA).

42. Gordon BM, Greene GH. Determination of nicotine, nornicotine and other secondary amine alkaloids in tobacco. *Tobacco Chem. Res. Conf. (TCRC), 35th*, 1981; Winston-Salem (NC).

43. Harvey WR, Handy BM. On-line generation of cyanogen chloride as a replacement for cyanogen bromide in the total alkaloids determination. *Tob. Sci.*, 1981; **20**: 131–3.

44. Muranaka H, Tamada T, Higashi E, Shunro I, Morimoto M. Microdetermination of nicotine and its metabolites in biological fluid. 3. Effects of smoking on plasma nicotine, plasma cotinine and other biochemical variables in human beings. *Rinsho Kagaku*, 1981; **10(2)**: 127–35.

45. Severson RF, McDuffie KL, Arrendale RF, Gwynn GR, Chaplin JF, Johnson AW. Rapid method for the analysis of tobacco nicotine alkaloids. *J. Chromatogr.*, 1981; **21(1)**: 111–21.

46. Kogan M, Verebey K, Jaffee J, Mule S. Simultaneous determination of nicotine and cotinine in human plasma by nitrogen detection gas–liquid chromatography. *J. Forensic Sci.*, 1981; **26**: 6–11.

47. Hollweg J, Schumacher HJ. Simultaneous determination of nicotine and water in tobacco smoke condensate by automatic gas chromatography. *Beitr. Tabakforsch.*, 1981; **11(1)**: 39–43.

48. Verebey KG, DePace A, Mule SJ, Kanzler M, Jaffe JH. A rapid, quantitative GLC method for the simultaneous determination of nicotine and cotinine. *J. Anal. Toxicol.*, 1982; **6(6)**: 294–6.

49. Thompson JA, Ho MS, Petersen DR. Analyses of nicotine and cotinine in tissues by capillary gas chromatography and gas chromatography–mass spectrometry. *J. Chromatogr. Biomed. Appl.*, 1982; **231(1)**: 53–64.

50. McClure WF, Giesbrecht FG, Weeks WW, Hamid A. Fourier analysis enhances NIR diffuse reflectance spectroscopy. *Tobacco Chem. Res. Conf. (TCRC) 36th*, 1982; Raleigh (NC).

51. Mizunuma H, Hirayama Y, Sakurai S, Ikekawa N. Microanalysis of nicotine in urine and plasma. *Esiei Kagaku*, 1982; **28(1)**: 13–17.

52. Hoffmann D, Adams JD, Vinchkowski N, Tso TC. On the fate of nicotine-[14]C during leaf curing and smoking. *CORESTA Symp.*, 1982; Winston-Salem (NC).

53. Higgins CE, Griest WH, Guerin MR. Sampling and analysis of cigarette smoke using solid adsorbents. *184th Amer. Chem. Soc.*, 1982; Kansas City (MO).

54. Kinnersley AM, Dougall DK. Variation in nicotine content of tobacco callus cultures. *Planta*, 1982; **154**: 447–453.

55. Heinzer F, Burrus FJ. Capillary gas chromatographic head space analysis of tobacco aroma. *25th Takak. Kolloquim.*, 1983; Boncourt, Switzerland.

56. Trundle J, Skellern G. Gas chromatographic determination of nicotine in human breast milk. *J. Clin. Host. Pharm.*, 1983; **8(3)**: 289–293.

57. Blume DE. Production of callus and suspension cultures derived from tobacco roots. *Tobacco Chem. Res. Conf. (TCRC) 37th*, 1983; Washington DC.

TABLE 1  Continued

58. Williams D, Whitaker J, Jennings W. Air monitoring for nicotine contamination. *J. Chromatogr. Sci.*, 1984; **22(6)**: 259–61.

59. Sakaki T, Sakuma H, Sugawara S. Analysis of the headspace volatiles of tobacco using an ether trap. *Agr. Biol. Chem.*, 1984; **48(11)**: 2719–2724.

60. Sakaki T, Niino K, Sakuma H, Sugawara S. Analysis of tobacco headspace volatiles using Tenax gas chromatography or active carbon. *Agr. Biol. Chem.*, 1984; **48(12)**: 3121–3128.

61. Dias E, Sznelwar R. Analytical profile of tobacco smokers urine by thin layer chromatography. *Rev. Farm. Bioquim. Univ. Sao Paulo*, 1984; **20(2)**: 201–210.

62. Sudan B, Brouillard C, Strehler C, Strub H, Sterboul J, Sainte-Laudy J. Determination of nicotine in allergenic extracts of tobacco leaf by high-performance liquid chromatography. *J. Chromatogr.*, 1984; **288(2)**: 415–22.

63. Atkinson W, Han S, Purdie N. Determination of nicotine in tobacco by circular dichroism spectropolarimetry. *Anal. Chem.*, 1984; **56(11)**: 1947–50.

64. Ayad MM, Khayyal SE, Farag NM. Atomic absoprtion determination of nicotine in tobacco *Nicotiana tabacum*. *Spectrochim. Acta, Pt. B. Atom. Spectrosc.*, 1985; **40(8)**: 1127–1131.

65. Yaegashi S, Suzuki S, Oikawa H. Nicotine pollution of mulberry farm caused by tobacco leaves and its control. I. Rapid detection of nicotine in mulberry contaminated by tobacco. *Iwate-Ken Sangyo Shikenjo Yoho*, 1985; **8**: 21–5.

66. McClure W. Status of near infrared technology in the tobacco industry. *NIR 84 Proc. Int. Symp. Near Infrared Reflectance Spectrosc.*, 1984; 127–33.

67. Murthy PS, Rao BV, Gopalalchari NC, Saunders J. Determination of alkaloid composition in Indian tobacco by HPLC. *Tob. Res.*, 1986; **12(2)**: 186–91.

68. Xiao X. Determination of nicotine in cigarettes by thin-layer chromatography. *Fenxi Huaxue*, 1986; **14(8)**: 626.

69. Li Z, Shang X, Zhou H, Song H. Extraction and spectrophotometric determination of nicotine in tobacco. *Fenxi Huaxue*, 1986; **14(1)**: 60–62.

70. Court WA. High-performance liquid chromatography of tobacco and tobacco smoke components. *Recent Adv. in Tobacco Sci.*, 1986; **(12)**: 143–184.

71. Long TM. Robotic processing of tobacco and tobacco smoke analysis. *Proceedings of CORESTA Symposium*, 1986; **I**: 82–87.

72. Tiburcio AF, Galston AW. Analysis of alkaloids in tobacco callus by HPLC. In: Linskens HF, Jackson JF (Eds.). Modern Methods for Plant Analysis, Vol. 5 High Performance Liquid Chromatography. Berlin: Springer-Verlag, 1987: 228–242.

73. Nyiredy S, Sticher O. Gradient possibilities with the "prisma" model. HPLC separation of tobacco alkaloids. *J. High Resolut. Chromatogr. (Chromatogr. Commun.)*, 1987; **10(4)**: 208–210.

74. Hassan SSM. New liquid membrane electrodes for selective determination of nicotine in tobacco products. *194th Amer. Chem. Soc.*, 1987; **194(0)**: Anyl 14.

75. Gottscho AM, Lin JL, Duck WN, Losty TA. Nicotine analysis of commercial tobacco products. *Tobacco Chem. Res. Conf. (TCRC) 41st*, 1987; Greensboro (NC).

76. Dai Z, Xie J, Su X, Li W. Nicotine graphite-coated-membrane and PVC membrane selective electrodes. *Gaodeng Xuexiao Huzxue Xuebao*, 1987; **8(10)**: 882–4.

77. Nel JG. An evaluation of two methods for the routine determination of nicotine and nornicotine in tobacco. *S. Afr. J. Plant Soil*, 1988; **5(1)**: 40–1.

78. Finster P, Hollweg J, Kausch E, Burmester U. Automated tobacco study with flow-injection analysis. *Beitr. Tabakforsch.*, 1988; **14(2)**: 105–18.

TABLE 1  Continued

79.  Chung HL, Blume DE. Identification of nicotine biosynthetic intermediates in tobacco roots by liquid chromatography–mass spectrometry. *J. Chromatogr.*, 1989; **474(1)**: 329–334.

80  Kang XJ, Wang CY. A new sensor for the determination of nicotine in tobacco products. *Electroanalysis (NY)*, 1990; **2(1)**: 81–4.

81.  Mahanwal JS, Singh V, Shukla SK. Determination and comparison of nicotine in tobacco products by derivative spectroscopy. *Indian J. Forensic Sci.*, 1991; **5(4)**: 176–87.

82.  Manceau F, Fliniaux MA, Jacquin-Dubreuil A. A high performance liquid chromatographic procedure for the analysis of tobacco alkaloids. Application to the evaluation of tobacco alkaloids in plants and cell suspension cultures. *Phytochem. Anal.*, 1992; **3(2)**: 65–68.

83.  Li H, Sun X, Yank F. Determination of nicotine and benzene in tobacco extracts by derivative uv-spectrophotometric method. *Yingyong Huaxue*, 1992; **9(5)**: 110–12.

84.  Coleman WM. The volatile and semivolatile components of supercritical fluid and methylene chloride extracts of selected tobaccos. *J. Essent. Oil Res.*, 1992; **4(2)**: 113–20.

85.  McCalley DV. Evaluation of reversed-phase columns for the analysis of very basic compounds by high-performance liquid chromatography. Application to the determination of the tobacco alkaloids. *J. Chromatogr.*, 1993; **636(2)**: 213–20.

86.  Lang H, Xie Z, Lei Z, Li H. Application of atomic absorption spectrometry to organic analysis, (IV) nonstandard determination of nicotine in tobacco. *Xibei Daxue Xuebao, Ziran Kexueban*, 1994; **24(3)**: 215–18.

87.  Barber TE, List MS, Haas JW, Wachter EA. Determination of nicotine by surface-enhanced Raman scattering (SERS). *Appl. Spectrosc.*, 1994; **48(11)**: 1423–7.

88.  Rai M, Ramachandran KN, Gupta VK. Spectrophotometric method for the determination of total tobacco alkaloids and nicotine. *Analyst (Cambridge, UK)*, 1994; **119(8)**: 1883–5.

89.  Munir C, Zaidi MI, Yousaf S. Zinc, cadmium and mercury as extractants of nicotine from tobacco leaves. *Main Group Met. Chem.*, 1994; **17(9)**: 673–7.

90.  Wooten JB. $^{13}$C CPMAS NMR of bright and burley tobaccos. *J. Agric. Food Chem.*, 1995 **43(11)**: 2858–2868.

91.  Zhang M, Phillips JB. Applications of multiple gas chromatography to the determination of organics in solid samples. *J. Chromatogr. A*, 1995; **689(2)**: 275–284.

92.  Hana M, McClure WF, Whitaker TB, White M, Bahler DR. Applying artificial neural networks, I. Estimating nicotine in tobacco from near infrared data. *J. Near Infrared Spectrosc.*, 1995; **(3)**: 133–142.

93.  Yang SS, Smetena I. Evaluation of capillary electrophoresis for the analysis of nicotine and selected minor alkaloids from tobacco. *Chromatogrpahia.* 1995; **40(7/8)**: 375–8.

94.  Ye Y, Qian G, Meng Y. Study on the indirect determination of nicotine by atomic-emission spectrometry. *Fenxi Shiyanshi*, 1995; **14(5)**: 63–5.

95.  Ravard A, Crooks PA. Chiral purity determination of tobacco alkaloids and nicotine-like compounds by $^1$H NMR spectroscopy in the presence of 1, 1′-binaphthyl-2, 2′-diylphosphoric acid. *Chirality*, 1996; **8(4)**: 295–2999.

96.  Yang SS, Smetena I, Goldsmith AI. Evaluation of micellar electrokinetic capillary chromatography for the analysis of selected tobacco alkaloids. *J. Chromatogr. A*, 1996; **746(1)**: 131–136.

97.  Guo W, Zhao Y, Yang W, Yuan H. Rapid, direct determination of nicotine in tobacco and cigar smoke by linear-sweep polarography. *Xibei Daxue Xuebao, Ziran Kexueban*, 1996; **26(6)**: 495–498.

98.  Ralapati S. Capillary electrophoresis as an analytical tool for monitoring nicotine in ATF regulated tobacco products. *J. Chromatogr. B*, 1997; **695(1)**: 117–129.

TABLE 1 Continued

99. Yang C, Jiao Y. Determination of nicotine in tobacco by non-aqueous titration method. *Fenxi Huaxue*, 1997; **25(3)**: 368.

100. Jamin E, Naulet N, Martin GJ. Multi-element and multi-site isotopic analysis of nicotine from tobacco leaves. *Plant Cell Environ.*, 1997; **20(5)**: 589–599.

101. Sheng L, Liu S, Yong G, Dai Y. Use of coprecipitation-potentiometric titration in determination of nicotine and potassium of tobacco. *Huaxue Chuanganqi*, 1997; **17(2)**: 153–155.

102. Yang SS, Smetena I. Determination of tobacco alkaloids using solid phase microextraction and GC-NPD. *Chromatographia*, 1998; **47(7/8)**: 443–448.

103. Perfetti TA, Coleman WM. Chiral-gas chromatography-selected ion monitoring-mass selective detection analysis of tobacco materials and tobacco smoke. *Beitr. Tabakforsch.*, 1998; **18(1)**: 15–33.

and other medical facilities [27]. Continuous flow systems provide a means to carry out a series of individual steps in an analytical procedure accurately, precisely and in a fully automated fashion. With such an instrument, samples can be automatically introduced into the system, diluted as appropriate, reagents added, reactions conducted at specific temperatures and for specific lengths of time, sample clean-up achieved by dialysis and reaction products of interest detected by colorimetry or other detection modes. A number of tobacco chemists have developed and refined chemistry compatible with continuous flow analysis equipment [18–20,28–33]. Continuous flow analysis is popular for the determination of nicotine because it is relatively simple and inexpensive, and it provides the capability to analyze more than one hundred samples per day.

CORESTA (Centre de Cooperation pour les Recherches Scientifiques Relatives au Tabac [Cooperation Center for Scientific Research Relative to Tobacco]) recommended method No. 35 describes the determination of tobacco total alkaloids by continuous flow analysis [34]. Results by this method, while expressed as 'nicotine', actually represent the combined response of nicotine and other alkaloids such as nornicotine and anabasine. It should be noted that the continuous flow analyzer is calibrated specifically with nicotine and that the reaction chemistry specified by CORESTA recommended method No. 35 has not been optimized for alkaloids other than nicotine. Nicotine is, therefore, the primary response observed with the method, although studies have indicated higher response with CORESTA recommended method No. 35 than with other methods, especially for tobaccos with significant levels of alkaloids other than nicotine.

The main features of continuous flow analysis methodology can be described as follows:

*Method scope.* An extract of ground tobacco is prepared and reacted with sulphanilic acid and cyanogen chloride. Total alkaloid content, expressed as nicotine, is determined based on absorbance at 460 nm. A continuous flow analyzer provides automated analysis of tobacco extracts.

TABLE 2

Standard/official methods for the determination of nicotine in tobacco and smoke

| Method | AOAC | ISO | TIOJ | CORESTA |
|---|---|---|---|---|
| **Tobacco:** | | | | |
| Alkaloids (Total as Nicotine) in Tobacco | 960.07 (1964) 960.08 (1964) | 2881–1992 | | 20–1968 35–1994 |
| **Smoke:** | | | | |
| Determination of Alkaloid Retention by Filters | | 3401–1991 | | 13–1968 |
| Determination of Alkaloid in Smoke Condensates | | 3400–1989 | | 12–1968 |
| Determination of Nicotine in Cigarette Filters | | | | 9–1989 |
| Nicotine on Cambridge Filter Pads, GC Method | 979.01 (1984) | 10315–1991 | 2nd edn.1993 | 7–1991 |
| **Others:** | | | | |
| Determination of the Purity of Nicotine and Nicotine Salts by Gravimetric Analysis – Silicotungstic Acid Method | 920.35 | 13276–1997 | | 39–1994 |
| Nicotine in Environmental Tobacco Smoke | 990.01 (1997) 991.50 (1997) | 11454–1997 | | 14–1993 |

**AOAC** = Association of Official Analytical Chemists
**ISO** = International Organization for Standardization
**TIOJ** = Tobacco Institute of Japan
**CORESTA** = Cooperation Center for Scientific Research Relative to Tobacco

*AOAC Methods:*

Official Methods of Analysis of (AOAC) International. 1998. (Ed.) P. Cunniff, 16th Ed., 4th Revision, Vol. 1, Alkaloids (Total as Nicotine) in Tobacco – Distillation Method. Final Action 1964. AOAC Official Method 960.07.

Official Methods of Analysis of (AOAC) International. 1998. (Ed.) P. Cunniff, 16th Ed., 4th Revision, Vol. 1, Alkaloids (Total as Nicotine) in Tobacco – Cundiff – Markunas Method. Final Action 1964. AOAC Official Method 960.08.

Official Methods of Analysis of (AOAC) International. 1998. (Ed.) P. Cunniff, 16th Ed., 4th Revision, Vol. 1, Nicotine in Environmental Tobacco Smoke – Gas Chromatographic Method. Final Action 1997. AOAC Official Method 990.01.

Official Methods of Analysis of (AOAC) International. 1998. (Ed.) P. Cunniff, 16th Ed., 4th Revision, Vol. 1, Nicotine in Environmental Tobacco Smoke – Gas Chromatography Method – Alternate Conditions. Final Action 1997. AOAC Official Method 991.50.

TABLE 2 Continued

Official Methods of Analysis of Association of Official Analytical Chemistry (AOAC) International. 1998. (Ed.) P. Cunniff, 16th Ed., 4th Revision, Vol. 1, Nicotine in Tobacco Products – Silicotungstic Acid Method. Final Action. AOAC Official Method 920.35.
Official Methods of Analysis of Association of Official Analytical Chemistry (AOAC) International. 1998. (Ed.) P. Cunniff, 16th Ed., 4th Revision, Vol. 1, Nicotine on Cambridge Filter Pads – Gas Chromatographic Method. Final Action 1964. AOAC Official Method 979.01.

*ISO Methods:*
International Standards Organization. 1991. Tobacco and tobacco products – Cigarettes – Determination of nicotine content in smoke condensates – Gas chromatographic method. ISO 10 315.
International Standards Organization. 1989. Cigarettes – Determination of alkaloids in smoke condensates – Spectrometric method. ISO 3400.
International Standards Organization. 1992. Tobacco and tobacco products – Determination of alkaloids content – Spectrometric method. ISO 2881.
International Standards Organization. 1991. Tobacco and tobacco products – Determination of alkaloids retention by filters – Spectrometric method. ISO 3401.
International Standards Organization. 1997. Tobacco and tobacco products – Cigarettes – Determination of vapour-phase nicotine in air – Gas chromatographic method. ISO 11 454.
International Standards Organization. 1997. Tobacco and tobacco products – Determination of nicotine purity – Gravimetric method using tungstosilic acid. ISO 13 276.

*TIOJ Method:*
Hayashi, Yoshiro, Minister of Finance (Financial Bureau). Standard method for the determination of tar (Nicotine-Free Dry Particulate Matter, NFDPM) and nicotine in mainstream smoke of cigarettes, Second Edition. Final Approval, April 1993. Memorandum, No. 4665, December 18, 1992. To: Tobacco Institute of Japan (TIOJ), From: Izumi Teramoto, Council for the Tobacco and Salt Industries, Financial Bureau, Ministry of Finance, Tokyo, Japan.

*CORESTA Methods:*
CORESTA (Co-operation Centre for Scientific Research Relative to Tobacco Association) Standard Method No. 7: Determination of nicotine in the mainstream smoke of cigarettes by gas chromatographic analysis. Final Approval, August 1991. CORESTA Inf. Bull. 1991(3): 112–116.
CORESTA (Co-operation Centre for Scientific Research Relative to Tobacco Association) Standard Method No. 9: Determination of nicotine in the cigarette filters by gas chromatographic anlaysis. Final Approval, October 1989. CORESTA Inf. Bull. 1989 (2): 15–17.
CORESTA (Co-operation Centre for Scientific Research Relative to Tobacco Association) Standard Method No. 12: Determination of alkaloids in cigarette smoke condensates. Final Approval, September 1968. CORESTA Inf. Bull. 1969 (2): 7–8.

TABLE 2

CORESTA (Co-operation Centre for Scientific Research Relative to Tobacco Association) Standard Method No. 13: Determination of alkaloid retention by cigarette filters. Final Approval, September 1968. CORESTA Inf. Bull. 1969 (2): 15–20.
CORESTA (Co-operation Centre for Scientific Research Relative to Tobacco Association) Standard Method No. 14: Determination of nicotine in ambient air by gas chromatographic analysis. Final Approval, September 1993. CORESTA Inf. Bull. 1993 (4): 108–112.
CORESTA (Co-operation Centre for Scientific Research Relative to Tobacco Association) Standard Method No. 20. Determination of alkaloids in manufactured tobacco. Final Approval, September 1968. CORESTA Inf. Bull. 1969 (2): 25–27.
CORESTA (Co-operation Centre for Scientific Research Relative to Tobacco Association) Standard Method No. 35: Determination of total alkaloids (as nicotine) in tobacco by continuous flow analysis. Final Approval November 1994. CORESTA Inf. Bull. 1994 (3–4): 39–46.
CORESTA (Co-operation Centre for Scientific Research Relative to Tobacco Association) Standard Method No. 39: Determination of the purity of nicotine and nicotine salts by gravimetric analysis – Tungstosilicic Acid Method. Final Approval, November 1994. CORESTA Inf. Bull. 1994 (3–4): 87–90.

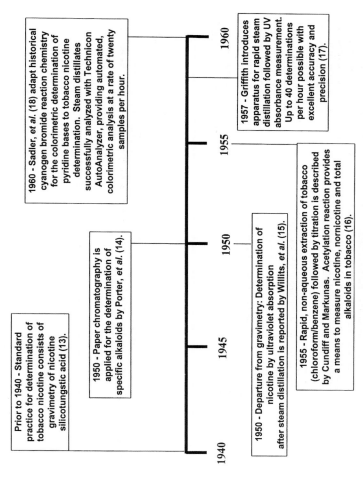

Prior to 1940 - Standard practice for determination of tobacco nicotine consists of gravimetry of nicotine silicotungstic acid (13).

1950 - Paper chromatography is applied for the determination of specific alkaloids by Porter, et al. (14).

1960 - Sadler, et al. (18) adapt historical cyanogen bromide reaction chemistry for the colorimetric determination of pyridine bases to tobacco nicotine determination. Steam distillates successfully analyzed with Technicon AutoAnalyzer, providing automated, colorimetric analysis at a rate of twenty samples per hour.

1950 - Departure from gravimetry: Determination of nicotine by ultraviolet absorption after steam distillation is reported by Willits, et al. (15).

1955 - Rapid, non-aqueous extraction of tobacco (chloroform/benzene) followed by titration is described by Cundiff and Markunas. Acetylation reaction provides a means to measure nicotine, nornicotine and total alkaloids in tobacco (16).

1957 - Griffith introduces apparatus for rapid steam distillation followed by UV absorbance measurement. Up to 40 determinations per hour possible with excellent accuracy and precision (17).

1940    1945    1950    1955    1960

Fig. 1 A simple chronology of tobacco nicotine analysis during the last fifty years.

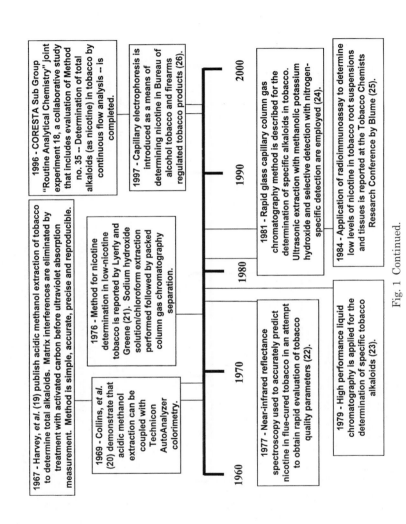

Fig. 1 Continued.

*Sample preparation.* Tobacco is ground to pass a 1 mm sieve and the moisture content of the tobacco is determined. The ground tobacco is extracted with either water or a 5% acetic acid solution. Tobacco extracts are filtered into an analyzer cup for subsequent determination.

*End determination.* A continuous flow analyzer is used to perform the necessary chemical reactions and to provide the end determination. A typical continuous flow analyzer consists of the following components: sampler, proportioning pump check dialyser for delay coils, colorimeter and recorder or digital data acquisition device. A commercially available system in use in a laboratory setting is pictured in Fig. 2. Total alkaloids, expressed as nicotine, are determined based on absorbance at 460 nm. Calibration with nicotine hydrogen tartrate is by peak height.

*Nicotine reporting requirements.* Nicotine concentrations are reported as a percentage of the original tobacco weight on a dry weight basis to an accuracy level of two decimal places.

Method precision for CORESTA recommended method No. 35 was investigated by international collaborative study under the auspices of a CORESTA Task Force between 1989 and 1993. Twelve laboratories

Fig. 2 A commercially available continuous flow analyzer in use in a tobacco laboratory.

evaluated tobacco samples from three different tobacco types; Oriental, flue cured and Burley. The method was found to be repeatable (individual measurements within a laboratory conducted during a short time interval varied less than 5% relative to the mean response for each tobacco type) and reproducible (measurements for a given tobacco type varied no more than 16% relative to the mean response for a given tobacco type when measurements were conducted in different laboratories).

Nineteen laboratories participated in a more recent collaborative study conducted in 1996 by the CORESTA Routine Analytical Chemistry Sub Group (Thomsen HV, personal communication). Two tobacco samples were analyzed in the study and results are summarized graphically in Fig. 3. The continuous flow analysis method was again found to be both repeatable and reproducible. It should be noted that collaborators were allowed to practice either the principal chemistry or alternative chemistry as described in CORESTA recommended method No. 35.

*Capillary gas chromatography method for the simultaneous determination of nicotine, nornicotine, myosmine, anabasine and anatabine in tobacco.* In 1958, a cooperative effort between tobacco growers, export tobacco leaf dealers, cigarette manufacturers, seed companies, USDA scientists, university researchers and agricultural extension personnel began to evaluate

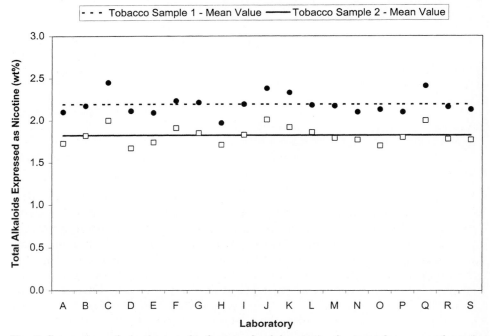

Fig. 3 Comparison of nicotine results from twelve laboratories for two tobacco samples using CORESTA recommended method No. 35 (data from International Collaborative Study).

new flue-cured (Virginia or Bright) tobacco cultivars before the new varieties are extensively planted and become part of an annual crop. The program is known as 'The Regional Minimum Standards Program for the release of Flue-Cured Tobacco Varieties in the United States' [35]. The minimum standards program was initiated in response to the release of undesirable varieties in the mid–1950s. Minimum standards for release of a new flue-cured tobacco cultivar originally included chemical, physical and smoke taste evaluations. Currently, emphasis is placed on chemical evaluation of tobacco samples because it was found that cultivars that were rejected based on taste were also rejected based on the chemical evaluation parameters.

The determination of individual alkaloids in tobacco is one requirement of the minimum standards program. Because significant levels of secondary alkaloids in flue-cured tobacco (i.e., nornicotine, myosmine, anabasine and anatabine) impart undesirable taste characteristics in smoke, there is a minimum standard for the relative amounts of secondary alkaloids in a tobacco cultivar. Total secondary alkaloids may not exceed 13% of the total alkaloids found in the tobacco if a new cultivar is to be accepted.

For many years nicotine, nornicotine, myosmine, anabasine and anatabine have been simultaneously determined by capillary gas chromatography in both flue-cured and Burley tobacco samples to support the development of new tobacco varieties. The main features of the capillary gas chromatography method can be described as follows:

*Method scope.* Ground tobacco samples are extracted with aqueous sodium hydroxide and partitioned into chloroform. The separation and quantitation of individual alkaloids is achieved by capillary gas chromatography with anethole as internal standard.

*Sample preparation.* Each ground tobacco sample is wetted with a small volume of water. After wetting, chloroform and 10% sodium hydroxide are each added and the tobacco sample is then extracted by agitation. Centrifugation follows the extraction step, from which an aliquot of the chloroform layer is analyzed by capillary gas chromatography.

*End determination.* A capillary gas chromatograph with temperature programming capability is used to separate the individual alkaloids. Gordon and Greene described capillary column specifications and chromatographic parameters to separate nicotine, nornicotine, myosmine, anabasine and anatabine in a tobacco extract in 1981 [36]. An apolar fused silica capillary column (SE–54 stationary phase) was used with flame ionization detection. Total separation time was approximately 40 min. The advent of 'fast chromatography' has greatly reduced separation times. Separation of the individual alkaloids is now possible in less than six minutes (Fig. 4) with a total separation time under ten minutes (Gordon BM, personal communication). Quantitation is based on an internal standard technique.

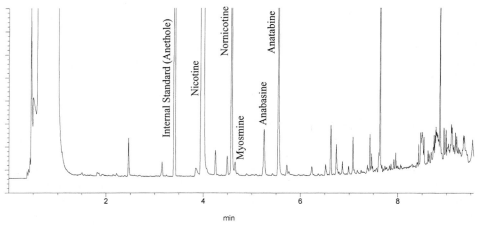

Fig. 4 Separation of nicotine and other alkaloids found in a tobacco sample by capillary gas chromatography. Fast GC analysis conditions were employed for this separation. Column: 10M × 180 μ O.D. with 0.4 μ DB-5 stationary phase. Oven temperature program: 60°C for 0.5 min, then programmed at 50°C/min to 100°C with no hold time, next programmed at 10°C/min to 165°C with no hold time, and then programmed at 40°C/min to a final temperature of 220°C with a 1 min hold time. Splitless injection. Helium carrier gas was employed with an average linear velocity of 60 cm/sec (19 psi column head pressure, 1.4 mL/min flow rate).

*Nicotine reporting requirements.* Individual alkaloid concentrations are reported as a percentage of the original tobacco weight on an 'as is' basis to an accuracy level of from two to four decimal places, depending on the specific alkaloid.

Studies reported by Gordon and Greene have shown the capillary gas chromatography method to be both accurate and precise for the determination of specific alkaloids. Results from tobacco samples analyzed simultaneously by the Cundiff and Markunas method [37] and the capillary gas chromatography method were in good agreement. Precision was improved with the chromatography method.

*The determination of nicotine in smokeless tobacco products*
In 1996, an initial collaborative investigation of a method for the determination of nicotine in smokeless tobacco was completed. Five corporate laboratories and one university laboratory conducted the investigation as part of a working group formed in response to federal regulations governing smokeless tobacco (Bennett, C.B. personal communication). The analytical method investigated for smokeless tobacco is similar in some ways to the method described earlier in this section for the determination of specific alkaloids in tobacco. The smokeless tobacco method requires that tobacco samples are ground, made basic with sodium hydroxide and partitioned with methyl-*t*-butyl ether. The organic layer is analyzed by capillary gas chromatography to determine the nicotine concentration in the smokeless

tobacco sample. Tobacco nicotine concentrations determined with the method represent nicotine response only, free of any additional response from other alkaloids such as nornicotine or anatabine. The method investigated is described in further detail in the Regulatory Aspects section of this chapter.

Results from an initial collaborative study to investigate the smokeless tobacco method are summarized in Table 3. The five tobacco types evaluated in the study (dry snuff, loose leaf, moist snuff, plug and twist) contained vastly different amounts of moisture. As measured by the oven volatiles method, 'moisture contents' ranged from approximately 10% (dry snuff) to in excess of 50% (moist snuff). For all tobacco types, it is clear from the results in Table 3 that repeatable results can be obtained within a given laboratory, and that reproducible results can be obtained between laboratories when using the smokeless tobacco method. In addition to the precision results in Table 3, nicotine recovery studies were performed in two laboratories to assess accuracy (Bennett C.B., personal communication). For the five smokeless tobacco types studied, one laboratory found nicotine recoveries of ~104–108%. The other laboratory found recoveries from 98% to 118%.

In summary, this section has attempted to provide an overview of analytical procedures for the determination of nicotine in tobacco. Emphasis has been placed on three examples of current analytical practice. However, many other analysis strategies beyond the examples presented are in use today. For example, in a recent CORESTA Routine Analytical Chemistry Sub Group joint experiment coordinated by Hans Thomsen to investigate the determination of nicotine in tobacco by gas chromatography, eighteen laboratories reported eleven separate combinations of sample preparation procedure and chromatographic conditions as described in Table 4 (Thomsen H.V., personal communication). Therefore, the choice of tobacco nicotine determination that is practiced in a particular laboratory will often be determined as much by practical considerations (e.g., instrument availability considerations) as it will be by scientific considerations such as the tobacco type of interest.

## IV. DETERMINATION OF NICOTINE IN TOBACCO PROCESSING ENVIRONMENTS

Tobacco is used in a variety of products (e.g., chewing tobacco, snuff, pipe tobacco and cigarettes). Growers and manufacturers process all tobacco in some way before or during its preparation into a variety of products for use by consumers. The flavor, taste and aroma, as well as the cost and consumer desires usually determine the choice and level of different tobacco types (Burley, flue-cured, Oriental, Maryland, cigar types, Latakia, Perique and various types of fire-cured and air-cured tobaccos) used in the production of different products. The taste and/or smoke flavor of tobacco leaf may vary

313

TABLE 3

Results from an interlaboratory investigation of a method to determine nicotine in smokeless tobacco

| lab | aliquot | Smokeless tobacco type | | | | | | | | | |
|---|---|---|---|---|---|---|---|---|---|---|---|
| | | moist snuff | | dry snuff | | plug | | twist | | loose leaf | |
| | | nicotine (% dry wt.) | oven volatiles (%) | nicotine (% dry wt.) | oven volatiles (%) | nicotine (% dry wt.) | oven volatiles (%) | nicotine (% dry wt.) | oven volatiles (%) | nicotine (% dry wt.) | oven volatiles (%) |
| A | 1 | 2.68 | 54.45 | 1.45 | 10.04 | 1.48 | 20.53 | 3.87 | 17.74 | 0.86 | 19.88 |
| A | 2 | 2.68 | 54.44 | 1.45 | 9.88 | 1.48 | 20.39 | 3.93 | 17.77 | 0.87 | 19.84 |
| A | 3 | 2.66 | 54.49 | 1.44 | 9.98 | 1.46 | 20.47 | 3.87 | 17.66 | 0.86 | 20.14 |
| B | 1 | 2.72 | 54.20 | 1.47 | 10.00 | 1.54 | 20.20 | 3.94 | 17.20 | 0.96 | 19.40 |
| B | 2 | 2.76 | 54.20 | 1.53 | 9.60 | 1.56 | 20.00 | 3.84 | 17.40 | 0.95 | 19.40 |
| B | 3 | 2.87 | 54.60 | 1.50 | 9.80 | 1.56 | 20.20 | 3.94 | 17.40 | 0.96 | 19.60 |
| C | 1 | 2.69 | 53.90 | 1.40 | 9.31 | 1.50 | 19.85 | 3.93 | 17.36 | 0.91 | 19.54 |
| C | 2 | 2.79 | 54.06 | 1.42 | 9.26 | 1.54 | 19.72 | 3.92 | 16.88 | 0.87 | 19.26 |
| C | 3 | 2.63 | 54.02 | 1.50 | 9.11 | 1.49 | 20.07 | 3.99 | 17.10 | 0.89 | 19.29 |
| D | 1 | 2.79 | 54.06 | 1.53 | 9.41 | 1.55 | 20.11 | 4.10 | 17.22 | 0.91 | 19.38 |
| D | 2 | 2.77 | 53.82 | 1.50 | 9.39 | 1.55 | 20.12 | 4.10 | 17.23 | 0.90 | 19.33 |
| D | 3 | 2.80 | 53.89 | 1.50 | 9.38 | 1.54 | 20.23 | 4.12 | 17.15 | 0.91 | 19.37 |
| E | 1 | 2.76 | 55.00 | 1.41 | 10.10 | 1.57 | 23.00 | 4.21 | 19.00 | 0.87 | 21.00 |
| E | 2 | 2.73 | 55.00 | 1.42 | 10.00 | 1.56 | 23.00 | 4.15 | 19.00 | 0.89 | 21.00 |
| E | 3 | 2.64 | 55.00 | 1.51 | 10.20 | 1.55 | 23.00 | 4.15 | 19.00 | 0.89 | 21.00 |
| F | 1 | 2.75 | 54.50 | 1.45 | 10.30 | 1.47 | 19.20 | 3.86 | 17.40 | 0.90 | 19.80 |
| F | 2 | 2.69 | 54.70 | 1.46 | 10.30 | 1.53 | 21.40 | 3.83 | 16.90 | 0.87 | 19.90 |
| F | 3 | 2.72 | 54.00 | 1.45 | 9.50 | 1.53 | 21.10 | 3.86 | 17.40 | 0.88 | 19.30 |
| *avg* | | *2.73* | *54.22* | *1.47* | *9.68* | *1.52* | *20.24* | *3.94* | *17.32* | *0.90* | *19.56* |
| *stdev* | | *0.06* | *0.29* | *0.04* | *0.38* | *0.04* | *0.53* | *0.12* | *0.27* | *0.03* | *0.28* |
| *rsd* | | *2.3%* | *0.5%* | *2.8%* | *4.0%* | *2.3%* | *2.6%* | *3.1%* | *1.5%* | *3.5%* | *1.4%* |

TABLE 4 (Part 1)

Sample extraction and GC analysis conditions practiced in different laboratories during CORESTA† joint experiment 23

| Laboratory Code | | B, E, F, H, AC, AJ | S | C, R | P | W | K | D |
|---|---|---|---|---|---|---|---|---|
| Column Type | | Stabilwax-DB | DB-1701 | CP WAX | CP WAX | CP-Sil (CB | Permabond CW20M | DB WAX |
| Dimension | m*mm | 15*0.25 | 15*0.53 | 25*0.53 | 10*0.32 | 25*0.25 | 25*0.53 | 15*0.53 |
| Film | um | 0.25 | | 2 | 0.2 | 1.2 | – | 1 |
| Split | | + | + | – | – | – | | + |
| Extraction solution | | MeOH | Water 5% NaOH | MeOH Ammonia | Chloroform | Methyl-ter-butyl-ether | MeOH KOH | Chloroform NaOH |
| Internal std | | Isoquinoline | n-heptadecane | | n-heptadecane | Quinoline | n-heptadecane | Anethole |
| Pre treatment | | 20% Ammonia | – | – | 20ml 0.5 m HCl | 2M NaOH | | |
| Pre treatment time | min | 30 | – | – | 15 | 15 | 30 | |
| Extraction time | | 60 | 60 | 60 | 30 | 120 | 270 | 20 |
| Shaking | | + | + | + | + | + | + | + |
| Reflux | | – | – | – | – | – | – | – |
| Ultra sound | | – | – | – | – | – | + | – |
| **Alternatives** | | | | | | | | |
| Internal std | | Quinoline Decanol | | | | | | |
| Column | | Carbowax+KOH Stabilwax – DB | | | | | | |

The laboratory procedures reported have been presented as accurately as possible. But some simplifications may have been necessary (Hans Valdemar Thomsen, 23-09-98).

† CORESTA = Cooperation Center for Scientific Research Relative to Tobacco.

TABLE 4 (Part 2)

Sample extraction and GC analysis conditions practiced in different laboratories during CORESTA† joint experiment 23

| Laboratory Code | | Z | R | I,V | AG |
|---|---|---|---|---|---|
| Column Type | | 10% carbowax | Versamid+KOH | 10% CP WAX | 20M+KOH |
| Dimension | m*mm | 1*4 | 2*4 | 2*4 | 1.5*4 |
| Film | um | | | | |
| Split | | – | – | – | – |
| Extraction solution | | Water 0.5N NaOH | EthOH KOH 10 mg/ml | Hexane 2.5 M NaOH | MeOH NaOH |
| Internal std | | Benzyl-OH | n-heptadecane | | n-heptadecane |
| Pre treatment | | | | | 32% NaOH |
| Pre treatment time | min | | | | 15 |
| Extraction time | | 24 | 120 | 60 | 60 |
| Shaking | | + | – | + | + |
| Reflux | | – | + | – | + |
| Ultra Sound | | – | – | – | – |
| **Alternatives** | | | | | |
| Internal std | | | | | |
| Column | | | Carbowax+KOH | | |
| Extraction konc. | | | | | |

The laboratory procedures reported have been presented as accurately as possible. But some simplifications may have been necessary (Hans Valdemar Thomsen, 23-09-98).

† CORESTA = Cooperation Center for Scientific Research Relative to Tobacco.

depending on the location in which it is grown, the climate, agricultural and post-agricultural practices employed, the location of the leaves on the tobacco stalk and the maturity of the leaf [38]. Because of the diversity in leaf characteristics and the inherent flavor in different tobaccos, scientists have explored the chemistry of tobacco in great detail. One of the chemical constituents often examined is nicotine. The nicotine content of tobacco lamina, stem and processed tobacco materials can often be used in conjunction with other chemical, physical and sensory measures to distinguish between different tobacco types, grades of tobacco and different processing means [39].

This section will describe and characterize the main types of tobacco used in the tobacco industry. Typical tobacco nicotine concentrations of those tobacco types will be presented. Tobacco processing techniques will then be discussed in terms of their effect on changes in tobacco nicotine concentrations. A description of the processing that occurs with tobacco from the field to finished products will be discussed. Tobacco processing within the cigarette industry will then be described with emphasis on how those processing steps affect the nicotine concentration in tobacco blends. Analytical techniques for direct and indirect, qualitative and quantitative determination of nicotine, where these determinations have been practiced and documented in the industry, will be discussed.

### Tobacco types

Flue-cured tobacco derives its name from the 'flues' used to heat the buildings (small barns) where the tobacco is packed to undergo its curing process [40]. During the flue-curing process stalk-ripened tobaccos are placed in barns and the temperature of the barns is slowly increased in a particular manner to wilt, yellow, fix and then dry the tobacco. During this process, enzymatic activity converts the majority of the tobacco starch to sugar and various tobacco pigments degrade to produce a golden or yellow color in the leaf [39,41]. Flue-cured tobacco is principally used in the cigarette industry. Its smoke is often described as having a sweet, aromatic character [38]. Flue-cured tobaccos contain significantly higher levels of reducing sugars than air-cured tobaccos such as Burley or Maryland. Flue-cured tobacco generally has a lower level of tobacco nicotine compared to Burley. Although the average nicotine level of flue-cured (~2 to 4%) is lower than Burley, the nicotine level of particular grades of flue-cured from different growing locations, having different climates, and employing varied agricultural practices can vary greatly [39]. Table 5 lists the mean percent tobacco nicotine levels for US flue-cured and Burley tobacco crops grown between 1967 and 1990. The data in Table 5 are archival in nature. Each set of data (flue-cured and Burley) for each year represents an average of multiple analyses conducted on inventoried tobaccos from all tobacco grades purchased during that year. Methods for determination of tobacco nicotine have varied and have been improved continuously over the period of 1967–1990. Spectrophotometric analysis of total tobacco alkaloids measured as nicotine was the method used in Table 5 from 1967 to ~1980. Gas chromatographic analysis of tobacco nicotine has been used as the method of choice since ~1980. During the period of 1967–1990 samples have been analyzed in different laboratories in R.J. Reynolds Tobacco Company (RJR) and by numerous analysts. All data were collected on dried and ground tobacco samples. All results are reported as either total tobacco alkaloids (measured as nicotine) or as nicotine on a dry weight basis. The historical data in Table 5 illustrates the year-to-year tobacco nicotine variability in flue-cured and Burley tobacco. Over the 23-year period, flue-cured tobacco nicotine ranged from 2.14–3.50%. Burley tobacco nicotine ranged from 2.43–4.63%. Although the percent tobacco nicotine for flue-cured is considered, in general, to be lower than the percent tobacco nicotine of Burley

TABLE 5

Percent nicotine levels for total US crop purchases*

| Crop Year | Tobacco type | |
|:---:|:---:|:---:|
| | Flue-cured | Burley |
| 1967 | 2.63 | 3.22 |
| 1968 | 2.53 | 3.63 |
| 1969 | 2.27 | 4.63 |
| 1970 | 2.76 | 4.16 |
| 1971 | 2.35 | 2.93 |
| 1972 | 2.41 | 3.46 |
| 1973 | 2.14 | 3.28 |
| 1974 | 2.63 | 3.12 |
| 1975 | 2.87 | 3.47 |
| 1976 | 3.29 | 3.62 |
| 1977 | 3.46 | 3.33 |
| 1978 | 2.76 | 3.08 |
| 1979 | 2.70 | 2.43 |
| 1980 | 3.50 | 3.92 |
| 1981 | 3.24 | 3.42 |
| 1982 | 2.26 | 3.14 |
| 1983 | 3.24 | 4.27 |
| 1984 | 2.84 | 3.40 |
| 1985 | 3.63 | 3.55 |
| 1986 | 3.15 | 3.40 |
| 1987 | 2.92 | 4.02 |
| 1988 | 2.98 | 3.80 |
| 1989 | 2.47 | 3.94 |
| 1990 | 3.50 | 3.40 |
| Average | 2.86 | 3.53 |
| Standard Deviation | 0.44 | 0.47 |

* Data is from R. J. Reynolds Tobacco Company

tobacco, there were years when this is not the case. In 1977, 1979 and 1985 the average nicotine level in flue-cured crop was higher than the Burley crop (Fig. 5).

Burley tobacco is considered a light, air-cured tobacco [38]. It is normally light-brown to reddish brown in color. It possesses a taste, smoke flavor and aroma described as having a 'chocolate, nutty, protein, winey, and ammoniacal' character [38]. Burley tobacco possesses excellent smoking characteristics when used in blended tobacco cigarettes and in pipe tobaccos. Smokers do not usually prefer smoking Burley tobacco alone because they are considered too 'harsh and strong'. Burley tobacco is low in sugar, which is metabolized from starch during the air curing process. The nicotine content of Burley is generally higher than that of flue-cured tobacco (~2 to 4%) depending on the type, leaf position on the stalk, growing season and growing location [39,42]. In smoking products, sugar based casings are normally applied to Burley tobaccos to mellow its smoking character. Burley tobacco is also often blended with flue-cured tobacco (that is high in sugar) to help 'smooth' the more 'alkaline' smoke of Burley tobacco [38].

Maryland tobacco is air-cured and is considered to have excellent smoking characteristics. It has characteristics similar to Burley tobacco but is much

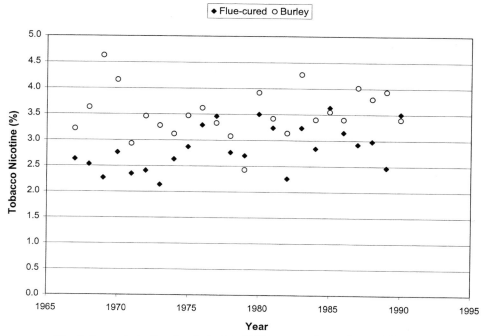

Fig. 5 Percent nicotine levels for total US crop purchases from 1967–1990.

lighter in taste. Straight Maryland cigarettes are popular in some European countries such as Switzerland [38]. The tobacco nicotine level of Maryland tobacco is about one to two and one-half percent by weight.

Oriental tobaccos are unique. Their small leaf size (three to six inches) relative to flue-cured and Burley leaf (twelve to sixteen inches) and their aromatic odor and taste characterize the uniqueness of Oriental tobaccos. Oriental tobaccos are usually used in blended cigarettes although straight Oriental cigarettes are produced and are popular in the regions where they are grown (most countries bordering the Aegean, eastern Mediterranean and Black seas). Oriental tobaccos have high levels of volatile flavor oils and provide for a rich smooth flavor effect when smoked. The smoke of Oriental tobaccos range from highly aromatic in character to a resinous-cedar character largely depending on the region where the tobacco is grown [38]. The nicotine content of Oriental tobaccos is normally low and ranges between one to two and one-half percent [39].

Cigar tobaccos are all air-cured tobaccos [43–46]. They are classified as a medium to heavy bodied tobacco and are medium to dark-brown in color. Cigars have a core made of the cigar filler. The filler is wrapped with a tobacco leaf or a reconstituted tobacco referred to as the cigar binder. The outer wrapper of the cigar is known as the cigar wrapper. Depending on the cost of the cigar this outer cigar wrapper may be a high-quality, fine, unblemished single tobacco leaf or a specially prepared reconstituted tobacco sheet prepared from cigar tobacco made to look like a cigar tobacco leaf. The cigar wrappers are usually grown under a cloth to shade them from the direct sunlight, temperature and wind that can damage the leaves. This type of tobacco is known as 'shade grown' tobacco. Not only is cigar tobacco air-cured, but it is also fermented under controlled conditions. To ferment the air-cured tobacco, the tobacco is moisturized, bulked and allowed to stand for varying periods of time. During the fermentation process, the tobacco reaches temperatures of 115 to 120°F. The fermentation process produces the characteristic color and aroma of cigar tobacco. Numerous chemical and biological changes occur during the fermentation process. Those changes produce a variety of color, flavor and odor differences (via Browning and Maillard Reactions) in the finished fermented tobacco. Much of the nicotine in the cigar tobacco is biologically degraded during the fermentation process. Air-cured cigar tobacco typically has a concentration of three and one-half to four percent tobacco nicotine prior to fermentation. After fermentation, as much as 80% of its original nicotine is lost [43].

There are other curing and fermenting processes that are used on tobacco [38]. Tobaccos can be fire-cured in which they are smoked over low burning fires during the curing process. The fire-cured tobacco is then packed for aging. Fire-cured tobaccos are not fermented. There is another unique tobacco called Latakia, which is an Oriental sun-cured variety grown principally in Syria that is fire-cured. These types of tobacco are generally used in pipe tobacco blends and give unique smoky notes to the pipe smoke. Perique tobacco is a fermented tobacco type that has a very unique flavor. Perique tobaccos are prepared by fermenting green tobacco leaf under pressure in wooden cases for several years. The finished leaf is black and has a sweet aroma; it is also used in pipe tobacco blends. The tobacco nicotine content of fire-cured tobaccos can vary depending on the nicotine content of the original leaf. There is not a substantial loss of nicotine from fire-curing. The nicotine content of Latakia is ~1% or lower, as it is an Oriental-type tobacco. The nicotine content of Perique tobacco is also very low (less than ~1%) due to the extensive fermentation that it undergoes.

*Processed tobaccos*

There are three principal types of processed tobacco; processed stems, reconstituted sheet materials and expanded tobacco.

*Processed stems.* After tobacco leaf is purchased – primarily flue-cured and Burley leaf – it is normally taken to a stemming operation where it is conditioned with water to a moisture content necessary to separate the lamina from the stem or midrib of the individual leaves. Only about 65 to 75 percent of the weight of the purchased leaf tobacco will end up as lamina strips that will be processed into tobacco cut filler. The rest of the material is classified as tobacco by-products and is used to prepare processed tobaccos.

As depicted in Fig. 6 [42], the stem and dust are separated from the broken leaf (tobacco scrap) and lamina. The lamina is dried to about 12% moisture and packed in bales. The bales are taken to storage where they are held for 18 to 36 months on average. The dust from the stemming operation and the drying operation are also dried to a low enough moisture, usually less than 10%, before they are boxed up and taken to storage. Likewise, the stems and tobacco scrap are dried and boxed for later use. When called for, the boxes of stems are sent to processing areas to prepare useful materials that can be added to or blended with the tobacco lamina. Burley or flue-cured stems can be processed. Processed stem materials are cut or crushed rolled stems (CRS) [42]; cut rolled stems can also be partially expanded. Stems, after being moistened, cut and sized to about one to two inches, are rolled under pressure and cut into disk shaped pieces of a size similar to tobacco cut filler. The cut stem pieces can also be partially expanded using steam [67]. The nicotine level of processed stems is low and is normally about 25% of the nicotine level found in the lamina [42].

Stemmery dust is not usually sent to the tobacco reconstitution process as it usually contains large amounts of sand and dirt (>10%). The sand content of stemmery dust is difficult to remove. As a result, some manufacturers simply dispose of this material and place it into land-fills.

*Reconstituted tobacco.* Preparation of reconstituted tobacco sheet is an excellent economic way that tobacco manufacturers have found to convert

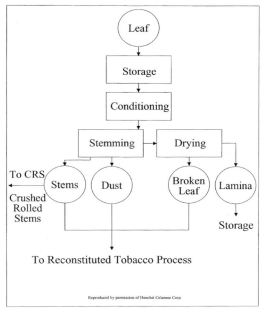

Fig. 6 Stemming process.

tobacco dust, fines, scrap and stem into useful tobacco filler. Tobacco reconstituted sheet materials can be of such good quality and appearance that they can be substituted for tobacco lamina in cigarettes and for cigar filler. As previously mentioned, reconstituted sheets are also used as cigar binders and in some cases, cigar and cigarillo outer wrappers [38].

There are at least five processes for making reconstituted tobacco sheet [47,48]. Browne [42] has described five different methods, but the two major methods employed today are the cast sheet, or slurry process, and the paper making process. Several reviews and articles have been written on the production and use of reconstituted tobacco which detail the physical, process engineering and smoking and health aspects of reconstituted sheet materials [49,50]. The books by Sittig and Halpern [51,52] review much of the patent literature on tobacco reconstituted sheet materials.

For the cast sheet process, tobacco by-products such as tobacco dust, fines, scrap and stems in various proportions are dried and ground to a very fine powder. The powder is mixed into an aqueous solution of binder. Typical binders are methylcellulose, carboxymethylcellulose, guar or locust-bean gum [42]. Additional fiber, such as cellulose, can be added. Additionally, combustion and ash modifiers, humectants and flavors can be added. The suspension is usually mixed thoroughly, sheered in a blender or put through a refiner to prepare a viscous slurry of fine particle size. The slurry is transferred to a head-box, and a thin film of the slurry is cast onto a stainless-steel belt, dried, reconditioned if necessary, doctored from the belt and either rolled up or cut into pieces the size of lamina. Certain binders can be used to prepare slightly expanded sheet materials. These binders are heat sensitive and expand during the drying process. Some cast sheets are prepared without the addition of a binder. In this type of cast sheet the natural pectin in tobacco is chemically released with the use of ammonia and diammonium phosphate as processing aids. The pectin is cross-linked during the drying process to produce an excellent tobacco reconstituted sheet material [66].

The nicotine content of cast sheet material is totally dependent on the initial nicotine content of the tobacco mix used to form the sheet. The cast sheet process is over 95% efficient in terms of material mass balance. Typically the nicotine content of cast sheet materials is between 1 and 4% nicotine [55]; but it is also true that about 5 to 20% of the nicotine can be lost in processing during the drying step. Therefore, the overall amount of tobacco nicotine that is found in the cast sheet material is always less than the amount of nicotine that was originally present in the starting material.

*Paper-processed reconstituted tobacco sheet.* To prepare reconstituted tobacco sheet by the paper process, a mix of tobacco by-products is used as the feedstock material (Fig. 7) [42]. Tobacco stems (flue-cured and Burley stems) are the largest component of the mix, usually at least 50% by weight. The rest of the mix is made of tobacco dust, tobacco scrap, by-products from the primary tobacco blending area, cigarette tobacco making by-products, and broken sheet from sheet operation (broke). The starting materials are called from tobacco storage. Often the materials are ground (scrap) or cut (stems) to reduce their size and to improve the extraction efficiency. The tobacco mix is then placed in extraction vessels with warm water and agitated to remove as much water-soluble material as possible. In the extraction process about 45 to 55% of the weight of the initial tobacco is soluble and can be separated from the tobacco pulp. The extract is then separated from the pulp. The pulp is washed with water and the combined dilute extracts are combined and concentrated in an evaporator. The 'concentrated' extract (20% concentration of tobacco solids and solubles) is still quite dilute. The washed pulp is then digested, beaten mechanically and/or refined to fibrillate the cellulose and reduce the size of the fibers comprising the biomass composition of the pulp.

The clean refined pulp is very dilute at this point in the process, usually less than 1% suspension in water. The dilute refined pulp is pumped to a head-box where it is formed into a web on a wire screen of a standard papermaking machine, dried by suction and heated drums or hot air to a moisture level of 40 to 60%. In a parallel operation, the extract was 'concentrated' as previously mentioned. Additional ingredients can be added to the extract at this point. Humectants, combustion and ash conditioners, sugars and flavors are often added to the extract to improve its material performance characteristics (friability, moisture retention, flexibility, etc.), burning properties and tobacco taste and smoke flavor. The 'concentrated' extract is then applied to the wet web, dried in ovens, conditioned to a moisture of about 10 to 12% and either rolled up or cut into pieces about the same size as tobacco lamina (3 inch by 5 inch rectangular shapes).

The paper process operation for the production of reconstituted tobacco sheet is reasonably efficient. Operating efficiencies range from about 80 to 95% at conversion costs that are quite dependent on the converter, the source of tobacco materials, the additives/processing aids used to prepare the sheet and the physical dimensions and properties of the reconstituted tobacco sheet produced. To be economical, the paper process requires a high throughput; initial capital investments are high. As a result only large tobacco manufacturers and national monopolies can justify their own facilities and equipment. Many smaller tobacco manufactures send their tobacco by-products to other converters for processing [42].

The nicotine content of reconstituted tobacco sheet material is totally dependent on the initial nicotine content of the tobacco mix used to form the sheet. The efficiency of the papermaking operation in terms of material mass

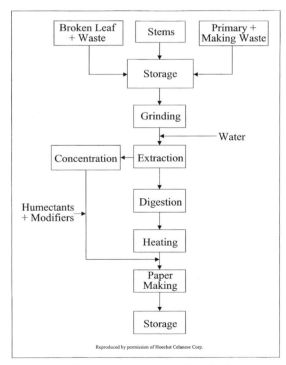

Fig. 7 Wet process for reconstituted tobacco.

balance is 80 to 95%. Typically the nicotine content of reconstituted tobacco sheet materials is between 1 and 4% nicotine [55]. But it is also true that about 5 to 20% of the original nicotine present in the feedstock can be lost in processing, especially during the drying step. Therefore, the overall amount of tobacco nicotine that is found in the reconstituted sheet material is always less than the amount of nicotine that was originally present in the starting material.

*Expanded tobacco.* Numerous processes have been developed over the years to expand tobacco and thus increase its volume and reduce the weight of the space it occupies [42,67,68]. The technique typically involves the impregnation of tobacco with an expansion agent which is a liquid or gas and then the release of that agent from the tobacco to increase the volume of the tobacco. Expansion agents have included water, steam, a wide variety of organic fluids and inorganic fluids. Volume expansions vary based on the mechanical/engineering reduction-to-practice and operating conditions, the impregnating liquid or gas employed, the tobacco type (e.g., flue-cured, Burley, Oriental, etc.) and tobacco form (lamina, stems). A schematic diagram for a typical expansion process is shown in Fig. 8. In this process, tobacco lamina is shredded into cut-filler form (lamina cut at 25 to 35 cuts per inch wide). The tobacco is moistened to 25 to 35% and then saturated (impregnated) with an expansion agent. The impregnated tobacco is then

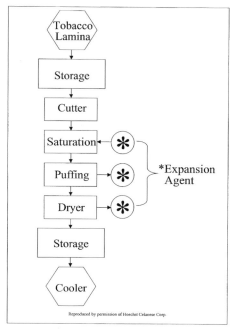

Reproduced by permission of Hoechst Celanese Corp.

Fig. 8 Tobacco expansion process.

contacted with a stream of steam to expand the tobacco; the expansion agent is most often recovered and reused. The expanded tobacco is then dried, cooled, conditioned to a set moisture content and either used immediately or packaged for storage. Today, the preferred expansion agent is carbon dioxide; use of expanded tobacco is cost effective and is widely used in the cigarette industry. The normal inclusion rate ranges from about 5 to 50%, with the normal level of incorporation in the range of 10 to 20%.

Expansion processing does not, in most cases, dramatically change the nicotine content of expanded tobacco compared to the infeed tobacco. Nicotine losses on the order of 5 to 10% are typical for most expansion processes used today. Factors that may effect these typical nicotine losses are the type of tobacco impregnated, any additives that might be applied to the tobacco, temperatures used to dry and condition the tobacco after expansion, and other design parameters that may be unique to each tobacco expansion facility.

*Tobacco processing*
Tobacco is a natural product and as such a certain amount of variability in the physical and chemical properties of tobacco is expected to occur for each type grown, for each growing season, and for each location where it is grown in the world. Year-to-year, crop-to-crop and location-to-location variations are due to a variety of circumstances; some controllable (e.g., certain kinds of agronomic practices), and some uncontrollable (e.g., weather-related variables). To compensate for tobacco variability, blends of different types of tobacco are usually used in smoking (pipe, cigar and cigarette tobacco blends) and smokeless (wet and dry snuff, chewing and plug tobacco) tobacco products, to ensure consistency in the final product mixture. For each different tobacco product a variety of different process steps take place due to manufacturing costs, tobacco availability, the different forms of products manufactured (e.g. dry snuff versus moist chewing tobacco) and the desired end use and preferences of the consumers (e.g. pipe tobacco vs. cigarettes). Depending on the particular finished tobacco product, various processing steps (e.g., conditioning, blending, casing, cutting, bulking, flavoring, etc.) can be used to compensate for certain variations in the chemical composition of the tobaccos used (relative amount of sugars, acids, volatile oils and alkaloids). In this way it is possible for a manufacturer to provide consistent products to consumers year after year. The following section will describe the processing steps that tobacco undergoes from when it is harvested to when it is prepared into finished products. Although several chemical constituents are known to change during the processing of tobacco, special emphasis will be placed on changes in tobacco nicotine.

*Effects of harvesting, curing methods and bulking on tobacco nicotine content.* The reported range of nicotine in different tobacco types used in the manufacture of tobacco products varies greatly (Table 6). Data in Table 6 are

TABLE 6

Tobacco handling practices and reported effects on tobacco nicotine * [13,17,19–21,28]

| Tobacco Type | Virginia | Burley | Burley | Maryland | Oriental | Cigar | Cigar |
|---|---|---|---|---|---|---|---|
| **Reported Ranges of Nicotine (%)** | 0.8–3.5 | 0.40–4.50 | 0.40–4.50 | 0.65–2.0 | 0.5–1.30 | 0.2–3.7 | 0.2–3.7 |
| ***Tobacco Before Sold*** | | | | | | | |
| **Tobacco Harvest Method** | | | | | | | |
| Nicotine Before Curing | Primed 1.10 | Stalk-cut 4.5 | Primed 4.5 | Stalk-cut NA | Primed 1.47 | Primed 0.35 | Stalk-cut 0.40 |
| **Tobacco Curing Method** | | | | | | | |
| Nicotine After Curing | Flue-cured 0.97 | Air-cured 4.25 | Barn-cured 3.99 | Air-cured NA | Sun-cured 0.91 | Air-cured 0.32 | Air-cured 0.40 |
| **Bulking Tobacco before Sale** | Nicotine losses are estimated to be on the order of 0.1% | | | | | | |
| ***Tobacco After Sold to Dealer/Manufacturer*** | | | | | | | |
| **Redrying of Tobacco** | | | | | | | |
| Before Redryer | 2.52 | 2.58 | | | | | |
| After Redryer | 2.45 | 2.48 | | | | | |
| **Natural Aging** | | | | | | | |
| Immediately After Curing | 2.74 | | 3.2 | | 1.00 | | |
| 4 Months Aging | | | 2.72 | | | | |
| 1 Year After Curing | 2.54 | | | | 0.92 | | |
| 2 Years After Curing | 2.46 | | | | 0.86 | | |
| 2.3 Years Aging | | | 2.14 | | | | |
| 30 Months Aging | 2.44 | | | | | | |
| **Fermentation** | | | | | | | |
| Before Fermentation | | | | | | | 0.73, 4.06 |
| After Fermentation | | | | | | | 0.33, 0.73 |
| Thermo-Vacuum | | | | | | | |
| Before | 2.03 | 3.47 | | | | | |
| After | 2.02 | 3.45 | | | | | |

* Note that data in this table come from numerous sources. The data represent different studies conducted by numerous researchers using different analytical methods on different tobacco types.

327

a compilation of literature data from numerous sources [39,43,45–47,54] and should not be considered a single data set. The data are presented to show relative change in tobacco nicotine that has been reported to occur at various stages in tobacco processing. The data under each tobacco type gives examples of data in the literature and are from different studies. The specific analytical methods used to report tobacco nicotine changes for the numerous studies cited within the table are not known. The particular varieties of each tobacco reported in Table 6 and the specific agronomic practices under which the tobaccos were grown, harvested and processed are also unknown. The reported ranges of nicotine for each tobacco type in Table 6 encompass normal cultivars and hybrid varieties of tobaccos as seen in Table 7. Table 8 illustrates the typical ranges of nicotine by stalk position that is normally seen in different tobacco types grown in the United States. At the time of harvest, green tobacco contains a level of nicotine that is governed in large part by the climatic, geographic and agronomic conditions that the tobacco experienced during its growing cycle. Data from the first part of Table 6 show

TABLE 7

Average total alkaloids of different tobacco cultivars [13]

| Cultivar | Total alkaloids (%) |
| --- | --- |
| **Flue-cured:** | |
| Coker 319 | 2.24 |
| VA 115 | 2.6 |
| NC 95 | 2.72 |
| McNair 12 | 2.62 |
| Speight G-7 | 2.64 |
| Hicks Broadleaf | 2.69 |
| NC 2326 | 2.62 |
| McNair 30 | 2.53 |
| Reams 266 | 1.69 |
| Golden Wilt | 2.75 |
| SC 58 | 3.58 |
| LN 38 | 0.1 |
| Coker 139 | 1.37 |
| MD 609 | 2.4 |
| Coker 129 | 1.5 |
| PD 33 | 1.6 |
| Hicks Broadleaf | 2.7 |
| NC402 | 2.9 |
| **Air-cured:** | |
| MD 609 | 0.8 |
| NC 2326 | 1.1 |
| Coker 319 | 1 |
| Burley 37 | 5.07 |
| Kentucky 14 | 5.07 |

TABLE 8

Average nicotine content of tobaccos by stalk position [13,28]

| Type of tobacco – curing | Stalk position | Nicotine (%) |
|---|---|---|
| Virginia, Flue-cured | Lower 1/3 | 1.87 |
| Virginia, Flue-cured | Middle 1/3 | 2.65 |
| Virginia, Flue-cured | Upper 1/3 | 3.26 |
| Burley, Air-cured | Lower 1/3 | 2.14, 2.02 |
| Burley, Air-cured | Middle 1/3 | 3.00, 2.69 |
| Burley, Air-cured | Upper 1/3 | 3.65, 2.60 |
| Maryland, Air-cured | Lower 1/3 | 1.79, 1.10 |
| Maryland, Air-cured | Middle 1/3 | 2.15, 1.68 |
| Maryland, Air-cured | Upper 1/3 | 2.96, 1.63 |
| Oriental, Sun-cured | Composite | 0.95 |
| Stem-Sheet | Composite | 0.85 |
| Virginia, Flue-cured | 1st Priming | 1.28 |
| Virginia, Flue-cured | 2nd Priming | 1.40 |
| Virginia, Flue-cured | 3rd Priming | 1.53 |
| Virginia, Flue-cured | 4th Priming | 1.82 |
| Virginia, Flue-cured | 5th Priming | 2.47 |
| Virginia, Flue-cured | 6th Priming | 3.35 |
| Virginia, Flue-cured | 7th Priming | 3.89 |

the change in nicotine content that those different tobacco types exhibit depending on the curing method employed by the grower. It should be emphasized that curing tobacco is an art. Different growers cure their tobaccos slightly differently depending on the maturity of the tobacco, the weather conditions during the time of curing, the overall climatic condition that the crop experienced during the season, etc. Typical nicotine losses that occur from curing are from 5 to 15% of the nicotine present in the green leaf. Data from this section of Table 6 illustrate that Oriental tobacco can lose nearly 40% of the nicotine in green leaf during sun-curing, although the nominal amount of nicotine loss (~0.5%) is similar to barn-cured Burley tobacco. Stalk-cut, fire-cured Burley tobacco can lose relatively little nicotine (5 to 15%) or nearly 30% [56] depending on how much heat the tobacco experiences during the curing process. Once the grower cures the tobacco it is ready to be sold at auction. The cured, dried tobacco is moistened or 'ordered' to prevent breakage during transport and handling. The conditioned tobacco is bulked before sale and is then either sold at auction or stored for a period before it is sold. The time that the tobacco is bulked often depends on the amount of tobacco the grower produced that year, the amount of stored tobacco the grower has from last year, the price of the leaf at market and other circumstances known to the grower or the market itself. Regardless, nicotine losses generally on the order of <0.1% are estimated to occur during the time that tobacco is bulked prior to sale [39].

Vacu-Dyne process. In this process tobacco was placed into a vessel, the vessel was evacuated, and steam was allowed to escape into the evacuated chamber via probes; the probes were pushed into the tobacco inside the vessel to furnish both heat and water to the tobacco. Several cycles were normally used to bring the dry tobacco to a pliable form that minimized tobacco breakage during processing of the leaf. The second type of process used today to remoisturize dry packaged tobacco from storage is an in-line process. Tobacco from inventory is brought to the processing area. The tobacco is removed from the bales employing 'kickers' that partially break up the tobacco bales, or in the case of the smaller burlap-wrapped Oriental tobacco, the Oriental bales are sliced into sections. The tobacco types are fed into direct conditioning cylinders and steamed at about 150 to 170°F. When the tobacco exits the direct conditioning cylinders, it is very pliable, moisturized and delaminated. The in-line process is used extensively today in large manufacturing areas, while the batch processing technique and equipment still finds a lot of application in smaller pilot plant operations. By employing either the batch process technique/equipment or the in-line conditioning process technique/equipment only a minimal amount of nicotine is lost. Typically less than 1% of the leaf nicotine is lost during this type of tobacco conditioning (Table 6).

*Effects of processing tobaccos/blends in the cigarette industry on tobacco nicotine content.* It was mentioned above that the first step in tobacco processing is to obtain tobaccos from a tobacco dealer or from a company's inventory. Questions such as what types of tobaccos are needed, how much of each type and what criteria should be used to determine quality and usability of the tobacco, are very important and need to be addressed. Before a blend can be put together, the blend specialist or cigarette designer must have a supply of different types of quality tobaccos and know information on the physical, chemical and organoleptic or sensory characteristics of those tobaccos. To obtain the necessary tobaccos, the tobacco manufacturers can contract grow the tobaccos needed based on specifications set by the manufacturers, buy tobacco at auction, or they can contract with one or more tobacco dealers to purchase specified types and grades of tobacco. Some tobacco dealers will stem and sometimes even store a manufacturer's tobacco until needed. A company's buying program and practices are usually based on a five-year or longer strategic plan. The buying plan is based on the needs that the company has in terms of its recent and projected market performance, its present tobacco blends, its present inventory of tobaccos, projections of future tobacco availability (domestic and off-shore) and new product development. The tobacco manufacturer sets minimum quality and uniformity standards for each type and grade of tobacco based on stalk position, tobacco maturity and consistency. Typical criteria for purchasing are tobacco maturity, color, odor and sensory attributes (when available).

Once the lots are purchased and stemmed the typical criteria of concern are particle size, nicotine level and total sugar content. Nicotine and sugar determinations can be obtained employing in-line or off-line near infrared (NIR) techniques [57–59] or by analysis of composite tobacco samples by colorimetric, gas chromatographic, ultraviolet spectroscopy or any number of other techniques known and available to the tobacco industry. The measurements for tobacco nicotine and sugars are done as a chemical check for the uniformity/consistency of each tobacco lot. Historically, the standard deviation expected in tobacco nicotine and sugar per lot of tobacco is 0.2% and 1.5%, respectively.

The cigarette industry in the United States (US) normally uses what has been termed the 'American Blend' which is comprised of varying ratios of flue-cured, Burley, Oriental, reconstituted sheet(s), expanded tobacco and a variety of tobacco by-products [42,48]. The majority of flue-cured and Burley tobaccos used in US cigarettes are grown and purchased domestically, although some manufacturers incorporate off-shore flue-cured and Burley tobaccos in their products. Oriental leaf is 100% off-shore tobacco. The majority of domestic flue-cured tobaccos are grown along the eastern coast of the United States. Flue-cured tobacco is grown and sold in four different tobacco belts (growing regions). These are Florida/Georgia, South Carolina, eastern North Carolina and the so-called 'Old Belt' that encompasses western North Carolina and Virginia [39]. Domestic Burley tobacco is grown and sold primarily in two different belts, Kentucky and eastern Tennessee. Both flue-cured and Burley tobaccos have similar growing seasons, but are harvested and cured differently, and are sold over several months. For example, the Florida/Georgia belt flue-cured tobacco matures, is harvested, cured and sold first (the tobacco market opens in early summer), while the Burley tobacco market opens early in the fall and often continues through early winter. At the beginning of the flue-cured market in the Florida/Georgia belt, lower stalk tobaccos are sold, stemmed, analyzed both chemically and physically for quality and placed in inventory. At the auction, prior to the sale, agents from the US Department of Agriculture (USDA) grade each lot of tobacco to be sold. For example, the USDA grade B3K indicates that the lot of tobacco is a flue-cured type, that it is a sample of leaf (B), that it is considered to have a uniform quality of 3rd (3) and that the leaf was ripe (K). A bundle or sheet of flue-cured tobacco can range from ~100–200 pounds. At the auction, the bundles of tobacco are sold quickly, are identified as purchased by the different tobacco companies or dealers, and often a company designation is placed on the different bundles of tobacco to specify a particular company grade. The tobacco is then shipped to a stemmery.

At the stemmery hundreds of bundles of tobacco are processed together in a sequential manner based largely on the sequence in which the different stalk positions ripen for harvest. Tobaccos from each belt are processed separately. Composite samples (five to ten) of each grade of tobacco from each belt are collected and analyzed for tobacco nicotine and sugar. An average tobacco nicotine and sugar value is assigned to each grade of tobacco for characterization of that grade and belt for a particular crop year. Additionally, other physical and often sensory characteristics are measured on the composite tobacco samples. As the sale of tobacco continues, the tobacco sold and processed at the stemmery comes from stalk positions higher up the stalk. Flue-cured and Burley tobaccos from different stalk positions and belts have different physical and chemical properties [13,28]. For example, lower stalk tobaccos are lower in nicotine compared to upper stalk tobaccos. Similarly, flue-cured tobaccos of similar stalk positions from the Florida/Georgia belt can be different in tobacco nicotine and sugar levels depending on the climatic and soil conditions compared to tobaccos grown in the Old Belt. During the tobacco market season, millions of pounds of tobacco will be graded, purchased, stemmed, analyzed (by composite samples) and placed in inventory. Each container of tobacco is designated by type, grade, belt and crop year, which constitutes a 'lot' of tobacco. After the last lot of Burley tobacco is placed into inventory, a complete picture of the results of the domestic tobacco-buying program for that year can be assessed and compared to the present inventory in light of the present and future needs of the company. The ultimate goal for the company's buying program is to obtain an inventory of sufficient quantities of high quality tobaccos that can be used to maintain product quality and consistency.

A company's tobacco utilization program and inventory management program are very important to the effective and efficient use of the millions of pounds of tobacco in inventory, while maintaining product quality and consistency. The objectives of these programs are to improve long-term blend consistency, minimize blend variability caused by disruptions in tobacco flow (e.g., unusual environmental conditions, catastrophic events, (fire, water damage, etc), political unrest in various regions of the world), and allow manufacturing operations to make tobacco grade substitutions without making dramatic changes in product specifications. One type of plan used in the tobacco industry to accomplish this objective involves the grouping and blending of different types of tobaccos (e.g., flue-cured upper stalk or lower stalk tobaccos) having similar chemical, physical, sensory properties to create common grouped blends of tobaccos. This approach can be accomplished by the use of computer simulation modeling techniques and statistical cluster analysis. Ideally, grouped lots of tobaccos, evaluated based on similarities (i.e., grade [quantified by stalk position, quality/uniformity and maturity], sensory ratings, chemical parameters and physical parameters) from the entire tobacco inventory are analyzed and placed into common group clusters. Each common group cluster of tobaccos contains thousands of containers of tobaccos, which are located in inventory and designated by a specific grade, belt and year. A typical common group cluster, for example, upper stalk flue-cured tobacco, would contain tobacco from several grades, belts and years. Based on the company's current blends, new product development needs, and future tobacco needs, the entire inventory can be clustered into several common group blends. Examples of common group blends might be: upper stalk flue-cured tobacco, lower stalk flue-cured tobacco, upper stalk Burley tobacco, lower stalk Burley tobacco, off-shore purchased flue-cured tobacco, off-shore purchased Burley tobacco, common blends of Oriental leaf, common blends of tobaccos used in the preparation of processed tobaccos, such as expanded or reconstituted tobacco sheet materials, or tobacco blends used in lower cost value cigarette brands. As the clusters are developed based on a set of criteria (grade [quantified by stalk position, quality/uniformity and maturity], sensory ratings, chemical parameters and physical parameters), it is possible to adjust the size (total available tobacco weight) and character of the common tobacco groups by weighting the different criteria. This is often important when tobacco crops differ greatly in criteria due to climatic and/or environmental changes that can occur in different growing seasons. Once adjustments to each common group cluster (in terms of size and common group character) are made to meet the needs of the company's production schedule, tobacco blends containing set percentages of different common group tobaccos can be developed to meet the product specifications. For a company to efficiently and effectively use its tobacco inventories and provide consistency in the physical, chemical and sensory parameters of its products, computer simulation models such as the one described can be employed (e.g., see Tables 9–11). Additionally, it is hoped that the use of such models may allow for better production efficiency and reduced waste.

When considering how a tobacco inventory is utilized, it is important first to understand the diversity of the marketplace for which cigarettes are being produced. For example, there are currently more than 1200 brand styles sold in the US that vary in their design, construction and tobacco blend. In the 12 months from June 1997 to June 1998, the US cigarette industry produced 475.35 billion cigarettes (Maxwell Report, [64]). About 98% of the cigarettes prepared and sold were filtered and only about 2% were unfiltered. Cigarettes are classified by the Federal Trade Commission (FTC) based on 'tar' yield per cigarette. The smoke nicotine and carbon monoxide (CO) yield per cigarette are also determined and reported by the FTC for each cigarette brand made and sold in the United States. Based on each cigarette brand's FTC yield of 'tar', it is classified as full flavored (FF, greater than 15 mg 'tar'), fuller flavored low 'tar' (FFLT, 7 to 15 mg 'tar') or ultra low 'tar' (ULT, 0 to 7 mg 'tar'). Within each brand category there are several hundred brands monitored by the FTC. Each company also monitors its brands for FTC 'tar' and CO yield to be sure that it falls within the values advertised. The cigarette market is highly fragmented and competition between brands is fierce. Each company tries to gain a 'competitive edge', by both developing an understanding of current and emerging consumer wants and by translating those consumer desires into product variations. To gain an understanding of the success of lead products in each cigarette category, companies conduct competitive brand analyses in efforts to better understand their competition, the competitor's product and reasons why smokers chose the competitor's product. Interpretation of the entire area of competitive brand analysis is outside the scope of this chapter and will not be discussed in detail. Suffice it to say that one of the studies normally conducted in competitive brand analysis is a breakdown of a competitor's product. An analysis of a competitor's product can reveal differences in its blend or cigarette construction parameters (cigarette length and circumference, cigarette papers, filter types, adhesives, filter ventilation type and level, cigarette draft, etc) compared to other products. Cigarette blend analyses are routinely conducted. A typical set of blend analyses includes an evaluation of the blend composition by tobacco type (percentage of each type of tobacco) and an evaluation of each blend component's level of tobacco nicotine and sugar. This information may provide clues to competitor's use of high vs. low stalk tobaccos in the blend. In formulating a blend to compete with the competitor's successful product difficulties can be encountered. There are numerous ways to blend tobaccos to match a particular tobacco blend's nicotine and sugar level. Not every blend of tobacco at a particular nicotine and sugar level will yield an acceptable smoke taste. The choice of tobaccos used in a blend is often directed by a company's tobacco inventory. It should be noted that in any particular growing season that the vast majority of the total US tobacco grown is purchased by the major US cigarette producers. The amount of tobacco that each manufacturer buys is proportional to the market share of each manufacturer. Therefore, each tobacco manufacturer is most likely to buy similar types of grades of tobaccos to ensure that its products remain consistent. If a company cannot prepare a sustainable blend that is consumer acceptable and that has a relatively constant tobacco nicotine and sugar level over long periods of time, its finished, marketed products will not be consistent to the consumer over time. The results of this inconsistency may well lead to the eventual rejection of product. Additionally, an analysis of the competitive product's FTC 'tar', smoke nicotine and CO is determined. From these types of data it is often possible to piece together valuable information as to whether the success of a competitor's product is product-based. It is possible to have a competitive product advantage based on other criteria such as name recognition or a particularly novel advertising or merchandising technique.

If a competitive advantage is product-based, it should be understood that the product was designed based on a specific FTC 'tar' category and that it has met the wants and acceptance criteria of smokers in that particular category. A company will often produce a prototype product for consumer evaluation that is based on product specifications that are similar to a competitor's product while trying to maintain the company's original product taste signature or product performance attributes. If successful, the company will try to market the new or improved product and will set product specifications for its manufacture. Typical product specifications for the new or improved product include: the type and percentage of each blend component based on common group clusters, flavoring and/or casings to be applied to the blend, the cigarette length, circumference, weight, level and type of filter ventilation, the cigarette paper and filter specification, cigarette tipping specifications, and packaging specifications. This total set of product specifications is necessary in order to compete with the taste and cigarette performance characteristics of the consumer-successful cigarette.

Neither tobacco blend nicotine content nor smoke nicotine yields are product specifications for cigarettes produced for sale in the United States. Tobacco nicotine is a result of the blend ratios of the common group blends of tobacco used to produce a consumer acceptable smoke flavor. Smoke nicotine is a consequence of smoking a consumer-acceptable combination of various levels of common group blended tobaccos placed into a specific cigarette configuration that are designed to generate a specific FTC 'tar' yield per cigarette. Tobacco blend nicotine content is not a product specification on exported cigarette products. Some cigarettes produced for export are required to have product specifications for smoke nicotine yields based on laws in the country to which they are being exported.

A typical cigarette manufacturing process based on a predetermined set of product specifications is illustrated in Fig. 9 [42]. The process can be divided into several steps. Based on product blend specifications exact levels of different common group blended flue-cured, different common group blended Burley, Oriental group blends, reconstituted tobacco materials and possibly group blends of Maryland tobaccos, are metered from common group blended tobacco bulkers and conditioned separately to moisten and delaminate the tobaccos. Each tobacco type is fed to a separate feeder line. Some of the flue-cured and/or Burley tobacco can be shunted to a tobacco expansion process for future inclusion in the blend or can be added via a separate tobacco expansion process that continuously produces expanded tobacco for later inclusion in different blends. The Burley tobacco itself or any part of the tobacco strip blend can be cased with sugars and humectants. Flavors can be added to the strip blend or components at this point. After the various specified tobaccos have been conditioned, delaminated and possibly cased and/or flavored, they are placed into blending silos or bulkers. From the blending silos or bulkers the tobacco strip blend can have additional casing materials applied, if specified. If casing is applied at this phase in the processing, the tobacco is run through a dryer and then to a bulker to reequilibrate the blend to a specified moisture for cutting (usually 18 to 24% moisture). The blended tobacco strip is then cut to a specified cut width, usually ~25 to 35 cuts per inch). The tobacco cut-filler is then delaminated and dried. Specified levels of processed tobacco stems and expanded tobacco filler are then added in a mixing drum. If specified, the final blend, which contains specified levels of cut tobacco strip materials, expanded tobaccos and stems can be top-dressed with flavors. The finished cut-filler tobacco blend is prepared at specified moisture and is sent to one or more bulkers for use in the preparation of cigarettes.

*Product consistency in terms of tobacco nicotine.* The ability for a company to produce a consistent product day after day is often dependent on the quality, quantity and consistency of a company's raw material and the company's ability to effectively and efficiently process the raw materials into finished

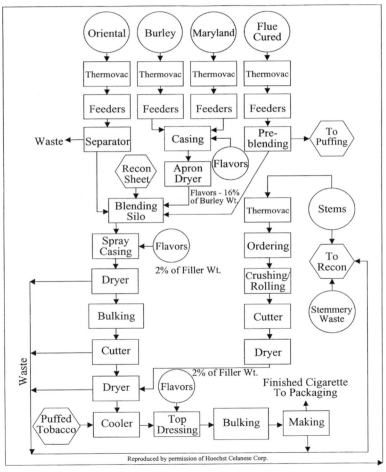

Fig. 9 Cigarette manufacturing process.

products. For the tobacco industry, a strategically directed tobacco purchasing plan and an effective tactically operated blending program are necessary.

R.J. Reynolds Tobacco Company (RJR) employs a tobacco blending system based on computer simulation modeling techniques and statistical clustering analysis. Its methods for effectively and efficiently utilizing their tobacco inventory are similar to those previously described. Estimates of blend nicotine variability have been calculated based on composite samples of tobacco collected and analyzed for tobacco nicotine during the stemming operation prior to placing the tobacco in inventory.

Target recipes for cigarette blends are composed of different percentages of common grouped RJR grades of tobacco strips. Within each common group grade of tobacco (for example, upper stalk flue-cured or lower stalk Burley strip tobaccos) there are various lots of tobacco differentiated by tobacco belt and crop year. While the lot percentages in a common group tobacco strip recipe can and do change due to crop availability, the common group grade percentages in a cigarette blend recipe remain almost constant. Each lot of tobacco is characterized by an averaged tobacco nicotine value, among other physical, chemical and sensory parameters. Thus, the nicotine variability within a common group of tobacco strips can be estimated. First, by determining the variance in nicotine of each grade across lots (weighting lots by the number of pounds in the lot) and then by combining the grades according to their target percentages to estimate the tobacco nicotine variability.

Similarly, tobacco blend cut-filler nicotine variability can be estimated by determining the weighted average percent tobacco nicotine variance of each common group of tobacco used in the final blend recipe. In the following example, the nicotine variability from sources other than flue-cured and Burley strip has been ignored.

Three major assumptions are included in the estimates that have significant effects: (1) Only one average nicotine determination is used for each lot of tobacco, thus ignoring within-lot variability. This underestimates the common group of tobacco strip nicotine variability. (2) The weighted average variance of lots within a common group of tobacco strip implies that tobacco from a common group of tobacco strips is randomly selected for inclusion in the bulked final blend recipe. As random selection of tobacco containers is difficult and not practical, the random selection criteria is not true and as a result this tends to overestimate the bulked final blend recipe nicotine variability, particularly in the short run. (3) Nicotine variability from other cut filler components is considered non-appreciable, thus underestimating cut filler variability. The following tables illustrate how the nicotine variability is calculated.

Table 9 shows an example of how the weighted variance and standard deviation in percent tobacco nicotine is calculated for one hypothetical

TABLE 9

Calculation of weighted standard deviation in percent tobacco nicotine in a common group of strip tobacco used in a hypothetical blend recipe*

| Common group of upper stalk flue-cured tobacco (FC1) | Percentage of each RJR tobacco grade in the common group | Standard deviation in tobacco nicotine by weight and size of each tobacco lot | Variance | Variance×% of each tobacco grade in FC1 weighted variance |
|---|---|---|---|---|
| Grade 1 | 17.0% | 0.2069 | 0.0428 | 0.0073 |
| Grade 2 | 47.0% | 0.2860 | 0.0818 | 0.0384 |
| Grade 3 | 15.5% | 0.4259 | 0.1814 | 0.0281 |
| Grade 4 | 10.5% | 0.5413 | 0.2930 | 0.0308 |
| Grade 5 | 10.0% | 0.3392 | 0.1151 | 0.0115 |
| **Sum of weighted variance** | | | | 0.1161 |
| **Weighted standard Deviation of tobacco nicotine in FC1** | | | | 0.3407 |

TABLE 10

Calculation of weighted standard deviation in tobacco nicotine in a common group of strip tobacco used in a hypothetical blend recipe

| Common groups of RJR* tobacco types | Weighted standard deviation of % tobacco nicotine in FC1 |
|---|---|
| FC1 (Flue-cured Group 1) | 0.3407 |
| FC2 (Flue-cured Group 2) | 0.3766 |
| FC3 (Flue-cured Group 3) | 0.2626 |
| FC4 (Flue-cured Group 4) | 0.3663 |
| B1 (Burley Group 1) | 0.3169 |
| B2 (Burley Group 2) | 0.2588 |
| B3 (Burley Group 3) | 0.9166 |
| B4 (Burley Group 4) | 0.4794 |
| Expanded Tobacco Type 1 | 0.3992 |
| Expanded Tobacco Type 2 | 0.4954 |

common group of flue-cured tobaccos (FC1). For this example five grades of flue-cured strip tobacco were selected. Each group is included into a common group of flue-cured tobaccos called FC1 at a particular percentage of the total weight of the common group of flue-cured strip. Each grade of flue-cured strip was previously analyzed for percent tobacco nicotine (along with other analyses) based on five to ten composite samples of tobacco collected during the stemming operation. A standard deviation in percent tobacco nicotine was determined taking into account the weight and size of each tobacco lot. A variance was calculated for each grade of flue-cured strip incorporated into the common group flue-cured blend of tobaccos. Finally, a weighted standard deviation of percent tobacco nicotine was calculated based on the sum of weighted variances from each grade used in the FC1 common group blend. In a similar way, Table 10 gives the calculated weighted standard deviation in percent tobacco nicotine for several other common group grades of tobaccos used in this example of how a blend recipe could be developed by RJR. Table 11 shows examples of two calculations of the total weighted standard deviation in percent tobacco nicotine for hypothetical blends of cut-filler tobacco based on the weighted standard deviation in percent tobacco nicotine of each common tobacco group used in the two hypothetical blends. From the calculations, the total weighted standard deviation in percent tobacco nicotine for blends of cut-filler tobacco are in the range of about 0.22% tobacco nicotine. If a tobacco company could maintain such a low variance in percent tobacco nicotine day after day their product consistency (at least in terms of tobacco nicotine) would be quite good.

Janjigian et al. [60] and Gordin et al. [61] have conducted studies on the level of tobacco nicotine necessary to elicit a change in sensory response by

TABLE 11

Calculated total weighted standard deviation in percent tobacco nicotine based on percent of each tobacco group in hypothetical blends

| Recipe | Cut-Filler Blend A % | Weighted Standard Deviation in % Tobacco Nicotine in Tobacco Groups | Weighted Standard Deviation in % Tobacco Nicotine Based on % of Each Tobacco Groups in the Blend |
|---|---|---|---|
| FC4 | 28.01 | 0.3663 | 0.1026 |
| B4 | 11.52 | 0.4794 | 0.0552 |
| Reconstituted Sheet | 15.14 | | |
| Oriental | 20.63 | | |
| Expanded Tobacco | 11.60 | 0.3992 | 0.0463 |
| Expanded Stems | 13.10 | | |
| | | Total Weighted Standard Deviation in % Tobacco Nicotine Based on % of Each Tobacco Groups in the Blend | 0.2041 |

| Recipe | Cut-Filler Blend B % | Weighted Standard Deviation in % Tobacco Nicotine in Tobacco Groups | Weighted Standard Deviation in % Tobacco Nicotine Based on % of Each Tobacco Groups in the Blend |
|---|---|---|---|
| FC1 | 8.63 | 0.3407 | 0.0294 |
| FC2 | 15.38 | 0.3766 | 0.0579 |
| FC3 | 2.55 | 0.2626 | 0.0067 |
| B1 | 4.35 | 0.3169 | 0.0138 |
| B2 | 10.13 | 0.2588 | 0.0262 |
| B3 | 4.05 | 0.9166 | 0.0371 |
| Reconstituted Sheet | 17.20 | | |
| Oriental | 14.48 | | |
| Casing | 2.25 | 0.0000 | 0.0000 |
| Expanded Tobacco Type 2 | 13.20 | 0.4954 | 0.0654 |
| Expanded Stems | 6.30 | | |
| Top-Dressing | 1.50 | 0.0000 | 0.0000 |
| | | Total Weighted Standard Deviation in % Tobacco Nicotine Based on % of Each Tobacco Groups in the Blend | |

a certain percentage of a smoker population. These studies employed the classic sensory methodology called 'just noticeable difference' (JND). The most conservative estimate of the JND for tobacco nicotine ($JND_{10}$, which is the just noticeable difference in a variable that can be detected by 10% of a population) was a $-0.27$ to $+0.22\%$ change in percent tobacco nicotine. They also found that there was a high correlation between tobacco nicotine and smoke nicotine for a single cigarette configuration (specific cigarette wrapper, blend, packing density, filter type, filter ventilation, etc.). Based on the calculated total weighted standard deviation in percent tobacco nicotine for blends of cut-filler tobacco it would appear that if a company could maintain such a low variance in percent tobacco nicotine, then its smoke nicotine within a single cigarette configuration would be expected to be consistent over time.

R.J. Reynolds Tobacco Company, like most companies, has a quality-monitoring program. Data in Table 12 shows the percent tobacco nicotine from four products made numerous times during 1996. The percent tobacco nicotine data of Table 12 represents only one of numerous analytes evaluated in their quality-monitoring program. The percent tobacco nicotine data are from a non-filtered 70 mm cigarette, a 85 mm filtered full flavor product, a 85 mm filtered low 'tar' product and a 85 mm filtered ultra low 'tar' product. The average percent tobacco nicotine and standard deviation for the four products were 2.37, 0.065; 1.92,0.078; 2.03,0.068 and 2.07,0.065, respectively. For each cigarette type there was a relatively narrow range in percent tobacco nicotine. The method for the analysis of percent tobacco nicotine for the data of Table 12 was a high volume colorimetric analysis and was used for all the data of Table 12. The blend recipes for these cigarette brands were all calculated using RJR's tobacco blending model. The model worked well in selecting tobaccos for each blend over one year. The blend nicotine for the four tobacco blends were statistically different (at the 95% confidence level) among all comparisons other than the blends for the 85 mm filtered low 'tar' product and the 85 mm filtered ultra low 'tar' product which were not statistically significantly different. Figure 10 is a graphical representation of the data.

Over an extended period of time, cigarette brand styles and configurations tend to change to meet changes in the market (changes in consumer wants, competition, changes in advertising, etc.) [69]. Additionally, the tobaccos employed in those brands vary based on environmental changes (hot-dry vs. cool-wet growing seasons). Table 13 gives data on the percent tobacco nicotine and smoke nicotine for a single RJR cigarette brand from 1978-1994. Each data entry is a yearly average of at least three or more data points for both percent tobacco nicotine and smoke nicotine obtained from RJR's quality-monitoring program. The method for the determination of percent tobacco nicotine was a high volume colorimetric analysis. Over the years, improvements have been made to both the equipment and methodologies for

TABLE 12

Tobacco nicotine analyses of a non-filtered and a full-flavored (FF), a low 'tar' (LT) and an ultra low 'tar' (ULT) filtered cigarette blends collected during one year (1996)*

| Date | Non-Filtered 70 mm | Colorimetric analysis of percent tobacco nicotine (%) | | | | | |
| | | Date | FF 85 mm | Date | LT 85 mm | Date | ULT 85 mm |
|---|---|---|---|---|---|---|---|
| 1/8/1996 | 2.37 | 1/8/1996 | 1.92 | 1/8/1996 | 2.09 | 1/15/1996 | 2.11 |
| 1/29/1996 | 2.40 | 1/22/1996 | 1.98 | 1/29/1996 | 2.04 | 3/18/1996 | 2.04 |
| 2/26/1996 | 2.33 | 2/12/1996 | 1.97 | 2/26/1996 | 1.97 | 5/13/1996 | 2.05 |
| 3/11/1996 | 2.41 | 2/26/1996 | 1.78 | 3/11/1996 | 2.06 | 8/5/1996 | 2.09 |
| 3/25/1996 | 2.33 | 3/18/1996 | 1.89 | 3/25/1996 | 1.95 | 11/4/1996 | 2.11 |
| 4/22/1996 | 2.33 | 4/8/1996 | 1.85 | 4/15/1996 | 2.03 | 11/25/1996 | 2.04 |
| 5/13/1996 | 2.41 | 4/29/1996 | 1.88 | 5/20/1996 | 2.11 | 1/22/1996 | 2.14 |
| 6/3/1996 | 2.39 | 5/20/1996 | 2.03 | 6/10/1996 | 2.05 | 3/4/1996 | 1.95 |
| 7/1/1996 | 2.37 | 6/10/1996 | 1.94 | 8/19/1996 | 2.05 | 4/8/1996 | 2.04 |
| 8/5/1996 | 2.40 | 7/29/1996 | 1.90 | 9/23/1996 | 2.15 | 8/19/1996 | 2.08 |
| 9/9/1996 | 2.52 | 8/26/1996 | 1.87 | 10/21/1996 | 1.95 | 11/11/1996 | 2.05 |
| 10/21/1996 | 2.28 | 9/23/1996 | 1.97 | 11/18/1996 | 1.92 | 4/15/1996 | 1.95 |
| 11/11/1996 | 2.27 | 10/28/1996 | 1.97 | 12/9/1996 | 1.94 | 5/20/1996 | 2.10 |
| 11/25/1996 | 2.43 | 11/25/1996 | 2.01 | 1/8/1996 | 2.04 | 9/9/1996 | 2.15 |
| 12/16/1996 | 2.29 | 12/16/1996 | 1.88 | 1/22/1996 | 2.10 | 11/11/1996 | 1.98 |
| 12/30/1996 | 2.42 | 1/8/1996 | 2.01 | 2/5/1996 | 2.16 | 1/15/1996 | 2.13 |
| Average | 2.37 | 1/22/1996 | 1.98 | 2/26/1996 | 1.99 | 3/18/1996 | 1.95 |
| standard deviation | 0.065 | 2/26/1996 | 1.92 | 3/11/1996 | 2.13 | 5/13/1996 | 2.04 |
| % coefficient of variation | 2.74 | 3/11/1996 | 2.09 | 4/1/1996 | 1.98 | 9/2/1996 | 2.08 |
| | | 4/1/1996 | 1.95 | 4/22/1996 | 2.00 | 12/9/1996 | 2.04 |
| | | 4/22/1996 | 1.98 | 5/13/1996 | 1.98 | 12/30/1996 | 2.17 |
| | | 5/20/1996 | 2.08 | 6/10/1996 | 2.03 | 4/22/1996 | 2.01 |
| | | 6/10/1996 | 1.89 | 7/1/1996 | 2.04 | 3/25/1996 | 2.05 |
| | | 7/1/1996 | 2.03 | 7/29/1996 | 1.94 | 9/30/1996 | 2.08 |
| | | 8/12/1996 | 1.86 | 9/2/1996 | 2.07 | 5/6/1996 | 2.19 |
| | | 9/16/1996 | 1.95 | 9/23/1996 | 2.02 | 11/18/1996 | 2.03 |
| | | 10/21/1996 | 1.85 | 10/14/1996 | 2.10 | 5/20/1996 | 2.14 |
| | | 11/11/1996 | 1.80 | 11/4/1996 | 2.08 | 4/22/1996 | 2.04 |
| | | 11/25/1996 | 1.93 | 11/18/1996 | 2.03 | 5/20/1996 | 2.15 |

TABLE 12 Continued

| Date | Non-Filtered 70 mm | Date | FF 85 mm | Date | LT 85 mm | Date | ULT 85 mm |
|---|---|---|---|---|---|---|---|
| | | | Colorimetric analysis of percent tobacco nicotine (%) | | | | |
| | | 12/16/1996 | 1.84 | 12/9/1996 | 1.95 | | 2.07 |
| | | 12/30/1996 | 2.00 | 12/30/1996 | 2.17 | | 0.065 |
| | | 1/8/1996 | 1.96 | 4/1/1996 | 1.93 | | 3.14 |
| | | 1/22/1996 | 1.89 | 2/26/1996 | 2.01 | | |
| | | 3/4/1996 | 1.95 | 4/8/1996 | 2.03 | | |
| | | 3/18/1996 | 1.81 | 5/13/1996 | 2.02 | | |
| | | 4/1/1996 | 1.79 | 12/2/1996 | 2.11 | | |
| | | 4/22/1996 | 1.86 | Average | 2.03 | Average | |
| | | 5/13/1996 | 1.88 | standard deviation | 0.068 | standard deviation | |
| | | 6/3/1996 | 1.77 | % coefficient of variation | 3.34 | % coefficient of variation | |
| | | 9/2/1996 | 1.94 | | | | |
| | | 10/7/1996 | 1.93 | | | | |
| | | 11/4/1996 | 1.92 | | | | |
| | | 11/18/1996 | 1.86 | | | | |
| | | 12/2/1996 | 1.98 | | | | |
| | | 1/8/1996 | 1.90 | | | | |
| | | 2/5/1996 | 1.88 | | | | |
| | | 3/11/1996 | 2.12 | | | | |
| | | 4/1/1996 | 1.81 | | | | |
| | | 5/6/1996 | 1.97 | | | | |
| | | 6/3/1996 | 1.99 | | | | |
| | | 7/8/1996 | 1.91 | | | | |
| | | 9/30/1996 | 1.89 | | | | |
| | | 11/11/1996 | 1.83 | | | | |
| | | 11/25/1996 | 1.96 | | | | |
| | | 12/16/1996 | 1.85 | | | | |
| | Average | | 1.92 | | | | |
| | standard deviation | | 0.078 | | | | |
| | % coefficient of variation | | 4.06 | | | | |

* Data is from R.J. Reynolds Tobacco Company

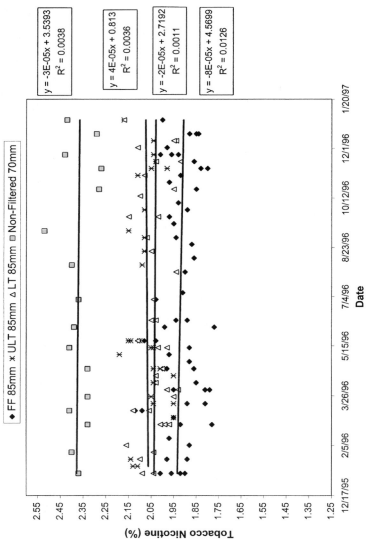

Fig. 10 Tobacco nicotine for four cigarette blends measured during 1996.

the determination of percent tobacco nicotine. The data of Table 13 although precise may have some bias due to the changes and improvements that were made over time. Smoke nicotine was determined by the official FTC methodology. The average and standard deviation for percent tobacco nicotine and smoke nicotine from data on Table 13 are 1.88, 0.19 and 0.77,0.09, respectively. The range in percent tobacco nicotine was 1.60 to 2.17%. The range in smoke nicotine was 0.92 to 0.64 mg/cigarette. Figures 11–13 show the relationships of percent tobacco nicotine over time, smoke nicotine over time and smoke nicotine to percent tobacco nicotine. Numerous blend and cigarette configuration changes were made to this brand over the 16-year period evaluated. As expected, the percent tobacco nicotine and the smoke nicotine varied. These changes are evident from the low linear correlation ($R^2$ values) seen in the data of Figs 11–13. Yet it is interesting that, on average, percent tobacco nicotine remained relatively flat, albeit highly variable. On average, smoke nicotine was reduced from 1978–1994. There is no linear relationship of percent tobacco nicotine and smoke nicotine over the period from 1978–1994.

Data on a 70 mm non-filtered RJR product was collected from numerous sources over a period of 44 years. Over that period of time, numerous cigarette design technologies were introduced into cigarette products [69]. In

TABLE 13

Tobacco nicotine and smoke nicotine for a low 'tar' cigarette brand 1978–1994

| Year | Tobacco nicotine (%) | Smoke nicotine (mg/cigarette) |
|---|---|---|
| 1978 | 2.05 | 0.92 |
| 1979 | 2.10 | 0.88 |
| 1980 | 2.09 | 0.79 |
| 1981 | 2.17 | 0.72 |
| 1982 | 2.13 | 0.77 |
| 1983 | 1.86 | 0.75 |
| 1984 | 1.74 | 0.73 |
| 1985 | 1.71 | 0.69 |
| 1986 | 1.66 | 0.66 |
| 1987 | 1.70 | 0.64 |
| 1988 | 1.76 | 0.64 |
| 1989 | N/A | 0.77 |
| 1990 | N/A | 0.77 |
| 1991 | 1.60 | 0.77 |
| 1992 | 1.78 | 0.79 |
| 1993 | 1.84 | 0.87 |
| 1994 | 1.97 | 0.90 |
| Average | 1.88 | 0.77 |
| standard deviation | 0.19 | 0.09 |

* Data is from R. J. Reynolds Tobacco Company.

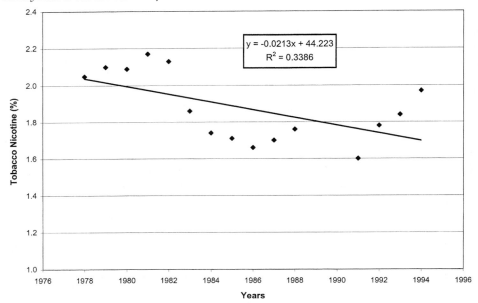

Fig. 11  Tobacco nicotine for a low 'tar' brand 1978–1994.

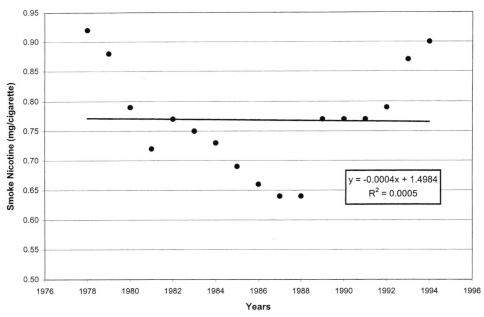

Fig. 12  Smoke nicotine for a low 'tar' brand 1978–1994.

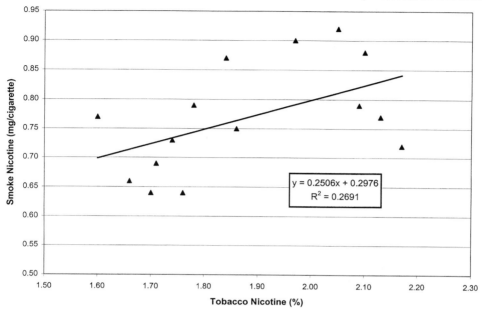

Fig. 13 Tobacco nicotine vs. smoke nicotine for a low 'tar' brand 1978–1994.

the early 1950s filter tips and reconstituted tobacco materials were introduced. In the late 1950s and early 1960s paper additives and porous paper were introduced. In the late 1960s and early 1970s expanded tobacco and filter ventilation were developed and used to varying degrees in cigarette products. All technologies used in the cigarette are interactive. The cigarette is a system of components that interact to produce changes in smoke chemistry. Inclusion of, or even changes in the level or type of technology, may require adjustments in the cigarette design to maintain certain smoke characteristics acceptable to consumers [62]. During this period, numerous improvements were also made in the analytical determination of nicotine in tobacco and smoke [63]. Since the data of Table 14 is on a non-filtered cigarette, filter-related technologies obviously have not impacted smoke deliveries, but the other paper and processed tobacco technologies have been implemented into the 70 mm product over time. Table 14 lists the 'tar' and smoke nicotine yields and methods of analysis used to determine both. References are cited that contained the data in Table 14. Prior to 1966, numerous analytical methods were used to determine the yields of 'tar' and nicotine in cigarette smoke. Some of the cited data are not actually 'tar', as we consider it today, but are measures of wet or dry particulate matter. After 1966 the data on Table 14 lists 'tar' and smoke nicotine values obtained employing the FTC method. Figure 14 is a plot of smoke nicotine vs. time. As can be seen there has been a large decrease in smoke nicotine over time. This

TABLE 14

Tar' and nicotine data (mg/cigarette) for a popular 70 mm cigarette found in the literature – 1953–1997

| Year | Tar' | Nicotine | Reference | Notes | Tar' | Nicotine |
|---|---|---|---|---|---|---|
| | | | | | Methods of analysis [13] | |
| 1953 | 18 | 1.9 | 2 | 2 | Dried chloroform soluble WTPM | UV spec. |
| 1953 | 15 | 1.9 | 5 | 2 | Dried chloroform soluble WTPM | UV spec. |
| 1954 | | | | 2 | | |
| 1955 | 20 | 3.3 | 2,6 | 2 | Dried chloroform soluble WTPM | UV spec. |
| 1956 | | | | 2 | | |
| 1957 | 31 | 2.8 | 1 | 2 | Dried TPM | Silicotungstic acid-nicotine precipatate |
| 1957 | 17 | 3.2 | 2 | 2 | Dried chloroform soluble WTPM | UV spec. |
| 1957 | 16.2 | 2.6 | 2 | 2 | Dried chloroform soluble WTPM | UV spec. |
| 1958 | 28.7 | 2.5 | 1 | 2 | Dried TPM | Silicotungstic acid-nicotine precipatate |
| 1958 | 22.4 | 1.94 | 3 | 2 | Dried chloroform soluble WTPM | UV spec. |
| 1958 | 13.4 | | | 1 | Modified dried chloroform soluble WTPM | |
| 1958 | 18 | 2.5 | 4 | 2 | Dried chloroform soluble WTPM | UV spec. |
| 1959 | 27.7 | 2.3 | 1 | 2 | Dried TPM | Silicotungstic acid-nicotine precipatate |
| 1959 | 40.3 | 2.31 | 7 | 2 | **Total WTPM** | UV spec. |
| 1960 | 23.5 | 1.76 | 10 | 3 | FTC method | GC |
| 1961 | 23.4 | 1.9 | 1 | 2 | Dried TPM | Silicotungstic acid-nicotine precipatate |
| 1961 | 24.2 | 1.77 | 10 | 3 | FTC method | GC |
| 1963 | 23.4 | 1.54 | 10 | 3 | FTC method | GC |
| 1964 | 23.8 | 1.51 | 10 | 3 | FTC method | GC |
| 1964 | 29.9 | 1.62 | 11 | | 1960 CORESTA method; T=WTPM-H2O-Nic. | UV spec. |
| 1965 | 23.5 | 1.61 | 10 | 3 | FTC method | GC |
| 1966 | 31.3 | 1.69 | 8 | 3 | Method described by Ogg [12] | Griffith still method, UV spec |
| 1967 | 24.2 | 1.39 | 9 | 3 | FTC method | Griffith still method, UV spec |
| 1968 | 25.7 | 1.4 | | 3 | FTC method | Official FTC method |
| 1968 | 25.6 | 1.55 | | 3 | FTC method | Official FTC method |
| 1969 | 25.1 | 1.6 | | 3 | FTC method | Official FTC method |
| 1969 | 24.9 | 1.56 | | 3 | FTC method | Official FTC method |
| 1970 | 25.1 | 1.51 | | 3 | FTC method | Official FTC method |
| 1970 | 23.5 | 1.45 | | 3 | FTC method | Official FTC method |
| 1971 | 24.7 | 1.45 | | 3 | FTC method | Official FTC method |
| 1971 | 24.6 | 1.49 | | 3 | FTC method | Official FTC method |

TABLE 14 Continued

| Year | 'Tar' | Nicotine | Reference | Notes | 'Tar' | Methods of analysis [13] | Nicotine |
|---|---|---|---|---|---|---|---|
| 1972 | 25.1 | 1.59 | | 3 | FTC method | | Official FTC method |
| 1972 | 24.8 | 1.59 | | 3 | FTC method | | Official FTC method |
| 1973 | 25.3 | 1.61 | | 3 | FTC method | | Official FTC method |
| 1973 | 23.4 | 1.51 | | 3 | FTC method | | Official FTC method |
| 1974 | 24.7 | 1.58 | | 3 | FTC method | | Official FTC method |
| 1974 | 24.7 | 1.61 | | 3 | FTC method | | Official FTC method |
| 1975 | 23.4 | 1.57 | | 3 | FTC method | | Official FTC method |
| 1975 | 23.7 | 1.48 | | 3 | FTC method | | Official FTC method |
| 1976 | 23.1 | 1.43 | | 3 | FTC method | | Official FTC method |
| 1976 | 24.6 | 1.58 | | 3 | FTC method | | Official FTC method |
| 1977 | 25 | 1.6 | | 3 | FTC method | | Official FTC method |
| 1978 | 26 | 1.8 | | 3 | FTC method | | Official FTC method |
| 1979 | 25 | 1.9 | | 3 | FTC method | | Official FTC method |
| 1981 | 21 | 1.4 | | 3 | FTC method | | Official FTC method |
| 1982 | 20 | 1.2 | | 3 | FTC method | | Official FTC method |
| 1983 | 21 | 1.4 | | 3 | FTC method | | Official FTC method |
| 1986 | 22 | 1.3 | | 3 | FTC method | | Official FTC method |
| 1987 | 22 | 1.4 | | 3 | FTC method | | Official FTC method |
| 1988 | 22 | 1.4 | | 3 | FTC method | | Official FTC method |
| 1989 | 22 | 1.4 | | 3 | FTC method | | Official FTC method |
| 1990 | 22 | 1.4 | | 3 | FTC method | | Official FTC method |
| 1991 | 22 | 1.5 | | 3 | FTC method | | Official FTC method |
| 1992 | 22 | 1.4 | | 3 | FTC method | | Official FTC method |
| 1993 | 22 | 1.5 | | 3 | FTC method | | Official FTC method |
| 1994 | 23 | 1.7 | | 3 | FTC method | | Official FTC method |
| 1994 | 22 | 1.45 | | 3 | FTC method | | Official FTC method |
| 1995 | 22.6 | 1.65 | | 3 | FTC method | | Official FTC method |
| 1996 | 24.1 | 1.66 | | 3 | FTC method | | Official FTC method |
| 1997 | 24 | 1.73 | | 3 | FTC method | | Official FTC method |

## TABLE 14 Continued

### Notes and References to Table 14

Note 1: Chloroform soluble fraction of WTPM.

Note 2: Depending on the method of analysis used for 'tar' and nicotine the values can vary significantly. 'Tar' values prior to ~1967 were both over-estimated and under estimated depending on the method used [13]. Nicotine values were in general over-estimated when using the silicotungstic method. Depending on which still method was employed, the value for smoke nicotine could vary. [14–18]

Note 3: Data from 1967–1997 come from Federal Trade Commission (FTC) Reports and the Tobacco Institute Testing Laboratory (TITL). These data were prepared by R.J. Reynolds Tobacco scientists L Lewis and PF Perfetti.

(1) Cigarettes: tests for nicotine and tar in the smokes of 33 brands [editorial]. Consumer Reports 1957 Mar; 22: 100–110. Miller Lois M, Monahan James. Facts we're not told about filter-tips. The Reader's Digest 1961 July; 79: 71–78.

(2) Cigarettes: tests for nicotine and tar in the smokes of 33 brands [editorial]. Consumer Reports 1957 March; 22(3); 100–110.

(3) Staberg EM. Content of tar and nicotine in cigarette smoke. Nordisk Hygienisk Tidskrift 1959; 40: 71–88.

(4) Tars and nicotine in cigarettes [editorial]. Consumer Reports 1958 April; 23: 224–225.

(5) Cigarettes: what cu's test showed... the industry and its advertising... and: how harmful are they? [editorial]. Consumer Reports 1953 February; 18(2): 59–74.

(6) Cigarettes: a three part report [editorial]. Consumer Reports 1955 Feb: 20(2): 56–73.

(7) Staberg EM. Quantities of tar and nicotine in cigarette smoke. Nordisk Hygienisk Tidskrift 1960; 41: 199–216.

(8) Moore GE, Bross I, Shamberger R, Bock FG. Tar and nicotine retrieval from fifty-six brands of cigarettes. Cancer 1967 March; 20(3): 323–332.

(9) Shea JW. Federal trade commission cigarettes testing for tar and nicotine content. Federal Register: 1967 Aug. Vol 32, No. 147, FR Doc 67–8978.

(10) T. A. Perfetti, FTC Analysis of frozen 70 mm products made between 1960 and 1965. Smoked on 11/11/97.

(11) Cederlof R, Edfors ML. Tar and nicotine in cigarettes. Nord Hyg Tidskrift 1964; 45: 83–88.

(12) Ogg CL. Tobacco determination of particulate matter and alkaloids (as nicotine) in cigarette smoke. Journal of the AOAC 1964; 47(2): 356–362.

(13) Rodgman A. FTC tar and nicotine in cigarette mainstream smoke: A retrospective. Recent Advances in Tobacco Science 1997; 23: 5–74.

(14) Pyriki C. Moldenhauer W. Contribution to the determination of nicotine in the mainstream and sidestream smoke of cigarettes. Ber. Inst. Tabakforsch., 1962; 9(1): 104–122.

(15) Morgareidge K. Evaluation of variables in cigarette smoke collection and analysis i. the use of a monitor sample. Tobacco Chem. Res. Conf., 13th: 1959: Lexington (KY).

(16) Morgareidge K. Observations on the determination of total solids in cigarette smoke. Tobacco Chemist's Research Conference, 12th: 1958 October 23: Durham (NC).

(17) Willits CO, Swain ML, Connelly JA, Brice BA. Spectrophotometric determination of nicotine. Anal Chem, 1950; 22: 430–433.

(18) Keith CH, Newsome JR. Quantitative studies on cigarette smoke i. an automatic smoking machine. Tobacco Sci, 1957; 1: 51–57.

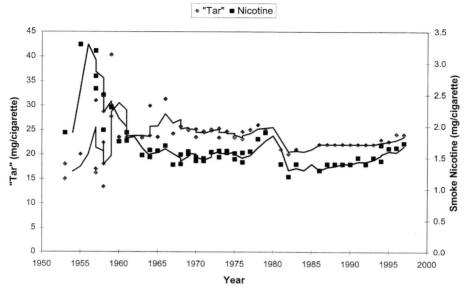

Fig. 14 'Tar' and smoke nicotine for 70 mm cigarette 1953–1997.

initial decrease was due to the implementation of reconstituted tobacco, use of porous papers and small changes in the size (circumference) of the cigarette rods. Additionally, there was competition among cigarette manufacturers to reduce 'tar' and smoke nicotine to satisfy the wants of their consumers' [69]. Since about 1960, the 'tar' and smoke nicotine for this 70 mm non-filtered cigarette has changed very little, comparatively. One obvious change that did occur from evaluation of the figure was an increase in smoke nicotine during the later half of the 1970s. This increase in smoke nicotine was due to the very high nicotine levels in both the flue-cured and Burley crops beginning in 1975 through 1979. In 1977 it is interesting to note that the mean nicotine for the flue-cured crop was higher than the Burley crop, 3.46 vs. 3.33%, respectively (Table 5). Figure 15 shows the FTC smoke nicotine data for the 70 mm non-filtered cigarette during the time period between 1966 and 1997. On average, the smoke nicotine has been relatively constant although wide swings in the data set have been noted (1977–1983).

Summarizing, it is critical for tobacco manufacturers to be able to produce consistent products for consumers. The raw material, tobacco, is a natural product and as such can vary considerable from year-to-year. To maintain consistent product quality, the manufacturers have limited means available to process or remove material inconsistencies. For the cigarette industry, the use of a strategic buying program coupled with inventory modeling programs has been an effective and efficient way to use tobacco inventories and

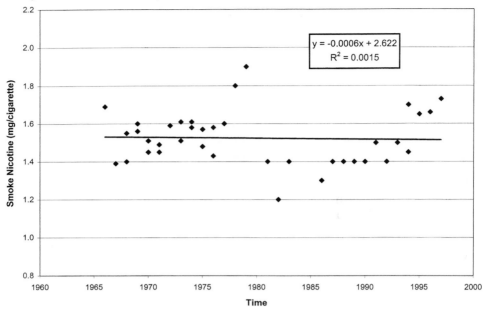

Fig. 15 FTC analyses of smoke nicotine for a 70 mm plain cigarette over time.

maintain consistent tobacco nicotine levels in their products from year to year. Additionally, cigarette design technologies have been used to provide a means for cigarette manufacturers to provide consistent yields of 'tar' and smoke nicotine. With the use of effective and efficient tobacco buying practices, tobacco use modeling efforts and use of advances in cigarette design technology, it is possible for a cigarette manufacturer to reproduce consistent levels of tobacco nicotine and smoke nicotine in their products year after year.

## V. REGULATORY ASPECTS

Tobacco is widely regulated both in the United States and abroad. Historically in the United States, tobacco has been regulated by the federal government to classify products for the purpose of taxation (Bureau of Alcohol, Tobacco and Firearms [70]), to provide cigarette smokers with information about cigarette smoke yields in advertising (Federal Trade Commission [71]), and to promote health education (Department of Health and Human Services [72,73]). Recently, regulations have been introduced in some states that require tobacco companies to conduct new testing methods and to report smoke yield classification schemes for cigarettes and other

354

tobacco products [74–76]. State regulations are based on the premise that smoke yields for 'an average smoker' can be determined. Therefore state regulations do not focus on issues of taxation or product advertising, but rather focus on issues within the broad umbrella of the smoking and health controversy. Specifically, current state regulations derive from nicotine addiction arguments and, tangentially, human smoking behavior research.

The determination of tobacco nicotine content has been a key element in regulatory schemes applied to tobacco products. Nicotine analyses conducted for this purpose can be divided into two general categories: (1) analyses conducted in government laboratories; and (2) analyses conducted in industry laboratories. Of note is the fact that the former are conducted at the expense of the taxpayer and the latter are incurred as a corporate expense. This section of the chapter will describe tobacco nicotine determinations used in the classification of tobacco products for federal taxation purposes. The analytical methods applied to classify products according to state guidelines and to meet state regulatory compliance will also be described and discussed.

*Analyses conducted in government laboratories: classification of tobacco products for tax purposes – bureau of alcohol, tobacco and firearms methods*

*Determination of nicotine in alcohol tobacco and firearms regulated tobacco products.* The Bureau of Alcohol, Tobacco and Firearms (ATF) was established as an independent agency within the US Department of the Treasury in 1972. Under Treasury Department Order 221, the functions, powers and duties of the Internal Revenue Service arising under laws relating to alcohol, tobacco and firearms were transferred to the newly created bureau where they presently reside. The primary responsibility of ATF is to regulate the alcohol, tobacco, firearms and explosives industries. ATF's mission in connection with the tobacco industry is as follows:

1. To secure voluntary compliance with law and regulations.
2. To exercise adequate controls to assure full and timely payment of the taxes imposed on tobacco products.
3. The revenue derived from these taxes is the primary consideration in regulating the tobacco industry.

The Federal taxes on manufactured tobacco products are among the oldest enacted legislation in the United States and are still in use today [77]. Historically, there has been nearly two hundred years of federal tobacco legislation, regulation and taxation. Table 15 illustrates key dates from 1791 to 1988 in the regulation and taxation of tobacco. Despite the current decrease in the quantity of cigarettes sold, federal excise taxes from cigarettes rose to a total of $4.7 billion for 1990. The Federal revenue

TABLE 15

Federal tax legislation regarding tobacco products – key historical events

| | | |
|---|---|---|
| 1791 | Hamilton's Excise Law (Repealed) | First tax on tobacco products in the United States |
| 1812 | Tax Law of 1812 (Repealed) | Tax levied on tobacco products to generate revenue during the war with England |
| 1862 | Internal Revenue (IR)Act of July 1, 1862 | Tax imposed on all tobacco products (domestic and imported). Tobacconist required to pay annual license fee ($10.00). Taxes collected by tax collectors |
| 1868 | IR Act of July 20, 1868, This act is the foundation of present IR laws concerning manufacture and tax payments on tobacco products | Tax collection discontinued, tax stamps issued and regulated by Federal government. Tax stamps used as payment of taxes |
| 1897 | IR Act of July 24, 1897 | First tax on cigars based on a weight per thousand cigars basis |
| 1917 | IR Act of October 3, 1917 | Act established a classification of cigar products into seven classes (Class A – G) for tax purposes |
| 1935 | Overhaul of the IR Code dealing with tobacco regulations | Tobacco regulations dealing with classification schemes for tobacco products |
| 1951 | Federal Excise | Federal excise tax of $4.00 per thousand levied on cigarette products |
| 1954 | IR Code of 1954 | Additional new tobacco regulations imposed on tobacco products, first major changes since 1935 |
| 1956 | Revision of stamp method of tobacco tax collection | Institution of optional daily tax return method for payment of taxes on cigars |
| 1959 | IR Act of June 24, 1959, Revamping of the stamp method for collection of taxes on tobacco | Today tobacco manufacturers are required to file semi-monthly tax returns on taxable products and make payments with the returns |
| 1982 | Tax Equity and Financial Responsibility Act (August 1982) | An additional Federal excise tax of $4.00 per thousand levied on cigarette products. Act also imposed an inventory tax on warehoused tobacco products |
| 1985 | Consolidated Omnibus Budget Reconciliation Act of 1985, Public Law 99-272 | This Act doubled the tax on cigarettes, it also set a tax on snuff of 24 cent per pound and 8 cent per pound on chewing tobacco |
| 1986 | The Comprehensive Smokeless Tobacco Health Education Act of 1986 | This law required manufacturers, packagers and importers of smokeless tobacco products to put health warnings on packaging and advertisements (FTC was given power to issue regulations) |
| 1988 | Technical and Miscellaneous Revenue Act of 1988 | Imposed a 45 cent per pound tax on pipe tobacco, and an inventory tax on pipe tobacco |

collected from the sale of all tobacco products amounted to $6 billion in 1997.

The mission of the ATF laboratories (ATF National Laboratory Center, Rockville, MD) is to produce the accurate and authoritative scientific information needed by law enforcement and compliance operations in protecting the public and collecting the revenue. The laboratory services provide scientific expertise to the bureau. Tobacco laboratory analysts perform a wide variety of tests to assist in consumer protection and proper tax classification of tobacco products.

Commercial tobacco products are defined (Table 16) and classified at different tax rates (Table 17) according to the Code of Federal Regulations [70,78]. ATF regulated tobacco products (Fig. 16) can be classified into two groups: *Smoking tobacco products* and *Smokeless tobacco products*. Smoking tobacco products include cigarettes, cigars, pipe tobacco and roll-your-own (RYO) tobacco. Smokeless tobacco products include chewing tobacco and snuff. Product classification is performed on a comparative basis: *cigars* vs. *cigarettes, pipe tobacco* vs. *roll-your-own tobacco, and chewing tobacco* vs. *snuff.* For example, the tax on cigarettes is approximately ten-fold greater than the tax on cigars. The current tax rate is $12.00 per thousand for small

TABLE 16

Definitions of ATF regulated tobacco products

| | |
|---|---|
| **CIGAR** | Any roll of tobacco wrapped in leaf tobacco or in any substance containing tobacco (other than any roll of tobacco which is a cigarette within the meaning of paragraph (2) of the definition for cigarette). |
| **SMALL CIGARS** | Cigars weighing not more than three pounds per thousand. |
| **LARGE CIGARS** | Cigars weighing more than three pounds per thousand. |
| **CIGARETTE** | (1) Any roll of tobacco wrapped in paper or in any substance not containing tobacco, and (2) Any roll of tobacco wrapped in any substance containing tobacco which, because of its appearance, the type of tobacco used in the filter, or its packaging and labeling, is likely to be offered to, or purchased by, consumers as a cigarette described in paragraph (1) of this definition. |
| **SMALL CIGARETTES** | Cigarettes weighing not more than three pounds per thousand. |
| **LARGE CIGARETTES** | Cigarettes weighing more than three pounds per thousand. |
| **PIPE TOBACCO** | Any tobacco which, because of its appearance, type, packaging, or labeling, is suitable for use and likely to be offered to, or purchased by, consumers as tobacco to be smoked in a pipe. |
| **CHEWING TOBACCO** | Any leaf tobacco that is not intended to be smoked. |
| **SNUFF** | Any finely cut, ground, or powered tobacco that is not intended to be smoked. |

It should be noted that a definition for **ROLL-YOUR-OWN TOBACCO** is not included since this is the only tobacco product that is exempt from tax as the regulations stand today.

TABLE 17

Excise tax rates effective January 1, 1993

| Tobacco product | Tax rate |
|---|---|
| Small cigars | $1.125 per thousand |
| Large cigars | 12.75% of price not to exceed $30 per thousand |
| Small cigarettes | $12.00 per thousand |
| Large cigarettes | $25.20 per thousand |
| Cigarette papers | 0.75 cent per 50 papers |
| Cigarette tubes | 1.5 cent per 50 tubes |
| Smokeless snuff | 36 cents per pound |
| Chewing tobacco | 12 cents per pound |
| Pipe tobacco | 67.5 cents per pound |

cigarettes compared to $1.125 per thousand for small cigars (Table 17). Identification of a tobacco product(s) relies on a wide variety of tests including a determination of nicotine.

The first step in defining and properly classifying an ATF regulated tobacco product is the determination of nicotine (Fig. 16). The rationale for

ATF REGULATED PRODUCTS

| Nicotine (present) | Nicotine (absent) |
|---|---|
| **Tobacco Product** | **Non-Tobacco Product** |
| Subject to Tax | Exempt from Tax |
| Subject to Classification | |

**Smoking Tobacco Product**
Cigar
Cigarette
Pipe Tobacco
Roll-Your-Own*

**Smokeless Tobacco Products**
Chewing Tobacco
Snuff

---

\* Roll-Your-Own products are not currently subject to tax. However, these products may be subject to tax in the future.

Fig. 16 Definition of an ATF regulated tobacco product.

analyzing nicotine in ATF regulated tobacco products, regardless of the type of tobacco product, is two-fold:

1. *Regulatory*: The presence, not the amount, of nicotine is the primary determinant of what constitutes a tobacco product.
2. *Abundance*: Nicotine is the most abundant tobacco alkaloid, comprising about 98% of the total alkaloid content.

Nicotine is one of the few liquid alkaloids. This principal tobacco alkaloid contains pyridine and pyrrolidine base rings [$C_{10}N_{14}N_2$; (S)-3-(1-methyl-2-pyrrolidinyl)pyridine]. First isolated from smoke in 1809 by Vauquelin [79], and its properties described in 1828 [80], nicotine is a colorless to pale yellow liquid, found in tobacco leaves and smoke, and which turns brown when exposed to light and air. Nicotine is a molecule that is strongly chromophoric in the ultraviolet region and is directly detected at about 260 nm [81]. Nicotine has been analyzed using spectrophotometry [15], gas chromatography [82], high-performance liquid chromatography [83,84] and more recently capillary electrophoresis [26,85].

*Historical development of analytical methods for analysis of nicotine in ATF regulated tobacco products.* Advances in the technology applied to the analysis of nicotine in ATF regulated tobacco products have occurred with time. The analytical methods for the analysis of nicotine historically practiced at ATF include spectrophotometry and gas chromatography. These two techniques will be discussed briefly in this section followed by a review of the current analytical method, the technique of capillary electrophoresis (CE).

*Spectrophotometric analysis of nicotine.* Historically, nicotine was first routinely analyzed in tobacco products at the ATF National Laboratory Center using spectrophotometry. The advantages of spectrophometry include ease of use, rapid analysis capability and well established, reliable instrumentation. Unfortunately, there are inherent disadvantages. The main disadvantage with using spectrophotometry, despite the simplicity of technique as well as instrumentation, is that structurally related tobacco alkaloids, for example nornicotine, and other UV absorbing compounds also absorb in the 260 nm region, thus interfering with the absorbance of nicotine. Nevertheless, prior analysis of nicotine in ATF regulated tobacco products reported total alkaloids as nicotine.

When working with pure nicotine (for example, nicotine standards), UV absorbance measurements can be very useful, providing valuable information regarding the chemistry of the nicotine molecule. Nicotine absorbs strongly in the UV at about 260 nm, the maximum UV absorbance of nicotine (Fig. 17). This UV absorbance is dependent on the pH of the solvent used. Figure 18 shows the UV spectra of nicotine in distilled, deionized water, (B), sodium phosphate buffer, pH 6.9 (A), and sodium phosphate buffer, pH 2.5

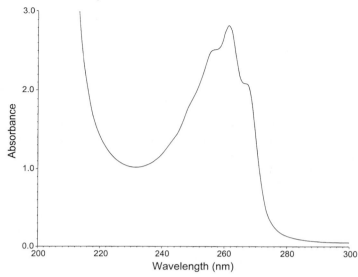

Fig. 17 UV absorbance spectrum of nicotine.

(C). It can be seen that the UV absorbance of nicotine at 260 nm in sodium phosphate buffer, is different at pH 2.5 and pH 6.9. This can be explained as follows.

When nicotine is dissolved in water, it can exist in one of three forms depending on the pH: unprotonated monoprotonated and diprotonated

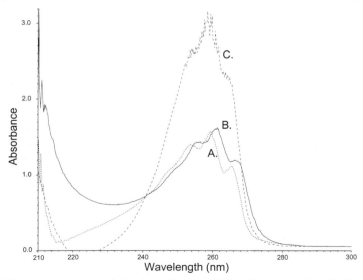

Fig. 18 UV absorbance spectra of nicotine: (A) in 10 mM sodium phosphate buffer, pH 6.9, (B) in distilled, deionized water, and (C) in 25 mM sodium phosphate buffer, pH 2.5.

[85,86]. Figure 19 taken from Morie [86] shows the fraction of these nicotine species as a function of pH. At pH 2.5, nicotine exists as a diprotonated species, while it is monoprotonated at pH 6.9. The UV absorbance of nicotine has been reported to vary as a function of pH [15,85]. As a diprotonated species (pH 2.5), the magnitude of UV absorbance of nicotine at 260 nm is approximately twice that for nicotine as a monoprotonated species (pH 6.9). From the known molar extinction coefficient of nicotine (depending on the solvent used), one can calculate the concentration of nicotine and determine the amount in a tobacco product. This pH dependence has a definite impact on the CE performance of nicotine as will be seen later.

The main features of the historical spectrophotometric method can be described as follows:

*Method scope.* An extract of tobacco is prepared by distillation and nicotine is determined by spectrophotometry.

*Sample preparation.* ATF regulated tobacco products are received into the laboratory, stored and physically prepared, prior to instrumental analysis of nicotine in the sample. Smoking and smokeless tobacco products are routinely provided by the Regulations Division and received by the Tobacco Laboratory at the ATF National Laboratory Center. These tobacco products are stored in a cold room (4°C) until they are analyzed.

Filler tobacco is separated from the wrappers when working with cigars, cigarettes, beedies and cheroots. In addition, cigars, beedies and cheroots are analyzed for their wrappers as well, in keeping with the ATF regulatory requirements for their subsequent classification. Pipe tobacco analysis requires some comminution in order to obtain a working sample size. Chewing tobacco, as a plug or loose leaf in a pouch, also requires comminution to obtain a working sample size. Snuff, both in its dry and moist form, is used as received.

Samples of tobacco products are used as is or ground depending on the type of tobacco product being analyzed. Cigar filler tobacco, cigarette filler tobacco, cigar wrapper tobacco are ground in a Wiley Mill equipped with a 20-mesh screen. Beedi filler tobacco and cheroot filler tobacco are used as is. Beedi wrapper and cheroot wrapper are cut into small pieces and used as is. Pipe tobacco (large pieces of tobacco are usually cut into small pieces) and used as is. Chewing tobacco, as a plug or in a pouch, is cut into small pieces using a pair of scissors or a razor blade and used as is; snuff, in its moist or dry form, is also used as is.

Nicotine in ATF regulated tobacco products is extracted for spectrophotometric analysis employing distillation techniques in the presence of an alkali-salt solution [87,88]. ATF Procedure 73–5 describes the method in detail. Briefly, a sample of tobacco (100 mg of filler tobacco or 400 mg of wrapper material) is extracted using steam distillation in the presence of a saturated NaCl solution in 30% NaOH. After acidifying the distillate with

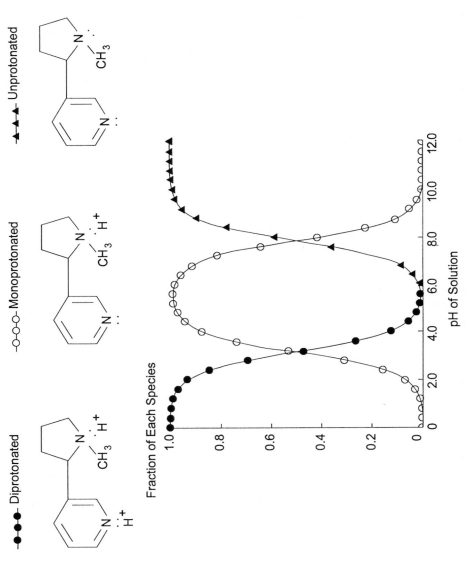

Fig. 19 Distribution of protonated and unprotonated nicotine in solution at various pH increments. (Reprinted from reference [86]).

HCl, nicotine in the acidified distillate is analyzed by UV spectrophotometry at 259 nm after correcting for the background absorbance.

*End determination.* UV absorbance measurements of nicotine are performed using a Shimadzu Model UV-2501 PC UV–VIS recording spectrophotometer equipped with a data system using UVPC Personal Spectroscopy Software vs. 3.9 (Shimadzu Scientific Instruments, Inc., Columbia, MD, USA). The UV spectral data are measured in the spectrum acquire mode at a wavelength range of 200–300 nm. After nicotine is extracted from the tobacco product, aliquots of the distillate are diluted, if necessary, with 0.05 N HCl such that the absorbance at 259 nm is between 0.5–0.8. Furthermore, the absorbance values at wavelengths 236, 259 and 282 nm are also read for background correction purposes.

UV absorbance measurements of nicotine are also obtained at times for capillary electrophoresis (CE) calibration studies. For quantitative CE studies, nicotine calibration standard concentrations are calculated using the molar extinction coefficient of $3016.92\ M^{-1}\ cm^{-1}$ for nicotine in water at 260 nm [15].

*Nicotine reporting requirements.* Current ATF regulatory classifications are based on qualitative information. As such, declaration of the presence or absence of nicotine in a particular sample is the only reporting requirement with this method. However, quantitative determination of nicotine in tobacco is also possible with the method. If quantitative results are reported, the percent total alkaloids, expressed as nicotine, are reported for samples on a dry weight basis. The percent total alkaloids (as nicotine) is calculated after correcting the absorbance at 259 nm as follows:

$$A_{259\ corr} = 1,059\ [A_{259} - \tfrac{1}{2}(A_{236} + A_{282})]$$

*Gas chromatographic (GC) analysis of nicotine.* Gas chromatography is very amenable to the analysis of nicotine in ATF regulated tobacco products, especially since the technique was originally applied to the analysis of nicotine in commercial tobacco products (vide supra). As with spectrophotometry described above, there are disadvantages associated with this technique as well. Despite the rapid analysis times, gas chromatography also suffers in that the extraction step employs organic solvents such as chloroform and hexane in relatively large volumes, is tedious, time consuming and is not as environmentally friendly as the spectroscopic method. The main features of the gas chromatographic method can be described as follows:

*Method scope.* Tobacco is extracted with organic solvent and nicotine is determined by capillary gas chromatography with a polar stationary phase.

*Sample preparation.* Nicotine in ATF regulated tobacco products is extracted using the method developed by Jablonski as described by Gottscho et al. [82]. A sample of tobacco product (1.0 g) is treated with 7.5 mL of 1N HCl and left to stand 10 min at room temperature. The sample is allowed to wet completely. The sample is then extracted with vigorous shaking on a shaker in 25 mL of extracting solution (prepared by mixing 600 mL of chloroform with 400 mL hexane containing 0.6 g octadecane as internal standard) and 5.0 mL of 36% NaOH for 20 min at room temperature. The extract is then treated with 3.0 g of diatomaceous earth, shaken to wet it completely, and filtered through a sintered glass funnel. A standard nicotine solution is prepared by weighing 0.06 g of nicotine into a 100 mL volumetric flask and made to volume with the extracting solution.

*End determination.* Aliquots of standard nicotine (1.0 μL) and filtered tobacco extracts are injected in tandem and analyzed by GC. The GC system typically used consists of a Hewlett-Packard (HP) GC Series 5890 fitted with a capillary column (0.53 mm×16 m coated with Supelcowax 10) and equipped with both a flame ionization detector (FID) and a HP Integrator Series 3393A integrator. Helium is used as the carrier gas at a flow rate of 11.6 mL/min, a head pressure of 5 KPa and a split ratio of 3:1. Column temperature is maintained at 190°C and injection temperature at 230°C. Samples are injected automatically using the HP Autoinjector Series 7736A. Figure 20 shows the gas chromatographic analysis of nicotine in an ATF regulated tobacco product.

*Nicotine reporting requirements.* Current ATF regulatory classifications are based on qualitative information. As such, declaration of the presence or absence of nicotine in a particular sample is the only reporting requirement with this method. However, quantitative determination of nicotine in tobacco is also possible with the method. If quantitative results are reported, the percent total alkaloids, expressed as nicotine, are reported for samples on a dry weight basis. Nicotine concentration in the starting tobacco is calculated based on an internal standard calculation.

*Current practices (capillary electrophoresis).* Capillary electrophoresis (CE) is rapidly gaining importance as an analytical technique, capturing the interest of analytical scientists in several areas [26,89,90]. The unique and powerful capabilities of CE include rapid analysis times, low sample consumption, high efficiency, high resolution and selectivity, thus making CE a powerful analytical tool that is complementary to other analytical techniques such as HPLC. Application of capillary electrophoresis to determine nicotine in ATF regulated tobacco products was first reported in 1997 [26]. The current analytical technique for the analysis of nicotine in ATF regulated tobacco products is capillary electrophoresis (CE).

*Method scope.* Nicotine is obtained by aqueous extraction of tobacco products. The aqueous extract is analyzed by CE using a pH 2.5 sodium phosphate buffer followed by UV detection at 260 nm.

*Sample preparation.* Nicotine in ATF regulated tobacco products is extracted using a method described by Ralapati [26,89]. A sample of tobacco (50–100 mg), used as is or ground, is extracted with occasional shaking in 10 mL of distilled, deionized water at room temperature for 1 h. The tobacco extracts obtained are used fresh, usually in less than 4 h. The aqueous tobacco extracts are diluted 1:10 or 1:20 with water to give a working sample. An aliquot of working sample is filtered directly into an autosampler vial for CE analysis.

*End determination.* CE is typically performed on a SpectraPHORESIS 500 System equipped with a UV absorbance detector, an autosampler and a data system using PC1000 System Software version 3.0.1A from Thermo Separation Products. The capillary is preconditioned prior to its first use and thereafter at the beginning of each day as follows: 10 min wash with 1.0 N NaOH at 60°C, 5 min wash with 0.1 N NaOH at 60°C, 5 min wash with distilled, deionized water at 60°C and 5 min wash with run buffer at 20°C. Electrolyte solutions are degassed under vacuum for 2 min prior to each run. Two buffer blank runs and two sample runs are performed prior to starting the actual analysis to allow the system to stabilize, and for the electrolyte

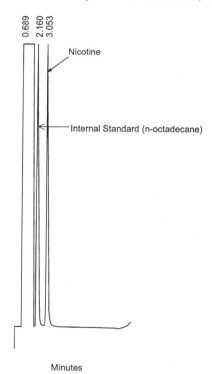

Fig. 20 Gas chromatogram of nicotine typically obtained with ATF specified methodology.

TABLE 18

CE separation conditions

| Capillary column | Fused silica |
|---|---|
| Capillary dimensions | 44 cm × 100 mm |
| Temperature | 20°C |
| Injection | Hydrodynamic, 1 second |
| Voltage | 10 kV for 10 minutes |
| Capillary buffer | 25 mM $NaH_2PO_4$, pH 2.5 |
| Detection | 260 nm |
| Instrument | SpectraPHORESIS 500 |

solutions, nicotine standards and tobacco sample extracts to equilibrate to the run conditions. The capillary is rinsed with 0.1 N NaOH and then with distilled, deionized water each for 1 min between runs in a postrun capillary wash. The capillary is rinsed and filled with the run buffer in a prerun capillary wash for 1 min. Nicotine is analyzed by CE using the separation conditions described in Table 18.

*Nicotine reporting requirements.* Current ATF regulatory classifications are based on qualitative information. As such, declaration of the presence or absence of nicotine in a particular sample is the only reporting requirement with this method. However, quantitative determination of nicotine in tobacco is possible with the method as described below.

Before establishing standard CE analysis conditions at the ATF National Laboratory, research was conducted to optimize CE analysis conditions. As mentioned above, nicotine in an aqueous solution can exist in three different ionic forms depending on the pH. This pH dependence has definite impact on the CE performance when analyzing nicotine [26]. CE analysis of nicotine in sodium phosphate buffer at pH 6.9 and pH 2.5 demonstrate two important pH effects, (1) impact on the net mobility of nicotine; and (2) detection sensitivity. As a monoprotonated species at pH 6.9, nicotine has a high electrophoretic mobility (migration time of 3.276 min) compared to nicotine as a diprotonated species at pH 2.5 (migration time of 6.200 min) (Fig. 21). Second, as a diprotonated species (pH 2.5), the molar extinction coefficient of nicotine at 260 nm is greater than that observed for nicotine as a monoprotonated species (pH 6.9) (Fig. 18). These experimental observations are in agreement with similar observations reported by others [15,85]. Based on these results, a pH 2.5 sodium phosphate buffer was established as the standard analysis condition for CE analysis of nicotine at the National Laboratory.

Method precision (repeatability) has been evaluated for the standard conditions used at the National Laboratory [26]. Repeated injections of nicotine standards typically yield coefficients of variation of 8 or less for corrected peak area response and 2.5 or less for migration times. Corrected peak area takes into account that in CE, analytes with high electrophoretic mobility pass the detection window faster than low mobility analytes. Identical concentrations of highly mobile analytes, therefore, will generate smaller peak areas when compared to less mobile analytes. Corrected peak area is calculated as the peak area of nicotine divided by the migration time of nicotine.

Many types of products are analyzed for classification purposes at the National Laboratory. Evaluation of a beedi provides an example of the classification process employing CE analysis. According to the Dictionary of Tobacco Terminology, a beedi (also known as *bidi, biri*) is 'a form of cigarette found in India; it consists of granulated tobacco rolled in a section of Indian ebony leaf and tied with thread' [79] (Fig. 22). Prior records have indicated that neither was this product analyzed for nicotine, nor a determination made to properly classify it for tax purposes according to the Code of Federal Regulations [78]. According to the Dictionary of Economic Plants, the species of Indian ebony leaf used for wrapping beedies in India is *Diospyros melanoxylon Rox* [91,92]. Furthermore, according to Hegnauer's Chemotaxonomie der Pflanzen, this and the other *Diospyros* species contain no

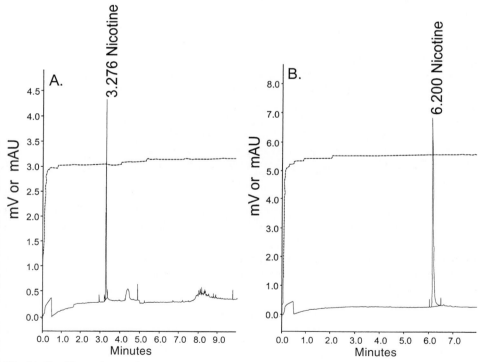

Fig. 21 Capillary electrophoresis analysis of nicotine in sodium phosphate buffer at (A) pH 6.9 and (B) pH 2.5 to demonstrate the effect of pH on the net mobility of nicotine.

alkaloids ([93], Ravindranath B, personal communication). Based on this information, evaluation of beedies provide both an example of the regulatory classification process for tobacco products and a challenging tobacco product substrate.

CE analysis results show the presence of nicotine in the beedi filler, indicating that the beedi is consistent with a tobacco product. On the other hand, CE analysis of the beedi wrapper demonstrates an absence of nicotine, indicating that the beedi is consistent with a cigarette and would be properly classified as such. Therefore, from CE nicotine analysis results, it is possible to show that the beedi is indeed a tobacco product, and beedies are properly classified as cigarettes. Results are shown in Fig. 23. Also, as can be seen in Fig. 23, the beedi wrapper did not analyze positive for nicotine which is in agreement with published reports ([93], Ravindranath B, personal communication).

It may come as a surprise to some, but many of the products analyzed by the National Laboratory do not have a dominant position in the US marketplace. Mandalay Cheroot is another product that has been analyzed for nicotine and subsequently classified according to ATF regulations (Fig. 24). Originally manufactured and imported from Myanmar (formerly,

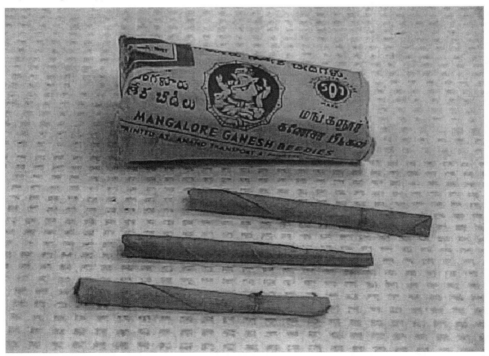

Fig. 22 A picture of beedies. A beedi is a form of cigarette found in India. It consists of granulated tobacco rolled in a section of Indian ebony leaf and tied with thread.

Burma), these Mandalay Cheroots consist of cut granulated tobacco wrapped in the leaf material from the Sebesten tree (*Cordia dichotoma Forst*), that belongs to the family Boraginaceae (Tana leaf). Like the beedies, CE analysis of the cheroot, shows the presence of nicotine in the filler material, indicating that the cheroot is consistent with a tobacco product. CE analysis of the wrapper did not demonstrate the presence of nicotine, indicating that the cheroot is consistent with a cigarette. Based on these CE data, Mandalay cheroot is indeed a tobacco product, and would be properly classified as a cigarette.

Capillary electrophoresis has also been applied to the analysis of nicotine in other ATF regulated tobacco products, including cigarettes, cigars, pipe tobacco, roll-your-own (RYO), chewing tobacco and snuff. For the different tobacco products analyzed, minor peaks, in addition to the major nicotine

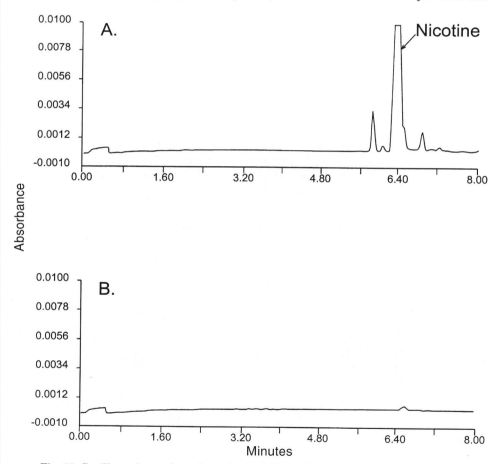

Fig. 23 Capillary electrophoresis analysis of (A) beedi filler and (B) beedi wrapper.

peak, are also detected with virtually no interference. The electrophero-grams obtained are generally clean and straightforward with all peaks well resolved. Minor alkaloids such as nornicotine, anabasine and anatabine have been reported to occur in tobacco and tobacco products [94]. Two of the minor alkaloids, nornicotine and anabasine, have been identified in a CE analysis of ATF regulated tobacco products [85]. Table 19 shows the application of CE to the analysis of nicotine in a wide variety of ATF regulated tobacco products.

*Quantitative analysis of nicotine by CE.* Despite the quantitative capability of the analytical techniques discussed above (spectrophotometry, gas chromatography and capillary electrophoresis), attempts to quantitate the amount of nicotine in ATF regulated tobacco products have not been made hitherto. The main reason for this is that the ATF regulation as it stands today calls for regulating the tobacco product based on the *presence*, not the *amount* of, nicotine.

In 1994, the Food and Drug Administration (FDA) held discussions to consider a proposal(s) to regulate the tobacco products based on nicotine content. Even though this proposal has yet to become law, scientists at the

Fig. 24 A picture of Mandalay Cheroots. They consist of cut granulated tobacco wrapped in the leaf material from the Sebesten tree.

National Laboratory Center have conducted CE studies to quantitatively analyze nicotine in ATF regulated tobacco products in anticipation of a possible FDA proposal to regulate tobacco products based on nicotine content. Method validation studies have been conducted to determine response linearity, limits of detection and limits of quantitation for nicotine

TABLE 19

CE analysis of nicotine in ATF regulated tobacco products

| Tobacco product | Sample # | Nicotine | |
| | | Wrapper | Filler |
|---|---|---|---|
| Cigarette | 16648 | NA | + |
| | 16649 | NA | + |
| | 16650 | NA | + |
| | 16651 | NA | + |
| | 16652 | NA | + |
| | 16653 | NA | + |
| | 16654 | NA | + |
| | 16655 | NA | + |
| | 16656 | NA | + |
| | 16657 | NA | + |
| | 16659 | NA | + |
| | 16660 | NA | + |
| | 16661 | NA | + |
| | 16662 | NA | + |
| | Marlboro | NA | + |
| | Newport | NA | + |
| Beedies | 16586 | - | + |
| | 16606 | - | + |
| | 16607 | - | + |
| | 16647 | - | + |
| Cigars | 16574 | + | + |
| | 16582 | + | + |
| | 16664 | + | + |
| | Winchester Little Cigar | + | |
| | Habana Cigar | + | + |
| Pipe tobacco | 16634 | NA | + |
| | 16635 | NA | + |
| Chewing tobacco | 16645 | NA | + |
| Roll-your-own | 16636 | NA | + |
| | 16637 | NA | + |
| Herbal smoking (non-tobacco) | 16608 | - | - |
| | 16658 | - | - |
| Herbal smokeless (non-tobacco) | 16665 | NA | - |

NA = Not applicable

TABLE 20

CE quantitative analysis of nicotine in ATF regulated products

| ATF regulated tobacco product | Nicotine (% dry wt) |
| --- | --- |
| (A) **Smoking tobacco products** | |
| Cigar 1 Wrapper | 0.712 |
| Cigar 1 Filler | 0.560 |
| Cigar 2 Wrapper | 0.840 |
| Cigar 2 Filler | 1.011 |
| Cigarette 1 Filler | 1.521 |
| Cigarette 2 Filler | 1.881 |
| Beedi Filler | 2.429 |
| Pipe Tobacco 1 | 1.785 |
| Pipe Tobacco 2 | 1.512 |
| (B) **Smokeless tobacco products** | |
| Chewing tobacco 1 | 0.417 |
| Chewing tobacco 2 | 3.483 |
| Snuff 1 | 3.141 |
| Snuff 2 | 1.535 |

when analyzing tobacco products [89]. Quantitative analysis results for selected ATF regulated tobacco products are summarized in Table 20.

Quantitative analysis results of nicotine in ATF regulated tobacco products determined by CE [89] (Table 20) compare reasonably well with the ranges reported for similar products in other studies [80,82,94–96]. However, some differences between CE results and values reported in the literature do exist.

As discussed earlier in this chapter, the nicotine content of tobacco is greatly influenced by the cultural practices employed. The nicotine content of the fully aged and cured tobacco leaf in the field is no indication of the nicotine content in the final commercially manufactured product. This is because in commercial tobacco products, processing parameters and conditions greatly impact final nicotine content. For example, gradual and mild curing and aging conditions, generally seen in the manufacture of cigarette products, show minor losses of nicotine. On the other hand, heavy aging and fermentation processes, commonly seen in the manufacture of cigar and smokeless tobacco products, show drastic losses of nicotine [80]. Nevertheless, the nicotine content of popular blended US cigarettes is generally about 1.75%. For cigars of all classes, the nicotine content is approximately ~0.5–1.5%. In case of pipe tobaccos containing the popular blended mixtures, the nicotine content is about the same as the blended cigarettes. The scrap type of chewing tobacco contains less than 1.0% nicotine while the plug type of chewing tobacco contains nicotine ranging from 1.5–2.5% [80,82,94–96]. Snuff is reported to contain nicotine ranging from <1.0–3.35% [82,93].

In summary, the analysis of nicotine in tobacco is the primary regulatory determinant that drives the science and regulation performed at the ATF National Laboratory Center. ATF regulations, as they are applied today, monitor the presence (without regard to the amount) of nicotine in tobacco products. Nevertheless, nicotine can be quantitated with techniques such as capillary electrophoresis in response to any future regulation.

*Tobacco nicotine analysis conducted in industry laboratories to comply with federal and state regulations*

Current regulatory requirements that include tobacco nicotine testing are: (1) federal regulations that require the determination of nicotine in smokeless tobacco samples [72]; and (2) state regulations that require the determination of nicotine in cigarettes and smokeless tobacco [74–76]. These regulations do not specify established standard methods for the determination of nicotine in tobacco (i.e., AOAC standard methods [97,98], CORESTA methods [99] and ISO methods [34] are not mandated). Rather, current regulations specify a single methodology that, at the time of this writing, has not been fully developed and validated for use in either federal or state regulation [100,101]. The application of a single analysis methodology is notable since federal and state regulations deal with different forms of tobacco (smokeless tobacco vs. cigarette cut filler and smokeless tobacco).

While the federal and state regulations designate a common tobacco nicotine testing methodology, a fundamental point of difference between the two types of regulation is the method development and validation approach that has been applied as regulations are formed. Development of a testing method to meet the federal smokeless tobacco regulations has, in principle, incorporated several of the key technical steps that must be followed when establishing a new standard method [102–104]. As of early 1999, the method development and validation process initiated in support of federal regulations has not been completed. Issues of method validation and interlaboratory method comparison have not been addressed in the same way during formation of state regulations. In an effort to promulgate state regulation as expeditiously as possible and rapidly provide data to consumers, state regulators have attempted to adapt 'published' testing procedures without the benefit of any experimentation to confirm the appropriateness of the 'adapted' procedures. Thus, although a common tobacco nicotine testing methodology has been proposed or adopted for federal and state regulations, differences in test purpose, the type of tobacco sample(s) evaluated, and the method development and validation processes followed are found when the two types of regulation are compared.

*The determination of nicotine in smokeless tobacco samples (Federal Regulation).* In 1986, the 99th Congress enacted into law the Comprehensive Smokeless Tobacco Health Education Act [72]. The stated purpose of this act is "to provide for public education concerning the health consequences of using smokeless tobacco products." The law calls upon the Department of Health and Human Services (HSS) to develop public education programs, the Federal Trade Commission to mandate smokeless tobacco warning labels and tobacco manufacturers to report ingredients added to smokeless tobacco. In addition to added ingredients, the federal regulation requires that the quantity of nicotine contained in fine cut, ground, powdered or leaf tobacco that is intended to be placed in the oral cavity be reported to the HSS Secretary on an annual basis.

*Method development and validation issues.* To meet the annual reporting requirement, the federal government has recognized that a standard methodology to measure the quantity of nicotine contained in smokeless tobacco is necessary. The federal government has also recognized that the annual reporting requirement cannot be implemented until a standard protocol can be developed and validated according to accepted scientific practices [100,101]. Further, the Department of Health and Human Services has affirmed the need for uniform testing methodology if understandable data are to be obtained. To achieve a technically sound method, the Department of Health and Human Services has determined that method development and validation should be completed by tobacco industry scientists, rather than mandating analytical methodology as a matter of policy without regard for scientific process.

While the draft methodology proposed for the determination of nicotine in smokeless tobacco has received substantial technical criticism from smokeless tobacco manufacturers and others [105,106], several aspects of the process that has been followed to establish the draft methodology, as described in the Federal Register, are worth noting:

- Prior to 1989 each smokeless tobacco company employed different methods for nicotine analysis.
- The Department of Health and Human Services requested that "a standard methodology be developed to ensure the accuracy and reliability of the information on nicotine and moisture."
- The Department of Health and Human Services recognized that it did not have the technical expertise and resources to develop a standardized methodology.
- A working group comprised of the ten major domestic manufacturers of smokeless tobacco was formed "to begin the development of uniform methods for analyzing smokeless tobacco products for nicotine and moisture content."
- The working group reached consensus on a technical alternative for subsequent method validation. The technical alternative was communicated to the Department of Health and Human Services.

- After revising the recommended technical alternative based on comments from peer reviewers, the Centers for Disease Control (CDC) published testing methodology for public comment.
- Nicotine reporting requirements for smokeless tobacco will not be implemented until the methodology has been finalized and published.

*Smokeless tobacco method synopsis.* This section will briefly summarize the main features of the draft analytical method proposed by federal regulators for the determination of nicotine in smokeless tobacco.

*Method scope.* Samples of smokeless tobacco are ground, made basic with sodium hydroxide and nicotine is partitioned into methyl-*t*-butyl ether. Sample extracts are analyzed by gas chromatography to determine the nicotine concentration in the smokeless tobacco sample. As first drafted [101], the method also specifies that nicotine recovery studies with a 'vegetable matrix' will be conducted with each set of samples.

*Sample preparation.* The method recognizes that smokeless tobacco is produced in various physical forms. Plug tobacco, twist tobacco and loose tobaccos are ground to a 4 mm particle size. Dry snuff, wet snuff and wet snuff portion packs are analyzed without grinding, although the packaging material or 'pouch' from the wet snuff portion pack is ground. For all tobacco types, a minimum sample size of 100 g is recommended to ensure a uniform and representative sample. Tobacco samples are analyzed 'as is' without drying. After grinding, 1 g samples of tobacco are wetted with 2 N sodium hydroxide and nicotine is partitioned into 50 mL of methyl-*t*-butyl ether while shaking the sample for two hours.

*End determination.* Nicotine is determined by capillary gas chromatography via the split injection mode using an apolar capillary column (5% phenyl/95% methyl silicone) and flame ionization detection. Quantitation is based on internal standard calibration employing quinoline as the internal standard. A typical separation is found in Fig. 25.

*Nicotine reporting requirements.* Nicotine concentrations are reported as mg/g tobacco to an accuracy level of two decimal places, along with summary statistics from repeated measurements. In addition to the total nicotine content found in the tobacco, a quantity termed 'free base nicotine' is estimated from a 'tobacco pH' measurement and the Henderson-Hasselbach equation. Nicotine concentrations are not corrected for the moisture content of the tobacco, but tobacco moisture is determined separately and reported with the other values.

*The determination of nicotine in cigarettes and smokeless tobacco (State Regulation).* In late 1996, legislation was passed by the Commonwealth of

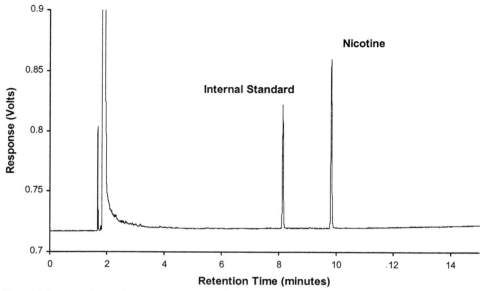

Fig. 25 Typical capillary gas chromatography separation using the smokeless tobacco method for the determination of nicotine in tobacco.

Massachusetts that mandated new testing and reporting regulations for cigarettes and smokeless tobacco products sold in Massachusetts [74]. In accordance with that law, draft regulations were proposed and a public hearing was announced for early 1997 [107–110]. Nicotine testing and reporting regulations were revised and promulgated in August 1997 [75,111]. The first nicotine testing results for cigarettes and smokeless tobacco were reported to the Massachusetts Department of Health on December 15, 1997 as required by the new regulations. Similar regulations were adopted the following year in Texas. The first testing results reported to the Texas Department of Public Health were initially to have been required on December 1, 1998 [76], however the reporting date was subsequently extended into the next year in view of uncertainties surrounding the ability to complete all necessary testing by the earlier deadline and other unresolved regulatory issues regarding cigarette additives and ingredients.

One of the stated purposes of the Massachusetts testing regulations is to provide 'cigarette nicotine yield ratings'. Those ratings are described as "a composite of information intended to show the range of nicotine that each cigarette brand can be expected to deliver to the average consumer . . .". The testing rationale adopted by the regulations assumes the premise that the range of smoke nicotine that each cigarette brand can be expected to deliver to the average consumer can be based on "machine testing parameters that seek to reflect actual smoking behavior." It is also assumed in the regulation that cigarette design features such as filter ventilation and cigarette tobacco nicotine content, as well as potential responses to cigarette design features such as 'smoke pH' can be used to predict average consumer smoke nicotine intake.

To comply with state regulations, smoke nicotine, filter ventilation, tobacco nicotine and 'smoke pH' are determined for all cigarette brand styles sold within the state. While four different cigarette tests are conducted, the nicotine yield rating is based on results from a single test rather than a composite of information: smoke nicotine results obtained with an exaggerated smoking regimen. The puffing regimen required consists of a 45 cc puff volume, a puff duration of 2 s, and puffs are taken once every 30 s. The smoking regimen specified also requires that cigarette filters are 50% 'blocked' by placing tape around half of the cigarette filter. In contrast to the smoking regimen specified by the Commonwealth of Massachusetts, the standard FTC smoking regimen consists of a 35 cc puff volume, a puff duration of 2 s, and puffs are taken once every 60 s. The FTC standard smoking regimen does not require any 'blocking' of the cigarette filter.

Smokeless tobacco manufacturers are required to test smokeless tobacco samples to determine tobacco pH, tobacco moisture content and the amount of nicotine in smokeless tobacco. In addition to this information, manufacturers must report the percentage of 'unionized (free) nicotine', the 'total unionized (free) nicotine' in units of mg/g and a 'smokeless tobacco nicotine delivery' rating based on the latter quantity.

As an era of increased tobacco product regulation begins, the choice of a draft method that is intended for analysis of a specific sample matrix (smokeless tobacco) as a state regulatory standard is scientifically troublesome. The endnotes section of the draft CDC method for smokeless tobacco states, "The comments and notes listed below can be described as Good Laboratory Practice guidelines; they are described in detail in this protocol to ensure minimal interlaboratory variability in the determination of nicotine, total moisture, and pH in *smokeless tobacco* [emphasis added]." Although the intended scope of application for draft methodology seems clear, state regulations exceed the intended method scope and require that cigarette cut filler tobacco shall also be analyzed with the smokeless tobacco methodology. As regulation of tobacco products increases, development of testing requirements based on sound scientific principles of method validation and interlaboratory testing will be a challenge, especially given the expedient alternative available to regulators of requiring published research methods without regard for any potential technical limitations.

*Method development and validation issues.* As discussed by Borgerding et al. [112], the major US cigarette manufacturers expressed numerous concerns to the Massachusetts Department of Public Health during the comment period in early 1997 as the testing regulations were being developed [113,114]. Central to the comments were three points. First, the underlying premise that such things as an 'average smoker' and an 'average smoke yield' exist and could provide useful information to individual consumers about their nicotine intake or 'average nicotine intake' was a flawed concept. Analytical smoking machines (i.e., the 20-port linear or rotary smoking machines currently used for standard methods such as the FTC, ISO, CORESTA, and TIOJ methods) cannot replicate actual human smoking behavior. These smoking machines are designed to take uniformly shaped puffs of a precisely set volume and duration when puffs are taken at a constant frequency. Human smoking behavior, in contrast, is variable. Puffing topography is not constant: (a) puff to puff as the smoker smokes a cigarette; (b) from cigarette to cigarette for a particular smoker; or (c) from individual to individual [115]. Second, without the benefit of scientifically accepted method development, method validation and interlaboratory collaborative evaluation studies, the 'testing methods' prescribed by the regulations would not necessarily produce comparable data from one laboratory to the next. Illustrations of the problems that can be and that were encountered follow. Third, the testing guidelines provided by the regulations did not provide sufficient technical detail with which to conduct meaningful experiments. The Massachusetts Department of Public Health incorporated some revisions to the proposed regulations, but did not address the major concerns of the manufacturers. The cigarette manufacturers were thus obliged to follow the prescribed test methods in order to report the required information by the deadline set by the state.

Table 21 compares the method development and validation process followed for the federal and state nicotine testing regulations. While neither regulation has fulfilled all of the steps generally expected in such a process, it is clear that the state regulations have implemented the draft methodology

TABLE 21

Comparison of the method development and validation process

| Steps completed in the method development and validation process | Federal regulation [100,101] | State regulation [75] |
|---|---|---|
| Define test purpose and scope | Yes | Yes |
| Conduct method development research | No | No |
| Identify best technical alternative to achieve test purpose | Yes | No |
| Complete validation process to determine method accuracy and precision | Pending | No |
|    Intra-laboratory validation | | No |
|    Inter-laboratory validation | | No |
| Finalize written protocol | Pending | Yes |

for smokeless tobacco as though it is an official, standardized analysis methodology, without the completion of any substantial method validation.

*Problems that can occur using non-standardized/non-official methods*
Testing conducted in 1997 to comply with state regulations provides some insight into the accuracy and precision of the tobacco nicotine method when applied to cigarette cut filler tobacco. For example, Fig. 26 depicts method quality control data and summarizes results for two different tobacco samples during a three-month period. During that time, more than 60 determinations each of a Burley tobacco and a typical American blend cut filler were conducted. Mean nicotine concentrations of 25.89±1.25 and 14.93±0.73 mg/g were found for the Burley tobacco and American blend cut filler tobacco samples, respectively, based on the quality control data. Thus, the coefficient of variation for repeated analysis of each tobacco type during the three-month period was less then 5 (i.e., 5% RSD).

While the method quality control data in Figure 26 provide an indication of the method precision that can be obtained by a single laboratory, data found in Figs 27 and 28 reveal information regarding interlaboratory accuracy and precision. As part of the1997 state compliance testing, one major tobacco manufacturer chose to have its cigarette brand styles tested by an independent contract laboratory. Although comparison of results from two laboratories is much more limited in scope than the typical eight laboratory comparison required when a research method is validated as an official standard method [102], data found in Figs 27 and 28 are the only example available to date in which more than one laboratory has used the

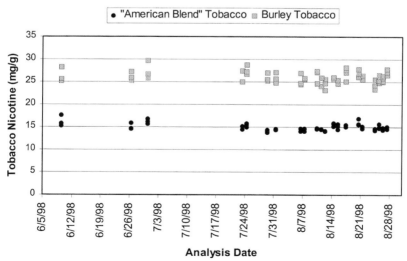

Fig. 26 Method quality control data for two 'monitor tobacco' samples when analyzed with smokeless tobacco methodology.

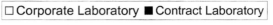

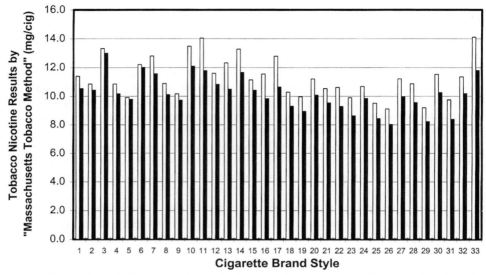

Fig. 27 Comparison of tobacco nicotine analysis results for cigarette cut filler samples when the samples are analyzed by the 'Massachusetts tobacco nicotine method' in two laboratories.

Massachusetts tobacco nicotine method to analyze the same cigarette samples.

Figure 27 compares tobacco nicotine analysis results determined by the corporate laboratory with the results obtained by the contract laboratory. Tobacco nicotine concentrations determined by the corporate laboratory

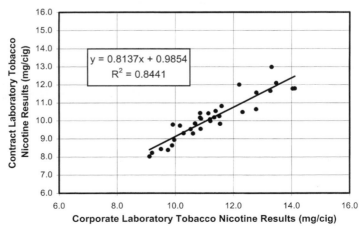

Fig. 28 Regression analysis of tobacco nicotine results for cigarette cut filler samples when the samples are analyzed by the 'Massachusetts tobacco method' in two laboratories.

380

ranged from approximately 9–14 mg/cig, while results from the contract laboratory were approximately 8–13 mg/cig. Individual cigarette brand style results when compared from the two laboratories differ by as little as a few tenths of a milligram per cigarette, and as much as roughly two milligrams per cigarette. Regression analysis (Fig. 28) suggests that results from the two laboratories are correlated, although there is both an offset (~1 mg/cig) and a bias (~19%) between the data sets. It is important to note that two separate laboratories following the same written protocol for identical samples obtained these results. Without a formal method validation process and an interlaboratory comparison study with a representative number of laboratories, it is not possible to assess which set of results (the corporate laboratory or the contract laboratory) are accurate. In fact, it is possible that neither set of results is correct.

The 1997 state compliance testing results also afford the opportunity to compare the results from two different tobacco nicotine methods when the methods are applied in the same laboratory. During 1997, analyses were conducted by the corporate laboratory using the Massachusetts tobacco nicotine method and a continuous-flow analysis method. The continuous-flow method is the standard method applied in the corporate laboratory. This method employs a colorimetric endpoint, rather than gas chromatography [34]. As such, the method is not specific for the determination of nicotine; rather total alkaloids are measured and reported as nicotine.

Figure 29 compares the results from the two analytical methods for thirty-three cigarette brand styles. Tobacco nicotine concentrations determined by the Massachusetts tobacco nicotine method ranged from approximately 9–14 mg/cig, while results from the company standard method were approximately 9–15 mg/cig. Comparison of individual cigarette brand style results yields absolute differences between the two methods of roughly a few tenths of a milligram per cigarette to approximately 1.7 mg/cig. This range of difference is similar to that observed for the between-laboratory comparison with a single method. Regression analysis (Fig. 30) suggests that results from the two methods are correlated, although there is both an offset (~1.6 mg/cig) and a bias (~17%) between the data sets. The continuous-flow method (standard method for company A), generally yields higher results consistent with the non-specific nature of the method, i.e.. the nicotine results obtained represent a summary response of 'all' tobacco alkaloids expressed as nicotine. Based on the type of data presented in Figs 29 and 30, the Massachusetts Department of Public Health has approved the continuous-flow method as an equivalent alternative for use in complying with nicotine testing and reporting requirements.

In summary, as regulation of tobacco and tobacco products increases based on the determination of nicotine in tobacco and smoke, cooperation between state agencies, federal agencies and the tobacco industry is essential. Without cooperation, it is likely that new regulations will require the generation of tremendous amounts of data that have little potential to become useful information. In fact regulations that require vast amounts of data from different laboratories without the benefit of a complete method validation process are more likely to misinform, rather than inform, the public. It is possible to provide unbiased information for public education concerning the use of tobacco and tobacco products. However, mutual trust and cooperation must be developed between all parties in order develop sound scientific data and ultimately provide the best possible information for the public as new questions regarding tobacco and tobacco products are posed. Ultimately, all parties should rely on the scientific method to provide unbiased answers to the technical questions that are posed.

## VI. CONCLUSIONS

The physical and molecular properties of nicotine have provided the analyst with a host of opportunities for qualitative and quantitative analysis. During the last fifty years, analytical methods reported for nicotine determination have evolved considerably, employing the newest instrumental analysis techniques of the day. In addition to its unique physical and molecular properties that provide many different qualitative analysis opportunities,

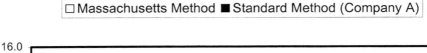

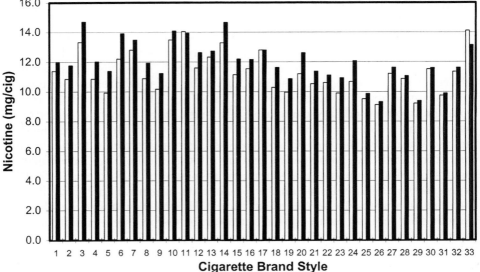

Fig. 29 Comparison of tobacco nicotine analysis results for cigarette cut filler samples when the samples are analyzed by two different methods in a single laboratory.

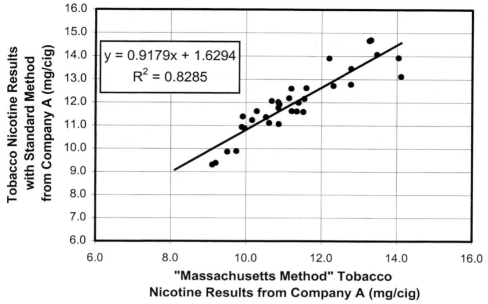

$$y = 0.9179x + 1.6294$$
$$R^2 = 0.8285$$

Fig. 30 Regression analysis of tobacco nicotine results for cigarette cut filler samples when the samples are analyzed by two different methods in a single laboratory.

nicotine is an ideal molecule for quantitative analysis. It is a significant component of the tobacco leaf, often representing more than 98% of the total alkaloid content in tobacco.

Nicotine analysis of tobacco has been conducted for many purposes. The chemical composition of tobacco crops varies from growing season to growing season and from one geographical region to the next. Tobacco nicotine analysis provides one means of characterizing new tobacco crops and understanding the raw material that will be included in final products. The research and development of new tobacco breeding lines also requires tobacco nicotine and secondary tobacco alkaloid analyses. The smoking quality of different tobacco varieties has been related to the alkaloid distribution found in the leaf. Another area in which the determination of tobacco nicotine has been of critical importance is the classification of tobacco products for subsequent taxation. The presence and location of nicotine within the components of a smoking article are critical to whether the smoking article is classified as a cigarette, cigar or non-tobacco product, and to the amount of tax that is levied by the federal government.

The legislative and regulatory history of tobacco control efforts is a substantial and varied topic that has been considered by all levels of government. Tobacco legislation and regulation has been enacted for many different purposes. There has been a steady progression of legislation and regulations dealing with tobacco products at the federal, state and local level. Tobacco control efforts have directly affected consumers in many ways. Taxation and restrictions on the use of tobacco products, particularly cigarettes, have greatly reduced the number of people that use tobacco products and the per capita use level. The determination of nicotine in tobacco products has been a specified requirement for a basis of taxation at the federal government level for many years. Since 1996, however, the determination of nicotine in tobacco has become a specific requirement in some state regulations. With this new requirement, the scope and limitations of analytical methods used for the determination of tobacco nicotine analysis are being carefully considered.

This chapter has dealt with three topics, the determinations of nicotine in tobacco, nicotine in tobacco processing environments (concentrating primarily with processes in the cigarette industry), and regulatory aspects dealing with nicotine in tobacco products. The first topic presented was an historical perspective of the progress of researchers in tobacco science and achievements in analytical chemistry made in the determination of alkaloids in tobacco (particularly nicotine) from 1826 to the present. Three examples of current laboratory practices for the determination of nicotine in tobacco were summarized. The second part of this chapter, on nicotine in tobacco processing environments, described tobacco processing that occur within the tobacco industry, mainly the cigarette industry, and the determinations of nicotine that have been made in the individual process streams. The third part of this chapter highlighted the current legislative and regulatory issues concerning the determinations of nicotine in tobacco and tobacco products.

## ACKNOWLEDGEMENTS

The authors would like to acknowledge the support of several colleagues during the preparation of this manuscript. Our appreciation is extended to Mr. L. A. Milhous, Jr. (R. J. Reynolds) and Mr. R. E. Davis (Philip Morris) for discussion of continuous flow analysis techniques, to Dr. C. B. Bennett (U.S. Tobacco) for insight into smokeless tobacco analysis, to Mr. H. V. Thomsen (House of Prince) for discussions of past CORESTA Routine Analytical Chemistry Sub Group collaborative studies, to Dr. B. M. Gordon for discussion of fast GC analysis and to Ms. H. S. Chung (R. J. Reynolds) for library services support. The authors would also like to thank Ms. L. A. Crumpler, Dr. W. M. Coleman, III and Dr. R. A. Lloyd, Jr. of R. J. Reynolds, and Mr. J. M. Burggraff and Dr. S. M. Dugar of BATF for editorial review and comment, and for their on-going support during this work.

## REFERENCES

1 Jacobson PD, Wasserman J, Anderson JR. Historical overview of tobacco legislation and regulation. *J. Soc. Issues*, 1997: **53**; 75–95.

2 Huber GL. Tobacco: its history, economics, and political influence. *Sem. Respir. Med.*, 1989: **10**(4); 278–296.

3 Huber GL. Physical, chemical, and biologic properties of tobacco, cigarette smoke, and other tobacco products. *Sem. Respir. Med.*, 1989: **10**(4); 297–332.

4 Schmeltz I. Nicotine and other tobacco alkaloids. *Nat. Occurring Insectic.*, 1971: **1971**; 99–136.

5 Holmstedt B, Rand MJ, Thuran K (Ed.). *The pharmacology of nicotine.* Holmstedt B. Toxicity of nicotine and related compounds. 61–88. Washington,DC: IRL Press, 1988.

6 Von Euler US (Ed.). *Tobacco alkaloids and related compounds.* Pailer M. Chemistry of nicotine and related compounds (including biosynthetic aspects). New York: The MacMillan Company, 1965, 15–36.

7 Bentley HR, Berry EGN (Ed.). *Research papers No.3: The constituents of tobacco smoke:an annotated bibliography.* London: The Imperial Tobacco Company, 1959.

8 Neurath G. Nitrogen compounds of tobacco smoke. *Beitrag. Tabakforsch.*, 1969: **5** (3); 115–133.

9 Jackson KE. Alkaloids of tobacco. *Chem. Rev.*, 1941: **29**; 123–197.

10 Markwood LN. *A review of information on nornicotine*; Pamphlet E–561 US Department of Agriculture, Bureau of Entomology and Plant Quarantine. Washington,DC: USDA, 1942.

11 Kuffner F, Schick K, Buhn H. Socratines, obeline, and other secondary alkaloids of cigar smoke. *Monat. Chem.*, 1956: **87**; 749–762.

12 William JJ. Alkaloids of tobacco: identification and determination. *Ind. Eng. Chem.*, 1952: **44**(2); 270–273.

13 U. S. Department of Agriculture, Bureau of Animal Industry, Bulletin 133, 1911.

14 Porter WA, Naghski J, Eisner A. Separation of alkaloids by chromatography. Tobacco Chem. Res. Conf. (TCRC) 3rd; 1949; Richmond (VA).

15 Willits CO, Swaim ML, Connelly JA, Brice BA. Spectrophotometric determination of nicotine. *Anal. Chem.*, 1950: **22**; 430–3.

16 Cundiff RH, Markunas PC. Determination of nicotine nornicotine and total alkaloids in tobacco. *Anal. Chem.*, 1955: **27**; 1650–3.

17 Griffith RB. The rapid determination of total alkaloids by steam distillation. *Tob. Sci.*, 1957: **1**; 130–7.

18 Sadler WW, Chesson RR, Schoenbaum AW. Automated procedure for determining the nicotine content of steam distillates. *Tob. Sci.*, 1960: **4**; 208–12.

19 Harvey W R, Badgett CE, Resnik F E. Nicotine alkaloids in tobacco leaf, cigarette filler, and particulate matter of smoke by acid-methanol extraction. *Tob. Sci.*, 1967: **11**; 84–6.

20 Collins P F, Sarji N M, Williams J F. Determination of nicotine alkaloids in tobacco using the Autoanalyzer. *Tob. Sci.*, 1969: **13**; 79–81.

21 Lyerly L A, Greene G H. Determination of nicotine in low-nicotine tobacco. *Beitr. Tabakforsch.*, 1976: **8**(6); 359–61.

22 McClure W F, Hamid A, Weeks W W, Whitaker, TB. Design and development of a computerized spectrophotometric instrument for measuring the chemical composition of tobacco. Tobacco Chem. Res. Conf. (TCRC) 31st; 1977; Greensboro (NC)

23 Sisson VA, Saunders JA. Qualitative and quantitative alkaloid screening of the USDA tobacco introduction by high performance liquid chromatography. Tobacco Chem. Res. Conf. (TCRC) 33rd; 1979; Lexington (KY).

24 Severson R F, McDuffie K L, Arrendale R F, Gwynn G R, Chaplin J F, Johnson A W. Rapid method for the analysis of tobacco nicotine alkaloids. *J. Chromatogr.*, 1981: **21**(1); 111–21.

25 Blume DE. Radioimmunoassay (RIA) for measuring low levels of nicotine in tobacco root suspension cultures and tobacco tissues. Tobacco Chem. Res. Conf. (TCRC) 38th; 1984; Atlanta (GA).

26 Ralapati S. Capillary electrophoresis as an analytical tool for monitoring nicotine in ATF regulated tobacco products. *J. Chromatogr. B*, 1997: **695**(1); 117–29.

27 Snyder L, Levine J, Stoy R. Conetta A. Automated chemical analysis: Update on continuous-flow approach. *Anal. Chem.*, 1976: **48**(12); 942A–956A.

28 Harvey WR, Stahr HM, Smith WC. Automated determination of reducing sugars and nicotine alkaloids on the same extract of tobacco leaf. *Tob. Sci.*, 1969: **13**; 13–5.

29 Charles JL, Stahr HM, Ikeda RM. Automated determination of nicotine in total particulate matter of cigarette smoke. *Tob. Sci.*, 1969: **13**; 54–8.

30 Harvey WR, Palmer AM. Field test estimation of reducing sugars and nicotine alkaloids in tobacco leaf. Tobacco Chem. Res. Conf. (TCRC) 24th; 1970; Montreal, CAN.

31 Derrick JC, Wilson WL, Hussey TA. The routine estimation of nicotine alkaloids and total sugars in flue-cured tobacco using dual channel autoanalyzers and electronic data handling systems. Tobacco Chem. Res. Conf. (TCRC) 24th; 1970; Montreal, CAN.

32 Davis RE. A combined automated procedure for the determination of reducing sugars and nicotine alkaloids in tobacco products using a new reducing sugar method. *Tob. Sci.*, 1976: **20**; 139–44.

33 Harvey WR, Handy BM. On-line generation of cyanogen chloride as a replacement for cyanogen bromide in the total alkaloids determination. *Tob. Sci.*, 1981: **20**; 131–3.

34 CORESTA (Co-operation Centre for Scientific Research relative to Tobacco Association) Standard Method No. 35: Determination of total alkaloids (as nicotine) in tobacco by continuous flow analysis. Final Approval, November 1994. CORESTA Inf. Bull. 1994 (3–4): 39–46.

35 Bowman DT. History of the regional minimum standards program for the release of flue-cured tobacco varieties in the United States. *Tob. Sci.*, 1996: **40**; 99–110.

36 Gordon B M, Greene G H. Determination of nicotine nornicotine and other secondary amine alkaloids in tobacco. Tobacco Chem. Res. Conf. (TCRC) 35th ; 1981; Winston-Salem (NC).

37 Cundiff RH, Markunas PC. Modification of the extraction procedure for the determination of alkaloids in tobacco. *J. Assoc. Off. Agr. Chem.*, 1960: **43**(3); 519–24.

38 Leffingwell JC, Young HJ, Bernasek E. Tobacco flavoring for smoking products. Winston-Salem,NC: RJ Reynolds Tobacco Co., 1972.

39 Tso TC. Interrelated factors among plant production, leaf usability and smoke characteristics. In: Tso TC. Production, physiology, and biochemistry of tobacco plant. Beltsville, MD: IDEALS, Inc., 1990: 3–134.

40 Voges E (Ed.). Tobacco Encyclopedia. Mainz, Germany: Tobacco Journal International, 1984.

41 Bacon CW, Wenger R, Bullock JF. Chemical changes in tobacco during flue-curing. *Ind. Eng. Chem.*, 1952: **44**(2); 292–6.

42 Browne CL. The design of cigarettes. 3rd rev. edn. Charlotte, NC, Hoechst Celanese Corp.,1990.

43 Frankenburg WG, Gottscho AM. Nitrogen compounds in fermented cigar leaves. *Ind. Eng. Chem.*, 1952: **44**(2); 301–5.

44 Jensen CO. Cigar tobaccos: chemical changes that occur during curing. *Ind. Eng. Chem.*, 1952: **44**(2); 306–9.

45 Frankenburg WG. Chemical changes in the harvested leaf : Part I. Chemical and enzymatic conversions during the curing process. In: Advances in enzymology, Vol. 6. New York: Interscience Publisher, Inc., 1946: 309–387.

46 Frankenburg WG. Chemical changes in the harvested leaf: Part II. Chemical and enzymatic conversions during fermentation and aging. In: Advances in enzymology, Vol.10. New York: Interscience Publisher, Inc., 1950: 325–441.

47 Moshy RJ. Reconstituted tobacco sheet. In: Wynder EL, Hoffmann D. Tobacco and tobacco smoke. New York: Academic Press, 1967: 47–83.

48 Wolf AW. Tobacco production and processing. In: Wynder EL, Hoffmann D. Tobacco and tobacco smoke. New York: Academic Press, 1967: 5–45.

49 Halter HM, Ito TI. Effect of tobacco reconstitution and expansion processes on smoke composition. In: Wernsman EA, Kuhn WH, Leger RG (Eds.). Recent advances in tobacco science: Vol 4, Physical parameters which affect the composition of cigarette smoke. Montreal, Canada: Tobacco Chemists' Research Conference, 1978: 113–31.

50 Halter HM, Ito TI. Reconstituted tobacco – Smoking and health possibilities. *J. Nat. Cancer Inst.*, 1972: **48**(6); 1869–83.

51 Sittig M. Tobacco substitutes – Chemical technology review No. 67. Park Ridge, NJ, Noyes Data Corp. 1976.

52 Halpern MG. Paper manufacture – Chemical technology review No.47. Park Ridge, NJ, Noyes Data Corp. 1975.

53 Dawson RF. Chemistry of green tobacco. *Ind. Eng. Chem.*, 1952: **44**(2); 266–70.

54 Darkis RF, Hackney EJ. Cigarette tobaccos: chemical changes that occur during processing. *Ind. Eng. Chem.*, 1952: **44**(2); 284–91.

55 Baskevitch N, Ferrer G. Modification of nicotine:tar ratio through the use of reconstituted tobacco. CORESTA Symposium, Winston-Salem NC, 1982.

56 Andersen RA, Fleming PD, Burton HR, Hamilton-Kemp TR, Hildebrand DF, Sutton TG. Levels of alkaloids and other derivatives in air- and fire-cured KY–171 dark tobacco during prolonged storage: Effects of temperature and moisture. *Tob. Sci.*, 1990: **34**; 50–56.

57 McClure W. Status of near infrared technology in the tobacco industry. In: Miskelly Diane, Law Donald, Clucas (Eds.). Proceedings of an International Symposium on Near Infrared Reflectance Spectroscopy; 1984 Oct. 15–16; Melbourne, Victoria, Australia. Cereal Chemistry Division: Royal Australian Chemical Institute, 1985.

58 Heckman RA, Diffee JT, Milhous LA. Transfer of near-infrared monochromator calibrations for tobacco constituents to tilting-filter instruments. *Analytica Chimica Acta*, 1987: **192**; 197–203.

59 Diffee JT. Tobacco analysis by NIR spectroscopy. In: Burns DA, Ciurczak EW (Eds.). Practical spectroscopy series, vol. 13. Handbook of near-infrared analysis. Marcel Dekker, 1992: 433–73.

60 Janjigian K, Perfetti TA, Green CR, Conrad FW, Rivers JM, Smith LA, Hawley RW. Nicotine just noticeable difference study of full flavor non-menthol 85 mm products. *Tob. Sci.*, 1986: **30**; 146–51.

61 Gordin HH, Perfetti TA, Hawley RW. Nicotine just noticeable difference study of full flavor low 'tar' and ultra low 'tar' non-menthol 85 mm products. *Tob. Sci.*, 1988: **32**; 62–5.

62 Rodgman A. FTC 'tar' and nicotine in cigarette mainstream smoke: A retrospective. *Recent Adv. Tob. Sci.*, 1997: **23**; 5–74.

63 Green CR, Rodgman A. The tobacco chemists' research conference: A half century forum for advances in analytical methodology of tobacco and its products. *Recent Adv. Tob. Sci.*, 1996: **22**; 131–304.

64 Maxwell Jr. JC, Fenstermacher SD. Second-quarter 1998 sales estimates for the cigarette industry. The Maxwell Consumer Report August 14, 1998.

65 Hind, inventor. Philip Morris Tobacco Company, assignee. Method for providing a reconstituted tobacco material. U. S. patent 3,386,449. June 1968.

66 Young HJ, Bernasek E, Lekwauwa AN, Young WRD, inventors. R. J. Reynolds Tobacco Company, assignee. Method for providing a reconstituted tobacco material. U. S. patent 5,339,838. August 1994.

67 Buchanan WM, Madures JW inventors. Philip Morris Tobacco Company assignee. Method for expanding tobaco stems. US patent 3,734,104. May 1973.

68 Ashburn G, inventor. R. J. Reynolds Tobacco Company, assignee. Method for providing an expanded tobacco material. U. S. patent 3,753,440. August 1973.

69 Jones EC, Norman AB. History of cigarette design. *Recent Adv. Tob. Sci.*, 1998: **24**; 3–64.

70 Title 27 United States Code Subchapter M (Alcohol Tobacco and Other Excise Taxes) Part 270 (Manufacture of Tobacco Products) Section 270.23 (Cigarette Tax Rates).

71 Title 15 United States Code Chapter 36 (Cigarette Labeling and Advertising) Section 1337 (Reports to Congress by the Secretary and Federal Trade Commission).

72 Title 15 United States Code Chapter 70 (Comprehensive Smokeless Tobacco Health Education) Section 4403 (Ingredient Reporting).

73 Title 15 United States Code Chapter 36 (Cigarette Labeling and Advertising) Section 1341 (Smoking, Research, Education and Information).

74 Annual Report: Massachusetts General Law Chapter 94, Section 307B (1996).

75 Cigarette and Smokeless Tobacco Products: Reports of Added Constituents and Nicotine Yield Ratings: 105 Code of Massachusetts Regulations 660.000 et seq. (as adopted August, 1997).

76 General requirements for annual reports by manufacturers. 25 T.A.C. secs. 101.1 et seq. Published May 29, 1998 at 23 Texas Register 5687–746.

77 U. S. Department of the Treasury, Bureau of Alcohol, Tobacco and Firearms. An Introduction to the bureau of Alcohol, Tobacco and Firearms and the Regulated Industries. Revised 11/91.

78 Title 27 United States Code Subchapter M (Alcohol Tobacco and Other Excise Taxes) Part 270 (Manufacture of Tobacco Products) Section 270.11 (Meaning of Terms).

79 DeBardeleben MZ. Dictionary of Tobacco Terminology. New York: Philip Morris, 2nd edn., 1987.

80 Garner WW. The Production of Tobacco. Philadelphia: The Blakiston Co., revised 1st edn., 1951.

81 Lewis LJ, Lamb JD, Eatough DJ, Hansen LD, Lewis EA. The determination of nicotine and cotinine by ion pair reversed-phase chromatography. *J. Chromatogr. Sci.*, 1990: **28**; 200–3.

82 Gottscho AM, Lin J.-L., Duck WN, Losty TA. Nicotine analysis of commercial tobacco products. *J. Assoc. Off. Anal. Chem.*, 1988: **71**; 1110–5.

83 Saunders JA, Blume DE. Quantitation of major tobacco alkaloids by high-performance liquid chromatography. *J. Chromatogr.*, 1981: **205**; 147–54.

84 Manceau F, Fliniaux M.-A., Jacquin-Dubreuil A. A high performance liquid chromatographic procedure for the analysis of tobacco alkaloids: Application to the evaluation of tobacco alkaloids in plants and cell suspension cultures. *Phytochem. Anal.*, 1992: **3**; 65–8.

85 Yang SS, Smetena I. Evaluation of capillary electrophoresis for the analysis of nicotine and selected minor alkaloids from tobacco. *Chromatographia*, 1995: **40**; 375–8.

86 Morie GP. Fraction of protonated and unprotonated nicotine in tobacco smoke at various pH values. *Tob. Sci.*, 1972: **16**; 167.

87 Griffith RB, Jeffrey RN. Improved steam-distillation apparatus: Application to determination of nicotine in green and dry tobacco. *Anal. Chem.*, 1948: **20**; 307–11.

88 Ogg CL, Bates, Jr. WW, Cogbill E, Harrow LS, Petersen EL. Determination of secondary, tertiary, and total alkaloids in tobacco. *J. Assoc. Off. Agr. Chem.*, 1960: **43**; 524–9.

89 Lu GH, Ralapati S. Application of high-performance capillary electrophoresis to the quantitative analysis of nicotine and profiling of other alkaloids in ATF regulated tobacco products. *Electrophoresis*, 1998: **19**; 19–26.

90 Proceedings of the Sixth Annual Frederick Conference on Capillary Electrophoresis. *J. Chromatogr. B*, 1996: **683**; 1–131.

91 Uphof JC. Dictionary of Economic Plants. New York: 3301 Lehre, Verlag Von J. Cramer, Stechert-Hafner Service Agency, 2nd edn. rev. and enl., 1968.

92 Akehurst BC. Tobacco. London: Longman Group Ltd., 1968.

93 Hegnauer R. Chemotaxonomie der Pflanzen, Band 4 (Dicotyledoneae: Daphniphyllaceae-Lythraceae). Basel, Stuttgart: Birkhauser, Verlag, 1966.

94 Murthy PSN, Kameswararao BV, Gopalachari NC, Saunders J. Determination of alkaloid composition in Indian tobacco by HPLC. *Tob. Res.*, 1986: **12**; 186–91.

95 Raisi A, Alipour E, Manouchehri S. Quantitative analysis of nicotine in several cigarette brands available in Iran. *Chromatographia*, 1986: **21**; 711–2.

96 Tilashalski K, Rodu B, Mayfield C. Assessing the nicotine content of smokeless tobacco products. *J. Am. Dent. Assoc.*, 1994: **125**; 590–4.

97 Official Methods of Analysis of Association of Official Analytical Chemistry (AOAC) International. 1998. (Ed.) P. Cunniff, 16th edn., 4th Rev., Vol. 1, Alkaloids

(Total as Nicotine) in Tobacco – Distillation Method. Final Action 1964. AOAC Official Method 960.07.

98 Official Methods of Analysis of Association of Official Analytical Chemistry (AOAC) International. 1998. (Ed.) P. Cunniff, 16th edn., 4th Revision, Vol. 1, Alkaloids (Total as Nicotine) in Tobacco – Cundiff - Markunas Method. Final Action 1964. AOAC Official Method 960.08.

99 International Standards Organization. 1992. Tobacco and tobacco products – Determination of alkaloids content – Spectrometric method. ISO 2881.

100 Department of Health and Human Services Centers for Disease Control and Prevention. Proposed data collection submitted for public comment and recommendations. *Federal Register*, 1997: **62**(85); 24115–6.

101 Department of Health and Human Services Centers for Disease Control and Prevention. Protocol to measure the quantity of nicotine contained in smokeless tobacco products manufactured, imported, or packaged in the United States. *Federal Register*, 1997: **62**(85); 24116–9.

102 Guidelines for collaborative study procedure to validate characteristics of a method of analysis. *J. Assoc. Off. Anal. Chem.*, 1989: **72**(4); 694–704.

103 International Standards Organization. 1981. Precision of test methods – Determination of repeatability and reproducibility by inter-laboratory tests. ISO 5725.

104 American Society for Testing and Materials. Standard guide for determination of precision and bias of methods of committee D–22. Annual Book of ASTM Standards 1996; 11.03:174–7.

105 Comments of United States Tobacco Company, Brown & Williamson Tobacco Corporation, Conwood Co., L.P., National Tobacco Co., L.P., The Pinkerton Tobacco Co. and Swisher International, Inc. entitled, "Proposed Regulations of the Department of Public Health 105 CMR §660.000 for implementation of M.G.L. Chapter 94, Section 307B." Before the Massachusetts Department of Public Health, February 21, 1997.

106 Comments of Brown & Williamson Tobacco Corporation, Conwood Co., L.P., House of Windsor, Inc., National Tobacco Co., L.P., The Pinkerton Tobacco Co., R.C. Owen Company, Fred Stoker & Sons, Inc., Swisher International, Inc., and United States Tobacco Company entitled, "Centers for Disease Control and Prevention Department of Health and Human Services: Proposed data collections submitted for public comment and recommendations." Before the Office on Smoking and Health, July 2, 1997.

107 *Request for Public Hearing Memorandum*: from G. N. Connolly and H. Saxner to Public Health Council, November 19, 1996, The Commonwealth of Massachusetts. Executive Office of Health and Human Services Department of Public Health. Boston, MA.

108 *Press Release*: "Massachusetts Department of Public Health. Massachusetts Seeks Truth About Tobacco Products." S. Fitzpatrick. Boston, MA. November 19, 1996.

109 Cigarette and Smokeless Tobacco Products: Reports of Added Constituents and Nicotine Yield Ratings: 105 Code of Massachusetts Regulations 660.000 et seq. (as proposed Dec. 1996).

110 Massachusetts Department of Public Health Hearing on Proposed Regulations to Implement Massachusetts General Law Chapter 94 Section 307c., January 30, 1997.

111 Cigarette and Smokeless Tobacco Products: Reports of Added Constituents and Nicotine Yield Ratings: 105 Code of Massachusetts Regulations 660.000 et seq. (as revised and proposed May 1997).

112 Borgerding MF, Norman V, Pannell WT, Whidby JF. Massachusetts regulations regarding cigarette nicotine yield ratings – A summary of 1997 testing results. In: Papers presented at the meeting of the smoke and technology groups. CORESTA Congress, Smoke and Technology Group. Brighton, England October 11–15, 1998: 19–38.

113 Comments of Philip Morris Incorporated, R.J. Reynolds Tobacco Company, Brown & Williamson Tobacco Corporation, and Lorillard Tobacco Company on proposed regulations entitled, "Cigarette and Smokeless Tobacco Products: Reports of Added Constituents and Nicotine Ratings (105 CMR 660.000)." Before the Massachusetts Department of Public Health, February 25, 1997.

114 Comments of Philip Morris Incorporated, R. J. Reynolds Tobacco Company, Brown & Williamson Tobacco Corporation, and Lorillard Tobacco Company on revised proposed regulations entitled, "Cigarette and Smokeless Tobacco Products: Reports of Added Constituents and Nicotine Ratings (105 CMR 660.000)." Before the Massachusetts Department of Public Health, June 27, 1997.

115 Borgerding MF. The FTC method in 1997 – What alternative smoking condition(s) does the future hold? *Recent Adv. Tob. Sci.*, 1997: **23**; 75–151.

*Chapter 10*

# Determination of nicotine in pharmaceutical products and dietary sources

Donald E. Leyden[1], Erich Leitner[2] and Barbara Siegmund[2]

*[1] Worldwide Scientific Affairs, Philip Morris International, CH-2003 Neuchâtel, Switzerland*
*[2] Institute of Bio- and Food Chemistry, Department of Food Chemistry, Technical University of Graz, A-8010 Graz, Austria*

## I. INTRODUCTION

There have been numerous publications concerning the intake of nicotine by smoking and exposure to environmental tobacco smoke (ETS); too numerous to attempt to reference them here. However, nicotine occurs in other sources as well. Some of these sources are naturally occurring and include edible nightshades (*Solanaceae*), their process products and teas. Of these, tomatoes, potatoes, eggplants (aubergines), certain peppers, and tea contribute significantly to the diet of many cultures. Diverse reports have appeared in the literature concerning the nicotine content of these foods, and the amount of nicotine that is ingested by the population from consumption of them.

Because nicotine serves as an insecticide in some parts of the world [1], it is possible that nicotine is ingested occasionally from products contaminated by this insecticide use. There is little definitive information on nicotine content of edible plants that result from insecticide usage, or of the probability of such nicotine reaching the consumer. Transportation, exposure to the climatic elements and pre-consumption preparation are likely to alter the superficial nicotine on plant surfaces. Clearly, the variability of this source of nicotine precludes obtaining valid estimates of ingestion. We do not give this topic further consideration in this chapter.

Other sources of nicotine are man-made and include a variety of pharmaceutical products primarily intended to aid smoking cessation. These products include chewing gum, inhalers, nasal spray, transdermal patches, tablets and roll-on applications. The availability of these products is growing worldwide. By the end of 1998, products were available in more than 50 countries involving approximately 40 firms. Even a commercially available bottled water containing nicotine for direct sale to the public is advertised on the Internet [2]. It has been suggested that use of smoking cessation products in combination with a recommended diet is more effective than use of the products alone [3]. At sufficiently high concentration, nicotine can be toxic [1]. Analytical methods for quality control during the manufacture of products containing nicotine and for studies of their efficacy are of considerable importance. Although reputable pharmaceutical houses practice stringent quality control, one must have some concern about start-up companies that sell products containing nicotine directly to the public.

The organization of this chapter is to first consider the literature on the methods and results for the analysis of pharmaceutical products followed by a similar treatment of dietary sources of nicotine. These sections will describe analytical methods applied to nicotine determination in the various materials and matrices. In some cases, studies were carried out using commercial products in which no analytical verification was conducted to determine the actual initial and final level of nicotine in the product. These studies will not be described in detail and such validation is highly recommended for future studies. Additionally, in some cases analytical information or methodology is reported in the literature as 'personal communication' from the manufacturer or other sources, and no details are given. In these cases little can be said about the methods, procedures or results.

## II. DETERMINATION OF NICOTINE IN PHARMACEUTICAL PRODUCTS

Most pharmaceutical products to address smoking cessation are designed to be nicotine replacement products and use of them is known as nicotine replacement therapy. Chewing gum containing nicotine and transdermal systems (patches) deliver small, steady doses of nicotine providing lower and less variable plasma nicotine concentration than smoking, which delivers nicotine in rapidly absorbed pulses. Nasal spray and inhalers attempt to mimic the rapid change in blood nicotine concentration associated with smoking.

Nicotine gum was first approved by the US Food and Drug Administration in 1984. The nicotine patch for transdermal delivery was approved in 1991/92, and nasal spray in 1996. Inhalers were approved in 1997. In the United States, gums and patches are available over the counter, whereas, nasal spray, and inhalers require a prescription from a physician. There are products in tablet form, but these do not contain nicotine and appear to act by a different mechanism.

An unfortunate characteristic in studies of the efficacy of nicotine replacement systems is that a variety of indirect and biological measures of nicotine uptake are made [4–14], but few direct measurements of the delivery of the system are conducted. Only a small percentage of such studies include direct measurement of the nicotine extracted from gum, depleted from transdermal delivery systems during use, or delivered by nasal spray. Assumptions that the distribution of nicotine in biological fluids is the same whether ingested by smoking or by use of nicotine replacement devices may not hold. For example, Lindell et al. [15] report that when nicotine is administered by transdermal patch it accumulates in the intragastric juices resulting in a salivary nicotine concentration 10.5 times that in plasma.

A recent report by Foulds et al. [16] is an interesting example illustrating when a direct analytical measurement is prudent. A single case was investigated of a male who had a history of chronic schizophrenia and who had been consuming eight packs of 30 per day of unlicensed, commercially available lozenges containing nicotine. The investigation included administering eight of these lozenges to non-smoker volunteers within a 2-hour period. Before and after consumption of the lozenges, samples were taken from the volunteers for the determination of blood nicotine. Several of the volunteers experienced dizziness and vomited after using the lozenges. The authors based the dose on the claims of the manufacturers of the various lozenges and administered them without confirming the nicotine levels. Because only one blood sample was taken, no estimate of the peak nicotine level is available. In view of the number of analytical measurements conducted in blood and saliva during the investigation, a determination of the nicotine content of the lozenges prior to administration would have seemed appropriate.

Considering the numerous variables associated with bioavailability, metabolism, and self-reported use of products that contain nicotine, and in light of modern analytical methods and instrumentation, to not characterize products before they are administered to volunteers is questionable. The absence of these data does not mean that the studies are not useful, but they could be improved and made more reliable by determining the mass balance between actual amount of nicotine delivered and the physiological levels.

This chapter is not intended as an exhaustive review of investigations of nicotine delivery systems, but attempts to provide a view of the state of knowledge related to analytical methods applied to their characterization. It is possible that a number of methods reside as proprietary property of the companies that manufacture these products.

## III. NICOTINE CHEWING GUM

Formulations of chewing gum containing nicotine and intended for use as smoking cessation aids are subject to approval by regulatory agencies. Some of these currently on the market contain Nicotine Polacrilex®, which is defined by the US Pharmacopeia [17] to be "a weak carboxylic cation-exchange resin prepared from methacrylic acid and divinylbenzene, in complex with nicotine. It contains not less than 95% and not more than 115% of the labeled amount of nicotine ($C_{10}H_{14}N_2$) calculated on the anhydrous basis." The US Pharmacopeia recommends infrared absorption for the identification of Nicotine Polacrilex®. A nicotine assay is performed by extraction of the nicotine in 1 M ammonium hydroxide, followed by addition of acetic acid to an aliquot of the extract. Liquid chromatography with UV detection at 254 nm is used for the analytical measurement. Nicotine released from Nicotine Polacrilex® is determined by extracting the nicotine into sodium chloride solution, filtration, and UV absorption at 259 nm for the analytical measurement, and 236 and 282 nm for background measurements.

The US Pharmacopeia states that Nicotine Polacrilex® Gum contains "an amount of Nicotine Polacrilex® equivalent to not less than 90% and not more than 120% of the labeled amount of nicotine" [18]. The US Pharmacopeia method for assay of Nicotine Polacrilex® Gum is similar to that for Nicotine Polacrilex®, except that the gum is first extracted using a mixture of *n*-hexane and a solvent consisting of water, acetonitrile and sodium 1-decanesulfonate. The solvent layer is collected, filtered and used for analysis. A liquid chromatographic method similar but not identical to that for the Nicotine Polacrilex® assay is used.

The liquid chromatographic method reported in the US Pharmacopeia must be modified for use with transdermal delivery systems and uses expensive solvents for the mobile phase. Dash and Wong [19] report the use of a buffered methanol mobile phase with lidocaine hydrochloride as an internal standard. Nicotine in both gum and transdermal patches was extracted into *n*-hexane and a filtered aliquot used for analysis. Although some shortcuts were taken in the validation of the method, some validation data are given. The limit of quantification at 10% relative standard deviation is reported to be $0.31 \pm 0.07$ µg/ml. Day-to-day replication (but only on the same solutions) ranged between 2.4 to 4.2%. Confidence in this method could have been enhanced with additional validation data.

Well before nicotine chewing gum was approved by the US Food and Drug Administration, studies began on its effects [20–22]. In one of the earlier studies of dose related effects of nicotine chewing gum on smoking behavior, initial nicotine levels were provided by the manufacturer, and levels in the chewed gum were determined by the manufacturer using HPLC techniques [23]. The initial levels were reported to be 0, 2 and 4 mg of nicotine per piece of gum. By combining two pieces of gum, 0, 2, 4, and 8 mg of nicotine were available. The authors report that 1.02 mg (51%) of the 2 mg, 2.39 mg (59%) of the 4 mg, and 5.2 (65%) of the 8 mg dose were extracted by chewing, and that these levels were all statistically significantly different from each other including adjacent pairs. The authors report that plasma nicotine levels increased with the amount of nicotine extracted by chewing the gum. The number of cigarettes smoked by the subjects after chewing the gum decreased, but not in a linear relationship to dose.

In a later study to determine the effects of gum chewing rate (number of seconds between chews) on subjective effects and plasma nicotine levels, no information was given on the analytical method [24]. The authors stated that the greatest amount of nicotine was extracted at the fastest chew rate, and the smallest at the lowest chew rate. The time allotted for the experiment was fixed. Therefore, a difference by a factor of eight in the number of chews between the slowest and the fastest chew rate would exist. The total number of chews was not taken into account. Both of these studies [23,24] included direct analytical measurements of the extraction from the gum, although few details were given.

A recent development in chromatographic column materials and solid-phase extraction materials is the use of molecularly imprinted substrates [25–27]. These materials are normally prepared by polymerization in the presence of the target molecule. The resulting specificity in affinity for small molecules has been favorably compared with corresponding immunoaffinity phases [25–27]. Zander et al. [28] have applied this approach to the solid-phase extraction of (S)-( − )-nicotine and its oxidation products (myosmine, cotinine, β-nicotyrine, nicotine-*trans*-N-oxide and nicotine-*cis*-N-oxide). Chewing gum was dissolved in a non-polar solvent and the nicotine compounds extracted by the molecularly imprinted solid-phase. Subsequent use of an acidic mobile phase eluted the analytes as a single sharp peak resulting in a clean-up of the sample. Three of the four target analytes were recovered using the imprinted phase, whereas, only nicotine was recovered using the non-imprinted phase. Zander et al. [28] reported some interesting chromatography, but there were no steps taken to offer a validated analytical method. Molecularly imprinted solid-phase extraction may hold promise for the future. However, the lack of commercially available and consistent materials will hinder implementation in routine analytical applications.

## IV. NICOTINE TRANSDERMAL SYSTEMS (PATCHES)

A review of the role of nicotine transdermal systems as an aid to smoking cessation has been published by Gora [29]. Transdermal patches consist of layers of membrane and a solid support medium for the nicotine. The driving force for the absorption of the nicotine is the concentration gradient between the patch and the adjacent skin. Carlise et al. [30] have reported that the analytical methods developed for the determination of nicotine in blood, urine and saliva are found to be inadequate for the analysis of transdermal patches. We will describe methods that have been developed for these products.

Dubois et al. [31] conducted a study of the pharmacokinetics and bioavailability of nicotine following administration of transdermal nicotine systems (patches). Dose was controlled by the surface area of the patch. The authors report that in the case of a single application of the transdermal delivery system the area under the curve for plasma nicotine and the amount of urinary nicotine excreted in 24 hours were linearly related to the nicotine dose. The amount of nicotine delivered was taken as "the difference between the verified declared nicotine content of the patch and the measured residual nicotine amount after 24 h application." Used patches were returned to the manufacturer for analysis. No information was given on the techniques or methodologies used for the determination of nicotine in the patches, or on the storage, transportation and sample preparation of the used patches. The authors report that the placebo delivered 1.9 mg of nicotine in 24 h, which they considered a pharmacologically inactive amount. This is roughly equivalent to the nicotine from two full flavored cigarettes. No serum or urine samples of subjects without patches appear to have been included, and the placebo data is not included in the plots of urinary cotinine vs. dose. Without the analytical data, one cannot determine how such an amount of nicotine can be delivered from a placebo that is only assumed to be nicotine free.

Sudan [32] has commented on the reports of studies in which placebos are stated to contain small amounts of nicotine that are assumed to be pharmacologically insignificant. Sudan reports dermatological reactions at the microgram level of nicotine and calls for the use of nicotine-free patches as placebos. These should be validated as nicotine-free by appropriate analytical measurements.

Carlisle et al. [30] reported a method for the determination of nicotine in transdermal dosage forms using high performance liquid chromatography. The method employs a multi-phase polymeric column that permits separation by a combination of reversed-phase and ion-exchange chromatography. Patches were prepared for analysis by removing them from the release liner, and sonicated in tetrahydrofuran. The sample was diluted to volume with absolute ethanol. An aliquot of the solution was added to 0.5 M perchloric acid and evaporated to about one third of its original volume under nitrogen flow at room temperature; water was then added. Sample size and dilution ratios resulted in a nominal concentration of nicotine in the range of 25–40 µg/ml. The method appears to provide good accuracy and precision, and prepared samples appeared to be stable for up to three days. It may not be particularly well suited for large numbers of samples because of the manipulations required. Interestingly, the authors investigated nicotine patches that had been stressed by being stored at 60°C for 24 days. One nicotine $N'$-oxide enantiomer (or both co-eluting), nicotine 1,1'-di($N$-oxide), and cotinine along with three unidentified compounds were detected in the chromatograms.

## V. DETERMINATION OF AIRBORNE NICOTINE IN MANUFACTURING SITES FOR NICOTINE TRANSDERMAL SYSTEMS

The manufacture of nicotine transdermal systems presents potential exposure opportunities for employees. Methods for the determination of airborne nicotine are well known [33]. Most of these are based on sampling for several hours with an absorbent that is later eluted with a solvent. The analytical measurement is conducted using gas chromatography with a nitrogen specific detector or with mass spectrometric detection. These methods easily meet the requirements for the determination of nicotine at the US OSHA standard of 0.5 mg/m$^3$ for an 8 h shift. However, even at this high level, as much as 30 min of sampling time are required, and the absorbent must be taken to a laboratory for analysis. Eiceman et al. [34] have proposed a method for the determination of nicotine in these environments using ion-mobility spectrometry. This technique is based on the drift rate (mobility) of ions formed by charge transfer at atmospheric pressure from ions such as $(H_2O)_n H^+$ to the analyte ion. The time required for the analyte ion to 'drift' from the inlet to the detector depends on the charge to mass ratio and instrumental parameters. Because no vacuum is used, these instruments can be constructed as small hand-held devices. Nicotine is well suited for ion mobility spectrometry because it is easily ionized. However, a monomer-dimer equilibrium of ions in the vapor phase complicates the linearity of response. Nevertheless, by summing the intensity of the monomer and dimer peaks, Eiceman et al. [34] were able to demonstrate an acceptable calibration between 0.01 and 0.25 mg/m$^3$. Developments that have occurred in hand-held ion mobility instrumentation would likely improve on this performance.

## VI. INHALATION AND NASAL SPRAY DEVICES

Gums and patches are alleged to have the disadvantage of low and slow delivery of nicotine to the body. A number of studies have suggested that cigarettes deliver a sharp rise in nicotine. Attempts have been made to imitate that delivery profile with a variety of approaches. A smokeless cigarette that contained a porous substrate impregnated with nicotine was evaluated [35]. Plasma nicotine levels were measured after subjects puffed on the cigarette. Only by maximal puffing in an unrealistic manner for several minutes were plasma nicotine levels reached that could compete with nicotine gum. It is interesting that extensive work was conducted on plasma nicotine measurement, but no results were described for the measurement of the actual vapor phase delivery of nicotine.

Cornaz-Gudet and Buri [36] reported a liquid chromatographic method using a C18 reversed-phase column (Supelcosil LC-ABZ) and UV detection for the quantification and evaluation of purity of nicotine in a nasal spray solution and in microspheres used in transnasal application [37]. The quantitative determination of nicotine was evaluated with respect to specificity, linearity, precision, and accuracy. A relative standard deviation of <2.5% was reported. Nicotine concentration in solution, a nasal spray and in Sephadex microspheres was followed for five months under a variety of storage conditions of temperature and light. The solution and spray showed a loss of <5% nicotine concentration after five months. Solutions exposed to light displayed a pale yellow color. The nicotine concentration in the microspheres did not vary significantly.

Davoli et al. [38] reported a solid-phase extraction method for the gas chromatographic-mass spectrometric determination of nicotine in plasma following administration with a 'low-dosage' nicotine chewing gum. No determinations were made of the nicotine content in the gum before or after use. A very basic solution of nicotine in ethylacetate was used and the authors reported a large loss during the concentration step. This was combated by the addition of isooctane to the solutions. There were no method validation data given.

## VII. ENANTIOMERIC PURITY

The determination of the enantiomeric purity of nicotine is of interest. The predominant naturally occurring nicotine in tobacco is the (S)-(−)-isomer. Until recently, chromatographic resolution of nicotine enantiomers was difficult. The reliability of optical methods to estimate enantiomeric purity has been questioned [39]. Because these methods depend on the difference between the optical properties (rotation, circular dichroism, etc.) of the enantiomers, the ability to characterize mixtures consisting of a high percentage of one enantiomer is limited.

Before chromatographic methods of enantiomeric resolution were adequately developed, Klus and Kuhn [40] and Crooks and Godin [41] employed chiral NMR shift reagents to detect (R)-(+)-nicotine in tobacco smoke condensate. It was suggested that the (R)-(+)-nicotine resulted from racemization of (S)-(−)-nicotine as a result of the process of smoking the tobacco. No measurements were reported for unaltered tobacco or tobacco products. Jaroszewski and Olsson [42] reported a method for this determination for nicotine in a variety of pharmaceutical preparations using $^{13}$C-NMR in the presence of a chiral lanthanide shift reagent. The nicotine was extracted with various solvents depending on the specific product. Although no data were given for accuracy or precision, the authors report that the method is applicable for the determination of the enantiomeric composition of mixtures containing ratios up to 100-fold excess of one isomer. For overnight spectrum acquisition, 1–3 mg of nicotine is adequate if the percent of one isomer is very low. In the case of pharmaceuticals that have adequate amounts of nicotine only 1–2 h are required for spectrum acquisition. The method was only applied to prepared mixtures and not to a wide variety of samples.

A microcolumn liquid chromatography method proved to be time consuming and not suitable for routine use [43]. Demetriou et al. [44] used a chiral stationary phase with gradient elution at 50°C to resolve the enantiomers of nicotine and nicotine-like compounds. Tang et al. [45] separated nicotine and nornicotine enantiomers via normal phase HPLC on derivatized cellulose chiral stationary phases. Armstrong et al. [46] used the method of Tang et al. [44] to investigate the enantiomeric composition of smokeless tobacco, tobacco leaf, pharmaceutical products and commercial reagents. The percentage of (R)-(+)-nicotine ranged from –0.1% to 1.2% of the total nicotine. The highest level of (R)-(+)-nicotine was found in a commercial transdermal patch. The results indicate that the extraction and purification processes used in obtaining commercial (S)-(–)-nicotine do not appear to decrease the percentage of (R)-(+)-nicotine.

## VIII. DETERMINATION OF NICOTINE IN DIETARY SOURCES

In recent years, there has been increased interest in the determination of nicotine in food. This is in part a result of suggestions that dietary nicotine intake could impact the level of nicotine metabolites otherwise attributed to exposure to environmental tobacco smoke (ETS). The concentration of biomarkers such as nicotine metabolites resulting from exposure to environmental tobacco smoke is expected to decrease with reduced exposure. One may ask whether at some point in reduced ETS exposure could the dietary contribution of nicotine intake become significant in comparison to that from ETS exposure.

It has been reported that nicotine has a wide distribution in flora. Leete [47] reported its presence in 12 families and 24 genera, including the nightshade-family (*Solanaceae*). Some common vegetables belong to the biological family of nightshades and throughout the world these plants are consumed in significant amounts. They include such basic foods as potatoes, tomatoes and aubergines (eggplants). The function of nicotine in these plants is still not fully known. However, it is assumed that nicotine serves as a natural defense against fungi, bacteria, insects and animals [48–50]. To a limited extent, nicotine is used as an insecticide in some parts of the world and could be a food contaminant [51,52].

Only a few citations can be found in the literature that address nicotine concentration in diverse foods and the ensuing dietary intake of nicotine.

Although no systematic investigations have been conducted, there is controversy on the subject. Nicotine is not only reported to be present in edible fruits from *Solanaceae*, but also in tea. The tea data are conflicting and controversial, especially as nicotine biosynthesis in teas has not been demonstrated [49]. Contamination from the use of nicotine as an insecticide has been proposed as the most likely source of nicotine in tea [49,56].

Even though the influence of dietary nicotine is not considered to be very important [49,51,53–56], discussion about its significance continues [51]. In some reports it is suggested that the contribution of dietary nicotine intake is significant when compared with exposure to ETS, or by active smoking of small numbers of cigarettes [49,51,52,55]. Others consider the dietary intake to be negligible unless inordinately large amounts of specific vegetables are consumed [53,54]. A confounding effect is discussed for people consuming high amounts of vegetables, in particular for vegans or vegetarians [52]. It has been suggested that dietary habits differ between smokers and non-smokers. A recently published study [57] proposed that non-smokers consume healthier diets than smokers, including a higher consumption of raw and cooked vegetables in combination with higher consumption of tea.

In general, foods represent difficult substrates for analysis because in most cases the matrix is very complex. The determination of nicotine in food matrices represents a demanding problem for the analytical chemist because of nicotine's basic properties. As almost all vegetables and fruits show acidic pH, nicotine is bound to the matrix as a salt and is therefore very hard to extract. A second issue related to nicotine determination at the low levels reported for fruits and vegetables is the high potential for contamination by nicotine present in the environment as part of environmental tobacco smoke. Therefore, carefully validated sample preparation and detection methods are required for selective and efficient determination of nicotine in dietary sources.

Only a small number of papers have been published dealing with the determination of nicotine in food material [49,51,55,56], but most of the methods used in these publications have not been fully validated. Nonetheless the publications provide interesting information on dietary nicotine. In 1986 Castro and Monji [55] were the first to provide reliable data on nicotine concentrations of edible nightshades in the order of μg/kg fresh fruits using radioimmunoassay (RIA). Although the procedure has been validated, the method is analytically selective but not specific. There are possibilities for cross-reactivity with the antibodies as well as non-consistent specificity based on differences of the antiserum used and the hapten analog that is employed for the preparation of the nicotine conjugate. Nevertheless, Castro and Monji [55] show some interesting results in this paper.

Davis et al. [49] presented an interesting study of nicotine in food using a sample preparation technique that has been validated for the extraction and detection of nicotine and cotinine in plasma [58] and was applied for the analysis of diverse vegetables. Concentrations reported by Davis et al. are in the range of µg/kg fresh fruits. Once again, no attempts were made to determine whether the detected nicotine concentrations were genuinely from the bulk food samples, or whether they were derived in part from possible surface contamination.

Sheen [56] investigated nightshades for their nicotine concentrations using gas chromatography. The results reported generate questions concerning the analytical approach. The nicotine concentrations reported are three orders of magnitude higher than values described in the two previously mentioned papers [49,55]. No attempts appear to have been made to check possible nicotine background contamination. For example, potato peel was reported to show large nicotine concentrations whereas in potato flesh no nicotine was detected. These large differences were interpreted as defense mechanism by the plant against fungi or bacteria. Possible surface contamination of the fruit was not taken into consideration. Additionally, the analytical techniques used were not validated.

In a more recently published study concerning the occurrence of nicotine in common vegetables [51] only rough descriptions were provided about the analytical techniques used. No detailed descriptions of the method, its validation, recoveries or attempts to avoid contamination by environmental tobacco smoke were discussed.

## IX. RESULTS FROM A SYSTEMATIC STUDY FOR THE DETERMINATION OF NICOTINE IN EDIBLE NIGHTSHADES AND IN TEA

Based on the limited information concerning nicotine concentrations in food material that is provided in the literature, we initiated a systematic study to estimate possible dietary nicotine intake. In the following sections a summary of this study will be given including a survey of analytical techniques, the nicotine concentrations found in various food materials and the use of these data to develop an estimation of the dietary nicotine intake.

### Analytical procedures

The aim of the presented study was to develop, validate, and apply an analytical method for routine analysis of nicotine in large numbers of samples. The method should be quick, but nevertheless rugged, reliable and sufficiently sensitive for the detection of small concentrations of nicotine in foods. Special emphasis was put on the analytical performance, sample preparation as well as on avoidance of possible contamination from any source. The development and validation of the analytical techniques used for this study have been described and discussed in detail elsewhere [59]. Therefore, only a short description will be given in the following section.

Gas chromatography-mass spectrometry (GC–MS) provides a very selective and sensitive method for the determination of nicotine. The use of the selected ion mode significantly increases the selectivity as well as the sensitivity of the analytical method. The danger of false positive results is eliminated as the identification of the compounds is based on relative retention indices, as well as on the ratio of the selected ions to one another. The separation of the analytes was performed on a standard high-resolution GC column.

Due to the polarity of the analytes, the performance of the GC–MS measurements is highly dependent on the condition of the injector system and the analytical column with special attention needed for the liner. Best results were obtained using a split/splitless injector and liners of single-taper geometry packed with deactivated glass wool in the splitless mode. We found that the surface purity of the liner plays a significant role for the sensitivity of the system. A properly cleaned and deactivated liner results in an improvement of the peak shape as well as in an increase of the signal-to-noise ratio by a factor of six.

For quantification of nicotine the use of an appropriate internal standard is required to compensate for the lack of a fully quantitative extraction of the analytes. The selection of suitable compounds as internal standards for nicotine analyses has been discussed widely [58,60,61]. For the quantification of nicotine, its deuterium labeled analog ($[^2H_3]$methylnicotine) is an excellent compound to use as an internal standard. Validation of the methods is described in detail in Siegmund et al. [59]. The limit of detection (LOD) is defined as the lowest concentration differing significantly from zero and the limit of quantification (LOQ) is the lowest concentration of the analyte that can be determined with a standard deviation $\leqslant 5\%$. The LOD and LOQ for nicotine were found to be 3.0 pg and 10.8 pg, respectively. These values correspond to 0.8 µg/kg (LOD) and 2.7 µg/kg (LOQ) for fresh fruit assuming a 3 g sample. A standard deviation of regression of 4.3% was obtained.

**Possible sources of contamination**

Because of the very low nicotine concentrations that are expected in food material, special emphasis is required to address possible nicotine contamination from chemicals and equipment, contamination from ETS, as well as surface contamination of the fruit samples. To minimize airborne nicotine contamination, smoking was not permitted in the laboratory area and airborne nicotine levels in the laboratory were measured using an ASTM method [33]. Nicotine concentrations in various indoor environments are reported to range from non-detectable to 70 µg/m$^3$, with values usually near the lower end of the range [62–64]. The measured nicotine concentrations show that contamination by airborne nicotine ($\leqslant 0.35$ µg/m$^3$ within the laboratory, $\leqslant 0.60$ μg/m$^3$ in adjoining hallways) did not represent a significant source of contamination.

Another concern was the history of possible exposure of the samples to nicotine, especially those samples purchased in supermarkets for which nothing is known about the history of the sample. As a consequence, surface contamination of the observed fresh fruits was studied very carefully. The values for nicotine derived from surface contamination can be estimated assuming that the surface of tomatoes are ideal spheres, and the surface of potatoes and aubergines are like ideal cylinders, respectively. Considering some of the previous reports [49,55], nicotine concentrations in diverse foods were expected to be in the order of 5–100 μg/kg fresh fruit. Nicotine concentration estimated from surface contamination (0.01–0.09 μg/kg fruits) is two to four orders of magnitude lower than the expected concentrations in the bulk vegetable and, therefore, out of statistical relevance for the nicotine concentration of the whole fruit.

## Extraction of the Analyte

Extraction of nicotine from food material represents a demanding problem. For the development of methodology a number of extraction techniques that are widely used for the extraction of organic compounds from plant or food material were investigated, but none of them gave acceptable results. Nevertheless, these methods are described very briefly in the following paragraph to show the difficulties that arise when extracting nicotine out of food material, and to allow future researchers to avoid unsuccessful approaches.

Three methods, that seemed to be very promising – (a) an AOAC method [65] designated for the determination of nicotine residues in apples, cabbage and spinach, (b) the simultaneous distillation-extraction according to Likens & Nickerson [66,67], which is widely used for the extraction of volatile organic compounds in aqueous systems [68], and (c) the microextraction according to DIN 38407 [69] – did not provide acceptable results. Either very stable emulsions were formed that could not be broken easily, or the recoveries were very poor ($\leqslant 50\%$) and inconsistent. The use of a rotation perforator was rejected mainly because of inconvenient manipulation of the apparatus and high danger of possible carry-over effects from prior extractions. Finally, a sample preparation technique that is widely used for the extraction of pesticides from plant material [70] was dismissed because under the necessary basic conditions a classical aldol condensation took place during the extraction procedure that resulted in a peak that interfered with nicotine.

A very simple sample preparation technique for the extraction of low concentrations of nicotine from plasma was described by Degen and Schneider [71]. This technique was adapted for the extraction of nicotine from food samples using simple liquid–liquid extraction with toluene at basic pH. Recoveries greater than 95% were obtained for the extraction of unprocessed fruits such as tomatoes, potatoes and eggplants. The main advantage of this method is its simplicity, which provides the possibility of processing large sample numbers simultaneously in a very simple way and which minimizes the risk of contamination. Investigations of processed foods showed that further clean-up steps were necessary for products containing ingredients in addition to vegetables such as spices or flavors. Interfering compounds showing similar m/z-ratios with nicotine or the internal standard made the identification and quantification impossible.

For many analytical applications, solid phase extraction (SPE) provides an excellent tool for simple and effective sample clean-up procedures. However, in this case we found that SPE using silica gel cartridges was not a suitable extraction technique mainly due to two reasons. Silica gel is a very good adsorbent for nicotine and, therefore, the blank concentration of nicotine on the silica material was high, and experiments with nicotine and labeled nicotine solutions resulted in very low and inconsistent recoveries.

An easy, but nonetheless highly effective liquid–liquid extraction procedure starting with the toluene extract was established making use of the polar and basic properties of nicotine and pH value (Fig. 1). Prior to

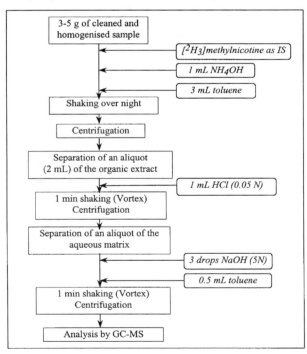

Fig. 1 Clean-up procedure for processed foods.

extraction fresh vegetable samples were washed with hot water followed with double distilled water, then homogenized. Processed foods were used as received. Figure 2 shows a comparison of the selected ion chromatograms of the extracts of commercially available tomato ketchup for nicotine and

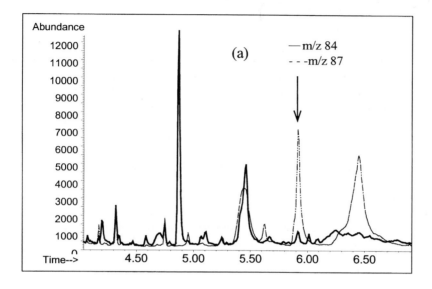

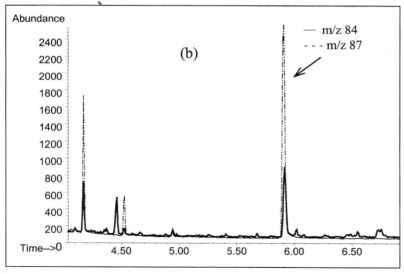

Fig. 2 Comparison of chromatograms of tomato ketchup (a) before and (b) after the clean-up procedure; the selected ion chromatograms for nicotine ($m/z$ 84) and for [$^2$H$_3$]methylnicotine ($m/z$ 87) are superimposed.

[$^2$H$_3$]methylnicotine before and after the clean-up procedure. In spite of the relatively high number of extraction steps, the losses of the analytes are very low with recoveries of 96% and relative standard deviation of 15–20% for the entire procedure.

## X. NICOTINE CONCENTRATION IN FOODS

As mentioned in the previous section, there are only a few papers published that deal with nicotine concentrations in food material [49,51,55,56]. The reports generally fail to use validated analytical methods, to consider potential contamination sources, or to perform adequate replicate determinations. Consequently, the concentrations reported are not comparable and not suitable for the estimation of dietary nicotine intake because the reported values show variations over three orders of magnitude.

The objective of the present study was to determine the feasibility that consumption of foods known to contain nicotine could contribute significantly to the nicotine intake via an estimation of the level of daily dietary nicotine intake. Hence, nicotine was determined in several fresh fruits (with emphasis on *Solanaceae* such as tomatoes, potatoes, eggplants and peppers) as well as on commercially available process foods originating from these fruits (e.g. tomato paste, ketchup, French fries, etc.). Based on early investigations for nicotine concentrations in tomatoes of different ripening stages (55), a systematic investigation of different varieties of tomatoes at various stages of ripening was conducted.

It was discussed earlier that diverging results are reported in the literature concerning the nicotine concentration in tea [49,56]. Consequently, the role of nicotine intake via tea consumption is also controversial [52,72]. To address this issue a selection of different types of tea was included in this study. As the consumption of green tea may increase due to public perception of advantages of its high anti-oxidative potential [73], green tea and black tea were investigated. A detailed description of the study as well as a comparison of the results to the values found in literature have been given elsewhere [82].

As will be described, by combining the nicotine concentrations found in foods with published values for daily dietary intake, these foods provide a means to estimate daily dietary nicotine intake. Using the mean values and precision estimates, a Monte Carlo simulation was conducted to show a distribution probability of daily dietary nicotine intake.

### Nicotine concentration in different matrices

Seven commonly grown varieties of tomato were investigated by following nicotine concentrations in the fruit during the course of the ripening process.

TABLE 1

Nicotine concentration in tomatoes of different ripening stages

| Degree of ripening (DR) | Mean nicotine conc. [μg/kg] ww | Conc. range [μg/kg] ww | $n$ |
|---|---|---|---|
| 1 | 8.0 | 4.4–16.1 | 7 |
| 3 | 2.9 | nq–5.6 | 7 |
| 7–8 | 2.9 | nq–3.9 | 7 |
| 9–12 | 2.6 | nq–3.9 | 10 |

nq   not quantifiable
ww   wet weight

This study was conducted for two reasons. First, in some countries the consumption of green unripe tomatoes in the form of fried or pickled tomatoes is very common. Secondly, frequent transportation of tomatoes after the harvest requires that the fruits are picked before they are fully ripe. The degree of ripening (DR) was determined using a classification system from the Sprenger Institute, Wageningen, Netherlands, that classifies the tomatoes into twelve ripening stages according to their color. DR 1 describes the unripe, green fruit, whereas, DR 12 describes the fully ripe stage. Nicotine was detected in all tomatoes up to a concentration of about 16 μg/kg fresh fruit (wet weight), but there were a few samples showing concentrations lower than the LOQ (Table 1). A clear relationship between nicotine concentrations and the degree of ripening was found. This can be seen best with the variety *Culina*, which is shown in Fig. 3. It is important to note that there is no significant difference in concentration between DR 7 to DR 12 because these fruits cover the major part of tomatoes on the market. Nevertheless, all investigated green unripe fruits show significantly higher nicotine values than tomatoes of higher ripening stages. These results could possibly confirm the assumption that the plants produce nicotine as a natural defense against insects or animals [48,49].

Potatoes, eggplants and peppers are other plants from the biological family of *Solanaceae* whose fruits play a significant role for human nutrition. Among them potatoes are the most important, as they belong to the most frequently consumed vegetables. Summarizing the results, nicotine was detected in all potato samples investigated, but the concentrations were rather low (Table 2). No significant differences could be observed between samples of different varieties or different harvesting years. Four different types of eggplants were investigated and nicotine could only be quantified in one sample. All varieties of peppers analyzed showed detectable, but low nicotine concentrations (Table 2).

Only two citations refer to nicotine concentrations in processed products of the vegetables [55,56]. In this investigation, a number of commercially available products (derived from tomatoes and potatoes) were analyzed for

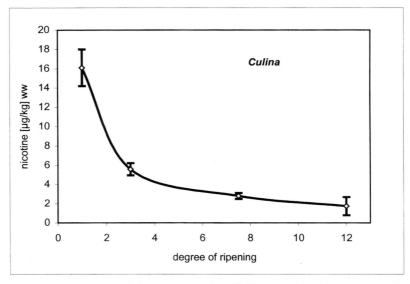

Fig. 3 Nicotine concentration of the tomato variety *Culina* showing dependence on the degree of ripening.

their nicotine concentration. These investigations do not cover the entire range of processed products available on the market. The purpose of conducting this part of the study was to determine (a) whether nicotine is present in processed products indicating its survival during processing, and (b) to check if the nicotine concentrations vary in a large amount within one group of product.

The nicotine concentrations found in the processed products are shown in Table 3. All products investigated showed detectable nicotine concentrations, even products that are thermally processed, which indicates that the analyte is thermally stable and does not degrade or evaporate during processing. Within these product groups the observed values are more or less consistent with the exception of concentrated tomato pulp. A slight increase

TABLE 2

Nicotine concentrations observed in potatoes, eggplants and peppers

| Vegetable | Mean nicotine conc. [μg/kg] ww | Conc. range [μg/kg] ww | *n* |
|---|---|---|---|
| Raw potatoes | 4.5 | nq–7.6 | 6 |
| Eggplants | 2.0 | nq–2.9 | 4 |
| Peppers | 6.5 | 3.7–9 0 | 7 |

nq not quantifiable
ww wet weight

TABLE 3

Nicotine concentration in various processed products

| Product | Mean nicotine conc. [μg/kg] ww | Conc. range [μg/kg] ww | n |
|---|---|---|---|
| Peeled tomatoes sold in cans | 2.8 | nq–4.4 | 4 |
| Homogenized tomatoes sold in cardboard boxes | 5.7 | 4.5–6.4 | 3 |
| 'Ready-to-eat' tomato sauce | 5.4 | 4.5–6.2 | 4 |
| Concentrated tomato pulp | 42.9 | 8.6–99.0 | 5 |
| Tomato ketchup | 7.3 | 5.8–10.3 | 6 |
| Cooked potatoes | 3.4 | nq–4.6 | 4 |
| French fries | 9.2 | 6.9–11.5 | 2 |

nq  not quantifiable
ww  wet weight

in nicotine concentration over fresh fruits was found in samples such as peeled tomatoes sold in cans, homogenized tomatoes, and commercially available tomato sauce as well as French fries. The observed increase in concentration within these three groups of products is related to the decrease of the water content associated with the thermal processing. Values for concentrated tomato pulp do not show definite correlation to the increasing dry matter, nor are the concentrations consistent within this group ranging over a factor of 10.

Conflicting results are found in the literature concerning nicotine concentrations in black tea [49,51,56]. Therefore, we analyzed tea samples for nicotine to address these contradictory results. The concentrations that were found in the dry tea-leaves (Table 4) were surprisingly high in a concentration range from 163–1600 μg/kg. Large variations were found within the types of black tea, whereas, the concentrations were more or less consistent within the green teas. For an estimation of the dietary nicotine intake from tea, the nicotine concentration of the tea-leaves is less relevant than that in brewed tea. Hence, tea was brewed using common amounts of tea-leaves and water. The results show that nicotine is not efficiently extracted by conventional brewing techniques. Even tea with very high nicotine concentrations in the leaves does not show high amounts in the

TABLE 4

Nicotine concentrations in dry leaves of green and black tea

| Type of tea | Mean nicotine conc. [μg/kg] ww | Conc. range [μg/kg] ww | n |
|---|---|---|---|
| Black tea | 737.4 | 163.8–1593.1 | 4 |
| Green tea | 370.4 | 317.2–469.8 | 4 |

brewed tea. If detectable, the extraction yield is in a range of 20–25%, yielding a mean nicotine concentration of 4.0 µg/l brewed tea.

## XI. DIETARY NICOTINE INTAKE

### Consumption data

The results obtained in this study provide an opportunity to estimate dietary intake of nicotine assuming that adequate food consumption data are available. Such data are available from a variety of governmental services, and are in a few cases summarized in publications. These data are most often based on crop production corrected for import/export, or on different consumption surveys, for example 24-hour recall questionnaire data, household surveys or dietary record methods. In most cases vegetables are grouped into categories and only vegetables consumed in large quantities are listed separately. Essentially no consumption information is available on processed products such as tomato ketchup or sauces. Any attempt to estimate consumption of vegetables and their products will be subjected to these limitations.

Table 5 shows the consumption data used to estimate daily dietary nicotine intake. These four vegetables make up the vast majority of vegetables consumed that are known to contain nicotine. The data were collected from a variety of sources [74–76] and represent mean and standard deviation of the daily per capita consumption of the vegetables and products shown for 13 European countries (Austria, Belgium, Denmark, Great Britain, Finland, France, Germany, Italy, Netherlands, Portugal, Spain, Sweden, Switzerland) and the United States. Concerning tea consumption data, Chappell and Gratt [72] reported an average consumption of brewed tea of 142 g/day with a very uneven distribution. An estimate of a standard deviation of 20% is difficult to justify. Nonetheless, we selected that value for use in the simulation described below. Table 5 also shows the average and

TABLE 5

Average consumption data for potatoes, tomatoes, tomato paste, eggplants, and brewed tea and the corresponding mean nicotine concentrations

| Product | Daily consumption average [g/day] | SD [g/day] | Averaged nicotine conc. nicotine conc. [ng/g] | SD [ng/g] |
|---|---|---|---|---|
| Potatoes | 157 | 80 | 4.5 | 1.9 |
| Tomatoes | 33 | 21 | 2.7 | 0.7 |
| Tomato pastes | 20 | 26 | 5.3 | 0.6 |
| Eggplants | 1.4 | 3.8 | 2.1 | 0.5 |
| Tea (brewed) | 142 | 28 | 4.0 | 0.3 |

standard deviation of nicotine content of the fruits, vegetables and their processed products as well as of brewed tea considered in this investigation.

## Calculation of the dietary nicotine intake

Using the data indicated in Table 5, an average daily dietary intake of 1.4 µg/day of nicotine is calculated for the countries for which data are available. To obtain this result, the average consumption of each vegetable for all countries in which data are available is multiplied by the average nicotine content found for this vegetable. It is beyond the scope of this report to describe in detail the average dietary intake for each country. The highest levels are calculated for Italy and Portugal because of high consumption of tomatoes, eggplants, and potatoes. Lower values are estimated for France because of lower consumption of these vegetables. As stated earlier, these estimates are subject to the variations of nicotine content, and the limitations of the consumption data that are available.

A point value such as the average daily nicotine intake for residents of the 14 countries represented in Table 5 does not provide information related to the probability distribution of the results. Given the mean and standard deviation, and assumptions concerning the upper and lower limits of the consumption data, a distribution can be simulated using a variety of mathematical approaches. One of these is Monte Carlo Simulation, which can be performed on a desktop computer using commercial software such as the one used in this study (Crystal Ball [77]).

In summary, a Monte Carlo approach begins with mean values of parameters, the standard deviation about those means, and some assumptions about the upper and lower boundaries of those values. In this case, the mean and standard deviation of the food consumption and nicotine content values are used. A lower limit of consumption of a food item is zero g/day indicating no consumption of that food. An upper limit of consumption was taken to be the value, plus three-times the standard deviation. A similar approach was taken for the nicotine content in each food substance. Considered together, these parameters and the statistical values associated with them make up a set of assumptions.

To perform a distribution simulation, one must either know or assume the nature of the distribution of the parameters used in the assumptions. For example, does the variation in the nicotine content in a particular vegetable follow normal, lognormal, or some other type of distribution? If sufficient data is available, this can be tested. In this case, without sufficient data to conduct such a test a lognormal distribution was assumed for both consumption and nicotine content of the vegetables. This type of distribution is frequently found when values cannot go below zero and have no practical upper limit. Based on the information available, it was decided that a lognormal distribution was the more appropriate for these simulations.

The Monte Carlo method randomly and independently selects a set of parameters consistent with the assumptions described above. In this case, a number of grams/day of each food substance and a value for the nicotine content are selected. The product of these parameters is calculated and summed for all vegetables considered. This process is repeated a large number of times (typically 10,000 to 20,000), eventually resulting in a simulation of the probability distribution daily dietary nicotine intake of the population.

A Monte Carlo simulation using the data in Table 5 is shown in Fig. 4. This simulation estimates that a mean daily dietary intake for the population of the 14 countries is 1.4 µg/day. The upper 95th percentile is estimated to consume 2.25 µg/day. A sensitivity analysis indicates that potato consumption combined with the nicotine content of potatoes account for about 91% of the variance in daily nicotine intake. This is not an unreasonable result when the values in are considered.

Benowitz [78] has reviewed the use of cotinine as a biomarker for nicotine intake from environmental tobacco smoke (ETS) exposure. Although the bioavailability and pharmacokinetics of dietary nicotine may differ from that ingested by ETS, relationships described by Benowitz provide a useful beginning to describe the conversion of dietary nicotine to salivary cotinine concentration. Using these relationships, one estimates a mean salivary cotinine concentration of 0.022 ng/ml with a standard deviation of 0.007 ng/ml from dietary sources. This value is smaller that those reported for subjects exposed to ETS.

However, in the past, the lower limits of detection (LOD) of the analytical methodology for the determination of salivary cotinine were 0.5 to 1.0 ng/ml and inadequate to quantify cotinine levels in non-smokers who were exposed to low levels of ETS. Improved methods [79–81] based on liquid chromatography coupled with atmospheric pressure mass spectrometry

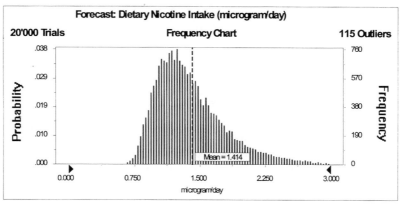

Fig. 4 Estimated daily dietary nicotine intake based on Monte Carlo simulation.

(LC/API/MS–MS)) has lowered the LOD to approximately 0.05–0.02ng/ml. The new analytical capability will provide increased opportunity to investigate the relationships between salivary cotinine concentration and low levels of ETS exposure.

## CONCLUSION

Determination of nicotine in pharmaceutical products intended for use as smoking cessation aids, and in foods that represent potential dietary intake of nicotine offer special challenges to the analyst. Patches, gums sprays and other manufactured materials provide a variety of matrices that preclude the use of many of the time-tested methods of nicotine determination. It may be for these reasons that many investigations concerning their use do not include determination of nicotine in the products either before or after their use. Many good methods likely reside in the laboratories of companies that manufacture these products.

Foods offer challenging matrices as well. Some of the results described in this chapter might reflect the problems of nicotine recovery, variability of nicotine concentration within and between samples, and the presence of interfering substances. There is little doubt that nicotine is present in some common foods, and that consumption of those foods does not contribute sufficient nicotine to explain cotinine levels using traditional methods of determination of cotinine in saliva. However, methods recently employed for determination of cotinine have reduced the limit of detection so that in the future this may not be the case.

## ACKNOWLEDGMENTS

The authors thank Philip Morris International and the Technical University of Graz for partial support of this work.

## REFERENCES

1 *The Merck Index*. 11 ed. Rahway, NJ: Merck & Co., Inc., 1989.
2 Nicotine Water. http://www.nicotinewater.com, 1997.
3 Høie LH. Use of a combination of an anti-smoking composition and a dietary composition in smoking cessation.1998. Norway patent WO 98/35669.
4 Gross J, Johnson J, Sigler L, Stitzer ML. Dose effects of nicotine gum. *Addict. Behav.*, 1995: **20(3)**; 371–381.
5 Gupta SK, Hwang SS, Causey D, Rolf CN, Gorsline J. Comparison of the nicotine pharmacokinetics of Nicoderm (nicotine transdermal system) and half-hourly cigarette smoking. *J. Clin. Pharmacol.*, 1995: **35**; 985–989.

6 Uematsu T, Mizuno A, Nagashima S, Oshima A, Nakamura M. The axial distribution of nicotine content along hair shaft as an indicator of changes in smoking behaviour: evaluation in a smoking-cessation programme with or without the aid of nicotine chewing gum. *Br. J. Clin. Pharmacol.*, 1995: **39**; 665–669.

7 Mizuno A, Uemastu T, Ishikawa T, Yoshimine N, Nakashima M. Clinical outcome of smoking-cessation trial of nicotine chewing gum evaluated by analysis of nicotine in hair. *Ther. Drug Monit.*, 1997: **19**; 407–412.

8 Molander L, Lunnell E, Andersson S-B, Kuylenstierna F. Dose released and absolute bioavailability of nicotine from a nicotine vapor inhaler. *Clin. Pharmacol. Ther.*, 1996: **59**; 394–400.

9 Paoletti P, Fornai E, Maggiorelli F, Puntoni R, Viegi G, Carrozzi L. Importance of baseline cotinine plasma values in smoking cessation: results from a double-blind study with nicotine patch. *Eur. Respir. J.*, 1996: **9**; 643–651.

10 Benowitz NL, Zevin S, Jacob III P. Sources of variability in nicotine and cotinine levels with use of nicotine nasal spray, transdermal nicotine and cigarette smoking. *Br. J. Clin. Pharmacol.*, 1997: **43**; 259–267.

11 Lawson GM, Hurt RD, Dale LC, Offord KP, Croghan IT, Schroeder DR, et al. Application of urine nicotine and cotinine excretion rates to assessment of nicotine replacement in light, moderate and heavy smokers undergoing transdermal therapy. *J. Clin. Pharmacol.*, 1998: **38**; 510–516.

12 Lawson GM, Hurt RD, Dale LC, Offord KP, Croghan IT, Schroeder DR, et al. Application of serum nicotine and plasma cotinine concentrations to assessment of nicotine replacement in light, moderate and heavy smokers undergoing trans-dermal therapy. *J. Clin. Pharmacol.*, 1998: **38**; 502–509.

13 Kochak GM, Sun JX, Choi RL, Piraino AJ. Pharmacokinetic disposition of multiple-dose transdermal nicotine in healthy adult smokers. *Pharm. Res.*, 1992: **9(11)**; 1451–1455.

14 Hegemann L, Forstinger C, Partsch B, Lagler I, Sabine K, Wolff K. Microdialysis in cutaneous pharmacology kinetic analysis of transdermall delivered nicotine. *J. Invest. Dermatol.*, 1995: **104**; 839–843.

15 Lindell G, Lunnell E, Graffner H. Transdermally administered nicotine accumulates in gastric juice. *Eur. J. Clin. Pharmacol.*, 1996: **51**; 315–318.

16 Foulds J, Russell MAH, Jarvis MJ, Feyerabend C. Nicotine absorption and dependence in unlicensed lozenges available over the counter. *Addiction*, 1998: **93(9)**; 1427–1431.

17 Nicotine Polacrilex. *The United States Pharmacopeia The National Formulary USP 23 NF 18*. Rockville MD: US Pharmacopeia, 1995: 2485.

18 Nicotine Polacrilex Gum. *The United States Pharmacopeia The National Formulary USP 23 NF 18*. Rockville MD: US Pharmacopeia, 1995: 2485–2486.

19 Dash AK, Wong S-T. Liquid-chromatographic method for the determination of nicotine in pharmaceutical formulations. *J. Chromatogr. A*, 1996: **749(1–2)**; 81–85.

20 Brantmark B, Ohlen P, Westling H. Nicotine-containing chewing gum as an anti-smoking aid. *Psychopharmacologia*, 1973: **31**; 191–200.

21 Russell MAH, Sutton SR, Reyerabend C, Cole PV, Saloojee Y. Nicotine chewing-gum as a substitute for smoking. *BMJ*, 1977: **280**; 1599–1602.

22 Herning RI, Jones T, Fischman P. The titration hypothesis revisted: Nicotine gum reduces smoking intensity. In: Grabowski J, Hall S, editors. *Pharmacological adjuncts in smoking cessation. National Institute on Drug Abuse Research*

*Monograph 53, (ADM) 85–1333.* Washington DC: US Government Printing Office, 1985.

23  Nemeth-Coslett R, Henningfield JE, OKeeffe MK, Griffiths RR. Nicotine gum: dose-related effects on cigarette smoking and subjective ratings. *Psychopharmacology*, 1987: **92**; 424–430.

24  Nemeth-Coslett R, Benowitz NL, Robinson N, Henningfield JE. Nicotine gum: chew rate, subjective effects and plasma nicotine. *Pharmacol. Biochem. Behav.*, 1988: **29**; 747–751.

25  Vlatakis G, I AL, Müller R, Mosbach K. Drug assay using antibody mimics made by molecular imprinting. *Nature*, 1993: **361**; 645–647.

26  Andersson LI, Müller R, Vlatakis G, Mosbach K. Mimics of the binding sites of opinoid receptors obtained by molecular imprinting of enkephalin and morphine. *Proc. Natl. Acad. Sci., USA*, 1995: **92(11)**; 4788–4792.

27  Andersson LI. Application of molecular imprinting to the development of aqueous buffer and organic solvent based radioligand binding assays for (S)-propanolol. *Anal. Chem.*, 1996: **68**; 111–117.

28  Zander A, Findlay P, Renner T, Sellergren B, Swietlow A. Analysis of nicotine and its oxidation products in nicotine chewing gum by a molecularly imprinted solid phase. *Anal. Chem.*, 1998: **70(15)**; 3304–3314.

29  Gora ML. Nicotine transderma system. *Ann. Pharmacother.*, 1993: **27**; 742–750.

30  Carlisle MR, Chicoine ML, Wygant MB. A stability-indicating high-performance liquid-chromatographic assay for nicotine in transdermal patches. *Int. J. Pharm.*, 1992: **80(2)**; 227–242.

31  Dubois JP, Sioufi A, Müller P, Mauli D, Imhof PR. Pharmacokinetics and bioavailability of nicotine in healthy volunteers following single and repeated administration of different doses of transdermal nicotine systems. *Meth. Find Exp. Clin. Pharmacol.*, 1989: **11(3)**; 187–195.

32  Sudan BJL. Transdermal nicotine and placebo. *The Lancet*, 1989 (February 11): 334.

33  American Society for Testing and Materials. D5075–96 Standard Test Method for Nicotine and 3-Ethenylpyridine in Indoor Air. © 1997, West Conshohocken, PA.

34  Eiceman GA, Sowa S, Lin S, Bell SE. Ion mobility spectrometry for continuous on-site monitoring of nicotine vapors in air during the manufacture of transdermal systems. *J. Haz. Mat.*, 1994: **43**; 13–30.

35  Russell MAH, Jarvis MJ, Sutherland G, Feyerabend C. Nicotine replacement in smoking cessation Absorption of nicotine vapor from smoke-free cigarettes. *JAMA*, 1987: **257**; 3262–3265.

36  Cornaz-Gudet A-L, Buri P. A simple and rapid high-performance liquid chromatography method to quantify and evaluate the purity of nicotine in solution and in microspheres. *S.T.P. Pharma. Sci.*, 1998: **8(2)**; 139–144.

37  Cornaz A-L, De Ascentis A, Colombo P, Buri P. In vitro characterization of nicotine microspheres for transnasal delivery. *Int. J. Pharm.*, 1996: **129**; 175–183.

38  Davoli E, Stramare L, Fanelli R, Diomede L, Salmona M. Rapid solid-phase extraction method for automated gas chromatographic-mass spectrometric determination of nicotine in plasma. *J. Chromatogr. B*, 1998: **707**; 312–316.

39  Zukowski J, Tang Y, Berthod A, Armstrong DW. Investigation of a circular dichroism spectrophotometer as a liquid chromatograph detector for enantiomers: sensitivity, advantages and limitations. *Anal. Chim. Acta*, 1992: **258**; 83–92.

40 Klus H, Kuhn H. Untersuchungen über die optische Aktivitat des Rauchnikotins. *Fachl. Mitt. Österr. Tabakregie.*, 1977: **17**; 331–336.

41 Crooks PA, Godin CS, F PW. Enatiomeric purity of nicotine in tobacco smoke condensate. *Med. Sci. Res.*, 1992: **20**; 879–880.

42 Jaroszewski JW, Olsson A. Determination of enantiomeric purity of nicotine in pharmaceutical preparations by carbon–13 NMR in the presence of a chiral lanthaninde shift reagent. *J. Pharm. Biomed. Anal.*, 1994: **12(3)**; 295–299.

43 Armstrong DW, Spino LA, Han SM, Seeman JI, Secor HV. Enantiomeric resolution of racemic nicotine and nicotine analogues by microcolumn liquid chromatography with β-cyclodextrin inclusion complexes. *J. Chromatogr.*, 1987: **411**; 490–493.

44 Demetriou D, Rustemeier K, Voncken P, Schepers G. HPLC separation of the enantiomers of nicotine and nicotine-like compounds. *Chirality*, 1993: **5**; 300–302.

45 Tang Y, Zielinski WI, Bigott HM. Separation of nicotine and nornicotine enantiomers via normal phase HPLC on derivatized cellulose chiral stationary phases. *Chirality*, 1998: **10**; 364–369.

46 Armstrong DW, Wang X, Ercal N. Enantiomeric Composition of nicotine in smokeless tobacco, medicinal products, and commercial reagents. *Chirality*, 1998: **10**; 587–591.

47 Leete E. Biosynthesis and metabolism of the tobacco alkaloids. In: *Alkaloids chemical and biochemical perspectives*. Pelletier SW (editor). New York: John Wiley and Sons; 1983; p. 86–139.

48 Ames BN. Dietary carcinogens and anticarcinogens. *Science*, 1983: **221**; 1256–64.

49 Davis RA, Stiles MF, deBethizy JD, Reynolds JH. Dietary nicotine: a source of urinary cotinine. *Food Chem. Toxicol.*, 1991: **29(12)**; 821–7.

50 Kuhn H. Tobacco Alkaloids and their pyrolysis products in smoke. In van Euler, U.S., editor. *Proceedings of the Fourth Wenner-Gren International Symposium*, Macmillan, New York. 1964; 37–51.

51 Domino EF, Hornbach E, Demana T. Relevance of nicotine content of common vegetables to the identification of passive tobacco smokers. *Med. Sci. Res.*, 1993: **21**; 571–572.

52 Idle RJ. Titrating exposure to tobacco smoke using cotinine - a minefield of misunderstandings. *J. Clinic Epidemiol.*, 1990: **43(4)**; 313–7.

53 Repace JL. Dietary nicotine - Won't mislead on passive smoking. *BMJ*, 1994: **308**; 61

54 Jarvis MJ. Dietary nicotine - . . . unless subjects eat 90 kg tomatoes a day. *BMJ*, 1994: **308**; 62.

55 Castro A, Monji N. Dietary nicotine and its significance in studies on tobacco smoking. *Biochem. Arch.*, 1986: **2(2)**; 91–7.

56 Sheen SJ. Detection of nicotine in foods and plant materials. *J. Food Chem.*, 1988: **53(5)**; 1572–3.

57 Osler M. The food intake of smokers and nonsmokers: the role of partner's smoking behaviour. *Prev. Med.*, 1998: **27**; 438–43.

58 Davis RA. The determination of nicotine and cotinine in plasma. *J. Chromatogr. Sci.*, 1986: **24**; 134–41.

59 Siegmund B, Leitner E, Pfannhauser W. Development of a simple sample preparation technique for the gas chromatographic-mass-spectrometric determination of nicotine in edible nightshades (*Solanaceae*). *J. Chromatogr. A* 1999: **840**; 249–260.

60 Jacob III P, Shulgin AT, Yu L, Benowitz NL. Determination of the nicotine metabolite trans–3'-hydroxycotinine in urine of smokers using gas chromatography with nitrogen-selective detection or selected ion monitoring. *J. Chromatogr.*, 1992: **583**; 145–54.

61 Jacob III P, Yu L, Wilson M, Benowitz NL. Selected ion monitoring method for determination of nicotine, cotinine and deuterium labeled analogs: absence of isotope effect in the clearance of (S)-nicotine–3',3'-d2 in humans. *Biol. Mass Spectrom.*, 1991: **20**; 247–52.

62 Holcomb LC. Indoor air quality and environmental tobacco smoke: concentration and exposure. *Environ. Int.*, 1993: **19**; 9–40.

63 Phillips K, Bentley MC, Howard DA, Alván G. Assessment of air quality in Paris by personal monitoring of nonsmokers for respirable suspended particles and environmental tobacco smoke. *Environ. Int.*, 1998: **24(4)**; 405–25.

64 Phillips K, Howard DA, Bentley M, Alván G. Measured exposures by personal monitoring for respirable suspended particles and environmental tobacco smoke of housewives and office workers resident in Bremen, Germany. *Int. Arch. Occup. Environ. Hlth*, 1998: **71**; 201–12.

65 AOAC Official methods of Analysis. Ed.: AOAC, Inc. Method 964.20 Nicotine Residues. Spectrophotometric Method. 1990; 15. edition, **1**, 307–308.

66 Likens ST, Nickerson GB. Detection of certain hop oil constituents in brewing products. *Am. Soc. Brew Chem. Proc.*, 1964: 5–13.

67 Nickerson GB, Likens ST. Gas chromatographic evidence for the occurrence of hop oil components in beer. *J. Chromatogr.*, 1966: 211–5.

68 Siegmund B, Leitner E, Mayer I, Farkas P, Sádecká J, Pfannhauser W, Kovác M. Untersuchungen zur Problematik der simultanen Destillation-Extraktion nach Likens-Nickerson. *Dtsch. Lebensm-Rdsch.*, 1996: **92(9)**; 286–90.

69 DIN 38407 (EN ISO 6468), part 1, Normenausschuß Wasserwesen im DIN Deutsches Institut für Normung e.V., 38. Lieferung, 1997.

70 Specht W, Tillkes M. Gas chromatographic determination of pesticide residues after clean-up by gel permeation and silica gel mini-column chromatography. III. Clean-up of foods and feeds of vegetable and animal origin for analysis of multiple residues of fat soluble and water soluble pesticides. *Fresenius Z. Anal. Chem.*, 1980: **301(4)**; 300–7.

71 Degen PH, Schneider W. Rapid and sensitive determination of low concentrations of nicotine in plasma by gas chromatography with nitrogen-specific detection. *J. Chromatogr. B*, 1991: **563**; 193–8.

72 Chappell WR, Gratt LB. Workplace indoor air quality: present day environmental tobacco smoke exposure for non-smokers. In: *The 7th international conference on indoor air quality and climate.* 1996 Jul 21; Nagoya, Japan: 1996: 399–404.

73 Zeyuan D, Bingyin T, Xiaolin L, Jinming H, Yifeng C. Effect of green tea and black tea on the blood glucose, the blood trigycerides, and antioxidation in aged rats. *J. Agric. Food Chem.*, 1998: **46(10)**; 3875–8.

74 TNO Nutrition and Food Research Institute. Survey of Vegetable and Fruit Consumption Data from 13 European Countries, Report No. V. 97.165, Zeist, the Netherlands, May 1997.

75 Department of Agriculture, Economic research Service. Food Consumption (on diskette), Springfield, VA, USA, NTIS Order Number ERS–89015B.

76  Grüter R, Schmid I, Sieber R. Bundesamt für Gesundheit BAG OFSP UFSP SFOPH, (Eds.). Vierter Schweizerischer Ernährungsbericht. Bern: Bundesamt für Gesundheit; 1998; Verbrauch an Lebensmitteln in der Schweiz in den Jahren *1994/95*. p. 5–16.

77  Crystal Ball, Denver CO. Forecasting & risk analysis for spreadsheet users, Version 4.0e, 1996.

78  Benowitz, NL. Cotinine as a biomarker of environmental tobacco smoke exposure. *Epidemiol. Rev.*, 1996: **18(2)**; 188–204.

79  Bernert Jr. JT, Turner W, E., Pirkle JL, S. SC, Akins JR, Waldrep MK, Ann, Q, et al. Development and validation of sensitive method for determination of serum cotinine in smokers and nonsmokers by liquid chromatography/atmospheric pressure ionization tandem mass spectrometry. *Clin. Chem.*, 1997;**43**; 2281–91.

80  Pirkle JL, Flegal KM, Bernert JT, Brody DJ, Etzel RA, Maurer KR. Exposure of the US population to environmental tobacco smoke. *JAMA*, 1996: **275(16)**; 1233–40.

81  Bentley MC, Abrar M, Kelk M, Cook J, Phillips K. Validation of an assay for the determination of cotinine and 3-hydroxycotinine in human saliva using automated solid-phase extraction and liquid chromatography with tandem mass-spectrometric detection. *J. Chromatogr. B*, 1999: **723(1–2)**; 185–94.

82  Siegmund B, Leitner E, Pfannhauser W. Determination of the nicotine content of various edible nightshades (solanaceae) and their products and estimation of the associated dietary nicotine intake. *J. Agric. Food Chem.* 1999: **47**; 3113–3120.

*Chapter 11*

# Nitrosamines derived from nicotine and other tobacco alkaloids

Stephen S. Hecht[1] and Anthony R. Tricker[2]

[1] *University of Minnesota Cancer Center, 420 Delaware Street S.E., Minneapolis, MN 55455, USA*
[2] *Philip Morris Europe, CH-2003 Neuchâtel, Switzerland*

## I. FORMATION AND OCCURRENCE OF TOBACCO-SPECIFIC N-NITROSAMINES

### Model studies

Model studies have shown that the reaction of nicotine with nitrous acid in aqueous solution at pH 3.4 and 20°C yields three nitrosation products; 4-(methylnitrosamino)-4-(3-pyridyl)butanal (NNA), N'-nitrosonornicotine (NNN) and 4-(methylnitrosamino)-1-(3-pyridyl)-1-butanone (NNK) in yields of 0.1–2.8%, with most of nicotine remaining unreacted after 17 hours [1]. Reaction at 90°C, with a five-fold excess of NaNO$_2$, results in the formation of NNN and NNK at higher yields (up to 13.5% and 4.3%, respectively). Under these conditions, extensive fragmentation and oxidation of the nicotine pyrrolidine ring occurs and both NNK and NNA give secondary products: NNK is nitrosated at the α-carbonyl position to give 4-(methyl-nitrosamino)-2-oximino-1-(3-pyridyl)-1-butanone, while NNA is unstable and decomposes to 1-methyl-5-(3-pyridyl)pyrazole. The kinetic mechanisms for the nitrosation of nicotine are complex and are first-order for both nicotine and nitrite [2]; typical for the nitrosation of tertiary amines [3].

Reduction of NNK and NNA yields 4-(methylnitrosamino)-1-(3-pyridyl)-1-butanol (NNAL) and 4-(methylnitrosamino)-4-(3-pyridyl)-1-butanol (*iso*-NNAL), respectively, while oxidation of NNA yields 4-(methyl-nitrosamino)-1-(3-pyridyl)butyric acid (*iso*-NNAC) [4]. Nitrosation of cotinine and cotinine acid both yield *iso*-NNAC, albeit in low yields [5]. N'-Nitrosoanabasine (NAB) and N'-nitrosoanatabine (NAT) are formed by the

421

nitrosation of anabasine [6] and anatabine [7], respectively. Collectively these *N*-nitroso compounds are referred to as tobacco-specific *N*-nitrosamines (TSNA) and their structures are shown in Fig. 1.

## Tobacco-specific *N*-nitrosamines in tobacco

The formation of TSNA in tobacco is significantly influenced by agricultural practice, tobacco type and nitrate content, and different curing conditions [8–11]. Although nitrosation of secondary amines is generally much more rapid than tertiary amines [3,12], nicotine rather than nornicotine has been suggested to be the major precursor of NNN in tobacco [6]. NNN could also be formed by nitrosation of nicotine-*N'*-oxide (13) and possibly acylnornicotines [11]. Pseudooxynicotine, formed by oxidation of nicotine during tobacco curing [14], undergoes rapid and irreversible nitrosation to NNK [15], and is probably the major precursor of NNK. Reduction of NNK probably accounts for the trace levels of NNAL occasionally detected in fermented snuff tobacco [16]. NNA, the major product of in vitro nicotine nitrosation [1,2], and 4-(methylnitrosamino)-2-oximino-1-(3-pyridyl)-1-butanone [1] have not been identified in tobacco. The apparent absence of NNA in tobacco is probably due to the reactivity of the aldehyde group, which could be either reduced to *iso*-NNAL or oxidized to *iso*-NNAC [4]. NAB and NAT are probably formed by the nitrosation of the corresponding alkaloid precursors. With the exception of NNA, all seven TSNA shown in Fig. 1 have been detected in cigarette tobacco filler [4,5,9,17,18], and at much higher concentrations in various smokeless tobacco products [4,16,19,20].

## Tobacco-specific *N*-nitrosamines in tobacco smoke

TSNA in mainstream cigarette smoke arise from two sources: *N*-nitrosamines pyrosynthetically formed during smoking, and transfer of endogenous *N*-nitrosamines present in the tobacco filler. The mechanisms involved in the pyrosynthesis of TSNA are poorly understood; however, it is apparent that the tobacco filler nitrate content is a significant determinant of TSNA yields in both mainstream and sidestream cigarette smoke [21,22]. Contrary to an early study reporting pyrosynthetic formation of NNK from [methyl-[14]C]nicotine added to cigarette tobacco [23], a later investigation reported that neither NNK nor NNN was formed from nicotine during tobacco combustion [22]. Nitrosation of nornicotine to NNN occurs during tobacco pyrolysis, albeit in small yields [24].

During smoking, transfer of TSNA [5,17,18,22] occurs almost exclusively into the particulate phase of mainstream cigarette smoke [25]. The presence of TSNA in mainstream cigarette smoke is well documented [4,18,26–28]. German data show mean mainstream smoke deliveries of $75\pm41$ (range 19–135) ng NNN/cigarette, $52\pm29$ (range 21–110) ng NNK/cigarette, and

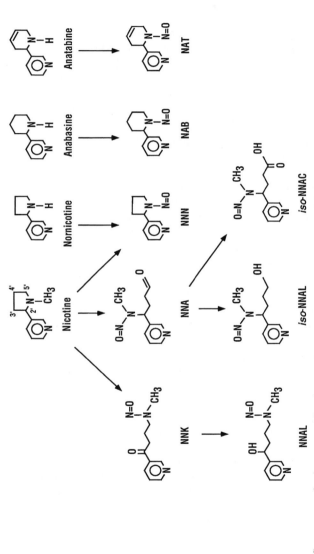

Fig. 1 Structures of tobacco-specific nitrosamines and tobacco alkaloid precursors. Abbreviations used: NNK, 4-(methylnitrosamino)-1-(3-pyridyl)-1-butanone; NNA, 4-(methylnitrosamino)-4-(3-pyridyl)butanal; NNN, $N'$-nitrosonornicotine; NAB, $N'$-nitrosoanabasine; NAT, $N'$-nitrosoanatabine; NNAL, 4-(methylnitrosamino)-1-(3-pyridyl)-1-butanol; *iso*-NNAL, 4-(methylnitrosamino)-4-(3-pyridyl)-1-butanol; *iso*-NNAC, 4-(methylnitrosamino)-4-(3-pyridyl)butanoic acid.

$73 \pm 47$ (range 18–155) ng NAB plus NAT/cigarette in filter cigarettes, and $431 \pm 304$ (range 79–855) ng NNN, $170 \pm 134$ (range 62–470) ng NNK, and $261 \pm 127$ (range 75–520) ng NAB/NAT in non-filter cigarettes [18]. American data show mainstream smoke deliveries (filter and non-filter cigarettes combined) of $162 \pm 104$ (range 40–278) ng NNN/cigarette, $109 \pm 66$ (range 17–156) ng NNK/cigarette, $17 \pm 11$ (range 4.5–30) ng NAB/cigarette, and $137 \pm 86$ (range 37–236) ng NAT/cigarette [4]. However, caution is required in interpretation of this data since no standardized methods exist for the measurement of these compounds, and these data may not necessarily be reproducible from one laboratory to another. In general, mainstream smoke TSNA deliveries are also lower in filter cigarettes with a Virginia tobacco filler compared to other tobacco types [29].

Both *iso*-NNAC [5,17] and *iso*-NNAL [18] occur rarely in mainstream cigarette smoke. Due to the absence of standardized smoking conditions for smoking articles other than cigarettes, mainstream smoke TSNA deliveries have been less seldomly reported for cigars [7] and other products [30].

Since TSNA are transferred to sidestream smoke of cigarettes [26] and cigars [7], and are presumably exhaled by smokers, they are also found in environmental tobacco smoke (ETS) present in indoor air [31,32]. Mean levels of 2.8 (range 0–6.0) ng/m$^3$ NNN and 4.9 (range 0–13.5) ng/m$^3$ NNK are reported in offices with poor ventilation conditions [31]. Similar levels of 0–23 ng/m$^3$ NNN and 3–24 ng/m$^3$ NNK, and in some cases 0–9 ng/m$^3$ NAT, have been reported in heavily smoke polluted bars and restaurants [32].

## Human exposure to tobacco-specific *N*-nitrosamines

Exposure to TSNA resulting from use of smokeless tobacco products is difficult to quantify and exposure estimates range between 39.6 μg/day TSNA for users of nasal snuff tobacco [19] to 11.5 mg/day NNK for toombak snuff-dippers in Sudan [33]. Based on the above data for German filter cigarettes [18], and the assumption that 70% retention of particle-bound TSNA occurs in the lung during oral inhalation [34], a smoker of 20 filter cigarettes per day would have a mean total exposure of about 3.0 μg/day TSNA: $1.1 \pm 0.6$ μg NNN, $0.7 \pm 0.4$ μg NNK and $1.0 \pm 0.7$ μg NAB/NAT. Based on American data, a smoker of 20 cigarettes per day would be exposed to about 6.0 μg TSNA: $2.3 \pm 1.5$ μg NNN, $1.5 \pm 0.9$ μg NNK, $1.9 \pm 1.2$ μg NAT, and $0.2 \pm 0.2$ μg NAB [4]. Similar exposure estimates have been calculated in other studies [27–29]. However, these exposure estimates are very crude as it has been shown that numerous factors also affect individual uptake of these compounds [4,22]. Estimates of TSNA exposure for users of other tobacco products have not been reported. Quantitation of TSNA metabolites in urine is potentially a better measure of individual uptake, as discussed below.

For non-smokers, the limited data for the presence of TSNA in ETS and uncertainty of the duration of exposure to ETS makes the estimation of exposure uncertain. However, maximum exposure to NNN and NNK in ETS of poorly ventilated offices has been estimated at 0.01 µg/h [31].

## II. IN VITRO METABOLISM OF TOBACCO-SPECIFIC *N*-NITROSAMINES

### 4-(Methylnitrosamino)-1-(3-pyridyl)-1-butanone (NNK)

NNK metabolic pathways are illustrated in Fig. 2. Five types of transformations have been observed: carbonyl reduction, pyridine *N*-oxidation, α-hydroxylation (hydroxylation of the carbons adjacent to the *N*-nitroso group), denitrosation, and adenosine dinucleotide phosphate (ADP) adduct formation. There is presently no evidence for β- or γ-hydroxylation. Metabolism of NNK has been examined in a large number of in vitro systems and these studies have been reviewed [35].

Most studies have focused on oxidative metabolism of NNK. All evidence indicates that these reactions are catalyzed by cytochrome P450s (Table 1), with little, if any, involvement of other enzyme systems such as flavin-containing monooxygenase (FMO) [36]. Possible exceptions are pulmonary lipoxygenase [37,38] and cyclooxygenase [39]. Cytochrome P450 (CYP) isoforms that catalyze NNK metabolism have been reviewed [35]. In summary, CYP1A2, CYP2A, CYP2B1 and CYP3A play a major role in oxidative metabolism of NNK in rats and mice, while CYP2E1 is minimally involved. In both species there appears to be considerable unidentified CYP activity. Much less is known about the involvement of specific CYP isoforms in hamster metabolism of NNK.

Human CYP1A2, CYP2A6 and CYP3A4 appear to be important for hepatic metabolic activation of NNK [40,41]. The possible involvement of CYP2D6 in lung metabolism has been studied [42]; however, both enzyme expression [43] and activity are low [41,42].

*Carbonyl reduction*

In many tissues, carbonyl reduction of NNK to NNAL is the predominant metabolic pathway in vitro (Table 2). This is the case in rodent liver [44–52] and rat pancreas [51,53]. An exception is the rat nasal mucosa in which minimal NNAL formation is observed [51,54,55]. Reduction of NNK to NNAL is also the major metabolic pathway in patas monkey liver, but not in the lung [56]. Reduction of NNK to NNAL was initially reported to account for more than 99% of total in vitro metabolism of NNK in human buccal mucosa, trachea, esophagus, bronchus, peripheral lung and urinary bladder tissues [57]. However, more recent studies report that human buccal mucosa predominantly reduces NNK to NNAL (95-99%) in addition to metabolism by α-hydroxylation (0.6-3.8%) and pyridyl *N*-oxidation (0.3–2.2%) [58]. Similar metabolic profiles are observed with isolated microsomes from the human esophagus (98.4% conversion of NNK to NNAL and 1.6% α-hydroxylation)

TABLE 2

Summary of extents of NNK metabolite formation in unpretreated rodents and in humans[a]

| Species | Tissue | Carbonyl Reduction | Pyridine-N-Oxidation | α-Methylene Hydroxylation | α-Methyl Hydroxylation | ADP Adduct |
|---|---|---|---|---|---|---|
| Rat | Liver | ++ | + | ++ | ++ | + |
| | Lung | ++ | +++ | ++ | ++ | ++ |
| | Nasal Mucosa | − | + | +++ | +++ | NA |
| | Oral Tissue | ++ | ++ | ++[b] | ++[b] | NA |
| | Pancreas | ++ | − | − | + | ++ |
| | Intestine | ++ | ++ | +[b] | +[b] | NA |
| Hamster | Liver | +++ | ++ | + | ++ | NA |
| | Lung | ++ | ++ | + | +++ | NA |
| | Intestine | ++ | +++ | ++[b] | ++[b] | NA |
| Mouse | Liver | +++ | − | ++ | ++ | NA |
| | Lung | ++ | +++ | ++ | ++ | NA |
| | Intestine | + | ++ | +++[b] | +++[b] | NA |
| Human | Liver | +++ | + | ++ | ++ | NA |
| | Lung | +++ | − | + | + | NA |

[a] Symbols refer to comparisons within a given tissue:
+++ Usually the predominant pathway
++ One of the major pathways
+ Minor pathway
− Not generally observed
NA Not analyzed
[b] Total α-hydroxylation

[59] and lung (96% conversion of NNK to NNAL and 2–4% α-hydroxylation) [37,41]. Human pancreas microsomes reduce NNK to NNAL in the absence of α-hydroxylation [60]. Reduction of NNK to NNAL accounts for 84–92% of total NNK metabolism in human liver microsomes; the remaining metabolic products result from α-hydroxylation (9–15%) with only a minor contribution by pyridyl N-oxidation (<1%) [60–62]. NNAL is further metabolized by human liver microsomes by α-hydroxylation, pyridine N-oxidation, and reconversion back to NNK [61]. Human red blood cells predominantly reduce NNK to NNAL and the keto acid is the only detectable product of α-hydroxylation [63]. In human liver microsomes, NNAL is further metabolized by α-hydroxylation and pyridine N-oxidation, and reconversion to NNK [61]. The absolute configuration of NNAL has only been recently elucidated and its formation appears to show considerable stereoselectivity depending on the system involved [64].

Further metabolism of NNAL is in many ways similar to that of NNK, with a notable exception being formation of [4-(methylnitrosamino)-1-(3-pyridyl)but-1-yl]-β-O-D-glucosiduronic acid (NNAL-Gluc) [51,64–67]. Rat hepatocytes produce mainly (S)-NNAL-Gluc from NNK [66], and liver microsomes from rats treated with phenobarbital (PB) give mainly (S)-NNAL-Gluc when incubated with racemic NNAL [68]. The major and minor NNAL-Gluc diasteriomers in the urine of patas monkeys and humans are formed from (R)-NNAL and (S)-NNAL, respectively [64]. NNAL-Gluc is a detoxification product of NNK/NNAL metabolism [35,65,69] and there is no evidence that NNAL is a substrate for sulfotransferase [35].

Enzymatic activity for NNK carbonyl reduction is widely distributed. In many studies CYP inducers or inhibitors modify the oxidative metabolism of NNK, but have little effect on the carbonyl reduction pathway [45,65,70–72]. Thus, CYP isoforms appear to play little role in NNAL formation. 11 β-Hydroxysteroid dehydrogenase 1 (11β-HSD 1; EC 1.1.1.146), a microsomal enzyme responsible for the conversion of active 11-hydroxyglucocorticoids to inactive 11-oxo forms, has been isolated from mouse liver and shown to be one carbonyl reductase involved in the reduction of NNK to NNAL [73]. 11β-HSD also reduces NNK to NNAL in mouse lung microsomes [74]. Whether 11β-HSD is the major enzyme responsible for NNK reduction in mammals is not known; however, it has been suggested that differences in tissue expression of this enzyme, as well as genetic polymorphisms, may influence NNK metabolism to NNAL-Gluc [75].

### α-Methylene hydroxylation

Hydroxylation of the methylene carbon adjacent to the N-nitroso group produces the unstable intermediate α-methylenehydroxy-NNK (Fig. 2, (3)), which spontaneously decomposes to methanediazohydroxide (7) (or the corresponding diazonium ion) and 4-(3-pyridyl)-4-oxobutanal (keto aldehyde). α-Methylene hydroxylation of NNAL similarly yields methanediazohydroxide (**7**) and 5-(3-pyridyl)-2-hydroxytetrahydrofuran (lactol). The lactol can be further oxidized to 4-(3-pyridyl)-4-hydroxybutanoic acid (hydroxy acid). NNAL is a poor substrate for α-methylene hydroxylation compared to NNK [76]. α-Methylene hydroxylation of both NNK and NNAL are considered to be metabolic activation pathways [35].

In microsomal systems, keto aldehyde can be quantified as its bisulfite adduct [48]. In the absence of bisulfite, further oxidation of keto aldehyde yields 4-(3-pyridyl)-4-oxobutanoic acid (keto acid), and there may be other uncharacterized reactions such as protein binding. Since bisulfite can inhibit metabolism under some conditions, care must be taken in choosing the appropriate in vitro conditions to trap the keto aldehyde [40,48,61,77].

α-Methylene hydroxylation of NNK is commonly observed in vitro (Table 2). Total α-methylene hydroxylation of NNK is more extensive than for NNAL, and occurs to a greater extent in Clara cells compared to alveolar Type II cells, small cells and alveolar macrophages in the rat lung [78]. α-Methylene hydroxylation is the predominant pathway of NNK metabolism in rat nasal mucosa, along with α-methyl hydroxylation [54,55,79–81]. Rat nasal mucosa generally has higher NNK α-hydroxylation activity than other tissues [82,83]. Antibody inhibition studies support the involvement of CYP1A2 and CYP2A1 in α-methylene hydroxylation of NNK in rat nasal mucosa [79,80], and CYP2B1 in rat lung [84]. Antibodies to CYP1A2 and CYP2A1 have significant inhibitory activity against keto aldehyde formation, but it is not clear which related forms are present in the lung [79]. Interestingly, in rat lung, but not liver, α-methylene hydroxylation is inhibited by chronic treatment with NNK, possibly due to CYP inactivation by keto aldehyde binding [66]. Inhibition of α-methyl hydroxylation is not observed under these conditions. CYP1A2 and CYP2A1 are present in the mouse lung and another related form may also be present since inhibition is seen with an antibody to CYP2F (P450$_{MP}$) [72]. In a reconstituted system, keto aldehyde is one of the minor products of NNK metabolism by rat hepatic CYP2B1 [85]. Antibody inhibition studies show some involvement of CYP1A2 and CYP3A in rat liver, and studies with enzyme inducers and inhibitors are generally consistent with a role of these two enzymes, as well as CYP2B1 [70,71].

Formation of keto aldehyde from NNK, and lactol from NNAL are highly correlated in human liver microsomes, indicating the involvement of the same enzyme [61]. Both CYP2A6 and CYP3A4 have been implicated in NNK α-methylene hydroxylation in human liver [40]; however, since NNK liver concentrations will be low, the low $K_m$ for CYP2A6 metabolism suggests that this is probably the more relevant enzyme.

*α-Methyl hydroxylation*
Hydroxylation of the NNK methyl group yields 4-(hydroxymethylnitrosamino)-1-(3-pyridyl)-1-butanone (Fig. 2) (α-hydroxymethyl-NNK). The

formation of this metabolite has been confirmed by characterization of its glucuronide [67]. α-Hydroxymethyl-NNK spontaneously decomposes to formaldehyde and 4-(3-pyridyl)-4-oxobutane-1-diazohydroxide (6) (or the corresponding diazonium ion). Reaction of diazohydroxide (6) with $H_2O$ gives 4-hydroxy-1-(3-pyridyl)-1-butanone (keto alcohol; also referred to as HPB). α-Methyl hydroxylation of NNAL produces a similar cascade of products, the major end product of which is 4-(3-pyridyl)butane-1,4-diol (diol). Rat liver and lung microsomes also yield 2-(3-pyridyl)-tetrahydrofuran (pyridyl-THF) [66]. Myosmine has also been detected following metabolism of NNK by liver microsomes from Aroclor pretreated rats [44]. Alternatively, myosmine could also be produced by denitrosation of NNK [86].

α-Methyl hydroxylation of NNK occurs to varying extents in vitro (Table 2). It is the predominant pathway, along with α-methylene hydroxylation, in rat nasal mucosa microsomes and cultures [54,55]. It is also a major pathway in rat lung, generally exceeding α-methylene hydroxylation [48,66,79]. On a per mg protein basis, total α-hydroxylation is greatest in the nasal mucosa [54,55], followed by lung [66,79,80], and liver [70]. α-Methyl hydroxylation of NNK also exceeds α-methylene hydroxylation in the mouse lung [87] and hamster lung [88]. An antibody to CYP1A2 strongly inhibits keto alcohol formation in rat nasal mucosa, while this reaction is inhibited by an antibody to CYP2A1 in lung [79]. α-Methyl hydroxylation is also inhibited by antibodies to CYP1A1 and CYP2B1 in rat lung [87], and CYP2A1 in mouse lung [72]. The effects of isothiocyanates on activities associated with CYP1A (ethoxyresorufin-*O*-dealkylase [EROD]) and CYP2B (pentoxyresorufin-*O*-dealkylase [PROD]) do not correlate with their effects on α-methyl hydroxylation of NNK in rat [71,89] or mouse lung [89]. Rat liver CYP2B1 catalyzes conversion of NNK to keto alcohol in a reconstituted system [85]. Keto alcohol formation is induced by pretreatment with PB; however, it is not inhibited by an antibody to CYP2B1 [65,70,85]. Antibodies to CYP1A2, CYP2A1, and CYP3A inhibit keto alcohol formation and inducers of these enzymes increase its formation [70].

α-Methyl hydroxylation of NNK and NNAL occur at approximately equal rates in human liver [61]. CYP1A2 appears to be important in the production of keto alcohol in human liver [41,90]; however, a variety of CYP isoforms catalyze this reaction including CYP2A6 (with a high $K_m$), CYP2D6, CYP2B7, CYP2E1, CYP2F1, CYP3A4, and CYP3A5 [40].

*Pyridine N-oxidation*
There are wide variations in the production of 4-(methylnitrosamino)-1-(3-pyridyl-*N*-oxide)-1-butanone (NNK-*N*-oxide) and 4-(methylnitros-amino)-1-(3-pyridyl *N*-oxide)-1-butanol (NNAL-*N*-oxide) depending on species and tissue (Table 2). NNK-*N*-oxide is generally the major metabolite of NNK in rodent lung microsomes [48,79,87,91], while pyridine-*N*-oxidation is a minor pathway, or absent, in rodent liver microsomes [46,49,67], and rat

nasal mucosa microsomes [54,55,70,79]. Similarly, NNK-*N*-oxide is the major metabolite formed in the isolated and perfused rat lung, but not in the isolated and perfused rat liver [92]. NNK-*N*-oxide is a major metabolite formed by patas monkey lung microsomes, but is not formed by liver microsomes [56]. In contrast to both rodents and the patas monkey, pyridine-*N*-oxidation of NNK and NNAL are observed in human liver microsomes [61], but not in human lung microsomes [37]. Pyridine *N*-oxidation of NNK is a detoxification pathway of NNK metabolism [47,93].

Pyridine-*N*-oxidation is strongly induced by pretreatment with PB, but not 3-methylcholanthrene (MC) in rat liver, suggesting the involvement of CYP2B1 [65,70]. CYP2B1 isolated from rat liver metabolizes NNK to NNK-*N*-oxide in a reconstituted system [85]. Collectively, there is strong evidence that CYP2B1 is one of the major rat hepatic CYP isoforms responsible for conversion of NNK to NNK-*N*-oxide.

In rat lung, where NNK-*N*-oxide is consistently the major metabolite [52,79,91], the picture is less clear. Although CYP isoforms appear to be the predominant enzymes involved in the formation of this metabolite, and there is some induction by PB treatment, studies with antibodies to CYP2B1 show no inhibition of NNK-*N*-oxide formation [70,79]. Moreover, there is little correspondence between the effects of phenethyl isothiocyanate (PEITC) on activity associated with CYP2B (PROD) and NNK-*N*-oxide formation. Similarly, in A/J mouse lung, antibodies to CYP isoforms do not inhibit NNK-*N*-oxide formation, and there is no correlation between the effects of isothiocyanates on EROD and PROD activities and NNK conversion to NNK-*N*-oxide [72,89].

Limited data are available on the role of CYP isoforms in pyridine-*N*-oxidation in human tissues. A correlation exists between formation of *N*-oxides of NNK and NNAL and CYP3A4 activity in human liver microsomes [61], while human kidney cells transfected with CYP2B1 cDNA convert NNK to NNK-*N*-oxide [94].

*Denitrosation*

Only one study has examined denitrosation of NNK [86]. By analogy to *N*-nitrosodimethylamine (NDMA), denitrosation of NNK would proceed by formation of an α-carbon radical followed by elimination of NO˙. The imines resulting from these reactions would hydrolyze to keto aldehyde and myosmine. Neither of these has been characterized specifically as a product of denitrosation, but enzymatic formation of nitrite from NNK by rat liver microsomes was observed.

*ADP adduct formation*

ADP adducts of NNK and NNAL, namely NNK(ADP)$^+$, NNK(ADPH), and NNAL(ADP)$^+$ have been identified in studies of NNK metabolism with rat pancreas [53], liver [53,66] and lung microsomes [66], NNK(ADP)$^+$ and

NNAL(ADP)$^+$ are formed with catalysis by NAD glycohydrolase, which also catalyzes similar reactions of nicotine, cotinine, and 3-acetylpyridine. These adducts have only been observed in vitro. Whether similar adducts are formed in vivo remains to be determined.

## N'-Nitrosonornicotine (NNN)

NNN metabolism pathways are illustrated in Fig. 3. Three types of reactions have been observed: pyridine-*N*-oxidation, hydroxylation of the pyrrolidine ring (including α-hydroxylation at the 2'- and 5'-positions and β-hydroxylation at the 3'- and 4'-positions), and norcotinine formation. Metabolism has been extensively studied in the rat liver [95–99], esophagus [50,99–101], oral mucosa [67,100,102], nasal mucosa [54,103] and lung [78]. Metabolism is less well documented in the hamster [98,104] and mouse [47], but has been studied in several human tissues [57,59,61,90,105–107].

### α-Hydroxylation

α-Hydroxylation of NNN at the 2'-position produces an unstable intermediate, 2'-hydroxy-NNN, which spontaneously loses HONO yielding myosmine, or ring opens to produce 4-(3-pyridyl)-4-oxobutane-1-diazohydroxide [97]. The same intermediate is formed by α-methyl hydroxylation of NNK (Fig. 2). The metabolites which ultimately result from this pathway are myosmine, keto alcohol, keto acid and diol; their sum indicates the total extent of NNN 2'-hydroxylation [47,54,95,100,101,108]. α-Hydroxylation at the 5'-position yields 5'-hydroxy-NNN which undergoes spontaneous ring opening to 1-(3-pyridyl)-4-oxobutane-1-diazohydroxide followed by hydrolysis to hydroxy aldehyde and cyclization to lactol [97,109].

The extent of 2'-hydroxylation and 5'-hydroxylation of NNN in the rat lung is cell-specific; small cells and alveolar macrophages predominantly metabolize NNN by 2'-hydroxylation while 5'-hydroxylation occurs preferentially in Clara cells and alveolar Type II cells [78]. The extent of 2'-hydroxylation in all cell types is less than that observed for NNK α-methyl hydroxylation [78]. The 2'- and 5'-hydroxylation pathways in rat liver are dependent on NADPH, inhibited by CO, and specifically induced or suppressed by pretreatment with MC, PB, and Aroclor [97–99]. Metabolism is similar in the hamster liver [98]. The CYP isoforms involved in NNN α-hydroxylation are poorly understood. A coumarin hydroxylase, possibly CYP2A6, appears to be involved in rat nasal mucosa [103]. An unidentified CYP isoform with a high affinity for α-hydroxylation of low concentrations of NNN is present in the rat esophagus, but appears to be absent in liver [99].

At equimolar doses, 4-(carbethoxynitrosamino)-1-(3-pyridyl)-1-butanone (CNPB), a stable precursor to the diazohydroxide formed by NNN 2'-hydroxylation, shows mutagenicity towards *S. typhimurium* strains TA 100 and TA 1535, but 4-(carbethoxynitrosamino)-1-(3-pyridyl)butanal, a stable precursor

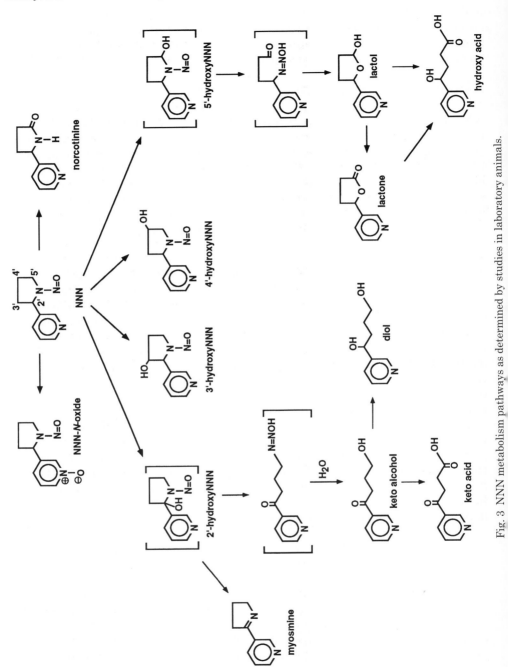

Fig. 3  NNN metabolism pathways as determined by studies in laboratory animals.

to the diazohydroxide formed by 5'-hydroxylation, is inactive [109]. Consequently, 2'-hydroxylation of NNN is regarded as an important route of NNN metabolic activation [97]. There is some consistency among ratios of 2'-hydroxylation to 5'-hydroxylation in different rodent tissues and their susceptibility to carcinogenesis by NNN. Thus, the 2'-hydroxylation to 5'-hydroxylation ratio is typically 2–4 in rat esophagus [99–101,108], and nasal mucosa [54], the main target tissues of NNN carcinogenicity in the rat [95,110–116]. The 2'-hydroxylation to 5'-hydroxylation ratio is similar (~ 3.8) in rat esophagus and oral tissue [100]; however, the oral mucosa is not a target tissue of NNN administered *per os* in the absence of chronic ethanol administration [117]. In liver, a non-target tissue, the ratios are 0.3–1.4 [95,99,101]. In the rat lung, another non-target tissue [95,110,113-116], the 2'-hydroxylation to 5'-hydroxylation ratio is approximately 1.1 in Clara cells, small cells and alveolar Type II cells, and 30 in alveolar macrophages [78]. Hamster trachea, a target tissue [104,111,118,119], metabolizes NNN by 2'-hydroxylation and 5'-hydroxylation to an equal extent [120]. In general, these results are consistent with a role for 2'-hydroxylation in the metabolic activation of NNN, and 5'-hydroxylation in detoxification [108].

Human CYP1A2 expressed in *Escherichia coli* and liver microsomes both catalyze 2'-hydroxylation and 5'-hydroxylation of NNN (90). Human liver microsomes catalyze predominantly 5'-hydroxylation of NNN [61,105,106]; the 2'-hydroxylation to 5'-hydroxylation ratio is about 0.02 [61]. Human liver microsomes metabolize NNN more rapidly than NNK, NNAL, or benzo[*a*]-pyrene (BaP) [61]. Convincing evidence for a major role of CYP2A6 in 5'-hydroxylation of NNN in human liver has been presented, whereas CYP3A4 is involved in 2'-hydroxylation [106]. CYP2E1 has also been suggested to be involved in human liver metabolism of NNN [107]. Human esophageal microsomes metabolize NNN by 2'-hydroxylation and 5'-hydroxylation to an equal extent, and α-hydroxylation appears to be mediated by CYP2E1 and CYP3A4, but not CYP2A6 [59]. Other extrahepatic human tissues predominantly metabolize NNN by 5'-hydroxylation [57].

### β-*Hydroxylation*

β-Hydroxylation of NNN has been reported in one study in which it was a minor metabolic pathway in rat liver microsomes [96].

### *Pyridine-N-oxidation*

NNN-*N*-oxide is a detoxification product of NNN metabolism [35]. It is observed in rat liver as a major metabolite [95–98], but appears to be a minor metabolite, if formed at all, in esophagus [50,99–101], lung [78], nasal mucosa [54,103], and oral tissue [67,100,102]. NNN-*N*-oxide is also a minor metabolite in the mouse lung [47]. It has not been reported as a metabolite of NNN in human liver [57,61,105,106], but is formed by the esophagus

[121]. The enzymology of NNN-*N*-oxide formation has not been investigated.

*Norcotinine*

Norcotinine is observed in cultured mouse lung [47] and, provisionally, in rat oral mucosa incubated with NNN [100]. The origin of norcotinine is unknown. It could result from denitrosation followed by oxidation of nornicotine, from oxidation of 5′-hydroxyNNN followed by denitrosation, or from loss of HNO from 5′-hydroxyNNN. Alternatively, norcotinine could be formed via CYP-mediated radical formation at the 5′-position followed by loss of NO˙ yielding *iso*-myosmine, which is oxidized to norcotinine.

## Other tobacco-specific *N*-nitrosamines

NAB metabolism has been subject to limited investigation. In cultured rat esophagus, α-hydroxylation of NAB occurs mainly at the 6′-position (equivalent to 5′-hydroxylation of NNN) [108]. The 2′-hydroxylation to 6′-hydroxylation ratio for NAB is 0.2–0.4, while the corresponding 2′-hydroxylation to 5′-hydroxylation ratio for NNN is 3 [108]. 2′-Hydroxylation of NAB and NNN are metabolic activation pathways whereas 5′-hydroxylation of NNN or 6′-hydroxylation of NAB are detoxification reactions. The metabolism of other TSNA have not been investigated.

## III. IN VIVO DISPOSITION AND EXCRETION IN LABORATORY ANIMALS

### 4-(Methylnitrosamino)-1-(3-pyridyl)-1-butanone (NNK)

Significant interspecies differences occur in the disposition and elimination of TSNA in laboratory animals. Administration of NNK results in rapid distribution to most tissues and rapid metabolism so that the NNK-NNAL equilibrium favors NNAL in all species [83,122–125].

In the rat, autoradiography studies using [carbonyl-$^{14}$C]NNK show tissue-bound radioactivity occurs extensively in nasal mucosa, lung and liver [82,126]. Accumulation of radioactivity also occurs in the stomach lumen, melanin of the eye, and hair follicles of the skin. Urine is the major route of excretion, with greater than 90% of the administered dose excreted in less than 24 h [44,51,52,67,127,128]. Most studies report that products of α-hydroxylation account for 50–60% of all metabolites [67,123,127,128]. Dose dependency of excretion has been studied by administration of 0.005–500 μmol/kg [5-$^3$H]NNK (0.1–103,500 μg/kg bw, i.p.) and collection of 48 h urine samples [123]. At high NNK doses (>103.5 μg/kg bw, i.p.), NNAL and a single diastereomer of (*S*)-NNAL-Gluc are minor excretion products (2–8% and 7–25%, respectively), while NNK-*N*-oxide and NNAL-*N*-oxide constitute 8–11% of urinary metabolites [123]. At lower doses, excretion of NNAL and NNAL-Gluc is reduced, and non-detectable at a dose of 0.1 µg/kg bw, while pyridine-$N$-oxides account for about 26% of urinary NNK metabolites. Chronic administration of NNK (2 ppm in drinking water for 28 days) results in a significantly longer terminal half-life (8.06 days) for NNAL-Gluc excretion [129]. 4-(Methylnitrosamino)-1-[3-(6-hydroxypyridyl)]-1-butanone has been identified as a minor urinary metabolite (~ 1%) of NNK [127]. In PB pretreated rats, the $O$-glucuronide conjugate of 4-hydroxymethylNNK is also a minor (< 1%) urinary metabolite [67]. NNAL-Gluc can undergo deconjugation to NNAL in vivo and further metabolism by both α-hydroxylation and pyridine $N$-oxidation [130]. Biliary excretion of NNAL-Gluc also occurs in the rat, but little if any reaches the feces [131].

In the mouse, administration of [carbonyl-$^{14}$C]NNK results in tissue-bound radioactivity in nasal mucosa, lung and liver [132]. A similar dose dependency of urinary NNK excretion is observed in the A/J mouse and F344 rat [123]. Excretion of products of α-hydroxylation usually account for 30-60% of all metabolites; no apparent excretion of either NNAL or NNAL-Gluc occurs at low doses (< 10.35 µg/kg bw, i.p.), while excretion of pyridine $N$-oxides are less than observed in rats at identical levels of administration.

In the hamster, binding of [carbonyl-$^{14}$C]NNK occurs primarily in the liver, lung, kidney, nasal cavity, lacrimal and adrenal glands [104]. Products of α-hydroxylation account for about 80% of all metabolites [104].

In the marmoset monkey, administration of [carbonyl-$^{14}$C]NNK results in accumulation of tissue-bound radioactivity in the liver, and to a lesser extent, in the nasal mucosa [133]. Accumulation of radioactivity is also observed in the melanin of the eyes, hair follicles of the skin, and ceruminous ear glands. The urinary metabolite profile is consistent with extensive metabolism by carbonyl reduction to NNAL resulting in excretion of NNAL (26%), hydroxy acid (56%), NNAL-$N$-oxide (7%), keto acid (7%), and keto alcohol (3%) [133].

In the patas monkey, metabolism of [5-$^3$H]NNK (16–18 µg/kg bw, i.v.) by α-hydroxylation is rapid and extensive and accounts for at least 50% of metabolites present in blood and urine (125). Other major urinary metabolites include NNK-$N$-oxide (4–16%), NNAL-$N$-oxide (8–16%), and two diastereomers of NNAL-Gluc (20–25%); minor metabolites include NNAL (< 2%) and 6-hydroxyNNK (1%). Excretion of NNK and metabolites in feces is negligible. The biological half-lives of NNK and NNAL in the patas monkey ($t_{1/2}$=5.5–18 min and 56–58 min, respectively) [125] are similar to those observed in the baboon ($t_{1/2}$=19.2–25.2 min and 60–85 min, respectively) [124].

In the rhesus monkey, metabolism of [5-$^3$H]NNK (4.6-9.8 µg/kg bw, i.v.) by α-hydroxylation is rapid and extensive and accounts for 20-40% of metabolites present in blood, and 53% of metabolites in 24 h urine (134). Other major urinary metabolites include NNAL (11%) and two diastereomers of

NNAL-Gluc (23%), NNAL-*N*-oxide (4.5%), NNK-*N*-oxide (5%); minor metabolites include diol (1.5%) and keto alcohol (2.5%). Decay of total urinary radioactivity is bi-exponential with half-lives of 1.7 h and 42 h; 86% of the dose being excreted in urine within the first 24 h. Metabolite patterns in urine from the first 6 h closely resembled those reported for the patas monkey [125], at latter time points NNAL and its *O*-glucuronide conjugates account for 60–70% of urinary NNK metabolites. Biliary excretion over 6 h (0.6%) is ten times less than predicted to occur in the rat [131].

### *N'*-Nitrosonornicotine (NNN)

In the F344 rat, [2'-$^{14}$C]NNN is rapidly distributed throughout the whole body followed by accumulation of tissue-bound radioactivity in the nasal mucosa, tracheobronchial mucosa, esophagus, liver and tongue [135,136]. Binding is higher in the nasal mucosa than in the liver and esophagus [136]. Following administration of [2'-$^{14}$C]NNN (3–300 mg/kg bw, i.p.), the major urinary metabolites result from 2'-hydroxylation to yield hydroxy acid (37–53%) and 5'-hydroxylation to yield keto acid (13-31%), NNN-*N*-oxide (7–11%), and norcotinine (3–5%) [137]. Excretion of hydroxy acid decreases with decreasing dose while excretion of keto acid increases suggesting total metabolic activation of NNN decreases with decreasing dose. 3'-Hydroxy-NNN and 4'-hydroxy-NNN resulting from β-hydroxylation have also been detected as very minor metabolites [96].

In the mouse, rapid tissue distribution of [2'-$^{14}$C]NNN occurs with binding in the tracheobronchial and nasal mucosa, liver, submaxillary and sublingual salivary glands, and esophagus [138]. Tissue binding of [2'-$^{14}$C]NNN occurs primarily in the hamster liver, lung, kidney and adrenal glands, and to a lesser extent in submaxillary salivary glands, lacrimal glands, pancreas and brain [104]. Based on limited experimental data, a similar urinary excretion profile for NNN occurs in F344 rats [137], strain A mice [137], and Syrian golden hamsters [104].

In the marmoset monkey, tissue binding of [2'-$^{14}$C]NNN occurs in the nasal mucosa and liver [133]. High levels of non-bound material also accumulate in the tracheobronchial and esophageal mucosa, melanin of the eyes, the hair follicles of the skin, and the ceruminous ear glands. Extensive 2'-hydroxylation of NNN occurs and hydroxy acid is the major metabolite detected in urine (81%), together with keto acid (7%) and NNN-*N*-oxide (6%) (133).

In the miniature pig, accumulation of tissue-bound [5-$^{3}$H]NNN is abundant in the nasal mucosa and liver, while unbound material accumulates in the mandibular and parotid salivary glands, Harder's gland, lacrimal glands, glands of the snout and respiratory part of the nasal cavity, melanin of the eyes and skin [139]. Hydroxy acid is the major metabolite detected in urine (78%) together with keto acid (14%) and norcotinine (7%) [139]. In contrast to rodents [137] and non-human primates [133], the miniature pig does not readily metabolize NNN by pyridine *N*-oxidation.

## Other tobacco-specific *N*-nitrosamines

The in vivo metabolism of other TSNA has not been subject to extensive study. The biological half-life of NAT in the rat ($t_{1/2}$=540 min) is much longer than that of NNK ($t_{1/2}$=25–37 min), NNAL ($t_{1/2}$=184–298 min) and NNN ($t_{1/2}$=184 min) [122]; however, no excretion data are available for any laboratory animal species. Limited data for the excretion of NAB in the F344 rat suggest that in vivo metabolism primarily occurs via pyridine *N*-oxidation and excretion of NAB-*N*-oxide (30%), and 6'-hydroxylation and excretion of 5-hydroxy-5-(3-pyridyl)pentanoic acid (10%) [108]. No evidence was found for in vivo metabolism by 2'-hydroxylation. *iso*-NNAC is rapidly excreted without further metabolism in rat urine and feces [17]. Lack of metabolism and rapid excretion is a common characteristic of *N*-nitrosamino acids and *N*-nitrosamines with carboxylic acid moieties.

## IV. DNA BINDING

### 4-(Methylnitrosamino)-1-(3-pyridyl)-1-butanone (NNK)

Both in vitro and in vivo data indicate that there are two major types of DNA adducts formed by NNK: methyl adducts formed by the α-methylene hydroxylation pathway and pyridyloxobutyl adducts formed by the α-methyl hydroxylation pathway (Fig. 4). The potential formation of adducts by these two pathways are consistent with expectations based on the well established α-hydroxylation metabolic activation pathway of *N*-nitrosamines.

α-Methylene hydroxylation of NNK yields methanediazohydroxide (7) and/ or the methyl diazonium ion (11), which can react with DNA producing 7-methylguanine (7-mG), $O^6$-methylguanine ($O^6$-mG), and $O^4$-methylthymidine ($O^4$-mT) adducts (Figure 4). DNA methylation by NNK is observed in a number of in vitro studies with different systems capable of its metabolic activation, including rat lung and lung cells [84,140], liver [140], oral tissue [100] and nasal mucosa [140], and hamster lung [141]. Mouse lung, rat lung and rat nasal microsomes catalyze a time- and protein-dependent DNA methylation in a microsome-mediated DNA alkylation system [80].

The chemistry of the intermediates resulting from α-methyl hydroxylation of NNK has been investigated in detail (Fig. 5). Since α-hydroxymethyl-NNK (2) is not very stable, this metabolite has been generated in situ by solvolysis of 4-(acetoxymethylnitrosamino)-1-(3-pyridyl)-1-butanone (NNKOAc) to produce diazohydroxide (6) (Fig. 5) [142]. CNPB is also a precursor to diazohydroxide (6) [109,143-147]. Diazohydroxide (6) yields diazonium ion (10) which has three fates: reaction with nucleophiles (Y:) producing (14),

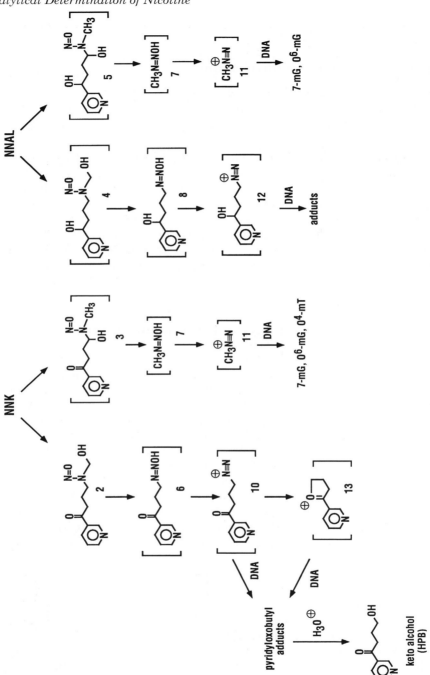

Fig. 4 DNA adduct formation from NNK and NNAL as determined by in vivo studies in laboratory animals. NNAL adducts can also be formed via metabolic oxidation of NNAL to NNK.

formation of the cyclic oxonium ion (**13**), or loss of $N_2$ and $H^+$ yielding the $\alpha,\beta$-unsaturated ketone (**15**). The latter two intermediates also react with nucleophiles to form products (**16**) and (**17**). Extensive in vitro and in vivo studies conclusively demonstrate that the major DNA adduct(s) formed by this pathway, accounting for at least 50% of the DNA binding, release HPB upon acid or neutral thermal, but not base hydrolysis

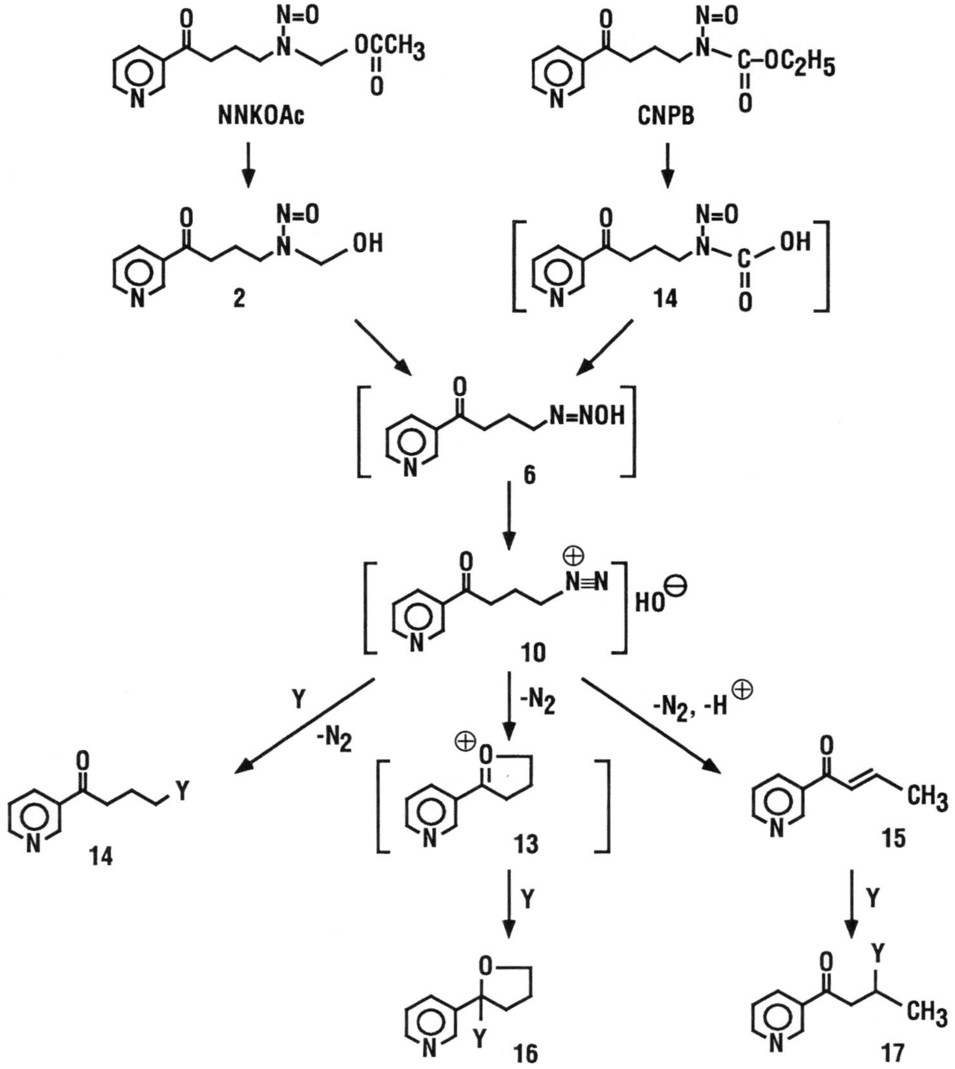

Fig. 5 Intermediates and products formed upon solvolysis of NNKOAc and CNPB. When Y=OH, (**16**) ring-opens to (**14**) [146].

[76,100,142,144,145,147–154]. This adduct(s) is produced via intermediates (**6**), (**10**) and/or (**13**), but not by HPB itself [148]. The HPB-releasing adduct(s) have differing stablilties in DNA, being released in a triphasic manner [144]. These adduct(s) do not appear to be stable at the nucleoside level, or under the conventional HPLC conditions used for adduct analysis [150]. The structure of the HPB-releasing adduct(s) has not been determined. The major HPB-releasing adduct appears to be a deoxyguanosine (dG) derivative (Hecht, unpublished data). Both the $N^2$ adduct (**18**) [150] and the $O^6$ adduct (**20**) [142] (Fig. 6) have been prepared, but neither of these has properties consistent with the major HPB-releasing adduct, since neither releases HPB upon neutral thermal hydrolysis, and both are stable under enzymatic hydrolysis and HPLC analysis conditions. Based on the chemistry described above, it appears likely that one HPB-releasing adduct may be a pyridyltetrahydrofuryl dG derivative (**21**) resulting from the reaction of $N^2$ with oxonium ion (**13**) and/or an acyclic 7-substituted pyridyloxobutyl dG adduct (**22**) (Fig. 6). Similar adducts from the reaction of oxonium ions with $N^2$ of dG have been characterized in reactions of $\alpha$-acetoxy-$N$-nitrosopyrrolidine with dG or DNA(155). The oxonium ion (**13**), an $S_N1$ type reactant, is likely to bind at $N^2$ and/or $O^6$ of dG, while the diazonium ion (**10**), an $S_N2$ type reactant, would be expected to bind preferentially at the 7-position of dG [156].

The identity of minor adducts formed by pyridyloxobutylation of DNA in vitro has been studied. The $O^6$ pyridyloxobutyl adduct (**20**) (Fig. 6), is formed by reaction of DNA with NNKOAc [142]. This is the only example of a structurally characterized DNA adduct resulting from methyl $\alpha$-hydroxylation of NNK. Another minor adduct, formed by incubation of NNK with cultured rat nasal mucosa, releases HPB upon acid hydrolysis, but is not identical to the $N^2$-pyridyloxobutyl adduct (**19**) (Fig. 6) [150]. Reactions of CNPB and the $\alpha,\beta$-unsaturated ketone (**15**) (Fig. 5), with dG yield the branched $N^2$ adduct (**19**) (Fig. 6), but this adduct has not been detected in tissues exposed to NNK in vitro [150,157] or in vivo [150].

Pyridyloxobutyl adducts inhibit $O^6$-alkylguanine-DNA-alkyltransferase (AGT), the enzyme responsible for repair of $O^6$-mG (142,158). This presumably occurs as a result of the transfer of the pyridyloxobutyl group from the $O^6$ position of guanine to the active site of AGT. Since $O^6$-mG is also formed upon metabolic activation of NNK, this phenomenon is likely to be important with respect to the persistence of $O^6$-mG in NNK-exposed tissues. Studies with NNKOAc and oligonucleotides demonstrate that only pyridyloxobutylated poly(dGdC), not poly(dAdT), is able to prevent repair of $O^6$-mG by rat liver AGT [159]. Furthermore, an oligomer containing dA, dT, and dG residues, but not its complement, reacts with NNKOAc to generate an AGT-reactive dG adduct. This dG adduct has a half-life of 1–2 weeks in DNA and represents a high percentage of the total HPB-releasing adducts in DNA.

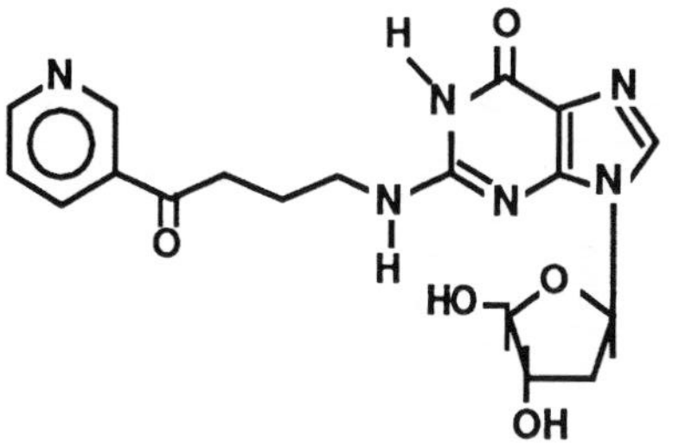

**18**

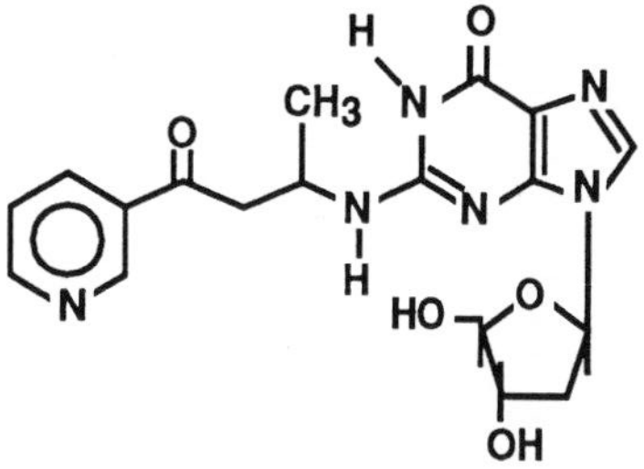

**19**

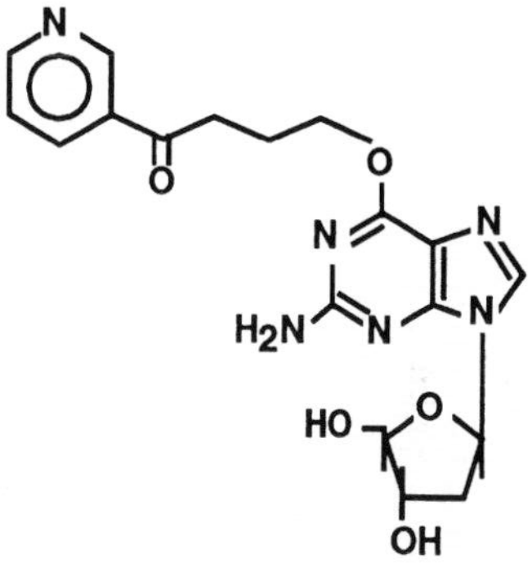

**20**

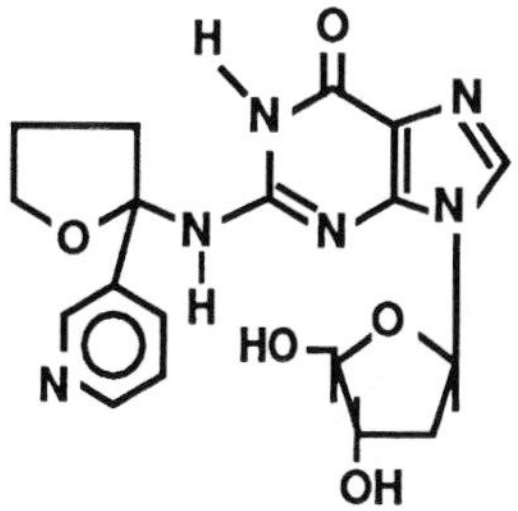

**21**

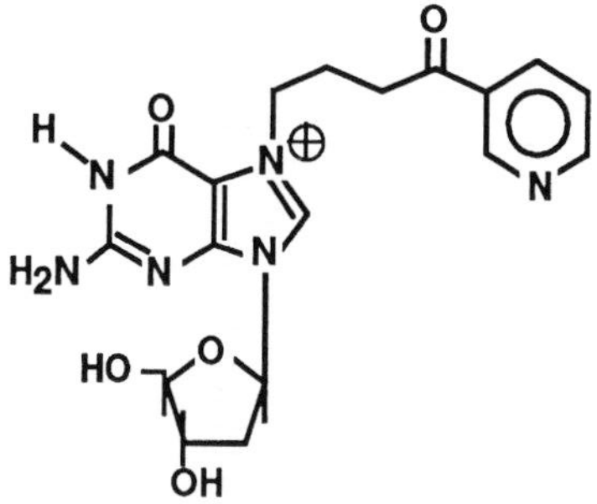

**22**

Fig. 6 Known and potential DNA adducts of NNK formed by α-methyl hydroxylation: (**18**), synthesized but not detected in hydrolysates of DNA from NNK-treated rats (150); (**19**), formed by reaction of CNPB or α,β-unsaturated ketone (**15**) (Fig. 5) with dG but not detected in DNA of NNK-treated rats [150,157]; (**20**), detected in DNA reacted with NNKOAc [142]; (**21,22**), potential structures of other unsynthesized HPB-releasing DNA adducts.

Since the first detection of $O^6$-mG and 7-mG in the liver and lung of NNK-treated F344 rats in 1984 [160], there has been substantial research on the in vivo occurrence and biological significance of methyl and pyridyloxobutyl adducts resulting from metabolic activation of NNK. With few exceptions, adduct formation occurs principally in target tissues of NNK carcinogenicity – lung, nasal mucosa, and liver. Adduct measurements provide some important information on the proposed mechanisms of carcinogenesis by NNK, as well as giving some new insights on the relationship of NNK exposure to DNA adduct dose [35]).

Treatment of rats with NNK gives slightly higher levels of 7-mG and $O^6$-mG in liver, lung and nasal mucosa DNA than after treatment with NNAL [76]. Most of the hepatic DNA methylation observed after treatment with NNAL appears to be due to its conversion to NNK. Following treatment with NNK, levels of 7-mG are 7.5–25 times higher than those of HPB released in whole rat lung, with lower ratios occurring at low doses [152]. HPB-releasing adducts are about twice the levels of $O^6$-mG, and levels of $O^6$-mG are about 10 times as great as those of $O^4$-mT in respiratory tissue of rats [154,161]. The highest levels of $O^6$-mG and HPB-releasing adducts are consistently found in the Clara cells of the rat lung, with lower amounts in Type II cells, macrophages, and small cells [84,154,162,163]. The dose-response for adduct formation in whole lung and lung cell types is non-linear (Fig. 7). The observed adduct levels in whole lung at the lowest doses are higher than

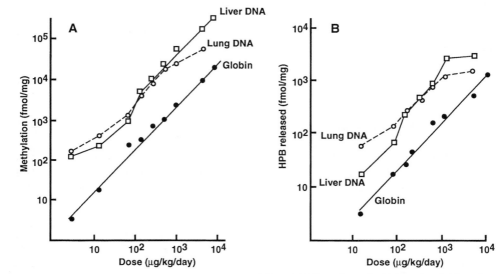

Fig. 7 Plots of (A) levels of 7-mG in lung and liver DNA and total tritium in globin after treatment of F344 rats with [methyl-$D_3$]NNK (3–10 000 μg/kg/day, i.p.) for 4 days or (B) HPB released from lung and liver DNA by acid hydrolysis or from globin by base hydrolysis [152].

would be expected by linear extrapolation from higher doses [84,152,154,162,164]. One interpretation of these data is the presence of CYP isoforms in the rat lung which efficiently catalyze α-hydroxylation of low doses of NNK [84,164]. Alternatively, reduced inhibition of AGT by pyridyloxobutyl adducts may occur at low NNK doses resulting in higher $O^6$-mG levels [158].

During semi-chronic treatment with high doses of NNK (100 mg/kg/day for 1, 2, 4, 6, 9, or 12 days, i.p.), $O^6$-mG increases progessively in the rat lung throughout the treatment period, while $O^6$-methylguanine-DNA transferase ($O^6$-mGT) decreases to less than 5% [161]. The concentration of $O^4$-mT in lung DNA reaches a steady state after 4 days of treatment. After NNK treatment is discontinued, $O^6$-mG is slowly repaired ($t_{1/2} \sim 7$ days), while $O^4$-mT is rapidly removed ($t_{1/2} \sim 18$ h) from the lung, suggesting that different repair pathways exist for the removal of these adducts in vivo. $O^6$-mG adducts are also formed in nasal mucosa, hepatocytes and non-parenchymal liver cells, but not in kidney and brain. The removal of both $O^6$-mG and $O^4$-mT occurs rapidly in hepatocytes ($t_{1/2} \sim 18$ h) after discontinuation of NNK treatment, but is much slower in non-parenchymal liver cells (4 and 2 days, respectively). The repair of $O^6$-mG is similar in both populations of liver cells ($t_{1/2} \sim 2$ days). At lower NNK doses (10 mg/kg/day for 4 days), $O^6$-mGT activity in macrophages is unaffected, but decreased by 57% and 84% in alveolar small cells and Type II cells, respectively (163). Removal of $O^6$-mG from DNA of macrophages follows first order kinetics ($t_{1/2} = 48$ h) while very little loss occurred in Clara cells over an 8 day period following cessation of NNK treatment. Even though $O^6$-mGT activity was reduced in alveolar small cells and Type II cells, ~90% of $O^6$-mG was removed within 8 days of cessation of NNK treatment. When NNK is administered over a 20 week period (1.76 mg/kg bw three times weekly for 4, 8, 12, 16, or 20 weeks, s.c.), $O^6$-mG adduct levels reach a plateau after 4 weeks in Clara cells, followed by a 82% decrease during the remaining treatment period (154). At the end of the 20-week treatment period, $O^6$-mG adduct levels are the highest in alveolar macrophages followed by Clara cells, Type II cells, and small cells. The decrease is due to inhibition of CYP catalyzed α-methylene hydroxylation by NNK or one of its metabolites, possibly the keto aldehyde, as demonstrated in a study of α-methylene hydroxylation by rat lung microsomes isolated during 20 weeks of NNK treatment [66]. Keto aldehyde has also been shown to inhibit $O^6$-mGT activity in rat hepatocytes, suggesting that this aldehyde may contribute to the carcinogenicity of NNK by inhibiting this repair enzyme [165].

Structure-activity studies suggest that both DNA methylation and pyridyloxobutylation are important in NNK-induced rat lung tumorigenesis. Neither NDMA [166], which only methylates DNA [84], nor NNN [95,110,113–116], which pyridyloxobutylates but does not methylate DNA [140], are effective lung carcinogens in the rat. NNK-induced rat lung tumors arise in the Type II cells [164]. NNK yields greater amounts of $O^6$-mG in Clara cells, but not Type II cells, than NDMA [84], and NNK is metabolized somewhat more effectively to a pyridyloxobutylating agent than NNN in different rat lung cell types [78]. Levels of HPB-releasing adducts in Type II cells of NNK-treated rats correlate with lung tumor incidence over a range of doses, suggesting that DNA pyridyloxobutylation is important in rat lung carcinogenesis (Fig. 8) [154]. The levels of $O^6$-mG in Clara cells also correlate with lung tumor incidence over a wide dose range, suggesting some role of this adduct in spite of the fact that Type II cells, and not Clara cells, are probably the cell of origin of the tumors (Fig. 8) [164]. The effects of PEITC on levels of HPB-releasing adducts in Type II cells and other cell types of the lung correlate well with inhibition of NNK-induced lung tumorigenesis by PEITC, providing further evidence for the importance of HPB-releasing adducts [154]. PEITC also inhibits $O^6$-mG levels in Clara cells to the same extent to which it inhibits lung tumorigenesis, but this is not seen in other cell types [154]. Collectively, these data indicate that HPB-releasing adducts and $O^6$-mG are both important in lung tumor induction by NNK in the rat.

Treatment of rats and hamsters with a single dose of NNK (80 mg/kg, s.c.) also results in high levels of 7-mG and $O^6$-mG in liver DNA, which reach a maximum 24 h after NNK treatment [165]. In the rat, $O^6$-mGT activity is depleted after 4 h but rapidly recovers resulting in repair of $O^6$-mG liver DNA adducts within 48 h. In the hamster, $O^6$-mGT activity is rapidly depleted with no recovery within 336 h resulting in highly persistant $O^6$-mG liver DNA adduct levels. Since NNK induces liver tumors in rats [114,164,166–172], but only rarely in hamsters [104,113,173–176], these

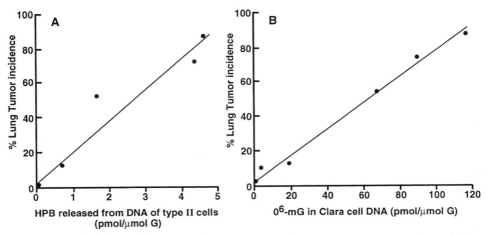

Fig. 8 Correlation between DNA adduct levels and lung tumor incidence in NNK-treated F344 rats: (A) HPB-releasing adducts in Type II cells vs. lung tumor incidence, $r^2 = 0.94$ (154); (B) $O^6$-mG in Clara cells verses lung tumor incidence, $r^2 = 0.98$ [164].

data suggest that factors other than $O^6$-mG DNA adduct levels may determine NNK-induced hepatocarcinogenesis in rats.

Elevated levels of 8-hydroxy-2'-deoxyguanosine (8-OH-dG), a promutagenic adduct and biomarker of oxidative DNA damage, have been measured in lung DNA of NNK-treated A/J mice [177–181] and F344 rats [177,178]. However, the role of 8-OH-dG and other types of oxidative damage in NNK-induced tumorigenesis have not been defined. While DNA methylation plays an impotant role in the initiation phase of NNK-induced lung tumorigenesis, oxidative damage may be important in the post-initiation phase [35].

## $N'$-Nitrosonornicotine (NNN)

2'-Hydroxylation of NNN is the only metabolic pathway currently known to induce DNA damage. This pathway generates 4-(3-pyridyl)-4-oxobutane-1-diazohydroxide (Fig. 3), as does methyl α-hydroxylation of NNK (Fig. 2). The fate of this intermediate and its role in DNA binding is discussed above. Studies using 2'-acetoxy-NNN suggest that 2'-hydroxylation of NNN could also yield an electrophilic nitrosiminium ion, but its relationship to DNA adduct formation is not known [143].

Little is known about 5'-hydroxylation of NNN and the fate of the resulting diazohydroxide. 5'-Acetoxy-NNN, a precursor to 5'-hydroxy-NNN, is mutagenic to *S. typhimurium* without activation, suggesting induction of DNA damage [97]. The diazohydroxide formed is structurally related to 4-oxobutane-1-diazohydroxide formed by α-hydroxylation of *N*-nitrosopyrrolidine [155,182]. Adduct formation from 4-oxobutane-1-diazohydroxide has been extensively characterized [155,182] suggesting that similar adducts may also be formed by 5'-hydroxylation of NNN.

Culture of rat nasal mucosa with NNN produces HPB-releasing DNA adduct(s) together with several unidentified adducts [150]. HPB-releasing adducts are also formed by culture of rat esophagus with NNN [100]. These results are consistent with the observed extensive 2'-hydroxylation of NNN in rat nasal mucosa [54] and esophagus [99–101,108]. In esophagus, approximately 10-fold higher HPB-releasing DNA adduct levels are formed from NNN than from NNK, and may be important in determining the carcinogenicity of NNN in this tissue [100]. The extent of 2'-hydroxylation of NNN is similar in rat esophagus and oral tissue; however, HPB-releasing adducts are not formed in oral cavity tissue, suggesting that factors, as yet unknown, protect the DNA from pyridyloxobutylation by NNN [100].

HPB-releasing adducts are present in both acid and enzymic hydrolysates of lung DNA from NNN-treated mice [147,150]. Acid hydrolysates of hepatic DNA, but not lung DNA, from NNN-treated rats contain HPB-releasing adducts [148]. HPB-releasing adducts are higher in the respiratory than in the olfactory parts of the nasal mucosa in rats after treatment with NNN [149]. Although pyridyloxobutylation of DNA occurs in the rat liver [148],

NNN rarely induces liver tumors in rats [95,110,111,113,114]. The chemical and physical properties of NNN-derived HPB-releasing adducts appear to be identical to those isolated from NNK-treated animals [76,148,149].

## V. HEMOGLOBIN BINDING

### 4-(Methylnitrosamino)-1-(3-pyridyl)-1-butanone (NNK)

NNK forms globin adducts in rats by both α-methylene hydroxylation leading to methylation of globin, and α-methyl hydroxylation resulting in pyridyloxobutylation of globin (183). Following treatment of rats with [5-$^3$H]NNK (6.2-807 μg/kg bw, i.p.), about 0.1% of NNK binds to globin as an HPB-releasing adduct(s) which decays with a half-life of 9.1 days [183]. Under identical conditions, treatment with [methyl-$^3$H]NNK also results in approximately 0.1% binding of the NNK methyl group to globin [183]. In rats treated with [$^{14}$C]NDMA, total radioactivity present in globin is about 7-fold higher compared to that found after treatment by an equivalent dose of [methyl-$^3$H]NNK [184]. Methyl globin adducts have not been investigated extensively, since they do not represent a specific biomarker of NNK metabolic activation. Depending on dose, species, and protocol employed, approximately 15–40% of the pyridyloxobutyl adducts are released as HPB upon mild base hydrolysis [63,145,152,183]. Less material is released upon acid treatment [183,185]. The HPB-releasing adduct(s) do not form upon treatment of hemoglobin or rodents with HPB, but are formed from NNK and stable precursors to α-hydroxymethyl-NNK [145]. Pyridyloxobutylation of rat globin by NNK is higher than that following treatment with NNAL, suggesting that alkylation by NNAL may result from metabolic reconversion to NNK [76]. PB treatment of rats reduces the formation of HPB-releasing hemoglobin adducts, possibly as a consequence of increased glucuronidation of α-hydroxymethyl-NNK, the metabolite thought to be responsible for adduct formation [65].

The HPB-releasing adduct(s) have been identified as carboxylic acid esters, most likely with glutamate or aspartate [185]. The reactivity of pyridyloxobutylating agents with *N*-acetylcysteine has been investigated to determine the potential role of cysteine in adduct formation (186): CNPB reacts with *N*-acetylcysteine to give cyclic, branched, and straight chain adducts with diazonium ion (10), oxonium ion (13), and α,β-unsaturated ketone (15) (Fig. 5). Hydrolysis of globin from NNK-treated rats shows that pyridyloxobutyl cysteine adducts are not formed in detectable amounts [186]. The structures of the non-HPB-releasing pyridyloxobutyl adduct(s) in globin have not been determined. Investigations of NDMA binding to human erythrocytes in vitro shows the formation of 3-methylhistidine, 1-methylhistidine, and *S*-methylcysteine (187), suggesting similar binding of NNK occurs.

Co-incubation of human red blood cells and rat hepatocytes, but not red blood cells alone, results in the formation of HPB-releasing adduct(s), as well as pyridyloxobutyl adducts that do not release HPB [63]. Similar adducts are observed in vivo. The HPB-releasing adducts are not formed in the absence of hepatocytes. Therefore, α-hydroxymethyl-NNK (**2**) or perhaps its glucuronide (**1**) (or possibly the corresponding diazohydroxide (**6**) or diazonium ion (**10**)), are stable enough to migrate out of the hepatocyte and into the red blood cell, where pyridyloxobutylation of globin occurs. α-Hydroxymethyl-NNK-Gluc (**1**) is a likely candidate for the transported metabolite leading to HPB-releasing adducts in globin, but this has not been investigated. Interestingly, red blood cells themselves are able to activate NNK producing globin adducts, but not HPB-releasing adducts [63]. Part of this activation process occurs by heme-catalyzed α-methylene hydroxylation of NNK resulting in formation of keto aldehyde, which probably forms lysine adducts in globin, and methanediazohydroxide (**7**), which methylates histidine and cysteine [187].

If hemoglobin adducts are to be useful surrogates for DNA adducts, there must be a predictable relationship between their formation. In NNK-treated rats, both methyl and pyridyloxobutyl hemoglobin adducts increase linearly over greater than a 3,000-fold dose range; however, methyl and pyridyloxobutyl-DNA adducts show a non-linear relationship to dose in liver and lung (Fig. 7) [152]. As discussed above, DNA adduct levels at low doses are higher than would be expected by extrapolation from higher doses. Thus, a complex relationship exists between NNK hemoglobin adducts and liver and lung DNA adducts in rats.

## *N'*-Nitrosonornicotine (NNN)

Treatment of rats with [5-$^3$H]NNN and [5-$^3$H]NNK results in the formation of an identical HPB-releasing globin adduct(s); the amount formed by [5-$^3$H]NNN is about 16% of the amount detected from [5-$^3$H]NNK [183].

## VI. CARCINOGENICITY OF TOBACCO-SPECIFIC *N*-NITROSAMINES

### 4-(Methylnitrosamino)-1-(3-pyridyl)-1-butanone (NNK)

The lung is the main target tissue for NNK carcinogenicity in laboratory animals [35]. Regardless of the route of administration, NNK preferentially induces systemic lung tumors in rats rather than local tumors; few if any oral cavity or esophageal tumors are observed upon oral swabbing [172] or administration in the drinking water [169], subcutaneous tumors are not seen upon s.c. injection [113] or topical administration [188], and bladder tumors are not found following intravesicular administration [171]. In dose-response studies, lung tumors are prevalent at lower doses, where they are often observed to the virtual exclusion of other tumor types [114,164]. The lung tumors are predominantly adenomas and adenocarcinomas, with lower incidences of adenosquamous and squamous cell carcinomas. Ultrastructural studies suggest that the tumors arise in the Type II cells of the rat lung [164].

Dose-response data for lung tumor induction from eleven studies of NNK carcinogenicity in F344 rats [114,164,166,167,169,170,189,190–193] are presented in Fig. 9. The eight studies using s.c. administration [114,164,166,167,170,191–193] were performed at two different institutions: two at the National Institute of Environmental Health Sciences [164,191] and six at the American Health Foundation [114,166,167,170,192,193]. The three drinking water studies performed at the American Health Foundation [169,189,190] gave similar tumor yields as the s.c. studies, in spite of the difference in protocols. In the drinking water studies, the NNK $TD_{50}$ (0.03 mmol; 62.1 mg/kg bw) corresponds to a single s.c. dose of 1 mg/kg bw. In the most extensive dose-response study, male F344 rats were treated with NNK (0, 0.03, 0.1, 0.3, 1.0, 10.0, and 50.0 mg/kg bw three times weekly for 20 weeks, s.c.) [164]. The incidence of tumors ranged from 6.7% at the lowest dose up to 87% in the high dose group. Tumor incidence was significantly different from vehicle-treated animals at all doses except at the lowest dose (0.03 mg/kg bw NNK). However, trend analysis of the lower three doses compared with the vehicle control indicated a significant induction of tumors

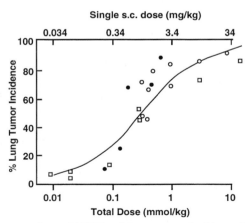

Fig. 9 Relationship between dose of NNK and lung tumor incidence in male F344 rats. Data combined from protocols using s.c. injection, 3 times weekly for 20 weeks, in studies carried out by investigators at the U.S. National Institute of Environmental Health Sciences (□) [164,191] or the American Health Foundation (○) [114,166,167,170,192,193] and by administration in drinking water (●) [169,189,190].

in this dose range. The dose-response curve for lung tumor incidence in this study showed a sharp decline in tumorigenicity as the NNK dose was decreased from 1.0 to 0.3 mg/kg bw. A significant incidence of alveolar hyperplasia was also evident in rats receiving 0.03 to 0.3 mg/kg bw NNK. The incidence of liver tumors increased from 7% in animals treated with 1.0 mg/kg bw NNK to 82% in the high dose group. In animals treated with 10.0 mg/kg bw NNK, the incidence of nasal tumors in the respiratory region and olfactory region was 53% and 26%, respectively. The majority of tumors observed in the respiratory region were benign polyploid adenomas, whereas a higher proportion of carcinomas occurred in the olfactory region. Lower doses of NNK treatment did not induce tumors of the nasal passages. In a later study, the same administration protocol and total dose (6 mg/kg bw, s.c.) induced only pulmonary hyperplasia in male F344/N rats [191]. A sharp increase in tumor induction occurs between total doses of approximately 0.1–0.3 mmol/kg bw, corresponding to single s.c. doses of 0.3–1 mg/kg body weight. A similar increase in $O^6$-mG formation in Clara cells occurs between single s.c. doses of 0.3–1 mg/kg bw, suggesting some critical role for $O^6$-mG in Clara cells in lung tumor induction, probably in combination with pyridyloxobutylation in Type II cells, as discussed above [154,164]. Collectively, these data demonstrate that the NNK rat lung tumor induction model is robust and reproducible.

Sensitive and resistant mouse strains develop lung tumors when treated with NNK, although tumor incidence and multiplicity in resistant strains are lower, and the tumors require a longer time to develop [194]. The A/J mouse is a sensitive strain that has been used extensively for studies of lung tumor induction by NNK. The most commonly used bioassay model is a single dose of NNK (10 μmol; 2.07 mg/mouse, i.p.) resulting in 7–12 lung tumors per mouse after 16 weeks [195]. Hyperplasia is present along the alveolar septa in lungs 14 weeks after treatment [196]. These appear to arise from the Type II cells. Progressively with time post treatment, the frequency of hyperplasia decreases with a subsequent increase in the frequency of adenomas, about 50% of which arise within the hyperplasias. Carcinomas increase in frequency 34 weeks after NNK treatment, eventually comprising 50% of the pulmonary lesions by 54 weeks. About 30% of these arise within adenomas. The model is rapid, reproducible, and lends itself to quantitation. Another advantage is the precision of the dose. The single dose bioassay [195] is a useful model for screening potential chemopreventive agents [35,197,198] and for mechanistic studies since the initiation and promotion/progression stages of carcinogenesis are clearly separated. A second A/J mouse bioassay model involves oral administration of NNK for seven weeks (total dose; 9.1 mg/mouse) [199,200]. A potential advantage of this assay is the use of semi-chronic NNK administration; however, NNK-treated mice lose body weight and the stages of carcinogenesis cannot be separated as readily as in the single-dose model. Other models have been suggested, but less frequently used to date [201–204]. Depending on the bioassay protocol, liver and forestomach tumors are occasionally seen [35,205].

Dose-response studies of NNK in A/J mice demonstrate that there is an inflection in the dose-tumor multiplicity curve which occurs at a total dose of about 3 μmol (0.62 mg)/mouse, with a sharp increase in lung tumor multiplicity at higher doses (Fig. 10) [147,195]. Persistent $O^6$-mG is detectable after a single dose of 2.5 μmol NNK, but not at lower doses, suggesting that the inflection in the dose-response curve is due to saturation of AGT [147]. The critical role of $O^6$-mG in NNK-induced lung tumors in the A/J mouse is strongly indicated by the correlation ($r^2$=0.96) between the levels of persistent $O^6$-mG in lung DNA and tumor multiplicity (Fig. 11) [147]. The central role of $O^6$-mG is also supported by the lower tumor-igenicity of [methylene-$D_2$]NNK than NNK or [methyl-$D_3$]NNK; [methylene-$D_2$]NNK also produces significantly lower $O^6$-mG levels in lung DNA than the other two compounds [206]. Lung tumors induced by NNK in strain A/J mice primarily have Ki-*ras* mutations (GC→AT transitions) at the second base of codon 12 (GGT→GAT) [196], although mutations (AT→GC transitions) have also been found at the second base of codon 61 (CAA→CGA) following treatment with high doses of NNK [207]. The specific GC→AT transitions at the second base of codon 12 are consistent with mispairing of $O^6$-mG with thymine during DNA replication [194,207]. Activation of this gene is not associated with lung tumor formation in the

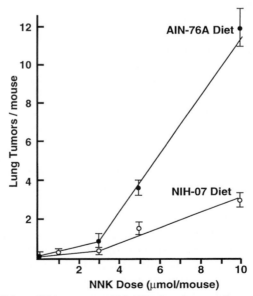

Fig. 10 Tumor multiplicity ±SE in groups of 30 A/J mice given various single i.p. doses of NNK in saline and were killed 4 months later. Mice were maintained on either a AIN-76A diet (●) or NIH-07 diet (○) [195].

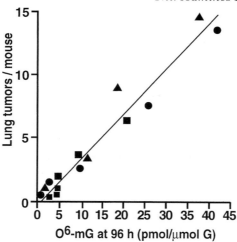

Fig. 11 Relationship between lung tumor multiplicity and concentration of $O^6$-mG in lung DNA 96 h after treatment with NNK (■), (acetoxymethyl)methylnitrosamine (AMMN) (▲), or AMMN plus NNKOAc (●), $r^2 = 0.96$ [147].

F344 rat [208]. In contrast to the rat, pyridyloxobutylation of DNA does not appear to play a direct role in lung tumor induction in the A/J mouse, but inhibits AGT, thus enhancing levels of $O^6$-mG [147,158]. NNK (total dose; 22.8 mg/mouse, i.p.) is a more potent lung carcinogen in the A/J mouse (37.6±11.8 tumors/mouse) than an equimolar dose of NNK-$N$-oxide (3.6±2.7 tumors/mouse) [47].

Although the A/J mouse model uses a precise dose of NNK, and is rapid and reproducible, it still has certain limitations. This model only results in lung adenocarcinomas since small cell and squamous cell carcinomas have not been observed following exposure to any test agent. Furthermore, differences in molecular changes in oncogenes and tumor suppressor genes are also apparent in A/J mouse and human lung adenocarcinomas [209].

In the Syrian golden hamster, lung, trachea, and nasal cavity are the main tumor target tissues of NNK administered s.c. [104,173–176] or applied to the cheek pouch (Hecht et al, unpublished). These respiratory tract tumors are induced following a number of different protocols [104,175,176] including a single dose of NNK (1 mg/kg bw, s.c.) [173]. The lung tumors are predominantly adenomas and adenocarcinomas, with some squamous and adenosquamous carcinomas [104,175,176]. Administration of NNK in the drinking water (1 or 3 ppm) to female hamsters, does not lead to tumors of any type [210]. Administration of NNK by gavage (0.25 mg/wk for 90 weeks) to male Syrian golden hamsters causes a slight increase in lung adenomas (10%), but no treatment-related tumors (Hecht and Lijinsky, unpublished). This contrasts to the sensitivity of the rat to tumor induction by orally administered NNK [169]. The reason for the apparent lower sensitivity of hamsters to NNK carcinogenicity is unknown, but could be related to higher clearance and hepatic metabolism of NNK to NNAL [45] and a low capacity for α-hydroxylation [211].

NNK is a weak transplacental carcinogen in the mouse inducing lung and liver tumors in A/J and C3B6F$_1$ mice (100 mg/kg bw on days 14, 16 and 18 of gestation, i.p.) [205], and outbred Swiss mice (100 mg/kg bw on days 15, 17 and 19 of gestation, i.p.) [205]. In outbred Swiss mice, Aroclor 1254 is a promoter of transplacental lung and liver tumorigenesis in male pups [212]. In neonatal Swiss mice, NNK (50 mg/kg bw on day 4, i.p.) is a weak lung and liver tumorigen, and tumors are promoted by treatment with Aroclor 1254 [212].

In contrast to the mouse, various administration protocols show that NNK is also a multi-organ transplacental carcinogen in the hamster [213–215]. Subcutaneous administration of NNK (50, 100, or 150 mg/kg bw on day 15 of gestation, or 50 and 100 mg/kg bw on days 13, 14 and 15 of gestation) results in respiratory tract tumors, ductular adenoma of the pancreas and hepatocellular carcinoma in hamster pups [213]. Lower doses of NNK (1–20 mg/kg bw on day 15 of gestation, s.c.) result in nasal cavity and adrenal gland tumors in pups [214]. Intratracheal administration of NNK (50 mg/kg bw on day 15 of gestation) induces adrenal tumors only in female pups; coadministration of ethanol (10% v/v) in drinking water on days 5–16 of gestation also resulted in a high incidence of ductular adenocarcinoma of the pancreas in female pups [215]. Lower doses of NNK (0.04–50 mg/kg bw on day 15 of gestation, i.t.) result in nasal cavity and adrenal gland tumors in addition to pancreas tumors in female pups [214].

## 4-(Methylnitrosamino)-1-(3-pyridyl)-1-butanol (NNAL)

NNAL and NNK administered in the drinking water (5.0 ppm for life) give almost identical incidences of 87 and 90% lung tumors in male F344 rats, respectively [169]. NNK, but not NNAL, also induces a significant incidence of nasal cavity and liver tumors, while NNAL induces a significant incidence of exocrine pancreas tumors. Depending on the protocol employed, racemic NNAL induces 30 to 70% as many lung tumors as an equimolar dose of NNK in A/J mice [47,204,206], and is about 10-fold more potent than NNK-*N*-oxide [47]. In the female Syrian golden hamster, no tumors of any type are induced by either NNAL (1 or 3 ppm) in drinking water [216] or similar doses of NNK [210]. Similarly, no treatment-related tumors are induced by either NNAL or NNK (0.25 mg/wk for 90 weeks, i.g.) in male hamsters (Hecht and Lijinsky, unpublished). This again contrasts to the sensitivity of the rat to tumor induction by orally administered NNK or NNAL [169]. NNAL is not a detoxification product of NNK in the rat, and the carcinogenic effects of NNAL could be due, in part, to reconversion to NNK.

## $N'$-Nitrosonornicotine (NNN)

The esophagus and nasal mucosa are the main target tissues of NNN carcinogenicity in the rat; trachea and nasal mucosa in the hamster; and lung in the mouse [35]. Other tumors are seldomly observed in any laboratory animal species.

The relative proportions of esophageal and nasal tumors are strongly affected by the route of administration in the F344 rat. Studies in which NNN administration occurs in drinking water or in a liquid diet result in esophageal and nasal tumors [95,110–112], while s.c. injection results exclusively or predominantly in nasal tumors, with few esophageal tumors [95,113–115]. NNN administration by gavage induces mainly nasal tumors, similar to s.c. treatment [116]. The esophageal tumors are mainly papillomas, but squamous cell carcinomas are sometimes observed [110,111]. Malignant nasal tumors are mainly olfactory neuroblastomas, although rhabdomyosarcomas are also observed [113,116]. NNN and NNK appear to have similar carcinogenic activities toward the rat nasal cavity [114,149]. NNN-$N$-oxide also induces squamous cell carcinomas of the rat nasal cavity, but few esophageal tumors compared to an equimolar dose of NNN (total dose; 120 mg/kg bw, p.o.) [111]. Unlike NNK, NNN very rarely induces lung tumors in the rat [113,114].

Only limited dose-response data are available in the rat. The lowest tested dose of NNN administered by s.c. injection (total dose; 177 mg/kg bw) induces approximately a 50% incidence of nasal tumors, mainly benign [114]. The lowest dose administered by gavage is approximately 0.8 mmol (142 mg/kg bw, p.o.), which induces about a 20% incidence of benign nasal tumors [116]. In a recently completed study, rats treated with NNN (5 ppm in drinking water) developed an esophageal tumor incidence of 71% with a multiplicity of 1.57 tumors/rat [112]. The total dose administered is about one third of the previously reported lowest total dose of 1 mmol administered in a liquid diet (177 mg/rat) [95]. Thus, NNN may be a more potent esophageal carcinogen in the rat than previously realized.

In mice, NNN consistently induces lung tumors regardless of the strain evaluated, although the effect is far weaker than observed with NNK [35]. In the A/J mouse, results are reproducible indicating that a total dose of 17.7 mg/mouse, administered over seven weeks, induces 1–2 lung tumors per mouse [47,217,218]. NNN (total dose; 21.2 mg/mouse, i.p.) is a more potent lung carcinogen in the A/J mouse (1.2±1.3 tumors/mouse) than equimolar doses of NNN-$N$-oxide (0.8±0.7 tumors/mouse) and 3'-hydroxy-NNN (0.9±1.4 tumors/mouse); but as potent as 4'-hydroxy-NNN (1.6±1.5 tumors/mouse) [47].

In hamsters, NNN bioassays also give consistent results for induction of tracheal and nasal tumors independent of the route of administration [104,111,118,119]. The tracheal tumors are squamous papillomas while the nasal tumors arise from the olfactory epithelium and can be invasive [119]. Tumor incidence is low at doses below 1 mmol (177 mg/hamster, i.p. or s.c.) administered over 25 weeks (104,119). NNN (total dose; 159 mg/hamster, p.o.), but not NNN-*N*-oxide (total dose; 173 mg/hamster, p.o.), induces nasal cavity and tracheal tumors [111].

Chronic swabbing of the rat oral mucosa with a mixture of NNN (total dose; 97 mg/rat) and NNK (total dose; 19 mg/rat) produces a significant incidence of oral tumors and some lung tumors [219]. Application of roughly the same dose of NNN and NNK in an aqueous snuff extract produced significantly fewer tumors. Oral swabbing of NNK (total dose; 539 mg/rat) does not produce oral cavity tumors, giving only lung and nasal cavity tumors tumors [172]. NNN has not been tested by oral swabbing in the rat.

## Other tobacco-specific *N*-nitrosamines

NAB (total dose; 1.5–2.2 g/rat in drinking water) is a relatively weak esophageal carcinogen in the Chester Beatty strain albino rat, producing mainly benign tumors [220]. NAB (total dose; 630 mg in drinking water) is also significantly less active than NNN (total dose 630 mg in drinking water) at inducing esophageal tumors in the F344 rat [110]. The weak carcinogenicity of NAB is probably due to its facile excretion in urine as NAB-*N*-oxide and detoxification by 6′-hydroxylation [108]. NAB and NNN have similar activity for lung adenomas in the A/J mouse model [204]. At similar total doses (~370 mg/hamster, s.c.), NNN induces papillary tumors of the trachea in the Syrian golden hamster while NAB is inactive [118].

Other tobacco-specific N-nitrosamines are apparently devoid of carcinogenic activity. NAT (total dose; 191–1,720 mg/kg bw, s.c.) is inactive in the F344 rat [114]. NNA is inactive as a lung tumorigen in the A/J mouse bioassay [217]. *iso*-NNAL shows no activity when applied topically to Sencar mice in a two-stage initiation-promotion model [188], and when administered to A/J mice (total dose; 41.8 mg/mouse, i.p.) [204]. *iso*-NNAC is inactive in the hepatocyte primary culture-DNA repair assay [221] and in the A/J mouse (total dose; 44.6 mg/mouse, i.p.) [204,221,222].

## VII. MODIFICATION OF TOBACCO-SPECIFIC *N*-NITROSAMINE METABOLISM AND CARCINOGENICITY IN LABORATORY ANIMALS

Strain A/J mice maintained on a semi-synthetic AIN-76A diet have significantly higher tumor multiplicities after a single dose of NNK than mice maintained on the cereal based NIH-07 diet (Fig. 10) [195]. However, a similar dietary effect is not observed in rats [192]. Observations of this kind have stimulated considerable interest in the identification of naturally occurring and synthetic chemopreventive agents for NNK-induced lung tumorigenesis in laboratory animals [197,223]. Results of these studies have been reviewed [35], and some of the major findings are presented below, together with studies reporting that cigarette smoke and other smoke components also modify the metabolism of TSNA.

## Diet and dietary constituents

F344 rats maintained on a high-fat diet (23.5% corn oil in AIN-76A diet) and treated with NNK (2 ppm in drinking water) show enhanced lung tumorigenesis compared to rats maintained on a 5% low-fat diet and treated with NNK [189]. Animals maintained on high-fat diets had significantly higher body weights and shorter life spans than animals on the low-fat diet. At 18 months, lung tumors occurred in 27% of rats on the high-fat diet compared to 5% of rats on the low fat diet. A prounounced effect of dietary fat on the pancreas tumor incidence was also observed: 18% of rats on the high-fat diet developed exocrine pancreas tumors compared to 2% of animals on the low-fat diet. Corn oil gavage to male F344 rats is known to increase the incidence of pancreatic tumors [224]. After three weeks under identical dietary conditions to the above, administration of NNK (81 mg/kg bw, s.c.) resulted in significantly enhanced formation of $O^6$-mG in lung DNA of rats maintained on the high-fat diet, but had no effect on $O^6$-mG and 7-mG formation in liver [52]. The significance of these observations is unclear since rats maintained on a high-fat diet do not show any difference compared to animals on a low-fat diet with regard to the distribution pattern of urinary NNK metabolites, or lung and liver microsome metabolism of NNK [52].

Decaffeinated green tea (prepared by supercritical $CO_2$ extraction of caffeine) as the sole source of drinking water, or black tea extract (0.6% in drinking water), fed to A/J mice during treatment with NNK (total dose; 103 mg/kg bw, i.p.) reduces tumor multiplicity by approximately 65%, [225]. When green tea was given after the NNK-treatment period until the end of the study at week 16, tumor incidence and multiplicity were reduced by 30 and 85%, respectively. Black tea extract reduced tumor multiplicity by about 63% but did not have a significant effect on tumor incidence. The inhibitory mechanism of green tea appears to result from suppression of NNK-induced 8-OH-dG in lung DNA, possibly by the tea antioxidant ($-$)-epigallocatechin-3-gallate (EGCG) [181]. EGCG is a potent inhibitor of several CYP isoforms including CYP1A, CYP2B1 and CYP2E1 [226]. Significant inhibition of NNK-induced $O^6$-mdG DNA adduct formation in the A/J mouse lung is not observed on treatment with EGCG, green tea or black tea [181,226]. The inhibitory effect of black tea may be due to theaflavin polyphenols which suppress NNK-induced lung hyperproliferation and tumorigenesis in the A/J mouse [227]. More recent studies have demonstrated that black tea (0.5, 1.0 and 2.0% in drinking water) has the beneficial effect of reducing lung and liver tumorigenesis induced by NNK (total dose; 90 mg/kg bw, s.c.) in a two-year bioassay in the F344 rat [193]. At the highest dose, lung and liver tumors were reduced from 47 to 19% and 34 to 12%, respectively. The tumor incidence in the nasal cavity was not affected. The inhibitory effect of black tea in the F344 rat appears to be attributed, to a significant extent, to caffeine as an active ingredient of tea.

Indole-3-carbinol (I3C), a metabolite of glucobrassicin present in cruciferous vegetables, supplemented to the diet at a concentration of 30 μmol/g, significantly reduces NNK-induced (total dose; 2.4 mg/kg bw, s.c.) DNA methylation in F344 rat lung and nasal mucosa, but increases 7-mG adduct levels in hepatic DNA [168]. I3C does not affect AGT activity in these tissues. Contrary to the effects of I3C, dietary sinigrin (3 μmol/g diet) which is a naturally occuring constituent of cruciferous vegetables decreases hepatic 7-mG adduct levels, but has no effect on 7-mG adduct levels in lung and nasal mucosa. In a two-year bioassay in the F344 rat, dietary sinigrin (3 μmol/g diet) had no effect on NNK (total dose; 105.6 mg/kg bw, s.c.) induced tumorigenesis in lung, liver and nasal mucosa, but induced a significant 16% incidence of pancreatic tumors, an otherwise rare occurrence in F344 rats [224].

Dietary PEITC (3 μmol/g diet) supplementation, before and during administration of NNK (total dose; 105.6 mg/kg bw, s.c.), significantly reduces lung tumor incidence from 80% to 43%, and causes a slight, but not statistically significant, reduction in both liver and nasal cavity tumors [170]. Using a different experimental protocol, PEITC (3 μmol/g diet) totally inhibits lung tumors in F344 rats chronically treated with NNK (total dose; 34.9 mg/kg bw, 2 ppm in drinking water) [190]. PEITC is a primary product of thioglucosidase-catalyzed hydrolysis of gluconasturtiin, a naturally occurring compound found in certain cruciferous vegetables [228]. In rats treated by subcutaneous injection of NNK, DNA 7-mG and pyridyloxobutylation were both decreased by 50% in lung of rats fed PEITC compared to rats maintained on control diet, but the levels of DNA 7-mG were not affected in liver and nasal mucosa [170]. In rats receiving dietary PEITC and chronic oral administration of NNK, a 4- to 6-fold increase in urinary NNAL and NNAL-Gluc excretion occurs together with a 50% reduction in HPB-releasing hemoglobin adducts [190]. PEITC reduces NNK α-hydroxylation in most rat tissues [45,67,91,170,229] except for the nasal mucosa (51); however, the effect appears to be greater in the lung [91,170] compared to the liver [66]. In rats treated with PEITC, lung microsomal CYP2B1 and CYP1A activities are not significantly affected, but they are decreased by 40–50% in nasal microsomes, while liver CYP2B1 and CYP1A activities are increased and CYP2E1 and UDP glucuronosyltransferase activity decreased [71]. Furthermore, treatment resulted in an increase of NAD(P)H: quinone oxidoreductase and glutathione S-transferase activities in liver, but not in lung and nasal mucosa. The progression of benign to malignant pancreatic tumors also appears to be inhibited by PEITC [190]. Further studies in which rats were treated with NNK (1.76 mg/kg bw three times a week, s.c.) for 4, 12 and 20 weeks and received dietary PEITC (3 μmol/g diet) confirmed that treatment with PEITC decreased the rates of NNK and NNAL α-hydroxylation in lung microsomes, but does not significantly affect liver oxidative metabolism of NNK and NNAL, or the rate of NNAL glucuronidation in liver microsomes [66]. PEITC treatment decreased pyridyloxobutyl DNA adducts by 57% in Clara cells, 51% in Type II cells, 40% in small cells, and 44% in whole lung, while $O^6$-mG DNA adducts decreased by 52% in Clara cells, 19% in Type II cells and small cells, and 36% in whole lung [154].

Treatment of F344 rats with the synthetic PEITC analog, 6-phenylhexyl isothiocyanate (PHITC; 1 μmol/g diet), also inhibits lung tumorigenesis induced by NNK (2 ppm in drinking water) [230]. Lung tumor incidence was reduced from 70% to 26% at week 111, but no effect on the total incidence of pancreatic tumors induced by NNK was observed. Significant reductions were also observed in HPB-releasing hemoglobin adduct levels and increased excretion of NNAL and NNAL-Gluc.

PEITC also inhibits NNK-induced lung tumorigenesis in the A/J mouse [89,197,198,231,232]. The mechanism of chemopreventive action in mice is unclear. PEITC modifies the kinetics of NNK metabolism by mouse lung microsomes resulting in no appreciable changes in the apparent $K_m$ and decreased apparent $V_{max}$ values, suggesting a mechanism of competitive inhibition of keto aldehyde, HPB and NNK-*N*-oxide formation, while NNK reduction to NNAL is not affected [72]. Consistent with this, PEITC inhibits NNK-induced $O^6$-mG DNA adduct formation in the lung [231].

A series of synthetic PEITC analogs including 3-phenylpropyl iso-thiocyanate (PPITC), 4-phenylbutyl isothiocyanate (PBITC), 5-phenylpentyl isothiocyanate (PPeITC), PHITC, and 4-(3-pyridyl)butyl isothiocyanate (PyBITC) have been investigated for their inhibition of NNK-induced lung tumorigenesis in the A/J mouse model [197]. PPITC, PBITC, PPeITC, and PHITC are all considerably more potent than PEITC as inhibitors of NNK-induced lung tumorigenesis, and PHITC appears to be the most potent inhibitor resulting in an 85% reduction in tumor multiplicity. PyBITC, an analog of both NNK and PBITC, is ineffective as an inhibitor. In A/J mouse lung microsomes, PEITC, PBITC, PHITC, but not benzyl isothiocyanate (BITC) decrease NNK α-hydroxylation by 40–85% [89]. PBITC and PHITC show the greatest inhibition of NNK α-hydroxylation. Structure-activity relationships of alkyl isothiocyanates demonstrate the essential role of the isothiocyanate functional group for inhibition of lung tumorigenesis [233], and a general trend of increasing inhibition with increasing alkyl chain length [198].

In the only reported study of chemoprevention of NNN-induced esophageal tumors in the F344 rat, administration of NNN (5 ppm in drinking water) and feeding AIN-76A diet containing PPITC (0, 1.0, or 2.5 μmol/g) reduced the 71% incidence of esophageal tumors to 2.9% (1.0 µmol/g PPITC in diet) and 5.9% (2.5 µmol/g PPITC in diet) [112]. In vitro incubation of NNN with esophageal explants from rats fed NIH-07 diet containing 2.5 µmol/g PPITC for one week resulted in over 75% reduction of both 2'- and 5'-hydroxylation of NNN. In the absence of NNN treatment, the incidence of testicular tumors (interstitial cell tumors) was increased from 44% in rats fed control AIN-76A diet to 93% in rats fed AIN-76A diet containing 2.5 µmol/g PPITC. However, the increase in testicular tumors is not believed to be related to treatment with PPITC, since this is a common spontaneous tumor in male F344 rats [224].

In contrast to the above studies, adverse effects of chronic administration of alkyl isothiocyanates have also been reported. PEITC and BITC, at twice the dose used for chemoprevention, induce papillary or nodular hyperplasia, and in some cases carcinoma of the urinary bladder in male F344 rats [234]. In addition to a weak carcinogenic potential, both PEITC and BITC are strong promoters of $N$-nitrosamine-induced bladder cancer. Similarly, PHITC promotes esophageal cancer in male F344 rats [235], and enhances experimental colon carcinogenesis [236]. Phenyl isothiocyanate (PITC), BITC and PEITC exhibit clastogenic effects in vitro in an SV40-transformed Indian muntjac cell line, suggesting that alkyl isothiocyanates may, potentially, have both beneficial and harmful consequences [237].

Various additional agents have been shown to inhibit NNK-induced lung tumorigenesis in the A/J mouse [35]. Of these, olive oil (19.6% in diet) and squalene (2% in diet) significantly decrease lung tumorigenesis induced by NNK (total dose; 103 mg/kg bw, i.p.) by 46 and 58%, respectively [238]. Squalene, a characteristic compound in olive oil, is an inhibitor of 3-hydroxy-3-methylglutaryl coenzyme A reductase activity and has been proposed to inhibit farnesylation of *ras* oncoproteins. Diallyl sulfide, a component of garlic oil, inhibits both methyl and methylene α-hydroxylation and pyridyl $N$-oxidation of NNK in lung and liver microsomes, and NNK-induced lung tumorigenesis [239]. β-Carotene has no effect on NNK-induced lung tumor incidence or multiplicity, but significantly retards malignant progression in the lung of NNK-treated A/J mice [202]. In contrast to this, large intervention trials testing the effect of β-carotene supplementation suggest an increased risk for lung cancer among smokers receiving β-carotene supplementation [240,241]. Vitamin E, at relatively high doses, suppresses proliferating nuclear cell antigen as a marker of cell proliferation, and is suggested to inhibit the development of NNK-induced lung tumorigenesis in the A/J mouse model [242].

The synthetic organoselenium compound, 1,4-phenylenebis(methylene)selenocyanate (*p*-XSC; 7.5 ppm in diet) significantly suppresses lung tumor multiplicity induced by NNK (2.07 mg/mouse, i.p.) [243]. Lower dietary levels (5 ppm in diet) are ineffective [232]. The chemopreventive effect of *p*-XSC is suggested to be partly due to inhibition of oxidative DNA damage and suppression of 8-OH-dG lung DNA adducts [178]. Dietary supplementation with *p*-XSC (5 ppm), PEITC (0.008%), I3C (0.18%), *d*-limonene (0.68%), and their admixture significantly reduces NNK-induced (total dose; 2.07 mg/mouse, i.p.) lung tumor multiplicity from $8.1 \pm 13.7$ tumors/mouse to $3.2 \pm 1.7$, $3.7 \pm 1.6$, $4.9 \pm 2.1$, $2.4 \pm 1.5$, and $2.5 \pm 1.7$ tumors/mouse, respectively [232]. However, neither individual agents nor their admixture had a measurable effect on lung tumor incidence.

## Pharmaceutical drugs

Non-steroidal anti-inflammatory drugs (NSAIDS) represented by acetyl-salicyclic acid (aspirin), naproxen, ibuprofen, piroxicam and sulindac account for about 5% of all prescribed medication in the United States. Lung tumorigenesis in the A/J mouse induced by NNK administered in drinking water for seven weeks (total dose; 9.1 mg/mouse, p.o.) is significantly inhibited by acetylsalicyclic acid (62%) and sulindac (52%) when administered in the diet at doses comparable to the maximum doses recommended for humans [200,244,245]. Inhibition of NNK-induced tumorigenesis by sulindac appears to result through both inhibition of NNK metabolism by CYP1A2 [244], and NNK-induced immunosuppression resulting in increased prostaglandin $E_2$ synthesis [180,245]. NSAIDS have no significant effect on endogenous or NNK-induced formation of 8-OH-dG in the lung, further suggesting that their action is more likely to be related to inhibition of prostaglandin synthesis rather than to direct inhibition of lipid peroxidation and oxidative DNA damage [180]. FGN-1 (sulindac sulfone), a sulindac derivative which lacks cyclooxygenase inhibitory action and associated toxic side effects, almost totally inhibits A/J mouse lung tumorigenesis compared to the 50% reduction seen following treatment with sulindac [246].

Lipoxygenase inhibitors which inhibit 5-lipoxygenase metabolism of arachidonic acid to leukotrienes inhibit NNK-induced lung tumorigenesis in the A/J mouse [38]. Under identical conditions, lipoxygenase inhibitors are as effective as NSAIDS at inhibiting NNK-induced lung tumors. Combination of acetylsalicyclic acid, a nonspecific cyclooxygenase (COX) inhibitor, and the investigational drug A-79715, a lipoxygenase inhibitor, is more effective than single treatment with either agent (38). NS-398, a specific COX-2 inhibitor, is a more effective immunosuppressive agent than either acetylsalicyclic acid or sulindac in preventing NNK-induced lung tumorigenesis in A/J mice [39,245]. NNK is metabolized in vitro by COX-1 and COX-2 suggesting that COX-2-specific inhibitors may be promising chemopreventive agents for future study in animal models [39].

Analogues of 4-ipomeanol, an investigational drug, which is metabolized in Clara cells to a reactive metabolite which covalently binds to tissue macromolecules resulting in toxicity and cell death, inhibit NNK-induced lung tumorigenesis in the A/J mouse [46,247]. These compounds inhibits oxidative metabolism of NNK in both A/J mouse lung and liver microsomes [46]. 4-Ipomeanol and less toxic analogs also strongly inhibit NNK α-hydroxylation and *N*-oxidation in human liver microsomes [62].

## VIII. TOBACCO SMOKE AND OTHER SMOKE CONSTITUENTS

It has been clearly demonstrated that tobacco smoke and tobacco smoke constituents also modify metabolism of TSNA in laboratory animals. Nicotine, NNN and NAT show dose-dependent inhibition of in vitro α-hydroxylation of NNK by rat oral tissue [102]. A 1,400-fold molar excess of either nicotine or cotinine over NNK reduces α-hydroxylation to 77% and 85% of control values in the isolated and perfused rat liver, and increases NNK-*N*-oxide and NNAL-Gluc formation 2.6- and 1.2-fold, respectively [248]. Under the same conditions, neither nicotine nor cotinine influence NNK metabolism in the isolated and perfused rat lung. A 500-fold molar excess of nicotine over NNK significantly reduces in vivo metabolic activation of NNK and excretion of α-hydroxylation metabolites (72% of control; $p < 0.001$) with a shift towards increased detoxification and increased excretion of *N*-oxides (155% of control; $p < 0.01$) and NNAL-Gluc (188% of control; $p < 0.01$) [128]. Acute (28 day) administration of [5-$^3$H]NNK (total dose; 7.2 nmol, 1.7 µg/kg bw) using osmotic mini-pumps, with or without nicotine or cotinine (total dose; 26 µmol/kg), causes a significant 50% reduction of binding of radioactivity (pyridyloxobutylation) to hemoglobin in both nicotine and cotinine treated rats [249].

NNK and coumarin, but not NDMA, competitively inhibit 2'- and 5'-hydroxylation of NNN by rat nasal mucosa microsomes [103]. The observation that coumarin, an inhibitor of CYP2A6, inhibits NNN metabolism suggests that other CYP2A6 substrates such as nicotine [250] and cotinine [251] should also be effective competitive inhibitors of α-hydroxylation of NNN.

A chronic 6-month bioassay in female A/J mice treated with NNK (total dose; 100 mg/kg bw, i.p.) followed by exposure to mainstream cigarette tobacco smoke (6 h/day, 5 days/week for 26 weeks at a mean total particulate matter [TPM] concentration of 248 mg/m$^3$) neither induced nor promoted NNK-induced lung tumorigenesis [252]. Exposure of A/J mice treated with NNK (0, 0.52, 1.04, 1.56 or 2.07 mg/mouse, i.p.) to cotinine (50 µmol/mouse) results in a significant NNK dose-related reduction in lung and liver $O^6$-mG DNA adducts [253]. Similarly, exposure of A/J mice treated with NNK (0, 0.77 or 1.56 mg/mouse, i.p.) nose-only to mainstream cigarette smoke (0, 0.4, 0.6, or 0.8 mg wet TPM/l for 2 hrs) also results in significant NNK and smoke exposure dose-related reductions in lung and liver $O^6$-mG DNA adducts [253]. Cigarette smoke exposure alone failed to yield detectable levels of $O^6$-

mG. In the cigarette smoke-exposed mice (0.6 mg wet TPM/l for 2 hrs), 24 h excretion of α-hydroxylation metabolites (hydroxy acid and keto acid) decreased significantly ($p < 0.05$) to 85% and 58% of control, respectively, while detoxification to the NNAL-Gluc increased significantly to 116% ($p < 0.0001$) of control [254]. Whether TSNA present in cigarette smoke are involved in the etiology of lung tumors in A/J mice exposed to cigarette smoke is doubtful since possible maximum exposure and uptake of NNK and NNN are three to six orders of magnitude below reported doses required to induce a single lung tumor [255].

In the Syrian golden hamster, smoke inhalation also fails to increase the incidence of respiratory tract tumors following administration of NNK (total dose; 1.0, 3.0 and 10.0 mg/kg bw, s.c.) [173]. Nicotine inhibits in vitro α-hydroxylation, pyridine *N*-oxidation and protein binding (pyridyloxo-butylation) in a concentration-dependent manner in hamster lung explants [256]. NNN and NDMA also significantly inhibit in vitro NNK-induced methylation and pyridyloxobutylation of hamster hepatic microsomal proteins, while nicotine inhibits only pyridyloxobutylation [49]. Nicotine also inhibits α-hydroxylation of NNK by rabbit nasal microsomes [81].

Consistent with the above results, nicotine and cotinine significantly reduce the induction of sister chromatid exchange in Chinese hamster ovary cells in vitro by NNK in the presence of metabolic activation [257]. Other in vitro studies show that the mutagenicity of NNK, but not NNN, to *S. typhimurium* strain TA1535 is reduced in a concentration-dependent manner by nicotine, cotinine, and aqueous cigarette smoke extract [257].

## IX. HUMAN STUDIES

### Biomonitoring of urinary excretion

The major NNK metabolites detected in urine of smokers are NNAL and NNAL-Gluc, and a significant correlation exists between total NNAL plus NNAL-Gluc and cotinine excretion [69,129,258–260]. Levels of NNAL-Gluc (0.27–19.0 pmol/mg creatinine) are generally higher than those of NNAL (0.08–7.2 pmol/mg creatinine), although the ratio of NNAL-Gluc:NNAL varies 16-fold among smokers [69]. Excretion of both NNAL and NNAL-Gluc appears to be higher in African-American smokers compared to Caucasian American smokers when adjusted for the number of cigarettes smoked; however, the reasons for this are unclear [258]. Excretion of NNAL and NNAL-Gluc both exceed that of NNAL-*N*-oxide [261]. NNK and NNK-*N*-oxide have not been detected in human urine [69,261].

Based on data for excretion of NNAL and NNAL-Gluc in 24 h urine samples from smokers after smoking cessation, the distribution half-life for NNAL and NNAL-Gluc is 3 to 4 days, with an elimination half-life of 40–50 days [129]. Both compounds show a relatively rapid initial decline in excretion rates followed by a slower terminal phase. Total body clearance of NNAL was estimated to be $61.4\pm35.4$ ml/min, and volume of distribution in the $\beta$-phase estimated to be $3,800\pm2,100$ litres, indicating substantial tissue distribution and the presence of a high-affinity compartment where NNK, NNAL and/or NNAL-Gluc are retained or sequestered and slowly released.

NNAL-Gluc is the major NNK metabolite detected in urine of nonsmokers experimentally exposed to sidestream cigarette smoke [262] and nonsmokers reporting exposure to ETS [263,264]. Total NNAL plus NNAL-Gluc excretion in nonsmokers reporting exposure to ETS ($42.3\pm44.9$ nmol/day) is almost 100-fold lower than that found in smokers of $20.9\pm4.6$ cigarettes/day ($3,242\pm1,954$ nmol/day) [263]. Total NNAL plus NNAL-Gluc excretion correlates significantly with urinary cotinine ($r=0.58$), but not with self-reported hours of ETS exposure ($r=0.08$). NNAL-Gluc has been detected in the urine of 22 of 31 (71%) newborns of smoking mothers (range 0.044–0.40 pmol/ml urine) while NNAL was detectable in four samples (0.031–0.066 pmol/ml urine) [265]. Neither compound was detected in 17 urines of newborns from non-smoking mothers. Thus, NNAL and NNAL-Gluc appear to be sensitive biomarkers for NNK uptake.

Excretion of NNAL, NNAL-$N$-oxide and NNAL-Gluc in urine of smokeless tobacco users is highly variable [266,267]. In extreme cases such as in the use of toombak, excretion of 0.16–0.95 nmol NNAL/mg creatinine and 0.37–1.63 nmol NNAL-Gluc/mg creatinine occurs [266].

Oral administration of I3C (total dose; 400 mg/day) to smoking women resulted in a significant reduction in excretion of both NNAL and NNAL-Gluc, and an increase in the NNAL-Gluc:NNAL ratio in 10 of 13 subjects [268], while dietary supplementation with watercress containing PEITC increased both NNAL and NNAL-Gluc excretion in most but not all smokers [269]. I3C treatment is associated with induction of human CYP1A1 and CYP1A2 [270], but its action on other human CYP isoforms involved in NNK metabolism is unknown. PEITC most likely inhibits NNK metabolism by inhibition of human CYP1A2 [90].

*iso*-NNAC excretion is seldom evident in smokers and there is no evidence for endogenous formation of this compound in humans administered either nicotine or cotinine [17]. Similarly, no evidence exists to support endogenous nitrosation of nicotine to NNK and excretion of NNAL and/or NNAL-Gluc in former smokers using nicotine patches during smoking cessation [129]. The excretion of metabolites derived from other TSNA have not been reported.

Metabolites potentially formed by $\alpha$-methylene hydroxylation of NNK and NNAL (keto acid and hydroxy acid, respectively), and either 2'-hydroxylation of NNN (keto acid) or 5'-hydroxylation (hydroxy acid) cannot be used as urinary biomarkers of NNK and NNN metabolic activation since these metabolites are also formed during nicotine metabolism [271]. Keto acid appears to be extensively converted to hydroxy acid in humans, in contrast to rats [272].

## DNA binding

Levels of $O^6$-mG and 7-mG in human lung have been examined in several studies [273–279]. The average level of $O^6$-mG adducts is reported to be 27 adducts/$10^8$ nucleotides in peripheral lung [273]. Four studies report average 7-mdG levels in lung as $2.1\pm0.9$ [274], $2.5\pm2.3$ [275], $2.7\pm1.4$ [276], and $12\pm10$ [277] adducts per $10^7$ dG. In one study, mean adduct levels were higher in smokers ($17.3/10^7$ dG) than in non-smokers ($4.7/10^7$ dG) [277]. A fifth study reported substantially higher 7-mdG levels in pulmonary alveolar cells ($42.7\pm146.3$ adducts per $10^7$ dG), with a higher prevalence in smokers than in non-smokers [278]. The results of the sixth study were questioned by its authors [279]. It is likely that sources other than tobacco smoke contribute to 7-mdG levels in the lung in both smokers and non-smokers; for example, NDMA is widely distributed in the environment and could be formed endogenously [280].

The average level of HPB-releasing DNA adducts in peripheral lung of smokers was found to be $0.5/10^8$ nucleotides [153]. Based on the results of the largest study measuring 7-mdG in lung, and providing information on smoking [274], the 7-mG:HPB ratio in smokers could be estimated as 41, somewhat higher than the 7.5–25 ratio observed in NNK-treated rats [152]. In another study, 7-mG, but not HPB-releasing adducts, were detected in various regions of different human lungs [275]. Therefore, the quantitative aspects of the relationship between methyl and pyridyloxobutyl adducts in the lungs of smokers are unclear. One study reports that human lung microsomes do not metabolize NNK by α-methyl hydroxylation to a pyridyloxobutylating species [41], while another found only limited metabolism of NNK by α-methyl hydroxylation in human lung tissues [57]. Metabolism of NNN by human lung predominantly results in 5′-hydroxylation, and not in 2′-hydroxylation to yield a pyridyloxobutylating species [57].

## Hemoglobin binding

Hemoglobin adducts of carcinogens are potentially useful as biomarkers of metabolic activation. Advantages of hemoglobin adducts over DNA adducts include the relative ease with which hemoglobin can be obtained in quantity, the lack of repair of adducts, and the relatively long lifetime of red blood cells in humans (120 days), potentially allowing adduct accumulation [281,282]. With these potential advantages in mind, determination of hemoglobin binding of NNK and NNN could provide a way for identifying tobacco- or tobacco smoke-exposed individuals who are particularly adept at activating NNK and NNN, and thus could be at a higher risk for cancer.

Initial biomonitoring studies reported elevated HPB-releasing hemoglobin adduct levels in American snuff dippers ($517\pm538$ fmol HPB/g globin) and smokers (80±189 fmol HPB/g globin) compared to non-smokers (29±26 fmol HPB/g globin) [283]. A large heterogeneity in adduct levels was confirmed in an extended study of smokers (mean 163 fmol HPB/g globin) and non-smokers (mean 68 fmol HPB/g globin) [284]. More recent studies from Germany report lower hemoglobin adduct levels in smokers (69±44 fmol HPB/g globin) and non-smokers (34±16 fmol HPB/g globin) [285], and in pregnant smoking (55±46 fmol HPB/g globin) and non-smoking (27±35 fmol HPB/g globin) women (286). Data obtained by the International Agency for Research on Cancer report even lower hemoglobin adduct levels in smokers (26±12 fmol HPB/g globin) and non-smokers (19±8 fmol HPB/g globin) [287]. No association is found between self-reported ETS exposure and HPB-releasing hemoglobin adduct levels in non-smokers [286]. Thus, HPB-releasing hemoglobin adducts have limited utility as biomarkers of exposure to TSNA in smokers because adduct levels are frequently not much higher than assay background amounts [35], and are unsuitable as biomarkers of ETS exposure in non-smokers [286].

TSNA-derived HPB-releasing hemoglobin adducts in smokers are considerably lower than adducts formed from other tobacco smoke and environmental carcinogens such as BaP [288] and 4-aminobiphenyl [285]. The comparatively low levels of HPB-releasing hemoglobin adducts is probably a consequence of the relative instability of α-hydroxymethylNNK compared to the reactive metabolites formed from BaP and 4-aminobiphenyl. The higher levels of HPB-releasing adducts in snuff-dippers [266,283] compared to smokers [266,283,285–287] may be due to differences in pharmacokinetics between orally absorbed NNK/NNN compared to inhaled substance. Alternatively, other constituents of tobacco smoke, not present in snuff, may inhibit α-methyl hydroxylation of NNK and/or 2'-hydroxylation of NNN.

## X. ANALYSIS OF TSNA AND THEIR METABOLITES

At present, no internationally recognized methodology has been adopted for the analysis of TSNA in either tobacco or tobacco smoke. However, several laboratory methods have been developed using similar techniques for the isolation of NAB, NAT, NNN and NNK from tobacco [9,10,16,18,20,289], mainstream cigarette smoke [4,18,26,28], sidestream cigarette smoke [26] and ETS [31,32]. Care is required using all these methods to avoid artifactual formation of N-nitrosamines during isolation and to incorporate appropriate internal standards. Analysis is mainly performed using gas chromatography (GC) interfaced to a Thermal Energy Analyzer (TEA) [4,9,10,16,18,20,26,28,31,32]. Depending on the matrix to be analyzed, additional modifications to the above methods are required for the isolation of NNAL, iso-NNAL and iso-NNAC which require chemical derivatization prior to GC-TEA analysis [5,17,18]. Gas chromatography – mass spectrometry (GC-MS) is seldom used for routine tobacco and tobacco smoke analysis, but can be used to provide additional analytical confirmation of results obtained by GC-TEA [289].

Metabolic studies in laboratory animals and in vitro systems are normally performed using radio-labeled TSNA which permit analysis by reversed-phase high-performance liquid chromatography (HPLC) using radioflow detection [51,61,91,123,125,128,229]. Stereoselective metabolism of NNK can be determined by HPLC [64] and by HPLC-MS-MS [272]. Highly sensitive methods have been developed for the biomonitoring of the NNK metabolites NNAL and NNAL-Gluc in human urine using both GC-TEA [129,260,265] and GC-MS-MS [264,265].

## SUMMARY AND CONCLUSIONS

While nicotine is not a carcinogen, several tobacco-specific nitrosamines derived from nicotine and other tobacco alkaloids are carcinogenic in laboratory animals, a property characteristic of over 200 nitrosamines [290]. NNK, NNAL, and NNN are strong rodent carcinogens while NAB, NAT, iso-NNAL, and iso-NNAC have little or no activity. The carcinogenicity of NNK, NNAL, and NNN leads to the hypothesis that they may play an important role in human cancer.

In support of this hypothesis, numerous analytical studies, summarized here, clearly demonstrate the presence of tobacco-specific nitrosamines in cured, unburned tobacco as well as in tobacco smoke. Virtually all marketed tobacco products contain these compounds. The uptake of NNK in people who use tobacco products, as well as in non-users exposed to environmental tobacco smoke, has been clearly demonstrated by analysis of NNAL and NNAL-Gluc in urine. A correlation is observed between the levels of these NNK metabolites in urine and the levels of urinary cotinine.

Numerous studies in rodents and primates, both in vitro and in vivo, demonstrate that NNK, NNAL, and NNN are extensively metabolized and form electrophilic intermediates that form covalent adducts with DNA and hemoglobin. The production of these intermediates occurs primarily through CYP catalyzed α-hydroxylation reactions which are major pathways of metabolism in rodents and primates. Characteristic DNA adducts are formed and persist in certain rodent tissues such as lung, nasal mucosa, and esophagus. These studies provide the mechanistic foundation for understanding the carcinogenic activities of tobacco-specific nitrosamines.

Extensive studies clearly demonstrate that NNK is a potent lung carcinogen in laboratory animals. In rats, NNK causes lung tumors independent of the route of administration, a property not shown by any other tobacco carcinogen. In mice, NNK induces lung tumors in both susceptible and resistant strains. Lung tumors are also caused by treatment of hamsters with NNK, and this nitrosamine is a particularly strong transplacental carcinogen in the hamster. The major NNK metabolite, NNAL, is also a strong pulmonary carcinogen in the rat and mouse. Both NNK and NNAL also produce pancreatic tumors in rats, while tumors of the nasal cavity and liver are also observed in NNK treated rats. NNN causes tumors of the esophagus and nasal cavity in rats, trachea and nasal cavity in hamsters, and lung in mice. The results of the carcinogenicity studies of NNK, NNAL, and NNN further support the hypothesis that these nitrosamines may be important in tobacco induced cancer.

The metabolism and carcinogenic activity of NNK and NNN can be strongly influenced by various modifiers including dietary fat; various components of vegetables such as isothiocyanates, indole-3-carbinol, and diallyl sulfide; drugs such as non-steroidal anti-inflammatory agents; tobacco smoke; and nicotine itself. Most studies to date have demonstrated inhibitory activities of such agents, and have stimulated interest in developing chemopreventive agents against NNK and NNN carcinogenesis. Isothiocyanates are particularly important in this respect. But some dietary components, such as fat, can enhance carcinogenicity. The complexities of NNK and NNN carcinogenesis in the context of mixtures, in particular in the presence of tobacco or tobacco smoke, have not been evaluated.

While the metabolism and carcinogenicity of NNK, NNAL, and NNN have been thoroughly studied in rodents, investigations in humans have been more limited, due in part to the inherent ethical difficulties of such studies. The available data demonstrate that most metabolic pathways operative in rodents are also observed in humans, although there are quantitative differences. Levels of α-hydroxylation metabolic activation pathways as well as levels of DNA adducts are generally less in human tissues than in rodent tissues. While the structural similarities of NNK, NNAL, and NNN to nicotine indicate that these nitrosamines, like nicotine, should be extensively metabolized in humans, this has been difficult to demonstrate so far, due in part to the identical structures of nicotine and tobacco-specific nitrosamine metabolites. Further studies in human systems are required before a more complete assessment can be made as to the involvement of NNK, NNAL, and NNN in the genesis of human cancer.

## ACKNOWLEDGEMENTS

Research in Dr. Hecht's laboratory is supported by grants CA-44377, CA-81301, and CA-46535 from The U.S. National Cancer Institute.

# REFERENCES

1 Hecht SS, Chen CB, Ornaf RM, Jacobs E, Adams JD, Hoffmann D. Reaction of nicotine with sodium nitrite: formation of nitrosamines and fragmentation of the pyrrolidine ring. *J. Org. Chem.*, 1978: **43**; 72–6.

2 Caldwell WS, Greene JM, Plowchalk DR, deBethizy JD. The nitrosation of nicotine: a kinetic study. *Chem. Res. Toxicol.*, 1991: **4**; 513–7.

3 Smith PAS, Loeppky RN. Nitrosative cleavage of tertiary amines. *J. Am. Chem. Soc.*, 1967: **89**; 1147–57.

4 Hoffmann D, Brunnemann KD, Prokopczyk B, Djordjevic MV. Tobacco-specific *N*-nitrosamines and *Areca*-derived *N*-nitrosamines: chemistry, biochemistry, carcinogenicity and relevance to humans. *J. Toxicol. Environ. Health*, 1994: **41**; 1–52.

5 Djordjevic MV, Sigountos CW, Brunnemann KD, Hoffmann D. Formation of 4-(methylnitrosamino)-1-(3-pyridyl)butyric acid in vitro and in mainstream cigarette smoke. *J. Agric. Food Chem.*, 1991: **39**; 209–13.

6 Mirvish SS, Sams J, Hecht SS. Kinetics of nornicotine and anabasine nitrosation in relation to *N′*-nitrosonornicotine occurrence in tobacco and to tobacco-induced cancer. *J. Natl. Cancer Inst.*, 1977: **59**; 1211–3.

7 Hoffmann D, Adams JD, Brunnemann KD, Hecht SS. Assessment of tobacco-specific *N*-nitrosamines in tobacco products. *Cancer Res.*, 1979: **39**; 2505–9.

8 Chamberlain WJ, Chortyk OT. Effect of curing and fertilization on nitrosamine formation in bright and Burley tobacco. *Beitr. Tabakforsch.*, 1992: **15**; 87–92.

9 Fischer S, Spiegelhalder B, Preussmann R. Preformed tobacco-specific nitrosamines in tobacco – role of nitrate and influence of tobacco type. *Carcinogenesis*, 1989: **10**; 1511–7.

10 Burton HR, Dye NK, Bush LP. Relationship between tobacco-specific nitrosamines and nitrite from different air-cured tobacco varieties. *J. Agric. Food Chem.*, 1994: **42**; 2007–11.

11 Andersen RA, Fleming PD, Burton HR, Hamilton-Kemp TR, Sutton TG. *N′*-Acyl and *N′*-nitroso pyridine alkaloids in alkaloid lines of Burley tobacco during growth and air curing. *J. Agric. Food Chem.*, 1989: **37**; 44–50.

12 Mirvish SS. Formation of *N*-nitroso compounds: chemistry, kinetics and in vivo occurrence. *Toxicol. Appl. Pharmacol.*, 1975: **31**; 325–51.

13 Klimisch H-J, Stadler L. Investigation of the formation of *N*-nitrosonornicotine from nicotine-*N*-oxide. *Talanta*, 1976: **23**; 614–6.

14 Maeda S, Uchida S, Kisaki T. Microbial degradation of nicotine-*N′*-oxide. I. Degradation products. *Agric. Biol. Chem.*, 1978: **42**; 1455–60.

15 Caldwell WS, Greene JM, Dobson GP, deBethizy JD. Intragastric nitrosation of nicotine is not a significant contributor to nitrosamine exposure. *Ann. NY Acad. Sci.*, 1993: **686**; 213–28.

16 Prokopczyk B, Wu M, Cox JE, Amin S, Desai D, et al. Improved methodology for the quantitative assessment of tobacco-specific *N*-nitrosamines in tobacco by supercritical fluid extraction. *J. Agric. Food Chem.*, 1995: **43**; 916–22.

17 Tricker AR, Scherer G, Conze C, Adlkofer F, Pachinger A, Klus H. Evaluation of 4-(*N*-methylnitrosamino)-4-(3-pyridyl)butyric acid as a potential monitor of endogenous nitrosation of nicotine and metabolites. *Carcinogenesis*, 1993: **14**; 1409–14.

18 Tricker AR, Dietrich C, Preussmann R. *N*-Nitroso compounds in cigarette tobacco and their occurrence in mainstream cigarette smoke. *Carcinogenesis*, 1991: **12**; 257–61.

19 Tricker AR, Preussmann R. Occurrence of and exposure to *N*-nitroso compounds in tobacco. In: *Relevance to Human Cancer of N-Nitroso Compounds, Tobacco Smoke and Mycotoxins*, O'Neill IK, Chen J, Bartsch H, (Eds.), International Agency for Research on Cancer, Lyon, France, 1991, pp. 493–5.

20 Hoffmann D, Djordjevic MV, Fan J, Zhang E, Glynn T, Connolly GN. Five leading U.S. commercial brands of moist snuff in 1994: assessment of carcinogenic *N*-nitrosamines. *J. Natl. Cancer Inst.*, 1995: **87**; 1862–9.

21 Adams JD, Lee SJ, Hoffmann D. Carcinogenic agents in cigarette smoke and the influence of nitrate on their formation. *Carcinogenesis*, 1984: **5**; 221–3.

22 Fischer S, Spiegelhalder B, Eisenbarth J, Preussmann R. Investigations on the origin of tobacco-specific nitrosamines in mainstream smoke of cigarettes. *Carcinogenesis*, 1990: **11**; 723–30.

23 Adams JD, Lee SJ, Vinchkoski N, Castonguay A, Hoffmann D. On the formation of the tobacco-specific carcinogen 4-(methylnitrosamino)-1-(3-pyridyl)-1-butanone during smoking. *Cancer Lett.*, 1983: **17**; 339–46.

24 Hoffmann D, Dong M, Hecht SS. Origin in tobacco smoke of *N*'-nitrosonornicotine, a tobacco-specific carcinogen. *J. Natl. Cancer Inst.*, 1978: **58**; 1841–4.

25 Hoffmann D, Hecht SS. Nicotine-derived *N*-nitrosamines and tobacco-related cancer: current status and future directions. *Cancer Res.*, 1985: **45**; 935–44.

26 Adams JD, O'Mara-Adams KJ, Hoffmann D. Toxic and carcinogenic agents in undiluted mainstream smoke and sidestream smoke of different types of cigarettes. *Carcinogenesis*, 1987: **8**; 729–31.

27 Atawodi SE, Preussmann R, Spiegelhalder B. Tobacco-specific nitrosamines in some Nigerian cigarettes. *Cancer Lett.*, 1995: **97**; 1–6.

28 Fischer S, Castonguay A, Kaiserman M, Spiegelhalder B, Preussmann. Tobacco-specific nitrosamines in Canadian cigarettes. *J. Cancer Res. Clin. Oncol.*, 1990: **116**; 563–8.

29 Spiegelhalder B, Bartsch H. Tobacco-specific nitrosamines. *Eur. J. Cancer Prev.*, 1996: **5(Suppl. 1)**; 33–8.

30 Nair J, Pakhale SS, Bhide SV. Carcinogenic tobacco-specific nitrosamines in Indian tobacco products. *Fd. Chem. Toxic.*, 1989: **27**; 751–3.

31 Klus H, Begutter H, Scherer G, Tricker AR, Adlkofer F. Tobacco-specific and volatile *N*-nitrosamines in environmental tobacco smoke of offices. *Indoor Environ.*, 1992: **1**; 348–50.

32 Brunnemann KD, Cox JE, Hoffmann D. Analysis of tobacco-specific *N*-nitrosamines in indoor air. *Carcinogenesis*, 1992: **13**; 2415–8.

33 Idris AM, Nair J, Friesen M, Ohshima H, Brouet I, et al. Carcinogenic tobacco-specific nitrosamines are present at unusually high levels in the saliva of oral snuff users in Sudan. *Carcinogenesis*, 1992: **13**; 1001–5.

34 Hinds W, First MW, Huber GL, Shea JW. A method for measuring respiratory deposition of cigarette smoke during smoking. *Am. Ind. Hyg. Assoc. J.*, 1983: **44**; 113–8.

35 Hecht SS. Biochemistry, biology, and carcinogenicity of tobacco-specific *N*-nitrosamines. *Chem. Res. Toxicol.*, 1998: **11**; 559–603.

36 Lamoureux J, Castonguay A. Absence of metabolism of 4-(methylnitrosamino)-1-(3-pyridyl)-1-butanone (NNK) by flavin-containing monoxygenase (FMO). *Carcinogenesis*, 1997: **18**; 1979–84.

37 Smith TJ, Stoner GD, Yang CS. Activation of 4-(methylnitrosamino)-1-(3-pyridyl)-1-butanone (NNK) in human lung microsomes by cytochromes P450, lipoxygenase, and hydroperoxides. *Cancer Res.*, 1995: **55**; 5566–73.

38 Rioux N, Castonguay A. Inhibitors of lipoxygenase: a new class of cancer chemopreventive agents. *Carcinogenesis*, 1998: **19**; 1393–400.

39 Rioux N, Castonguay A. Prevention of NNK-induced lung tumorigenesis in A/J mice by acetylsalicylic acid and NS–398. *Cancer Res.*, 1998: **58**; 5354–60.

40 Patten CJ, Smith TJ, Murphy SE, Wang M-H, Lee J, et al. Kinetic analysis of the activation of 4-(methylnitrosamino)-1-(3-pyridyl)-1-butanone by heterologously expressed human P450 enzymes and the effect of P450-specific chemical inhibitors on this activation in human liver microsomes. *Arch. Biochem. Biophys.*, 1996: **333**; 127–38.

41 Smith TJ, Guo ZY, Gonzalez FJ, Guengerich FP, Stoner GD, Yang CS. Metabolism of 4-(methylnitrosamino)-1-(3-pyridyl)-1-butanone in human lung and liver microsomes and cytochromes P–450 expressed in hepatoma cells. *Cancer Res.*, 1992: **52**; 1757–63.

42 Crespi CL., Penman, B. W., Gelboin, H. V., and Gonzalez, F. A tobacco smoke-derived nitrosamine, 4-(methylnitrosamino)-1-(3-pyridyl)-1-butanone, is activated by multiple human cytochrome P450s including the polymorphic human cytochrome P4502D6. *Carcinogenesis*, 1991: **12**; 1197–201.

43 Lo Guidice J-M, Marez D, Sabbagh N, Legrand-Andreoletti M, Spire C, et al. Evidence for CYP2D6 expression in human lung. *Biochem. Biophys. Res. Comm.*, 1997: **241**; 79–85.

44 Hecht SS, Young R, Chen CB. Metabolism in the F344 rat of 4-(*N*-methyl-*N*-nitrosamino)-1-(3-pyridyl)-1-butanone, a tobacco specific carcinogen. *Cancer Res.*, 1980: **40**; 4144–50.

45 Hamilton SM, Teel RW. Effects of isothiocyanates on the metabolism of 4-(methylnitrosamino)-1-(3-pyridyl)-1-butanone (NNK) and benzo[*a*]pyrene by hamster and rat liver microsomes. *Anticancer Res.*, 1994: **14**; 1089–94.

46 Lin J-M, Desai DH, Morse MA, Amin S, Hecht SS. Inhibition of 4-(methylnitrosamino)-1-(3-pyridyl)-1-butanone pulmonary metabolism and tumorigenicity in mice by analogues of the investigational chemotherapeutic drug 4-ipomeanol. *Chem. Res. Toxicol.*, 1992: **5**; 674–9.

47 Castonguay A, Lin D, Stoner GD, Radok P, Furuya K et al. Comparative carcinogenicity in A/J mice and metabolism by cultured mouse peripheral lung of *N'*-nitrosonornicotine, 4-(methylnitrosamino)-1-(3-pyridyl)-1-butanone and their analogues. *Cancer Res.*, 1983: **43**; 1223–9.

48 Peterson LA, Mathew R, Hecht SS. Quantitation of metabolic α-hydroxylation of the tobacco-specific nitrosamine, 4-(methylnitrosamino)-1-(3-pyridyl)-1-butanone. *Cancer Res.*, 1991: **51**; 5495–500.

49 Castonguay A, Rossignol G. Modulation of the activation of 4-(methylnitrosamino)-1-(3-pyridyl)-1-butanone by hamster lung microsomes to protein alkylating species. *Toxic in Vitro*, 1992: **6**; 397–404.

50 Carmella SG, Hecht SS. High-performance liquid chromatographic analysis of metabolites of the nicotine-derived nitrosamines, *N'*-nitrosonornicotine and

4-(methylnitrosamino)-1-(3-pyridyl)-1-butanone. *Anal. Biochem.*, 1985: **145**; 239–44.

51 Staretz ME, Hecht SS. Effects of phenethyl isothiocyanate on the tissue distribution of 4-(methylnitrosamino)-1-(3-pyridyl)-1-butanone and metabolites in F344 rats. *Cancer Res.*, 1995: **55**; 5580–8.

52 El-Bayoumy K, Prokopczyk B, Peterson LA, Desai D, Amin S et al. Effects of dietary fat content on the metabolism of NNK and DNA methylation induced by NNK. *Nutr. Cancer.*, 1996: **26**; 1–10.

53 Peterson LA, Ng DK, Stearns RA, Hecht SS. Formation of NADP(H) analogs of tobacco specific nitrosamines in rat liver and pancreatic microsomes. *Chem. Res. Toxicol.*, 1994: **7**; 599–608.

54 Brittebo EB, Castonguay A, Furuya K, Hecht SS. Metabolism of tobacco specific nitrosamines by cultured rat nasal mucosa. *Cancer Res.*, 1983: **43**; 4343–8.

55 Hong J-Y, Smith T, Lee M-J, Li W, Ma B-L et al. Metabolism of carcinogenic nitrosamines by rat nasal mucosa and the effect of diallyl sulfide. *Cancer Res.*, 1991: **51**; 1509–14.

56 Smith TJ, Liao AM, Liu Y, Butler-Jones A, Anderson LM et al. Enzymes involved in the bioactivation of 4-(methylnitrosamino)-1-(3-pyridyl)-1-butanone in patas monkey lung and liver microsomes. *Carcinogenesis*, 1997: **18**; 1577–84.

57 Castonguay A, Stoner GD, Schut HAJ, Hecht SS. Metabolism of tobacco-specific N-nitrosamines by cultured human tissues. *Proc. Natl. Acad. Sci., USA*, 1983: **80**; 6694–7.

58 Liu Y, Sundqvist K, Belinsky SA, Castonguay A, Tjälve H, Grafström RC. Metabolism and macromolecular interaction of the tobacco-specific carcinogen 4-(methylnitrosamino)-1-(3-pyridyl)-1-butanone in cultured explants and epithelial cells of human buccal mucosa. *Carcinogenesis*, 1993: **14**; 2383–8.

59 Smith TJ, Liao A, Wang L-D, Yang G, Starcic S et al. Characterization of xenobiotic-metabolizing enzymes and nitrosamine metabolism in the human esophagus. *Carcinogenesis*, 1998: **19**; 667–72.

60 Anderson KE, Hammons GJ, Kadlubar FF, Potter JD, Kaderlik KR et al. Metabolic activation of aromatic amines by human pancreas. *Carcinogenesis*, 1997: **18**; 1085–92.

61 Staretz ME, Murphy SE, Nunes MG, Koehl W, Amin S et al. Comparative metabolism of the tobacco smoke carcinogens benzo[*a*]pyrene, 4-(methylnitrosamino)-1-(3-pyridyl)-1-butanone, 4-(methylnitrosamino)-1-(3-pyridyl)-1-butanol, and *N'*-nitrosonornicotine in human hepatic microsomes. *Drug Metab. Dispos.*, 1997: **25**; 154–62.

62 Nunes MG, Desai D, Koehl W, Spratt TE, Guengerich FP, Amin S. Inhibition of 4-(methylnitrosamino)-1-(3-pyridyl)-1-butanone (NNK) metabolism in human hepatic microsomes by ipomeanol analogs – an exploratory study. *Cancer Lett.*, 1998: **129**; 131–8.

63 Murphy SE, Coletta KA. Two types of 4-(methylnitrosamino)-1-(3-pyridyl)-1-butanone hemoglobin adducts, from metabolites which migrate into or are formed in red blood cells. *Cancer Res.*, 1993: **53**; 777–83.

64 Hecht SS, Spratt TE, Trushin N. Absolute configuration of 4-(methylnitrosamino)-1-(3-pyridyl)-1-butanol (NNAL) formed metabolically from 4-(methylnitrosamino)-1-(3-pyridyl)-1-butanone (NNK). *Carcinogenesis*, 1997: **18**; 1851–5.

65  Murphy SE, Nunes MG, Hatala MA. Effects of phenobarbital and 3-methylcholanthrene induction on the formation of three glucuronide metabolites of 4-(methylnitrosamino)-1-(3-pyridyl)-1-butanone, NNK. *Chem-Biol. Interact.*, 1997: **103**; 153–66.

66  Staretz ME, Koenig L, Hecht SS. Effects of long term phenethyl isothiocyanate treatment on microsomal metabolism of 4-(methylnitrosamino)-1-(3-pyridyl)-1-butanone and 4-(methylnitrosamino)-1-(3-pyridyl)-1-butanol in F344 rats. *Carcinogenesis*, 1997: **18**; 1715–22.

67  Murphy SE., Heiblum S, King PG, Bowman D, Davis WJ, Stoner GD. Effect of phenethyl isothiocyanate on the metabolism of tobacco-specific nitrosamines by cultured rat oral tissue. *Carcinogenesis*, 1991: **12**; 957–61.

68  Murphy SE, Spina DA, Nunes MG, Pullo DA. Glucuronidation of 4-(hydroxymethyl)nitrosamino-1-(3-pyridyl)-1-butanone, a metabolically activated form of 4-(methylnitrosamino)-1-(3-pyridyl)-1-butanone, by phenobarbital treated rats. *Chem. Res. Toxicol.*, 1995: **8**; 772–9.

69  Carmella SG, Akerkar SA, Richie Jr JP, Hecht SS. Intra-individual and inter-individual differences in metabolites of the tobacco-specific lung carcinogen 4-(methylnitrosamino)-1-(3-pyridyl)-1-butanone in smokers' urine. *Cancer Epidemiol Biomarkers & Prev.*, 1995: **4**; 635–42.

70  Guo Z, Smith TJ, Thomas PE, Yang CS. Metabolism of 4-(methylnitrosamino)-1-(3-pyridyl)-1-butanone by inducible and constitutive cytochrome P450 enzymes in rats. *Arch. Biochem. Biophys.*, 1992: **298**; 279–86.

71  Guo Z, Smith TJ, Wang E, Sadrieh N, Ma Q et al. Effects of phenethyl isothiocyanate, a carcinogenesis inhibitor, on xenobiotic-metabolizing enzymes and nitrosamine metabolism in rats. *Carcinogenesis*, 1992: **13**; 2205–10.

72  Smith TJ, Guo Z, Li C, Ning SM, Thomas PE, Yang CS. Mechanisms of inhibition of 4-(methylnitrosamino)-1-(3-pyridyl)-1-butanone bioactivation in mouse by dietary phenethyl isothiocyanate. *Cancer Res.*, 1993: **53**; 3276–82.

73  Maser E, Richter E, Friebertshauser J. The identification of 11β-hydroxysteriod dehydrogenese as carbonyl reductase of the tobacco-specific nitrosamine 4-(methylnitrosamino)-1-(3-pyridyl)-1-butanone. *European J. Biochem.*, 1996: **238**; 484–9.

74  Maser E. 11β-hydroxysteroid dehydrogenase responsible for carbonyl reduction of the tobacco-specific nitrosamine 4-(methylnitrosamino)-1-(3-pyridyl)-1-butanone in mouse lung microsomes. *Cancer Res.*, 1998: **58**; 2996–3003.

75  Maser E. Stress, hormonal changes, alcohol, food constituents and drugs: factors that advance the incidence of tobacco smoke-related cancer? *TIPS*, 1997: **18**; 270–5.

76  Hecht SS, Trushin N. DNA and hemoglobin alkylation by 4-(methylnitrosamino)-1-(3-pyridyl)-1-butanone and its major metabolite 4-(methylnitrosamino)-1-(3-pyridyl)-1-butanol in F344 rats. *Carcinogenesis*, 1988: **9**; 1665–8.

77  Jorquera R, Castonguay A, Schuller HM. Effects of pregnancy and ethanol treatment on the metabolism of 4-(methylnitrosamino)-1-(3-pyridyl)-1-butanone by hamster liver and lung microsomes. *Drug Metab. Dispos.*, 1992: **20**; 510–7.

78  Belinsky SA, White CM, Trushin N, Hecht SS. Cell specificity for the pulmonary metabolism of tobacco-specific nitrosamines in the Fisher rat. *Carcinogenesis*, 1989: **10**; 2269–74.

79 Smith TJ, Guo Z, Hong J-Y, Ning SM, Thomas PE, Yang CS. Kinetics and enzyme involvement in the metabolism of 4-(methylnitrosamino)-1-(3-pyridyl)-1-butanone (NNK) in microsomes of rat lung and nasal mucosa. *Carcinogenesis*, 1992: **13**; 1409–14.

80 Guo Z, Smith TJ, Thomas PE, Yang CS. Metabolic activation of 4-(methylnitrosamino)-1-(3-pyridyl)-1-butanone as measured by DNA alkylation in vitro and its inhibition by isothiocyanates. *Cancer Res.*, 1991: **51**; 4798–803.

81 Hong J-Y, Ding X, Smith TJ, Coon MJ, Yang CS. Metabolism of 4-(methylnitrosamino)-1-(3-pyridyl)-1-butanone (NNK), a tobacco-specific carcinogen, by rabbit nasal microsomes and cytochrome P450s NMa and NMb. *Carcinogenesis*, 1992: **13**; 2141–4.

82 Castonguay A, Tjälve H, Hecht SS. Tissue distribution of the tobacco-specific carcinogen 4-(methylnitrosamino)-1-(3-pyridyl)-1-butanone, and its metabolites in F344 rats. *Cancer Res.*, 1983: **43**; 630–8.

83 Tjälve H, Castonguay A. The in vivo tissue disposition and in vitro target-tissue metabolism of the tobacco-specific carcinogen 4-(methylnitrosamino)-1-(3-pyridyl)-1-butanone in Syrian golden hamsters. *Carcinogenesis*, 1983: **4**; 1259–65.

84 Devereux TR, Anderson MW, Belinsky SA. Factors regulating activation and DNA alkylation by 4-(N-methyl-N-nitrosamino)-1-(3-pyridyl)-1-butanone and nitrosodimethylamine in rat lung and isolated lung cells, and their relationship to carcinogenicity. *Cancer Res.*, 1988: **48**; 4215–21.

85 Guo Z, Smith TJ, Ishizaki H, Yang CS. Metabolism of 4-(methylnitrosamino)-1-(3-pyridyl)-1-butanone (NNK) by cytochrome P450IIB1 in a reconstituted system. *Carcinogenesis*, 1991: **12**; 2277–82.

86 Castonguay A, Pepin P, Briere N. Modulation of 4-(methylnitrosamino)-1-(3-pyridyl)-1-butanone demethylation and denitrosation by rat liver microsomes. *Cancer Lett.*, 1991: **59**; 67–74.

87 Smith TJ, Guo Z, Thomas PE, Chung F-L, Morse MA et al. Metabolism of 4-(N-methyl-N-nitrosamino)-1-(3-pyridyl)-1-butanone in mouse lung microsomes and its inhibition by isothiocyanates. *Cancer Res.*, 1990: **50**; 6817–22.

88 Zhang Z, Huynh H, Teel RW. Effects of orally administered capsaicin, the principal component of capsicum fruits, on the in vitro metabolism of the tobacco-specific nitrosamine NNK in hamster lung and liver microsomes. *Anticancer Res.*, 1997: **17**; 1093–8.

89 Guo Z, Smith TJ, Wang E, Eklind KI, Chung F-L, Yang CS. Structure-activity relationships of arylalkyl isothiocyanates for the inhibition of 4-(methylnitrosamino)-1-(3-pyridyl)-1-butanone metabolism and the modulation of xenobiotic-metabolizing enzymes in rats and mice. *Carcinogenesis*, 1993: **14**; 1167–73.

90 Smith TJ, Guo Z, Guengerich FP, Yang CS. Metabolism of 4-(methylnitrosamino)-1-(3-pyridyl)-1-butanone (NNK) by human cytochrome P450 1A2 and its inhibition by phenethyl isothiocyanate. *Carcinogenesis*, 1996: **17**; 809–13.

91 Doerr-O'Rourke K, Trushin N, Hecht SS, Stoner GD. Effect of phenethyl isothiocyanate on the metabolism of the tobacco-specific nitrosamine 4-(methylnitrosamino)-1-(3-pyridyl)-1-butanone by cultured rat lung tissue. *Carcinogenesis*, 1991: **12**; 1029–34.

92 Schrader E, Hirsh-Ernst KI, Richter E, Foth H. Metabolism of 4-(methylnitrosamino)-1-(3-pyridyl)-1-butanone (NNK) in isolated rat lung and liver. *Naunyn-Schmiedeberg's Arch. Pharmacol.*, 1998: **357**; 336–43.

93  Liu L, Alaoui-Jamali A, El Alami N, Castonguay A. Metabolism and DNA single strand breaks induced by 4-(methylnitrosamino)-1-(3-pyridyl)-1-butanone and its analogues in primary culture of rat hepatocytes. *Cancer Res.*, 1990: **50**; 1810–6.

94  Lacroix D, Desrochers M, Castonguay A, Anderson A. Metabolism of 4-(methyl-nitrosamino)-1-(3-pyridyl)-1-butanone (NNK) in human kidney epithelial cells transfected with rat CYP2B1 cDNA. *Carcinogenesis*, 1993: **14**; 1639–42.

95  Castonguay A, Rivenson A, Trushin N, Reinhardt J, Spathopoulos S et al. Effects of chronic ethanol consumption on the metabolism and carcinogenicity of *N'*-nitrosonornicotine in F344 rats. *Cancer Res.*, 1984: **44**; 2285–90.

96  Hecht SS, Chen CB, Hoffmann D. Metabolic β-hydroxylation and *N'*-oxidation of N-nitrosonornicotine. *J. Med. Chem.*, 1980: **23**; 1175–8.

97  Chen CB, Hecht SS, Hoffmann D. Metabolic α-hydroxylation of the tobacco specific carcinogen, *N'*-nitrosonornicotine. *Cancer Res.*, 1978: **38**; 3639–45.

98  McCoy GD, Chen CB, Hecht SS. Influence of mixed function oxidase inducers on the in vitro metabolism of *N'*-nitrosonornicotine by rat and hamster liver microsomes. *Drug Metab. Dispos.*, 1981: **9**; 168–9.

99  Murphy SE, Spina DA. Evidence for a high affinity enzyme in rat esophageal microsomes which α-hydroxylates *N'*-nitrosonornicotine. *Carcinogenesis*, 1994: **15**; 2709–13.

100  Murphy SE, Heiblum R, Trushin N. Comparative metabolism of *N'*-nitrosonor-nicotine and 4-(methylnitrosamino)-1-(3-pyridyl)-1-butanone by cultured rat oral tissue and esophagus. *Cancer Res.*, 1990: **50**; 4685–91.

101  Hecht SS, Reiss B, Lin D, Williams GM. Metabolism of *N'*-nitrosonornicotine by cultured rat esophagus. *Carcinogenesis*, 1982: **3**; 453–6.

102  Murphy SE, Heiblum R. Effect of nicotine and tobacco-specific nitrosamines on the metabolism of *N'*-nitrosonornicotine and 4-(methylnitrosamino)-1-(3-pyridyl)-1-butanone by rat oral tissue. *Carcinogenesis*, 1990: **11**; 1663–6.

103  Patten C, Peterson LA, Murphy SE. Evidence for metabolic activation of *N'*-nitrosonornicotine and N-nitrosobenzylmethylamine by rat nasal coumarin hydroxylase. *Drug Met. Dispos.*, 1998: **26**; 177–80.

104  Hoffmann D, Castonguay A, Rivenson A, Hecht SS. Comparative carcinogenicity and metabolism of 4-(methylnitrosamino)-1-(3-pyridyl)-1-butanone and *N'*-nitrosonornicotine in Syrian golden hamsters. *Cancer Res.*, 1981: **41**; 2386–93.

105  Hecht SS, Chen CB, McCoy GD, Hoffmann D, Domellöf L. α-Hydroxylation of N-nitrosopyrrolidine and *N'*-nitrosonornicotine by human liver microsomes. *Cancer Lett.*, 1979: **8**; 35–41.

106  Patten C, Smith T, Friesen M, Tynes R, Yang C, Murphy S. Evidence for cytochrome P450 2A6 and 3A4 as major catalysts for *N'*-nitrosonornicotine α-hydroxylation by human liver microsomes. *Carcinogenesis*, 1997: **18**; 1623–30.

107  Yamazaki H, Inui Y, Yun C-H, Guengerich FP, Shimada T. Cytochromes P450 2E1 and 2A6 enzymes as major catalysts for metabolic activation of N-nitrosodialk-ylamines and tobacco-related nitrosamines in human liver microsomes. *Carcinogenesis*, 1992: **13**; 1789–94.

108  Hecht SS, Young R. Regiospecificity in the metabolism of the homologous cyclic nitrosamines, *N'*-nitrosonornicotine and *N'*-nitrosoanabasine. *Carcinogenesis*, 1982: **3**; 1195–9.

109 Hecht SS, Lin D. Comparative mutagenicity of 4-(carbethoxynitrosamino)–4-(3-pyridyl)butanal and 4-(carbethoxynitrosamino)-1-(3-pyridyl)-1-butanone, model compounds for α-hydroxylation of *N'*-nitrosonornicotine. *Carcinogenesis*, 1986: **7**; 611–4.

110 Hoffmann D, Raineri R, Hecht SS, Maronpot R, Wynder EL. Effects of *N'*-nitrosonornicotine and *N'*-nitrosoanabasine in rats. *J. Natl. Cancer Inst.*, 1975: **55**; 977–81.

111 Hecht SS, Young R, Maeura Y. Comparative carcinogenicity in F344 rats and Syrian golden hamsters of *N'*-nitrosonornicotine and *N'*-nitrosonornicotine-1-*N*-oxide. *Cancer Lett.*, 1983: **20**; 333–40.

112 Stoner GD, Adams C, Kresty LA, Hecht SS, Murphy SE, Morse MA. Inhibition of *N'*-nitrosonornicotine-induced esophageal tumorigenesis by 3-phenylpropyl isothiocyanate. *Carcinogenesis*, 1998: **19**; 2139–43.

113 Hecht SS, Chen CB, Ohmori T, Hoffmann D. Comparative carcinogenicity in F344 rats of the tobacco specific nitrosamines, *N'*-nitrosonornicotine and 4-(*N*-methyl-*N*-nitrosamino)-1-(3-pyridyl)-1-butanone. *Cancer Res.*, 1980: **40**; 298–302.

114 Hoffmann D, Rivenson A, Amin S, Hecht SS. Dose-response study of the carcinogenicity of tobacco-specific *N*-nitrosamines in F344 rats. *J. Cancer Res. Clin. Oncol.*, 1984: **108**; 81–6.

115 Hecht SS, Young R, Rivenson A, Hoffmann D. On the metabolic activation of *N*-nitrosomorpholine and *N'*-nitrosonornicotine: effects of deuterium substitution. In *N-Nitroso Compounds: Occurrence and Biological Effects* (Bartsch H, O'Neill IK, Castegnaro M, Okada M, Davis W, Eds.) pp. 499–507, International Agency for Research on Cancer, Lyon, France, 1982.

116 Griciute L, Castegnaro M, Bereziat JC, Cabral JRP. Influence of ethyl alcohol on the carcinogenic activity of *N*-nitrosonornicotine. *Cancer Lett.*, 1986: **31**; 267–75.

117 Nachiappan V, Mufti SI, Chakravarti A, Eskelson CD, Rajasedkharan R. Lipid peroxidation and ethanol-related tumor promotion in Fischer–344 rats treated with tobacco-specific nitrosamines. *Alcohol*, 1994: **29**; 565–74.

118 Hilfrich J, Hecht SS, Hoffmann D. Effects of *N'*-nitrosonornicotine and *N'*-nitrosoanabasine in Syrian golden hamsters. *Cancer Lett.*, 1977: **2**; 169–76.

119 McCoy GD, Hecht SS, Katayama S, Wynder EL. Differential effect of chronic ethanol consumption on the carcinogenicity of *N*-nitrosopyrrolidine and *N'*-nitrosonornicotine in male Syrian golden hamsters. *Cancer Res.*, 1981: **41**; 2849–54.

120 McCoy GD, Katayama S, Young R, Wyatt M, Hecht S. Influence of chronic ethanol consumption on the metabolism and carcinogenicity of tobacco-related nitrosamines. In: *N-Nitroso Compounds: Occurrence and Biological Effects* (Bartsch H, O'Neill IK, Castegnaro M, Okada M, Davis W, Eds.) pp. 635–642, International Agency for Research on Cancer, Lyon, France, 1982.

121 Chakradeo PP, Nair J, Bhide SV. Metabolism of *N'*-nitrosonornicotine by adult and fetal human oesophageal cultures. *Cell Biol. International*, 1995: **19**; 53–8.

122 Adams JD, LaVoie EJ, Hoffmann D. On the pharmacokinetics of tobacco-specific *N*-nitrosamines in Fischer rats. *Carcinogenesis*, 1985: **6**; 509–11.

123 Morse MA, Eklind KI, Toussaint M, Amin SG, Chung F-L. Characterization of a glucuronide metabolite of 4-(methylnitrosamino)-1-(3-pyridyl)-1-butanone (NNK) and its dose-dependent excretion in the urine of mice and rats. *Carcinogenesis*, 1990: **11**; 1819–23.

124 Adams JD, LaVoie EJ, O'Mara-Adams KJ, Hoffmann D, Carey KD et al. Pharmacokinetics of *N'*-nitrosonornicotine and 4-(methylnitrosamino)-1-(3-pyridyl)-1-butanone in laboratory animals. *Cancer Lett.*, 1985: **28**; 195–201.

125 Hecht SS, Trushin N, Reid-Quinn CA, Burak ES, Jones AB et al. Metabolism of the tobacco-specific nitrosamine 4-(methylnitrosamino)-1-(3-pyridyl)-1-butanone in the patas monkey: pharmacokinetics and characterization of glucuronide metabolites. *Carcinogenesis*, 1993: **14**; 229–36.

126 Tjälve H, Castonguay A, Rivenson A. Microautoradiographic localization of bound metabolites in the nasal cavities of F344 rats treated with the tobacco-specific carcinogen 4-(methylnitrosamino)-1-(3-pyridyl)-1-butanone. *J. Natl. Cancer Inst.*, 1985: **74**; 185–9.

127 Desai D, Kagan SS, Amin S, Carmella SG, Hecht SS. Identification of 4-(methylnitrosamino)-1-[3-(6-hydroxypyridyl)]-1-butanone as a urinary metabolite of 4-(methylnitrosamino)-1-(3-pyridyl)-1-butanone in rodents. *Chem. Res. Toxicol.*, 1993: **6**; 794–9.

128 Richter E, Tricker AR. Nicotine inhibits the metabolic activation of the tobacco-specific nitrosamine 4-(methylnitrosamino)-1-(3-pyridyl)-1-butanone in rats. *Carcinogenesis*, 1994: **15**; 1061–4.

129 Hecht SS, Carmella SG, Chen M, Koch JFD, Miller AT et al. Quantitation of urinary metabolites of a tobacco-specific lung carcinogen after smoking cessation. *Cancer Res.*, 1999: **59**; 590–6.

130 Atawodi SE, Michelsen K, Richter E. Metabolism of a glucuronide conjugate of 4-(methylnitrosamino)-1-(3-pyridyl)-1-butanone in rats. *Arch. Toxicol.*, 1994: **69**; 14–7.

131 Schulze J, Richter E, Binder U, Zwickenpflug W. Biliary excretion of 4-(methylnitrosamino)-1-(3-pyridyl)-1-butanone in the rat. *Carcinogenesis*, 1992: **13**; 1961–5.

132 Castonguay A, Tjälve H, Trushin N, Hecht SS. Perinatal metabolism of the tobacco-specific carcinogen 4-(methylnitrosamino)-1-(3-pyridyl)-1-butanone in C57BL mice. *J. Natl. Cancer Inst.*, 1984: **72**; 1117–26.

133 Castonguay A, Tjälve H, Trushin N, d'Argy R, Sperber G. Metabolism and tissue distribution of tobacco-specific *N*-nitrosamines in the marmoset monkey (*Callithrix jacchus*). *Carcinogenesis*, 1985: **6**; 1543–50.

134 Meger M, Richter R, Zwickenpflug W, Oehlmann C, Hargaden MB et al. Metabolism and disposition of 4-(methylnitrosamino)-1-(3-pyridyl)-1-butanone (NNK) in rhesus monkeys. *Drug Met. Dispos.*, 1999: **27**; 471–8.

135 Brittebo EB, Tjälve H. Formation of tissue-bound *N'*-nitrosonornicotine metabolites by the target tissues of Sprague-Dawley and Fisher rats. *Carcinogenesis*, 1981: **2**; 959–63.

136 Löfberg B, Brittebo EB, Tjälve H. Localization and binding of *N'*-nitrosonornicotine metabolites in the nasal region and in some other tissues of Sprague-Dawley rats. *Cancer Res.*, 1982: **42**; 2877–83.

137 Hecht SS, Lin D, Chen CB. Comprehensive analysis of urinary metabolites of *N'*-nitrosonornicotine. *Carcinogenesis*, 1981: **2**; 833–8.

138 Brittebo EB, Tjälve H. Autoradiographic observations on the distribution and metabolism of *N'*-/$^{14}$C/nitrosonornicotine in mice. *J, Cancer Res, Clin, Oncol,*, 1980: **98**; 233–42.

139 Domellöf L, Andersson M, Tjälve H, Veals S, Trushin N, Hecht SS. Distribution and metabolism of *N'*-nitrosonornicotine in the minature pig. *Carcinogenesis*, 1987: **8**; 1741–7.

140 Castonguay A, Foiles PG, Trushin N, Hecht SS. Study of DNA methylation by tobacco-specific *N*-nitrosamines. *Environ. Health Perspect.*, 1985: **62**; 197–202.

141 Rossignol G, Alaoui-Jamali MA, Castonguay A, Schuller HM. Metabolism and DNA damage induced by 4-(methylnitrosamino)-1-(3-pyridyl)-1-butanone in fetal tissues of the Syrian golden hamster. *Cancer Res.*, 1989: **49**; 5671–6.

142 Wang L, Spratt TE, Liu X-K, Hecht SS, Pegg AE, Peterson LA. Pyridyloxobutyl adduct, $O^6$-[4-oxo–4-(3-pyridyl)butyl]guanine is present in 4-(acetoxymethyl-nitrosamino)-1-(3-pyridyl)-1-butanone-treated DNA and is a substrate for $O^6$-alkylguanine-DNA alkytransferase. *Chem. Res. Toxicol.*, 1997: **10**; 562–7.

143 Hecht SS, Chen CB. Hydrolysis of model compounds for α-hydroxylation of the carcinogens, *N*-nitrosopyrrolidine and *N'*-nitrosonornicotine. *J. Org. Chem.*, 1979: **44**; 1563–6.

144 Peterson LA, Mathew R, Murphy SE, Trushin N, Hecht SS. In vivo and in vitro persistence of pyridyloxobutyl DNA adducts from 4-(methylnitrosamino)-1-(3-pyr-idyl)-1-butanone. *Carcinogenesis*, 1991: **12**; 2069–72.

145 Peterson LA, Carmella SG, Hecht SS. Investigations of metabolic precursors to hemoglobin and DNA adducts of 4-(methylnitrosamino)-1-(3-pyridyl)-1-butanone. *Carcinogenesis*, 1990: **11**; 1329–33.

146 Spratt TE, Peterson LA, Confer WL, Hecht SS. Solvolysis of model compounds for α-hydroxylation of *N'*-nitrosonornicotine and 4-(methylnitrosamino)-1-(3-pyridyl)-1-butanone: evidence for a cyclic oxonium ion intermediate in the alkylation of nucleophiles. *Chem. Res. Toxicol.*, 1990: **3**; 350–6.

147 Peterson LA, Hecht SS. $O^6$-Methylguanine is a critical determinant of 4-(methyl-nitrosamino)-1-(3-pyridyl)-1-butanone tumorigenesis in A/J mouse lung. *Cancer Res.*, 1991: **51**; 5557–64.

148 Hecht SS, Spratt TE, Trushin N. Evidence for 4-(3-pyridyl)–4-oxobutylation of DNA in F344 rats treated with the tobacco specific nitrosamines 4-(methyl-nitrosamino)-1-(3-pyridyl)-1-butanone and *N'*-nitrosonornicotine. *Carcinogenesis*, 1988: **9**; 161–5.

149 Trushin N, Rivenson A, Hecht SS. Evidence supporting the role of DNA pyridyloxobutylation in rat nasal carcinogenesis by tobacco specific nitrosamines. *Cancer Res.*, 1994: **54**; 1205–11.

150 Spratt TE, Trushin N, Lin D, Hecht SS. Analysis for $N^2$-(pyridyloxobutyl)deox-yguanosine adducts in DNA of tissues exposed to tritium-labeled 4-(methylnitrosamino)-1-(3-pyridyl)-1-butanone and *N'*-nitrosonornicotine. *Chem. Res. Toxicol.*, 1989: **2**; 169–73.

151 Foiles PG, Peterson LA, Miglietta LM, Ronai Z. Analysis of mutagenic activity and ability to induce replication of polyoma DNA sequences by different model compounds of the carcinogenic tobacco-specific nitrosamine 4-(methylnitros-amino)-1-(3-pyridyl)-1-butanone. *Mutat. Res.*, 1992: **279**; 91–101.

152 Murphy SE, Palomino A, Hecht SS, Hoffmann D. Dose-response study of DNA and hemoglobin adduct formation by 4-(methylnitrosamino)-1-(3-pyridyl)-1-butanone in F344 rats. *Cancer Res.*, 1990: **50**; 5446–52.

153 Foiles PG, Akerkar SA, Carmella SG, Kagan M, Stoner GD et al. Mass spectrometric analysis of tobacco-specific nitrosamine-DNA adducts in smokers and non-smokers. *Chem. Res. Toxicol.*, 1991: **4**; 364–8.

154 Staretz ME, Foiles PG, Miglietta LM, Hecht SS. Evidence for an important role of DNA pyridyloxobutylation in rat lung carcinogenesis by 4-(methylnitrosamino)-1-(3-pyridyl)-1-butanone: effects of dose and phenethyl isothiocyanate. *Cancer Res.*, 1997: **57**; 259–66.

155 Wang M, Young-Sciame R, Chung F-L, Hecht SS. Formation of $N^2$-tetrahydrofuranyl and $N^2$-tetrahydropyranyl adducts in the reactions of α-acetoxy-$N$-nitrosopyrrolidine and α-acetoxy-$N$-nitrosopiperidine with DNA. *Chem. Res. Toxicol.*, 1995: **8**; 617–24.

156 Latif F, Moschel RC, Hemminki K, Dipple A. Styrene oxide as a stereochemical probe for the mechanism of arylalkylation at different sites on guanosine. *Chem. Res. Toxicol.*, 1988: **1**; 364–9.

157 Hecht SS, Lin D, Chuang J, Castonguay A. Reactions with deoxyguanosine of 4-(carbethoxynitrosamino)-1-(3-pyridyl)-1-butanone, a model compound for α-hydroxylation of tobacco specific nitrosamines. *J. Am. Chem. Soc.*, 1986: **108**; 1292–5.

158 Peterson LA, Liu X-K, Hecht SS. Pyridyloxobutyl DNA adducts inhibit the repair of $O^6$-methylguanine. *Cancer Res.*, 1993: **53**; 2780–5.

159 Liu X-K, Spratt TE, Murphy SE, Peterson LA. Pyridyloxobutylation of guanine residues by 4-[(acetoxymethyl)nitrosamino]-1-(3-pyridyl)-1-butanone generates substrates of $O^6$-alkylguanine-DNA alkyltransferase. *Chem. Res. Toxicol.*, 1996: **9**; 949–53.

160 Castonguay A, Tharp R, Hecht SS. Kinetics of DNA methylation by the tobacco specific carcinogen 4-(methylnitrosamino)-1-(3-pyridyl)-1-butanone in F344 rats. In: *N-Nitroso Compounds: Occurrence, Biological Effects and Relevance to Human Cancer* (O'Neill IK, Von Borstel RC, Miller CT, Long J, Bartsch H, Eds.) pp. 805–810, International Agency for Research on Cancer, Lyon, France. 1984.

161 Belinsky SA, White CM, Boucheron JA, Richardson FC, Swenberg JA, Anderson M. Accumulation and persistence of DNA adducts in respiratory tissue of rats following multiple administrations of the tobacco specific carcinogen 4-($N$-methyl-$N$-nitrosamino)-1-(3-pyridyl)-1-butanone. *Cancer Res.*, 1986: **46**; 1280–4.

162 Belinsky SA, White CM, Devereux TR, Swenberg JA, Anderson MW. Cell-selective alkylation of DNA in rat lung following low dose exposure to the tobacco specific carcinogen 4-($N$-methyl-$N$-nitrosamino)-1-(3-pyridyl)-1-butanone. *Cancer Res.*, 1987: **47**; 1143–8.

163 Belinsky SA, Dolan ME, White CM, Maronpot RR, Pegg AE, Anderson MW. Cell specific differences in $O^6$-methylguanine-DNA methyltransferase activity and removal of $O^6$-methylguanine in rat pulmonary cells. *Carcinogenesis*, 1988: **9**; 2053–8.

164 Belinsky SA, Foley JF, White CM, Anderson MW, Maronpot RR. Dose-response relationship between $O^6$-methylguanine formation in Clara cells and induction of pulmonary neoplasia in the rat by 4-(methylnitrosamino)-1-(3-pyridyl)-1-butanone. *Cancer Res.*, 1990: **50**; 3772–80.

165 Liu L, Castonguay A, Gerson SL. Lack of correlation between DNA methylation and hepatocarcinogenesis in rats and hamsters treated with 4-(methylnitrosamino)-1-(3-pyridyl)-1-butanone. *Carcinogenesis*, 1992: **13**; 2137–40.

166 Hecht SS, Trushin N, Castonguay A, Rivenson A. Comparative tumorigenicity and DNA methylation in F344 rats by 4-(methylnitrosamino)-1-(3-pyridyl)-1-butanone and *N*-nitrosodimethylamine. *Cancer Res.*, 1986: **46**; 498–502.

167 Hecht SS, Lin D, Castonguay A, Rivenson A. Effects of α-deuterium substitution on the tumorigenicity of 4-(methylnitrosamino)-1-(3-pyridyl)-1-butanone in F344 rats. *Carcinogenesis*, 1987: **8**; 291–4.

168 Morse MA, Wang C-X, Amin SG, Hecht SS, Chung F-L. Effects of dietary sinigrin or indole-3-carbinol on $O^6$-methylguanine-DNA-transmethylase activity and 4-(methylnitrosamino)-1-(3-pyridyl)-1-butanone-induced DNA methylation and tumorigenicity in F344 rats. *Carcinogenesis*, 1988: **9**; 1891–5.

169 Rivenson A, Hoffmann D, Prokopczyk B, Amin S, Hecht SS. Induction of lung and exocrine pancreas tumors in F344 rats by tobacco-specific and *Areca*-derived *N*-nitrosamines. *Cancer Res.*, 1988: **48**; 6912–7.

170 Morse MA, Wang C-X, Stoner GD, Mandal S, Conran PB et al. Inhibition of 4-(methylnitrosamino)-1-(3-pyridyl)-1-butanone-induced DNA adduct formation and tumorigenicity in lung of F344 rats by dietary phenethyl isothiocyanate. *Cancer Res.*, 1989: **49**; 549–53.

171 Lijinsky W, Thomas BJ, Kovatch RM. Local and systemic carcinogenic effects of alkylating carcinogens in rats treated by intravesicular administration. *Jpn. J. Cancer Res.*, 1991: **82**; 980–6.

172 Prokopczyk B, Rivenson A, Hoffmann D. Comparative carcinogenicity of 3-(methylnitrosamino)propionitrile and 4-(methylnitrosamino)-1-(3-pyridyl)-1-butanone upon local application to mouse skin and rat oral mucosa. *Cancer Lett.*, 1991: **60**; 153–7.

173 Hecht SS, Adams JD, Numoto S, Hoffmann D. Induction of respiratory tract tumors in Syrian golden hamsters by a single dose of 4-(methylnitrosamino)-1-(3-pyridyl)-1-butanone (NNK) and the effect of smoke inhalation. *Carcinogenesis*, 1983: **4**; 1287–90.

174 Schüller HM, Witschi H-P, Nylen E, Joshi PA, Correa E, Becker KL. Pathobiology of lung tumors induced in hamsters by 4-(methylnitrosamino)-1-(3-pyridyl)-1-butanone and the modulating effect of hyperoxia. *Cancer Res.*, 1990: **50**; 1960–5.

175 Oreffo VIC, Lin H-W, Padmanabhan R, Witschi H. K-*ras* and *p53* point mutations in 4-(methylnitrosamino)-1-(3-pyridyl)-1-butanone-induced hamster lung tumors. *Carcinogenesis*, 1993: **14**; 451–5.

176 Sunday ME, Willet CG, Graham SA, Oreffo VIC, Linnoila RI, Witschi H. Histochemical characterization of non-neuroendocrine tumors and neuroendocrine cell hyperplasia induced in hamster lung by 4-(methylnitrosamino)-1-(3-pyridyl)-1-butanone with or without hyperoxia. *Am. J. Pathol.*, 1995: **147**; 740–51.

177 Chung F-L, Xu Y. Increased 8-oxodeoxyguanosine levels in lung DNA of A/J mice and F344 rats treated with the tobacco-specific nitrosamine 4-(methylnitros-amino)-1-(3-pyridyl)-1-butanone. *Carcinogenesis*, 1992: **13**; 1269–72.

178 Rosa JGV, Prokopczyk B, Desai DH, Amin SG, El-Bayoumy K. Elevated 8-hydroxy–2'-deoxyguanosine levels in lung DNA of A/J mice and F344 rats treated with 4-(methylnitrosamino)-1-(3-pyridyl)-1-butanone and inhibition by dietary 1,4-phenylenebis(methylene)selenocyanate. *Carcinogenesis*, 1998: **19**; 1783–8.

179 Sipowicz MA, Amin S, Desai D, Kasprzak KS, Anderson LM. Oxidative DNA

damage in tissues of pregnant female mice and fetuses caused by the tobacco-specific nitrosamine, 4-(methylnitrosamino)-1-(3-pyridyl)-1-butanone (NNK). *Cancer Lett.*, 1997: **117**; 87–91.

180 Bilodeau J-F, Wang M, Chung F-L, Castonguay A. Effects of nonsteroidal antiinflammatory drugs on oxidative pathways in A/J mice. *Free Rad. Biol. Med.*, 1995: **18**; 47–54.

181 Xu Y, Ho C-T, Amin SG, Han CH, Chung F-L. Inhibition of tobacco-specific nitrosamine-induced lung tumorigenesis in A/J mice by green tea and its major polyphenol as antioxidants. *Cancer Res.*, 1992: **52**; 3875–9.

182 Young-Sciame R, Wang M, Chung F-L, Hecht SS. Reactions of α-acetoxy-*N*-nitrosopyrrolidine and α-acetoxy-*N*-nitrosopiperidine with deoxyguanosine: formation of $N^2$-tetrahydrofuranyl or $N^2$-tetrahydropyranyl adducts. *Chem. Res. Toxicol.*, 1995: **8**; 607–16.

183 Carmella SG, Hecht SS. Formation of hemoglobin adducts upon treatment of F344 rats with the tobacco-specific nitrosamines 4-(methylnitrosamino)-1-(3-pyridyl)-1-butanone and *N'*-nitrosonornicotine. *Cancer Res.*, 1987: **47**; 2626–30.

184 Pereira MA, Chang LW. Binding of chemical carcinogens and mutagens to rat hemoglobin. *Chem-Biol. Interact.*, 1981: **33**; 301–5.

185 Carmella SG, Kagan SS, Hecht SS. Evidence that a hemoglobin adduct of 4-(methylnitrosamino)-1-(3-pyridyl)-1-butanone is a 4-(3-pyridyl)-4-oxobutyl carboxylic acid ester. *Chem. Res. Toxicol.*, 1992: **5**; 76–80.

186 Carmella SG, Kagan SS, Spratt TE, Hecht SS. Evaluation of cysteine adduct formation in rat hemoglobin by 4-(methylnitrosamino)-1-(3-pyridyl)-1-butanone and related compounds. *Cancer Res.*, 1990: **50**; 5453–9.

187 Kim S, Paik WK, Choi J, Lotlikar PD, Magee PN. Microsome-dependent methylation of erythrocyte proteins by dimethylnitrosamine. *Carcinogenesis*, 1981: **2**; 179–82.

188 LaVoie EJ, Prokopczyk G, Rigotty J, Czech A, Rivenson A, Adams JD. Tumorigenic activity of the tobacco-specific nitrosamines 4-(methylnitrosamino)-1-(3-pyridyl)-1-butanone (NNK), 4-(methylnitrosamino)-1-(3-pyridyl)-1-butanol (*iso*-NNAL) and *N'*-nitrosonornicotine (NNN) on topical application to Sencar mice. *Cancer Lett.*, 1987: **37**; 277–83.

189 Hoffmann D, Rivenson A, Abbi R, Wynder E. Effect of the fat content of the diet on the carcinogenic activity of 4-(methylnitrosamino)-1-(3-pyridyl)-1-butanone in F344 rats. *Cancer Res.*, 1993: **53**; 2758–61.

190 Hecht SS, Trushin N, Rigotty J, Carmella SG, Borukhova A et al. Complete inhibition of 4-(methylnitrosamino)-1-(3-pyridyl)-1-butanone induced rat lung tumorigenesis and favorable modification of biomarkers by phenethyl isothiocyanate. *Cancer Epidemiol Biomarkers & Prev.*, 1996: **5**; 645–52.

191 Boorman GA, Hailey R, Grumbein S, Chou BJ, Herbert RA et al. Toxicology and carcinogenesis studies of ozone and 4-(*N*-nitrosomethylamino)-1-(3-pyridyl)-1-butanone in Fischer–344/N rats. *Toxicol. Pathol.*, 1994: **22**; 545–54.

192 Chung F-L, Kelloff G, Steele V, Pittman B, Zang E et al. Chemopreventive efficacy of arylalkyl isothiocyanates and *N*-acetylcysteine for lung tumorigenesis in Fischer rats. *Cancer Res.*, 1996: **56**; 772–8.

193 Chung F-L, Wang M, Rivenson A, Iatropoulos MJ, Reinhardt JC et al. Inhibition of lung tumorigenesis by black tea in Fischer rats treated with a tobacco-specific carcinogen: caffeine as an important constituent. *Cancer Res.*, 1998: **58**;

4096–101.

194 Devereux TR, Belinsky SA, Maronpot RR, White CM, Hegi ME et al. Comparison of pulmonary $O^6$-methylguanine DNA adduct levels and Ki-*ras* activation in lung tumors from resistant and susceptible mouse strains. *Mol. Carcinog.*, 1993: **8**; 177–85.

195 Hecht SS, Morse MA, Amin S, Stoner GD, Jordan KG et al. Rapid single-dose model for lung tumor induction in A/J mice by 4-(methylnitrosamino)-1-(3-pyri-dyl)-1-butanone and the effect of diet. *Carcinogenesis*, 1989: **10**; 1901–4.

196 Belinsky SA, Devereux TR, Foley JF, Maronpot RR, Anderson MW. Role of the alveolar type II cell in the development and progression of pulmonary tumors induced by 4-(methylnitrosamino)-1-(3-pyridyl)-1-butanone in the A/J mouse. *Cancer Res.*, 1992: **52**; 3164–73.

197 Morse MA, Eklind KI, Hecht SS, Jordan KG, Choi C-I et al. Structure-activity relationships for inhibition of 4-(methylnitrosamino)-1-(3-pyridyl)-1-butanone lung tumorigenesis by arylalkyl isothiocyanates in A/J mice. *Cancer Res.*, 1991: **51**; 1846–50.

198 Morse MA, Eklind KI, Amin SG, Hecht SS, Chung F-L. Effects of alkyl chain length on the inhibition of NNK-induced lung neoplasia in A/J mice by arylalkyl isothiocyanates. *Carcinogenesis*, 1989: **10**; 1757–9.

199 Boukharta M, Jalbert G, Castonguay A. Biodistribution of ellagic acid and dose-related inhibition of lung tumorigenesis in A/J mice. *Nutr. and Cancer*, 1992: **18**; 181–9.

200 Castonguay A, Rioux N. Inhibition of lung tumorigenesis by sulindac: comparison of two protocols. *Carcinogenesis*, 1997: **18**; 491–6.

201 Hecht SS, Isaacs S, Trushin N. Lung tumor induction in A/J mice by the tobacco smoke carcinogens 4-(methylnitrosamino)-1-(3-pyridyl)-1-butanone and benzo[α]-pyrene: a potentially useful model for evaluation of chemopreventive agents. *Carcinogenesis*, 1994: **15**; 2721–5.

202 Conaway CC, Jiao D, Kelloff GJ, Steele VE, Rivenson A, Chung F-L. Chem-opreventive potential of fumaric acid, *N*-acetylcysteine, *N*-(4-hydroxyphenyl) retinamide and β-carotene for tobacco-nitrosamine induced lung tumors in A/J mice. *Cancer Lett.*, 1998: **124**; 85–93.

203 Belinsky SA, Stefanski SA, Anderson MW. The A/J mouse lung as a model for developing new chemointervention strategies. *Cancer Res.*, 1993: **53**; 410–416.

204 Hoffmann D, Djordjevic MV, Rivenson A, Desai D, Amin S. Relative potencies of tobacco-specific *N*-nitrosamines as inducers of lung tumors in A/J mice. *Cancer Lett.*, 1993: **71**; 25–30.

205 Anderson LM, Hecht SS, Dixon DE, Dove LF, Kovatch RM et al. Evaluation of the transplacental tumorigenicity of the tobacco-specific carcinogen 4-(methylnitros-amino)-1-(3-pyridyl)-1-butanone in mice. *Cancer Res.*, 1989: **49**; 3770–5.

206 Hecht SS, Jordan KG, Choi C-I, Trushin N. Effects of deuterium substitution on the tumorigenicity of 4-(methylnitrosamino)-1-(3-pyridyl)-1-butanone and 4-(methylnitrosamino)-1-(3-pyridyl)-1-butanol in A/J mice. *Carcinogenesis*, 1990: **11**; 1017–20.

207 Belinsky SA, Devereux TR, Maronpot RR, Stoner GD, Anderson MW. The relationship between the formation of promutagenic adducts and the activation of the K-*ras* proto-oncogene in lung tumors from A/J mice treated with nitrosamines. *Cancer Res.*, 1989: **49**; 5305–11.

208 Belinsky SA, Anderson MW. Detection of non-*Ras* oncogenes in 4-(methyl-nitrosamino)-1-(3-pyridyl)-1-butanone (NNK) induced lung tumors from F344 rats. *Proc. Amer. Assoc. Cancer Res.*, 1990: **30**; 778.

209 Malkinson AM. Molecular comparison of human and mouse pulmonary adeno-carcinomas. *Expt. Lung Res.*, 1998: **24**; 541–55.

210 Furukawa F, Nishikawa T, Yoshimura H, Mitsui M, Imazawa T et al. Effects of 4-(methylnitrosamino)-1-(3-pyridyl)-1-butanone (NNK) on *N*-nitrosobis-(2-oxopro-pyl)amine (BOP)-initiated carcinogenesis in hamsters. *Cancer Lett.*, 1994: **86**; 75–82.

211 Miller CH, Zhang Z, Hamilton SM, Teel RW. Effects of capsaicin on liver microsomal metabolism of the tobacco-specific nitrosamine NNK. *Cancer Lett.*, 1993: **75**; 45–52.

212 Beebe LE, Kim YE, Amin S, Riggs CW, Kovatch RM, Anderson LM. Comparison of transplacental and neonatal initiation of mouse lung and liver tumors by *N*-nitrosodimethylamine (NDMA) and 4-(methylnitrosamino)-1-(3-pyridyl)-1-butanone (NNK) and promotability by a polychlorinated biphenyls mixture (Aroclor 1254). *Carcinogenesis*, 1993: **14**; 1545–8.

213 Correa E, Joshi PA, Castonguay A, Schüller HM. The tobacco-specific nitrosamine 4-(methylnitrosamino)-1-(3-pyridyl)-1-butanone is an active transplacental carcin-ogen in Syrian golden hamsters. *Cancer Res.*, 1990: **50**; 3435–8.

214 Schüller HM, Jorquera R, Lu X, Riechert A, Castonguay A. Transplacental carcinogenicity of low doses of 4-(methylnitrosamino)-1-(3-pyridyl)-1-butanone administered subcutaneously or intratracheally to hamsters. *J. Cancer Res. Clin. Oncol.*, 1994: **120**; 200–3.

215 Schüller HM, Jorquera R, Reichert A, Castonguay A. Transplancental induction of pancreas tumors in hamsters by ethanol and the tobacco-specific nitrosamine 4-(methylnitrosamino)-1-(3-pyridyl)-1-butanone. *Cancer Res.*, 1993: **53**; 2498–501.

216 Furukawa F, Nishikawa T, Enami T, Mitsui M, Imazawa T et al. Promotional effects of 4-(methylnitrosamino)-1-(3-pyridyl)-1-butanol (NNAL) on *N*-nitrosobis (2-oxo-propyl)amine (BOP)-initiated carcinogenesis in hamsters. *Food Chem. Toxicol.*, 1997: **35**; 387–92.

217 Hecht SS, Chen CB, Hirota N, Ornaf RM, Tso TC, Hoffmann D. Tobacco specific nitrosamines: formation from nicotine in vitro and during tobacco curing and carcinogenicity in strain A mice. *J. Natl. Cancer Inst.*, 1978: **60**; 819–24.

218 Hecht SS, Abbaspour A, Hoffmann D. Bioassay in A/J mice of some structural analogues of tobacco-specific nitrosamines. *Cancer Lett.*, 1988: **42**; 141–5.

219 Hecht SS, Rivenson A, Braley J, DiBello J, Adams JD, Hoffmann D. Induction of oral cavity tumors in F344 rats by tobacco-specific nitrosamines and snuff. *Cancer Res.*, 1986: **46**; 4162–6.

220 Boyland E, Roe F JC, Gorrod JW, Mitchley BCV. The carcinogenicity of nitrosoanabasine, a possible constituent of tobacco smoke. *Br. J. Cancer*, 1964: **18**; 265–70.

221 Djordjevic MV, Brunnemann KD, Hoffmann D. Identification and analysis of a nicotine-derived *N*-nitrosamino acid and other nitrosamino acids in tobacco. *Carcinogenesis*, 1989: **10**; 1725–31.

222 Rivenson A, Djordjevic MV, Amin S, Hoffmann D. Bioassay in A/J mice of some *N*-nitrosamines. *Cancer Lett.*, 1989: **47**; 111–4.

223 Hecht SS. Approaches to cancer prevention based on an understanding of *N*-

nitrosamine carcinogenesis. *Proc. Soc. Expt. Biol. Med.*, 1997: **216**; 181–91.

224 Haseman JK, Hailey JR, Morris RW. Spontaneous neoplasm incidences in Fischer 344 rats and B6C3F$_1$ mice in two-year carcinogenicity studies: a National Toxicology Program update. *Toxicologic Path.*, 1998: **26**; 428–41.

225 Wang ZY, Hong J-Y, Huang M-T, Reuhl KR, Conney AH, Yang CS. Inhibition of *N*-nitrosodiethylamine- and 4-(methylnitrosamino)-1-(3-pyridyl)-1-butanone-induced tumorigenesis in A/J mice by green tea and black tea. *Cancer Res.*, 1992: **52**; 1943–7.

226 Shi ST, Wang Z-Y, Smith TJ, Hong J-Y, Chen W-F et al. Effects of green tea and black tea on 4-(methylnitrosamino)-1-(3-pyridyl)-1-butanone bioactivation, DNA methylation, and lung tumorigenesis in A/J mice. *Cancer Res.*, 1994: **54**; 4641–7.

227 Yang G, Liu Z, Seril DN, Liao J, Ding W et al. Black tea constituents, theaflavins, inhibit 4-(methylnitrosamino)-1-(3-pyridyl)-1-butanone (NNK)-induced lung tumor igenesis in A/J mice. *Carcinogenesis*, 1997: **18**; 2361–2365.

228 Van Etten CH, Daxenbichler ME, Williams PH, Kwolek WF. Glucosinolates and derived products in cruciferous vegetables. Analysis of the edible part from twenty-two varieties of cabbage. *J. Agric. Food Chem.*, 1976: **24**; 452–5.

229 Schulze J, Malone A, Richter E. Intestinal metabolism of 4-(methylnitrosamino)-1-(3-pyridyl)-1-butanone in rats: sex-difference, inducibility and inhibition by phenethylisothiocyanate. *Carcinogenesis*, 1995: **16**; 1733–40.

230 Hecht SS, Trushin N, Rigotty J, Carmella SG, Borukhova A et al. Inhibitory effects of 6-phenylhexyl isothiocyanate on 4-(methylnitrosamino)-1-(3-pyridyl)-1-buta-none metabolic activation and lung tumorigenesis in rats. *Carcinogenesis*, 1996: **17**; 2061–7.

231 Morse MA, Amin SG, Hecht SS, Chung F-L. Effects of aromatic isothiocyanates on tumorigenicity, $O^6$-methylguanine formation, and metabolism of the tobacco-specific nitrosamine 4-(methylnitrosamino)-1-(3-pyridyl)-1-butanone in A/J mouse lung. *Cancer Res.*, 1989: **49**; 2894–7.

232 El-Bayoumy K, Upadhyaya P, Desai DH, Amin SG, Hoffmann D, Wynder EL. Effects of 1,4-phenylenebis(methylene)selenocyanate, phenethyl isothiocyanate, indole–3-carbinol, and *d*-limonene individually and in combination on the tumorigenicity of the tobacco-specific nitrosamine 4-(methylnitrosamino)-1-(3-pyr-idyl)-1-butanone in A/J mouse lung. *Anticancer Res.*, 1996: **16**; 2709–12.

233 Jiao D, Smith TJ, Kim S, Yang CS, Desai D et al. The essential role of the functional group in alkyl isothiocyanates for inhibition of tobacco nitrosamine-induced lung tumorigenesis. *Carcinogenesis*, 1996: **17**; 755–9.

234 Hirose M, Yamaguchi T, Kimoto N, Ogawa K, Futakuchi M et al. Strong promoting activity of phenethyl isothiocyanate and benzyl isothiocyanate on urinary bladder carcinogenesis in F344 male rats. *Int. J. Cancer*, 1998: **77**; 773–7.

235 Stoner GD, Siglin JC, Morse MA, Desai DH, Amin SG et al. Enhancement of esophageal carcinogenesis in male F344 rats by dietary phenylhexyl iso-thiocyanate. *Carcinogenesis*, 1995: **16**; 2473–6.

236 Rao CV, Rivenson A, Simi B, Zang E, Hamid R et al. Enhancement of experimental colon carcinogenesis by dietary 6-phenylhexyl isothiocyanate. *Cancer Res.*, 1995: **55**; 4311–8.

237 Musk SRR, Johnson IT. The clastogenic effects of isothiocyanates. *Mutation Res.*, 1993: **300**; 111–7.

238 Smith TJ, Yang G, Seril DN, Liao J, Kim S. Inhibition of 4-(methylnitrosamino)-1-(3-pyridyl)-1-butanone-induced lung tumorigenesis by dietary olive oil and squalene. *Carcinogenesis*, 1998: **19**; 703–6.

239 Hong J-Y, Wang ZY, Smith TJ, Zhou S, Shi S et al. Inhibitory effects of dially sulfide on the metabolism and tumorigenicity of the tobacco-specific carcinogen 4-(methylnitrosamino)-1-(3-pyridyl)-1-butanone (NNK) in A/J mouse lung. *Carcinogenesis*, 1992: **13**; 901–4.

240 Omenn GS, Goodman GE, Thornquist MD, Balmes J, Cullen MR et al. Risk factors for lung cancer and for intervention effects in CARET, the Beta-Carotene and Retinol Efficacy Trial. *J. Natl. Cancer Inst.*, 1996: **88**; 1550–9.

241 The Alpha-Tocopherol, Beta Carotene Cancer Prevention Study Group. The effect of vitamin E and beta carotene on the incidence of lung cancer and other cancers in male smokers. *N. Engl. J. Med.*, 1994: **330**; 1029–35.

242 Kishimoto M, Yano Y, Yajima S, Otani S, Ichikawa T, Yano T. The inhibitory effect of vitamin E on 4-(methylnitrosamino)-1-(3-pyridyl)-1-butanone-induced lung tumorigenesis in mice based on regulation of polyamine metabolism. *Cancer Lett.*, 1998: **126**; 173–8.

243 Prokopczyk B, Amin S, Desai DH, Kurtzke C, Upadhyaya P, El-Bayoumy K. Effects of 1,4,-phenylenebis(methylene)selenocyanate and selenomethionine on 4-(methylnitrosamino)-1-(3-pyridyl)-1-butanone-induced tumorigenesis in A/J mouse lung. *Carcinogenesis*, 1997: **18**; 1855–7.

244 Duperron C, Castonguay A. Chemopreventive efficacies of aspirin and sulindac against lung tumorigenesis in A/J mice. *Carcinogenesis*, 1997: **18**; 1001–6.

245 Rioux N, Castonguay A. Recovery from 4-(methylnitrosamino)-1-(3-pyridyl)-1-butanone-induced immunosuppression in A/J mice by treatment with nonsteroidal anti-inflammatory drugs. *J. Natl. Cancer Inst.*, 1997: **89**; 874–80.

246 Malkinson AM, Koski KM, Dwyer-Nield LD, Rice PL, Rioux N et al. Inhibition of 4-(methylnitrosamino)-1-(3-pyridyl)-1-butanone-induced mouse lung tumor formation by FGN-1 (sulindac sulfone). *Carcinogenesis*, 1998: **19**; 1353–6.

247 Desai D, Chang L, Amin S. Synthesis and bioassay of 4-ipomeanol analogs as potential chemopreventive agents against 4-(methylnitrosamino)-1-(3-pyridyl)-1-butanone (NNK)-induced tumorigenicity in A/J mice. *Cancer Lett.*, 1996: **108**; 263–70.

248 Schulze J, Schrader E, Foth H, Kahl GF, Richter E. Effect of nicotine or cotinine on metabolism of 4-(methylnitrosamino)-1-(3-pyridyl)-1-butanone (NNK) in isolated rat lung and liver. *Naunyn-Schmiedeberg's Arch. Pharmacol.*, 1998: **357**; 344–50.

249 Kutzer C, Richter E, Oehlmann C, Atawodi SE. Effect of nicotine and cotinine on NNK metabolism in rats. In: *Effects of Nicotine on Biological Systems II*, Clarke PBS, Quik M, Adlkofer F, Thurau K (Eds.), Birkhäuser Verlag, Basel. 1994, pp. 385–90.

250 Nakajima M, Yamamoto T, Nunoya K-I, Yokoi T, Nagashima K et al. Role of human cytochrome CYP2A6 in C-oxidation of nicotine. *Drug Metab. Dis.*, 1996: **24**; 1212–7.

251 Nakajima M, Yamamoto T, Nunoya K-I, Yokoi T, Nagashima K et al. Characterization of CYP2A6 involved in 3'-hydroxylation of cotinine in human liver microsomes. *J. Pharmacol. Exp. Ther.*, 1996: **227**; 1010–5.

252 Finch GL, Nikula KJ, Belinski SA, Barr EB, Stoner GD, Lechner JF. Failure of

cigarette smoke to induce or promote lung cancer in the A/J mouse. *Cancer Lett.*, 1996: **99**; 161–7.

253  Brown BG, Chang C-JG, Ayres PH, Lee CK, Doolittle DJ. The effect of cotinine or cigarette smoke co-administration on the formation of $O^6$-methylguanine adducts in the lung and liver of A/J mice treated with 4-(methylnitrosamino)-1-(3-pyridyl)-1-butanone (NNK). *Toxicol. Sci.*, 1999: **87**; 33–9.

254  Brown BG, Richter E, Tricker AR, Doolittle DJ. The effect of cigarette smoke on the metabolism of the tobacco-specific nitrosamine 4-(methylnitrosamino)-1-(3-pyridyl)-1-butanone (NNK) in the A/J mouse. *Proc. Amer. Assoc. Cancer Res.*, 1999: **40**; 250.

255  Witschi H, Espiritu I, Maronpot RR, Pinkerton KE, Jones AD. The carcinogenic potential of the gas phase of environmental tobacco smoke. *Carcinogenesis*, 1997: **18**; 2035–42.

256  Schüller HM, Castonguay A, Orloff M, Rossignol G. Modulation of the uptake and metabolism of 4-(methylnitrosamino)-1-(3-pyridyl)-1-butanone by nicotine in hamster lung. *Cancer Res.*, 1991: **51**; 2009–14.

257  Lee CK, Fulp C, Bombick BR, Doolittle DJ. Inhibition of mutagenicity of *N*-nitrosamines by tobacco smoke and its constituents. *Mutat. Res.*, 1996: **367**; 83–92.

258  Richie Jr JP, Carmella SG, Muscat JE, Scott DG, Akerkar SA, Hecht SS. Differences in the urinary metabolites of the tobacco-specific lung carcinogen 4-(methylnitrosamino)-1-(3-pyridyl)-1-butanone in black and white smokers. *Cancer Epidemiol Biomarkers & Prev.*, 1997: **6**; 783–90.

259  Carmella SG, Akerkar S, Hecht SS. Metabolites of the tobacco-specific nitrosamine 4-(methylnitrosamino)-1-(3-pyridyl)-1-butanone in smokers' urine. *Cancer Res.*, 1993: **53**; 721–4.

260  Meger M, Meger-Kossien I, Dietrich M, Tricker AR, Scherer G, Adlkofer F. Metabolites of 4-(methylnitrosamino)-1-(3-pyridyl)-1-butanone in urine of smokers. *Eur. J. Cancer Prev.*, 1996: **5(Suppl. 1)**; 121–4.

261  Carmella SG, Borukhova A, Akerkar SA, Hecht SS. Analysis of human urine for pyridine-*N*-oxide metabolites of 4-(methylnitrosamino)-1-(3-pyridyl)-1-butanone, a tobacco-specific lung carcinogen. *Cancer Epidemiol. Biomarkers & Prev.*, 1997: **7**; 113–20.

262  Hecht SS, Carmella SG, Murphy SE, Akerkar S, Brunnemann KD, Hoffmann D. A tobacco-specific lung carcinogen in the urine of men exposed to cigarette smoke. *N. Engl. J. Med.*, 1993: **329**; 1543–6.

263  Meger M, Meger-Kossien I, Scherer G. Metabolites of 4-(methylnitrosamino)-1-(3-pyridyl)-1-butanone (NNK) in urine of smokers and non-smokers. *Proc. Amer. Assoc. Cancer Res.*, 1998: **39**; 333–4.

264  Parsons WD, Carmella SG, Akerkar S, Bonilla LE, Hecht SS. A metabolite of the tobacco-specific lung carcinogen 4-(methylnitrosamino)-1-(3-pyridyl)-1-butanone in the urine of hospital workers exposed to environmental tobacco smoke. *Cancer Epidemiol. Biomarkers & Prev.*, 1998: **7**; 257–60.

265  Lackmann GM, Salzberger U, Töllner U, Chen M, Carmella SG, Hecht SS. Metabolites of a tobacco-specific carcinogen in the urine of newborns. *J. Natl. Cancer Inst.*, 1999: **91**; 459–65.

266  Murphy SE, Carmella SG, Idris AM, Hoffmann D. Uptake and metabolism of carcinogenic levels of tobacco-specific nitrosamines by Sudanese snuff dippers.

*Cancer Epidemiol. Biomarkers & Prev.*, 1994: **3**; 423–8.

267 Kresty LA, Carmella SG, Borukhova A, Akerkar SA, Gopalakrishnan R et al. Metabolites of a tobacco-specific nitrosamine, 4-(methylnitrosamino)-1-(3-pyridyl)-1-butanone (NNK), in the urine of smokeless tobacco users: relationship of urinary biomarkers and oral leukoplakia. *Cancer Epidemiol. Biomarkers & Prev.*, 1996: **5**; 521–5.

268 Taioli E, Garbers S, Bradlow HL, Carmella SG, Akerkar S, Hecht SS. Effects of indole–3-carbinol on the metabolism of 4-(methylnitrosamino)-1-(3-pyridyl)-1-butanone in smokers. *Cancer Epidemiol. Biomarkers & Prev.*, 1997: **6**; 517–22.

269 Hecht SS, Chung F-L, Richie Jr JP, Akerkar SA, Borukhova A et al. Effects of watercress consumption on metabolism of a tobacco-specific lung carcinogen in smokers. *Cancer Epidemiol Biomarkers & Prev.*, 1995: **4**; 877–84.

270 Tiwari RK, Guo L, Bradlow HL, Telang NT, Osborne MP. Selective responsiveness of human breast cancer cell lines to indole–3-carbinol, a chemopreventive agent. *J. Natl. Cancer Inst.*, 1994: **86**; 126–31.

271 Hecht SS, Hatsukami DK, Bonilla LE, Hochalter JB. Quantitation of 4-oxo–4-(3-pyridyl)butanoic acid and enantiomers of 4-hydroxy–4-(3-pyridyl)butanoic acid in human urine: a substantial pathway of nicotine metabolism. *Chem. Res. Toxicol.*, 1999: **12**; 172–9.

272 Trushin N, Hecht SS. Stereoselective metabolism of nicotine and tobacco-specific *N*-nitrosamines to 4-hydroxy–4-(3-pyridyl)butanoic acid in rats. *Chem. Res. Toxicol.*, 1999: **12**; 164–71.

273 Wilson VL, Weston A, Manchester DK, Trivers GE, Roberst D et al. Alkyl and aryl carcinogen adducts detected in peripheral lung. *Carcinogenesis*, 1989: **10**; 2149–53.

274 Kato S , Bowman E, Harrington AM, Blomeke B, Shields PG. Human lung carcinogen-DNA adduct levels mediated by genetic polymorphisms in vivo. *J. Natl. Cancer Inst.*, 1995: **87**; 902–7.

275 Blömeke B, Greenblatt MJ, Doan VD, Bowman ED, Murphy SE et al. Distribution of 7-alkyl–2′-deoxyguanosine adduct levels in human lung. *Carcinogenesis*, 1996: **17**; 741–8.

276 Kato S, Petruzzelli S, Bowman ED, Turteltaub KW, Blomeke B et al. 7-Alkyldeoxyguanosine adduct detection by two-step HPLC and the $^{32}$P-postlabelling assay. *Carcinogenesis*, 1993: **14**; 545–50.

277 Mustonen R, Schoket B, Hemminki K. Smoking-related DNA adducts: $^{32}$P-postlabeling analysis of 7-methylguanine in human bronchial and lymphocyte DNA. *Carcinogenesis*, 1993: **14**; 151–4.

278 Petruzzelli S, Tavanti LM, Celi A, Giuntini C. Detection of $N^7$-methyldeoxyguanosine adducts in human pulmonary alveolar cells. *Am. J. Respir. Cell Mol. Biol.*, 1996: **15**; 216–23.

279 Shields PG, Povey AC, Wilson VL, Weston A, Harris CC. Combined high performance liquid chromatography/$^{32}$P-postlabelling assay of N7 methyldeoxyguanosine. *Cancer Res.*, 1990: **50**; 6580–4.

280 Tricker AR. *N*-Nitroso compounds and man: sources of exposure, endogenous formation and occurrence in body fluids. *Eur. J. Cancer Prev.*, 1997: **6**; 226–68.

281 Ehrenberg L, Osterman-Golkar S. Alkylation of macromolecules for detecting mutagenic agents. *Teratog. Carcinog. Mutag.*, 1980: **1**; 105–27.

282 Skipper PL, Tannenbaum SR. Protein adducts in the molecular dosimetry of

chemical carcinogens. *Carcinogenesis*, 1990: **11**; 507–18.

283 Carmella SG, Kagan SS, Kagan M, Foiles PG, Palladino G et al. Mass spectrometric analysis of tobacco-specific nitrosamine hemoglobin adducts in snuff dippers, smokers, and non-smokers. *Cancer Res.*, 1990: **50**; 5438–45.

284 Foiles PG, Murphy SE, Peterson LA, Carmella SG, Hecht SS. DNA and hemoglobin adducts as markers of metabolic activation of tobacco-specific carcinogens. *Cancer Res.*, 1992: **52(Suppl.)**; 2698s–701s.

285 Falter B, Kutzer C, Richter E. Biomonitoring of hemoglobin adducts: aromatic amines and tobacco-specific nitrosamines. *Clin. Investig.*, 1994: **72**; 364–71.

286 Branner B, Kutzer C, Zwickenpflug W, Scherer G, Heller W, Richter E. Hemoglobin adducts from aromatic amines and tobacco-specific nitrosamines in pregnant smoking and non-smoking women. *Biomarkers*, 1998: **3**; 35–47.

287 Atawodi SE, Lea S, Nyberg F, Mukeria A, Constantinescu V et al. 4-Hydroxy-1-(3-pyridyl)-1-butanone-hemaglobin adducts as biomarkers of exposure to tobacco smoke: evaluation of a method to be used in multicenter studies. *Cancer Epidemiol. Biomarkers & Prev.*, 1998: **7**; 817–22.

288 Melikian AA, Sun P, Pierpont C, Coleman S, Hecht SS. Gas chromatography-mass spectrometric determination of benzo[a]pyrene and chrysene diol epoxide globin adducts in humans. *Cancer Epidemiol. Biomarkers & Prev.*, 1997: **6**; 833–9.

289 Song S, Ashley DL. Supercritical fluid extraction and gas chromatography/mass spectrometry for the analysis of tobacco-specific nitrosamines in cigarettes. *Anal. Chem.*, 1999: **71**; 1303–8.

290 Preussman R, Steward BW. *N*-Nitroso carcinogens. In: *Chemical Carcinogens* 2nd Edn (Searle CE, Ed.), ACS Monographs Series No. 182, American Chemical Society, Washington, DC, 1984, pp. 643–823.

*Chapter 12*

# Determination of nicotine in mainstream and sidestream cigarette smoke

Klaus Rustemeier[1] and Jean-Jacques Piadé[2]

[1] *INBIFO Institut für biologische Forschung GmbH, Fuggerstrasse 3, D-51149, Cologne, Germany*
[2] *Philip Morris Europe, Fabriques de Tabac Réunies, CH-2003 Neuchâtel, Switzerland*

## I. FORMATION OF SMOKE AND NICOTINE TRANSFER

### Mainstream smoke

The processes governing cigarette smoke formation and nicotine transfer have been studied extensively. In particular, Baker and his team investigated the temperature and gas flux distributions inside a burning cigarette. The conclusions of this massive research effort have been reviewed [1]. A detailed summary of the current understanding of smoke formation and dynamic evolution is also available [2]. With a focus on the transfer of nicotine into smoke, these processes can be summarized as follows.

The heated part of a burning cigarette can be divided into two regions, an exothermic combustion zone and an endothermic zone immediately downstream where distillation and pyrolysis take place. The high temperature in the combustion zone, reaching 700 to 950°C, carbonizes the tobacco and makes the glowing cone almost impermeable to air. Consequently, most of the air that is puffed enters the cigarette near the char line. The oxygen causes combustion of the carbonized tobacco and the release of heat sustains the whole burning process. As a result, the hot gases in the heated part of a burning cigarette are depleted of oxygen but rich in hydrogen. Very high temperature gradients and heating rates up to 500°C/s are produced during a puff.

The distillation zone starts approximately 2 mm behind the char line with temperatures of 600°C and is characterized by temperatures ranging down

with no filter ventilation is very close to that reported by Houseman (see above).

With an increase in filter ventilation, the total nicotine accountability decreases. The authors speculate that this could be due to nicotine traveling with the smoke stream down the rod during "... the dynamic smolder (smolder during active smoking)". This is known to happen in highly ventilated cigarettes and results in more distillation and pyrolysis of nicotine, thus, lowering the total nicotine accountability.

## IV. STANDARD SMOKING

### History of regulatory smoking

Human cigarette smoking varies to a large extent with respect to the frequency of puffs, and the amount of smoke and the way it is drawn from the cigarette. The conditions under which a cigarette is smoked have a significant influence on both mainstream and sidestream smoke delivery. In order to ensure that the results of analytical determinations obtained in different laboratories are comparable, standard methods have been developed for the investigation of mainstream smoke yields. At present the most commonly used ones are the closely related Federal Trade Commission (FTC) and International Standards Organization (ISO) methods.

Before the standardization of cigarette testing for 'tar' and nicotine in 1967 [15], various methods were used by cigarette manufacturers when smoking cigarettes to obtain specifications for their products. As a result, values that were not comparable were used by the various tobacco companies in their advertising. The reaction of the FTC to this situation was to prohibit the use of 'tar' and nicotine values in advertising. In the years following, the industry took steps to develop standardized and reliable methods to generate these values. The adequacy of these methods when performing routine testing was demonstrated by comparing results obtained for a carefully designed monitor cigarette in different laboratories [16]. At the end of the standardization process, the FTC recommended a method based on the procedure developed by Ogg et al. [17], which became known as the 'Cambridge Filter Method'. The FTC clearly stated that this test method was "... not designed to determine the amount of 'tar' and nicotine inhaled by any human smoker, but rather to determine the amount of 'tar' and nicotine generated when a cigarette is smoked by a machine in accordance with the prescribed method" [18]. The purpose of testing for 'tar' and nicotine was solely for "... the cataloging or sequential listing of the cigarette brands for their mainstream smoke yields of the two smoking entities generated under standard and reproducible conditions" [19].

Basically, an identical method then became part of the CORESTA (Centre de Cooperation pour les Recherches Scientifiques Relatives au Tabac)

Recommended Method [20], part of various national standards [21,22], and finally part of the international standard ISO, 3308 [23]. All of these standards differ in some specific aspects. Today, the ISO standard derived from the CORESTA standard has gained general acceptance worldwide. In the US, however, cigarette testing for domestic brands is still performed according to the recommendations of the FTC.

The first method recommended by the FTC for determination of nicotine from a Cambridge filter pad was spectrophotometric. In 1980 the FTC recommended a gas chromatographic procedure described by Wagner [24]. This gas chromatographic procedure made it possible to perform a fast and reproducible analysis of Cambridge filter pad extracts that was free of interference, especially from other amines. CORESTA [25]/ISO [26] have recommended basically the same gas chromatographic procedures.

In the case of sidestream smoke, in which interest has been relatively recent and mainly at a research level, no standard method has yet been defined. For consistency, smoke generation for sidestream analysis usually follows the same rules as for mainstream analysis. However, with regard to smoke collection, the most critical step in sidestream smoke analysis, a standard method remains to be defined.

Between 1987 and 1990, a group of laboratories within a CORESTA task force carried out a collective and systematic examination of the most promising approaches, and investigated the influence of the different experimental options on the results. The essence of the conclusions of this cooperative work was reported in 1990 [27].

Quite recently, Health Canada issued a draft of a compendium of analytical test protocols, a document that was, as of the end of 1998, at the public examination stage [28]. Prepared under contract by Labstat Incorporated (Kitchener Ontario, Canada N2C 1L3), this document contains a proposed procedure for the simultaneous determination of 'tar', nicotine, and carbon monoxide in sidestream smoke. At this time, however, none of these methods has been validated through suitable inter-laboratory tests nor been accepted as a standard.

## Smoking machines

To date, the operating principles of the smoking machines used for regulatory purposes are the same as those of the syringe smoking machine described by O'Keeffe and Lieser in 1958 [29]. An improved version of this machine, the linear 20-port syringe machine (initially developed by Philip Morris and further developed and produced by Phipps & Bird and by Filtrona), was recommended as the standard in 1969 [30].

An alternative construction to these multi-syringe machines is a smoking machine with a rotating carrousel (Fig. 1). Rotary machines can be more easily automated because they need only one fixed cigarette insertion device

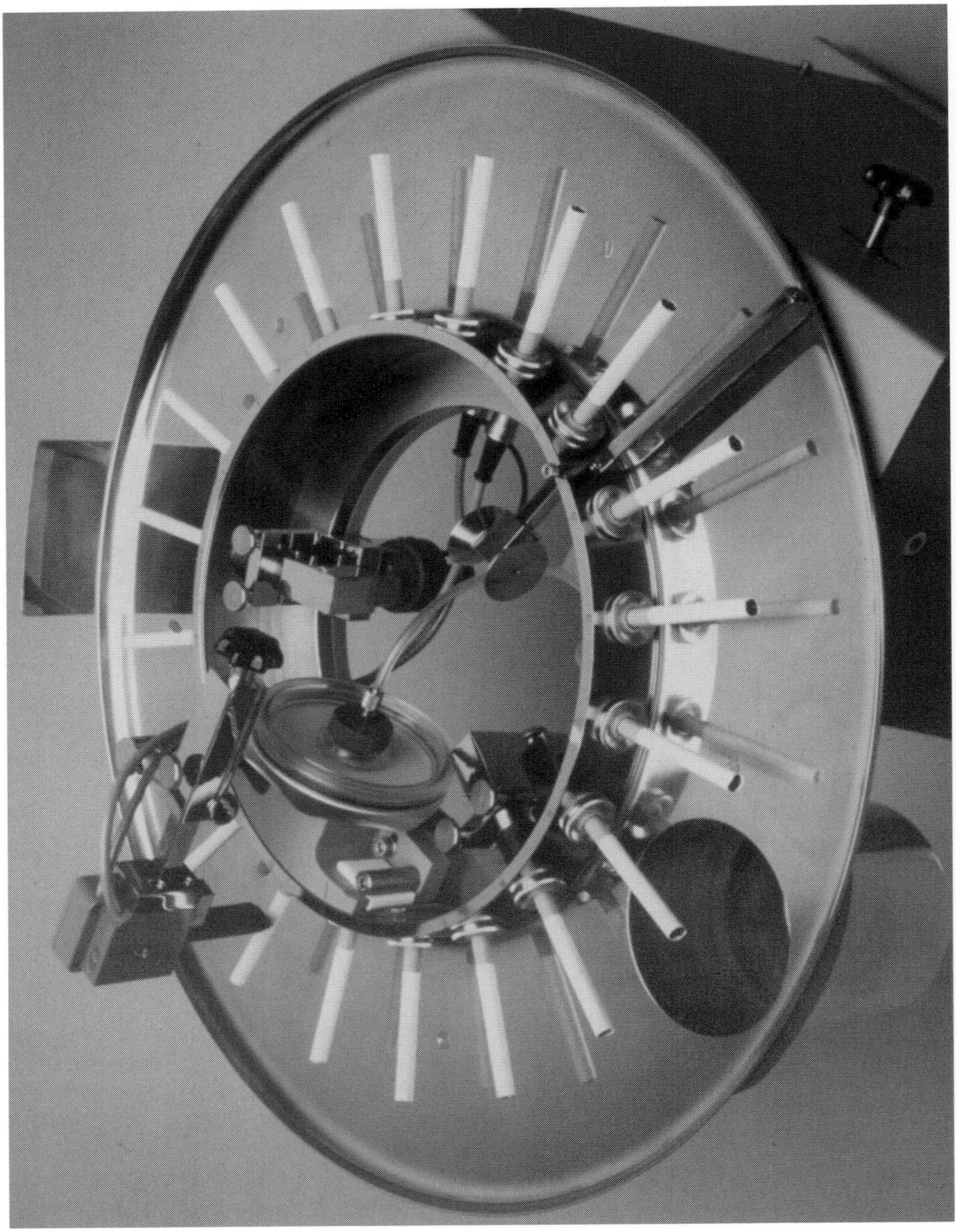

Fig. 1 Top view of a 20-port rotary smoking machine equipped with a central collection device.

and one fixed ejection device. In these machines, the airflow around the cigarettes can also be controlled more precisely and kept uniform by means of a circular hood above the cigarette carrousel. Rotary smoking machines normally collect the smoke from all the cigarettes smoked together for one sampling, (usually 20) on one central collection device, usually a Cambridge filter, and cannot easily accommodate multiple smoke collection devices. The carrousel containing the burning cigarettes must be connected to the static collection device via a sealed smoke duct that switches from one cigarette to another. This may be a source of losses due to deposition or leakage, which does not happen in linear machines where the cigarette is connected directly to the Cambridge filter holder. Differences in mainstream smoke yields between the two types of smoking machines have not been reported in the literature. In any case, these differences appear to be no greater than typical inter-laboratory variations.

The use of a piston pump on both machines ensures that puff volume and duration remain constant independent of the changing pressure drop of the burning cigarette. The machines are therefore referred to as constant volume/constant time machines.

## Smoking parameters

The parameters that define a puff are puff volume, puff duration, puff frequency, and puff profile. The FTC recommended the use of the parameters from Ogg's Cambridge filter method, which originated from a selection by Bradford et al. [31] in 1936. Thus, the standard puff became a puff of 35 ml drawn over 2 s and generated once every minute. Because a syringe type of smoking machine was used, the standard puff profile became bell-shaped.

Another important parameter for machine smoking is butt length. Butt length requirements for non-filter cigarettes were set by both the FTC and ISO at 23 mm. For filter cigarettes, the FTC specified length of overwrap plus 3 mm while ISO specified length of overwrap plus 3mm or filter length plus 8 mm, whichever is longer (see Table 1 for comparison of the two standards).

Ambient conditions for cigarette equilibration and smoking were also specified in the FTC standard and set at 75°F (23.9°C) and 60% relative humidity (RH). The ISO standard differs slightly from the FTC standard in terms of these recommended conditions, i.e., temperature at 22°C. One important difference is that the ISO standard specifies that the air velocity around the cigarette should be adjusted to 200 mm/s [23], while the FTC only requires removal of the sidestream smoke, the air velocity being adjusted to yield results for a monitor cigarette in agreement with accepted values.

## Mainstream smoke collection

The collection of smoke particles for analytical purposes became reproducible with the introduction of the Cambridge filter as a smoke trap in 1959 by

*Analytical Determination of Nicotine*

TABLE 1

Differences between FTC and ISO smoking procedures

| Parameter | FTC | ISO |
|---|---|---|
| Ambient conditions: | | |
| Temperature (°C) | 23.9±1 | 22±2 (22±1)[a] |
| Relative humidity (%) | 60±2 | 60±5 (60±2)[a] |
| Equilibration time | minimum 24 h | minimum 48 h |
| | maximum 14 d | maximum 10 d |
| Butt Length (mm): (select the option with the longest butt length) | | |
| Non-filtered cigarettes | 23 | 23 |
| Filtered cigarettes | minimum 23 | minimum 23 |
| | tipping paper length plus 3 | tipping paper length plus 3 |
| | | filter length plus 8 |
| Ashtray position | not specified | 20–60 mm below cigarette |
| Smoke trap | Cambridge filter | Cambridge filter |
| | holder assembly not specified | 2-piece reusable holder with washer that partly obstructs the butt end |
| Airflow around the cigarette | just high enough to remove sidestream smoke | 200 mm/s |
| Filter holder wiped after use | with separate filter pad | with back of used pad |

[a] Values in brackets are conditions during cigarette equilibration

Wartman et al. [32]. The Cambridge filter (Cambridge Filter Corporation, East Syracuse, NY) is a glass fiber filter capable of trapping 99.9% of particles with diameters larger than 0.3 μm. Due to the high sampling capacity of this filter, the flow resistance increases only slightly during smoking and does not affect the puff parameters. With the authorized use of Ogg's Cambridge filter method by the FTC, the Cambridge filter became part of the recommended 'tar' and nicotine testing routines. However, it took some time until it replaced the method of electrostatic precipitation of the smoke particulate phase, an equally reliable method that had been part of some of the standards valid in Europe.

The standardization of cigarette testing methods was facilitated by the use of reference and monitor cigarettes. Monitor cigarettes are used exclusively for quality control purposes and the most common ones are listed below.

The CM3 CORESTA monitor is a test piece manufactured under the auspices of CORESTA in 1999 under tightly controlled conditions in order to ensure uniformity within the lot. It is calibrated yearly through interlaboratory collaborative tests performed among CORESTA members.

The Industry Monitor (IM) is routinely used by US tobacco companies for quality assurance purposes. It is produced on a regular basis under strict, uniform conditions, and is used to monitor the performance of routine smoking laboratories.

Reference cigarettes were produced that provide quantitative results for research purposes, thus enabling comparisons over the years or between locations. They are used to assess new analytical or biological procedures, or to provide a reference when evaluating cigarette prototypes [33]. The most common ones are listed below.

The C20 and C50 research cigarettes, essentially used in Europe at the end of the 1980s

The Kentucky Reference Cigarettes are a series of research cigarettes available from the University of Kentucky [33]. The different types available represent the range of cigarettes on the U.S. market when the reference cigarettes were produced. A large amount of data are available on the Kentucky Reference Cigarette 1R4F, a 10mg 'tar' American-blend cigarette with a ventilated filter that was introduced in 1983.

Although the only goal of the recommendation of standard methods for smoke analysis was to provide comparable 'tar' and nicotine ratings to permit the ranking of cigarettes, the meaning of these ratings has been frequently questioned over the past several years. In particular, for smokers switching to cigarettes with lower 'tar' and nicotine yields, compensatory smoking like taking larger and more puffs and the blocking of ventilation holes was discussed [34,35,36]. To address these differences in smoking behavior, a puffing procedure was proposed by the FTC [37]. It is intended to be used in addition to the current procedure. This proposed puffing regimen differs from the current procedure in that a puff of 2 s in duration and 55 ml in volume is taken every 30 s. In 1998 the British Columbian government in Canada specified another puffing regimen for analytical determinations in both mainstream and sidestream smoke, according to which a puff of 2 s in duration and 56 ml in volume is taken every 20 s, with the ventilation holes of the filter blocked.

Altering the puffing regimen can result in substantial changes in the measured smoke deliveries. It was generally observed, however, that such a change had little influence on the ranking of cigarettes.

## V. NICOTINE QUANTIFICATION IN SMOKE

The procedures detailed below are targeted at the determination of the nicotine yield in mainstream or sidestream smoke. Quite often, however, this determination has to be combined with that of the yields of particulate matter, water, and carbon monoxide. Most of the procedures for nicotine determination have thus been designed to be fully compatible with this. Performing the determination of the other analytes requires additional equipment, the addition of another internal standard for water determination in the trapping solution, and additional analytical steps. Although these additional determinations are out of the scope of this chapter, they will be mentioned briefly at the relevant steps.

## VI. EARLY METHODS (GRAVIMETRY, TITRATION)

For a long time, the most frequently used method for the isolation of nicotine from the smoke matrix was steam-distillation. Various types of distillation apparatus were used, among which were Kjeldall-type instruments and the Griffith still [38]. Originally designed for the analysis of nicotine in tobacco, the Griffith still became very popular and was also used for the analysis of nicotine in smoke. Instead of the mixture of tobacco and MgO used in the case of tobacco analysis, the alkalized trapping solution, or the Cambridge filter and an alkaline saline solution [17] were placed in the distillation chamber to achieve both extraction and distillation. The purified nicotine was then collected in a receiver containing aqueous hydrochloric acid.

Early methods for the determination of nicotine in aqueous solution were based on the formation of precipitating nicotinium salts. These salts could be generated using the acidic receiver solution from the steam distillation apparatus or by using the acidic smoke-trapping solution directly [39]. To remove neutral smoke constituents, the acidic trapping solution is generally washed with an organic solvent before the precipitate-forming acid is added. For precipitation of the nicotine salts, one of the following two reagents was most frequently used:

Silicotungstic acid was usually added to the acidic receiver solution as a 12% aqueous solution. After heating followed by crystallization overnight in a refrigerator, the precipitate was washed, filtered, and ignited at 650°C. The amount of nicotine was then calculated from the weight of the remainder. This procedure became an AOAC method [40] in 1955. The drawback to this method is that other basic smoke constituents co-precipitate.

Picric acid forms a dipicrate with nicotine specific for compounds with two basic nitrogens. According to Pfyl [39], this reaction can be used for the determination of nicotine directly from the trapping solution. The determination of the amount of the isolated dipicrate is performed by titration of the picric acid with sodium hydroxide solution or in a revised procedure in a second titration step after isolation of the precipitated nicotine picrate by its titration with hydrochloric acid [41]. This method has the drawback that nicotinium dipicrate is, to a small degree, soluble in water. This fact becomes especially crucial for low nicotine concentrations in the trapping solution, which cause erroneously low nicotine results.

Both of these methods are tedious and time consuming. Therefore another, simpler method was developed: the titration of a benzene-chloroform extract, which contains nicotine, with perchloric acid. This method was originally described by Cundiff and Markunas [42,43] for tobacco extracts. It has the advantage of making possible the determination of both nicotine and nornicotine by titrating the extract with and without acetylation of nornicotine. The use of this method for nicotine in smoke led to highly variable results, obviously due to its low specificity for nicotine. With the development of the more reliable spectrophotometric methods, however, 'wet chemistry' methods were rapidly abandoned.

## Photometric methods

Nicotine in acidic aqueous solution has a characteristic absorption spectrum with a maximum at approximately 260 nm [44]. Willits et al. [45] originally suggested a procedure that uses this absorption spectrum for the quantification of nicotine in tobacco samples.

Different methods were proposed for achieving efficient clean-up and correction of the absorbance from interfering components in the extract (background correction).

Willits et al. measured absorbance at three wavelengths, i.e. 236, 259, and 282 nm, and performed a correction assuming a linear background in this spectral region. This procedure was applied to smoke samples, generally by photometrically analyzing the acidic receiver solution from the steam distillation apparatus. The baseline correction, however, caused problems due to the much higher background interference in smoke extracts. As demonstrated by Laurene et al. [46], this interference could be considerably reduced by introducing an additional steam distillation step to distill the acidic smoke solution prior to alkalization for the distillation of nicotine. The evaluation of the absorbance measurements was then performed as described by Willits. In a similar steam distillation/photometric procedure, a graphic evaluation of the ultraviolet absorbance spectrum is suggested by Pleasants et al. [47].

Harvey et al. [48] demonstrated that the two steam distillation steps can be replaced by one extraction step with a hydrochloric acid-methanol mixture followed by clean up with an activated carbon suspension in 50% aqueous glycerol. The background evaluation was again performed graphically by drawing straight lines between the two minima. The results were reported to be close to the steam distillation values.

Another procedure for the elimination of background interference was described by Barkemeyer and Seehofer [49] and later recommended by the

Verband der Cigarettenindustrie [50]. This procedure consisted of only one steam distillation step; the distillate, however, was trapped in water not in acid. Two photometric measurements were performed: one of the aqueous distillate and one of the acidified distillate. The difference between the absorption measurements is proportional to the nicotine concentration.

## Derivatization methods

Compounds like nicotine that contain a pyridine moiety form colored derivatives under the conditions of the Koenig reaction [51]. These derivatives can be used for quantitative determinations as described by Asmus et al. [52]. For the determination of nicotine, the absorption at 509 nm was used, corrected for the absorption of pyridine at 578 nm. The authors demonstrated that their procedure agrees closely with the picrate method.

The same reaction principle was utilized by Sadler et al. [53]. For the determination of nicotine in the steam distillate, he used cyanogen bromide and aniline instead of KCN, chloramin T, and barbituric acid. This method is stable and fast and can be used in a flow-injection analyzer to determine nicotine in a Cambridge filter extract as was described by Charles et al. [54]. The same extract can also be used for the determination of water in TPM, thus saving a second extraction step. The method was compared to the steam distillation/spectrophotometric method and found to agree closely with it. The system can analyze 35 samples per hour.

An alternative flow-injection analyzer method was published by Bourlas et al. [55], who used citric acid and bromomaleic acid, both in acetic anhydride, to form a red nicotine derivative absorbing at 505 nm from the isopropanol extract. An advantage of this method over that of Charles et al. is the use of less toxic chemicals and the fact that no dialyzer step has to be included. A series of compounds were investigated for interference, showing that only tertiary aliphatic amines form an interfering red derivative. Therefore, the method is not responsive to alkaloids like nornicotine.

## Gas chromatographic methods

One of the first chromatographic determinations of nicotine was reported by Quin in 1959 [56]. He generated a benzene extract from a TPM-loaded Cambridge filter. This extract was analyzed by repetitive gas chromatographic runs combined with paper chromatography for nicotine and other alkaloids using polypropylene glycol columns.

Later, in 1967, Lyerly reported on the "Vapor Chromatographic Determination of Menthol, Propylene Glycol, Nicotine, and Triacetin in Cigarette Smoke" [57]. The TPM-loaded filter in a brass filter holder was connected directly to the injection port of the vapor chromatograph. The filter holder was heated to 200°C and the nicotine was purged with the carrier gas on the separation column consisting of copper tubing packed with 14% diethylene glycol succinate on Chromosorb W.

The breakthrough for analysis by gas chromatography (GC) came in 1974 with the publications of Wagner and Thaggard [58], and Randolph [59]. Both groups used the isopropanol extract used for water determination to analyze for nicotine. Wagner and Thaggard used a 4 ft column containing 20% Carbowax 20M on Gas-Chrom R. Randolph used a 6 ft column containing 7% Carbowax 20M, 3% Polyphenyl Ether, 2% KOH phase on Gas-Chrom Q. The methods run isothermically and take approximately three minutes per analysis. The results from the GC procedures and the standard procedures (cyanogen bromide method and steam distillation/photometry) agree closely.

By the end of the 70s, many labs used GC procedures to determine nicotine in mainstream smoke. Several different methods were being used operating with different chromatographic phases. Two collaborative studies were conducted under the sponsorship of the Analytical Methods Committee of the Tobacco Chemists' Research Conference in order to harmonize the procedures and to find the optimum conditions. The results of the two studies were discussed comprehensively by Wagner and Thaggard in 1979 [24]. This discussion resulted in the definition of a procedure recommended to the AOAC. The GC procedure recommended by Wagner and Thaggard became the official method of the FTC in 1980 and is still part of the actual CORESTA [25] and ISO [26] methods.

Meanwhile, packed columns have been substituted in many labs, including those of regulatory agencies, by capillary columns because of their higher efficiency and better reproducibility. Liquid phases similar to those used in packed columns are also used in capillary columns like Carbowax or equivalent phases, or phenyl-methyl siloxane phases. Nicotine is usually detected by means of a flame ionization detector or a thermionic detector. A broad spectrum of different compounds are used as internal standards: from anethole in the AOAC method, to n-heptadecane or quinaldine in the CORESTA and ISO methods, to carvone or quinoline. Table 2 compares the conditions of the FTC/ISO/CORESTA packed column method with the conditions of the capillary GC method used successfully in our laboratories for several years. Figure 2 represents a typical chromatogram obtained with this method for the Kentucky Reference Cigarette 1R4F. The nicotine yield determined by the Tobacco and Health Research Institute under FTC conditions is reported to be 0.90 mg per cigarette [33].

Frequently, the GC instruments used for the determination of nicotine in smoke are also used for the determination of water in smoke. For this purpose, the chromatographs are equipped with a dual injector and an additional thermal conductivity detector for the detection of water. The use of a single injector fitted with a flow-divider provides an alternative for the

501

TABLE 2

Comparison of packed column method described by ISO [26] and capillary GC method

| Parameter | Packed column | Capillary GC |
|---|---|---|
| Extraction solvent containing internal standard: | isopropanol containing *n*-heptadecane or quinaldine | isopropanol containing isoquinoline |
| Injector: | | |
| Type | packed column | split/splitless |
| Temperature | 250°C | 220°C |
| Injection volume | 2 µl | 1 µl |
| Split ratio | — | 1:50 |
| Septum purge | — | 2 to 3 ml/min |
| Column: | | |
| Phase | 10% PEG 20,000 plus 2% potassium hydroxide on acid washed silanized support | (5%-Phenyl)-methylpolysiloxane (DB-5) |
| Length | 1.5 to 2.0 m | 15 m |
| Internal diameter | 2.0 to 4.0 mm | 0.25 mm |
| Temperature | 170°C (isothermal) | 140°C (isothermal) |
| Carrier gas | He or $N_2$ at 30 $cm^3$/min | He at 1.4$cm^3$/min |
| Detector: | | |
| Type | FID | FID or NPD |
| Temperature | 250°C | 250 or 300°C |
| Make-up gas | none | $N_2$ |
| Analysis Time | 6 to 8 min | 2.5 min |

determination of nicotine and water on a single chromatograph. An example of this technique was presented by Deutsch and Jeffords [60] in 1994. They used a 0.53 mm splitter to transfer a single injection onto a PLOT column for water analysis and onto a Carbowax-type column for nicotine analysis. The drawback to this technique is that the gas flows through the two columns cannot be controlled independently.

## VII. QUANTIFICATION OF NICOTINE IN CIGARETTE SIDESTREAM SMOKE

### Sidestream smoke sampling

Sidestream smoke is usually defined as all the smoke issuing from a burning cigarette that does not exit through the mouth-end [61]. This definition excludes the material that might exit through the mouth-end between puffs, and thus makes it possible to employ commonly used smoking machines for the determination.

Sidestream smoke sampling is in effect a two-step process: One first needs to collect all the smoke generated from the space surrounding the burning

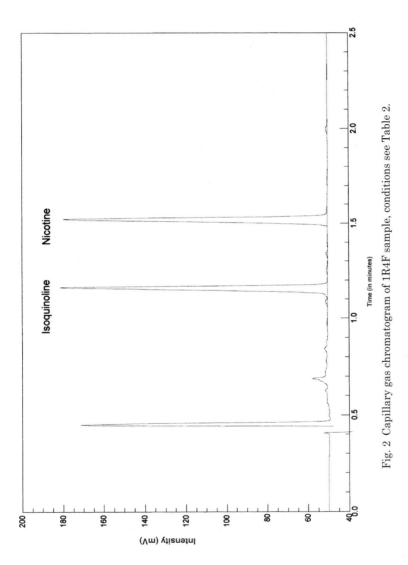

Fig. 2 Capillary gas chromatogram of 1R4F sample, conditions see Table 2.

cigarette, and the component under investigation must then be channeled to a suitable trapping system where it is collected for quantitative analysis. Compared to mainstream smoke analysis, this raises two additional issues. The first one is the design of the trapping system, which will likely have to handle a higher and warmer flow than when sampling mainstream smoke. The second, even more challenging one, resides in performing an efficient and non-disturbing smoke collection.

Over the years, a number of devices have been proposed for collecting sidestream smoke, and their use is discussed in several reviews [2,62,63]. The numerous approaches that are found in the literature reflect in part the differing aims and priorities of the authors.

Three issues should be taken into consideration when assessing the performance of an analytical set-up for the optimization of its operation:

The collection apparatus should not alter the smoking of the cigarette, and in particular should not change mainstream yields or puff count [2,64].

The collection of the sidestream smoke should be complete.

Alterations in smoke composition while collecting the sidestream smoke may happen due to the increased temperature next to the lit cigarette, to interaction with the apparatus walls, or as a result of reactions within the smoke. These changes should be minimized by careful design of the collection and trapping systems, and taken into account when reporting.

In practice the above issues present the experimenter with very demanding challenges, making any sidestream smoke determination more difficult and less precise than its mainstream counterpart. This also implies that at least as many cigarettes should be smoked for the determination of sidestream smoke constituents as for the determination of mainstream smoke constituents (20 to 40 for nicotine).

## VIII. COMPARATIVE EVALUATION OF DIFFERENT COLLECTION PROCEDURES

A number of different experimental approaches have been proposed for sidestream smoke analysis. Some authors smoked cigarettes in a sealed test chamber (static), while others used a small collection device continuously purged with air (dynamic). Many early collection systems were aimed at sampling the smoke from a small number of cigarettes. As methods for trapping and quantifying target components were developed, however, the need for smoking higher numbers of cigarettes with limited time and labor resources led to new developments in the smoke collection procedures.

These different options are reviewed and discussed in this chapter with specific reference to nicotine sampling and analysis.

## Enclosed test chamber

An obvious alternative to capturing sidestream smoke from the immediate vicinity of the cigarette is to smoke a number of cigarettes in a sealed chamber of known volume. Provided that the mainstream smoke is not released into the chamber, measuring the concentration of the analyte being investigated in the chamber air provides a means to calculate the sidestream smoke yield. This approach has merits and has been followed by many teams, principally in the early phases of research on the generation and dynamics of environmental tobacco smoke [65]. It suffers from a major drawback, however, in that the sampling cannot be initiated before completion of the smoke run and homogenization of the enclosed atmosphere. As a result, an important influence of aging effects within the smoke can be expected. Chemical reactivity and sorption effects may have a very considerable impact since a high smoke density is often needed to ensure precise analytical determination. In the specific case of a nicotine determination, it has been shown that the chemical stability of the compound was not an issue, but that sorption phenomena to the walls are massive and very fast [66–68]. Their effect is so important and their kinetics so complex [69] that they will preclude any valid yield determination from being obtained under these conditions. This means that the continuous collection of sidestream smoke from the immediate vicinity of the cigarette at the time when it is being produced is the only acceptable method for an accurate assessment.

## Early collection flasks

Over the years, a number of devices have been proposed for direct collection of the sidestream smoke from a burning cigarette [2,62,63]. The respective merits of the different approaches proposed in the literature were discussed extensively by Proctor et al. [64], McRae [63] and Klus [27].

In the first generation of chambers, the cigarette was enclosed in a flask with an internal volume of a few hundred cubic centimeters designed to receive one cigarette at a time. The flasks were fitted with a port for lighting the cigarette and with an air inlet, often fitted with a diffuser to ensure a more even distribution of the incoming air, and sometimes with a water jacket to cool the walls. The air was drawn from a top outlet through a Cambridge filter, usually at flow rates of 0.5 to 1.5 l/min.

This kind of apparatus design and operation did not fulfill the requirements for smoke collection. One reason for this failure is that the temperature and humidity inside the flask are extremely high. One of the most frequently cited devices, the sidestream chamber developed by Neurath and modified by Brunnemann et al. [70], can be taken as an example (Fig. 3). It was reported that the temperature of the stream exiting the chamber could reach up to 90°C [71]. In addition, due to the amount of water

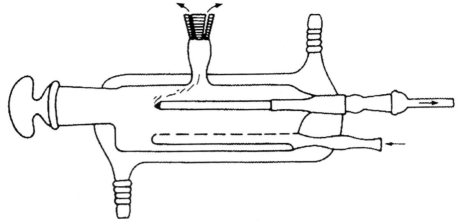

Fig. 3 Sidestream collection device according to Brunnemann et al. (1969) [70].

generated by the burning cigarette, the relative humidity was at saturation, resulting in visible condensation on the walls. A second cause for the failure is that the smoke remains for a substantial time within the enclosure before it is cleared. For some compounds, reactions and losses occur. Smoking the cigarette in an atmosphere of concentrated sidestream smoke also influences mainstream smoke composition. The importance of these effects was clearly demonstrated, again in the case of the Neurath-type device, where sidestream smoke yields in dimethylnitrosamine varied seven-fold when the flow rate through the flask was changed from 0.25 to 1.5 l/min [72].

The reason for the failure of these devices to properly clear the smoke from the vicinity of the cigarette has been identified. A detailed determination of the velocity distribution of the plume in the open was made by Robinson using a laser Doppler velocimeter [73]. A mathematical model was found to fit the results by treating the system as a density-driven natural convection plume [74]. Through this experimental study, Robinson showed that the upward velocity increases with distance above the burning zone, reaching more than 60 cm/s at a vertical distance of 7 cm as air is drawn radially into the convection plume. Measured 5 cm above the cigarette, this translates into an upward, laminar mass flow of 76 mg/s. Computer modeling predicts that this value will roughly double at twice this distance [74]. Consequently, a crude calculation shows that in a flask sufficiently wide to allow some convection to take place, flow rates in excess of 6 l/min would be needed to avoid recirculation of the smoke back to the cigarette.

To accommodate these flow rates, a second generation of chambers was introduced. Browne et al. [75] reported satisfactory smoking conditions using a flow rate of 5.9 l/min through a vertical cylinder 9 cm in diameter (Fig. 4). Kaneki et al. [76] investigated the effect of varying the flow up to 15 l/min through a vertical cylinder 8cm in diameter. Their data showed that the

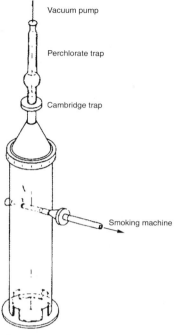

Fig. 4 Sidestream collection device according to Browne et al. (1974) [75].

highest flow rates provided mainstream smoke yields and puff-counts that were the closest to the values obtained without the chamber.

This experimental approach is very sound, but only permits the smoking of one cigarette at a time. As a consequence, it is not ideal for research studies on minor components because the collection of large amounts of smoke would be tedious.

## Systems allowing a high sample throughput

Several teams have investigated the possibility of collecting the sidestream smoke from many cigarettes at the same time. Several systems based on rotary multi-port machines were proposed. Harris et al. used a set of small glass chimneys fitted over each port of a 20-port carrousel machine [77]. The operation was delicate in that the hoods needed to be constantly repositioned. Chortyk et al. proposed modifying a 30-port machine by adding a large glass funnel to collect the sidestream smoke from the cigarettes and bent glass cigarette holders to fit 15 cigarettes underneath [78]. The impact of the collection system on the mainstream yields was not presented and in addition, the reproducibility of the results was questioned [79]. In their reply [80], the authors stated that the airflow through the system was 5 l/min, which would appear to be too low to properly draw the smoke away from the cigarette. To generate high amounts of fresh sidestream smoke and feed it into a test room, Ueno et al. used a modified multiport machine in a sealed enclosure [81]. The puffing could thus be performed by applying a positive pressure and the sidestream smoke could be directly flushed. This procedure is not suitable for nicotine analysis because of the impossibility of properly accounting for losses to the internal surfaces.

## The 'fishtail chimney' collection device

### Concept of a narrow chimney

A solution to the problem of collecting the sidestream smoke from several cigarettes at one time was proposed by Proctor et al. [64]. The cigarette is enclosed in a bottomless, narrow chamber with a long vertical neck, often referred to as a "fishtail chimney" (Fig. 5). The walls are only 4 mm from each side of the cigarette and the total width is about 18 mm, which allows a series of these chambers to be mounted side by side on most non-rotating smoking machines. As an example, Wingate et al. reported on a 20-port linear smoking machine that was modified to include a sidestream smoke collection capability with fishtail chimneys [82]. The results obtained with

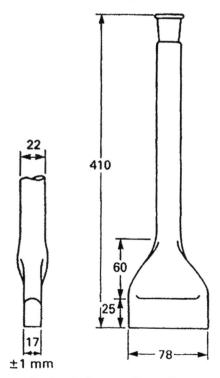

Fig. 5 Sidestream collection device according to Proctor et al. (1963) [64].

the device were reported to compare favorably to those obtained with a traditional collection system. The vessel has no bottom and just needs to be lowered into place to enclose the cigarette immediately after it has been lit. The simplicity of this manipulation makes it possible to automate part of the operation. The smoke losses through the open bottom of the flask can be prevented by placing a plate underneath the cigarette, leaving only a small gap between it and the chimney when it is in the lowered position. The air velocity through this restriction is such that no smoke diffuses out of the apparatus. The influence of the gap width is discussed below.

There is an additional benefit in adopting a narrow configuration for the collection flask: the sidestream smoke can be cleared efficiently from the cigarette at relatively low flow rates because the restriction at the level of the cigarette increases the air velocity around the burning cone. Operating with a low collection flow rate is an obvious advantage for the subsequent trapping step.

*Trapping procedures*
Cambridge glass fiber filters are a proven and effective means of retaining the particulate matter in cigarette smoke at room temperature. They are thus an obvious choice as the first element in the sampling train. The first investigations on fishtail chimneys have shown the influence of the distance between the cigarette and the filter pad on the temperature of the smoke as it reaches the filter pad. Depending on this distance and on the flow rate of air through the system, a variable amount of gas phase nicotine will pass through the filter. It is necessary therefore to add a back-up trap behind it. A wash-bottle with a solvent such as isopropanol is sufficient to ensure a complete trapping of the nicotine that passes through the filter. Finally, a small amount of particulate matter and some nicotine can be found as a deposit on the chimney walls. To ensure a complete collection, each flask should be rinsed with a small amount of the solvent used in the wash-bottle.

Perfetti et al. [13] published the amounts of nicotine found in the different trapping elements when a fishtail chimney identical to the one described by Proctor et al. [64] is used. Sixty-eight to 78% of the nicotine in sidestream smoke was found on the Cambridge filter, 9 to 17% in the flask washings, and 11 to 18% in the impinger.

*Recent applications*
This kind of collection system has been widely used since the initial publication of the method, and its operation has been investigated and discussed by several authors. All used a flask geometry that was similar to that proposed by Proctor et al. [64], the variations reflecting only the different lengths of the cigarettes to be analyzed. In this chapter, what is referred to as a fishtail chimney is the one described by Proctor et al., unless

otherwise specified. As reported by Klus [27], the use of a fishtail chimney was investigated through a cooperative method development within a CORESTA task force. This collection device is also the one proposed by Health Canada as part of a draft method for the quantification of nicotine in sidestream smoke. The methodological prescriptions in this chapter are centered on the use of fishtail chimneys.

## IX. SUGGESTED METHOD FOR SIDESTREAM SMOKE ANALYSIS

The sidestream smoke from a pre-conditioned cigarette smoked under a defined procedure for mainstream smoke generation is collected by means of a fishtail chimney at a flow rate of 2.5 to 3 l/min. The particulate matter generated by 3 cigarettes is retained on a Cambridge filter and the gas phase components are trapped by means of a wash-bottle containing cooled isopropanol. The chimney walls are rinsed and the filter is extracted with isopropanol. The solutions are combined and nicotine is quantified by capillary GC using anethole as an internal standard (for details see addendum).

### Effect of operating parameters

*Number of cigarettes smoked*
As reported by Proctor et al. [64], the amounts of sidestream smoke particulate matter recovered on a Cambridge filter were found to be additive when up to three cigarettes were smoked on the same filter pad. It is thus recommended to smoke three cigarettes on the same port without changing the filter pads.

Taking into account the variability between individual cigarettes, at least three or four determinations performed on three cigarettes each should be made for the result to be derived from a statistically acceptable sample.

*Relative humidity*
Tobacco moisture content has a marked influence on the transfer of nicotine into sidestream smoke. Neurath et al. [83] showed that this transfer changed from 3.9 mg/cig. at 3% tobacco moisture content to 4.3 mg/cig. at 22% tobacco moisture content. Equilibration of the cigarette in ambient conditions with a fixed relative humidity (RH) before smoking is therefore a critical step.

The relative humidity of the air at the time of smoking is also very important. In a communication to the CORESTA task force on sidestream smoke (reviewed by Klus [27]), Green presented the results obtained when 1R4F cigarettes, previously equilibrated at 60% RH, are smoked in a room where the RH is varied. The results available from Baker [84] show that the sidestream smoke nicotine yield of this cigarette varied under these conditions from 5.71 mg at 13% RH to 6.18 mg at 80% RH. This emphasizes the need for good control of the climatic conditions in the smoking laboratory where the determination is made, even when the cigarettes have been properly conditioned.

*Flow rate*

The airflow rate through the chimney is probably the parameter that has been studied the most in fishtail operation. It is an interesting feature of the fishtail collection devices, that changes in the flow rate have a smaller influence on the mainstream smoke yields than they do in similar but larger sidestream smoke collection devices (chimneys 3 cm wide or larger), as was demonstrated by Piadé et al. [7]. These authors noted, however, that the influence of the flow rate on static burn rate was still important in all devices, and particularly in the fishtail chimney.

In their original work, Proctor et al. [64] operated at a flow rate of 2 l/min. According to the studies by Piadé et al. [7], video observation under contrasted lighting demonstrated that a flow of about 3 l/min was in fact needed to effectively pull the smoke into the tubular neck of the chimney without any convection. Moreover, the conditions producing mainstream smoke yields that would best match those found without the collection system require flow rates above 2.5 l/min.

A limitation to increasing the flow is brought about by the resulting increase in temperature of the cigarette cone, as reported by Piadé et al. who measured this temperature in different collection flasks with a high-resolution infrared camera. They observed the largest increase with the fishtail chimney.

For a fishtail chimney with a geometry close to that proposed by Proctor et al., the best compromise appears to be in the 2.5 to 3 l/min range.

Finally, it should be kept in mind that with any collection system, increasing the flow may lead to losses of nicotine if the trapping system is not adapted. This effect explains some of the discrepancies that can be found in the literature.

*Width of the gap to horizontal plate*

Haut [85] calculated the impact on air velocity of varying the gap between the plate and the chimney. This parameter is not critical, but a wide gap will lead to losses, whereas a very narrow gap will result in air being drawn mostly from the vicinity of the cigarette filter instead of over the full cigarette length. The recommended range of operation is 3 to 4 mm.

*Distance to the pad and nicotine retention by the filter*

The TPM retention efficiency of the Cambridge filter was systematically investigated by Kaneki et al. [76]. In their study they varied both the distance between the cigarette and the Cambridge filter and the air velocity through the pad, but they did not measure nicotine breaking through the filter. This issue was also studied by Proctor et al. when they assessed their fishtail chimney prototype [64], initially without a wash-bottle behind the filter. They tested a series of chimneys with different neck lengths at 2 l/min, and found that a distance between the cigarette and the filter pad of more than 36 cm was needed to ensure that the amount of nicotine recovered on the Cambridge filter would be constant. At this distance, the smoke temperature is within 2°C of ambient temperature.

The conclusions from these studies can be rationalized as follows: Sidestream smoke temperature rapidly drops to ambient as it is diluted with the incoming air, causing a decrease in the gas phase concentration of the semi-volatile compounds such as nicotine, which then start to evaporate from the particles. The retention of nicotine on a Cambridge filter is thus going to be influenced by the distance between the cigarette and the filter pad and by the smoke temperature as it reaches the filter pad. A shorter distance or a higher smoke temperature will result in an increase in the amount of gas phase nicotine that passes through the filter. Since nicotine may evaporate from the material deposited on the filter pad, the air velocity through the filter material will also play an important role. A higher airflow will result in an increase in the amount of nicotine that passes through the filter. This sensitivity of the Cambridge filter retention to sampling conditions is one of the main reasons why adding a wash-bottle behind the filter is strongly recommended.

*Atmospheric pressure*
The effect of atmospheric pressure on sidestream smoke yields obtained with the fishtail chimney was investigated by Gonzàlez et al. [86]. In their study, the barometric pressure was varied between 690 and 755 Torr, values within the acceptable range specified by the ISO norm. It was reported that a lower pressure increased the yield of nicotine in the particulate phase by about 10%. These variations may be within the scope of analytical uncertainty.

*Effect of changing the cigarette smoking regimen*
Sidestream smoke yields depend heavily on the smoking regimen used. This is because sidestream smoke generation is influenced by the occurrence of a puff. Altering the smoking regimen will change the puff count, and thus the ratio of mainstream to sidestream smoke.

As an example, Rickert [87] investigated a set of conditions that he described as 'intense smoking'. This consisted of occluding half of the ventilation holes and drawing a 48 ml puff for 2.4 s with an interval between puffs of 44 s. He found that the yield of nicotine in sidestream smoke under these conditions was lowered by 7% whereas the yield of nicotine-free dry particulate matter was lowered by 20%. A larger effect is expected if the smoking regimen specified by the British Columbian government (see page 497) is selected. From a survey of sidestream smoke yields reported for a series of commercial products on its web-site [88], a very consistent 21% (range 17 to 25%) reduction is observed when comparing the 'intense smoking' to the 'standard smoking' nicotine sidestream deliveries.

## Validation and performance of the method

Methods for sidestream smoke analysis that are based on the utilization of fishtail chimneys have been practiced in different laboratories for over a decade. The validity of their use as a means for collecting the sidestream smoke of a cigarette has been repeatedly challenged. Published reports on this subject are scarce, however.

The original paper by Proctor et al. [64] that first described the fishtail sampling of sidestream smoke (1988) provides an estimate for the coefficient of variation (CV) found in the determination of the total yield of nicotine in sidestream smoke. According to these authors, the CV would be better than 6% for successive smoke runs.

Using the same device and flow rate, Evans [89] reported a CV of 12.4% for successive determinations of nicotine in sidestream smoke for a 1R4F Kentucky Reference Cigarette. For the same cigarette, operating in about the same way but without rinsing the chimney walls, Haut [85] reported a CV of 7.5%.

The Kentucky Reference Cigarettes [34] have become a benchmark for analytical method development and standardization. A collection of the published values for the nicotine yields in mainstream and sidestream smoke of the 1R4F cigarette is given in Table 3 as a reference for assessing the accuracy of a determination. These yields were all determined employing a fishtail chimney, and the flow rate used for the sidestream collection is reported.

TABLE 3

Collection of published analytical data obtained for the 1R4F Reference Cigarette

| Reference | MS 'tar' (mg/cig.) | MS nicotine (mg/cig.) | SS nicotine (mg/cig.) | Flow rate (l/min) | Remarks |
|-----------|--------------------|-----------------------|-----------------------|-------------------|---------|
| [88] | 10.5 | 0.84 | 4.5 | 2.0 | — |
| [85] | 12.1 | 0.83 | 4.5 | 2.5 | no chimney rinse |
| [86] | 11.6 | 0.95 | 5.6 | 3.0 | no backup trap |
| [13] | 9.0 | 0.87 | 6.07 | unspecified | — |
| [2] | — | — | 5.6 | — | unspecified method |

Note: Data obtained with the fishtail chimney for sidestream smoke collection, with backup trap and chimney rinse, unless otherwise noted.

For comparison, the values reported by Guerin et al. [2] for the sidestream smoke yields of nicotine for the 1R4F are included.

Additional information can be found among the collection of tobacco research documents posted on Internet sites for public inspection. A 1989 report [84] provides a set of results regarding the analysis of this same 1R4F Reference Cigarette by different laboratories using fishtail chimneys, albeit with slightly different procedures.

The relevant data have been consolidated in Table 4. They show that the variability among the laboratories is significant, probably due to differences within the cigarettes, their equilibration, or ambient conditions. Within the same laboratory, however, puff counts and mainstream smoke yields are little affected by the sidestream smoke collection when a fishtail chimney is in place. This was demonstrated by Perfetti et al. [13] who showed that, with the possible exception of CO and $CO_2$, the difference caused by the use of a fishtail chimney is of no practical significance and usually not statistically significant.

From the results in Table 4, the average value for the sidestream smoke nicotine yield is 5.4 mg/cig.

# X. DETERMINATION OF NICOTINE ENANTIOMERS IN CIGARETTE SMOKE

## Stereochemistry of nicotine

The nicotine molecule contains a chirogenic center at the 2'-position of the pyrrolidine moiety. Thus, nicotine exists as two enantiomers: R- and S-nicotine.

TABLE 4

Results obtained for the 1R4F Reference Cigarette with a fishtail-type chimney, from data found among the collection of tobacco research documents posted on Internet sites for public inspection [84]

| Parameter | Laboratory | | | | | | | |
| --- | --- | --- | --- | --- | --- | --- | --- | --- |
| | 1 | | 2 | 3 | | | 4 | |
| | | | | Chimney | | | | |
| | no | yes | no | yes | no | yes | no | yes |
| Flow rate (l/min) | — | 3 | — | 2 | — | 3 | — | 2 |
| Puff count | 9.3 | 9.0 | 8.8 | 9.2 | 9.0 | 9.4 | 9.4 | 9.4 |
| MS 'tar' (ms/cig.) | 12.9 | 12.4 | 9.4 | 9.6 | 8.7 | 8.5 | 8.3 | 8.5 |
| MS nicotine (mg/cig.) | 1.1 | 1.0 | 0.8 | 0.8 | 0.8 | 0.7 | 0.7 | 0.7 |
| SS nicotine (mg/cig.) | — | 4.9 | — | 6.0 | — | 5.5 | — | 5.4 |

R-(+)-Nicotine          S-(-)-Nicotine

S-Nicotine is reported to be the only naturally occurring enantiomeric form of nicotine [90]. Since the metabolism and biokinetics of the two enantiomers are different [91–94], it would be of great interest to know if R-nicotine is formed under the conditions of cigarette smoking by the racemization of S-nicotine. Published results concerning the amount of R-nicotine in mainstream smoke vary from no R-nicotine [95] to significant amounts of R-nicotine [90,96]. A prerequisite for the determination of nicotine racemization in smoke is a method for the determination of the enantiomeric ratio. Only a few methods have been described.

**Methods for the determination of the enantiomeric ratio of nicotine**

*Polarimetry*
From the measurement of the optical rotation of a solution of nicotine enantiomers, the excess amount of one of the enantiomers can be calculated directly. For this determination, a sufficient amount of nicotine of high purity is needed. Even small amounts of impurities may influence the results, e.g., small amounts of acid will result in erroneously high values for R-nicotine because the S-nicotinium ion is dextrorotatory whereas the free base is levorotatory.

*[1]H-NMR spectroscopy*
Crooks et al. [96] demonstrated that [1]H-NMR spectroscopy in conjunction with a chiral lanthanide shift reagent can be used to determine the enantiomeric purity of nicotine samples. The authors used *tris*(3-(hepta-fluoropropylhydroxymethylene)-d-camphorato) europiumIII (EU(hfc)$_3$) and obtained signal splitting for the protons in the 2- and 2'-position, and the N-methyl protons. The splitting due to the formation of a diastereomeric complex was largest for the protons in the 2-position. Therefore, these signals were used for the analysis of mixtures of the two stereoisomers. A similar method was published earlier by Chavdarian et al. [97] for the determination of the enantiomeric purity of various nicotinoids.

## Gas chromatography

A gas chromatographic method for determining the enantiomeric purity of nicotine was described by Jacob et al. [91]. Nicotine was demethylated to nornicotine via nicotine $N'$-oxide followed by conversion to diastereomeric amides by derivatization with ($-$)-camphanic acid chloride. The two diastereomers can be baseline separated on an achiral GC column (25 m × 0.2 mm SE-54), but the derivation steps are time consuming.

The direct gas chromatographic separation of the nicotine enantiomers was described by Perfetti and Coleman [98]. They used a combination of two different cyclodextrin-based columns each of 30 m length and a mass selective detector. The nicotine enantiomers eluted after approximately 160 min and were nearly baseline separated. A detection limit of 2.5% S-nicotine of total nicotine was reported.

## High performance liquid chromatography

The enantiomeric resolution of racemic nicotine by microcolumn liquid chromatography using a chiral mobile phase was demonstrated by Armstrong et al. [99]. The column was a 1 m × 250 μm I.D. microcolumn packed with 5 μm $C_1$ stationary phase. The mobile phase was acetonitrile/water saturated with β-cyclodextrin. With the flow rate kept at 1.3 μl/min, the nicotine enantiomers eluted with baseline separation after 3 to 4 h.

A direct method for the separation of nicotine enantiomers using a chiral stationary phase was reported by Demetriou et al. [100] who employed a commercially available $\alpha_1$-acid glycoprotein (AGP) column. Under the conditions described, i.e. mobile phase pH 8.2, decanoic acid and methanol as modifiers and elevated temperature, the enantiomers were baseline separated within 5 min.

## Racemization of nicotine during smoking

Most of the methods capable of differentiating between nicotine enantiomers have been used in order to determine the racemization of nicotine during the smoking process.

Generally, additional extensive clean-up procedures have to be employed because the large number of interfering smoke constituents have to be separated prior to the last step of chiral differentiation. This is particularly important for those methods that have no or a limited ability to differentiate between nicotine and compounds from the smoke matrix, like polarimetric methods or $^1$H-NMR spectroscopy.

Crooks et al. [96] used liquid-liquid extraction followed by two vacuum distillation steps or one vacuum distillation step combined with preparative high-performance liquid chromatography (HPLC) on a cation exchanger. As a validation of the method, they took an S-nicotine standard through the total purification process without observing any racemization. The purified smoke nicotine was analyzed by ¹H-NMR spectroscopy with the chiral shift reagent. The different types of Kentucky Reference Cigarettes investigated yielded a broad range of values for the amount of R-nicotine in the total nicotine: from 3.6% R-nicotine for the 2R1 cigarette to 11.8% R-nicotine for the 1A4 cigarette.

With the cyclodextrin-based GC method, Perfetti et al. [13] determined the amount of R-nicotine in the smoke of 6 Kentucky Reference cigarettes to be around 3% of the total nicotine for all cigarettes with the exception of 1R1, which was found to be below the detection limit of 2.5%. Smoking according to an alternative puffing regimen did not significantly change these results.

When the method was applied to sidestream smoke, again about 3% R-nicotine was found [98]. The detection limit for this method did not facilitate the detection of R-nicotine in tobacco.

The racemization of nicotine in the smoke of the 2R1 has also been investigated in our laboratories employing the HPLC method of Demetriou et al. [100]. To prevent artificial racemization, we used an extremely mild clean-up procedure for the nicotine avoiding all heating steps. The procedure (Fig. 6) consisted of a liquid-liquid extraction followed by two solid phase extractions and preparative HPLC. The purpose of the additional final solid phase extraction step was to concentrate the eluate from the HPLC column. The complete procedure was validated in triplicate:

First, an S-nicotine standard was taken through the total purification process.

Second, the same was done with a pure R-nicotine standard. Both standards showed no racemization under the clean-up conditions.

Third, a mixture of radio-labeled S- and R-nicotine was added to the total particulate matter collected on a Cambridge filter. The eluate of the final HPLC was monitored using a radio-activity detector. Under the clean-up conditions applied, the ratio of the added mixture of enantiomers did not change.

The third validation step was found to be very important because it showed that with the large excess of S-nicotine, the nicotine enantiomers separated partly on the achiral cation exchange column. A similar effect had already been reported by Cundy and Crooks [101]. It is important to collect all the nicotine from the preparative column, otherwise the ratio of the nicotine enantiomers may change during the clean-up procedure.

Our procedure yielded a significantly lower value than that of Crooks and of Perfetti. The mean of five determinations showed R-nicotine to be 1.64±0.04% of the total nicotine. The amount of R-nicotine in the tobacco filler of the 2R1 was determined with the same method and found to be 0.11±0.01% of the total nicotine. Assuming the nicotine in the tobacco filler to be the only source for R-nicotine in smoke, the conversion of S-nicotine into R-nicotine during smoking is about 1.5% for the Reference Cigarette 2R1.

*Analytical Determination of Nicotine*

TPM/toluene extract

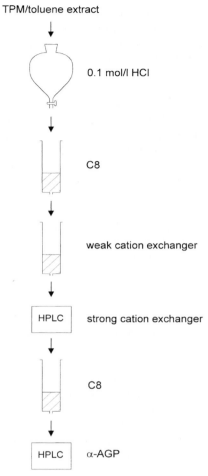

Fig. 6 Clean-up procedure for the determination of the enantiomeric ratio of nicotine in TPM of reference cigarette 2R1.

## XI. ADDENDUM

### Equipment and setting for sidestream smoke collection

*Laboratory*
The laboratory in which the determination is to take place should have its air and humidity controls set exactly as is done for mainstream smoke determinations. The exhaust from the pumping system should be evacuated outside of the laboratory.

518

*Smoking machine and ancillary equipment*

The determination can be made using any machine that can fulfill the requirements for the standard procedure of mainstream smoke analysis. Rotary machines can only be used at the cost of rather cumbersome adaptations and increased complexity of use, so single port or linear machines are normally preferred. It was reported by Allman [102] that an automated machine dedicated to sidestream smoke collection had been designed on the basis of the Filtrona SM302. Blueprints were produced, but no machine; however, all the newer Filtrona smoking machines are designed to be easily adapted for sidestream smoke collection.

The availability of equipment and the number of samples to be processed dictates the choice of the smoking machine. When numerous determinations are required, a multi-port linear machine is preferred, such as the 4-port Borgwaldt RM-4, or, from Filtrona, the 8-port (SM302 or 342) or the 20-port (SM300, 350 or 400) smoking machines. Wingate et al. [82] reported the modification of a 20-port linear smoking machine that was fitted with fishtail chimneys. Note that in laboratories where a 20-port smoking machine is used, it is often the case that 1 port out of 2 or 4 is effectively equipped with a chimney.

*Additional equipment*

An adjustable holder should be available for each fishtail chimney. It should allow an easy vertical movement of the chimney with sufficient clearance to permit lighting and removing the cigarette when the device is in the upper position.

A plate should also be fitted horizontally below the cigarettes in such a way that when the chimneys are lowered into position over the cigarettes, the bottom rim is 3 to 4 mm above the plate. The purpose of this plate is to ensure that no smoke leaks out of the enclosure through the bottom. Once selected, the width of the gap should be kept constant.

*Several further modifications may be considered in some instances*

The adjustment of the chimneys must be both precise, so that the flask is always correctly positioned with respect to the cigarette and the bottom plate, and sufficiently quick to ensure that smoke is sampled immediately after lighting, keeping losses to a minimum. In practice, a manually operated system is acceptable for a single port machine. It can even be sufficient for handling a 4-port machine such as the RM-4, in which the puffs are taken out of phase at 15-s intervals. It is impractical to operate a multi-port smoking machine that takes simultaneous puffs on all ports without a motorized raising and lowering of the chimneys with pre-set stopping positions.

Another useful improvement is the automated detection of the time when the burn line has reached the selected butt length. A sensing device can be used to cause the puff to be terminated and to end the smoking of the cigarette. Fishtail chimneys are compatible with the widespread use of a cotton thread, which breaks when reached by the burn line. A yet more convenient solution consists in the use of an infrared sensor that can easily be mounted on the plate under the cigarettes.

### Fishtail chimneys

Fishtail chimneys were first described by Proctor [64]. Made of glass, they can be obtained from many glass-blowing operations. They have been made to various sizes to accommodate the different cigarette lengths and diameters. For instance, some researchers have used a longer model for collecting the sidestream smoke from very long cigarettes (e.g. 120 mm) [103]. Minor variations due to the availability of specific profiled glass rods were also reported.

In practice, the model represented in Fig. 3 will fit most of the cigarettes commonly analyzed, and it is the one that is referred to in this section.

### Collection train

The chimneys are fitted with a ground glass joint at the top, to which a filter holder can be connected.

A filter is normally used in order to collect the particulate fraction of the sidestream smoke. It is the same as for the mainstream smoke determinations – a Cambridge filter with a 44 mm diameter.

The nicotine that breaks through this filter is collected in a wash-bottle containing 50ml of trapping solution. The dipping tube can be fitted with a very coarse fritted glass diffuser to increase the interface between the gas and liquid phases, but equivalent results are obtained with an open tube and no diffuser. Because of the high flow-rate through the flask, care should be taken to leave a sufficient overhead volume over the liquid to avoid liquid being sucked into the piping. It is also a recommended practice to cool the flask in an ice bath to minimize evaporation.

When nicotine is the only analyte to be determined, a sampling train of two 100 ml gas wash-bottles containing isopropanol could be used and the filter omitted, or alternatively an acid impregnated filter can be used with no back-up gas wash-bottle. These approaches have been used in some laboratories but no real validation has been published.

When a determination of carbon monoxide yields is to be performed, the exhausts from the pumps can be directed to an analyzer and the response integrated, or they can be collected in a bag.

### Pumping system

A single pump can be used with a proper manifold when cigarettes are smoked on several ports at the same time. However, because resistance to flow may vary differently in the different sampling lines over the duration of the sidestream smoke collection, provision should be made for a reading of the actual airflow rate through each chimney and for means to adjust the flow separately on each port.

## Procedure for cigarette smoking and smoke collection

### Cigarettes
The cigarettes are marked with a pencil on the paper lap seam for insertion depth and butt length. They are sampled and conditioned exactly as is done for mainstream smoke determinations. In addition, it is advisable to further select them by weight and resistance to draw. The cigarettes selected are those within one standard deviation of the mean of the sample (typically about 50 cigarettes).

### Extraction, stock and standard solutions
The nicotine that breaks through the Cambridge filter is trapped in isopropanol. Anethole is often used as an internal standard for the GC analysis at a typical concentration of 200 µl/l. Some authors [86] prefer to use a nitrogen-specific detector in their GC quantification and use quinaldine as an internal standard.

A nicotine stock solution is prepared by weighing about 100 mg of pure nicotine into a volumetric flask (wear gloves and operate under a fume hood) and filling up to volume with extraction solution.

Typical concentrations for the calibration standards for GC analysis are obtained by diluting the stock solution with extracting solution in ratios ranging from 0.5:100 to 5:100.

If a water determination is foreseen on the extracts, anhydrous conditions should be maintained and an internal standard for the water determination should be added.

The solutions should be stored in the dark, sealed with paraffin film.

### Preparation of the equipment
Before each smoke run, the following parameters are controlled:

*Calibration of the smoking machine.* After warming up the smoking machine, place the chimneys in the upper position, then check puff duration, puff frequency and puff volume as for mainstream smoke determination.

*Preparation of the filter holders.* Surgical gloves should be worn at all times when handling filter holders and pads. Place a Cambridge filter pad into each holder with the rough side facing the smoke inlet of the holder. If a gravimetric determination of particulate matter is desired, the Cambridge filter pads should be equilibrated by storing them in the smoking room overnight and tare weights should be obtained with a resolution of 0.1 mg.

*Preparation of the gas wash-bottle.* Pour exactly 50 ml of extraction solution into each gas-wash-bottle, and connect the bottle inlet to the outlet of the CF holder.

*Analytical Determination of Nicotine*

*Cigarette positioning.* Insert the cigarettes to the mark into the filter holders connected to the smoking machine. Lower each chimney to check that the cigarettes are horizontal and properly centered with respect to the collection flask.

*Adjustments.* Adjust the sensing device used to terminate the smoking when the burn line reaches the required butt length. If the collection flask has been removed, its positioning should be checked and the distance to the horizontal plate adjusted to 4mm. The lighter operation should be checked at this stage whenever it is not a manual operation on an individual channel.

*Airflow adjustment.* With the chimneys in the lower position and the trapping system in place, check that the flow rate is adjusted to the required flow with a flow meter.

It is advised to operate with a flow rate between 2.5 and 3 l/min. The airflow can be fine-tuned within this range to reflect local conditions so as to obtain the best match of mainstream smoke yields compared to those of a run performed without the sidestream sampling equipment in place. The setting should not be altered once selected.

*Smoking and smoke collection*
The chimneys are put in the upper position and the pumps for the sidestream smoke collection are started immediately before lighting the cigarette. The cigarette is lit and the chimney is lowered as soon as possible thereafter. When the butt-mark is reached, the chimney is raised and the cigarette is removed with forceps and extinguished. The chimney volume is about a third of a liter. The flow on the corresponding channel is thus maintained for 15 s after the time the butt mark is reached to ensure that the chimney volume has been cleared.

If a cigarette fails to light or self extinguishes, it is improper to re-light it and the analytical determination on the corresponding channel should be discarded.

Three cigarettes are smoked on each port on the same set of Cambridge filters. It is good practice to include a monitor cigarette in the smoking schedule. A check on the analytical results obtained for this monitor will allow for the screening out of those smoke runs in the course of which some anomaly occurred. These results should be discarded and the whole set of determinations repeated.

*Rinsing and extraction*
When the three cigarettes have been smoked, the Cambridge filter holders are opened and the filters are removed with forceps, folded in four (condensate inside) and used to wipe the inside surface of the holder. The pads are then extracted, as is done for mainstream smoke analysis, in a conical flask containing exactly 2 ml of extracting solution for 45 minutes on a rotary shaker at ambient temperature. Do not disintegrate the pad.

The chimney is freed from the holder, held upside down over a collection vessel and rinsed twice with isopropanol.

If the nicotine yield is the only analyte to be determined in the smoke run, the filter pad extract, the wash-bottle contents, and the chimney washings are pooled. An aliquot is filtered and analyzed by gas chromatography as is done for mainstream smoke analysis.

When the yield in nicotine-free dry particulate matter is also to be measured, the procedure should be altered to allow the separate determination of nicotine in the filter pad. The amount of particulate matter deposited on the chimney wall should also be estimated. This is done by determining the amount of particulate matter and nicotine found on the filter, and by applying a proportion factor to this value. This factor can be derived from the ratio of either spectroscopic absorbance of extracts or of the nicotine amounts found in the respective locations.

# REFERENCES

1 Baker RR, Robinson DP. Tobacco Combustion – the Last Ten Years. In: Durocher DF, Bell JW, Sisson VA (Eds.). Symposium of the 44th Tobacco Chemists' Research Conference; 1990 Sept 30–Oct 3; Winston-Salem, North Carolina: Tobacco Chemists' Research Conference, 1990: 3–71.

2 Guerin MR, Jenkins RA, Tomkins BA. Mainstream and sidestream cigarette smoke. in: Eisenberg M. (Ed.). *The Chemistry of Environmental Tobacco Smoke, Composition and Measurement.* Lewis Publishers (Boca Raton) 1992: 75–85.

3 Baker RR. The release of nicotine and semi-volatile components inside a burning cigarette. Proceedings of the CORESTA Symposium (Taormina Italy) 1986: **2**; 522–573.

4 Johnson WR, Hale RW, Nedlock JW, Grubbs HJ, Powel DH. The distribution of products between mainstream and sidestream smoke. *Tob. Sci.*, 1973: **17**; 141–144.

5 Creamer R. The light transmission of cigarette smoke and its contribution to the study of smoke kinetics, particle sizes, and number of particles. Paper presented at the 33rd Tobacco Chemists' Research Conference (Lexington KY) Oct. 1979.

6 Townsend, DE. The effect of tobacco moisture on the removal of cigarette smoke by the tobacco rod. Paper presented at the 37th Tobacco Chemists' Research Conference (Washington DC) Oct. 1983.

7 Piadé J-J, Bindler GN. Design optimization for sidestream smoke collection systems: influence of flask geometry and airflow patterns within the enclosure. Paper presented at the 43rd Tobacco Chemists' Research Conference (Richmond VA) 1989.

8 Baker RR. Mechanisms of Smoke Formation and Delivery. In: Martin RH, Davis DL, Litzinger EF, editors. Symposium of the 34th Tobacco Chemists' Research Conference; 1980 Oct 27–29; Richmond, Virginia: Tobacco Chemists' Research Conference, 1980, p.184–224.

9 Baker RR. Variation of sidestream gas formation during the smoking cycle. *Beitr. Tabakforsch. Int.*, 1982: **11**; 181–193.

10  Dittmann R, Feld H-J, Müller B-H, Schneider W. Time-resolved emission of sidestream smoke particles. *Beitr. Tabakforsch. Int.*, 1992: **15**; 53–57.

11  Jenkins RW, Comes RA, Bass RT. The Use of Carbon–14 Labeled Compounds in Smoke Precursor Studies – A Review. In: Woltz WG, Rodgman A, Morie GP, editors. Symposium of the 29th Tobacco Chemists' Research Conference; 1975 Oct 8–10; Beltsville-College Park, Maryland: Tobacco Chemists' Research Conference, 1976. p. 1–30.

12  Houseman TH, Studies of cigarette smoke transfer using radioisotopically labelled tobacco constituents. *Beitr. Tabakforsch.*, 1973: **7**; 142–147.

13  Perfetti TA, Coleman WM, Smith WS. Determination of mainstream and sidestream cigarette smoke components for cigarettes of different tobacco types. *Beitr. Tabakforsch. Int.*, 1998: **18**; 95–113.

14  Eatough DJ, Lee VC, Benner CL, Bayona JM, Galen R, Lamb JD, Lee ML, Lewis EA, Hansen LD. Chemical composition of environmental tobacco smoke, 1. Gas-phase acids and bases. *Environ. Sci. Technol.*, 1989: **23**; 679–687.

15  Federal Register 1967: **32**; no. 147, 1178.

16  Morgareidge K. Evaluation of variables in cigarette smoke collection and analysis. I. The use of a monitor sample. Paper presented at the 13th Tobacco Chemists' Research Conference, 1959 Oct. 29–30, Lexington KY; paper no. 21.

17  Ogg CL. Determination of particulate matter and alkaloids (as nicotine) in cigarette smoke. *J. Assoc. Off. Agr. Chem.*, 1964: **47**; 356–362.

18  Federal Trade Commission, press release September 9, 1997.

19  Federal Trade Commission 'Tar' and nicotine of the smoke of 59 varieties of cigarettes November, 1967.

20  CORESTA, Machine smoking of cigarettes, determination of crude and dry smoke condensate. CORESTA Inf. Bull. 1969; CORESTA method no. 10, 24–33.

21  British Standard Methods, Routine analytical cigarette smoking machine part 1, specification and standard conditions. BS 5668, Part 1, 1992.

22  Deutsches Institut für Normung. Maschinelles Abrauchen von Zigaretten und Bestimmung des Rauchkondensats, DIN 10240. Beuth Verlag (Berlin) 1978.

23  International Standards Organization, Cigarettes – Routine analytical cigarette-smoking machine – definitions and standard conditions. ISO 3308, 1991.

24  Wagner JR, Thaggard NA. Gas-liquid chromatographic determination of nicotine contained on cambridge filter pads: collaborative study. *J. Assoc. Off. Anal. Chem.*, 1979: **62**; 229–236.

25  CORESTA. Determination of nicotine in the mainstream smoke of cigarettes by gas chromatographic analysis. CORESTA Inf Bull 1986; CORESTA method no. 7, 39–41.

26  ISO, Cigarettes – Determination of nicotine content in smoke condensates gas chromatographic method. ISO 10315, 1989.

27  Klus H. Distribution of mainstream and sidestream smoke components. . In: Durocher DF, Bell JW, Sisson VA (Eds.). Symposium of the 44th Tobacco Chemists' Research Conference; 1990 Sept 30–Oct 3; Winston-Salem, North Carolina: Tobacco Chemists' Research Conference, 1990. 189–232.

28  Health protection branch's information letter no. 819, proposed tobacco (reporting) regulations. Prepared for Health Canada by Labstat Inc. under contract no. H4097-7-008, drafted November 1997, edited March 1998.

29  O'Keeffe AE, Lieser RC. An improved smoking machine. *Tob. Sci.*, 1958: **2**; 73–76.

30 Pillsbury HC, Bright CC, O'Connor KJ, Irish FW. Tar and nicotine in cigarette smoke. *J. Assoc. Off. Anal. Chem.*, 1969: **52**; 458–462.

31 Bradford JA, Harlan WR, Hammer HR. Nature of cigarette smoke: technique of experimental smoking. *Ind. Eng. Chem.*, 1936: **28**; 836–839.

32 Wartman WB, Cogbill EC, Harlow ES. Determination of particulate matter in concentrated aerosols. Application to analysis of cigarette smoke. *Anal. Chem.*, 1959: **31**; 1705–1709.

33 Diana JN, Vaught A. The Research Cigarette. The Tobacco and Health Research Institute (THRI), University of Kentucky (Lexington KY) 1990.

34 Benowitz NL, Hall SM, Herning RI, Jacob P, Jones RT, Osman A-L. Smokers of low-yield cigarettes do not consume less nicotine. *N. Engl. J. Med.*, 1983: **309**; 139–142.

35 Kozlowski LT, Frecker RC, Khouw V, Pope MA. The misuse of 'less-hazardous' cigarettes and its detection: hole-blocking of ventilated filters. *Am. J. Pub. Health*, 1980: **70**; 1202–1203.

36 Hoefer I, Nil R, Baettig K. Ultralow-yield cigarettes and type of ventilation: The role of ventilation blocking. *Pharmacol. Biochem. Behav.*, 1991: **40**; 907–914.

37 Federal Trade Commission, press release September 9, 1997.

38 Griffith RB, Jeffrey RN. Improved steam-destillation apparatus: application to determination of nicotine in green and dry tobacco. *Anal. Chem.*, 1948: **20**; 307–311.

39 Pfyl B, Schmitt O. Zur Bestimmung des Nikotins im Tabak und Tabakrauch. Ii., *Z. Unters Lebensm.*, 1927: **54**; 60–77.

40 Assoc. Offic. Agr. Chemists, Official methods of analysis (8th Ed.) 1955; 66.

41 Pfyl B. Zur Bestimmung des Nikotins im Tabakrauch. *Z. Unters Lebensm.*, 1933: **66**; 510–524.

42 Cundiff RH, Markunas PC. A rapid titrimetric procedure for the determination of nicotine, nornicotine, and total alkaloids in tobacco and tobacco products. 7th Tob. Chem. Res. Conf. 1953 Oct. 1–2, Winston-Salem NC; paper no. 03.

43 Cundiff RH, Markunas PC. Determination of nicotine, nornicotine, and total alkaloids in tobacco. *Anal. Chem.*, 1955: **27**; 1650–1653.

44 Swain LM, Eisner A, Woodward CF, Brice BA. Ultraviolet absorption spectra of nicotine, nornicotine and some of their derivatives. *J. Am. Chem. Soc.*, 1949: **71**; 1341–1345.

45 Willits CO, Swain ML, Connelly JA, Brice BA. Spectrophotometric determination of nicotine. *Anal. Chem.*, 1950: **22**; 430–433.

46 Laurene AH, Harrell TG. Improved spectrophotometric method for determination of nicotine in tobacco smoke. *Anal. Chem.*, 1958: **30**; 1800–1802.

47 Pleasants SW, Haynes CM, Harrow LS. A rapid spectrophotometric determination of total nicotine alkaloids in tobacco smoke. *J. Assoc. Off. Agr. Chem.*, 1959: **42**; 424–427.

48 Harvey WR, Badgett CE, Resnik FE. The determination of nicotine alkaloids in tobacco leaf, cigarette filler, and TPM by acid-methanol extraction. 20th Tob. Chem. Res. Conf. (Winston-Salem) 1966; paper no. 15.

49 Barkemeyer H, Seehofer F. Zur spektrophotometrischen Bestimung des Nicotins in Tabak und Tabakrauchkondensaten. *Z. Lebensm. Unters Forsch.*, 1960: **112**; 50–52.

50 Verband der Cigarettenindustrie, Vci, Einheitliche Vorschriften für die Analyse von Tabak und Tabakrauch. 4. Vorschrift für die Bestimmung des Nikotins im Tabak, im Tabakrauchkondensat und in Filtern. *Beitr. Tabakforsch.*, 1965: **3**; 87–89.

51 Koenig W. Untersuchungen aus dem organischen Laboratorium der technischen Hochschule zu Dresden. Zur Kenntnis der Pyridinfarbstoffe. *J. Prakt. Chem.*, 1904: **70**; 19–56.

52 Asmus E, Hoehne R, Kraetsch J. Photometrische Bestimmung von Nicotin und Pyridin im Tabakrauch. *Fresenius Z. Anal. Chem.*, 1962: **187**; 33–37.

53 Sadler WW, Chesson RR, Schoenbaum AW. Automated procedure for determining the nicotine content of steam distillates. *Tob. Sci.*, 1960: **4**; 208–212.

54 Charles JL, Stahr HM, Ikeda RM. Automated determination of nicotine in total particulate matter of cigarette smoke. *Tob. Sci.*, 1969: **13**; 54–58.

55 Bourlas MC, Thomas CE, Esperdy RC, Kuhn WF. The determination of nicotine in particulate matter using citric acid and bromomaleic acid. *Beitr. Tabakforsch.*, 1976: **8**; 354–358.

56 Quin LD. Alkaloids of tobacco smoke. I. Fractionation of some tobacco alkaloids and of the alkaloid extract of burley cigarette smoke by gas chromatography. *J. Org. Chem.*, 1959: **24**; 911–916.

57 Lyerly LA. Direct vapor chromatographic determination of menthol, propylene glycol, nicotine, and triacetin in cigarette smoke. *Tob. Sci.*, 1967: **11**; 49–51.

58 Wagner JR, Thaggard NA. An automated method for the gas chromatographic determination of nicotine in smoke. *Tob. Sci.*, 1974: **18**; 65–68.

59 Randolph HR. A gas chromatographic determination of nicotine in an isopropyl alcohol extract of smoke particulate matter. *Tob. Sci.*, 1974: **18**; 137–139.

60 Deutsch LJ, Jeffords AL. Simultaneous gas chromatographic nicotine/water analysis in smoke via split injection on dual capillary columns. 48th Tob. Chem. Res. Conf. (Greensboro) 1994; paper no. 35.

61 Seehofer F, Schulz W. *Beitr. Tabakforsch.*, 1965: **3**; 151–156.

62 Dube M.F., Green C.R. Methods of collection of smoke for analytical purposes. In: Durocher DF, Bell JW, Sisson VA (Eds.). Symposium of the 36th Tobacco Chemists' Research Conference; 1982 Oct 24–27; Raleigh, North Carolina: Tobacco Chemists' Research Conference, 1982. 42–102.

63 McRae DD. The physical and chemical nature of tobacco smoke. In: Durocher DF, Bell JW, Sisson VA (Eds.). Symposium of the 44th Tobacco Chemists' Research Conference; 1990 Sept 30–Oct 3; Winston-Salem, North Carolina: Tobacco Chemists' Research Conference, 1990. 233–323.

64 Proctor CJ, Martin C, Beven JL, Dymond HF. Evaluation of an apparatus designed for the collection of sidestream tobacco smoke. *Analyst*, 1988: **113**; 1509–1513.

65 Jermini CA, Weber A, Grandjean E. Quantitative determination of various gas-phase components of the sidestream smoke of cigarettes in the room air as a contribution to the problem of passive smoking. *Environm. Health*, 1976: **36**; 169–181.

66 Van Loy MD, Lee VC, Gundel LA, Daisey JM, Sextro RG, Nazaroff WW. Dynamic behavior of semivolatile organic compounds in indoor air. 1. Nicotine in a stainless steel chamber. *Environ. Sci. Technol.*, 1997: **31**; 2554–2561.

67 Baker RR, Proctor CJ. *Environ. Int.*, 1990: **10**; 231–245.

68  Nelson PR, Heavner DL, Collie BB, Maiolo KC, Ogden MW. Effect of ventilation and sampling time on environmental tobacco smoke components ratios. *Environ. Sci. Technol.*, 1992: **26**; 1909–1915.

69  Piadé JJ, D'Andrès S, Sanders EB. Sorption phenomena of nicotine and ethenylpyridine vapors on different materials in a test chamber. *Envir. Sci. & Techn.*, 1999: **33**; 2046–2052.

70  Brunnemann KD, Hoffmann D. The pH of tobacco smoke. *Fd Cosmet. Toxicol.*, 1974: **12**; 115–124.

71  Klus H, Kuhn H. Verteilung verschiedener Tabakrauchbestandteile auf Haupt- und Nebenstromrauch (eine Übersicht). *Beitr. Tabakforsch. Int.*, 1982: **11**; 229–265.

72  Brunnemann KD, Adams JD, Ho DPS, Hoffman D. The influence of tobacco smoke on indoor atmospheres II. Volatile and tobacco specific nitrosamines in main- and sidestream smoke and their contribution to indoor pollution. in: Proceedings of the ACS 4th Joint Conf. Sens. Environ. Pollution (Washington DC) 1977; 876–880.

73  Robinson DP. Aerodynamic characteristics of the plume generated by a burning cigarette. Proceedings of the International Conference on the Physical and Chemical Processes Occurring in a burning Cigarette. Wake Forest University (Winston Salem NC) 1987; 115–150.

74  Robinson DP. Prediction of the aerodynamic characteristics of the cigarette sidestream plume. Paper presented at the 42nd Tobacco Chemists' Research Conference (Lexington KY) 1988.

75  Browne CL, Keith CH, Allen RE. The effect of filter ventilation on the yield and composition of mainstream and sidestream smokes. *Beitr. Tabakforsch. Int.*, 1980: **10**; 81–90.

76  Kaneki K, Masuo Y, Okada T. Determination of particulate matter and carbon oxides in sidestream smoke. *Beitr. Tabakforsch. Int.*, 1989: **14**; 155–162.

77  Harris JL, Hayes LE. A 20-port sidestream smoke collection system. Paper presented at the 32nd Tobacco Chemists' Research Conference (Montreal Canada) Oct. 1978.

78  Chortyk OT, Schlotzhauer WS. Modification of an automatic cigarette smoking machine for sidestream smoke collection. *Tob. Sci.*, 1986: **30**; 122–126.

79  Green CR. A discussion of smoke generation and analytical procedure in "The contribution of low tar cigarettes to environmental tobacco smoke". Letter to the editor, *J. Anal. Toxicol.*, 1990: **14**; 261–262.

80  Chortyk OT, Schlotzhauer WS. A discussion of smoke generation and analytical procedure in "The contribution of low tar cigarettes to environmental tobacco smoke". Letter to the editor, authors' reply, *J. Anal. Toxicol.*, 1990: **14**; 262.

81  Ueno Y, Peters LK. Size and generation rate of sidestream cigarette smoke particles. *Aerosol. Sci. Technol.*, 1986: **5**; 469–476.

82  Wingate DE, Beard KA, Jordan FD, Morrison CC, Risner CH, Smith WS, Rogers JC, Smith DH. Utilization of the BAT fishtail chimney system for collection and quantification of sidestream smoke constituents. Paper presented at the 49th Tobacco Chemists' Research Conference (Lexington KY) 1995.

83  Neurath G, Ehmke H, Horstmann H. Einfluss des Feuchtigkeitsgehaltes von Cigaretten auf die Zusammensetzung des Rauches III. *Beitr. Tabakforsch.*, 1964: **2**; 361–369.

84 Baker RR. Summary of main points. Minutes of the CORESTA SS & ETS Task Force (Interlaken, Switzerland) Sept. 1989; retrieved February 12, 1999 from the World Wide Web: http://documents.rjrt.com/

85 Haut SA. A simple apparatus for the simultaneous collection and measurement of chemical components of mainstream and sidestream cigarette smoke. Paper presented at the 42nd Tobacco Chemists' Research Conference (Lexington KY) 1988.

86 Gonzalez JM, Sarabia J. Influence of barometric pressure on the sidestream smoke of cigarettes. Paper presented at the Smoke Group Meeting, CORESTA (Kallithea Greece) 1990.

87 Rickert WS, Robinson JC, Collishaw N. Yields of tar, nicotine, and carbon monoxide in the sidestream smoke from 15 brands of Canadian cigarettes. *Am. J. Public Health*, 1984: **74**; 228–231.

88 British Columbia Ministry of Health and Ministry Responsible for Seniors. 1998 Reports on cigarette additives and smoke constituents. Analysis of cigarette smoke. Results by smoking condition and smoke constituent; retrieved February 26, 1999 from the World Wide Web: http://www.asalvin.com/poisons/chemlookup.asp

89 Evans WH, Sefton GV. Predictive relationships for sidestream smoke cigarette yields. *Sci. Tot. Environ.*, 1992: **116**; 269–280.

90 Klus H, Kuhn H. Untersuchungen über die optische Aktivität des Rauchnikotins. *Fachl. Mitt. Oesterr. Tabak. R.*, 1977: **17**; 331–336.

91 Jacob P, Benowitz NL, Copeland JR, Risner ME, Cone EJ. Disposition kinetics of nicotine and cotinine enantiomers in rabbits and beagle dogs. *J. Pharm. Sci.*, 1988: **77**; 396–400.

92 Nwosu CG, Godin CS, Houdi AA, Damani LA, Crooks PA. Enantioselective metabolism during continuous administration of S-(−)- and R-(+)-nicotine isomers to guinea-pigs. *J. Pharm. Pharmacol.*, 1988: **40**; 862–869,

93 Nwosu CG, Crooks PA. Species variation and stereoselectivity in the metabolism of nicotine enantiomers. *Xenobiotica*, 1988: **18**; 1361–1372.

94 Risner ME, Cone EJ, Benowitz NL, Jacob P. Effects of the stereoisomers of nicotine and nornicotine on schedule-controlled responding and physiological parameters of dogs. *J. Pharmacol. Exp. Ther.*, 1988: **244**; 807–813.

95 Yoshida D, Tomita H. Optical rotatory power of nicotine in the smoke of cigarette. *Jpn. Tob. Salt Public Corp. Cent. Res. Inst. Sci. Pap.*, 1975: **117**; 125.

96 Crooks PA, Godin SC, Pool WF. Enantiomeric purity of nicotine in tobacco smoke condensate. *Med. Sci. Res.*, 1992: **20**; 879–880.

97 Chavdarian CG, Sanders EB, Bassfield RL. Synthesis of optical active nicotinoids. *J. Org. Chem.*, 1982: **47**; 1069–1073.

98 Perfetti TA, Coleman WM. Chiral-Gas Chromatography-Selected Ion Monitoring-Mass Selective Detection Analysis of Tobacco Materials and Tobacco Smoke. *Beitr. Tabakforsch. Int.*, 1998: **18**; 15–33

99 Armstrong DW, Spino LA, Han SM, Seeman JT, Secor HV. Enantiomeric resolution of racemic nicotine and nicotine analogs by microcolumn liquid chromatography with beta-cyclodextrin inclusion complexes (technical note). *J. Chromatogr.*, 1987: **411**; 490–493.

100 Demetriou D, Rustemeier K, Voncken P, Schepers G. HPLC separation of the enantiomers of nicotine and nicotine-like compounds. *Chirality*, 1993: **5**; 300–302.

101 Cundy KC, Crooks PA. Unexpected phenomenon in the hplc analysis of racemic [14]C labelled nicotine: separation of enantiomers in a totally archiral system. *J. Chromatogr.*, 1983: **281**; 17–33.

102 Allman DR. A proposed smoking machine modification to facilitate the collection of sidestream cigarette smoke. Paper presented at the Technology Group Meeting, CORESTA (Interlaken Switzerland) 1989.

103 Proctor CJ. Evaluation of fishtail chimneys designed to accommodate 100 mm and 120mm-long cigarettes. Report presented at the CORESTA SS and ETS Task Force meeting (Guang-zhou) October 1988; retrieved February 12, 1999 from the World Wide Web: http://documents.rjrt.com

*Chapter 13*

# Nicotine in environmental tobacco smoke

Michael W. Ogden[1] and Roger A. Jenkins[2]

[1] *R.J. Reynolds Tobacco Co., Research & Development, P.O. Box 1236, Winston-Salem, NC 27102-1236, USA*
[2] *Chemical and Analytical Science Division, Oak Ridge National Laboratory, Building 4500S, P.O. Box 2008, Mail Stop 6120, Oak Ridge, TN 37831-6120, USA*

## I. INTRODUCTION

The tobacco leaf, as with many natural products, is chemically complex, drawing its complexity from both genetic and environmental sources. More than 60 genetic species of plants belong to the genus *Nicotiana*; however, only two (*N. tabacum* and *N. rustica*) are widely cultivated for use as a consumer product. The tobacco of commerce, *N. tabacum*, is a hybrid plant capable of a high degree of variability as evidenced by the different major commercial types (flue-cured, Burley, Maryland, and Oriental) and the numerous agronomic varieties within each type [1]. Each of these types, in addition to others, is used currently in the manufacture of cigarettes, cigars and pipe tobaccos [2].

Superimposed on the chemical complexity of the tobacco leaf is an additional dimension of complexity resulting from combustion of these tobaccos on smoking. In a 1982 review, Dube and Green estimated 3875 individual components had been identified in tobacco smoke [3]. Of these, 2740 were reported to be unique to smoke with the remaining 1135 compounds being common to both tobacco leaf and smoke. Of the 3875 compounds known, 921 are *N*-heterocycles [3].

As background for discussing nicotine in environmental tobacco smoke (ETS), it appears useful to describe briefly some of the major processes that occur during the burning of tobacco. This, combined with the dilution of ETS in indoor air, will hopefully give the reader added insight into the issue of measuring nicotine (or anything else) in ETS and help frame the discussion of how these measurements can, and cannot, be interpreted.

531

Pyrolysis, pyrosynthesis, and distillation are three general processes that occur simultaneously during tobacco combustion. Pyrolysis refers to thermal decomposition of the organic leaf constituents into smaller molecules. As a result of being partially unstable chemically, some of these newly formed fragments can recombine to form additional components (pyrosynthesis). Both pyrolysis and pyrosynthesis result in the presence of chemical constituents in tobacco smoke that were not originally present in the tobacco. The third process, distillation, results in the intact transfer of certain compounds directly from tobacco to smoke. The principal alkaloid of tobacco, nicotine, contributes to the complexity of tobacco smoke through all three processes.

Depending on the circumstances surrounding its generation, the smoke from burning tobacco is usually categorized as either mainstream smoke (MS), sidestream smoke (SS), or environmental tobacco smoke (ETS). Mainstream smoke is considered to be the smoke emerging through the mouth end of the tobacco product during puffing and is primarily inhaled by the smoker. Sidestream smoke (SS) is generally defined as the combination of the smoke emanating from the lit end of a cigarette during the smolder period between puffs, and the smoke emanating from the lit end of a cigarette during the puff, including vapor phase constituents diffusing through the cigarette paper.

ETS, or second-hand smoke as it is sometimes referred to, can be defined simply as the smoke existing in an air space as the result of people smoking tobacco. More precisely, it is the combination of aged and diluted SS (approximately 85%) with aged and diluted exhaled MS (approximately 15%). The chemical compositions of SS, inhaled MS, exhaled MS, and ETS are qualitatively the same but quantitatively different [4–8].

In both the scientific literature and popular press, the terms SS and ETS are sometimes (although erroneously) used interchangeably. Due to its composition, concentration, and temperature of formation, SS is a complex, dynamic, and chemically reactive aerosol. In discussing SS, it is important to differentiate SS as it transforms into a component of ETS, and SS as it is typically studied in the laboratory. As studied in the laboratory, SS is an artificial measurement; that is, the SS collection process itself influences both the qualitative and quantitative composition determined. SS which ultimately contributes to ETS is generated under unencumbered, free-burning conditions. SS, as studied in the laboratory, is confined in small vessels, typically made of glass [3,9,10].

As fresh SS transforms into a component of ETS, rapid aging and *extensive* dilution occur. Aging of SS is, conceptually, a very important phenomenon for appreciating the quantitative differences in chemical composition (and differences in potential biological activity) between SS and ETS. Many of the known and suspected animal carcinogens identified in SS [11] are chemically unstable species (e.g., aromatic amines, nitrosamines, etc.) [12]. The potential for the diminution or removal of these compounds from ETS at typical, real-world concentration levels due to chemical reactions with oxygen, ozone, ultraviolet (UV) light, etc. is high but, at present, has not been studied. (In fact, exposure to UV light is the mechanism by which nitrosamines are deactivated for disposal purposes.) This is because methods of analysis in analytical chemistry are not yet of sufficient sensitivity to investigate the presence and concentration in ETS of most of the chemicals identified in SS in a laboratory.

These distinctions are of significant importance as one attempts to assess the impact of ETS on indoor air quality (IAQ) in an enclosed environment and, further, to assess an individual's exposure to ETS as they occupy that environment. Due to the chemical complexity, dynamics, and extreme dilution of ETS in indoor air, and the complexity of the indoor air background, it is impossible to ascertain the concentration level of ETS per se. Attempts are made to put ETS concentrations into perspective by measuring selected ETS components that can then be used as markers or tracers for ETS as a whole. Obviously then, any useful marker compound must behave in a predictable and verifiable way with regard to the bulk of ETS. The best summary of this concept, put forth by the National Research Council (NRC) in 1986 [11], lists the following four attributes of an ideal ETS marker:

1. unique to, or at least highly selective for, ETS;
2. easily measurable in air, even at low smoking rates;
3. similar in emission rates for a variety of tobacco products;
4. "In a fairly consistent ratio to the individual contaminant of interest or category of contaminants of interest (e.g., suspended particulates) under a range of environmental conditions encountered and for a variety of tobacco products".

No single ETS component has yet been demonstrated to successfully meet all these criteria. In fact, we are not aware of the systematic assessment of any ETS constituent against these criteria. Therefore, it is our intention to evaluate nicotine as a quantitative marker of ETS using the NRC guidelines as a framework. Further, we will also include brief discussion of other commonly used ETS and IAQ markers as appropriate. These commonly used markers include: 3-ethenylpyridine; solanesol; respirable suspended particles (RSP); ultraviolet particulate matter (UVPM); fluorescent particulate matter (FPM); and scopoletin. In so doing, it is our aim to provide an understanding of the characteristics of nicotine as an ETS marker in both an absolute sense (against the NRC criteria) and in a relative sense (against other commonly used ETS markers).

## II. NRC CRITERION 1 – UNIQUENESS FOR TOBACCO

During smoking, approximately two-thirds of the nicotine present in tobacco is distilled directly into the smoke [13,14] and the remaining one-third is pyrolyzed. Accordingly, nicotine is the predominant tobacco alkaloid in smoke, including environmental tobacco smoke. However, the pyrolysis products of nicotine play an increasingly important role in the chemical composition of SS and ETS.

The principal pyrolysis products of nicotine and related alkaloids are myosmine, nicotyrine, cotinine, bipyridine and a series of simpler pyridine derivatives [2]. Besides pyridine itself, the major derivatives are alkyl-substituted pyridines which include: 3-ethenylpyridine (3-EP, or 3-vinylpyridine), 3-picoline and the various isomeric lutidines [13]. Because of the extreme dilution of ETS in most indoor environments, many of these alkaloids (with the exception of 3-EP) are virtually non-detectable in real-world settings.

Nicotine is, by far, the most commonly used indicator of ETS in indoor air. However, many authors state erroneously that nicotine is unique to tobacco. Nicotine is present in a surprisingly large number of species other than *Nicotiana* [15]. (See also Chapters 2 and 10.) It has also been detected at trace levels in a variety of common foods [16–18]. Although more widely dispersed in nature than commonly thought, the presence of nicotine in indoor air should be uniquely attributable to tobacco smoke and/or tobacco smoking (the distinction will be discussed later in the section on ratios of nicotine to other constituents). The same is also believed to hold true for the related alkaloids and the more unique pyrolysates (e.g., 3-EP and myosmine).

Of all the commonly measured ETS constituents, only the nicotine alkaloids (including nicotine, 3-EP, and myosmine) and solanesol appear to offer the required specificity for use as a quantitative tracer of the ETS aerosol [2]. Solanesol is a C45 isoprenoid alcohol found in the chloroplasts of tobacco leaves [19,20]. Solanesol, like nicotine, is not believed to be unique to tobacco but similarly distributed in nature among the *Solanaceae* family (which includes *Nicotiana*). Moreover, also like nicotine, solanesol has not been shown to be (nor is it suspected of being) present in indoor air through any means other than burning tobacco. In general, the same can be said for scopoletin, even though it has been detected in oak leaves [21].

However, the sporadic presence at trace levels of any of these compounds in outdoor air cannot be ruled out. Such instances might include, by way of example, the following: nicotine (because of its volatility) in the immediate vicinity of tobacco-processing facilities (curing barns, storage sheds, manufacturing plants, etc.); scopoletin as a result of uncontrolled or controlled fires (e.g., forest fires or residential leaf burning); or nicotine and/or solanesol resulting from controlled post-harvest burn off of commercial tomato fields (an historical agricultural practice that may still be used today).

The remaining commonly used ETS and IAQ indicators (RSP, UVPM, and FPM) are not as selective for tobacco smoke in indoor air as are nicotine, 3-EP, myosmine, solanesol, and scopoletin. RSP simply connotes particles of 'respirable' size (typically defined as those particles passing through a size-selection device having a median cut-point particle diameter of 4.0 μm [22]) without regard for their origin or chemical makeup. ETS particles have a mass median diameter in the range of 0.2 μm [23]. Virtually all combustion sources, and many non-combustion sources, also produce airborne particles in this size range. Examples include: kerosene and oil smoke, candles, internal combustion engines (automobiles, lawnmowers, etc.), carbon black, talc, bacteria, atmospheric dust, etc. There is no question that tobacco smoking in indoor spaces contributes to the overall RSP background; however, the contribution of ETS to total RSP generally is much less than 50% [24–37].

Realizing this potentially serious limitation of RSP as an indicator of ETS, several workers exploited the generally-known phenomena that solutions of tobacco smoke particles exhibited broad-band ultraviolet absorbance and fluorescence. Taking the position that sensitivity and selectivity for ETS particles in the presence of other RSP could be enhanced via wavelength optimization, two methods have emerged. Conner and co-workers [38] in the mid-1980s developed the so-called UVPM method based on ultraviolet absorbance and Ogden and co-workers [25] in the late 1980s developed the so-called FPM method based on fluorescence. The predictions of the originators of these methods have been borne out: both methods offer sensitivity and selectivity enhancement over RSP for determining airborne ETS particles but both methods suffer from the presence of non-tobacco interferences (generally presumed to be RSP from other combustion sources) [25–37]. Overall, FPM has been shown to be more selective for ETS RSP than UVPM. This too was also predicted at the inception of the FPM method based on the naturally more selective process of fluorescence (compared to UV absorption).

Summarizing the commonly used ETS and IAQ markers against only this first NRC criterion (uniqueness) yields the following ranked comparison.

For Uniqueness to ETS:
Nicotine=3-EP=Myosmine=Solanesol=Scopoletin>FPM>UVPM>>RSP

## III. NRC CRITERION 2 – EASILY MEASURABLE

By today's standards, nicotine (and other commonly measured ETS markers) is fairly easily measured. However, this was not the case in the infancy of ETS research and considerable effort has been expended over the last 15 or so years in developing and refining methods of collection and analysis. Prior to presenting the current state-of-the-art in nicotine measurement, it is necessary to discuss one important area that impacts it: the particle/vapor distribution of nicotine in the ETS aerosol.

## Particle/vapor distribution and sample collection

Tobacco smoke is an aerosol, meaning it is composed of both vapor and particulate (small droplets suspended in the vapor) phases. Besides this physical difference between the aerosol components, there are chemical differences as well. In the case of many smoke constituents, the chemical differences are not constant but vary dynamically with the dilution and aging of the aerosol.

As discussed earlier, the vast dilution, typically 100,000- to 1,000,000-fold, is perhaps the most readily obvious distinction between SS and ETS. Not only is the dilution factor important for understanding the concentration differences between SS and ETS, it is fundamentally responsible for many other observed differences as well. Because of the concentrated nature of the SS aerosol, most semivolatile and many volatile chemicals are associated with the smoke particles or are distributed between the smoke particles and the smoke vapor phase. As SS ages and dilutes, both the size and chemical composition of the smoke particles change. Water, and many volatile and semivolatile chemicals (such as nicotine) evaporate from the particles of SS as it transforms into a component of ETS [4,5,39]. Concurrently with the evaporation phenomenon, the average particle size decreases slightly (by about a factor of 2 from ~0.3–0.4 microns to ~0.15–0.2 microns). Within 10 minutes of generation, the mass median aerodynamic diameter (MMAD) of SS particles is indistinguishable from ETS (at about 0.2 microns) [23]. The aerosol vapor phase is composed of both the more volatile smoke constituents and many (if not most) of the semivolatile compounds that have evaporated from the particles of concentrated smoke. This phenomenon is illustrated best with nicotine, the major alkaloid component of tobacco smoke.

A semivolatile compound, nicotine is found almost exclusively in the particulate phase of MS and is distributed between the two aerosol phases of SS (to a varying degree depending on the nature of the SS collection procedure). However, due to evaporation, nicotine in ETS is found almost exclusively in the vapor phase (95+%) [40–45]. Along this same line of reasoning, the more non-volatile components of ETS such as solanesol [19] reside in the particulate fraction of the aerosol.

During the early years of ETS research, this behavior of nicotine was unknown and it was assumed, erroneously, that ETS nicotine was a particulate phase component (as had been assumed for many years from knowledge of mainstream and sidestream smoke). It wasn't until the advent of real-time monitoring of ETS alkaloids with an atmospheric pressure chemical ionization mass spectrometer coupled to an environmentally controlled test chamber that this phenomenon was discovered in the mid-1980s [46].

Capable of determining airborne pyridine, ethenylpyridines, picolines, lutidines, nicotine, and ammonia in real-time at parts-per-trillion levels, this instrumentation was of paramount importance in the infancy of ETS research. Two discoveries were first made with this instrumentation in the mid-1980s which have proven to be cornerstones in the present understanding of nicotine in ETS. First was the discovery by Eudy et al. [40] that ETS nicotine was virtually completely contained in the vapor fraction of the ETS aerosol and not in the particles. Eatough and coworkers have subsequently verified this finding [41–44]. The second discovery was that the decay rate of nicotine was dramatically different from the other measurable constituents of ETS [46].

Both the phase residence of nicotine and its adsorptive tendencies (giving rise to its unusual decay) are key, fundamental properties that must be understood prior to contemplating the usefulness of nicotine as a marker for ETS. Not only do they impact the means by which air is sampled for nicotine, they impact further the methods of analysis and the interpretation of the nicotine measurement (for example, in its relationship to other markers and the ETS aerosol as a whole). The impact of these phenomena will be discussed in greater detail in later sections.

Historically, some investigators have employed filter-based sampling systems that mimic collection systems used in the laboratory for MS and SS collection [47], on the erroneous assumption that ETS nicotine would be particle-bound. More recently, others have used filter-based sampling systems with sorbent beds located downstream in an attempt to study the vapor-particulate phase dynamics of ETS nicotine. However, due to adsorptive properties of the filters and the predominance of vapor phase nicotine in ETS, these systems, employing chemically untreated filters, are fraught with uncertainties and errors [45] and cannot be used for studying nicotine phase distribution.

It is now generally accepted that approximately 97% of indoor air nicotine exists in the vapor phase [40–45]. As a result, most sampling systems in use today focus on this major fraction. Vapor phase nicotine can be effectively sampled with both so-called active and passive monitoring systems. Active systems derive their name based on the 'active' transport of air to (and through) the collection surface. This is accomplished by air sampling pumps that operate at a known sampling rate, typically characterized by the air volume displacement measured in liters per minute. Passive systems, also known as diffusive systems, derive their name from the absence of a mechanical air movement device (pump). Delivery of gas phase analytes in the air medium to the collection surface is controlled by molecular diffusion through porous barriers. Several passive samplers for nicotine have been developed and described in the literature but none are commercially available [44,48–50].

Whether molecular transport to the collection surface is controlled by active or passive mechanisms, trapping of nicotine has been accomplished by a variety of means. In general terms, these include granular polymeric resins (sorbents) and chemically treated filters. Most commonly used are sorbent sampling systems that collect vapor phase nicotine (and related constituents) on either XAD-4 resin for solvent extraction [2,50–53] or on other porous polymer sorbents (such as Tenax) for thermal desorption [54–59].

Regardless of the collection means, passive (or diffusive) samplers collect predominantly vapor phase nicotine. Moreover, regardless of the transport means (active or passive), granular sorbent beds also collect predominantly vapor phase nicotine. Granular beds, in general, have a very low efficiency for aerosol particles in the size range of ETS particles [50].

Filter sampling systems can be used effectively to collect ETS nicotine if they are acidified so that they trap the semivolatile nicotine, which is slightly alkaline, and prevent it from re-volatilizing [41,43,44,50,60–62]. Typical acidifying agents include benzenesulfonic acid [43,44,60], citric acid [62], and sodium bisulfate [50,61]. However, in general, sampling systems employing acid-treated filters as the collection medium collect total ETS nicotine (i.e., vapor plus particulate phase) and not just vapor phase nicotine. Because of the relatively small amount of particulate phase nicotine present in most indoor air sampling situations, there should not be a significant discrepancy in results between systems for vapor phase nicotine and systems for total nicotine [50]. This may not hold true for airborne nicotine in situations that might perturb the vapor-particle equilibrium (outdoor aerosols, concentrated smoke, etc.) [42,45].

Diffusion denuder sampling systems (typically followed by acid-treated filters) are most commonly used for definitive separation of vapor and particulate phase nicotine. Acidification of the denuder surface is usually required and is accomplished by the same acids as listed above as well as by phosphoric acid [40]. Recently, an annular denuder system for nicotine has been described which employs ground XAD-4 resin coated onto the surface [63]. In diffusion denuder systems under pumped air flow, vapor phase nicotine diffuses radially to the collection surface while particles (including particulate phase nicotine) exit the denuder axially (and are collected downstream). Several of the sampling systems described in this section have been compared for nicotine determination [44,50,59,62].

## Analytical (chromatographic) methods

Due to the typically small amounts of nicotine present in indoor air and the chemical composition of the typical indoor air background, analytical selectivity and sensitivity are prerequisites for successful measurement. Gas chromatography (GC) with open-tubular (or capillary) columns coupled with *N*-thermionic detection, using a nitrogen-phosphorus detector (NPD) is the most often used technique. This is due to the widespread availability of GC instrumentation, the high efficiency and reproducibility available with modern-day capillary columns, the high sensitivity and selectivity in detection offered by NPDs, and the relative ease in data collection and interpretation.

Chromatographic techniques other than GC/NPD have been, and continue to be used. GC/FID (flame ionization detection) can be used for samples generated in laboratories and environmental test chambers where concentration levels are high and the chemical background is controlled. Problems with analytical sensitivity and specificity occur when trying to use GC/FID in ETS/IAQ research.

GC/MS (mass spectrometry) is sometimes used for nicotine determination. GC/MS systems (particularly bench top mass selective detector, MSD, systems) are becoming much more readily available in laboratories. Although moderately superior to GC/NPD in terms of selectivity and sensitivity, GC/MS systems tend to be more complex systems, requiring a higher level of operator skill, larger maintenance costs, and slightly more complex data analysis. They are commonly used, however, coupled with thermal desorption systems that attempt to measure a broad range of volatile organic chemicals (VOCs) in the indoor atmosphere [56,64–66]. This is due to the more variable nature of retention times in thermal desorption systems and the desire to quantify many compounds with different functionalities using a single instrument.

Other chromatographic systems, such as high performance liquid chromatography (HPLC) and ion chromatography (IC) have also been used [67]. Generally coupled with ultraviolet (UV) absorbance detectors, these systems are the exception rather than the rule for determining nicotine in ETS/IAQ investigations and research.

## Methods

As illustrated above, there are a variety of different approaches to each step in the overall determination of ETS nicotine. Accordingly, there are numerous combinations of sampling/desorption/analysis conditions that either have been, or could be used. Of all these methods, we will present slightly more detailed information on one of them: the method for collection of vapor phase nicotine on XAD-4 sorbent with extraction in triethylamine-modified ethyl acetate with endpoint determination by GC/NPD and quantitation by the method of internal standards using quinoline. This method is by far the most widely documented, validated, and used method for ETS nicotine determination. This methodology has been validated by ruggedness testing [52] and two international collaborative studies [52,68], and has been adopted as the official or recommended method by the following organizations: Association of Official Analytical Chemists [69,70], American Society for Testing and Materials [53], International Organization for

Standardization [71], CORESTA [72] and the U.S. Environmental Protection Agency [73].

## Sample collection

Air is sampled through a glass sorbent tube containing 20/40 mesh XAD-4 (styrene-divinylbenzene copolymer, part no. S2-0361, SKC Inc., Eighty Four, PA). The current tube in use consists of a single section of XAD-4 with glass wool plugs on either end. Previous tubes (still available) contained two sections of sorbent separated by a third plug of glass wool. The two-section tubes offer the built-in capability of determining collection efficiency (or breakthrough) by analyzing the XAD sections separately. After more than 10 years' experience, it has been shown that breakthrough of nicotine onto the backup section of XAD resin is not a problem, and the new tube was configured (to simplify extraction and analysis).

Air flow through the tube is controlled (typically at < 1.5 L/min) with a personal air sampling pump. Flow rate can be adjusted depending on the anticipated concentration of nicotine in the air and/or the expected duration of sampling (for example, flow rates of <0.5 L/min are often used in laboratory test chambers). Flow rates greater than 1.5 L/min are possible for nicotine, but may cause breakthrough of other analytes collected simultaneously (such as 3-EP). After sampling, the tube ends are sealed with plastic caps (supplied with the tubes) and the tubes are stored in a laboratory freezer until they can be analyzed. Collected samples appear to be stable indefinitely; that is, no degradation of collected nicotine has ever been observed.

Sample preparation consists of breaking the tubes open and transferring the entire tube contents to a vial. An extremely convenient means of sample prep employs transferring the tube contents directly to a standard $12 \times 32$ mm GC autosampler vial (approximate internal volume, 2.0 mL). The extraction solvent consists of the following: ethyl acetate containing 0.01% (v/v) triethylamine (TEA) and ca. 10 µg/mL quinoline (internal standard). The addition of a stronger base than nicotine (i.e., TEA) is necessary to prevent adsorptive losses of nicotine from solution onto the surface of the glass vials [51]. The XAD resin is extracted in this solution (1.25 mL in the GC vial) for at least 30 min prior to analysis.

Sample analysis is performed by capillary GC/NPD on a 30 m $\times$ 0.32 mm id capillary column with a 1.0 µm film of a relatively non-polar stationary phase (e.g., CP-Sil 8 CB from Chrompack Inc., Raritan, NJ). Sample injection is typically in the splitless mode followed by temperature programming from 50°C (hold 1 min) to 215°C at 10°C/min followed by 20°C/min to 275°C (hold 2 min). Helium is the carrier gas (15 psig head pressure) and the NPD is operated according to manufacturer's recommendations (typically ca. 15 mL/min helium makeup gas, 75 mL/min air, and 3 mL/min hydrogen). These same method conditions (including sample collection and extraction) are also suitable for the coincident determination of other ETS constituents (e.g., 3-EP and myosmine). A representative chromatogram is shown in Fig. 1. Complete details of this method are available elsewhere [2,53,70,74].

Quantitation is performed by the internal standard method with quinoline as the internal standard. The use of an internal standard in a GC/NPD method is essential due to the inherent fluctuation and drift in the NPD response [51]. Method detection limits for nicotine are in the range of ca. 0.01 μg/sample [53,75]. A typical calibration range with this method is 0.1–5 μg/sample. After laboratory quantitation, the amount of nicotine determined in the extract solution is converted to the average airborne concentration in the following manner. Amount of nicotine in the sample (typically in μg) is divided by the air volume sampled (typically in cubic meters) to yield air nicotine concentration in units of μg/m$^3$. The air volume sampled is determined by the pump flow rate (in L/min) multiplied by the sampling duration (in minutes) divided by 1000 (conversion from L to cubic meters) [53].

Not only is this method the most thoroughly validated and widely used for nicotine, it is one of the easiest to use as well. This is due to the thorough literature documentation (referenced above) and the use of the most

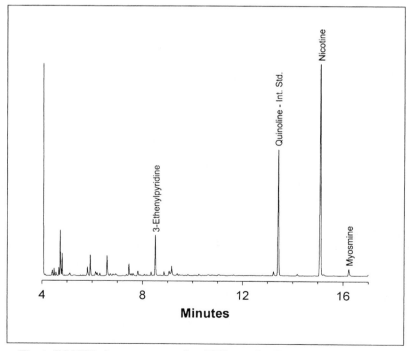

Fig. 1  GC/NPD chromatogram of an ETS sample. For conditions, see text.

commonly available equipment and supplies (starting with the commercially available sorbent sampling tubes). In order to conclude this section with a comparison of nicotine to other commonly used ETS and IAQ markers, a brief description of methods for the other markers follows.

The method of determination for the two other commonly measured ETS alkaloids (3-ethenylpyridine and myosmine) is the same as that for nicotine. Therefore, from an analytical standpoint, they have the same ease of determination as nicotine. This is complicated slightly by the lower concentrations of these two markers in most environments, which can result in a larger fraction of samples being closer to, or below, method detection limits. In very general terms, 3-ethenylpyridine is 25–50% of the nicotine level and myosmine is 5–10% of the nicotine level [2,26,60,76,77].

For the particulate phase markers (RSP, UVPM, FPM, solanesol, scopoletin), the general starting point is the gravimetric determination of RSP collected on pre-weighed, membrane filters [22,25,38,78]. In principle, this technique is simple and straightforward. Suitable filters (typically 37-mm o.d. PTFE membrane filters with 1 μm pore size) are weighed and sample is collected by drawing a known amount of air through the filter for a known period of time. Afterwards, the filter is reweighed and the collected sample weight is determined by weight difference (filter weight after sampling minus filter weight before sampling).

In practice, this method can be quite tedious due to several reasons. First, the very low weights typically encountered (usually less than 100 μg and very often less than 30 μg) require a microbalance with readability of 1 μg or less. In general, such balances are affected by air flow, minor surface vibrations, and atmospheric conditions such as relative humidity and temperature. In addition, the PTFE membrane filter media are inherently static prone which, at the levels often encountered, is a most serious problem. And finally, weights in this range can show a marked sensitivity to atmospheric moisture, so this must usually be controlled. These problems, and effective solutions, are described in detail elsewhere [22,38,78]. As a result, limits of detection (LOD) for collected particle weights are, in the best of cases, generally in the order of 10–15 μg per filter but, in the case of inexperienced operators or inadequate equipment or controls, LODs can often be 3–5 times higher (i.e., ca. 50 μg).

Determination of the other common ETS particulate phase markers is performed on a methanol extract of the filter used to collect RSP. Extraction is simple, generally the filter is placed in a 4-mL HPLC autosampler vial and extracted in 3 mL methanol. UVPM and FPM determination are best performed by what has been described as 'columnless HPLC' or a variant of flow-injection analysis where HPLC components (pump, autosampler, detector) are used to automate the determination. The isocratic 'mobile phase' is methanol and detection is by UV absorbance at 325 nm (for UVPM) and by fluorescence at 300 nm excitation, 420 nm emission (for FPM)

[22,25,38,75]. Of all the methods used in ETS marker determinations, these are the easiest.

Solanesol is determined on the same methanol extract by traditional HPLC using, typically, a low-carbon loading C18 column with an isocratic mobile phase consisting of 5% methanol in acetonitrile with UV detection at 205 nm [75,79,80]. This analysis is also relatively simple; much more so than the method first used to identify solanesol in ETS [19,79,81] which relied on GC determination of solanesol as its trimethylsilyl derivative.

For scopoletin determination, an aliquot of the original methanol filter extract is further diluted with acetic acid and analyzed by gradient HPLC on a C18 column with fluorescence detection (342 nm excitation; 464 nm emission) [21]. Gradient mobile phases consist of aqueous acetic acid and acetonitrile. In comparison to UVPM, FPM, and solanesol, scopoletin determination is more difficult but certainly well within the capabilities of any moderately-equipped HPLC laboratory.

Prior to concluding with a formal comparison of these markers against the criterion 'ease of measurement', it is necessary to more fully consider this issue. 'Ease of measurement' has two components: (1) the 'ease' of laboratory measurement (including operator skill level, necessary laboratory equipment, complexity of operations, etc.) which has been discussed above, and (2) the ability to detect this particular ETS component or marker in the air. This second consideration must include relative comparisons of the marker amounts in ETS against the ability to detect their presence (i.e., against the method LOD). This can be accomplished in the following manner.

Limits of detection of all these markers have been published [75]. In addition, average airborne concentrations of all these markers are available for the 50 leading brand-styles of US cigarettes [76] smoked in a controlled environment. These original data are reproduced in Table 1, along with a combined variable called 'Detectability' which is simply a ratio of the LOD (in μg/m$^3$) to the average airborne marker concentration (also in μg/m$^3$) expressed as a percentage. Thus, 'Detectability' can be thought of as the minimum percentage of the ETS level that can be ascertained by analysis of that particular marker compound. The choice of ETS level is arbitrary, as are the absolute magnitudes of the percentages calculated therefrom. The only requirement is that all analytes must be determined at the same time, in the same environment (which they were in this example). Thus, the 'Detectability' percentages shown are useful only for rank-ordering purposes. Smaller 'Detectability' indicates a more sensitive measure.

Summarizing the commonly used ETS and IAQ markers against only this second NRC criterion (ease of measurement) yields two comparisons. The first, ease of laboratory analysis, is in some respects, a subjective assessment.

For Ease of Laboratory Analysis:
UVPM = FPM > Solanesol > Nicotine = 3-EP = Myosmine > Scopoletin = RSP

The second comparison (ability to detect the marker in ETS using state-of-the-art methodology) is more objective and results in the following rank ordering (based on the data in Table 1).

For Detectability:

Scopoletin > Solanesol > FPM > Nicotine = 3-EP = UVPM > Myosmine >> RSP

## IV. NRC CRITERION 3 – SIMILAR EMISSION RATES

Very limited data are available on this topic in the literature. Virtually all of the tobacco products studied for determining emission rates have been cigarettes, although Nelson et al. [82] have recently reported some data for cigars. Since the vast majority of tobacco smoked is in the form of cigarettes (at least in the US), this is not a serious limitation. To our knowledge, only one study has characterized true ETS for a large number of commercial cigarette brand-styles and for a large number of analytes [76]. ETS from the leading 50 brand-styles of US cigarettes was characterized by measuring approximately 50 individual compounds. Sales-weighted average emission

TABLE 1

Relative comparison of analyte limit of detection (LOD) against mean airborne concentration in a test chamber

| Marker compound | Mean airborne concentration ($\mu$g/m$^3$)[1] | LOD ($\mu$g/m$^3$)[2] | Detectability (%)[3] |
|---|---|---|---|
| Nicotine | 90.84 | 0.083 | 0.1 |
| 3-Ethenylpyridine | 37.14 | 0.042 | 0.1 |
| Myosmine | 5.42 | 0.017 | 0.3 |
| RSP | 1439 | 36 | 3 |
| UVPM | 1441[4] | 1.4 | 0.1 |
| FPM | 1458[5] | 0.72 | 0.05 |
| Solanesol | 42.41 | 0.018 | 0.04 |
| Scopoletin | 1.89 | 0.00036 | 0.02 |

[1] Average data for 50 leading US cigarette brand-styles smoked separately in an environmental test chamber [76]. Concentrations are not typical of real-world environments. In each experiment, two cigarettes of one brand-style were smoked in an 18-m$^3$ chamber with no ventilation.
[2] LOD calculated from $\mu$g/sample data [75] for 2-h sampling at 2.3 L/min for filter and 1.0 L/min for XAD samplers. LODs are reported to two significant figures.
[3] Percentage of the test chamber concentration corresponding to the individual method LOD. For rank comparison purposes only. Data are reported to one significant figure.
[4] Raw UVPM data (192.45 $\mu$g/m$^3$) from Martin et al. [76] converted to ETS RSP equivalents using correction factor of 7.49 [22]. Corresponding LOD is in ETS RSP equivalents.
[5] Raw FPM data (37.39 $\mu$g/m$^3$) from Martin et al. [76] converted to ETS RSP equivalents using correction factor of 38.99 [22]. Corresponding LOD is in ETS RSP equivalents.

TABLE 2

ETS emission rates for sales-weighted average US cigarette

| Marker Compound | Emission rate (μg/cigarette)[1] | Standard error (n=50)[1] | Relative standard error (%) |
|---|---|---|---|
| Nicotine | 1585.08 | 42.21 | 2.7 |
| 3-Ethenylpyridine | 333.74 | 7.75 | 2.3 |
| Myosmine | 49.04 | 1.20 | 2.4 |
| RSP | 13674 | 410.65 | 3.0 |
| UVPM[2] | 13667 | 405.66 | 3.0 |
| FPM[3] | 13678 | 404.33 | 3.0 |
| Solanesol | 414.33 | 14.76 | 3.6 |
| Scopoletin | 18.17 | 0.74 | 4.1 |

[1] Sales-weighted average data for 50 leading US cigarette brand-styles smoked separately in an environmental test chamber [76].
[2] Raw UVPM data (1824.70±54.16 μg/cigarette) from Martin et al. [76] converted to ETS RSP equivalents using correction factor of 7.49 [22].
[3] Raw FPM data (350.81±10.37 μg/cigarette) from Martin et al. [76] converted to ETS RSP equivalents using correction factor of 38.99 [22].

rates (measured in μg/cigarette) were presented along with corresponding standard errors; these are reproduced in Table 2. For this third criterion (similar emission rates), the data for US cigarettes in Table 2 can be rank ordered as follows.

For Similarity of Emission Rates:
3-EP > Myosmine > Nicotine > FPM = UVPM = RSP > Solanesol > Scopoletin

In this context, the symbol '>' signifies a relatively more similar emission rate among cigarette brand-styles (although differences among this range of values may not have much practical significance considering larger differences in the other criteria).

## V. NRC CRITERION 4 – FAIRLY CONSISTENT RATIO

Of the four criteria presented for discussion, all are important but this fourth criterion is the one most centrally critical to the concept of a good marker compound. Unfortunately, however, it also is one which is virtually impossible to verify (at least according to the exact wording offered by NRC [11], ". . . consistent ratio to the individual contaminant of interest or category of contaminants of interest . . ."). Even though a variety of compounds and classes of compounds have been hypothesized to be "contaminants of interest", there exists no consensus opinion. Not only is this true for ETS (where non-smokers are exposed to very minute quantities of smoke compared to smokers), it is also true for the mainstream smoke inhaled by smokers. For MS, this has been summarized by Rodgman [5]:

Attempts to attribute the carcinogenicity of MS to a particular component are questionable. Over the past four decades, numerous scientists and organizations have stated that no single MS component or class of components acting either individually or in concert can explain epidemiological observations in human smokers or biological observations in laboratory animals treated with heroic doses of MS.

Therefore, for ETS, the best that can be hoped for at present is to identify marker compounds that stay in a consistent ratio to other markers and to the bulk of the ETS aerosol (either gas phase, particulate phase, or both).

Prior to summarizing the literature on this subject, we must first address the important issue of marker stability. There are two properties critical in this regard. One is chemical stability; that is, the tendencies of the marker to change into a different chemical species (either airborne, in the condensed state, or both). The other is behavioral stability; that is, the tendency of the marker to interact differentially or non-differentially (with respect to other compounds in the ETS aerosol) with the environment through means other than chemical degradation.

## Chemical stability

Most airborne chemicals are not stable indefinitely; they degrade chemically through a number of pathways including reaction with hydroxyl radicals, photooxidation with ozone, and direct photolysis [12]. Depending upon numerous variables, an individual chemical's stability in the air (as measured by half-life) can range from minutes to months. This important distinction among chemicals is very often overlooked. Two examples of this oversight at its most obvious extreme are the following: (1) the prediction of ETS particles existing in the outdoor urban atmosphere based on very stable alkanes as markers [83]; and (2) the quantitative prediction of highly reactive smoke components (which have never been measured in ETS) from data on chemically stable smoke components [84,85].

It is erroneous to assume that a good marker for ETS must be chemically stable in the air. Reflecting on the fourth NRC criterion reveals that the only necessary attribute for chemical stability of an ideal ETS tracer is that it have approximately the same stability as the contaminant of interest. However, the marker must remain stable in the ETS sample that has been obtained for analysis. That is, once a representative sample of the air has been concentrated for analysis, no further chemical changes should take place before the final determination is made (generally by laboratory analysis). Otherwise, the laboratory data will not accurately reflect the nature of ETS in the sampled air space.

Both of these aspects of marker stability have been studied (albeit in a limited way) for several of the commonly used ETS and IAQ marker compounds. Eatough and co-workers have investigated the effect of intense

UV radiation in a test chamber on the concentration of various ETS alkaloids and solanesol. In two experiments, nicotine concentrations fell by an average of 75% after irradiation while 3-EP concentrations dropped only about 25%. Results for myosmine were equivocal; concentrations increased 47% in the first experiment and dropped 19% in the second [60]. In a separate experiment, solanesol was shown to be stable over four hours in the test chamber without, but reactive in the presence of intense UV radiation [20]. The intense UV radiation was chosen to simulate midday solar radiation; however, as noted by the authors, this situation was not expected to be relevant to indoor environments.

Of seemingly more relevance are studies investigating the stability of collected samples. Nicotine collected on XAD-4 has been shown to be stable at room temperature in the dark for at least two weeks and in a freezer for at least four weeks [52]. In another experiment, nicotine and ethenylpyridine were spiked onto XAD-4 and subjected to continuous flow of various gases for four days. Recovery of the spiked amount was at least 95% in separate experiments with air, nitrogen, and oxygen, suggesting very good stability of the collected samples [50].

By comparison, collected samples of ETS particles were evaluated for stability under conditions of exaggerated sunlight and indoor light (both fluorescent and incandescent) [86]. Results showed the potential for UVPM, FPM, and solanesol to degrade in the presence of direct sunlight and moderately intense fluorescent light. However, there was no evidence of stability problems with samples subjected to typical indoor light conditions. In addition, collected samples were shown to be stable for up to six weeks in a freezer (UVPM, FPM, and solanesol) and up to one week at room temperature in the dark (for UVPM and solanesol, but not for FPM).

Thus, while the potential for instability of airborne nicotine and other markers has been demonstrated under extreme conditions, it is not known whether this is a deleterious attribute (i.e., changes in a manner inconsistent with the 'contaminant of interest') or an advantageous one (i.e., changes in a manner consistent with the 'contaminant of interest'). The only relevant conclusion that can be made from the data in hand is that it appears all of the markers studied (nicotine, 3-EP, UVPM, FPM, and solanesol) are sufficiently stable in their collected state.

## Behavioral stability

Behavioral stability of the airborne ETS markers is the most critical attribute impacting on the fourth NRC criterion. On the positive side, the behavioral stability of nicotine has been studied more than any other potential marker in research spanning the last 15 or so years. On the negative side, the reason for this scrutiny is due to the now firmly established anomalous behavior of airborne nicotine. To establish the foundation for this, we will briefly recap the historical evolution over this time period.

As mentioned earlier, two important discoveries were made in the mid 1980s with the real-time mass spectrometric monitoring of ETS. One was the finding that ETS nicotine resided in the aerosol gas phase [40], the other was the unusual adsorptive tendencies of nicotine [46]. While ethenylpyridine and pyridine decayed with nearly first-order kinetics, nicotine was shown to decay very rapidly initially and then to approach first-order decay at a considerable time after smoking. Further, in the terminal phase of decay, nicotine decays more slowly than other compounds measured. Thus, while typical indoor ventilation takes other ETS components to not-detectable levels, nicotine asymptotically approaches a non-zero (and still detectable) background level. Based on these findings, Eudy et al. were the first to recognize that measured exposure to airborne nicotine (and even body burden of nicotine or cotinine) may have no quantifiable relationship to ETS exposure, as it may reflect exposure to simply nicotine and nothing else [40].

Due to these unpredictable decay kinetics, nicotine does not appear to be the ideal tracer once thought. Readily adsorbing to building materials and room furnishings, nicotine is depleted from ETS at a rate faster than most other components. Even on stainless steel surfaces, nicotine can readily adsorb [46,57,63,66]. Under more commonly encountered conditions, nicotine can adsorb on carpeting, painted walls, heating and air conditioning system components, [44,87,88] and human clothing [89]. This has resulted in the suggestion that nicotine concentrations underestimate concentrations of other ETS constituents [90]. This is undoubtedly true in many environments during the generation of smoke; however, the converse is true in environments with a recent past history of smoking. As demonstrated by Nelson and co-workers [87,91,92], the adsorbed nicotine slowly desorbs over time and results in an overestimate of ETS exposure in such environments. That is, as the source strength of nicotine diminishes, nicotine will begin to desorb from these surfaces. For example, nicotine levels measured in a non-occupied, stationary Boeing 767 aircraft in which smoking had been permitted were found to range from 1.1–8.7 $\mu g/m^3$ [91]. Presumably, this was due to the volatilization of nicotine from fabric and cigarette butts. Nicotine has been determined at a level of 0.09 $\mu g/m^3$ in the den of a non-smoker's home two days after smoking had occurred in the room [91]. The same investigation has also demonstrated that nicotine can desorb from a smoker's clothing and can contribute significant levels of nicotine to an otherwise pristine environment. This phenomenon has been confirmed by Piadé, et al. [89]. In these situations, ETS constituents other than nicotine decay to zero while nicotine decays rapidly to some standing background level with further decay occurring more slowly. Thus, measured concentrations of nicotine precisely assess only airborne nicotine and indicate only that smoking has taken place; they do not necessarily indicate the presence, and certainly not the levels, of other constituents.

Whether or not these biases are significant will depend on many factors, including the duration of sampling, the level of smoke in the air, the nature of the ventilation system, and the frequency of smoking activity. Such phenomena, and the magnitude of their impact, are discussed in detail elsewhere [63,87–89,91–93]. However, not taking these phenomena into account can cause incorrect assessment of actual ETS levels. For example, Hammond and co-workers [84] reported the 7-day use of passive samplers for nicotine in workplace environments. As later discovered, the nicotine concentration data from this study were incorrectly calculated and reported. Samplers were exposed continuously for seven days (i.e., 168 hours) yet, in the calculation of concentration, Hammond et al. used only 45 hours. Not only does this practice ignore the well-established phenomenon of nicotine being re-emitted from surfaces, it also violates one of the first principles of air sampling (that you must use the actual air volume sampled in calculations of airborne concentration). As a result, the nicotine concentrations *reported* by Hammond et al. are 3.7 times higher (i.e., 168/45) than the actual nicotine concentrations *measured* [66,94].

### Emission consistency

The composition of the cigarette may have an important impact on the relative amounts of material emitted into the surrounding air, and therefore, in the potential relative ratios of components in ETS. For example, the nitrate content of the tobacco may impact upon the SS yield of oxides of nitrogen ($NO_x$), without affecting the nicotine yield as much. Thus, the relationship of the nicotine to $NO_x$ in SS would be different for different cigarette brands or brand-styles, leading to potential differences in the ratios in ETS. Nelson et al. [95,96] have reported substantial differences in the relative solanesol content of the ETS respirable suspended particulate matter (RSP) for sales weighted composites of cigarette brands from foreign nations. For example, for Canadian cigarettes, solanesol comprised ca. 1.5% of the ETS RSP, compared with 4% for Argentinian cigarettes. Differences in the relative composition of tobacco blends utilized in different countries can translate into major differences in ratios of common ETS markers. For example, in Table 3 are compared the ratio of nicotine to four other markers of ETS, 3-EP, UVPM, FPM, and solanesol (converted into ETS particulate units – Sol-PM) for 18 nations and the United States. Both studies were conducted in controlled chambers, where outside contamination from other sources was considered unlikely. Within each study, there is considerable variation in the ratios of two tobacco specific constituents: nicotine and solanesol. In Study I [95], the ratio varies from 13.7–21. For Study II [96], the same ratio varies from 8.4 to 18. The reported ratios for a US sales-weighted composite are substantially lower in Study II than in Study I. This general trend seems to exist for all the determinations in Study II, and may point to the difficulty of obtaining reproducible data for nicotine, due to its adsorptive behavior, even in controlled atmosphere chambers. This may be due to the substantial interactions between nicotine and stainless steel surfaces noted above.

An additional challenge with attempting to quantify relationships between nicotine and other tobacco-combustion-derived chemical species in real environments is that, with a few exceptions, most of the other commonly determined species have major sources other than tobacco smoke. As such, the relative contribution from other sources may dominate that from ETS. Consequently, large variations in marker ratios determined in real-world environments are to be expected. For example, a variety of consumer products contribute to the overall levels of formaldehyde [97]. Wood burning stoves, candles, and other sources, as well as ETS, can contribute a variety of polycyclic hydrocarbons, UVPM, oxides of nitrogen, carbonyls, benzene, etc. [19,64,65,98,99]. Shah and Singh [100] have reviewed an extensive database of volatile organic chemicals in indoor and outdoor air, many of which have been reported to be in tobacco smoke. As a result of these confounding sources, and given the wide variety of micro-environments

TABLE 3

ETS constituent ratios determined in chamber. Studies sales-weighted market share composite (Adopted from Nelson et al. [95,96])

|  | Country | 3-EP:Nicotine | UVPM:Nicotine | FPM:Nicotine | Sol-PM:Nicotine |
|---|---|---|---|---|---|
| Study I [95] | Czech Republic | 0.386 | 17.3 | 17.2 | 17.2 |
|  | France | 0.431 | 17.8 | 17.9 | 18.0 |
|  | Germany | 0.447 | 18.4 | 18.7 | 17.8 |
|  | Hong Kong | 0.457 | 20.5 | 20.3 | 21.1 |
|  | Italy | 0.398 | 16.4 | 16.5 | 16.8 |
|  | Malaysia | 0.459 | 16.9 | 16.6 | 16.7 |
|  | Portugal | 0.451 | 16.9 | 17.2 | 17.3 |
|  | Spain | 0.406 | 15.7 | 15.8 | 16.2 |
|  | Sweden | 0.431 | 13.9 | 13.8 | 13.7 |
|  | Switzerland | 0.423 | 19.3 | 19.3 | 19.4 |
|  | United States I | 0.557 | 19.8 | 19.6 | 19.7 |
| Study II [96] | Argentina | 0.307 | 11.1 | 11.3 | 11.1 |
|  | Australia | 0.281 | 11.3 | 11.4 | 11.5 |
|  | Brazil | 0.341 | 11.2 | 11.2 | 11.4 |
|  | Canada | 0.264 | 10.0 | 10.2 | 10.3 |
|  | China | 0.307 | 18.1 | 17.8 | 18.4 |
|  | England | 0.268 | 9.3 | 9.3 | 9.3 |
|  | Japan | 0.260 | 8.5 | 8.5 | 8.4 |
|  | Korea | 0.310 | 9.6 | 9.7 | 9.4 |
|  | United States II | 0.281 | 10.4 | 10.3 | 10.4 |

which individual non-smokers are likely to inhabit [101], it is not surprising that there is wide variation in the ratio of nicotine to other ETS-derived components.

Given these properties of nicotine and other ETS components, as well as the behavior of nicotine in controlled environments, it is expected that the relationship between nicotine and other ETS components is variable and complex in realistic settings. For example, the relationship between nicotine and other common markers of ETS in selected field studies is summarized in Table 4. It is important to note here that data are not reported for a comparison of RSP with nicotine. It is the authors' assessment that RSP is simply not a marker of ETS. There are so many sources other that ETS in most indoor environments – combustion derived and otherwise – that, except at the highest concentrations of ETS, tobacco smoke is not the predominant source of airborne particles [4,26]. This has been confirmed by a number of additional studies [27–37]. Because of this, the utility of RSP as an indicator of ETS level in specific environments is non-existent. For example, in Fig. 2 is portrayed the nicotine and RSP data obtained from a large personal monitoring study in the United States. All the data pairs represent 8-hour personal exposure samples obtained in smoking workplaces where smoking was directly observed by the subjects. Note that both axes are logarithmic scales. While the correlation between RSP and nicotine is statistically significant ($R^2=0.188$), at any given RSP level, associated nicotine levels can vary over at least two orders of magnitude.

For the data presented in Table 4, most derive from personal monitoring studies where subjects were free to move through a number of micro-environments throughout the course of the sampling duration. As might be expected, the relationship between nicotine and a gas-phase combustion derivative of nicotine, 3-EP, is strongest. With one exception, the coefficients of determination for 3-EP vs. nicotine are greater than 0.55, indicating a strong statistical relationship. While the relationship may be statistically significant, the predictive utility is very limited, and for any given concentration of one marker, the other can usually vary by at least one order of magnitude. Mean ratios reported in Table 4 ranged from 0.19 to 0.68 for the studies reported. This suggests that in a given society or situation, nicotine may be a good general predictor of 3-EP levels (or vice versa), but that it would be first necessary to establish expected ratios in a given class of environments. However, as described above, many standard sampling and analysis methods provide both 3-EP and nicotine levels in the same analysis. Thus, it is more straightforward merely to report concentrations of both constituents, rather than use one to estimate the value of the other. For general combustion-derived particulate phase indicators, such as UVPM and FPM, or a tobacco specific component, such as solanesol, the relationship with nicotine is more variable, and even less quantitative. Coefficients of determination ($R^2$) were generally lower, predictive values even less, and the

TABLE 4

Mean ratios[1] and coefficients of determination (R²) for concentrations of nicotine and common markers of ETS from selected field studies. Except where noted, all data are from personal monitoring studies

| Study | Sample duration hrs | No. of data points | Constituent | | | | | | | |
|---|---|---|---|---|---|---|---|---|---|---|
| | | | 3-EP | | UVPM | | FPM | | Sol-PM | |
| | | | R² | Ratio | R² | Ratio | R² | Ratio | R² | Ratio |
| Phillips et al. 1998 Hong Kong [34] | 24 | 77–102 | 0.968 | 0.191 | 0.012 | 0.639 | 0.875 | 4.75 | 0.670 | 14.9 |
| Phillips et al. 1998 Lisbon [32] | 24 | 86–134 | 0.769 | 0.309 | 0.311 | 6.4 | 0.347 | 6.19 | 0.393 | 7.37 |
| Phillips et al. 1998 Paris [33] | 24 | 204–314 | 0.767 | 0.379 | 0.583 | 7.19 | 0.444 | 7.22 | 0.512 | 22.6 |
| Phillips et al. 1997 Turin [30] | 24 | 290–292 | 0.874 | 0.380 | 0.637 | 9.87 | 0.633 | 11.05 | 0.645 | 15.8 |
| Phillips et al. 1998 Prague [36] | 24 | 201–229 | 0.824 | 0.269 | 0.441 | 12.6 | 0.389 | 10.3 | 0.276 | 8.88 |
| Philips et al. 1998 Kuala Lumpur [35] | 24 | 82–139 | 0.590 | 0.190 | 0.010 | 0.59 | 0.010 | 0.43 | 0.010 | 0.16 |
| LaKind et al., 1999 [119] US Smoking workplaces[2] | 8 | 170–256 | 0.627 | 0.616 | 0.716 | 14.7 | 0.716 | 11.9 | 0.627 | 5.62 |
| Sterling et al., 1996 [107] Smoking workplace in Richmond VA | 8 | 24–25 | 0.143 | 0.678 | N/A | N/A | 0.000 | 6.16 | 0.006 | 5.57 |
| Crouse et al., 1988 [120] Restaurants in Greensboro, NC[3] | 1 | 36 | N/A | N/A | 0.133 | 6.46 | N/A | N/A | N/A | N/A |
| Oldaker et al., 1990 [121] Restaurants in Winston-Salem, NC[3,4] | 1 | 41 | N/A | 0.238 | N/A | 6.19 | N/A | 5.14 | N/A | 5.13 |

N/A=Not Available

[1] Ratios Given as X:Nicotine, in unitless values, where X is constituent. All data pairs at levels above reported limit of quantification

[2] Original data derived from Jenkins et al. [26].

[3] Area monitoring data.

[4] 3-EP data from Ogden and Nelson [2]. FPM and Sol-PM data from Ogden et al. [25]. Also, ratios obtained from summary means, rather than individual data points.

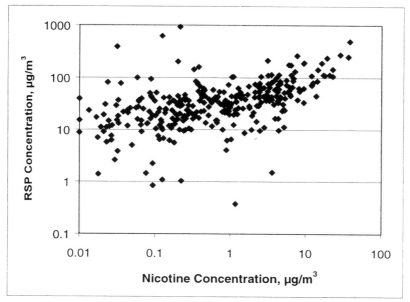

Fig. 2 Respirable suspended particulate matter (RSP) as a function of nicotine concentration for personal exposure measurements in smoking workplaces where smoking has been directly observed. Data from LaKind et al. [119] and Jenkins et al. [26]. Note logarithmic scale on both axes.

ratios variable over a somewhat wider range. This may be due in part to the expected differences in behavior between particle phase and vapor phase constituents and to suspected lower pre-collection stability of the particle phase species in lighted environments [86].

The extent to which nicotine concentrations may be correlated with or predictive of other species in ETS has not been studied extensively. This is the case for common species (e.g., benzene, toluene, phenol) as well as those of potentially greater toxicological interest. While a variety of chemical species are postulated to be responsible for the alleged toxicological properties of ETS, greater attention lately has been directed toward the tobacco specific nitrosamine 4-(*N*-methylnitrosamino)-1-(3-pyridyl)-1-buta-none (NNK). Unfortunately, due to the extreme difficulty in measuring NNK, there have been few reports of simultaneous measurement of nicotine and NNK or other tobacco specific nitrosamines. An exception is the work of Klus et al. [102], who reported individual levels of nicotine and NNK, among other species, in office environments which were deliberately polluted with artificially high levels of ETS. In that study, the mean nicotine : NNK ratio was ca. 24,000 : 1. However, the correlation between the two constituent levels was very low, with $R^2 = 0.103$. These findings suggest that nicotine would not be a good predictor of NNK values.

where the subjects were employed during the period in which they worked. On the basis of statistical groupings, the differences between area and personal monitoring levels encountered were not great. For example, the median area level of FPM for bartenders was 39 μg/m³ vs. 45 μg/m³ for the personal samples. The median area 3-EP level for restaurant servers was 0.58 μg/m³ vs. the median personal level of 0.59 μg/m³. Such a comparison suggests that area measurements are good surrogates for personal monitoring measurements in this type of environment. And while this may be true for groups of subjects, graphical examination of the data indicates that on an individual basis, area samples can only estimate personal exposure to within a factor of 5–10. For example, in Fig. 3 are portrayed graphical comparisons of area vs. personal ETS levels of nicotine, for all subjects in the restaurant/tavern servers study described above for which direct comparisons could be made. While there is general correlation among the data, the variation of personal levels, for example, at a given area concentration is considerable. This is likely due to individual activity patterns of the subjects, who occupy a variety of micro-environments over the course of their work shift. The results of all these studies indicate that area samples are more likely to be useful for assessing ETS exposures of individuals who are relatively stationary within their environment.

Guerin et al. [4] have extensively reviewed the 25–30 field studies of nicotine levels in non-manipulated environments reported in 1991 and earlier. A review of these data will not be repeated here, other than to

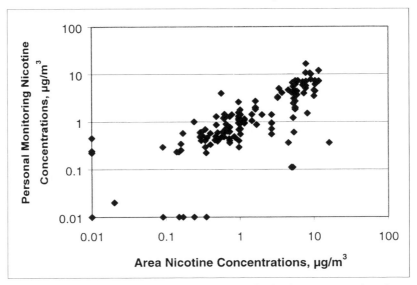

Fig. 3 Comparison of area and personal monitoring ETS nicotine concentrations in a study of restaurant and tavern servers. Data from Jenkins and Counts [109] and Maskarinec et al. [108]. Note logarithmic scale on both axes.

TABLE 5

Summary of concentrations of ETS nicotine ($\mu$g/m$^3$) in non-manipulated environments* prior to 1992. Includes area monitoring and limited personal exposure measurements

| Exposure category | Background[1] | Common[2] | Heavy[3] | Extreme[4] |
|---|---|---|---|---|
| Lower Exposure venues<br>Public Buildings<br>Offices<br>Work Situations<br>Restaurants | <2–5 | <2–10 | 5–25 | >25 |
| Higher Exposure Venues<br>Smoking Lounges<br>Transportation<br>Bars and Taverns | <2 | <2–35 | 20–70 | >70 |

* Abstracted from Guerin et al. [4].
[1] Occasionally found in non-smoking environments. Also factors in the limits of detection or quantification for various studies.
[2] Encountered in most situations (lower 70% of reported concentrations).
[3] Encountered in the ~70th–95th percentiles of nicotine concentrations.
[4] Encountered in the upper 95th percentile of reported nicotine concentrations.

contrast with more recent results. The majority of the studies reported by 1991 relied on fixed location monitoring of airborne nicotine levels. Guerin et al. summarized the levels found in these studies as being 'background', 'common', 'heavy', or 'extreme' for higher and lower exposure venues (see Table 5). According to the authors, at the time, ambient concentrations of nicotine up to 10 $\mu$g/m$^3$ were considered to be 'common'.

In Table 6 are summarized the findings of more recent studies reporting airborne nicotine concentrations determined from area monitoring data. There are several key observations from the compilation. First, compared with the number of large studies reporting levels of nicotine from personal exposure measurements, the number of area monitoring studies is limited. In fact, of the nine studies reported in Table 6, four are actually summaries of area monitoring data collected in conjunction with personal exposure studies reported in Table 7. Secondly, based on median data, it appears that nicotine levels as high as 10 $\mu$g/m$^3$ are higher than concentrations commonly encountered in most settings, with the exception of higher-end exposure settings, such as nightclubs and taverns. The exception to this is the data reported by Hammond et al. [84] for offices in facilities where smoking was unrestricted. However, the calculation of concentration levels in that study has been called into question [66,94]. The data reported in Hammond et al. [84] may overestimate the actual concentrations by as much as a factor of 3.7. It seems clear that, by today's standards, commonly encountered

TABLE 6

Nicotine concentrations as determined from area sampling in selected recent studies. Concentrations in μg/m³

| Study | Sample duration, hrs | Number of Samples | Venue | Median | Geometric Mean | Arithmetic mean | 80th percentile | 90th percentile | 95th percentile | Maximum |
|---|---|---|---|---|---|---|---|---|---|---|
| Scherer et al. 1995 Germany [110] | 3 | 20 | Smoking homes | 3.45 | | 7.33 | | | | 35.2 |
| Scherer et al. 1995 Germany [110] | 3 | 10 | Non-smoking Homes | 0.01 | | <0.01 | | | | 0.04 |
| Scherer et al. 1995 Germany [110] | 168 | 20 | Smoking homes | 1.04 | | 5.09 | | | | 34.6 |
| Scherer et al. 1995 Germany [110] | 168 | 10 | Non-smoking Homes | 0.02 | | 0.02 | | | | 0.06 |
| Williams et al. 1993 Research Triangle Park NC [122] | 48 | 34 | Home only | | 2.24 | 8.43 | | | | 94 |
| Jenkins and Counts, 1999 Knoxville TN [109] | 7 | 32 | Restaurants | 0.76 | | 5.8 | 4.4 | | 36 | |
| Jenkins and Counts, 1999 Knoxville TN [109] | 7 | 53 | Bars and bar areas within restaurants | 5.8 | | 14.4 | 29.6 | | 49.6 | |
| Hammond et al. 1995 Massachusetts workplaces [84] | 45* | 13 | Closed smoking offices | 9.1* | | 8.3* | | | | 22* |
| Hammond et al. 1995 Massachusetts workplaces [84] | 45* | 61 | Open offices unrestricted smoking | 8.6* | | 14.4* | | 34* | | |
| Hammond et al. 1995 Massachusetts workplaces [84] | 45* | 35 | Open offices restricted smoking | 1.3* | | 3.4* | 9* | | | |

TABLE 6 Continued

| Study | Sample duration, hrs | Number of Samples | Venue | Median | Geo-metric Mean | Arith-metic mean | 80th percen-tile | 90th percen-tile | 95th percen-tile | Maximum |
|---|---|---|---|---|---|---|---|---|---|---|
| Hammond et al. 1995 Massachusetts workplaces [84] | 45* | 29 | Open offices: smoking banned | 0.3* | | 0.7* | | 1.7* | | |
| Hammond et al. 1995 Massachusetts Workplaces [84] | 45* | 114 | Shop production areas and fire stations unrestricted smoking | 2.3* | | 4.4* | | 7.2* | | |
| Hammond et al. 1995 Massachusetts Workplaces [84] | 45* | 54 | Shop production areas and fire stations: restricted smoking | 0.7* | | 2.2* | | 4* | | |
| Hammond et al. 1995 Massachusetts Workplaces [84] | 45* | 53 | Shop production areas and fire stations: smoking banned | 0.2* | | 0.2* | | 0.6* | | |
| Collett et al. 1992 Vancouver BC [111] | 1 | 31 | Taverns and nightclubs | | | 38.6–58.0 | | | | 63.9–119.5 |
| Ogden et al. 1993 Columbus OH [105] | 120 | 48 | Smoking homes – heavy use area | 2.45 | | | | | | |
| Ogden et al. 1993 Columbus OH [105] | 120 | 48 | Smoking homes – light use area | 0.56 | | | | | | |
| Sterling et al. 1996 Richmond VA [107] | 8 | 16 | Office building where smoking was unrestricted | 2.1 | | 2.2 | 2.4 | | 3.9 | |
| Lambert et al. 1993 Albuquerque NM [123] | 12 | 7 | Restaurant smoking sections | 3.2 | | | | | | 3.8 |
| Lambert et al. 1993 Albuquerque NM [123] | 12 | 7 | Restaurant non-smoking sections | 1 | | | | | | 2.8 |

TABLE 6 Continued

| Study | Sample duration, hrs | Number of Samples | Venue | Median | Geo-metric Mean | Arith-metic mean | 80th percen-tile | 90th percen-tile | 95th percen-tile | Maximum |
|-------|---------------------|-------------------|-------|--------|-----------------|------------------|------------------|------------------|------------------|---------|
| Oldaker et al. 1995 Winston-Salem NC [124] | 8 | 40 | Two office buildings where smoking was unrestricted | 1.8–2.3 | | | | | | 9.9–11.7 |

* Passive sampling over the course of one week (ca. 168 h) was used in this study. However, concentrations were calculated based on a sampling duration of 45 h. Reported values are overestimates of true concentration levels by as much as a factor of 3.7 [66,94].

TABLE 7

Nicotine concentrations as determined from personal exposure sampling in selected recent studies. Concentrations in $\mu g/m^3$

| Study | Criteria conditions | Sample duration, hrs | Number of subjects | Subject type | No of pumps or samplers | Exposure venue | Median | Geo-metric mean | Arith-metic mean | 80th %ile | 90th %ile | 95th %ile | Maximum |
|---|---|---|---|---|---|---|---|---|---|---|---|---|---|
| Phillips et al. 1998 Sydney [37] | Smoking homes | 24 | 30 | Housepersons | 1 | Home only | 0.3 | 0.3 | 0.67 | | | | 1.6 |
| Phillips et al. 1998 Sydney [37] | Non-smoking homes | 24 | 46 | Housepersons | 1 | Home only | 0.05 | 0.05 | 0.06 | | | | 0.1 |
| Phillips et al. 1998 Sydney [37] | Smoking workplaces | 24 | 20 | Workers | 1 | Workplace only | 0.22 | 0.42 | 1.1 | | | | 2.9 |
| Phillips et al. 1998 Sydney [37] | Non-smoking workplaces | 24 | 57 | Workers | 1 | Workplace only | 0.15 | 0.15 | 0.15 | | | | 0.19 |
| Phillips et al. 1998 Sydney [37] | Smoking homes | 72 | 25 | Anyone | 1 | Outside of home or work | 0.32 | 0.47 | 1.8 | | | | 6.1 |
| Phillips et al. 1998 Sydney [37] | Non-smoking homes | 72 | 53 | Anyone | 1 | Outside of home or work | 0.16 | 0.22 | 0.41 | | | | 0.94 |
| Phillips et al. 1998 Sydney [37] | Smoking homes | 24 | 27 | Workers | 1 | Home, work, and elsewhere | 0.39 | 0.33 | 0.66 | | | | 1.4 |
| Phillips et al. 1998 Sydney [37] | Non-smoking homes | 24 | 51 | Workers | 1 | Home, work, and elsewhere | 0.05 | 0.07 | 0.14 | | | | 0.16 |
| Phillips et al. 1998 Lisbon [32] | Smoking homes | 24 | 24 | Housepersons | 1 | Home only | 0.19 | 0.24 | 0.59 | | | | 1.2 |
| Phillips et al. 1998 Lisbon [32] | Non-smoking homes | 24 | 53 | Housepersons | 1 | Home only | 0.05 | 0.07 | 0.14 | | | | 0.31 |
| Phillips et al. 1998 Lisbon [32] | Smoking home smoking work | 24 | 27 | Workers | 2 | Home, work, and elsewhere | 0.58 | 0.55 | 1.2 | | | | 2.8 |
| Phillips et al. 1998 Lisbon [32] | Smoking home non-smoking work | 24 | 7 | Workers | 2 | Home, work, and elsewhere | 0.5 | 0.46 | 0.88 | | | | 2.1 |
| Phillips et al. 1998 Lisbon [32] | Non-smoking home smoking work | 24 | 58 | Workers | 2 | Home, work, and elsewhere | 0.32 | 0.37 | 0.84 | | | | 2.1 |
| Phillips et al. 1998 Lisbon [32] | Non-smoking home non-smoking work | 24 | 20 | Workers | 2 | Home, work, and elsewhere | 0.09 | 0.16 | 0.37 | | | | 0.83 |

TABLE 7 Continued

| Study | Criteria conditions | Sample duration, hrs | Number of subjects | Subject type | No of pumps or samplers | Exposure venue | Median | Geo-metric mean | Arith-metic mean | 80th %ile | 90th %ile | 95th %ile | Maximum |
|---|---|---|---|---|---|---|---|---|---|---|---|---|---|
| Scherer et al. 1995 Germany [110] | Smoking homes | 168 | 43 | Anyone | 1 | Home, work, and elsewhere | 0.3 | | 0.76 | | | | 4.57 |
| Scherer et al. 1995 Germany [110] | Non-smoking homes | 168 | 39 | Anyone | 1 | Home, work, and elsewhere | 0.07 | | 0.14 | | | | 1.1 |
| Jenkins et al. 1996 USA [26] | Smoking home smoking work | 24 | 157 | Workers | 2 | Home, work, and elsewhere | 1.47 | | 2.98 | 3.93 | 9 | | |
| Jenkins et al. 1996 USA [26] | Smoking home non-smoking work | 24 | 234 | Workers | 2 | Home, work, and elsewhere | 0.473 | | 1.21 | 1.84 | 4.28 | | |
| Jenkins et al. 1996 USA [26] | Non-smoking home smoking work | 24 | 281 | Workers | 2 | Home, work, and elsewhere | 0.107 | | 0.543 | 0.585 | 2.07 | | |
| Jenkins et al. 1996 USA [26] | Non-smoking home non-smoking work | 24 | 808 | Workers | 2 | Home, work, and elsewhere | 0.031 | | 0.12 | 0.087 | 0.35 | | |
| Phillips et al. 1996 Stockholm [28] | Smoking homes | 24 | 9 | Housepersons | 1 | Homes | 1.1 | | 3.1 | | | 7.5 | |
| Phillips et al. 1996 Stockholm [28] | Non-smoking homes | 24 | 33 | Housepersons | 1 | Homes | 0.05 | | 0.34 | | | | 3.2 |
| Phillips et al. 1996 Stockholm [28] | Smoking homes | 16 | 10 | Workers | 1 | Home and elsewhere | 0.15 | | 0.3 | | | | 1.6 |
| Phillips et al. 1996 Stockholm [28] | Non-smoking homes | 16 | 119 | Workers | 1 | Home and elsewhere | 0.07 | | 0.19 | | | | 9.3 |
| Phillips et al. 1996 Stockholm [28] | Smoking workplaces | 8 | 53 | Workers | 1 | Workplace only | 0.2 | | 0.48 | | | | 3.1 |

TABLE 7 Continued

| Study | Criteria conditions | Sample duration, hrs | Number of subjects | Subject type | No of pumps or samplers | Exposure venue | Median | Geo-metric mean | Arith-metic mean | 80th %ile | 90th %ile | 95th %ile | Maximum |
|---|---|---|---|---|---|---|---|---|---|---|---|---|---|
| Phillips et al. 1996 Stockholm [28] | Non-smoking workplaces | 8 | 82 | Workers | 1 | Workplace only | 0.15 | | 0.23 | | | | 1.8 |
| Williams et al. 1993 RTP NC area [122] | Smoking homes | 48 | 42 | Children | 1 | Home only | | 2.52 | 6.43 | | | | 47 |
| Phillips et al. 1998 Hong Kong [34] | Smoking homes | 24 | 31 | Housepersons | 1 | Home only | 0.07 | 0.12 | 0.26 | | 0.51 | | |
| Phillips et al. 1998 Hong Kong [34] | Non-smoking homes | 24 | 33 | Housepersons | 1 | Home only | 0.05 | 0.07 | 0.1 | 0.27 | | | |
| Phillips et al. 1998 Hong Kong [34] | Smoking home non-smoking work | 24 | 18 | Workers | 2 | Home, work, and elsewhere | 0.44 | 0.74 | 3.2 | | 7 | | |
| Phillips et al. 1998 Hong Kong [34] | Smoking home non-smoking work | 24 | 25 | Workers | 2 | Home, work, and elsewhere | 0.27 | 0.32 | 0.62 | | 1.2 | | |
| Phillips et al. 1998 Hong Kong [34] | Non-smoking home smoking work | 24 | 22 | Workers | 2 | Home, work, and elsewhere | 0.24 | 0.25 | 0.32 | | 0.66 | | |
| Phillips et al. 1998 Hong Kong [34] | Non-smoking home non-smoking work | 24 | 43 | Workers | 2 | Home, work, and elsewhere | 0.1 | 0.13 | 0.17 | | 0.22 | | |
| Phillips et al. 1998 Prague [36] | Smoking homes | 24 | 51 | Housepersons | 1 | Home only | 0.72 | 0.63 | 1.3 | | 3.1 | | |
| Phillips et al. 1998 Prague [36] | Non-smoking homes | 24 | 38 | Housepersons | 1 | Home only | 0.15 | 0.16 | 0.31 | | 0.51 | | |
| Phillips et al. 1998 Prague [36] | Smoking home smoking work | 24 | 57 | Workers | 2 | Home, work, and elsewhere | 1.6 | 1.5 | 2.3 | | 4.9 | | |
| Phillips et al. 1998 Prague [36] | Smoking home non-smoking work | 24 | 12 | Workers | 2 | Home, work, and elsewhere | 0.45 | 0.57 | 1.3 | | 4.6 | | |

TABLE 7 Continued

| Study | Criteria conditions | Sample duration, hrs | Number of subjects | Subject type | No of pumps or samplers | Exposure venue | Median | Geo-metric mean | Arith-metic mean | 80th %ile | 90th %ile | 95th %ile | Maximum |
|---|---|---|---|---|---|---|---|---|---|---|---|---|---|
| Phillips et al. 1998 Prague [36] | Non-smoking home smoking work | 24 | 45 | Workers | 2 | Home, work, and elsewhere | 0.49 | 0.54 | 1.1 | | | | 2.2 |
| Phillips et al. 1998 Prague [36] | Non-smoking home non-smoking work | 24 | 19 | Workers | 2 | Home, work, and elsewhere | 0.16 | 0.2 | 0.25 | | | | 0.52 |
| Phillips et al. 1998 Kuala Lumpur [35] | Smoking homes | 24 | 40 | Housepersons | 1 | Home only | 0.18 | 0.17 | 0.65 | | | | 1.3 |
| Phillips et al. 1998 Kuala Lumpur [35] | Non-smoking homes | 24 | 51 | Housepersons | 1 | Home only | 0.05 | 0.08 | 0.24 | | | | 0.24 |
| Phillips et al. 1998 Kuala Lumpur [35] | Smoking home smoking work | 24 | 29 | Workers | 2 | Home, work, and elsewhere | 0.3 | 0.34 | 0.66 | | | | 1.2 |
| Phillips et al. 1998 Kuala Lumpur [35] | Smoking home non-smoking work | 24 | 27 | Workers | 2 | Home, work, and elsewhere | 0.23 | 0.24 | 0.4 | | | | 1.1 |
| Phillips et al. 1998 Kuala Lumpur [35] | Non-smoking home smoking work | 24 | 44 | Workers | 2 | Home, work, and elsewhere | 0.15 | 0.22 | 1.1 | | | | 1.3 |
| Phillips et al. 1998 Kuala Lumpur [35] | Non-smoking home non-smoking work | 24 | 43 | Workers | 2 | Home, work, and elsewhere | 0.1 | 0.12 | 0.17 | | | | 0.27 |
| Phillips et al. 1997 Turin [30] | Smoking homes | 24 | 36 | Housepersons | 1 | Home only | 1.1 | 1.1 | 1.9 | | | | 4.9 |
| Phillips et al. 1997 Turin [30] | Non-smoking homes | 24 | 47 | Housepersons | 1 | Home only | 0.14 | 0.14 | 0.32 | | | | 0.6 |
| Phillips et al. 1997 Turin [30] | Smoking home smoking work | 24 | 21 | Workers | 2 | Home, work, and elsewhere | 1.3 | 1.2 | 1.6 | | | | 3.6 |
| Phillips et al. 1997 Turin [30] | Smoking home non-smoking work | 24 | 8 | Workers | 2 | Home, work, and elsewhere | 1 | 1.1 | 1.9 | | | | 4.7 |

TABLE 7 Continued

| Study | Criteria conditions | Sample duration, hrs | Number of subjects | Subject type | No of pumps or samplers | Exposure venue | Median | Geo-metric mean | Arith-metic mean | 80th %ile | 90th %ile | 95th %ile | Maximum |
|---|---|---|---|---|---|---|---|---|---|---|---|---|---|
| Phillips et al. 1997 Turin [30] | Non-smoking home smoking work | 24 | 51 | Workers | 2 | Home, work, and elsewhere | 0.38 | 0.43 | 0.61 | | 1.3 | | |
| Phillips et al. 1997 Turin [30] | Non-smoking home non-smoking work | 24 | 24 | Workers | 2 | Home, work, and elsewhere | 0.25 | 0.31 | 0.41 | | 0.87 | | |
| Phillips et al. 1998 Bremen [31] | Smoking homes | 24 | 21 | Housepersons | 1 | Home only | 0.49 | 0.31 | 0.63 | | 1.5 | | |
| Phillips et al. 1998 Bremen [31] | Non-smoking homes | 24 | 59 | Housepersons | 1 | Home only | 0.05 | 0.07 | 0.1 | | 0.22 | | |
| Phillips et al. 1998 Bremen [31] | Smoking home smoking work | 24 | 17 | Workers | 2 | Home, work, and elsewhere | 0.69 | 0.66 | 1.2 | | 2.1 | | |
| Phillips et al. 1998 Bremen [31] | Smoking home non-smoking work | 24 | 6 | Workers | 2 | Home, work, and elsewhere | 0.23 | 0.23 | 0.32 | | 0.63 | | |
| Phillips et al. 1998 Bremen [31] | Non-smoking home smoking work | 24 | 44 | Workers | 2 | Home, work, and elsewhere | 0.2 | 0.26 | 0.52 | | 1.1 | | |
| Phillips et al. 1998 Bremen [31] | Non-smoking home non-smoking work | 24 | 35 | Workers | 2 | Home, work, and elsewhere | 0.1 | 0.13 | 0.16 | | 0.24 | | |
| Phillips et al. 1998 Paris [33] | Smoking homes | 24 | 48 | Housepersons | 1 | Home only | 0.52 | 0.55 | 0.93 | | 2.4 | | |
| Phillips et al. 1998 Paris [33] | Non-smoking homes | 24 | 42 | Housepersons | 1 | Home only | 0.13 | 0.13 | 0.18 | | 0.29 | | |
| Phillips et al. 1998 Paris [33] | Smoking home smoking work | 24 | 41 | Workers | 2 | Home, work, and elsewhere | 1.4 | 1.3 | 1.8 | | 4.1 | | |
| Phillips et al. 1998 Paris [33] | Smoking home non-smoking work | 24 | 11 | Workers | 2 | Home, work, and elsewhere | 0.72 | 0.55 | 0.78 | | 1.2 | | |
| Phillips et al. 1998 Paris [33] | Non-smoking home smoking work | 24 | 58 | Workers | 2 | Home, work, and elsewhere | 0.42 | 0.45 | 0.58 | | 1.1 | | |
| Phillips et al. 1998 Paris [33] | Non-smoking home non-smoking work | 24 | 7 | Workers | 2 | Home, work, and elsewhere | 0.27 | 0.29 | 0.3 | | 0.42 | | |
| Phillips et al. 1997 Barcelona [29] | Smoking homes | 24 | 43 | Housepersons | 1 | Home only | 0.74 | 0.72 | 1.4 | | 2.8 | | |

TABLE 7 Continued

| Study | Criteria conditions | Sample duration, hrs | Number of subjects | Subject type | No of pumps or samplers | Exposure venue | Median | Geo-metric mean | Arith-metic mean | 80th %ile | 90th %ile | 95th %ile | Maximum |
|---|---|---|---|---|---|---|---|---|---|---|---|---|---|
| Phillips et al. 1997 Barcelona [29] | Non-smoking homes | 24 | 40 | Housepersons | 1 | Home only | 0.11 | 0.14 | 0.33 | | 0.46 | | |
| Phillips et al. 1997 Barcelona [29] | Smoking homes | 16 | 28 | Workers | 1 | Home and elsewhere | 0.86 | 0.83 | 1.8 | | 4.4 | | |
| Phillips et al. 1997 Barcelona [29] | Non-smoking homes | 16 | 41 | Workers | 1 | Home and elsewhere | 0.17 | 0.21 | 0.48 | | 0.61 | | |
| Phillips et al. 1997 Barcelona [29] | Smoking workplaces | 8 | 61 | Workers | 1 | Workplace only | 2.4 | 2.3 | 4.1 | | 9 | | |
| Phillips et al. 1997 Barcelona [29] | Non-smoking workplaces | 8 | 8 | Workers | 1 | Workplace only | 0.71 | 0.56 | 0.91 | | 2 | | |
| Phillips et al. 1994 UK [27] | No venue discrimination | 24 | 249 | Anyone | 1 | Home, work, and elsewhere | 0.5 | | 1.7 | | | | 26 |
| Trout et al. 1998 Atlantic City NJ [112] | Casinos | 8 | 18 | Workers | 1 | Workplace only | 9.5 | | 9.4 | 12 | 14.2 | | |
| Ogden 1993 Columbus OH [105, 125] | Female non-smokers married to non-smokers | 120 | 48 | Anyone | NA | Anywhere | 0.12 | | | | | | |
| Ogden 1993 Columbus OH [105, 125] | Female non-smokers married to smokers | 120 | 47 | Anyone | NA | Anywhere | 1.61 | | | | | | |
| Jenkins and Counts 1999 Knoxville TN [109] | Restaurant servers | 4–8 | 82 | Workers | 1 | Workplace only | 1.22 | | 5.9 | 6.1 | | 28.9 | |

TABLE 7 Continued

| Study | Criteria conditions | Sample duration, hrs | Number of subjects | Subject type | No of pumps or samplers | Exposure venue | Median | Geo-metric mean | Arith-metic mean | 80th %ile | 90th %ile | 95th %ile | Maximum |
|---|---|---|---|---|---|---|---|---|---|---|---|---|---|
| Jenkins and Counts 1999 Knoxville TN [109] | Bartenders | 5–9 | 80 | Workers | 1 | Workplace only | 4.45 | | 14.1 | 27.1 | 43.6 | | |
| O'Conner et al. 1995 New Haven CT [126] | Pregnant women reporting exposure to ETS | 168 | 131 | Women | NA | Anywhere | 0.1 | | | | 0.6 | | |
| Bergman et al. 1996 Oklahoma City OK [113] | Nightclub musicians | ?? | 9 | Workers | NA | Workplace only | 38.8 | | 37.1 | | | | 50.3 |

concentrations (i.e., median levels) of ambient nicotine in all but the highest exposure venues will be less than 5 $\mu g/m^3$.

Some additional features of the data reported in Table 6 are interesting. In the study reported for German homes [110], longer duration samples resulted in lower levels of nicotine. This presumably reflects sampling during those time periods in the home when smoking was more limited. In general, there is a tendency for shorter duration sampling periods to result in higher concentrations than longer term measurements, even when the latter are restricted to one 'shift'. This may reflect 'peak' concentrations vs. 'average' concentrations. It also may reflect an unconscious bias on the part of the field team to collect samples when smoking rates are highest in order to insure a measurable amount of material is collected. For example, the Vancouver, BC nightclub study [111] yielded mean levels of ca. 40–60 $\mu g/m^3$ nicotine from samples collected over one hour. In contrast, the Knoxville, TN bar and bar area study [109] yielded a mean of ca. 14 $\mu g/m^3$ from samples collected over the course of 7 h. However, this difference may also reflect the difference in smoking rates in a large vs. small city, differences in societal smoking rates, and/or changes in smoking practices during the 6 years between the two studies.

In contrast to determination of area measurements of airborne nicotine over the past 7–8 years, there has been a massive effort to characterize personal exposure to ETS. In studies conducted in the United States, Europe, and Asia, the exposure of more than 4000 subjects to environmental tobacco smoke has been characterized. The results of these studies are summarized in Table 7. Many of the studies reported had similar experimental designs. That is, subjects would wear two pumps over the course of 24 hours; one sampling pump at their workplace for 8–9 h, and another away from work for 15–16 h. In studies conducted in Europe and Asia, an additional effort was made to characterize exposure of subjects who did not work outside the home, by providing them with a sampling pump capable of collecting air samples for a 24-h period. The usual experimental design was to classify the subjects according to the smoking status of their spouse or partner. The strength of the experimental designs in these studies was that it was possible, with multiple sampling pumps, to characterize the relative importance of exposure from inside and outside the workplace. Virtually all the studies conducted in this manner indicated that exposure (the product of concentration and time) outside of work (with home being the predominant source) is greater than workplace exposure.

Another general feature of these data sets is that for the most part, subjects who spend more time in smoking environments encounter higher time-averaged concentrations of ETS nicotine, relative to those who spend less time in smoking environments. For example, subjects who work and live in smoking environments are exposed to higher levels of nicotine than those who may live in a smoking home, but work in a non-smoking workplace, etc.

However, the magnitudes of time-averaged concentrations are typically lower than those measured by fixed location area sampling. A review of the data reported in Table 7 indicates that with the exception of certain occupational venues where greater levels of ETS would be anticipated, 24 hour time-weighted average (TWA) personal exposure levels of nicotine were all less than 2 $\mu g/m^3$. Interestingly, for all the countries in which measurements were made, subjects who lived and worked in smoking environments in the United States [26] were exposed to higher levels of ETS nicotine than all of the other locations, other than Prague [36]. Given the distribution of ETS nicotine levels reported in all of these studies, it is clear there are some individuals who are exposed to nicotine levels that exceed 10 $\mu g/m^3$ on a 24-hr TWA basis. However, the vast majority of non-smokers appear to be exposed to levels much lower than that.

In an initial assessment, such a conclusion may run counter to that which could be derived by an assessment of the area monitoring data. However, while individuals may live or work in smoking environments, stationary monitors can not take into account changes in smoke exposure resulting from changes in an individual's micro-environment. In these micro-environments, they may be closer to or farther away from various sources of ETS. It seems unlikely that many non-smokers remain in constant ETS concentration environments throughout their day or night. Even when smoking is unrestricted throughout a facility or home, clearly there will be micro-environments which have higher and lower concentrations of ETS. It seems nearly impossible that individual personal exposure to ETS can be described by long-term area monitoring. Area monitoring may overestimate or underestimate actual personal exposure because it fails to account for individual activity patterns. Among some non-smokers, aversion to ETS is a likely driving force in controlling exposure when individuals have some freedom of movement in their working or living environment.

There have been a few studies of personal exposure to ETS for workers in occupations associated with entertainment (dining, gambling, nightclubs, taverns) which are thought to involve higher exposure to ETS. Jenkins and Counts [109] have reported a study of ca. 160 restaurant and tavern servers. The median shift duration level reported for bartenders was ca. 4.5 $\mu g/m^3$ nicotine. Trout et al. [112] reported shift duration nicotine levels for casino workers of 9.5 $\mu g/m^3$. The highest median TWA nicotine levels reported for any occupational setting are those reported by Bergman et al. [113] for nine nightclub musicians playing in a total of three clubs (different musicians in different clubs). The median reported nicotine level was ca. 39 $\mu g/m^3$. It was reported that for two of the three facilities in which the subjects played, there was no fresh air provided by the ventilation system, and that in general, the fraction of smokers present in the facility (30–40%) was greater than that in the surrounding population. It seems clear that nicotine at TWA levels

## VII. CONCLUSIONS

If nicotine in environmental tobacco smoke is evaluated for its utility as a marker of ETS according to the National Research Council criteria, the assessment is complex. Airborne nicotine itself is uniquely attributable to tobacco smoking (although not to tobacco). Given the sampling and analytical technology available today, in addition to the effort expended by a number of investigators to understand and describe its analysis, it can be determined in most situations where ETS is present. Furthermore, nicotine can also be measured where its presence is due to off-gassing of adsorbed nicotine, rather than real ETS being present. Under controlled conditions in experimental chambers, the consistency with which it is emitted into ETS is reasonably good. However, differences in tobacco blends, especially when assessing ETS in a variety of countries, can contribute to considerable variation in emission ratios.

It is in the fourth NRC criterion – fairly consistent ratio with other components of interest in ETS – where nicotine has difficulties. First, 'components of interest' is a broad category, whose individual constituents change frequently. Second, due to the chemical structure and properties of nicotine, it tends to adsorb much more readily on surfaces than other measured components of ETS. Also, it can desorb back into room air long after true ETS is no longer present. Differences in chemical instability among other species (which includes photosensitivity of them as well as nicotine) also contribute to wide potential variation in the ratios of nicotine to other ETS components. Indeed, many field studies have shown that while there may be statistically significant correlation between nicotine and other ETS components in air, the predictive value of nicotine as a quantitative indicator of other ETS components is very limited. Given these limitations, it seems best to measure both nicotine and other ETS components (e.g., solanesol, 3-ethenylpyridine, etc.) to describe ETS concentrations and exposures.

That said, nicotine has been used in dozens of studies to determine both area concentrations of ETS, as well as personal exposures of individuals. Studies conducted prior to the mid-1990s focused more on area measurements than personal exposure determinations. This may have been due to the relative complexity and expense of involving human subjects in the latter-type studies. Probably the most important finding of the more recent, larger personal monitoring studies has been that for the most part, they indicate that the levels to which subjects are exposed tend to be lower than those estimated from smaller studies employing area measurements or short duration personal monitoring measurements. Some comparisons which have been performed do suggest that for a given micro-environment, area samples may be useful for describing potential exposure levels, but only for those groups of subjects confined to that particular environment, and not necessarily useful for estimating personal exposure for individuals in that environment. The latter distinction is important because it acknowledges avoidance practices among certain groups of non-smokers. Additionally, these personal exposure studies have indicated that while salivary cotinine may be useful for describing overall categories of ETS nicotine exposure for large groups of individuals, it is not predictively useful for estimating individual exposures.

Advances in analytical methodology for nicotine and cotinine will likely improve the capability for characterizing the magnitude of relative exposure to ETS. However, improved sensitivity in the analytical measurement of these species seems unlikely to improve their quantitative predictive capability. Thus, it appears that significant improvement in quantifying actual ETS exposure can only come about by developing and utilizing airborne markers that more fully satisfy the NRC criteria. In this regard, 3-ethenylpyridine and solanesol appear to be better quantitative markers for ETS exposure than is nicotine.

## REFERENCES

1 Wolf WA. Tobacco Production and Processing. In: *Tobacco and Tobacco Smoke*, Wynder EL, Hoffmann D (Eds.), New York: Academic Press, 1967: pp. 5–45.

2 Ogden MW, Nelson PR. Detection of Alkaloids in Environmental Tobacco Smoke. In: *Alkaloids*, Linskens HF, Jackson JF (Eds.), Berlin: Springer-Verlag, 1994: pp. 163–189.

3 Dube MF, Green CR. Methods of Collection of Smoke for Analytical Purposes. In: *Recent Advances in Tobacco Science: Formation, Analysis and Composition of Tobacco Smoke*. Raleigh, NC (Ed.), 36th Tobacco Chemists Research Conference, 1982: pp. 42–102.

4 Guerin MR, Jenkins RA, Tomkins BA. *The Chemistry of Environmental Tobacco Smoke: Composition and Measurement*. Chelsea, MI: Lewis, 1992.

5 Rodgman A. Environmental Tobacco Smoke. *Regul. Toxicol. Pharmacol.*, 1992: **16**; 223–244.

6 Smith CJ, Sears SB, Walker JC, DeLuca PO. Environmental Tobacco Smoke: Current Assessment and Future Directions. *Toxicol. Pathol.*, 1992: **20**; 289–305.

7 Roe FJC, Perry R. The Physical and Chemical Characteristics of Environmental Tobacco Smoke with Special Reference to Exposure Dose. In: *Other People's Tobacco Smoke*. East Yorkshire, England: Galen Press, 1991.

8 Steenland K. Passive Smoking and the Risk of Heart Disease. *JAMA*, 1992: **267**; 94–99.

9 Guerin MR, Higgins CE, Griest WH. The Analysis of the Particulate and Vapor Phases of Tobacco Smoke. In: *Environmental Carcinogens: Methods of Analysis and Exposure Measurement*, O'Neill IK (Ed.), New York: Oxford University Press, 1987: pp. 115–140.

10 Proctor CJ, Martin C, Beven JL, Dymond HF. Evaluation of an Apparatus Designed for the Collection of Sidestream Tobacco Smoke. *Analyst.*, 1988: **113**; 1509–1513.

11  National Research Council. *Environmental Tobacco Smoke: Measuring Exposures and Assessing Health Effects.* Washington, D.C.: National Academy Press, 1986.

12  Howard PH, Boethling RS, Jarvis W, Meylan WM, Michalenko EM. *Handbook of Environmental Degradation Rates.* Chelsea, Michigan: Lewis Publishers, 1991.

13  Kuhn H. Tobacco Alkaloids and their Pyrolysis Products in the Smoke. In: *Tobacco Alkaloids and Related Compounds: Proceedings of the Fourth International Symposium,* von Euler US (Ed.), Stockholm: Macmillan, 1965: pp. 37–51.

14  Wynder EL, Hoffmann D. Certain Constituents of Tobacco Products. In: *Tobacco and Tobacco Smoke,* Wynder EL, Hoffmann D (Eds.), New York: Academic Press, 1967: pp. 317–501.

15  Leete E. Biosynthesis and Metabolism of the Tobacco Alkaloids. In: *Alkaloids: Chemical and Biological Perspectives,* Pelletier SW (Ed.), New York: John Wiley, 1983: pp. 85–152.

16  Castro A, Monji N. Dietary Nicotine and its Significance in Studies on Tobacco Smoking. *Biochem. Arch.,* 1986: **2**; 91–97.

17  Sheen S. Detection of Nicotine in Foods and Plant Materials. *J. Food Sci.,* 1988: **53**; 1572–1573.

18  Davis RA, Stiles MF, deBethizy JD, Reynolds JH. Dietary Nicotine: A Source of Urinary Cotinine. *Food Chem. Toxicol.,* 1991: **29**; 821–827.

19  Ogden MW, Maiolo KC. Collection and Determination of Solanesol as a Tracer of Environmental Tobacco Smoke in Indoor Air. *Environ. Sci. Technol.,* 1989: **23**; 1148–1154.

20  Tang H, Richards G, Benner CL, Tuominen JP, Lee ML, Lewis EA, et al. Solanesol: A Tracer for Environmental Tobacco Smoke Particles. *Environ. Sci. Technol.,* 1990: **24**; 848–852.

21  Risner CH. The Determination of Scopoletin in Environmental Tobacco Smoke by High-Performance Liquid Chromatography. *J. Liq. Chromatogr.,* 1994: **17**; 2723–2736.

22  American Society for Testing and Materials. D5955-Standard Test Methods for Estimating Contribution of Environmental Tobacco Smoke to Respirable Suspended Particles Based on UVPM and FPM. In: *Annual Book of ASTM Standards.* West Conshohocken, PA: American Society for Testing and Materials, 1997: pp. 640–647.

23  Ingebrethsen BJ, Sears SB. Particle Evaporation of Sidestream Tobacco Smoke in a Stirred Tank. *J. Colloid Interface Sci.,* 1989: **131**; 526–536.

24  Spengler JD, Treitman RD, Tosteson TD, Mage DT, Soczek ML. Personal Exposures to Respirable Suspended Particulates and Implications for Air Pollution Epidemiology. *Environ. Sci. Technol.,* 1985: **19**; 700–707.

25  Ogden MW, Maiolo KC, Oldaker III GB, Conrad Jr. FW. Evaluation of Methods for Estimating the Contribution of ETS to Respirable Suspended Particles. In: *Indoor Air '90. Precedings [sic] of the 5th International Conference on Indoor Air Quality and Climate.* Toronto, Canada: Canada Mortgage and Housing Corporation, 1990: pp. 415–420.

26  Jenkins RA, Palausky A, Counts RW, Bayne CK, Dindal AB, Guerin MR. Exposure to Environmental Tobacco Smoke in Sixteen Cities in the United States as Determined by Personal Breathing Zone Air Sampling. *J. Expo. Anal. Environ. Epidemiol.,* 1996: **6**; 473–502.

27 Phillips K, Howard DA, Browne D, Lewsley JM. Assessment of Personal Exposures to Environmental Tobacco Smoke in British Smokers. *Environ. Int.*, 1994: **20**; 693–712.

28 Phillips K, Bentley MC, Howard DA, Alvan G. Assessment of Air Quality in Stockholm by Personal Monitoring of Non-smokers for Respirable Suspended Particles and Environmental Tobacco Smoke. *J. Work Environ. Health*, 1996: **22**; 1–24.

29 Phillips K, Bentley MC, Howard DA, Alvan G, Huici A. Assessment of Air Quality in Barcelona by Personal Monitoring of Non-smokers for Respirable Suspended Particles and Environmental Tobacco Smoke. *Environ. Int.*, 1997: **23**; 173–196.

30 Phillips K, Howard DA, Bentley MC. Assessment of Air Quality in Turin by Personal Monitoring of Non-smokers for Respirable Suspended Particles and Environmental Tobacco Smoke. *Environ. Int.*, 1997: **23**; 851–871.

31 Phillips K, Howard DA, Bentley M, Alvan G. Measured Exposures by Personal Monitoring for Respirable Suspended Particles and Environmental Tobacco Smoke of Housewives and Office Workers Resident in Bremen, Germany. *Int. Arch. Occup. Environ. Health*, 1998: **71**; 201–212.

32 Phillips K, Howard DA, Bentley MC, Alvan G. Assessment of Environmental Tobacco Smoke and Respirable Suspended Particle Exposure of Nonsmokers in Lisbon by Personal Monitoring. *Environ. Int.*, 1998: **24**; 301–324.

33 Phillips K, Bentley MC, Howard DA, Alvan G. Assessment of Air Quality in Paris by Personal Monitoring of Nonsmokers for Respirable Suspended Particles and Environmental Tobacco Smoke. *Environ. Int.*, 1998: **24**; 405–425.

34 Phillips K, Howard DA, Bentley MC, Alvan G. Assessment of Environmental Tobacco Smoke and Respirable Suspended Particle Exposures for Non-smokers in Hong Kong using Personal Monitoring. *Environ. Int.*, 1998: **24**; 851–870.

35 Phillips K, Bentley MC, Howard D, Alvan G. Assessment of Environmental Tobacco Smoke and Respirable Suspended Particle Exposures for Non-smokers in Kuala Lumpur Using Personal Monitoring. *J. Expo. Anal. Environ. Epidemiol.*, 1998: **8**; 519–541.

36 Phillips K, Bentley M, Howard D, Alvan G. Assessment of Environmental Tobacco Smoke and Respirable Suspended Particle Exposures for Non-smokers in Prague using Personal Monitoring. *Int. Arch. Occup. Environ. Health*, 1998: **71**; 379–390.

37 Phillips K, Howard DA, Bentley MC, Alvan G. Assessment by Personal Monitoring of Respirable Suspended Particles and Environmental Tobacco Smoke Exposure for Non-Smokers in Sydney, Australia. *Indoor Built Environ.*, 1998: **7**; 188–203.

38 Conner JM, Oldaker III GB, Murphy JJ. Method for Assessing the Contribution of Environmental Tobacco Smoke to Respirable Suspended Particles in Indoor Environments. *Environ. Technol.*, 1990: **11**; 189–196.

39 Ramsey RS, Moneyhun JH, Jenkins RA. Generation, Sampling and Chromatographic Analysis of Particulate Matter in Dilute Sidestream Tobacco Smoke. *Anal. Chim. Acta*, 1990: **236**; 213–220.

40 Eudy LW, Thome FA, Heavner DL, Green CR, Ingebrethsen BJ. Studies on the Vapor-Particulate Phase Distribution of Environmental Nicotine by Selective Trapping and Detection Methods. In: *Proceedings of the 79th Annual Meeting of the Air Pollution Control Association*. Pittsburgh: Air Pollution Control Association, 1986: paper 86–38.7.

41 Eatough DJ, Benner C, Mooney RL, Bartholomew D, Steiner DS, Hansen LD, et al. Gas and Particle Phase Nicotine in Environmental Tobacco Smoke. In: *Proceedings of the 79th Annual Meeting of the Air Pollution Control Association.* Pittsburgh: Air Pollution Control Association, 1986: paper 86–68.5.

42 Benner CL, Bayona JM, Caka FM, Tang H, Lewis L, Crawford J, et al. Chemical Composition of Environmental Tobacco Smoke. 2. Particulate-Phase Compounds. *Environ. Sci. Technol.*, 1989: **23**; 688–699.

43 Eatough DJ, Benner CL, Bayona JM, Caka FM, Tang H, Lewis L, et al. Sampling for Gas Phase Nicotine in Environmental Tobacco Smoke with a Diffusion Denuder and a Passive Sampler. In: *Proceedings of the 1987 EPA/APCA Symposium on Measurement of Toxic and Related Air Pollutants*; Air Pollution Control Association, 1987: pp. 132–139.

44 Eatough DJ, Wooley K, Tang H, Lewis EA, Hansen LD, Eatough NL, et al. Sampling Gaseous Compounds in Environmental Tobacco Smoke. In: *Proceedings of the 1988 EPA/APCA International Symposium on Measurement of Toxic and Related Air Pollutants*; Air Pollution Control Association, 1988: 739–749.

45 Ogden MW, Maiolo KC, Nelson PR, Heavner DL, Green CR. Artefacts in Determining the Vapour-Particulate Phase Distribution of Environmental Tobacco Smoke Nicotine. *Environ. Technol.*, 1993: **14**; 779–785.

46 Thome FA, Heavner DL, Ingebrethsen Bradley J, Eudy LW, Green CR. Environmental Tobacco Smoke Monitoring with an Atmospheric Pressure Chemical Ionization Mass Spectrometer/Mass Spectrometer Coupled to a Test Chamber. In: *Proceedings of the 79th Annual Meeting of the Air Pollution Control Association.* Pittsburgh: Air Pollution Control Association, 1986: paper 86–37.6.

47 Hinds WC, First MW. Concentrations of Nicotine and Tobacco Smoke in Public Places. *N. Engl. J. Med.*, 1975: **292**; 844–845.

48 Hammond SK, Leaderer BP. A Diffusion Monitor to Measure Exposure to Passive Smoking. *Environ. Sci. Technol.*, 1987: **21**; 494–497.

49 Ogden MW, Nystrom CW, Oldaker III GB, Conrad Jr. FW. Evaluation of a Personal Passive Sampling Device for Determining Exposures to Nicotine in Environmental Tobacco Smoke. In: *Proceedings of the 1989 EPA/A&WMA International Symposium Measurement of Toxic and Related Air Pollutants.* Pittsburgh: Air & Waste Management Association, 1989: pp. 552–558.

50 Ogden MW, Maiolo KC. Comparative Evaluation of Diffusive and Active Sampling Systems for Determining Airborne Nicotine and 3-Ethenylpyridine. *Environ. Sci. Technol.*, 1992: **26**; 1226–1234.

51 Ogden MW, Eudy LW, Heavner DL, Conrad Jr. FW, Green CR. Improved Gas Chromatographic Determination of Nicotine in Environmental Tobacco Smoke. *Analyst*, 1989: **114**; 1005–1008.

52 Ogden MW. Gas Chromatographic Determination of Nicotine in Environmental Tobacco Smoke: Collaborative Study. *J. Assoc. Off. Anal. Chem.*, 1989: **72**; 1002–1006.

53 American Society for Testing and Materials. D5075-Standard Test Method for Nicotine and 3-Ethenylpyridine in Indoor Air. In: *Annual Book of ASTM Standards.* West Conshohocken, PA: American Society for Testing and Materials, 1997: 400–407.

54 Muramatsu M, Umemura S, Okada T, Tomita H. Estimation of Personal Exposure to Tobacco Smoke with a Newly Developed Nicotine Personal Monitor. *Environ. Res.*, 1984: **35**; 218–227.

55 Muramatsu M, Umemura S, Fukui J, Arai T, Kiro S. Estimation of Personal Exposure to Ambient Nicotine in Daily Environment. *Int. Arch. Occup. Environ. Health*, 1987: **59**; 545–550.

56 Proctor CJ, Warren ND, Bevan MAJ. An Investigation of the Contribution of Environmental Tobacco Smoke to the Air in Betting Shops. *Environ. Toxicol. Lett.*, 1989: **10**; 333–338.

57 Thompson CV, Jenkins RA, Higgins CE. A Thermal Desorption Method for the Determination of Nicotine in Indoor Environments. *Environ. Sci. Technol.*, 1989: **23**; 429–435.

58 Tang H, Richards G, Gunther K, Crawford J, Lee ML, Lewis EA, et al. Determination of Gas Phase Nicotine and 3-Ethenylpyridine and Particulate Phase Nicotine in Environmental Tobacco Smoke with a Collection Bed-Capillary Gas Chromatography System. *J. High Resolut. Chromatogr. & Chromatogr. Commun.*, 1988: **11**; 775–782.

59 Jenkins RA, Moody RL, Higgins CE, Moneyhun JH. Nicotine in Environmental Tobacco Smoke: Comparison of Mobile Personal and Stationary Area Sampling. In: *Proceedings of the 1991 EPA/APCA Symposium on Measurement of Toxic and Related Air Pollutants*. Raleigh, NC, 1991: pp. 437–442.

60 Eatough DJ, Benner CL, Bayona JM, Richards G, Lamb JD, Lee ML, et al. Chemical Composition of Environmental Tobacco Smoke. 1. Gas-Phase Acids and Bases. *Environ. Sci. Technol.*, 1989: **23**; 679–687.

61 Hammond SK, Leaderer BP, Roche AC, Schenker M. Collection and Analysis of Nicotine as a Marker for Environmental Tobacco Smoke. *Atmos. Environ.*, 1987: **21**; 457–462.

62 Caka FM, Eatough DJ, Lewis EA, Tang H, Hammond SK, Leaderer BP, et al. An Intercomparison of Sampling Techniques for Nicotine in Indoor Environments. *Environ. Sci. Technol.*, 1990: **24**; 1196–1203.

63 Van Loy MD, Lee VC, Gundel LA, Daisey JM, Sextro RG, Nazaroff WW. Dynamic Behavior of Semivolatile Organic Compounds in Indoor Air. 1. Nicotine in a Stainless Steel Chamber. *Environ. Sci. Technol.*, 1997: **31**; 2554–2561.

64 Heavner DL, Morgan WT, Ogden MW. Determination of Volatile Organic Compounds and ETS Apportionment in 49 Homes. *Environ. Int.*, 1995: **21**; 3–21.

65 Heavner DL, Morgan WT, Ogden MW. Determination of Volatile Organic Compounds and Respirable Suspended Particulate Matter in New Jersey and Pennsylvania Homes and Workplaces. *Environ. Int.*, 1996: **22**; 159–183.

66 Daisey JM, Mahanama KRR, Hodgson AT. Toxic Volatile Organic Compounds in Simulated Environmental Tobacco Smoke: Emission Factors for Exposure Assessment. *J. Expo. Anal. Environ. Epidemiol.*, 1998: **8**; 313–334.

67 Lewis LJ, Lamb JD, Eatough DJ, Hansen LD, Lewis EA. The Determination of Nicotine and Cotinine by Ion Pair Reversed-Phase Chromatography. *J. Chromatogr. Sci.*, 1990: **28**; 200–203.

68 Ogden MW. Equivalency of Gas Chromatographic Conditions in Determination of Nicotine in Environmental Tobacco Smoke: Minicollaborative Study. *J. AOAC Int.*, 1992: **75**; 729–733.

69 Association of Official Analytical Chemists. Method 990.01-Nicotine in Environmental Tobacco Smoke: Gas Chromatographic Method. In: *Official Methods of Analysis*, Helrich K (Ed.), Arlington, VA: Association of Official Analytical Chemists, 1990.

70 AOAC International. Method 991.50-Nicotine in Environmental Tobacco Smoke: Gas Chromatographic Method-Alternative Conditions. In: *Official Methods of Analysis*, Helrich K (Ed.), Arlington, VA: AOAC International, 1991.

71 International Organization for Standardization. *ISO 11454-Determination of Vapour-Phase Nicotine in Air–Gas Chromatographic Method*. Geneva: International Organization for Standardization, 1997.

72 CORESTA. Method 14-Determination of Nicotine in Environmental Tobacco Smoke by Gas Chromatographic Analysis. In: *CORESTA Information Bulletin*. Paris: CORESTA, 1990.

73 United States Environmental Protection Agency. Method IP–2A-Determination of Nicotine in Indoor Air Using XAD–4 Sorbent Tubes. In: *Compendium of Methods for the Determination of Air Pollutants in Indoor Air*, Winberry WT, Forehand L, Murphy NT, Ceroli A, Phinney B, Evans A (Eds.), Research Triangle Park, NC: United States Environmental Protection Agency, 1989.

74 Ogden MW. Use of Capillary Chromatography in the Analysis of Environmental Tobacco Smoke. In: *Capillary Chromatography–The Applications*, Jennings W, Nikelly JG (Eds.), Heidelberg: Hüthig, 1991: pp. 67–82.

75 Ogden MW, Heavner DL, Foster TL, Maiolo KC, Cash SL, Richardson JD, et al. Personal Monitoring System for Measuring Environmental Tobacco Smoke Exposure. *Environ. Technol.*, 1996: **17**; 239–250.

76 Martin P, Heavner DL, Nelson PR, Maiolo KC, Risner CH, Simmons PS, et al. Environmental Tobacco Smoke (ETS): A Market Cigarette Study. *Environ. Int.*, 1997: **23**; 75–90.

77 Eatough DJ, Benner CL, Tang H, Landon V, Richards G, Caka FM, et al. The Chemical Composition of Environmental Tobacco Smoke III. Identification of Conservative Tracers of Environmental Tobacco Smoke. *Environ. Int.*, 1989: **15**; 19–28.

78 Ogden MW, Fix RJ, Thompson JW. A Robotic System for Microgram-Level Filter Weighing. *Am. Lab.*, 1996: **28**; 13–19.

79 Ogden MW, Maiolo KC. Comparison of GC and LC for Determining Solanesol in Environmental Tobacco Smoke. *LC–GC*, 1992: **10**; 459–462.

80 American Society for Testing and Materials. D6271-Standard Test Methods for Estimating Contribution of Environmental Tobacco Smoke to Respirable Suspended Particles Based on Solanesol. In: *Annual Book of ASTM Standards*. West Conshohocken, PA: American Society for Testing and Materials, 1998.

81 Ogden MW, Maiolo KC. Gas Chromatographic Determination of Solanesol in Environmental Tobacco Smoke (ETS). *J. High Resolut. Chromatogr. & Chromatogr. Commun.*, 1988: **11**; 341–343.

82 Nelson PR, Kelly SP, Conrad FW. Cigars as a Source of Environmental Tobacco Smoke. *J. Aerosol. Sci.*, 1998: **29**; S1307-S1308.

83 Rogge WF, Hildemann LM, Mazurek MA, Cass GR, Simonelt BRT. Sources of Fine Organic Aerosol. 6. Cigarette Smoke in the Urban Atmosphere. *Environ. Sci. Technol.*, 1994: **28**; 1375–1388.

84  Hammond SK, Sorenson G, Youngstrom R, Ockene JK. Occupational Exposure to Environmental Tobacco Smoke. *JAMA*, 1995: **274**; 956–960.

85  Hammond SK. Evaluating Exposure to Environmental Tobacco Smoke. In: *Sampling and Analysis of Airborne Pollutants*, Winegar ED, Keith LH (Eds.), Boca Raton, Florida: Lewis Publishers, 1993: pp. 319–338.

86  Ogden MW, Richardson JD. Effect of Lighting and Storage Conditions on the Stability of Ultraviolet Particulate Matter, Fluorescent Particulate Matter, and Solanesol. *Tob. Sci.*, 1998: **42**; 10–15.

87  Nelson PR, Heavner DL, Collie BB, Maiolo KC, Ogden MW. Effect of Ventilation and Sampling Time on Environmental Tobacco Smoke Component Ratios. *Environ. Sci. Technol.*, 1992: **26**; 1909–1915.

88  Nelson PR, Conrad FW. Interaction of Environmental Tobacco Smoke Components with a Ventilation System. *Tob. Sci.*, 1997: **41**; 45–52.

89  Piadé JJ, D'Andres S, Sanders EB. Sorption Phenomena of Nicotine and Ethenylpyridine Vapours on Different Materials in a Test Chamber. *Environ. Sci. Technol.*, 1999: **33**; 2046–2052.

90  Eatough DJ, Hansen LD, Lewis EA. The Chemical Characterization of Environmental Tobacco Smoke. *Environ. Technol.*, 1990: **11**; 1071–1085.

91  Nelson PR, Heavner DL, Oldaker III GB. Problems with the use of Nicotine as a Predictive Environmental Tobacco Smoke Marker. In: *EPA/A&WMA International Symposium: Measurement of Toxic and Related Air Pollutants*. Pittsburgh: Air & Waste Management Association, 1990: pp. 550–555.

92  Nelson PR, Ogden MW, Maiolo KC, Heavner DL, Collie BB. Predictive Value of Nicotine as an Environmental Tobacco Smoke Marker. In: *Indoor Air '90. Precedings [sic] of the 5th International Conference on Indoor Air Quality and Climate*. Toronto, Canada: Canada Mortgage and Housing Corporation, 1990: pp. 367–372.

93  Van Loy M, Nazaroff WW, Daisey JM. Nicotine as a Marker for Environmental Tobacco Smoke: Implications of Sorption on Indoor Surface Materials. *J. Air Waste Manage. Assoc.*, 1998: **48**; 959–968.

94  Ogden MW. Occupational Exposure to Environmental Tobacco Smoke. *JAMA*, 1996: **275**; 441.

95  Nelson PR, Conrad FW, Kelly SP, Maiolo KC, Richardson JD, Ogden MW. Composition of Environmental Tobacco Smoke (ETS) from International Cigarettes and Determination of ETS-RSP:Particulate Marker Ratios. *Environ. Int.*, 1997: **23**; 47–52.

96  Nelson PR, Conrad FW, Kelly SP, Maiolo KC, Richardson JD, Ogden MW. Composition of Environmental Tobacco Smoke (ETS) from International Cigarettes Part II: Nine Country Follow-Up. *Environ. Int.*, 1997: **24**; 251–257.

97  Kelly TJ, Smith DL, Satola J. Emission Rates of Formaldehyde from Materials and Consumer Products Found in California Homes. *Environ. Sci. Technol*, 1999: **33**; 81–88.

98  Traynor GW, Apte MG, Carruthers AR, Dillworth JF, Grimsrud DT, Gundel LA. Indoor Air Pollution Due to Emissions from Wood-Burning Stoves. *Environ. Sci. Technol.*, 1987: **21**; 691–697.

99  Rickert WS, Kaiserman MJ. An Assessment of the Relative Contributions of Burning Wood, Candles, Lamps, Incense and Cigarettes to Levels of Particulates, Nicotine, Benzo[a]pyrene, Carbonyls, Solanesol, HCN, Benzene, NO and CO in

Ambient Air. Presented at the 49th Tobacco Chemists' Research Conference. Lexington, KY, 1995.

100 Shah JJ, Singh HB. Distribution of Volatile Organic Chemicals in Outdoor and Indoor Air: A national VOC data base. *Environ. Sci. Technol.*, 1988: **22**; 1381–1388.

101 Jenkins PL, Phillips TJ, Mulberg EJ, Hui SP. Activity Patterns of Californians: Use and Proximity to Indoor Pollutant Sources. *Atmos. Environ.*, 1992: **26A**; 2141–2148.

102 Klus H, Begutter H, Scherer G, Tricker AR, Adlkofer F. Tobacco-Specific and Volatile *N*-Nitrosamines in Environmental Tobacco Smoke of Offices. *Indoor Environ*, 1992: **1**; 348–350.

103 Eatough DJ. Assessing Exposure to Environmental Tobacco Smoke. In: *Modeling of Indoor Air Quality and Exposure*, Nagda NL (Ed.), Philadelphia, PA: American Society of Testing and Materials, 1993: pp. 42–63.

104 Hodgson AT, Daisey JM, Mahanama KRR, Brinke JT, Alevantis LE. Use of Volatile Tracers to Determine the Contribution of Environmental Tobacco Smoke to Concentrations of Volatile Organic Compounds in Smoking Environments. *Environ. Int.*, 1996: **22**; 295–307.

105 Ogden MW, Davis RA, Maiolo KC, Stiles MF, Heavner DL, Hege RB, et al. Multiple Measures of Personal ETS Exposure in a Population-Based Survey of Nonsmoking Women in Columbus, Ohio. In: *Indoor Air '93. Proceedings of the 6th International Conference on Indoor Air Quality and Climate*, Jaakkola JJK, Ilmarinen R, Seppänen O (Eds.), Helsinki, Finland: Indoor Air '93, 1993: pp. 523–528.

106 Crouse WE, Oldaker GB. Comparison of Area and Personal Sampling Methods for Determining Nicotine in Environmental Tobacco Smoke. In: *Proceedings of the 1990 EPA/AWMA Conference on Toxic and Related Air Pollutants*. Raleigh, NC, 1990: pp. 562–566.

107 Sterling EM, Collett CW, Ross JA. Assessment of Non-Smokers' Exposure to Environmental Tobacco Smoke Using Personal-Exposure and Fixed-Location Monitoring. *Indoor Built Environ.*, 1996: **5**; 112–125.

108 Maskarinec MP, Jenkins RA, Counts RW, Dindal AB. Comparison of Area Samplers With Personal Samplers for the Determination of Markers of ETS in Bars and Restaurants. Presented at the 51st Tobacco Chemists' Research Conference. Winston-Salem, NC, 1997.

109 Jenkins RA, Counts RW. Occupational Exposure to Environmental Tobacco Smoke: Results of Two Personal Exposure Studies. *Environ. Health Perspect.*, 1999: **107**; 341–348.

110 Scherer G, Ruppert T, Daube H, Kossien I, Riedel K, Tricker AR, et al. Contribution of Tobacco Smoke to Environmental Benzene Exposure in Germany. *Environ. Int.*, 1995: **21**; 779–789.

111 Collett CW, Ross JA, Levine KB. Nicotine, RSP, and CO2 Levels in Bars and Nightclubs. *Environ. Int.*, 1992: **18**; 347–352.

112 Trout D, Decker J, Mueller C, Bernert JT, Pirkle J. Exposure of Casino Employees to Environmental Tobacco Smoke. *J. Occup. Environ. Med.*, 1998: **40**; 270–276.

113 Bergman TA, Johnson DL, Boatright DT, Smallwood KG, Rando RJ. Occupational Exposure of Nonsmoking Nightclub Musicians to Environmental Tobacco Smoke. *Am. Ind. Hyg. Assoc. J.*, 1996: **57**; 746–752.

114 Etzel RA. A Review of the Use of Saliva Cotinine as a Marker of Tobacco Smoke Exposure. *Prev. Med.*, 1990: **19**; 190–197.

115 Ogden MW, Morgan WT, Heavner DL, Davis RA, Steichen TJ. National Incidence of Smoking and Misclassification Among the U.S. Married Female Population. *J. Clin. Epidemiol.*, 1997: **50**; 253–263.

116 Pirkle JL, Flegal KM, Bernert JT, Brody DJ, Etzel RA, Maurer KR. Exposure of the US Population to Environmental Tobacco Smoke: The Third National Health and Nutrition Examination Survey, 1988 to 1991. *JAMA*, 1996: **275**; 1233–1240.

117 Davis RA, Steichen TJ, Nelson PR, Kelly SP, Gentry GB, Li LC, et al. Body Fluid Concentrations of Nicotine and Cotinine after Controlled Exposures to Environmental Tobacco Smoke. In: *Indoor Air '96: Proceedings of the 7th International Conference on Indoor Air Quality and Climate*, 1996: pp. 15–20.

118 Proctor CJ, Warren ND, Bevan MAJ, Baker-Rogers J. A Comparison of Methods of Assessing Exposure to Environmental Tobacco Smoke in Non-Smoking British Women. *Environ. Int.*, 1991: **17**; 287–297.

119 LaKind JS, Jenkins RA, Naiman DQ, Ginevan ME, Graves CG, Tardiff RG. Use of Environmental Tobacco Smoke (ETS) Constituents as Markers for ETS Exposure. *Risk Anal.*, 1999: **19**; 359–373.

120 Crouse WE, Ireland MS, Johnson JM, Striegel RM, Williard CS, DePinto RM, et al. Results From A Survey of Environmental Tobacco Smoke (ETS) in Restaurants. In: *Transactions of an International Specialty Conference-Combustion Processes and the Quality of the Indoor Environment*, Harper JP (Ed.), Pittsburgh: Air and Waste Management Association, 1988: pp. 214–222.

121 Oldaker III GB, Ogden MW, Maiolo KC, Conner JM, Conrad Jr. FW, DeLuca PO. Results from Surveys of Environmental Tobacco Smoke in Restaurants in Winston-Salem, North Carolina. In: *Indoor Air '90. Precedings [sic] of the 5th International Conference on Indoor Air Quality and Climate*. Toronto, Canada: Canada Mortgage and Housing Corporation, 1990: pp. 281–285.

122 Williams R, Collier A, Lewtas J. Environmental Tobacco Smoke Exposure of Young Children as Assessed Using a Passive Diffusion Device for Nicotine. *Indoor Environ*, 1993: **2**; 98–104.

123 Lambert WE, Samet JM, Spengler JD. Environmental Tobacco Smoke Concentrations in No-Smoking and Smoking Sections of Restaurants. *Am. J. Public Health*, 1993: **83**; 1339–1341.

124 Oldaker GB, Taylor WD, Parrish KB. Investigations of Ventilation Rate, Smoking Activity and Indoor Air Quality at Four Large Office Buildings. *Environ. Technol.*, 1995: **16**; 173–180.

125 Ogden M. Environmental Tobacco Smoke Exposure of Smokers Relative to Nonsmokers. *Anal. Commun.*, 1996: **33**; 197–198.

126 O'Conner TZ, Holford TR, Leaderer BP, Hammond SK, Bracken MB. Measurement of Exposure to Environmental Tobacco Smoke in Pregnant Women. *Am. J. Epidemiol.*, 1995: **142**; 1315–1321.

*Chapter 14*

# Determination of nicotine and its metabolites in biological fluids: in vivo studies

Riley A. Davis[1] and Margareta Curvall[2]

[1] *Research and Development, Bowman Gray Technical Center, R.J. Reynolds Tobacco Company, P.O. Box 1236, Winston-Salem, NC 27102-1236, USA*
[2] *Swedish Match, North Europe Division, Research and Analysis, SE-118 85 Stockholm, Sweden*

## I. INTRODUCTION

Presently there are numerous tobacco products available for use. While product form and its end use may vary, a common ingredient in all tobacco is nicotine. Within the past decade nicotine therapeutic products have become available; the primary use of these products is to slowly wean the tobacco user from tobacco. Nicotine may become more important as a therapeutic drug due to its alleged benefits as a regimen for Alzheimer's dementia, Parkinson's disease, ulcerative colitis, and attention deficit disorder.

The human interaction with nicotine is a topic of discussion in this chapter. The absorption, distribution, metabolism, and elimination of nicotine in humans are areas of focused attention. Sources of nicotine are examined including tobacco products, nicotine therapeutic products used for smoking cessation, and an incidental source, environmental tobacco smoke.

The evolution of analytical methods that progressed from the bioassay to very sophisticated instruments with modes of detection that ensure both sensitivity and exquisite selectivity is another area of discussion. The relationship between the analytical method and the question posed is highlighted by discussions of the matrix, sample collection, and storage. Finally, the results of comprehensive methods of analysis illustrate their utility in estimating nicotine intake by humans regardless of source.

583

## II. DISPOSITION OF NICOTINE IN HUMANS

### Pharmacokinetics of nicotine

*Absorption*

Nicotine is an alkaloid and a tertiary amine and comprises both a pyridine and pyrrolidine ring. It is a weak base and strongly alkaline with pKa values of about 3.2 and 7.9 (see also Chapter 5 for review of literature) and is soluble both in water and in lipids. In its ionized form, nicotine does not rapidly cross membranes. In its non-ionized form, however, nicotine is rapidly absorbed through the mucous membranes of the mouth and the bronchial tree. The extent of mucosal absorption varies with pH [1]. Smoke from cigarettes containing flue-cured tobacco is acidic (pH around 5.5); there is almost no buccal absorption of nicotine even when the smoke is held in the mouth [2]. Smoke from air-cured tobacco, however, which is the predominant tobacco in pipe tobaccos, cigars and a few cigarettes, is alkaline (pH >6.5). Nicotine in smoke from these products is well absorbed via the buccal mucosa [3,4].

The absorption of nicotine through the buccal mucosa is less rapid than the almost instant absorption by inhaling tobacco smoke into the lungs. The absorption is facilitated by increased pH, and in order to facilitate the absorption of nicotine, most moist snuff products and nicotine gums are buffered to an alkaline pH. With use of moist snuff or chewing tobacco, there is a rapid increase in blood nicotine level during the first 10 min after exposure [5]. The absorption continues at a slower rate for up to 60 min after the tobacco has been removed from the mouth [6]. The average blood nicotine levels are similar in habitual smokers and smokeless tobacco users [7,8]. Smokeless tobacco users, however, seem to absorb a greater total amount of nicotine through the gastrointestinal mucosa by swallowing, followed by a first-pass metabolism of nicotine to pharmacologically inactive metabolites in the liver [5,9]. This results in equal levels of blood nicotine, but somewhat higher levels of cotinine in smokeless tobacco users than in smokers. Nicotine is poorly absorbed from the stomach due to the acidity of gastric fluid, but is well absorbed in the small intestine [10]. The absorption of nicotine from finely ground nasal snuff via the nasal mucosa is much more rapid than through the buccal mucosa. Within ten min after taking a pinch of snuff, blood nicotine concentrations have been found to be comparable to those obtained after finishing a cigarette [7].

Alternative nicotine delivery systems, such as chewing gum, patch, nasal spray and inhaler are clinically important products, which have been developed as an adjunct to smoking cessation therapy. Nicotine levels and intake have been shown to be much lower while chewing nicotine gum than during cigarette smoking. Benowitz et al. [8] have estimated the total dose of nicotine absorbed by abstinent smokers while chewing 12 pieces per day of 2- and 4-mg gum. They found that extraction of nicotine from the gum was incomplete and averaged 53% and 72% for the 2-mg and 4-mg gum, respectively. Since the cotinine to nicotine ratio was higher in chewers than in smokers, it was assumed that some nicotine was swallowed and underwent first-pass metabolism. The systemic dose from a 2-mg gum was on average 0.86 mg and from a 4-mg gum 1.2 mg [11].

In a clinical study involving 11 subjects it was estimated that the average transdermal bioavailability of nicotine was 82%. Ten percent of the nicotine was absorbed after the patch was removed indicating that skin can act as a reservoir for nicotine [12]. The nasal route for systemic administration of nicotine has the pharmacokinetic advantages of both a rapid absorption and no gastrointestinal effects. Johansson et al. [13] has found that compared to an intravenous infusion of nicotine, the average bioavailability of nicotine applied to different nasal regions is within the range 60 to 75%. Mean plasma profiles showed that the rate of absorption is fast and $t_{max}$ values are obtained between 11 and 13 min after nasal administration [13].

## Distribution in body fluids and tissues

After absorption nicotine enters the blood stream. Distribution of nicotine into tissues cannot be studied in humans; this is why tissue levels of nicotine have been simulated using blood perfusion models with human organ weight and blood flow and partition coefficients from animal studies. When cigarette smoking was simulated by intrapulmonary injection, it was shown that concentrations in arterial blood, lung and brain were high, while concentrations in tissues such as muscle and adipose were low [6]. The disappearance of nicotine from blood is bi-exponential; an initial rapid distributional phase which is followed by a slower terminal phase representing metabolism and excretion. In man the distribution half-life is about eight min and the elimination half-life is about two h [6]. Nicotine, which is a highly lipid soluble alkaloid is extensively distributed to body tissues with a steady state volume of distribution ranging from 2 to 3 l/kg body weight [14]. The large variation in volume of distribution seen among subjects after intravenous infusion of nicotine could not be explained by variable protein binding since the nicotine binding is only 5% [14]. Nicotine is extensively metabolized which is reflected in a high hepatic clearance ranging from 1.3 to 2.5 l/min, which is close to normal hepatic blood flow. Renal clearance of nicotine in humans has been found to vary with urinary pH. Under alkaline conditions, renal clearance of nicotine is approximately 17 ml/min, while the renal clearance increases to 245 ml/min on urinary acidification [15,16].

Cotinine, which occurs in much higher concentrations than nicotine in body fluids of tobacco users, has a $t_{1/2}\beta$ (terminal half-life) ranging from 10 to 17 h [17]. The volume of distribution is on average 1.1 l/kg, which is lower than for nicotine due to lower lipophilicity and higher polarity than the parent compound [18]. The mean plasma clearance has been reported to be 0.9 ml/min/kg. *Trans*-3′-hydroxycotinine has been found in plasma, saliva, and urine of cigarette smokers [19–21]. Plasma pharmacokinetic parameters for $t_{1/2}\beta$, volume of distribution, and total plasma clearance of *trans*-3′-hydroxycotinine obtained after intravenous infusion in cigarette smokers were 5.9 h, 0.9 l/kg and 1.8 ml/min/kg [22].

*Metabolism* (see also Chapter 3)

Despite the fact that the use of tobacco has been worldwide for several centuries, the human metabolism of nicotine has not been elucidated until recently. Nicotine is extensively metabolized primarily in the liver, but also in the lungs and in the kidneys [23,24]. Nicotine is excreted partially unchanged by the kidney, but largely in the form of 20 or more different metabolites, which contain an intact pyridine ring. Not all of these have been isolated and identified in human urine. Renal excretion of nicotine depends on urinary pH and urine flow and accounts for 5–10% of total nicotine elimination in humans. The primary transformation of nicotine follows six different metabolic pathways; α-carbon oxidation of the pyrrolidine ring to form cotinine, $N$-oxidation of the pyrrolidine ring to form nicotine-1′-$N$-oxide, $N$-demethylation of the pyrrolidine ring to form nornicotine, conjugation to form nicotine glucuronide, $N$-methylation of the pyridine moiety to form the $N$-methylnicotinium ion and peroxidation to form nicotine-$\Delta^{4',5'}$-enamine [17]. There is a great variation in contribution of each metabolic pathway both between and within species. In humans four pathways seem to be predominant, i.e. α-carbon oxidation, $N$-oxidation, $N$-demethylation and glucuronidation. Great inter- and intraindividual variations in metabolic excretion profiles have been found in humans which could be attributed to different factors such as age, sex, tobacco habits, dietary habits, medications and physical fitness.

One of the major metabolites of nicotine in most species is cotinine. The conversion of nicotine to cotinine is a two-step reaction, the first step involving C-oxidation of the five position of the pyrrolidine ring by a cytochrome P-450 enzyme to nicotine-$\Delta^{1',(5')}$-iminium ion [25], which is in equilibrium with 5′-hydroxynicotine. This ion is then metabolized by cytosolic aldehyde oxidases to cotinine [26,27]. Further C-oxidation of cotinine results in two metabolites; 3′-hydroxycotinine and 5′-hydroxycotinine (allohydroxycotinine). The conversion of cotinine to 3′-hydroxycotinine is highly stereospecific and yields almost exclusively the *trans* isomer (>98%) [28]. Mass balance studies have shown that *trans*-3′-hydroxycotinine is the major metabolite of nicotine in humans and guinea pigs [19–21,29].

Another important metabolic pathway of nicotine is $N$-oxidation. Oxidation of the nitrogen atom in the pyrrolidine ring is catalyzed by the flavin-containing monooxygenase system. The main $N$-oxide metabolite, nicotine-1′-$N$-oxide, has been identified in human urine of tobacco users and occurs as two diastereoisomers, i.e. *cis*-1′R, 2′S- and *trans*-1′S, 2′S-nicotine-

1'-*N*-oxide [30]. Nicotine-1-*N*-oxide and nicotine -1,1'-di-*N*-oxide have also been postulated as *N*-oxide products of nicotine, but at most would account for less than 5% of the *N*-oxidation pathway [17]. The nicotine-*N*-oxides are not further metabolized since they are quantitatively recovered after intravenous administration [31].

Oxidative *N*-dealkylation is an important route in xenobiotic metabolism. Nicotine is demethylated to nornicotine, which is a major tobacco alkaloid and has been found in smokers urine. Its presence as a nicotine metabolite per se was determined by intravenous administration of stable isotope-labeled nicotine to smokers. Only a small amount of the nicotine dose (~0.3%) was converted to nornicotine [32]. Demethylcotinine (norcotinine) has been found in urine of both tobacco users and subjects given nicotine intravenously. Since norcotinine has not been found in urine after intravenous administration of cotinine, it could be assumed that norcotinine is formed via demethylation to nornicotine, which is then further oxidized to norcotinine [33,34]. The demethylation of nicotine has been postulated to proceed via a nicotine methylene-iminium ion, which is mediated by a P-450 cytochrome enzyme. The nicotine iminium ion, which is in equilibrium with its α-carbinolamine, spontaneously breaks down to nornicotine [35,36].

Besides the phase I metabolic pathways described above, nicotine, cotinine and *trans*-3'-hydroxycotinine also undergo phase II metabolism [37], since conjugates of these alkaloids are found in urine from tobacco users and from subjects after intravenous injection and transdermal application of nicotine [37–41]. Incubation in vitro of urine samples from tobacco users with β-glucuronidase has provided evidence that these conjugates of nicotine, cotinine and *trans*-3'-hydroxycotinine are glucuronides. The conjugate of cotinine has been synthesized and identified as an *N*-glucuronide [42]. It has been suggested that similar conjugating enzymes are involved in the conjugation of nicotine and cotinine, since individuals who extensively conjugate nicotine also extensively conjugate cotinine, while the conjugation of *trans*-3'-hydroxycotinine is not thought to be related to that of nicotine and cotinine. The conjugates of nicotine and cotinine can be deconjugated both by alkaline hydrolysis and by β-glucuronidase, while *trans*-3'-hydroxycotinine can only be hydrolyzed by the enzyme. Since *N*-glucuronides, but not *O*-glucuronides, are readily cleaved by basic hydrolysis, it can be assumed that both nicotine and cotinine form pyridine $N^+$-glucuronides, while *trans*-3'-hydroxycotinine forms an *O*-glucuronide [41].

*Elimination – excretion profiles*
The metabolic pattern of a xenobiotic compound may be influenced by its route of administration. The quantitative aspects of nicotine metabolism have been investigated in smokers, smokeless tobacco users and users of transdermal nicotine. Urinary excretion of nicotine and eight of its main metabolites were measured under steady-state conditions in 12 subjects

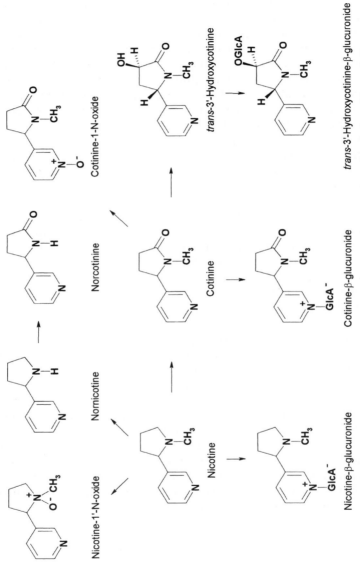

Fig. 1 An abbreviated scheme of nicotine metabolism showing major phase I and phase II metabolites.

while smoking cigarettes or while receiving transdermal nicotine. The systemic intake of nicotine was calculated from circadian plasma nicotine levels and intravenous clearance data. From these data the average urinary excretion of each metabolite as molar percentage was calculated. During cigarette smoking, on average 98% of the nicotine dose was accounted for in the urine, while, a little less, on average 88% was accounted for in the urine after transdermal administration. For the major nicotine metabolites similar metabolic patterns were obtained when nicotine was inhaled or applied transdermally. After transdermal application of nicotine the percentage of dose recovered as total free and conjugated *trans*-3′-hydroxycotinine averaged 41%; total free and conjugated cotinine, 26%; and total free and conjugated nicotine, 14%. The minor metabolites, nicotine-1′-*N*-oxide, cotinine-*N*-oxide and nornicotine accounted for 8%, 3%, and 8.9%, respectively. More nornicotine and nicotine-1′-*N*-oxide were excreted after smoking, while more cotinine-*N*-oxide was excreted after transdermal uptake of nicotine (Table 1) [41]. Excretion profiles have been studied in a group of Swedish smokeless tobacco users. Recruited for the study were 54 habitual users of smokeless tobacco; twenty two users of Swedish moist snuff (snus), 23 users of portion-bag packed snus and nine users of chewing tobacco. Nicotine and seven of its main metabolites were measured in 24-h urine and the amount of each metabolite expressed as nicotine equivalents was calculated. All three groups showed the same metabolic pattern. Nicotine together with the glucuronide accounted for on average 11%, cotinine and its glucuronide for 17%, and *trans*-3′-hydroxycotinine and its glucuronide for 61%, nicotine-1′-*N*-oxide for 9% and cotinine-*N*-oxide for 3% (38). In a similarly designed study, the average amounts of nicotine and metabolites excreted in 24-h urine were measured in a group of 91 cigarette smokers [40]. The excretion profile for this group of smokers was similar to the metabolic profile found by Benowitz et al. [41]. Nicotine and its glucuronide accounted for 14%, cotinine and its glucuronide for 23%, *trans*-3′-hydroxycotinine and its glucuronide for 59%, nicotine-1′-*N*-oxide for 3% and cotinine-*N*-oxide for 0.9%. The total of free plus conjugated cotinine and *trans*-3′-hydroxycotinine accounted for, on average, 82% in smokers and 78% in smokeless tobacco users (Table 1). Since similar excretion profiles have been found in smokers and smokeless tobacco users and users of transdermal nicotine, it can be concluded that the metabolism of nicotine is independent of the route of administration.

## Nicotine intake studies

*Tobacco smoke*
Different biomarkers have been used both in experimental and field studies in order to quantify the exposure to tobacco smoke, smokeless tobacco and environmental tobacco smoke. Measurement of nicotine in biological samples has the advantage of being specific to tobacco and tobacco smoke. Blood

TABLE 1

Distribution of urinary nicotine metabolites as percentage of the total amount of metabolites excreted during 24 h (average and standard deviation)

| Nicotine metabolites | Smokers[a] (*n* = 91) | Smokeless tobacco users[a] (*n* = 54) | Smokers[b] (*n* = 12) | Transdermal nicotine[b] (*n* = 12) |
|---|---|---|---|---|
| Nicotine | 9.4 (5.7) | 8.3 (5.7) | 10.4 (4.4) | 11.1 (4.3) |
| Nicotine-GlcA | 4.5 (2.5) | 3.0 (1.8) | 4.6 (2.9) | 5.3 (3.3) |
| Cotinine | 9.2 (2.6) | 7.9 (2.2) | 13.3 (3.1) | 14.9 (4.6) |
| Cotinine-GlcA | 14.0 (5.4) | 8.9 (4.6) | 15.8 (7.8) | 15.4 (7.9) |
| *Trans*-3'-hydroxycotinine | 36.1 (10.6) | 41.6 (10.6) | 39.1 (12.5) | 37.0 (10.8) |
| *Trans*-3'-hydroxycotinine-GlcA | 22.8 (10.0) | 19.4 (11.0) | 7.8 (5.9) | 7.9 (4.7) |
| Nicotine-1'-N-oxides | 3.0 (2.1) | 8.6 (6.9) | 3.7 (0.9) | 2.7 (1.2) |
| Cotinine-1-N-oxide | 0.9 (0.9) | 2.5 (2.3) | 4.5 (1.5) | 5.2 (1.5) |
| Nornicotine | | | 0.65 (0.15) | 0.41 (0.12) |

[a] Published by Andersson *et al.* (1997)
[b] Published by Benowitz *et al.* (1994)

or plasma nicotine levels peak at the end of smoking, the increment in plasma concentration ranging from 5 to 30 ng/mL depending on how the cigarette is smoked [43–45]. Generally, the afternoon samples from smokers have blood or plasma nicotine concentrations ranging from 10 to 50 ng/mL. In a study by Russell et al. [46] comprising 330 smokers with blood nicotine concentrations ranging from 4–72 ng/mL, there was a poor correlation between nicotine yields of the cigarettes and the blood concentrations. Similar results were obtained by Herning et al. [45], who found that only 25% of the individual differences in blood nicotine levels were attributable to the machine delivered nicotine yield, while 50 to 60% were attributable to individual differences in smoking behavior.

Due to great inter-individual variation in the metabolism and renal excretion of nicotine, blood nicotine concentrations could be used to estimate nicotine exposure, rather than the actual amount of nicotine absorbed from cigarette smoke [47]. In order to determine the actual daily intake of nicotine from plasma nicotine levels during tobacco consumption, both blood nicotine concentrations and nicotine elimination rate must be measured. Benowitz and Jacob [48] have estimated the daily intake of nicotine from metabolic clearance data obtained after intravenous infusion of nicotine in combination with blood and urinary nicotine concentration data obtained over 24 h when the subjects were exposed to cigarette smoke. The daily intake of nicotine averaged 37.6 mg, but varied widely among subjects (10.5 to 78.6 mg). Intake correlated with the number of cigarettes smoked but not with the smoking machine yields of nicotine.

Cotinine is present in much higher concentrations ($\sim 10$ times) in blood of tobacco users than nicotine. Blood or plasma cotinine levels average about 250–300 ng/mL in groups of smokers [49–51]. Since cotinine has a longer elimination half-life than nicotine ($\sim 15$ h compared with 2 h), it is more suitable for estimating nicotine exposure during a certain period of time [49]. Since there is a great intraindividual variation in the amount of nicotine converted to cotinine, only average nicotine exposure data for groups of smokers and not for individual smokers could be estimated from cotinine levels in different body fluids. Intravenous infusion studies have shown that the steady-state levels of plasma or saliva cotinine significantly correlate with nicotine intake both in smokers [52] and in non-smokers exposed to nicotine [53]. Saliva cotinine concentrations closely reflect those in plasma with a mean saliva-to-plasma ratio of 1.2 [53].

Different biomarkers have been used in order to distinguish smokers from non-smokers. In a comparison of different tests, it was found that the concentration of cotinine, whether measured in plasma, saliva or urine, was the best indicator of smoking with a sensitivity of 96% and a specificity of 100%. The average cotinine concentrations in plasma, saliva and urine were 1.5, 1.7 and 4.8 ng/mL for a group of true non-smokers ($n=100$), while the corresponding values for cigarette smokers ($n=75$) were 294, 330 and 1448 ng/mL [54]. Determinations of cotinine concentrations in plasma and saliva are valid to differentiate smokers from non-smokers [55,56]. Salivary cotinine is considered to be the marker of choice in evaluating smoking prevalence since saliva collection is preferable to venipuncture in most field studies. [57–59]. In subjects with relatively constant daily smoking habits, the daily elimination of cotinine is the same as the daily conversion of nicotine to cotinine. Using average data for cotinine clearance (Clcot = 72 ml/min) and the average fractional conversion for nicotine to cotinine ($F = 0.80$), the average 24 h dose of nicotine can be estimated from salivary cotinine data using the following equation:

$$\text{Dnic} = (\text{CLcot} \times \text{Ccot})/F.$$

The average smoker with a salivary cotinine level of 300 ng/mL will be predicted to consume about 25 mg nicotine per day, which is consistent with experimental data [60].

While the model above may be used to approximate daily nicotine intake in active smokers, application of the model to non-smokers exposed to nicotine from environmental tobacco smoke (ETS) may not be appropriate. A recent study [61] measured the actual nicotine exposure in the breathing zone of a large population ($n = 1486$) of non-smokers for 24 h. Saliva cotinine concentrations were measured also. When both the 24-h time-weighted average (TWA) nicotine levels and saliva cotinine concentrations whose measurement was unequivocal (>upper 95% confidence bound above the mean limit of detection for the two indicators) were taken, a population subset of 263 subjects was selected. For these subjects individually the relationship between nicotine exposure and saliva cotinine concentrations was poor, yielding an $R^2 = 0.105$. Cluster analysis of the median 24-h TWA nicotine levels and the median cotinine concentrations in groups of 25 subjects resulted in a vastly improved correlation ($R^2 = 0.915$). However, the study data indicated that the amount of inhaled nicotine was only a small fraction of the systemic dose of nicotine estimated from salivary cotinine levels. In fact the median ratio of systemic dose to absorbed dose of nicotine differed by nearly a factor of 10. These investigators [61] concluded that this large disparity was due to metabolic differences between smokers and non-smokers.

Analyses of scalp hair from smokers have demonstrated that inhaled nicotine becomes fairly concentrated in the hair shaft. Nicotine concentrations in the hair of unwashed hair samples of 10 smokers and 10 non-smokers averaged 15.8 and 2.42 µg/g, respectively. Cotinine averaged 0.5 µg/g in samples from smokers and could not be detected in seven of the non-smokers [50]. The usefulness of nicotine in hair to determine cigarette smoking status has been questioned, since inconsistent results could be obtained due to contamination of the exterior of the hair shaft by environmentally derived nicotine and also due to hair colour and hair treatments [62,63].

The infants of smokers are exposed to nicotine *via* the milk of the nursing mother. A wide range of nicotine concentrations have been found in the milk of smoking mothers (<20–512 ng/mL) [64]. There is a linear correlation between nicotine and cotinine concentrations in serum or plasma samples and milk samples of the smoking mother [65,66]. In samples obtained within 30 min after the mother had smoked, a strong correlation was found between both nicotine and cotinine concentrations in milk and plasma samples; the concentrations of nicotine in milk being about three times higher than in plasma, while the corresponding ratio for cotinine was found to be about 1.2. Cotinine, but not nicotine concentrations, in milk samples from nursing smokers have been shown to correlate with the number of cigarettes smoked by the mother [64,66]. The average nicotine and cotinine concentrations found in milk samples collected at all nursing periods within 24 h intervals from three groups of smokers ($n=34$), who smoked 1–10, 11–20 and 21–40 cigarettes per day were 18 and 76, 28 and 125 and 48 and 230 ng/mL, respectively. In a study comprising 22 smoking mothers and their newborn infants, it was found that the daily dose of nicotine via the mothers' milk was 6 μg per kg infant body weight. There was no correlation between nicotine and cotinine concentrations in the infant's urine and the amount of nicotine given to the infant via the mother's milk [66]. The concentration of nicotine in breast milk is such that the dose of nicotine given to the baby is small and considered unlikely to have physiological consequences [6].

Nicotine and cotinine may be transferred to the human fetus since detectable amounts have been found in amniotic fluid, placental tissues and umbilical vein blood of smoking mothers [67]. Nicotine concentrations in the placenta (range 3.3–28 ng/g), in amniotic fluid (range 1.5–23 ng/mL) and in fetal serum (range 0.5–25 ng/mL) have been found to be higher than the corresponding maternal serum values. The cotinine values however, were lower than or similar to the corresponding maternal serum values and were in the range of 10–131 ng/g in placental tissues, 5–188 ng/mL in amniotic fluid and 15–233 ng/mL in fetal serum [67]. In neonates, the urine concentrations of cotinine have been found to be significantly higher in children of smoking mothers than in children of non-smoking mothers. During the first day of life, a median cotinine/creatinine ratio level of 1233 ng/mg(14–3891 ng/mg) was found in neonates of smokers, while the corresponding value for neonates of non-smokers was 14.5 ng/mg (0–40.6 ng/mg) [68]. These urinary cotinine values did not correlate with the mothers cigarette consumption during 24 h before delivery. In another study on 22 neonates aged 3.7 d, similar urinary cotinine values were obtained, but in this group of children, maternal smoking during pregnancy correlated with the urinary cotinine/creatinine ratio of the infants [66].

Both nicotine and cotinine have been found in breast fluid of non-lactating smoking women [69,70]. Thirty minutes after smoking two cigarettes, similar concentrations of nicotine were found in breast fluid and plasma, while the concentrations of cotinine were only half of those found in plasma.

## Oral snuff

At pH 7.4 only 24% of nicotine is in its non-ionized form that can cross lipophilic cell membranes. Since oral moist snuff has an alkaline pH, the nicotine is easily absorbed via the buccal mucosa. The nicotine release from a given smokeless tobacco product is determined by different factors: the nicotine concentration of the product, the pH of the product and the particle size of the in-going tobacco. The amount absorbed by the snuff user is not only affected by the availability of nicotine in the product, but also by the consumption pattern such as the size of the pinch, the site of the placement in the mouth and the oral treatment of the pinch. Nicotine is absorbed more slowly from smokeless tobacco than from tobacco smoke, but peak venous concentrations of nicotine after single doses of oral moist snuff or chewing tobacco are similar to those seen after smoking a cigarette. After smoking the nicotine level rapidly declines, while during and after use of smokeless tobacco, the nicotine concentration plateaus. This prolonged elimination has been suggested to be attributable to continued absorption by release of nicotine from the mucous membranes or to absorption of nicotine that has been swallowed [8]. Similar peak levels and circadian blood nicotine pattern were seen in smokeless tobacco users and smokers in a cross-over study in which eight cigarette smokers in a balanced order used cigarettes, oral snuff, chewing tobacco and abstained from tobacco [9]. By use of plasma cotinine as an indicator of daily nicotine intake, different populations of tobacco users can be compared. Most studies have shown that the level of cotinine in moist snuff users and chewers is similar to those found in cigarette smokers [5,9,71,72]. There is a tendency towards higher cotinine levels in smokeless tobacco users compared to smokers in some studies, which most likely is due to some nicotine being swallowed and therefore undergoing first pass metabolism to cotinine before reaching the systemic circulation [5,9].

Steady-state levels of nicotine and cotinine have been measured in a group of habitual Swedish moist snuff users. They were male college students with an average consumption of 12 pinches of moist snuff per day or 150 g of snuff per week. From each subject an afternoon sample of plasma, saliva and urine was collected for two consecutive weeks. In plasma the average steady-state value of nicotine was found to be 29.8 ng/mL and the corresponding value for cotinine was 10 times higher (304 ng/mL). In saliva, the average steady-state value for cotinine was 531 ng/mL, the saliva to plasma ratio being 1.8, and in urine, means of steady-state values for nicotine and cotinine were 1.05 and 2.49 μg/ml, respectively. The steady-state concentrations of nicotine and cotinine in the different body fluids did not correlate with any of the consumption factors (habit duration, number of pinches per day and amount of moist snuff per day). For this group of Swedish moist snuff users, the steady-state levels of nicotine and cotinine in plasma, saliva, and urine were similar to those found in habitual smokers [73]. Since the metabolic conversion of nicotine to cotinine differs among people, the concentration of cotinine in different body fluids is a poor measure of the actual intake of nicotine. The best estimate, however, is obtained by recording the total amount of nicotine and its metabolites excreted in the urine after nicotine exposure. At steady-state, the rate of excretion reflects the intake rate. The daily nicotine intake by groups of Swedish users of moist snuff (snus) has been estimated by measuring 24 h excretion of nicotine and seven of its major metabolites, e.g. cotinine, *trans*-3'-hydroxycotinine, the glucuronides of nicotine, cotinine and *trans*-3'-hydroxycotinine, nicotine-1'-*N*-oxide and cotinine-*N*-oxide. The study comprised 54 habitual users of smokeless tobacco; 22 loose moist snuff users, 23 users of portion-bag packed moist snuff, and 9 users of chewing tobacco. The average daily intake of nicotine, measured as nicotine and its seven metabolites excreted in the urine during 24 h and expressed as nicotine equivalents, was 35 mg for both users of loose and portion-bag packed snus although the daily consumption was lower for users of portion-bag (14 g/day) than for users of loose (21 g/day) snus. The daily nicotine intake by the chewers was 50% higher and averaged 54 mg. The average steady-state saliva cotinine concentration was about 300 ng/mL for both categories of snus users, while the average concentration found in users of chewing tobacco was 50% higher [38]. The short-term effects on consumption and nicotine intake of switching from regular moist snuff (pH 8.2–8.5, nicotine 0.8–0.9%) to a low-nicotine snus (pH 7.8–8.2, nicotine 0.4–0.5%) were studied in a group of 24 habitual male users of Swedish snus. Consumption and nicotine intake data were compared with a control group, who had used the low-nicotine moist snuff product for at least one year. There was a significant reduction in nicotine intake when switching to the low-nicotine product. The individual average salivary cotinine levels decreased from 336 ng/mL to 153 ng/mL and the amount of nicotine equivalents in 24 h urine samples decreased from 25.2 mg to 14.6 mg, reaching about the same levels as found in the control group (159 ng/mL and 14.3 mg, respectively). Although a slight increase (+15%) in moist snuff consumption after short-term switching, both short- and long-term switching to a product containing half the concentration of nicotine resulted in a corresponding reduction in nicotine intake [74].

*Nasal snuff*
While the absorption of nicotine from oral snuff via the mucous membrane is relatively slow and pH dependent, the absorption of nicotine from nasal snuff is rapid. Within 10 min after taking a pinch of snuff, the increase in plasma nicotine concentration shown by habitual snuff takers (12.6 ng/mL) was

comparable to the average increase (10.1 ng/mL) obtained from a single cigarette by a group of heavy smokers. In a group of 11 daily users of nasal snuff, the peak plasma concentrations of nicotine and cotinine were 36 and 412 ng/mL, respectively [7].

*Nicotine in food* (see also Chapter 10)
Several foods contain small amounts of nicotine [75–77]. Repace (78) estimated a daily nicotine intake from diet of 0.7 µg/day, while Davis et al. [76] calculated a daily intake of 8.8 µg/day. Even in the latter case, the resulting steady-state urine cotinine level would be less than 0.7 ng/mL, which is only 10% of that seen in the urine of a person exposed to significant amounts of ETS (6 ng/mL). For most people, food is an insignificant source of nicotine exposure compared to ETS [79].

*Nicotine medications* (see also Chapter 10)
There are several forms of nicotine replacement therapy such as nasal spray, inhaler, nicotine gum, and nicotine patch. A nasal nicotine solution has been proposed as a potential cessation aid for cigarette smokers, since nicotine is rapidly absorbed by the nasal route [80]. When 2 mg of nicotine was administered hourly as a nasal nicotine solution eight times a day to five subjects, the plasma nicotine concentrations rose from a near-zero baseline to an average peak value of 16.3 ng/mL after the sixth dose, which is higher than that normally obtained from 2 mg nicotine chewing gum. There was great interindividual variation in plasma nicotine peak values with one subject obtaining 33.3 ng/mL, which is similar to that found in heavy smokers [81]. The administration of a nasal nicotine solution could be a problem during repeated self-administration since it is difficult to retain the droplets in the nose. The dosing accuracy and convenience are probably better when nicotine sprays and aerosols are used as nasal administration devices. Similar plasma nicotine concentrations are obtained after administration of a nasal dose of 2 mg nicotine over 30 s (by taking eight shots each equivalent to 0.25 mg nicotine base) from a nasal nicotine spray as from a nasal nicotine aerosol. Blood nicotine concentrations rose rapidly after nasal dosage. During the first 2.5 min the average rise in blood nicotine concentration was 8.6 ng/mL followed by a small rise during the next 2.5 min, which resulted in a total average peak increase of about 11 ng/mL after the dose of 2 mg nicotine base from both devices. After ad libitum use of the nasal nicotine spray during four weeks, the blood nicotine levels averaged 14.1 ng/mL in nine subjects, which was only 44% of their prior smoking level [82].

Nicotine could be supplied *via* an inhaler through inhalation from a plastic enclosure containing a soft plug impregnated with nicotine. Hjalmarson et al. [83] have measured salivary cotinine concentrations in 247 smokers taking part in a 1-year, randomized double-blind, placebo-controlled smoking cessation study. A nicotine inhaler was given to 123 participants and a placebo inhaler to 124 participants. The nicotine inhaler, which contained 10 mg of nicotine, released approximately 13 µg of nicotine in a 50 mL puff. The mean salivary cotinine concentrations obtained after three and six weeks use of the nicotine inhaler were only 30% (144+128 ng/mL) and 20% (116+116 ng/mL), respectively, of the concentrations obtained during smoking. There was a weak positive correlation between the average number of nicotine inhalers used per day and salivary cotinine levels at three weeks [83].

Nicotine polacrilex gum has been widely used as an aid to giving up smoking. In order to estimate the dose of nicotine obtained from the gum, plasma levels of nicotine have been measured in abstinent smokers while chewing 2- and 4-mg nicotine gum [11,46,84–86]. Chewing 2- and 4-mg pieces of nicotine gum hourly during a day produced mean steady-state plasma nicotine levels of 11.8 and 23.2 ng/mL, respectively. Smoking a cigarette with a machine-smoke nicotine yield of 1.1 mg hourly over the same time period resulted in a mean plasma nicotine level of 18.3 ng/mL [84]. In order to characterize intake of nicotine from gum, circadian blood nicotine concentrations, daily intake of nicotine, and extraction of nicotine from the gum were investigated in smokers switched experimentally to 2- or 4-mg nicotine gum, respectively. It was found that the extraction of nicotine from gum by the chewer was incomplete and averaged 53% and 72% for 2- and 4-mg gum. The daily intake of nicotine was estimated from intravenous clearance data and $AUC_{nic}$ during smoking (ad libitum) and chewing 2- and 4-mg gum (12 pieces per day). Only 29% and 42% of the nicotine dose obtained during smoking were obtained during chewing of the 2- and 4-mg gum. Average afternoon blood nicotine concentrations were 26.7, 7.9 and 14.3 ng/mL for smoking, 2-mg and 4-mg gum, respectively [11].

Since transdermal systems are designed to deliver nicotine at a more constant rate than other forms of administration more stable plasma nicotine concentrations are achieved during the application period. Studies of the pharmacokinetics of transdermal nicotine following single application of different doses have reported dose-related increases in plasma nicotine concentrations [87,88]. In a bioavailability study in 24 smokers, similar $C_{max}$ values were obtained when using a transdermal nicotine system delivering 22 mg nicotine/24 h as when using a 2 mg piece of nicotine gum chewed for 30 min every hour for 15 h. After 9 h use the $C_{max}$ values were 13.8 ng/mL and 11.9 ng/mL for the transdermal and gum formulations, respectively [89]. Steady-state concentrations were reached within 2 to 4 days after application, when transdermal nicotine patches are administered at 24-hours intervals [90,91]. Within 8 h after application plasma nicotine peak concentrations are obtained. Bannon et al. [87] have reported concentrations of 16 ng/mL on day 1 and 17 ng/mL on day seven with the use of a system delivering nicotine 30 mg/24h, while Dubois et al. [90] have found Cmax plasma nicotine levels of 10.4 ng/mL on day 1 and 11.6 ng/mL on day 10 with a system that delivered 13.5 mg/24 h. Since nicotine is delivered over 16 to

24 h, the peak concentrations of nicotine achieved following the use of a transdermal delivery system are considerably less than those achieved after smoking cigarettes. Single application of a transdermal nicotine system delivering 21 mg/day resulted in plasma nicotine concentrations of about half those of ad libitum smoking over 15 h [88].

*Environmental tobacco smoke* (see also Chapter 13)
The presence of nicotine and its metabolites in body fluids of non-tobacco users is highly specific to environmental tobacco smoke (ETS) exposure. Since nicotine has a short initial half-life of 8 min and a relatively short terminal half-life of 2 h, measurements of nicotine concentrations in body fluids could only be used to reflect recent or acute exposure to ETS. Neither saliva nor plasma nicotine concentrations in non-smokers exposed to ETS under natural conditions differentiate self-reported exposed and non-exposed non-smokers [92]. However, since urinary nicotine concentrations differ substantially between exposed and non-exposed non-smokers, they may provide an estimate of incidental exposure to ETS. The average urinary nicotine concentration in 188 urban non-smokers exposed to ETS under average natural conditions was found to be 10.8 ng/mL, which was 0.7% of the average of 1471 ng/mL found in a sample of 229 cigarette smokers [93]. However, urinary nicotine concentration is affected by urine pH and urine flow, which makes it an unreliable measure of intake in individual cases unless these factors are controlled. In order to estimate the daily nicotine intake from ETS exposure, six subjects were given a slow infusion of 1 mg nicotine at a steady rate over 1 h. The resulting plasma nicotine concentration – time curves were compared with those obtained after exposure to ETS. The plasma and urinary concentrations were much higher than those found in non-smokers even after heavy exposure to ETS. Using these infusion data, it was estimated that the nicotine intake from ETS exposure averages 0.014 mg/h among urban non-smokers living their normal life, 0.23 mg/h among non-smokers visiting a smoked-filled public house and 0.36 mg/h during maximum exposure in an unventilated room [94].

Since cotinine has a longer half-life (10–20 h), it is considered to better reflect long-term exposure to ETS. Although they differ, the concentrations in urine, saliva, and plasma are highly correlated ($r=0.82$–$0.90$) even at the low levels found in non-smokers [94]. The daily life exposure to other people's smoke has been assessed in a number of publications by relating cotinine concentrations in body fluids to self-reports of recent exposure [92,95–103]. Most studies have found increasing cotinine levels with increasing levels of self-reported ETS exposure.

The relationship between reports of recent exposure to ETS and urinary cotinine concentrations was investigated in 663 non-smokers who attended a cancer screening clinic in New York, USA. Cotinine was found in the urine of 91% of the subjects although only 76% reported recent exposure to ETS.

The cotinine levels ranged from 0 to 85 ng/mL with a mean of 8.84 ng/mL and 92% of the cotinine values were less than 20 ng/mL [97]. There are many epidemiological studies in which living with a smoker has been used as an index of exposure to ETS. Since the interpretation and interpretability of epidemiological studies of ETS depend largely on the validity of self-reported exposure, it is of great importance to investigate to what extent questionnaires can indicate exposure levels. In a 10-country collaborative study, 1369 non-smoking women were interviewed about their daily exposure to ETS and spot samples of urine were collected. About 20 percent of the subjects had cotinine values below the detection limit (>2 ng of absolute concentration) of the radioimmunoassay used for quantification of urinary cotinine. Mean cotinine/creatinine levels showed a linear increase from the group of women ($n=627$) not exposed either at home or at work (2.7 ng/mg creatinine) to the group ($n=124$) exposed both at home (10.0 ng/mg creatinine). For the group of women ($n=210$) exposed at work, but not at home the urinary cotinine values (4.8 ng/mg creatinine) were significantly lower than for those ($n=359$) exposed at home, but not at work (9.0 ng/mg creatinine) [103].

A US study [98] assessed ETS exposure in ca. 1000 confirmed non-smokers by continuously monitoring ETS markers in the breathing zone of these individuals for a 24-h period. Personal breathing zone monitoring for ETS markers is superior to either self-reported exposure or area monitoring. Separate personal systems monitored ETS exposure in the work environment and in the environment away from work. In each of the 12 cities participants were categorized into one of four groups depending on smoking activity at work and away from work. Cell 1 participants were exposed to ETS both at work and away from work. Cell 2 participants were exposed to ETS away from work, but not at work. Cell 3 participants were ETS exposed at work, but not away from work. Cell 4 participants were not ETS exposed in either venue. The highest ETS exposures were reported to be in the 'away from work' environment. The median salivary cotinine concentrations were related to the median nicotine concentrations in the ETS exposure venues. In cell 1, salivary cotinine was 1.62 ng/mL with a nicotine exposure of 42.3 μg; in cell 2, salivary cotinine was 0.959 ng/mL with a nicotine exposure of 15.2 μg; in cell 3, salivary cotinine was 0.360 ng/mL with a nicotine exposure of 2.85 μg; and in cell 4, salivary cotinine was 0.162 ng/mL with a nicotine exposure of 0.682 μg. Regression of these data results in an $R^2$ value of 0.958 ($p=0.021$). These authors further report that an important difference exists between the perceived ETS exposure and the actual measured ETS exposure for the study participants. The rate of smoker misclassification (3.6%) was similar to that reported by Riboli, et al. [103].

When serum cotinine levels were measured in a large US population comprising 10642 subjects aged 4 years and older, it was found that 88% of the non-tobacco users had levels of serum cotinine above the detection limit

(0.050 ng/mL) of the LC-MS method used for quantification of cotinine. The average serum cotinine values of adults and children not exposed at home or outside the household were in the range 0.11–0.13 ng/mL. Increased serum cotinine values were observed both in the group that was exposed at work only (mean 0.32 ng/mL) and in the group that was exposed at home only (mean 0.65–1.14 ng/mL). The mean serum cotinine value for those subjects who were exposed both at home and at work was 0.93 ng/mL. No dietary contribution to serum cotinine levels were found in this study [102].

It has been shown that schoolchildren with parents who smoke have measurable levels of cotinine through ETS exposure. In a study comprising 569 non-smoking schoolchildren, it was found that their saliva cotinine concentrations were strongly related to the smoking habits of their parents. The average concentration of cotinine in saliva rose in a dose related fashion from 0.44 ng/mL in children with non-smoking parents to 3.38 ng/mL when both parents smoked. Mothers' smoking had a stronger influence than did fathers' [96]. Similar results were obtained in another English study, in which 4043 schoolchildren aged 5–7 years participated. Half of the children were exposed to cigarette smoke at home or by an outside carer. When no exposure was identified, the average salivary cotinine level was 0.29 ng/mL. In households where both parents smoked the average cotinine level was 4.05 ng/mL and rose to 9.03 ng/mL if both parents smoked more than 20 cigarettes per day [101].

## III. METHODS OF ANALYSIS FOR NICOTINE AND ITS METABOLITES

### Body fluids

*Matrices, collection, and sample handling*
Presently blood serum (plasma), urine, and/or saliva are the preferred body fluids for the analysis of nicotine and its primary metabolite, cotinine. Training and tradition, as well as the goals of the study, are certainly influencing factors as to the choice of these body fluids, especially among basic life science and clinical research investigators. The size of the subject pool can influence the choice of body fluid for analysis, hence epidemiology studies normally choose a sample matrix that is easily obtained and causes little or no discomfort to the subject. Most important in the choice of the sample matrix should be answers to the questions: (1) For what are the analytical data to be used?; (2) Is the effect to be measured acute or chronic?; and (3) Is the choice of sample matrix consistent with the method of analysis, including sample pre-treatment?

Today there is virtually no body fluid or tissue that has not been examined for the presence of nicotine or one or more of its metabolites to investigate the effects of nicotine or to indicate its presence as it relates to presumed health effects or exposure. Nicotine and/or cotinine concentrations have been observed in hair [103–109], meconium [110,111], seminal fluid [112,113], cervical fluid [114–116], breast milk [64,65], and amniotic fluid [117]. Even bone and hair of mummies have been examined for nicotine and cotinine [118,119]. Each of these matrices comes with its own attendant problems regarding the separation of the analyte. Fortunately the state of technology provides various means for analysts to perform this daunting task including liquid–liquid extraction, solid phase extraction, supercritical fluid extraction, and, if all else fails, distillation from the alkalinized fluid into a weak acid solution. Immunological methods of analysis allow liquid matrices to be used directly without prior extraction in most cases. Modern analytical techniques have made isolation of the analyte(s) less of a burden for the analyst.

*Whole blood, serum, and plasma*

Obtaining up to three or four samples of whole blood by multiple venipunctures is certainly possible, but not very comfortable for the subject. In pharmacokinetic studies [121–123] or smoking behavioral studies [124] when numerous blood samples must be obtained, it is best to not only reduce the sample volume, but to insert an indwelling catheter into the vein. Whether serum or plasma is required for analysis, the erythrocytes must be separated from the liquid phase by centrifugation. While plasma may be separated immediately from the red blood cells, to procure serum one must mix the serum tubes thoroughly and allow clot formation for about 45 min before centrifuging the samples. Serum or plasma is aliquoted into cryogenic vials and frozen at $-20°C$ or $-80°C$ until analysis. Although it is common practice to freeze serum samples that must be analyzed later, apparently nicotine and cotinine are stable for as long as 12 days in unfrozen serum/plasma [125].

Losses of nicotine or cotinine due to protein binding are minimal. Although Schievelbein [126] estimated the protein binding of nicotine to be about 20%, equilibrium dialysis studies demonstrated a protein binding of these analytes at 4%–5% [127]. A procedure for acquiring serum for analysis of nicotine and cotinine was reported by Feyerabend [128].

Separation of the analyte(s) from the serum or plasma matrix is accomplished by liquid–liquid extraction of the alkalinized sample or by solid phase extraction. The removal of the protein from plasma and serum can be a problem for the analyst. In liquid–liquid extraction the organic solvent(s) will denature the protein and promote emulsion formation during the necessary mixing step(s). Alkalinization of the samples with saturated $K_2CO_3$ instead of NaOH can aid in breaking the emulsion during centrifugation of the samples and ensures that pH>9 is achieved for the extraction of non-polar analytes into the organic solvent [129–131]. The use of solid phase extraction techniques [132,133] allows the separation of the analytes through selective adsorption. Protein removal occurs simultaneously with this technique; during the elution of the analytes, the protein remains on the column solid phase. While liquid–liquid extraction of the samples allows analyte(s) recovery >90% in most instances, analyte recovery using solid phase extraction technique is generally 80%. Internal standards are usually added to the samples as part of the sample work-up.

*Saliva*
As a body fluid for analysis of drugs, hormones, antigens, and antibodies saliva is nearly ideal. Factors that contribute to the movement of drugs from the systemic circulation into saliva include molecular weight and size, lipid solubility, pH and the pKa of the drug [134]. Both nicotine and cotinine move easily from the systemic circulation into saliva. Cotinine concentrations exist at a saliva:plasma ratio of 1.2 to 1.4 [18] with a high correlation ($r=0.99$) between plasma and saliva. For a number of reasons, including its short half-life, nicotine was found to be less desirable for saliva monitoring [18]. Following nicotine infusion or smoking, saliva collected from the sub-mandibular gland and the parotid gland selectively, nicotine concentrations were found to be 20-fold or 60-fold higher than plasma concentrations, respectively [135]. Differences in the concentration of cotinine were negligible. Rose [136] achieved results similar to these authors while examining transdermal nicotine systems. Nicotine concentrations in stimulated salivary flow were about eight times the plasma concentration; cotinine concentrations were highly correlated with plasma concentrations ($r=0.94$). In examining serum, saliva, and urine with 11 other tests for biochemical means to separate active smokers from non-smokers, Jarvis et al. [92] determined that saliva cotinine provided the best means of discriminating smoking status and yielded more accurate data than self-report.

There are several methods for the collection of saliva many of which are discussed in 'Saliva as a Diagnostic Fluid' [137] and by Caddy [138]. Salivary flow may be stimulated mechanically by chewing a rubber band, a square of Teflon®, or a dental dam. Citric acid, candy, or a packet containing sucrose [139] stimulate saliva flow by chemical means. Rose employed both methods to stimulate saliva flow in his investigation of nicotine transdermal systems [136]. Generally, mechanical stimulation is preferred for collection of mixed saliva for analysis of nicotine and/or cotinine. Van Vunakis [59] reported that saliva cotinine concentrations are independent of saliva flow; this is contrary to the findings of Schneider et al. [140]. The system that has been used at R.J. Reynolds is primarily the Salivette™ [141] system which uses a dental dam for saliva collection and results in a sample free of mouth debris and mucoid substances.

Since saliva contains little protein compared to serum (<1%), protein removal from the matrix is usually not necessary. In some liquid chromatographic methods the saliva is treated with an equal volume of acetonitrile, centrifuged, and used directly for analysis. Generally saliva may be used directly for solid phase extraction of analytes. Liquid–liquid extraction of saliva is similar to extracting analytes from an aqueous solution.

*Urine*

Urine is the body fluid that receives both natural compounds that result from normal or abnormal body metabolism and xenobiotic compounds and their metabolites as a result of clearance from the systemic circulation. Therefore urine is both an ideal body fluid for analysis and, simultaneously, one of the most difficult for the analyst due to the number of interfering substances and high salt concentrations. While the promise of liquid chromatography with mass spectrometric determination (LC/MS/MS) offers potential for the so-called 'dilute and shoot' technique, application of this technique of sample introduction and analysis has not been fulfilled in many instances. For exacting quantitation the analyte(s) must be separated from the urine matrix. The use of liquid–liquid extraction, solid phase extraction, and/or analyte resolution via a reversed-phase chromatographic column are techniques available to recover non-polar and polar analytes from urine. However, the end determination and its sensitivity will impact the choice of extraction method. Haley and O'Neill [142] reported a method for urine collection and preservation for the determination of nicotine and its metabolites. Recently the necessity for low temperature storage of urine samples has received emphasis due to large variations in the cotinine concentration as storage temperature increased [143]. The source of the variation in cotinine concentrations was degradation of cotinine-*N*-glucuronide to cotinine [143].

**Methods of analysis**

As can be seen from Fig. 2, there are a number of methods for the determination of nicotine and its metabolites in virtually any biological matrix. Both the choice of analytes and the sensitivity required usually dictate the method of determination. One may speculate that the predominant method of analysis is gas chromatography with nitrogen-specific detection (thermionic detection) if only nicotine and/or cotinine are the analytes. This method of analysis affords the analyst with excellent selectivity and sufficient sensitivity for most applications. Micro-samples of serum (0.1 ml) have been analyzed using this technique [128,129]. Volume-limited samples may be analyzed by RIA as this analytical technique requires only 0.1 ml for analysis, however, selectivity is diminished especially for cotinine. From a practical point of view, one should have at least 0.5 ml sample volume for RIA determination of nicotine and cotinine since two separate RIAs must be done. When selectivity is the factor that predominates the analytical procedure, then a gas or liquid chromatographic method with mass spectrometric detection is the method of choice.

## Analytical Methods for the Determination of Nicotine and its Metabolites in Body Fluids

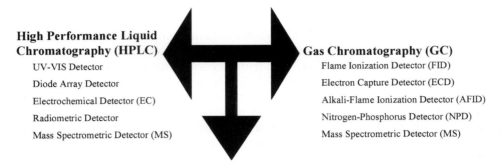

**High Performance Liquid Chromatography (HPLC)**

UV-VIS Detector

Diode Array Detector

Electrochemical Detector (EC)

Radiometric Detector

Mass Spectrometric Detector (MS)

**Gas Chromatography (GC)**

Flame Ionization Detector (FID)

Electron Capture Detector (ECD)

Alkali-Flame Ionization Detector (AFID)

Nitrogen-Phosphorus Detector (NPD)

Mass Spectrometric Detector (MS)

**Immunological Methods**

Radioimmunoassay (RIA)

Fluorescence Immunoassay (FIA)

Enzyme Linked Immunosorbent Assay (ELISA)

Fig. 2 Methods for the analysis of nicotine and its metabolites in biological fluids. Modified from Curvall and Kazemi-Vala (51) with kind permission of Kluwer Academic Publishers.

Obviously as one's requirements for selectivity and sensitivity increase, so do the analytical costs. A modest gas chromatographic system may range in cost between $25,000–$50,000. Typical costs for a HPLC system range from $30,000–$75,000. An advantage of the chromatographic systems is the automated, unattended sample analysis and data acquisition. If one considers that the RIA requires either a gamma counter or a liquid scintillation counter for end determination, then the instrumental cost approaches a modest gas chromatographic system. Costs per sample may range from $25 up to $100 for contract laboratories performing analyses for nicotine and cotinine depending upon the number of samples, the matrix, the information required (qualitative vs. quantitative), the type of instrumentation used for end determination, and the skill level of laboratory personnel.

While there are referenced methods of analysis for nicotine and cotinine by either RIA or gas chromatography [128,144,145], none of these methods represents a standard method of analysis. An assessment was made of the chromatographic and RIA methods for the determination of cotinine in persons exposed to ETS at a workshop [146]. Apparently most attendees favored GC/MS as a reference method against which all other analytical methods would be compared, but no action toward establishing GC/MS as the reference method was taken. A standard reference material [SRM 8444] has been prepared by the National Institute for Standards and Technology [147] composed of freeze-dried human urine with three concentrations of cotinine: (a) blank urine (<1 ppb), (b) a low level of 50 ppb; and (c) a high level of 500 ppb. Agreement was reached concerning the manner of data reporting for urine cotinine values in two formats, namely, as ng/mL and as ng/mg creatinine [148]. The use of urinary creatinine to normalize cotinine concentrations between individuals has been adopted by most analytical laboratories, particularly for 'spot' samples. Precautions are warranted for this use of creatinine especially in neonates and young children [149–151].

*Spectrophotometric methods of determination*
Qualitative and semi-quantitative colorimetric methods for the determination of nicotine metabolites in urine samples were based upon the König reaction using barbituric acid or diethylthiobarbituric acid (DETB) as condensation reagents [152]. A variation of the method included the extraction of the colored products of DETB into ethyl acetate. Subjective observation of colored products from the three tests allowed the discrimination of smokers and non-smokers. However, attempts to be more quantitative via colorimetric determinations were disappointing although a high correlation ($r=0.85$) was obtained for the direct barbituric acid method and the number of cigarettes smoked. Significant overlap was demonstrated for the smoker subjects who smoked between 6 cigarettes/day to greater than 26 cigarettes/day. This colorimetric method allowed discrimination of smokers from non-smokers at about the same variance as measurements of thiocyanate, expired carbon monoxide, and carboxyhemoglobin at a cost of $0.01 to $0.03 per sample.

Further investigation and optimization of the direct barbituric acid (DBA) method for the determination of nicotine metabolites in urine [153] was modified for automated analysis of urine samples [154]. Comparison with an [125]I-cotinine RIA yielded a high correlation between the two analytical methods ($r=0.91$), but the DBA produced analytical results that were consistently higher than the RIA. A false-positive rate of 1.5% was observed for both methods with a detection rate of 98% for self-reported cigarette smoking.

The discovery that cotinine only accounted for between 2% to 30% of the nicotine metabolites in urine provided impetus for the development of a high performance liquid chromatography (HPLC) method that employed pre-column derivatization with diethylthiobarbituric acid (DETB) [155]. The colored derivatives were separated by isocratic reversed-phase HPLC. Nicotine, cotinine, and five unidentified metabolites of nicotine were separated. Urines samples from 20 smokers and two non-smokers, who chewed a single piece of nicotine gum (4 mg), confirmed that the colored derivatives resulted from nicotine metabolism. One of the metabolites accounted for 40% of the total metabolite concentration; nicotine and cotinine accounted for 5% and 15%, respectively. Since neither nicotine nor cotinine was sensitive enough to investigate exposure to ETS, an investigation was carried out using the unidentified metabolite [156]. Preliminary results suggested that the unidentified metabolite was comprised of 3-pyridylcarbinol and 3'-hydroxycotinine as a single co-eluting peak. Identification of 3-pyridyl-carbinol and 3'-hydroxycotinine as urinary metabolites of nicotine was accomplished through analysis of a purified preparation by gas chromatography with mass spectrometric (GC/MS) detection and nuclear magnetic resonance studies [157]. This study independently confirmed that 3'-hydroxycotinine was a major urinary metabolite of nicotine [20] in tobacco users.

Smith and Cooke [158] evaluated the DETB method for the determination of nicotine and related compounds as a means to screen samples prior to gas chromatographic determinations. These authors identified numerous deficiencies of the DETB method for quantitative analysis of urinary nicotine and its metabolites including: (a) departure from the Beer-Lambert law; (b) the variation of color density with time and reagent additions; and (c) interference of structurally related compounds such as methylxanthines, nicotinic acid, and therapeutic drugs.

The ultimate application of the DETB method of analysis allowed the determination of nicotine and 12 of its metabolites in urine with $N'$-ethylnorcotinine as internal standard [159]. The HPLC separation of the nicotine metabolites was accomplished using a ternary solvent system. The mobile phase gradient was programmed so that the ion-pairing buffer (0.06M 1-pentanesulfonic acid sodium salt) was gradually replaced with organic solvent. Polar metabolites were eluted with methanol while acetonitrile was used to elute the non-polar basic metabolites in the latter portion of the chromatographic run. Interestingly, little or no 3-pyridylcarbinol was found in the human urine. Glucuronides of *trans*-3'-hydroxycotinine, cotinine, and nicotine were determined indirectly following urine treatment with β-glucuronidase. Method sensitivity was increased by injecting volumes ranging from 50 μL to 250 μL. At 250 μL injection volume the detection limit for the majority of nicotine metabolites ranged between 0.5 nM to 1.5 nM.

### Gas chromatographic methods of determination

Consistent with advances in more sophisticated instruments, column technology, and detector selectivity the gas chromatographic determination of nicotine and cotinine has undergone significant changes. Isaac and Rand [160] investigated plasma levels of nicotine in response to smoking cigarettes. The determination of nicotine involved four extraction steps, the use of an internal standard (modaline), and an alkali flame ionization detector that was selective for phosphorus and nitrogen compounds. Manual injections of 2 μL of the extract were sufficient for nicotine quantitation with sensitivity of 1 ng/mL. Feyerabend et al. [161], using an extensive extraction and sample clean-up procedure, were able to obtain a sensitivity of 0.1 ng/mL for nicotine. Quinoline was used as an external standard for quantitation. Nicotine recovery was reported at 75% from plasma samples with known quantities of added nicotine and quinoline. Application of this method of analysis allowed the differentiation of method blank samples and the low nicotine concentrations found in non-smokers exposed to cigarette smoke. A micro-method for the determination of nicotine in plasma, saliva, urine, and breast milk by gas chromatography was reported [162] with improvements in sample extraction and clean-up. The determination of plasma nicotine concentrations that resulted from intravenous administration, cigarette smoking, and nicotine gum on separate occasions were used to demonstrate the utility of this modified method. Additional improvements were made to this gas chromatographic method including the use of a capillary column, a single internal standard (5-methylcotinine) for the determination of both nicotine and cotinine simultaneously, and more efficient sample preparation [161]. The determination of nicotine and cotinine concentrations routinely ranged from 1 to 100 ng/mL, however, the sensitivity of the method was maintained at 0.1 ng/mL. Applications of the improved method allowed these determinations: (a) cotinine in saliva samples of children exposed to ETS; (b) plasma nicotine and cotinine concentrations that resulted from inhalations from a nicotine vaporizer; and (c) plasma nicotine and cotinine concentrations achieved by a non-smoker from a transdermal nicotine patch.

The use of solvent selectivity specific for the analyte and structural analogues, N-ethylnornicotine and N-(2-methoxyethyl)norcotinine, as internal standard compounds advanced the gas chromatographic determination of nicotine and cotinine [130]. Sensitivities of 1ng/mL and 5 ng/mL were achieved for nicotine and cotinine, respectively, from sample volumes of 1 mL. Simultaneous determination [32] of nicotine and cotinine via the use of a capillary column and a mixed solvent system for extracting the analytes from blood, saliva, and urine were additional modifications to this method

The simultaneous extraction into dichloromethane, the use of a capillary column, the use of structural analogues of nicotine and cotinine, and the use of a nitrogen-sensitive detector allowed the analysis of nicotine and cotinine from plasma samples with 1 mL or 0.1 mL volumes [129]. While the linearity of the calibrated range extended from 5 ng/mL to 100 ng/mL for nicotine and 5 ng/mL to 500 ng/mL for cotinine, the limits of detection were 5 pg for nicotine and 20 pg for cotinine at a signal-to-noise ratio of four. These extraordinary limits of detection were achieved by placing the capillary column into position directly under the rubidium bead of the nitrogen-sensitive detector. This allowed the column effluent to pass to the detector without restriction or adsorption in the detector port inlet. Dichloromethane was evaporated and replaced with ethanol that was concentrated about five-fold. The concentrated ethanolic solutions were manually injected into separate gas chromatographs maintained at different isothermal temperatures during the chromatographic runs.

## Internal standards

Numerous internal standard compounds have been chosen by analysts as shown by Curvall and Kazemi-Vala [51] (Fig. 3). Compounds with similar structure and physico-chemical behavior to the analytes serve as better internal standards and result in more accurate quantitation of the analyte(s). *N*-Ethylnornicotine, *N*-(2-methoxyethyl) norcotinine [130], *N*-ethylnornicotine and *N*-ethylnorcotinine [131], and *N*-methylanabasine and *N*-ethylnorcotinine [129] and deuterated nicotine and cotinine have been used as internal standards for the determination of nicotine and cotinine. Other internal standard compounds frequently used include quinoline [160], β-nicotyrine [133)], and lidocaine [145]. Deficiencies of this latter group of

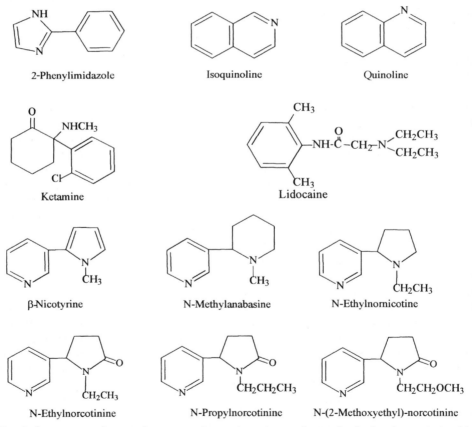

Fig. 3 Structures of several compounds used as internal standards in chromatographic methods for the determination of nicotine and its metabolites.

internal standard compounds lead to loss of sensitivity and inaccurate quantitiation due to selective losses during work-up or during the chromatographic end determination [129,130].

*Contamination*

When determining nicotine and cotinine at trace levels, contamination from any source can lead to both imprecision and inaccuracy in the end determination [164]. Sources of contamination may include the sample matrix, reagents, atmosphere, and the glassware. Reagents used during the extraction procedure were routinely checked for nicotine contamination by extraction and examination via chromatography [129,164]. Distilled water was contaminated either by the atmosphere or by a smoker touching the apparatus; consequently all samples were diluted with tap water. All glassware was thoroughly washed and placed into an oven at 200°C–450°C. The entire analytical procedure was conducted in a sealed room with all air passing through an active carbon filter. Smoking was expressly forbidden in the immediate area and the analyst was a non-smoker. Sample extraction and evaporation was conducted in a chemical hood.

*Extraction procedures*

The lengthy extractions required in most of the gas chromatographic methods for the analysis of nicotine and cotinine are performed: (1) to increase the recovery of the analytes by using several small volumes of extracting solvent; (2) to remove weakly acidic or neutral interfering compounds via back-extractions; and (3) to take advantage of solvent selectivity for each of the analytes, i.e., diethyl ether is preferred for nicotine extraction and dichloromethane for cotinine extraction. As previously discussed dichloromethane extraction affords sufficient recovery for both nicotine and cotinine [129,131]. Solid-phase column chromatography has been employed as a means to extract nicotine, cotinine, and other metabolites from biological fluids [132,133].

The extraction of nicotine and cotinine from urine, saliva, and plasma in a single step using an ethyl-silica sorbent (C2) has been reported [133]. After conditioning the column with 2 mL methanol and 2 ml borate buffer (0.05M disodium tetraborate), pH 9, the sample matrix containing internal standard compounds is applied as a mixture with borate buffer to the column. With the aid of vacuum the liquid passes through the column and the analytes partition into the bonded stationary phase. Following a washing step with distilled water and 80% aqueous methanol and a drying step, nicotine, cotinine, and the internal standard compounds, β-nicotyrine and nikethamide, were eluted with 500 μL methanol. Propyl acetate (20 μL) was added to the methanolic eluant and the sample concentrated to a final volume of 10 μL. The end determination was achieved through the use of fused silica capillary chromatography. Splitless injection technique, temperature programming, and nitrogen-sensitive detection were used to enhance method sensitivity. Thus the limits of detection for nicotine and cotinine in plasma, saliva, and urine were 0.2 ng/mL for nicotine and 0.5 ng/mL for cotinine. The recovery of nicotine, cotinine, and the internal standard compounds was >90% when the final volume of the evaporation/concentration step was 50 μL. Decreases in the recovery of nicotine and β-nicotyrine to ca. 75% were observed when the volume of the concentrated sample diminished to 10 μL, but cotinine and nikethamide did not exhibit any losses. This method of analysis for nicotine and cotinine demonstrates: (a) that solid phase extraction is suitable for quantitative removal of the analytes from plasma, saliva, and urine using a single extraction procedure; and (b) that the use of high resolution capillary gas chromatography with splitless sample injection and nitrogen-sensitive detection provides accuracy, sensitivity, and specificity.

Other metabolites of nicotine that are amenable to determination by gas chromatography include nicotine-1′-*N*-oxide, nornicotine, and *trans*-3′-hydroxycotinine. As early as 1971, Beckett et al. [165] determined nicotine-1′-*N*-oxide in urine through its reduction with titanium trichloride and subsequently analyzing for nicotine. Urinary excretion rates for nicotine, cotinine, and nicotine-1′-*N*-oxide were investigated under the influence of acidic, alkaline, and normal urinary pH. Diastereoisomers of nicotine-1′-*N*-oxide were detected by paper chromatography, but the *trans*-diastereoisomer was excreted to a larger extent than the *cis*-diastereoisomer. Sequential extraction of nicotine and cotinine followed by the reduction of nicotine-1′-*N*-oxide was applied to urine samples. A novel approach to the determination of nicotine, cotinine, and their *N*-oxides was taken by Stehlik et al. [132]. The analytes and the internal standard compounds were adsorbed onto a silica column, selectively eluted, and submitted to gas chromatography for end determination. Following a 30 min adsorption phase, nicotine, cotinine, lidocaine, and quinoline were eluted with a solvent mixture composed of 1:4 (v/v) diethyl ether and dichloromethane for direct determination. Saturation of the column with $SO_2$ reduced the *N*-oxides. After elution in the manner described, nicotine and cotinine were determined along with phendimetrazine as internal standard, as an indirect measure of the concentrations of the *N*-oxides. Recovery of nicotine and cotinine from urine and plasma was ca. 80%–85%. Recovery of the *N*-oxides was less in both matrices, being 60% in urine and 30% in plasma. While the limits of detection for nicotine and cotinine were quite sufficient, the limit of detection for the *N*-oxides was quite high (≥1000 ng/mL) in the urine matrix. Vast improvement to the limits of detection was observed in plasma with nicotine and cotinine at 5 ng/mL and the *N*-oxides at 15 ng/mL. nicotine-1′-*N*-oxide is thermally labile and undergoes a rearrangement to form an oxazine compound, 2-methyl-6-(3-pyridyl) tetrahydro-1,2-oxazine. Jacob et al. [166] used this oxazine formation to their advantage and developed a gas chromatographic method for the direct determination of nicotine-1′-*N*-oxide based on this derivative. The analytical scheme for the determination of nicotine-1′-*N*-oxide was developed incorporating derivative formation followed by a series of solvent extractions for clean-up and final analysis by gas chromatography. 5-Methylnicotine-1′-*N*-oxide was prepared and used as the internal standard. Examination of urine samples (*n* = 12) by packed column and capillary column chromatography for the presence of nicotine-1′-*N*-oxide yielded similar analytical data. These results gave evidence that the integrity of the analyte was maintained throughout a storage time of two years.

Nornicotine has been reported as a minor component of smoker's urine [167] with its daily excretion accounting for 0.5%–2% of nicotine intake. Direct determination of nornicotine can not be achieved with gas chromatography, therefore an *N*′-alkyl derivative of nornicotine was formed by reductive alkylation. The internal standard, 5-methylnornicotine, was converted to its *N*′-propyl derivative similarly. Following extraction from a one mL urine sample and concentration, 1 μL of the organic phase was injected into the gas chromatograph using the splitless injection technique and a fused silica capillary column. Detection was with a nitrogen-sensitive detector. Recovery of nornicotine from the urine matrix was 90% or better. Accuracy of the method ranged from 96% to 106% with coefficients of variation of 1% to 4.5%. The average daily excretion rate of nornicotine from a group of smoker subjects (*n* = 25) was 151 μg/day with a range from 12 μg to 361 μg. The excreted nornicotine concentration could not be apportioned into nornicotine derived from mainstream cigarette smoke and nicotine metabolism.

Dagne and Castagnoli [168] provided information both on the structure and the synthetic mechanisms for preparation of *trans*-3′-hydroxycotinine in 1972. Neurath and co-workers [19,20] reported two gas chromatographic methods for the determination of *trans*-3′-hydroxycotinine. In a brief communication [19], *trans*-3′-hydroxycotinine was determined in plasma and urine of smokers by forming the acetate derivative with detection provided by nitrogen-sensitive detector. Subsequently a second method of analysis employing derivatization reactions with heptafluorobutyric anhydride and electron-capture detection was reported [20]. Impetus for method development was the relatively low recovery of nicotine and its metabolites (<20%) in urine from nicotine intake studies with active smokers. *Trans*-3′-hydroxycotinine concentration was found to be the most abundant nicotine metabolite in the urine of smokers and second only to cotinine concentration in serum. With the addition of *trans*-3′-hydroxycotinine concentration to the other urinary nicotine metabolites, expressed as nicotine equivalent, the recovery of nicotine and its metabolites was raised to approximately 40% [20]. *Trans*-3′-hydroxycotinine accounted for 50% to 60% of the nicotine metabolites excreted. Elimination of the derivatization step and simultaneous determination with nicotine and cotinine provided for a simple, straightforward means of measuring *trans*-3'-hydroxycotinine using *N*-ethylnornicotine for nicotine quantitation and *N*-ethylnorcotinine, for quantitation of cotinine and *trans*-3'-hydroxycotinine [21]. Alkaline extraction into either *n*-butyl acetate or dichloromethane allowed the quantitation of nicotine, cotinine, and *trans*-3'-hydroxycotinine in urine or serum via fused silica capillary chromatography with nitrogen-sensitive detection. Recoveries of *trans*-3'-hydroxycotinine, nicotine, and cotinine from plasma into dichloromethane was only 49%, 77%, and 69%, respectively. Increased recoveries were observed when the urine was extracted with *n*-butyl acetate, thus for nicotine, 86%; for cotinine, 81%; and for *trans*-3'-hydroxycotinine, 82%. The limit of detection was reduced for *trans*-3'-hydroxycotinine in urine about five-fold with the use of a mass selective detector in the single-ion monitoring mode, thus 10 ng/mL could be detected at a signal-to-noise ratio of 3 : 1. Limits of detection for nicotine and cotinine in this mode of detection were 2 ng/mL and 10 ng/mL, respectively. Blood samples yielded a higher limit of detection (50 ng/mL) for *trans*-3'-hydroxycotinine with nicotine and cotinine remaining about the same. *Cis*-3'-hydroxycotinine was also identified as a urinary metabolite of nicotine using essentially the same method of determination [169], however this isomer of 3'-hydroxycotinine amounted for only 0.21% of the excreted 3'-hydroxycotinine in human smokers. The highly stereoselective metabolic formation of 3'-hydroxycotinine was confirmed by GC/MS [170] and demonstrated that between 95%–98% of the 3'-hydroxycotinine excreted in humans is the *trans*-isomer. The determination of urinary *trans*-3'-hydroxycotinine as the *tert*-butyldimethylsilyl derivative by gas chromatography with nitrogen-sensitive or mass spectrometric detection was developed as a means to study the metabolic formation and pharmacokinetics of this principal metabolite of nicotine [171]. Generally the agreement between the two methods of determination was good except at the lower concentrations where interfering substances promoted inaccuracy in the gas chromatographic method using nitrogen-sensitive detection. The concentration of *trans*-3'-hydroxycotinine in urine samples from 22 smoking subjects averaged 3028 ng/mL and 2910 ng/mL when measured by gas chromatography with nitrogen-sensitive detection and mass spectrometric detection, respectively. Furthermore the daily excretion of *trans*-3'-hydroxycotinine, as the nicotine equivalent mass, represented about 40% of the total urinary nicotine metabolites measured. This is fairly consistent with the data reported by Neurath and co-workers [19,20]. Other urinary metabolites measured included nicotine, cotinine, nicotine-1'-*N*-oxide, cotinine-1-*N*-oxide, and nornicotine. However, the determination of these major metabolites of nicotine in urine presents a skewed and imperfect picture of nicotine uptake in smokers. Only 60% to 65% of the nicotine intake could be accounted for by these measures of nicotine metabolism [172]. Curvall et al. [37] described their investigation of phase II metabolism of nicotine and developed an indirect method for the determination of the glucuronides of nicotine, cotinine, and *trans*-3'-hydroxycotinine. Undeniably method development for the determination of both phase I and phase II nicotine metabolites would allow more comprehensive investigations of nicotine pharmacodynamics, pharmacokinetics, and metabolism in both users and non-users of tobacco products and/or therapeutic nicotine containing medications such as are used in smoking cessation programs.

## High-performance liquid chromatography

For the determination of nicotine and cotinine in biological matrices high performance liquid chromatography (HPLC) offers no clear advantage compared with gas chromatography. The potential for HPLC to provide the analyst with a means to determine nicotine and several of its metabolites simultaneously is its contribution to this area of investigation. The abilities of HPLC to separate polar and non-polar compounds, to eliminate thermal degradation of the compounds, and to provide analyte detection without compound destruction are clear advantages of this analytical technique compared to gas chromatography. With few exceptions the sensitivity of HPLC methods is less than gas chromatographic methods. For the determination of only nicotine and cotinine the increased system costs are hardly justified.

Numerous HPLC methods for nicotine and cotinine have been developed over the past 15–20 years. Early methods for the determination of nicotine and cotinine used normal phase columns to resolve the analytes [173,174], but the majority of HPLC methods use reverse phase chromatographic columns. The analysis of nicotine and cotinine in plasma [175] and urine [176] employed extensive sample pre-treatment and allowed the detection of nicotine and cotinine at 1 ng/mL and 3 ng/mL, respectively. Isolation of the analytes from the matrices was accomplished by liquid–liquid extraction. Solid phase extraction was used to remove additional interfering substances. Nicotine, cotinine, and internal standard compounds were separated on a minibore reverse phase C18 column with a mobile phase consisting of a citrate-phosphate buffer containing 50 mL/L acetonitrile and 1 mM sodium heptanesulfonate at pH 6.1–6.5. The isocratic separation of the compounds was complete in 10 min. Relative recoveries of nicotine and cotinine exceeded 90% for both plasma and urine. Comparison of urinary concentrations of nicotine and cotinine by GC-NPD and HPLC yielded correlation coefficients well above 0.9.

Cotinine concentrations in plasma and saliva were investigated by Machacek and Jiang [177] to discriminate smokers and non-smokers. The method used ion-pair reverse phase chromatography with 2-phenylimidazole as the internal standard. The resulting analytical data for cotinine was compromised by interference from caffeine that eluted between cotinine and the internal standard. Elimination of caffeine interference in the determination of cotinine was the emphasis of other HPLC determinations [63,178].

Restriction of caffeine intake for 2–3 days prior to sample collection was a pragmatic way to eliminate this concern [179].

The direct determination of cotinine-1-*N*-oxide in urine samples obtained from smokers and nonsmokers was demonstrated using HPLC as the end determination [180]. Quantitative extraction of cotinine-1-*N*-oxide from urine was effected via the use of 25% aqueous base (NaOH) and acetonitrile. Two structural analogues were synthesized as internal standard compounds for quantitation. HPLC fractions containing cotinine-1-*N*-oxide were captured, reduced to cotinine, and confirmed by GC-NPD analysis [130]. Smokers' urines contained cotinine-1-*N*-oxide in concentrations ranging from 20 ng/mL to 1170 ng/mL with a mean concentration of 400 ng/mL. Cotinine-1-*N*-oxide was not present in non-smokers' urine samples above the detection limit (10 ng/mL).

To study the distribution and elimination kinetics of nicotine in smokers and non-smokers an HPLC method was developed to measure both radioactive and unlabeled nicotine, cotinine, and nicotine-1'-*N*-oxide in urine and plasma [179]. Nicotine, nicotine-1'-*N*-oxide, and cotinine were isolated from the matrices by solid phase extraction employing a silica gel column. Timed fractions (0.3 min) were collected into glass scintillation vials for counting when radiometric detection was necessary and data reported as corrected disintegrations per minute (dpm). Consequently the HPLC method is applicable to both [14]C-labeled nicotine and derived metabolites, [14]C-nicotine-1'-*N*-oxide, and [14]C-cotinine. Recoveries were 86% and 87% for nicotine, 82% and 84% for nicotine-1'-*N*-oxide, and 73% and 82% for cotinine in urine and plasma, respectively. For the radiometric assay method sensitivity was 60 dpm; sensitivity was 5 ng/mL for unlabeled compounds. Pharmacokinetic analysis of the analytical data obtained with this method led these investigators to conclude that the terminal half-life for nicotine and cotinine was shorter in smokers than non-smokers. However, no significant difference in nicotine or cotinine clearance was observed. The differences observed in the disposition of nicotine between smokers and non-smokers were due to induced metabolism of nicotine in smokers according to these authors.

Extensive investigation into the various factors that determine compound resolution by sorption phenomena in the chromatographic column used in HPLC allowed the determination of nicotine and twelve of its metabolites in rat plasma and urine [181]. Separation of these nicotine metabolites using HPLC serves to emphasize the utility of this analytical technique to resolve compounds with varying charge and polarity. With some minor modifications to mobile phase composition and the use of a single cyano column, nicotine and eight metabolites in both urine and plasma samples from smokers and nonsmokers were determined after intravenous administration of a single dose of [14]C-nicotine (190 μg). The compounds isolated with this radiometric HPLC method include nicotine, nicotine-1'-*N*-oxide, nornicotine, cotinine, demethylcotinine, cotinine-1-*N*-oxide, 3'-hydroxycotinine, 3'-hydroxycotinine glucuronide, and demethylcotinine $\Delta^{4',5'}$-enamine. These latter two metabolites were confirmed by enzymatic hydrolysis with β-glucuronidase or mass spectrometry. The formation of demethylcotinine $\Delta^{4',5'}$-enamine is an artifact of the MS determination of 5'-hydroxynorcotinine (182). As in an earlier report from this laboratory (181), the pharmacokinetic disposition of nicotine differed between the smoker and non-smoker groups of subjects. More importantly the urinary excretion of both 3'-hydroxycotinine glucuronide and the enamine compound exceeded that of cotinine. These compounds were suggested as potential markers of tobacco smoke exposure.

A ternary gradient solvent system was used in ion-pair reverse phase ($C_{18}$) HPLC for the separation of 29 nicotine metabolites as reference compounds with detection by UV and radioactivity [183]. Additionally, these authors modified the cation exchange HPLC method described by Cundy and Crooks [29] so that 17 reference substances could be separated using a binary gradient solvent system. Application of both the ion pair reverse phase and the cation exchange HPLC systems was used for the determination of nicotine metabolites in urine from rats injected with 2'-$^{14}$C-nicotine. Of the 40 radioactive peaks detected 18 were identified. A high correlation was found for major urinary nicotine metabolites separated by the two chromatographic systems with the exception of cotinine, which co-eluted with other metabolites in the ion-exchange HPLC system [183].

*Gas chromatography – mass spectrometry* (see also Chapter 6)
Because the problems of interfacing chromatographic instruments to the mass spectrometric detector were significantly less with gas chromatography than with HPLC, methods employing gas chromatography as a means of sample introduction into the mass selective detector were the first to show significant growth. Many of the various problems associated with the use of HPLC and interfacing to the mass spectrometric detector have yielded to research efforts. For instance, the reduction of column size in HPLC makes possible a reduction in mobile phase flow thereby eliminating the need for moving belt and particle beam interfaces. Presently, it appears that thermospray, electrospray, and atmospheric pressure chemical ionization are the interfaces most frequently in use. An additional advantage of these interfaces is the soft ionization technique that they have in common. A limited number of GC and HPLC methods for the determination of nicotine and its metabolites is offered as examples of both applications and analytical techniques.

An extensive review of nicotine and its metabolites obtained from synthesis or extraction, isolation, and purification from biological matrices resulted from a collaborative effort of investigators in Sweden and the U.S. [184]. A number of compounds were examined by mass spectroscopy using either gas chromatography or direct probe as modes of sample introduction.

All of the compounds examined were metabolites of nicotine or were present in tobacco or tobacco smoke. While the purpose of this investigation was to provide spectral and gas chromatographic data on some 30 compounds for future studies in plants and animals, the utility of the method of analysis was demonstrated by the determination of plasma nicotine concentrations in two smokers by mass fragmentography using deuterated nicotine as the internal standard. Subsequently high resolution selected ion monitoring (SIM) was used in conjunction with capillary gas chromatography to determine nicotine in plasma at a concentration <100 femtograms [185]. This highly sensitive method used deuterated nicotine as the internal standard and a modified interface that allowed the direct introduction of the capillary column effluent into the ion source of the mass spectrometer. Breast fluid, aspirated from non-lactating women, was shown to contain nicotine and cotinine at on-column concentrations as low as 25 picograms through the use of GC-MS and selected ion monitoring [69,186]. Deuterated nicotine and cotinine were used both as internal standards and as carriers in an inverse isotope dilution method of analysis. Breast fluid nicotine content ranged from 50–200 ng/mL in women smokers while the plasma concentrations ranged from 10–20 ng/mL. Elimination of sources of nicotine contamination was necessary to achieve an estimate of nicotine concentration in the body fluids. The use of selected ion monitoring for nicotine determination was demonstrated earlier by Dow and Hall [187]. Two modes of ionization within the ion source of the mass spectrometer were investigated by Skarping, et al. [188] in order to determine the concentrations of cotinine in urine samples from a smoker and nonsmokers. Three modes of cotinine detection were compared: (a) GC with nitrogen-sensitive detection; (b) SIM of cotinine and trideuterated cotinine with electron impact ionization; and (c) SIM of cotinine by chemical ionization using isobutane as the reagent gas. The sensitivity of gas chromatography with nitrogen-sensitive detection and SIM with chemical ionization was similar, however both these modes of detection were an order of magnitude more sensitive than SIM in the EI mode. In each of the modes of detection the response to cotinine was linear. Calibrated ranges of cotinine concentrations were: 1–100 ng/mL in the EI mode, 0.2–10 ng/mL in the CI mode, and 5–50,000 ng/mL for the nitrogen-sensitive detector. Limits of detection were calculated for both modes of ionization in the GC-MS determination and found to be 2.2 ng/mL and 0.2 ng/mL for EI and CI, respectively.

The application of GC-MS with SIM was used in a number of studies investigating the disposition and metabolism of nicotine [32,189–191]. The unique specificity of mass spectrometric analysis was demonstrated through the administration of a deuterated nicotine analogue to human volunteers to determine if its metabolism and clearance differed from natural nicotine obtained from the use of tobacco products [32]. Both GC and GC-MS were used for the determination of plasma nicotine-$d_0$, cotinine-$d_0$, nicotine-$d_2$, and cotinine-$d_2$. Intra-day accuracy and precision of all four compounds ranged from 91%-103% and 0.8%–5.5%, respectively, by GC-MS analysis. Comparison GC and GC-MS determinations of nicotine and cotinine from pooled plasma samples yielded small deviations from the mean concentrations of 2.6%–4.7% for nicotine and 6%–7% for cotinine. The limits of quantitation for nicotine and cotinine were 1 ng/mL and 10 ng/mL, respectively. The pharmacokinetic evaluation of nicotine-$d_0$ and nicotine-$d_2$ was conducted in abstinent active smokers by intravenous infusion of a 50:50 mixture at a dosage of 2 µg/Kg/min over a 90 min period. Plasma samples were collected during the infusion and for 6 h after the infusion. Total clearance, terminal half-life, and apparent volume of distribution for natural and deuterated nicotine were similar. Consequently the use of deuterated analogues of nicotine for pharmacokinetic studies was validated. Functionally this work eliminated the two to three day period of abstinence preceding most pharmacokinetic studies of nicotine and makes possible these types of studies in a more natural manner.

This same laboratory investigated the urinary levels of nornicotine, anabasine, anatabine, nicotine, and cotinine in subject groups of smokers, tobacco chewers, and oral snuff users [191]. GC-MS-SIM was used to determine the compounds after reductive alkylation with propionaldehyde and phenylacetaldehyde to form the propyl and the phenethyl derivatives. The compounds were extracted from urine using $C_{18}$ solid phase extraction columns, eluted with methanol, and the acidified residues concentrated to dryness. The residue was derivatized with proprionaldehyde or phenyl-acetaldehyde reagent solutions in the presence of sodium borohydride. Sample clean-up and concentration were accomplished with liquid–liquid extraction and partitioning into acidic or basic solutions. Injection volumes of 2 µL were analyzed by GC-MS-SIM. Derivatives of structural analogues were used as internal standards. Concentrations as low as 1 ng/mL could be determined for these compounds. Other compounds of interest such as metanicotine, dihydrometanicotine, and 2,3′-bipyridyl were at or below the limit of detection in the urine samples. This is not surprising since all of the compounds are minor components of tobacco or tobacco smoke. Daily excretion of nicotine and cotinine accounted for 43.8% and 52.4% of these alkaloids in cigarette smokers, 31.7% and 55.4% in chewers, and 31.7% and 59.5% in snuff users. Nornicotine excretion ranged from 3.1% to 9.3%. Anabasine and anatabine excretion were minimal in cigarette smokers (<0.5%), about 1% in oral snuff users, and maximal in the chewers (1%–3%). This was the first report of anabasine and anatabine in the urine of tobacco users.

*HPLC – mass spectrometry* (see also Chapter 7)
HPLC with mass spectrometric detection was used for the determination of plasma nicotine and cotinine for the investigation of the pharmacokinetics of

a transdermal nicotine patch [193]. The stated goals of the method were adequate sensitivity and rapid sample analysis. Other analytical methods were eliminated on the basis of the complicated extraction procedure, lack of sensitivity, the necessity for pre-derivatization, long run times, low recovery of analytes, or influence of contamination. The method employs a simple extraction procedure and liquid chromatography with tandem mass spectrometry that results in a highly sensitive, specific method with high sample throughput. Plasma (1 mL) and internal standards were made alkaline with ammonium hydroxide and extracted into dichloromethane. Centrifugation was used to separate the phases and the aqueous phase was discarded. The organic phase was transferred into a clean tube and concentrated to dryness under a gentle stream of nitrogen. The residue was dissolved in 100 μL mobile phase consisting of acetonitrile-methanol-10 mM ammonium acetate, and 10 μL was injected into the LC-MS-MS system. Reported recoveries for nicotine and cotinine were 106% and 86.2%, respectively. Resolution of nicotine, cotinine, and their deuterated internal standards was accomplished with a $C_{18}$ reverse phase column at a mobile phase (acetonitrile-methanol-10 mM ammonium acetate, 53:32:15, v/v) flow rate of 1.4 mL/min in the isocratic mode. Run time was about 1 min. An atmospheric pressure chemical ionization (APCI) interface was used to introduce the sample into the tandem mass spectrometer. The protonated molecular ions (MH$^+$) for nicotine ($m/z$ 163.2), cotinine ($m/z$ 177.2), nicotine-methyl-$d_3$ ($m/z$ 166.2), and cotinine-methyl- $d_3$ ($m/z$ 180.2) passed from the first quadropole into the collision chamber (2nd qaudropole) where fragmentation occurred using argon as the collision gas. The daughter ions of nicotine ($m/z$ 84), cotinine ($m/z$ 98), and the deuterated internal standard compounds (nicotine-$d_3$, $m/z$ 87; cotinine-$d_3$, $m/z$ 101) were monitored in the third quadropole by multiple reaction monitoring. The calibrated ranges for nicotine and cotinine were 1–50 ng/mL and 10–500 ng/mL, respectively. Method sensitivity was determined to be 1 ng/mL for nicotine and 10 ng/mL for cotinine. Coefficients of variation for both calibration standards and control samples were <8%. Sample stability and chromatographic stability were excellent in that no degradation occurred during chromatography, extraction, or sample storage. Daily sample analysis was reported as 400 samples per day. Certainly the application of LC-MS-MS for analysis of serum samples resulted in a rapid, sensitive, and specific method of determination. However, the expense of the instrument is presently a deterrent to many laboratories.

Using a similar system and approach, but with sensitivity being the focus of method development, the limit of quantitation for cotinine in serum was reported to be 50 pg/ml [194]. A high performance liquid chromatography system was coupled to a tandem mass spectrometer via an APCI interface. Salient features of this method include elimination of potential interference due to the specificity of MS/MS, preliminary sample prescreening with an enzyme immunosorbent assay (EIA) into high and low categories, a lengthy extraction method, and analytic validation through quality control samples. Sample analysis required a 10 µL injection onto the HPLC system consisting of a base-deactivated $C_{18}$ reverse phase column (4.6 mm × 3 cm, 3-µm particle size) operated with a mobile phase (methanol-10 mM ammonium acetate, 80:20, v/v) at an isocratic flow rate of 1 ml/min. Mass spectrometric detection was carried out using a corona discharge ionization in an APCI interface with heated nebulizer. Multiple reaction monitoring was used for quantitation of cotinine using the $m/z$ 177 → 80 transition ion of cotinine, the $m/z$ 177 → 98 transition ion as confirmatory ion for cotinine, and $m/z$ 180 → 80 transition ion for deuterated cotinine. Peak area ratio of $(m/z$ 177 → 80)/$(m/z$ 180 → 80) was plotted against cotinine concentration. The range of cotinine calibration was 0 to 25 ng/mL. Cotinine retention time was <1 min in the system with a total run time of 2 minutes per sample. Two batches of 50 samples per day were analyzed in this manner. Sample injection, analysis, and data reporting were under computer control. Owing to losses during centrifugal vacuum evaporation and protein binding, the recovery of cotinine ranged between 60%–70%. The method of cotinine determination appears to be quite robust and rugged in that >32,000 samples have been analyzed over four years to investigate exposure to environmental tobacco smoke (ETS) in individual participants in the Third National Health and Nutrition Examination Survey (NHANES III). The applicability of this method of analysis to other body fluids remains to be demonstrated.

Recently a method for the determination of cotinine and *trans*-3′-hydroxycotinine was validated for the saliva matrix [195]. The method is based upon an HPLC interfaced into a tandem mass spectrometric system. Cotinine and *trans*-3′-hydroxycotinine along with deuterated internal standards were extracted from the saliva matrix using solid phase extraction (SPE) in a semi-automated mode. Following concentration of the SPE eluant to dryness, the residue was re-dissolved in 200 µL 12.5 mM ammonium formate-methanol-formic acid (80:20:0.5, v/v/v) and transferred into a microvial where 100 µL was injected onto the system. Cotinine and 3′-hydroxycotinine were eluted in about two minutes from an Omnipac PCX-500 50 × 4.6 mm column by a mobile phase consisting of methanol-30mM ammonium formate-formic acid (50:50:5, v/v/v) with a flow rate of 1 ml/min. Mass spectrometric detection was provided by a triple quadropole mass spectrometer with a positive ion electrospray interface; five mass/charge transitions were monitored by multiple reaction monitoring. Transition ions for cotinine included $m/z$ 177 → 80 for quantitation and $m/z$ 177 → 98 for confirmation. The three additional transition ions monitored included cotinine-$d_3$, $m/z$ 180 → 101; 3′-hydroxycotinine, $m/z$ 193 → 80; and 3′-hydroxycotinine-$d_3$, $m/z$ 196 → 80. The method calibration range was 0.020–10 ng/mL prepared from standards in water. The lower limits of quantitation for cotinine and 3′-hydroxycotinine were 0.05 ng/mL and 0.10 ng/mL, respectively. Saliva cotinine and 3′-hydroxycotinine values correlated well with the self-reported exposure to ETS. The most obvious advantages of this method for the determination of cotinine and 3'-hydroxy-cotinine in saliva are an increase of sensitivity beyond the majority of reported methods of analysis and the simultaneous determination of 3'-hydroxycotinine, a metabolite of cotinine. Both of these nicotine metabolites have been suggested as biomarkers for exposure to ETS. If levels of nicotine and especially its metabolites in the urine matrix could approach the sensitivity of this method [195] or the method of Bernert et al. [194], a more valid estimate of nicotine intake would be achieved whether from active smoking or exposure to ETS.

A thermospray LC-MS method for nicotine and 17 of its urinary metabolites has been reported [196]. The method utilized the direct injection of urine into the instrumental system following a two-step ultra-filtration procedure that used centrifugal force to aid in the filtration process. Chromatographic separation was effected by means of a $4.6 \text{ mm} \times 50 \text{ mm}$ polymeric column preceded by a polymer guard column. A linear gradient of methanol and water (70:30, v/v) was used to elute the compounds in 10 min at a flow rate of 0.5 mL/min. Positive chemical ionization was aided by the post-column addition of 50 mM ammonium acetate at a flow rate of 0.9 ml/min. Mass spectral data were collected in the selected ion monitoring mode for the protonated molecular ions and quantitation was done using peak area ratio technique with cotinine-$d_3$ as internal standard. While nicotine and 17 of its metabolites could be determined by this method from a standard solution, when smokers' urine samples were analyzed only nicotine, cotinine, 3'-hydroxycotinine, and demethylcotinine (norcotinine) were detected. These compounds and nicotine-1'-$N$-oxide were used for method validation. For nicotine and these metabolites a linear response was obtained from 20–8000 ng/mL. The ease of protonating cotinine, 3'-hydroxycotinine, and de-methylcotinine provided for the limit of detection of 200 pg on column. The limit of detection for nicotine and nicotine-1'-$N$-oxide was 900 pg on column owing to a decrease in proton affinity. Similar quantities of urinary cotinine were determined by LC-MS and GC-NPD methods of analysis in samples obtained from five smoker subjects. After a slight modification of the method described by McManus et al. [196], Byrd et al. [197] examined the urines of 11 smokers for nicotine and its metabolites that demonstrated the presence of glucuronides of nicotine, cotinine, and *trans*-3'-hydroxycotinine as phase II metabolites. Method modifications included a change in the mobile phase gradient profile such that 100% water began the gradient. Methanol was added to the mobile phase in a linear profile to provide for 30% methanol at 3.1 min, 45% methanol at 5 min, and 90% methanol at 8 min. Additionally the post-column addition of 0.1 M ammonium acetate was added to the mobile phase in a reverse gradient beginning at 55 nM concentration and increasing to a final concentration of 95 nM to provide for a more stable baseline and to improve the reproducibility of the response to the analytes.

Protonated molecular ions monitored for quantitation were as follows: $m/z$ 163, nicotine and demethylcotinine; $m/z$ 177, cotinine; $m/z$ 179, nicotine-1'-$N$-oxide; $m/z$ 180, cotinine-$d_3$, internal standard; $m/z$ 193, cotinine-$N$-oxide and *trans*-3'-hydroxycotinine at a dwell time of 250 msecs for each ion. Sample aliquots were treated with β-glucuronidase to release the glucuronic acid moiety. The aliquots treated with β-glucuronidase were analyzed in the same manner as the untreated urine aliquots. Concentrations of nicotine, cotinine, and *trans*-3'-hydroxycotinine glucuronides were determined by difference. The distribution of the metabolites in the urines of the 11 smokers was assessed by converting all the metabolites to nicotine equivalents according to the equation:

$$\text{Nicotine equivalent} = \mu g \text{ per } 24 \text{ hr} \times 162/\text{mol. wt.}$$

The sum of the nicotine equivalents should provide an estimate of nicotine intake for the individual smoker. While it is clearly recognized that the nicotine yield per brand of cigarette, as determined by the FTC (US Federal Trade Commission) method, does not present an accurate value for nicotine intake in an individual smoker, it was of interest to assess the urinary excretion of nicotine in relation to the FTC nicotine yield and the number of cigarettes consumed on the day of urine collection. The total daily nicotine yield for the usual brands consumed by these smokers ranged from 12.3 mg to 44.2 mg. The ratio of nicotine intake:urine nicotine ranged from 0.58 to 1.44 with a mean value of 0.96±0.29 [197]. However, the demonstration that an estimate of nicotine intake may be calculated from the total urinary metabolite profile expressed as nicotine equivalents provides a more accurate means to investigate this relationship.

An additional application of LC-MS to the determination of nicotine and its metabolites, include the determination of cotinine-$N$-glucuronide directly in urine (198). Comparison of the concentrations of cotinine-$N$-glucuronide by both the direct and indirect (β-glucuronidase treatment to release the aglycon) yielded differences $\leqslant 14\%$. This study confirmed that cotinine-$N$-glucuronide is a significant urinary metabolite of nicotine. Only *trans*-3'-hydroxycotinine was excreted in greater amounts. There was significant inter-individual variability in the amount of cotinine-$N$-glucuronide excreted. This was interpreted by these authors to indicate the presence of competition between phase I and phase II metabolism of nicotine

*Immunological methods of analysis* (see also Chapter 8)
The determination of nicotine and cotinine using immunological techniques involves the use of either polyclonal or monoclonal antibodies to the ligands in a variety of analytical formats. Generally the method formats are described by the means of end determination, for instance, radio-immunoassay (RIA), enzyme immunoassay (EIA), and enzyme-linked immunosorbent assay (ELISA).

The principle underlying the immunological methods is the establishment of a dynamic equilibrium between antibody binding to labeled ligand and the competition for restricted antibody binding sites by unlabeled ligand present in the calibration standard or sample. The extent of competitive binding, i.e., binding inhibition, is used for quantification of the unlabeled ligand present in the sample or standard by comparison with a calibration curve.

Of the immunological methods available to the analyst for the determination of nicotine and cotinine the radioimmunoassay is most familiar. The RIA for nicotine and cotinine was reported initially by Langone et al. [199] who employed *O*-succinyl-3'-hydroxymethylnicotine and *trans*-4'-carboxy-cotinine as hapten molecules conjugated to keyhole limpet hemocyanin or human serum albumin. Six to 18 moles of hapten molecule were coupled to the protein molecules to form the functionalized hapten. Antisera were prepared by immunizing rabbits. The antisera were evaluated in competitive binding assays with the aid of $^3$H-nicotine and I$^{125}$-cotinine as labeled ligands. The selectivity of anti-nicotine and anti-cotinine antisera for unlabeled ligands was demonstrated; for 50% binding inhibition ($I_{50}$), 3.2 ng nicotine and 1.2 ng cotinine were required in the respective RIAs. Ligand selectivity was excellent in these RIAs; cotinine was 2500 times less effective at inhibiting nicotine binding and nicotine was 2000 times less effective as an inhibitor of cotinine binding than was cotinine. The limit of detection for nicotine was reported as 0.35 ng/mL. This level of assay sensitivity exceeds the majority of chromatographic methods of analysis.

Other RIAs were developed that used hapten molecules with the linkage to the macromolecule at positions on the pyridine ring of nicotine or cotinine that differed from Langone et al. (199). Thus Haines et al. [200] used *N*-succinyl-6-aminonicotine for conjugate formation. Matsukura, et al. [201] also used the 6 position of the pyridine ring to form 6-(*p*-aminobenzamido)-nicotine as conjugates to protein molecules to form a complete hapten molecule. Castro [202] developed RIAs based on antisera raised against conjugates of *N*-succinyl-6-aminonicotine and 6-(Σ-aminocapramido)-nicotine. Matsukura et al. [201] developed a RIA utilizing antibodies raised in rabbits to $\ell$-1-(β-aminoethyl)-cotinine coupled to bovine serum albumin that proved to be problematic. Since the radio-labeled antigen was similar in structure to the conjugated antigen, the antisera reacted strongly with both the cotinine antigen and the labeled derivative of cotinine. Accordingly, the antiserum was selective not only for the pyridine and pyrrolidine rings, but also to the side chain attached to the pyridine ring. Antisera that recognized the sidechain to a tyramine derivative of *N*-(*p*-hydroxyphenethyl)-4'-*trans*-cotinine carboxamide were problematic for Knight et al. [203] in their initial RIA procedure. Partial improvement in RIA performance was achieved by adsorbing the bridge antibodies to a structurally related compound (nicotine) prepared with the identical sidechain. To circumvent bridge recognition by antisera in a RIA, Jones and Amatayakul [204] simply changed the structure of the sidechain so that the antisera no longer expressed binding affinity for the sidechain. Both of these approaches only slightly improved these $I^{125}$-cotinine RIAs in that the slope of the standard inhibition curve remained shallow and binding inhibition in the presence of excess cotinine remained incomplete [203]. Each of the RIAs developed from the variety of functionalized haptens claims excellent sensitivity, high selectivity for the analyte ligand, and high titers of antibody production. A table describing the characteristics of several RIA procedures has been compiled by Curvall and Kazemi-Vala [51].

Compared to the chromatographic methods of analysis the RIA offers several advantages. Among these advantages are: (a) a reduction in sample preparation prior to quantitation; (b) ease of application; (c) lower cost per sample; (d) higher analytical efficiency (throughput), thus applicability in studies composed of large sample numbers; (f) reduced sample volume; and (g) the reduction of skill level required from laboratory technical personnel. Simple dilution of samples is often the only preparation necessary prior to submitting the sample to the RIA for quantitation of nicotine and/or cotinine.

A number of studies directly compared RIA and chromatographic methods of analysis for the determination of nicotine and/or cotinine and found that the correlation ($r > 0.95$) of the values obtained by these two methods of analysis was high [71,199,205,206]. Some laboratories have found large discrepancies between analytical values obtained by RIA and chromatographic methods [208–211]. A variety of reasons account for the lack of agreement between the RIA and chromatographic methods of analysis for the determination of nicotine and/or cotinine.

In examining this issue, Letzel et al. [210] concluded that interlaboratory variation accounted for about 60% of the observed variance thus casting doubt on the direct comparison of analytical results reported from two or more laboratories performing analyses by RIA. High variability was observed in an inter-laboratory study reported by Biber et al. [209] where the objective was to determine the concentrations of nicotine and cotinine in serum and urine. Eleven laboratories, experienced with either gas chromatographic or RIA determinations for the analytes, were chosen for participation in the study. Yet the interlaboratory coefficient of variation ranged between 4% to 59%. Intralaboratory variance was high when a cotinine RIA on urine samples was performed by two individuals using a common sample pool. The manner in which samples are collected may affect analytical integrity and variability. These observations of analytical variability in the RIA serve to emphasize that meticulous attention to details of analysis does not always promote precise and accurate data. However, the presence of nicotine metabolites that cross-react with the antisera cannot be overcome easily and significant sample pre-treatment is required to eliminate this source of variability.

To avoid inaccurate quantitation due to cross-reactivity with structurally similar nicotine metabolites, some investigators have extracted nicotine and/ or cotinine from the matrix prior to analysis by RIA [200]. This practice does not appear justified in the majority of cases. There are two instances when sample extraction might be critical to ensure the validity of the analytical data; the measurement of nicotine and/or cotinine concentration in a urine matrix and examination of serum, saliva, or urine from non-smokers exposed to environmental tobacco smoke. The extraction of nicotine and cotinine from the matrix ensures an accurate determination of the analyte due to the absence of other nicotine metabolites that could cross react with the antisera. One may concentrate the extract to dryness and reconstitute to a lesser volume thus achieving an analyte concentration within the sensitivity of the RIA. Despite the examination of 50 to 100 compounds structurally related to nicotine and/or cotinine by Langone and co-workers [144,199] only nornicotine, nicotine-1'-*N*-oxide, and anabasine were demonstrated to possess significant cross-reactivity toward the nicotine antisera. Cotinine-*N*-oxide, demethylcotinine, *N'*-nitrosonornicotine cross-reacted with the cotinine antisera. Inexplicably, both metyrapone and reduced metyrapone cross-reacted with the cotinine antisera. These studies did not investigate the predominant urinary metabolite of nicotine, *trans*-3'-hydroxycotinine, reported by Neurath and co-workers [19,20]. Others [211–213] investigated the cross-reactivity of *trans*-3'-hydroxycotinine. These investigators reported that cross-reactivity due to *trans*-3'-hydroxycotinine in both a RIA [211] and an ELISA for cotinine ranged between 30% to 35% [212]. *Cis*-3'-hydroxycotinine cross-reacted with anti-cotinine antibody in the ELISA at a 12% level. Zuccaro et al. [211] found no cross-reactivity of cotinine glucuronide in the cotinine RIA. Through the use of Extrelut-3 column chromatography and direct addition of *trans*-3'-hydroxycotinine to blank urine, this nicotine metabolite was demonstrated to cross-react with the cotinine antisera at an efficiency of 40% [213]. The glucuronides of nicotine, cotinine, and *trans*-3'-hydroxycotinine apparently did not significantly cross-react with the cotinine antisera [213]. We at RJRT have confirmed that the glucuronides of nicotine, cotinine, and *trans*-3'-hydroxycotinine are not the source of significant interference in either the nicotine or cotinine RIAs through direct addition of known concentrations of the *N*-glucuronides (unpublished data).

Recently, *trans*-3'-hydroxycotinine has been shown to be present in both serum [20] and saliva [214,215]. The concentration of *trans*-3'-hydroxycotinine in both serum and saliva is about one-third the level of cotinine. We (RJRT) have observed a positive bias in cotinine RIAs performed in these matrices of about 8% to 12% in samples obtained from active smokers. In non-smoker saliva samples no bias in the cotinine concentration was detected [61,216].

Development of ELISAs for nicotine and cotinine that employ polyclonal [217] or monoclonal antibodies (McAb) [218] eliminated two significant problems attendant to the RIAs. End determinations that used either fluoresence or enzymic oxidation to a colored product effectively eliminated the need for using radioisotopic ligands. Additionally, the use of McAb with high selectivity and minimal cross-reactivity practically eliminated concerns about cross-reactivity in urine, saliva, and serum. (R)-(+) nicotine and 6-hydroxynicotine exhibited cross-reactivity against the nicotine McAb at $I_{50}$ doses five and 13 times greater than (S)-(−)-nicotine ($I_{50}$=0.25 ng), respectively. (S)-(−)-Cotinine had an $I_{50}$ dose of 0.12 ng against the anti-cotinine McAb. At $I_{50}$ doses, desmethylcotinine, (−)-cotinine-N-oxide, and N′-nitrosonornicotine cross-reacted against the anti-cotinine McAb at 1.6, 4.6, and 6.5 nanograms, respectively. In typical body fluid analyses for nicotine and/or cotinine by ELISA the cross-reactivity of these compounds is of little concern. The range of the ELISAs is 5 ng/mL to 500 ng/mL; only at the lowest concentration does the coefficient of variation exceed 10% (nicotine at 10 ng/mL, 16.4%; cotinine at 5 ng/mL, 13.6%). Antibody titer, reagent additions, and coating concentration of poly-L-lysine nicotine or cotinine conjugate were optimized to yield an absorbance reading of 1.0 at 490 nm after color development.

Monoclonal antibodies to cotinine were used to maximize the ELISA for determination of cotinine in physiological fluids [206]. These assays were 5–20 times more sensitive than six rabbit antisera RIAs. Thus $I_{50}$ doses ranged between 0.024–0.063 ng cotinine with as little as 0.005 ng yielding a 15% inhibition. The use of an $I^{125}$-labeled tyramine derivative of cotinine inhibited maximal binding at 100%. ELISAs that used $^3$H-cotinine were superimposable with ELISAs that used horseradish peroxidase labeled staphylococcal protein A for end determination, consequently there was no need to use isotopically labeled ligand. The analysis of serum samples for cotinine by both the ELISA and the RIA yielded a high correlation (r=0.944). Similar correlation values were obtained when either the ELISA or the RIA was compared with gas chromatographic determinations of cotinine.

Further advances were made in assay selectivity for cotinine with the development of a solid phase ELISA based upon idiotype-anti-idiotype ligand recognition [219]. Under optimal assay conditions as little as 0.9 ng cotinine inhibited binding by 50%. The limit of detection for this type ELISA was 0.04 ng. Amounts of *trans*-3′-hydroxycotinine and nicotine necessary to inhibit binding by 50% were 61 ng and >1000 ng, respectively. Comparison of the idiotype-anti-idiotype ELISA and the anti-cotinine ELISA for the determination of cotinine in saliva yielded values that were highly correlated (r=0.994).

Chang investigated the use of anti-nicotine and anti-cotinine ELISAs for the determination of plasma nicotine and cotinine in rats exposed to aged and diluted sidestream smoke [220]. Confirmation of smoking status by the measurement of urinary cotinine with ELISA was another application [221]. Due to increased selectivity of the anti-nicotine and anti-cotinine McAb, these ELISAs were used for the analysis of nicotine and cotinine in urine of smokers and non-smokers [222].

## Comparison of analytical methods

An inter-laboratory comparison of the various methods for the determination of nicotine and its metabolites was done using pooled smokers' urine as the matrix [223]. Methods of analysis included LC-MS [196,197], GC-MS [21], HPLC [159], GC-NPD [38], RIA [144,205], and ELISA [220]. With the exception of RIA and ELISA, the instrumental analyses were performed on urine samples that were hydrolyzed enzymatically to release the conjugated forms of nicotine, cotinine, and *trans*-3'-hydroxycotinine (Table 2). Following conversion of the urinary nicotine metabolites to nicotine equivalents as described earlier, the total nicotine excreted averaged 12.61±0.78 μg/mL yielding a coefficient of variation of 6.51% for the instrumental modes of analysis. The immunological methods (RIA and ELISA) could only analyze for nicotine and cotinine concentrations in the urine, consequently only about 50% of the total nicotine determined by the more comprehensive methods of analysis could be accounted for by these methods. Amounts of nicotine in urine did not differ between the RIA and ELISA (2.29±0.33 μg/mL vs. 2.31±0.39 μg/mL). The RIA yielded a value for cotinine (4.06±1.05 μg/mL) that was 15% higher than the cotinine concentration obtained by the ELISA (3.43±0.41 μg/mL). The observed difference between these results was not significant due to the small sample size, but the difference may reflect cross-reactivity from other urinary compounds in the RIA.

Application of the comprehensive method of analysis provided by LC/MS was used to investigate the presence of nicotine and its metabolites in pooled samples of blood serum, saliva, and urine from smokers by direct injection of sample aliquots after filtration [214]. Limits of detection ranged from 2 ng/mL to 40 ng/mL. Analytes included nicotine, cotinine, *trans*-3'-hydroxy-cotinine, the oxides of nicotine and cotinine, demethylcotinine, and nornicotine. Conjugated forms of nicotine, cotinine, and *trans*-3'-hydroxycotinine were not included in these analyses. Precision of the analyses was <10% at concentrations above 250 ng/mL and 22% for all detected analytes. The results of these analyses are shown in Table 3. The determination of nicotine and all six of these metabolites was possible only in the urine matrix. In both serum and saliva matrices the concentrations of a number of the analytes were below their determined limit of detection. Since these matrices were collected from different groups of smokers, it is inappropriate to make comparisons of analyte concentrations across the matrices. Comparisons of analyte concentrations within the serum and saliva matrices indicate similar profiles with the exception of nicotine. Cotinine is the most abundant nicotine metabolite in both serum and saliva. The observed nicotine concentration in saliva can reflect both nicotine adsorption in the oral cavity

TABLE 2

Interlaboratory analysis of standard smokers' urine: summary data

| Lab. | Method SD† | NIC | NIC-GlcA | COT | COT-GlcA | t3HC | t3HC-GlcA | NNO | CNO | DMC | NNIC | POMBA | PAA | c3HC |
|---|---|---|---|---|---|---|---|---|---|---|---|---|---|---|
| | | URINARY NICOTINE METABOLITES*, µg/mL | | | | | | | | | | | | |
| GRI | HPLC | 1.89 | 0.54 | 1.45 | 2.65 | 4.46 | 1.80 | 1.12 | 0.43 | 0.32 | 0.09 | 0.24 | 0.03 | 0.04 |
| | SD | 0.05 | 0.09 | 0.04 | 0.14 | 0.13 | 0.10 | 0.02 | 0.09 | 0.01 | 0.01 | 0.02 | 0.00 | 0.00 |
| GRI | GC-MS | 1.91 | ‡ | 1.62 | ‡ | 4.65 | ‡ | ‡ | ‡ | 0.19 | 0.08 | 0.12 | | 0.07 |
| | SD | 0.05 | | 0.03 | | 0.09 | | | | 0.00 | 0.01 | 0.00 | | 0.01 |
| RES | GC-NPD | 1.48 | 0.47 | 1.42 | 2.10 | 4.43 | 2.22 | 1.59 | 0.25 | | | | | |
| | SD | 0.13 | 0.07 | 0.18 | 0.22 | 0.34 | 0.30 | 0.15 | 0.02 | | | | | |
| RJR | LC-MS | 2.04 | 0.82 | 1.59 | 2.64 | 4.03 | 1.11 | 1.14 | 0.55 | 0.14 | | | | |
| | SD | 0.28 | 0.23 | 0.11 | 0.25 | 0.34 | 0.21 | 0.08 | 0.03 | 0.04 | | | | |
| RJR | GC-MS§ | 2.64 | | 4.06 | | 6.28 | | | | | | | | |
| | SD | 0.10 | | 0.19 | | 0.90 | | | | | | | | |
| RJR | RIA¶ | 2.29 | | 4.06 | | | | | | | | | | |
| | SD | 0.33 | | 1.05 | | | | | | | | | | |
| RJR | ELISA¶ | 2.31 | | 3.43 | | | | | | | | | | |
| | SD | 0.39 | | 0.41 | | | | | | | | | | |
| | Mean | 1.83 | 0.61 | 1.52 | 2.47 | 4.39 | 1.71 | 1.28 | 0.41 | 0.22 | 0.09 | 0.18 | 0.03 | 0.06 |
| | SD | 0.24 | 0.18 | 0.10 | 0.31 | 0.26 | 0.56 | 0.26 | 0.15 | 0.09 | | | | |

* Metabolite abbreviations: NIC, nicotine; NIC-GlcA, nicotine glucuronide; COT, cotinine; COT-GlcA, cotinine glucuronide; t3HC, *trans*-3′-hydroxycotinine; t3HC-GlcA, *trans*-3′-hydroxycotinine glucuronide; NNO, nicotine-1′-N-oxide; CNO, cotinine-1-N-oxide; DMC, demethylcotinine (norcotinine); NNIC, nornicotine; POMBA, 4-(3-pyridyl)-4-oxo-N-methylbutyramide; PAA, 3-pyridylacetic acid; c3HC, *cis*-3′-hydroxycotinine. Laboratories: GRI, German Research Institute; RES, Swedish Match; RJR, R.J. Reynolds Tobacco Co.

† SD = standard deviation of mean value shown directly above; mean value represents at least five determinations.

‡ Not determined

§ Analysis of hydrolyzed urine only; data values include free and conjugated forms. These values are not included in calculation of the mean of the analytical values from the three locations.

¶ Values not included in calculation of the mean of the analytical values from the three locations.

** Used with permission of Byrd [223]

TABLE 3

Nicotine metabolites detected in biological fluids of smokers

| Analyte | Biological matrices, * ng/mL | | |
| | Urine† | Serum† | Saliva† |
| --- | --- | --- | --- |
| Nicotine | 1220±20 | < 40 | 1380±60 |
| *trans*-3'-hydroxycotinine | 4720±430 | 77±1 | 131±10 |
| Cotinine | 1450±70 | 190±6 | 303±9 |
| Nicotine-*N*'-oxide | 640±50 | <28 | <28 |
| Cotinine-*N*-oxide | 502±27 | <2 | <2 |
| Demethylcotinine | 228±31 | 10±1 | 19±1 |
| Nornicotine | 60±13 | <6 | <6 |

* Data reported are the mean±standard deviation of triplicate determinations.
† Limit of detection based on response to blank sample+3σ. The limit of detection was analyte specific and ranged from 2 to 40 ng/mL
Used with permission by Byrd [214]

and transfer from systemic circulation. As a reflection of the brief half-life of nicotine in blood its concentration was below the limit of quantitation although the estimated nicotine level was 8.5 ng/mL. The concentration of *trans*-3'-hydroxycotinine is approximately 2.5 times less than cotinine in both serum and saliva. Demethylcotinine is present in both serum and saliva, but at levels that are probably inconsequential so far as nicotine disposition in the body is concerned. This study demonstrates that it is possible to obtain information on the disposition of nicotine and its metabolites in common matrices to determine the active or passive exposure to nicotine and tobacco products.

*Quo vadis?*
The multitude of analytical methods for the determination of nicotine and cotinine in biological fluids presents a dilemma for analysts. The dilemma is an admixture of analyte(s) of interest, sensitivity, specificity, and costs. The choice of analytical method may be selected quickly by answering this question: What am I going to do with the data? For instance, to dichotomize smokers and non-smokers, the determination of cotinine in serum, saliva, and/or urine will accomplish that very easily at minimal costs. Both gas chromatographic and RIA methods readily provide the analytical results. More comprehensive analytical methods are required for protracted pharmacokinetic studies and investigations of nicotine intake in smokers where nicotine and its major metabolites in urine are the focus of attention. Of the two methods (GC/MS and LC/MS) available for the comprehensive determination of nicotine and its metabolites, LC/MS/MS appears to be the preferred method of analysis. Virtually all of the major metabolites of nicotine may be resolved from a single sample injection *via* this method including the direct determination of phase II metabolites. Specificity and sensitivity are

effectively addressed using this analytical technique. With respect to cotinine and *trans*-3'-hydroxycotinine in saliva [195] and cotinine in serum [194], sufficient sensitivity was achieved to allow determinations in non-smokers. High sample through-put can be achieved [193] at reduced sensitivity. Additional efforts by investigators will likely provide increased-sensitivity for other nicotine metabolites as necessary. The ability to characterize urinary excretion of nicotine and its metabolites in non-smokers exposed to environmental tobacco smoke with high sensitivity and specificity would be a tremendous step forward in better estimating the exposure dose of nicotine received by these persons. System automation, composed of sample pretreatment, sample analysis, data acquisition and reduction, and report generation, would be less onerous with LC/MS/MS.

## SUMMARY

Information in this chapter has reviewed the analytical methods used to study the interaction of humans with various tobacco products and nicotine containing medications. Both the pharmacodynamics and pharmacokinetics of nicotine absorbed from tobacco products and nicotine therapeutic products have been elucidated. Due to its prolonged half-life cotinine may be used as a biomarker compound, especially related to environmental tobacco smoke exposure, although there are deficiencies with its use for this purpose. Analytical methods have been discussed as integral parts of these investigations. Factors concerning the choice of matrix, sample collection and storage, and analytical method selection were discussed from the perspective of the application of the acquired data. The advantages and disadvantages of numerous analytical methods, as well as method applications have been discussed. While in some cases it may be appropriate to restrict analysis to one or more analytes, it should be clear that estimates of nicotine intake are best determined by comprehensive methods of analysis that provide information on nicotine and its major metabolites simultaneously. Costly and sophisticated instruments provide the means to perform these measurements in serum, saliva, and urine with high sample efficiency. Increasing method sensitivity for the analytes of interest for studies of low level nicotine or tobacco exposure remains a challenge, although progress toward this goal has been made for cotinine in serum and for cotinine and *trans*-3'-hydroxycotinine in saliva.

## REFERENCES

1 Armitage A, Turner DM. Absorption of nicotine in cigarette and cigar smoke through the oral mucosa. *Nature*, 1970: **226**; 1231–2.
2 Gori GB, Benowitz NL, Lynch CJ. Mouth vs. deep airways absorption of nicotine in cigarette smokers. *Pharmacology*, 1986: **26**; 1181–4.

629

3  Armitage A, Dollery C, Houseman T, Kohner E, Lewis PJ, Turner D. Absorption of nicotine from small cigars. *Clin. Pharmacol. Ther.*, 1978: **23**; 143–51.

4  Russell MAH, Raw M, Jarvis MJ. Clinical use of nicotine chewing-gum. *Br. Med. J.*, 1980: **280**; 1599–602.

5  Holm H, Jarvis MJ, Russel MAH, Feyerabend C. Nicotine intake and dependence in Swedish snuff takers. *Psychopharmacology*, 1992: **108**; 507–11.

6  Benowitz NL, Porchet H, Jacob P III. Pharmacokinetics, metabolism and pharmacodynamics of nicotine. In: Wonnacott S, Russell MAH, Stolerman IP, editors. *Nicotine Psychopharmacology Molecular, Cellular and Behavioural Aspects*. Oxford, New York, Tokyo: Oxford University Press, 1990: pp. 112–57.

7  Russell M, Jarvis M, Devitt G, Feyerabend C. Nicotine intake by snuff users. *Br. Med. J.*, 1981: **2**; 814–7.

8  Benowitz N, Porchet H, Sheiner L, Jacob P III. Nicotine absorption and cardiovascular effects with smokeless tobacco use: Comparison with cigarettes and nicotine gum. *Clin. Pharmacol. Ther.*, 1988: **44**; 23–8.

9  Benowitz N, Jacob P III, Yu L. Daily use of smokeless tobacco: systemic effects. *Ann. Intern. Med.*, 1989: **111**; 112–6.

10  Jenner P, Gorrod JW, Beckett AH. The absorption of nicotine–1'-N-oxide and its reduction in the gastrointestinal tract in man. *Xenobiotica*, 1973: **3**; 341–9.

11  Benowitz NL, Jacob P, Savanapridi C. Determinants of nicotine intake while chewing nicotine polacrilex gum. *Clin. Pharmacol. Ther.*, 1987: **41**; 467–73.

12  Benowitz NL, Chan K, Denaro CP, Jacob P. Stable isotope method for studying transdermal drug absorption: the nicotine patch. *Clin. Pharmacol. Ther.*, 1991: **50**; 286–93.

13  Johansson C-J, Olsson P, Bende M, Carlsson T, Gunnarsson PO. Absolute bioavailability of nicotine applied to different nasal regions. *Eur. J. Clin. Pharmacol.*, 1991: **41**; 585–8.

14  Benowitz NL, Jacob P, Jones RT, Rosenberg J. Interindividual variability in the metabolism and cardiovascular effects of nicotine in man. *J. Pharmacol. Exp. Ther.*, 1982: **221**; 368–372.

15  Svensson CK. Clinical pharmacokinetics of nicotine. *Clin. Pharmacokin.*, 1987: **12**; 30–40.

16  Rosenberg J, Benowitz NL, Jacob P, Wilson KM. Disposition kinetics and effects of intravenous nicotine. *Clin. Pharmacol. Ther.*, 1980: **28**; 517–22.:

17  Kyerematen GA, Vesell ES. Metabolism of nicotine. *Drug Metab. Rev.*, 1991: **23**; 3–41.

18  Curvall M, och Elwin C-E, Kazemi-Vala E, Enzell CR. The pharmacokinetics of cotinine in plasma and saliva from non-smoking healthy volunteers. *Eur. J. Clin. Pharmacol.*, 1990; **38**; 281–7.

19  Neurath GB, Dunger M, Orth D, Pein FG. *Trans*-3'-hydroxycotinine as a main metabolite in urine of smokers. *Int. Arch Occup. Environ. Health*, 1987: **59**; 199–201.

20  Neurath GB, Pein FG. Gas chromatographic determination of *trans*-3'-hydroxy-cotinine, a major metabolite of nicotine in smokers. *J. Chromatogr. B Biomed. Appl.*, 1987: **415**; 400–6.

21  Voncken P, Schepers G, Schäfer K-H. Capillary gas chromatographic determination of *trans*-3'-hydroxycotinine simultaneously with nicotine and cotinine in urine and blood samples. *J. Chromatogr.*, 1989: **479**; 410–8.

22 Scherer G, Jarczyk L, Heller W-D, Biber A, Neurath GB, Adlkofer F. Pharmacokinetics of nicotine, cotinine and 3′-hydroxycotinine in cigarette smokers. *Klin-Wochenschr.*, 1988: **66**; 5–11.

23 Gorrod JW, Jenner P. *Essays in Toxicology.* NY, San Francisco, London: Academic Press, 1975.

24 Turner DM, Armitage AK, Briant RH, Dollery CT. Metabolism of nicotine by the isolated perfused dog lung. *Xenobiotica*, 1975: **5**; 539–51.

25 Peterson LA, Trevor A, Castagnoli N. Stereochemical studies on the cytochrome P–450 catalyzed oxidation of (S)-nicotine to the (S)-nicotine $\Delta^{1'(5')}$-iminium species. *J. Med. Chem.*, 1987: **30**; 249–54.

26 Brandänge S, Lindblom L. Synthesis, structure and stability of nicotine delta 1′(5′) iminium ion, an intermediary metabolite of nicotine. *Acta Chemica Scandinavica*, 1979: **B33**; 187–91.

27 Gorrod JW, Hibberd AR. The metabolism and excretion of nicotine $\Delta^{1'(5')}$ iminium ion, in vivo and in vitro. *Eur. J. Drug Metabol. Pharmacol.*, 1982: **7**; 243–8.

28 Jacob P, Benowitz NL, Shulgin AT. Recent studies of nicotine metabolism in humans. *Pharmacology Biochemistry and Behavior*, 1988: **30**; 249–53.

29 Cundy KC, Crooks PA. Biotransformation of primary nicotine metabolites II. Metabolism of [³H]-S-(−)-cotinine in the guinea pig: determination of in vivo urinary metabolites by high-performance liquid-radiochromatography. *Xenobiotica*, 1987: **17**; 785–92.

30 Booth J, Boyland E. Enzymatic oxidation of (−)-nicotine by guinea-pig tissues in vitro. *Biochem. Pharmacol.*, 1971: **20**; 407–15.

31 Beckett AH, Gorrod JW, Jenner P. Absorption of (−)-nicotine–1′-N-oxide in man and its reduction in the gastrointestinal tract. *J. Pharm. Pharmacol.*, 1970: **22**; 722–3.

32 Jacob P, Yu L, Wilson M, Benowitz NL. Selected ion monitoring method for determination of nicotine, cotinine and deuterium-labelled analogs: absence of an isotope effect in the clearance of (S)-nicotine–3′–3′-d$_2$ in humans. *Biol. Mass Spectrom.*, 1991: **20**; 247–52.

33 Bowman ER, Turnbull LB, McKennis H. Metabolism of nicotine in the human and excretion of pyridine compounds by smokers. *J. Pharmacol. Exp. Ther.*, 1959: **127**; 92–5.

34 Bowman ER, McKennis H. Studies on the metabolism of (−)-cotinine in the human. *J. Pharmacol. Exp. Ther.*, 1962: **135**; 306–11.

35 Nguyen T, Gruenke LD, Castagnoli H. Metabolic oxidation of nicotine to chemically reactive intermediates. *J. Med. Chem.*, 1979: **22**; 259–63.

36 Shigenaga MK, Trevor AJ, Castagnoli N. Metabolism-dependent covalent binding of (S)-[5–³H]nicotine to liver and lung microsomal macromolecules. *Drug Metabolism and Disposition*, 1988: **16**; 397–402.

37 Curvall M, Kazemi Vala E, Englund CR. Conjugation pathways in nicotine metabolism. In: Adlkofer F, Thurau K, editors. *Advances in Pharmacological Sciences, Effects of Nicotine on Biological Systems.* Basel, Boston, Berlin: Birkhäuser Verlag, 1991: pp. 69–75.

38 Andersson G, Björnberg G, Curvall M. Oral mucosal changes and nicotine disposition in users of Swedish smokeless tobacco products: a comparative study. *J. Oral Pathol. Med.*, 1994: **23**; 161–7.

631

39 Byrd GD, Robinson JH, Caldwell WS, Debethizy JD. Comparison of measured and FTC-predicted nicotine uptake in smokers. *Psychopharmacology (Berl.)*, 1995: **122**; 95–103.

40 Andersson G, Kazemi Vala E, Curvall M. The influence of cigarette consumption and machine yields of tar and nicotine on the nicotine uptake and oral mucosal lesions in smokers. *J. Oral Pathol. Med.*, 1997: **26**; 117–23.

41 Benowitz NL, Jacob P, Fong I, Gupta S. Nicotine metabolic profile in man: Comparison of cigarette smoking and transdermal nicotine. *J. Pharmacol. Exp. Ther.*, 1994: **268**; 296–303.

42 Caldwell WS, Greene JM, Byrd GD, Chang KM, Uhrig MS, deBethizy JD. Characterization of the glucuronide conjugate of cotinine: a previously unidentified major metabolite of nicotine in smokers' urine. *Chem. Res. Toxicol.*, 1992: **5**; 280–5.

43 Isaac PF, Rand MJ. Blood levels of nicotine and physiological effects after inhalation of tobacco smoke. *Eur. J. Pharmacol.*, 1969: **8**; 269–83.

44 Armitage AK, Dollery CT, George CF, Houseman TH, Lewis PJ, Turner DM. Absorption and metabolism of nicotine from cigarettes. *Br. Med. J.*, 1975: **4**; 313–6.

45 Herning RI, Jones RT, Benowitz NL, Mines AH. How a cigarette is smoked determines blood nicotine levels. *Clin. Pharmacol. Ther.*, 1983: **33**; 84–90.

46 Russell MA, Jarvis M, Iyer R, Feyerabend C. Relation of nicotine yield of cigarettes to blood nicotine concentrations in smokers. *Br. Med. J.*, 1980: **280**; 972–6.

47 Benowitz NL, Kuyt F, Jacob P. Circadian blood nicotine concentrations during cigarette smoking. *Clin. Pharmacol. Ther.*, 1982: **32**; 758–64.

48 Benowitz NL, Jacob P. Daily intake of nicotine during cigarette smoking. *Clin. Pharmacol. Ther.*, 1984: **35**; 499–504.

49 Benowitz NL, Hall SM, Herning RI. Smokers of low-yield cigarettes do not consume less nicotine. *N. Engl. J. Med.*, 1983: **309**; 139–42.

50 Haley NJ, Hoffmann D. Analysis for nicotine and cotinine in hair to determine cigarette smoker status. *Clin. Chem.*, 1985: **31**; 1598–600.

51 Curvall M, Kazemi Vala E. Nicotine and metabolites: analysis and levels in body fluids. In: Gorrod JW, Wahren J, editors. *Nicotine and related alkaloids. Absorption, distribution, metabolism, excretion.* London: Chapman&Hall, 1993: pp. 147–79.

52 Galeazzi RL, Daenens P, Gugger M. Steady-state concentration of cotinine as a measure of nicotine-intake by smokers. *Eur. J. Clin. Pharmacol.*, 1985: **28**; 301–4.

53 Curvall M, Kazemi Vala E, Enzell CR Wahren J. Simulation and evaluation of nicotine intake during passive smoking: Cotinine measurements in body fluids of non-smokers given intravenous infusions of nicotine. *Clin. Pharmacol. Ther.*, 1990: **47**; 42–9.

54 Jarvis MJ, Tunstall-Pedoe H, Feyerabend C, Vesey C, Saloojee Y. Comparison of tests used to distinguish smokers from non-smokers. *Am. J. Public Health*, 1987: **77**; 1435–8.

55 Pojer R, Whitfield JB, Poulos V, Eckhard IF, Richmond R, Hensley WJ. Carboxyhemoglobin, cotinine and thiocyanate assay compared for distinguishing smokers from non-smokers. *Clin. Chem.*, 1984: **30**; 1377–80.

56 Ogden MW, Morgan WT, Heavner DL, Davis RA, Steichen TJ. National incidence of smoking and misclassification among the U.S. married female population. *J. Clin. Epidemiol.*, 1997: **50**; 253–63.

57 Haley NJ, Axelrad CM, Tilton KA. Validation of self-reported smoking behavior: Biochemical analyses of cotinine and thiocyanate. *Public Health*, 1983: **73**; 1204–7.

58 Giusto ED, Eckhard I. Some properties of saliva cotinine measurements in indicating exposure to tobacco smoking. *Am. J. Public Health*, 1986: **76**; 1245–6.

59 Van Vunakis H, Tashkin DP, Rigas B, Simmons M, Gjika HB, Clark VA. Relative sensitivity and specificity of salivary and serum cotinine in identifying tobacco-smoking status of self-reported non-smokers and smokers of tobacco and/or marijuana. *Arch of Environmental Health*, 1989: **44**; 53–8.

60 Benowitz NL. Importance of nicotine metabolism in understanding the human biology of nicotine. In: Adlkofer F, Thurau K, editors. *Advances in Pharmacological Sciences, Effects of Nicotine on Biological Systems.* Basel, Boston, Berlin: Birkhäuser Verlag, 1991: pp. 19–24.

61 Jenkins RA and Counts RW. Personal exposure to environmental tobacco smoke: salivary cotinine, airborne nicotine, and non-smoker misclassification. *J. Expo. Anal. Environ. Epidemiol.*, 1999: **9**; 352–363.

62 Seaton MJ, Vesell ES. Variables affecting nicotine metabolism. *Pharmacol. Ther.*, 1993: **60**; 461–500.

63 Pichini S, Altieri I, Pellegrini M, Pacifici R, Zuccaro P. Hair analysis for nicotine and cotinine: evaluation of extraction procedures, hair treatments, and development of reference material. *Forensic Sci. Int.*, 1997: **84**; 243–52.

64 Luck W, Nau H. Nicotine and cotinine concentrations in serum and milk of smoking mothers: influence of cigarette consumption and diurnal variation. *Eur. J. Pediatr.*, 1987: **146**; 21–6.

65 Luck W, Nau H. Nicotine and cotinine concentrations in serum and milk of nursing smokers. *Br. J. Clin. Pharmacol.*, 1984: **18**; 9–15.

66 Dahlström A, Lundell B, Curvall M, Thapper L. Nicotine and cotinine concentrations in the nursing mother and her infant. *Acta Paediatr. Scand.*, 1990: **79**; 142–7.

67 Luck W, Nau H, Hansen R, Steldinger R. Extent of nicotine and cotinine transfer to the human fetus, placenta and amniotic fluid of smoking mothers. *Dev. Pharmacol. Ther.*, 1985: **8**; 384–95.

68 Etzel RA, Greenberg RA, Haley NJ, Loda FA. Urine cotinine excretion in neonates exposed to tobacco smoke products in utero. *J. Pediatr.*, 1985: **107**; 146–8.

69 Petrakis NL, Gruenke LD, Beelen TC, Castagnoli N, Craig JC. Nicotine in breast fluid of nonlactating women. *Science*, 1978: **199**; 303–5.

70 Hill P, Wynder EL. Nicotine and cotinine in breast fluid. *Cancer Lett.*, 1979: **6**; 251–4.

71 Gritz E, Baer-Weiss V, Benowitz N, Van Vunakis H, Jarvik M. Plasma nicotine and cotinine concentrations in habitual smokeless tobacco users. *Clin. Pharmacol. Ther.*, 1981: **30**; 201–9.

72 Wennmalm Å, Benthin G, Granström EF, Persson L, Petersson A-S, Winell S. Relation between tobacco use and urinary excretion of thromboxane A2 and prostacyclin metabolites in young men. *Circulation*, 1991: **83**; 1–7.

73 Larsson I, Curvall M, Enzell CR. Disposition of nicotine and cotinine in plasma, saliva and urine of snuff-users. Proceedings from Third European Congress of Biopharmaceutics and Pharmacokinetics 1987: pp. 318–24.

74 Andersson G, Axéll T, Curvall M. Reduction in nicotine intake and oral mucosal changes among users of Swedish oral moist snuff after switching to a low-nicotine product. *J. Oral Pathol. Med.*, 1995: **24**; 244–50.

75 Domino EF, Hornbach E, Demana T. The nicotine content of common vegetables. *N. Engl. J. Med.*, 1993: **329**; 437.

76 Davis RA, Stiles MF, DeBethizy JD, Reynolds JH. Dietary nicotine: A source of urinary cotinine. *Food Chem. Toxicol.*, 1991: **29**; 821–7.

77 Castro A, Monji N. Dietary nicotine and its significance in studies on tobacco smoking. *Biochem. Arch.*, 1986: **2**; 91–7.

78 Repace JL. Dietary nicotine. *Br. Med. J.*, 1994: **308**; 61–2.

79 Benowitz NL. Cotinine as a biomarker of environmental tobacco smoke exposure. *Epidemiol. Rev.*, 1996: **18**; 188–204.

80 Russell MAH, Jarvis MJ, Feyerabend C. Nasal nicotine solution: a potential aid to giving up smoking? *Br. Med. J.*, 1983: **286**; 683–4.

81 West RJ, Jarvis MJ, Phil M, Russell MAH, Feyerabend C. Plasma nicotine concentrations from repeated doses of nasal nicotine solution. *Br. J. Addict.*, 1984: **79**; 443–5.

82 Sutherland G, Russell MAH, Stapleton J, Feyerabend C, Ferno O. Nasal nicotine spray: a rapid nicotine delivery system. *Psychopharmacology*, 1992: **108**; 512–8.

83 Hjalmarson A, Nilsson F, Sjöström L, Wiklund O. The nicotine inhaler in smoking cessation. *Arch Intern. Med.*, 1997: **157**; 1721–8.

84 McNabb ME, Ebert RV, McCusker K. Plasma nicotine levels produced by chewing nicotine gum. *JAMA*, 1982: **248**; 865–8.

85 Russell MAH, Feyerabend C, Cole PV. Plasma nicotine levels after cigarette smoking and chewing nicotine gum. *Br. Med. J.*, 1976: **1**; 1043–6.

86 Ebert RV, McNabb ME, Snow SL. Effect of nicotine chewing gum on plasma nicotine levels of cigarette smokers. *Clin. Pharmacol. Ther.*, 1984: **35**; 495–8.

87 Bannon YB, Corish J, Corrigan OI, Devane JG, Kavanagh M, Mulligan S. Transdermal delivery of nicotine in normal human volunteers: a single dose and multiple dose study. *Eur. J. Clin. Pharmacol.*, 1989: **37**; 285–90.

88 Gorsline J, Benowitz NL, Rolf CN, Gupota SK, Garaud JJ. Comparison of plasma nicotine concentrations for nicotine transdermal system (NTS), cigarette smoking and nicotine polacrilex (nicotine gum). *Clin. Pharmacol. Ther.*, 1992: **51**; 129.

89 Palmer KJ, Buckley MM, Faulds D. Transdermal nicotine. A review of its pharmacodynamic and pharmacokinetic properties and therapeutic efficacy as an aid to smoking cessation. *Drugs*, 1992: **44**; 498–529.

90 Dubois JP, Sioufi A, Muller P, Mauli D, Imhof PR. Pharmacokinetics and Bioavailability of nicotine in healthy volunteers following single and repeated administration of different doses of transdermal nicotine systems. *Meth. Find Exp. Clin. Pharmacol.*, 1989: **11**; 187–95.

91 Ross HD, Chan KKH, Piraino AJ, John VA. Pharmacokinetics of multiple daily transdermal doses of nicotine in healthy smokers. *Pharm. Res.*, 1991: **8**; 385–8.

92 Jarvis M, Tunstall-Pedoe H, Feyerabend C, Vesey C, Salloojee Y. Biochemical markers of smoke absorption and self reported exposure to passive smoking. *J. Epidemiol. Community Health*, 1984: **38**; 335–9.

93 Russell MAH, Jarvis MJ, West RJ. Use of urinary nicotine concentrations to estimate exposure and mortality from passive smoking in non-smokers. *Br. J. Addict.*, 1986: **81**; 275–81.

94 Russell MAH. Estimation of smoke dosage and mortality of non-smokers from environmental tobacco smoke. *Toxicol. Lett.*, 1987: **35**; 9–18.

95 Wald NJ, Boreham J, Bailey A, Ritchie C, Haddow JE, Knight G. Urinary cotinine as marker of breathing other people's tobacco smoke. *Lancet*, 1984: **1**; 230–1.

96 Jarvis MJ, Russell MAH, Feyerabend C, Eiser JR, Morgan M, Gammage P et al. Passive exposure to tobacco smoke: saliva cotinine concentrations in a representative population sample of non-smoking schoolchildren. *Br. Med. J.*, 1985: **291**; 927–9.

97 Cummings KM, Markello SJ, Mahoney M, Bhargava AK, McElroy PD, Marshall JR. Measurements of current exposure to environmental tobacco smoke. *Arch. Environ. Health*, 1990: **45**; 74–9.

98 Jenkins RA, Palausky MA, Counts RW, Guerin MR, Dindal AB, Bayne CK. Determination of personal exposure of non-smokers to environmental tobacco smoke in the United States. *Lung Cancer*, 1996: **14** (Suppl. 1): S195–213.

99 Greenberg RA, Haley NJ, Etzel RA, Loda FA. Measuring the exposure of infants to tobacco smoke. *N. Engl. J. Med.*, 1984: **310**; 1075–8.

100 Tunstall-Pedoe H, Brown CA, Woodward M, Tavendale R. Passive smoking by self report and serum cotinine and the prevalence of respiratory and coronary heart disease in the Scottish heart health study. *J. Epidemiol. Community Health*, 1995: **49**; 139–43.

101 Cook DG, Whincup PH, Jarvis MJ, Strachan DP, Papacosta O, Bryant A. Passive exposure to tobacco smoke in children aged 5–7 years: Individual, family and community factors. *Br. Med. J.*, 1994: **308**; 384–9.

102 Pirkle JL, Flegal KM, Bernert JT, Brody DJ, Etzel RA, Maurer KR. Exposure of the US population to environmental tobacco smoke. *JAMA*, 1996: **275**; 1233–40.

103 Riboli E, Preston-Martin S, Saracci R, Haley NJ, Trichopoulos D, Becher H et al. Exposure of non-smoking women to environmental tobacco smoke: a 10-country collaborative study. *Cancer Causes Control*, 1990: **1**; 243–52.

104 Kintz, P. Gas chromatographic analysis of nicotine and cotinine. *J. Chromatogr. B Biomed. Sci. Appl.*, 1992; **580**; 347–53.

105 Kintz, P. Drug testing in addicts: a comparison between urine, sweat, and hair. *Ther. Drug Monit.*, 1996; **18**; 450–55.

106 Dimich-Ward H, Gee H, Brauer M, Leung, V. Analysis of nicotine and cotinine in the hair of hospitality workers exposed to environmental tobacco smoke. *J. Occup. Environ. Med.*, 1997: **39**; 946–48.

107 Haley NJ, Hoffmann, D. Analysis for nicotine and cotinine in hair to determine cigarette smoker status. *Clin. Chem.*, 1985: **31**; 1598–1600.

108 Zahlsen K, Nilsen T, Nilsen, OG. Interindividual differences in hair uptake of air nicotine and significance of cigarette counting for estimation of environmental tobacco smoke exposure. *Pharmacol. Toxicol.*, 1996: **79**; 183–90.

109 Mizuno A, Uematsu T, Oshima A, Nakamura M, Nakashima, M. Analysis of nicotine content of hair for assessing individual smoking behavior. *Ther. Drug Monit.*, 1993: **15**; 99–104.

110 Mizuno A, Uematsu T, Ishikawa T, Yoshimine N, Nakashima, M. Clinical outcome of smoking cessation trial of nicotine chewing gum evaluated by analysis of nicotine in hair. *Ther. Drug Monit.*, 1997: **19**; 407–12.

111 Baranowski J, Pochopien G, Baranoska, I. Determination of nicotine, cotinine, and caffeine in meconium using high performance liquid chromatography. *J. Chromatogr. B, Biomed. Sci. Appl.*, 1998: **707**; 317–21.

112 Moore C, Negrusz A, Lewis, D. Determination of drugs of abuse in meconium. *J. Chromatogr. B, Biomed. Sci. Appl.*, 1998: **713**; 137–46.

113 Vine MF, Hulka BS, Margolin BH, Truong YK, Hu P-C, Schramm MM, Griffith JD, McCann M, Everson, RB. Cotinine concentrations in semen, urine, and blood of smokers and non-smokers. *Am. J. Public Health*, 1993: **83**; 1335–38.

114 Pacifici R, Altieri I, Gandini L, Lenzi A, Pichini S, Rosa M, Zucczro P, Dondero, F. Nicotine, cotinine and *trans*-3′-hydroxycotinine levels in seminal plasma of smokers: effects on sperm parameters. *Ther. Drug Monit.*, 1993: **15**; 358–63.

115 Schiffman MH, Haley NJ, Felton JS, Andrews AW, Kaslow RA, Lancaster WD, Kurman RJ, Brinton LA, Lannom LB, Hoffman D. Biochemical epidemiology of cervical neoplasia: measuring cigarette smoke constituents in the cervix. *Cancer Res.*, 1987: **47**; 3886–88.

116 Hellberg D, Nilsson S, Haley NJ, Hoffman D, Wynder E. Smoking and cervical intraepithelial neoplasia: nicotine and cotinine in serum and cervical mucus in smokers and non-smokers. *Am. J. Obstet. Gynecol.*, 1988: **158**; 910–13.

117 Poppe WA, Peeters R, Daenans P, Ide PS, Van Assche FA. Tobacco smoking and the uterine cervix: cotinine in blood, urine, and cervical fluid. *Gynecol. Obstet. Invest.*, 1995: **39**; 110–14.

118 Hibberd AR, O'Connor V, Gorrod JW. Detection of nicotine, nicotine–1′-*N*-oxide, and cotinine in maternal and foetal body fluids. In: Gorrod JW, editor. *Biological oxidation of nitrogen*. Amsterdam, New York, Oxford: Elsevier/North Holland Biomedical Press, 1978: pp. 353–61.

119 Parsche F, Balabanova S, Pirsig W. Drugs in ancient populations [letter]. *Lancet*, 1993: **341**; 503.

120 Balabanova S, Rosing FW, Buhler G, Schoetz W, Scherer G, Rosenthal J. Nicotine and cotinine in prehistoric and recent bones from Africa and Europe and the origin of these alkaloids. *Homo*, 1997: **48**; 72–77.

121 Feyerabend C, Ings RMJ, Russell MAH. Nicotine pharmacokinetics and its application to intake from smoking. *Br. J. Clin. Pharmacol.*, 1985: **19**; 237–47.

122 DeBethizy JD, Robinson JH, Davis RA, Doolittle DJ, Burger GT, Reynolds JH, Fletcher RG, Hayes AW. Absorption of nicotine from a cigarette that does not burn tobacco. *Pharmacology*, 1988: **37**; 328–32.

123 Benowitz NL, Jacob III P. Nicotine and cotinine elimination pharmacokinetics in smokers and non-smokers. *Clin. Pharmacol. Ther.*, 1993: **53**; 316–23.

124 Robinson JH, Pritchard WS, Davis RA. Psychopharmacological effects of smoking a cigarette with typical 'tar' and carbon monoxide yields but minimal nicotine. *Psychopharmacology*, 1992: **108**; 466–72.

125 Foulds J, Feyerabend C, Stapleton J, Jarvis MJ, Russell MAH. Stability of nicotine and cotinine in unfrozen plasma. *J. Smoking-Related Dis.*, 1994: **5**; 41–44.

126 Schievelbein, H. Nicotine, resorption and fate. In: Balfour DJK, editor. *Nicotine and the tobacco smoking habit*. Oxford: Pergamon, 1984: pp. 1–12.

127 Benowitz NL, Jacob P, Jones RT, Rosenberg J. Interindividual variability in the metabolism and cardiovascular effects of nicotine in man. *J. Pharmacol. Exp. Ther.*, 1982: **221**; 368–72.

128 Feyerabend C. Determination of nicotine in physiological fluids by gas chromatography. In: O'Neill IK, Brunnemann KD, Dodet B, Hoffmann D, editors. *Environmental carcinogens methods of analysis and exposure measurement, (Volume 9, Passive Smoking)*. Lyon: International Agency for Research on Cancer, 1987: pp. 299–307.

129 Curvall M, Kazemi-Vala E, Enzell CR. Simultaneous determination of nicotine and cotinine in plasma using capillary column gas chromatography with nitrogen-sensitive detection. *J. Chromatogr. B Biomed. Sci. Appl.*, 1982: **232**; 283–93.

130 Jacob P, Wilson M, Benowitz NL. Improved gas chromatogrphic method for the determination of nicotine and cotinine in biologic fluids. *J. Chromatogr. B Biomed. Sci. Appl.*, 1981: **222**; 61–70.

131 Davis RA. The determination of nicotine and cotinine in plasma. *J. Chromatogr. Sci.*, 1986: **24**; 134–41.

132 Stehlik G, Kainzbauer J, Tausch H, Richter O. Improved method for routine determination of nicotine and its metabolites in biological fluids. *J. Chromatogr. B Biomed. Appl.*, 1982: **232**; 295–303.

133 Teeuwen HWA, Aalders RJW, Van Rossum JM. Simultaneous estimation of nicotine and cotinine in biological fluids using high-resolution capillary-column gas chromatography combined with solid phase extraction work-up. *Mol. Biol. Rep.*, 1989: **13**; 165–75.

134 Jusko JJ, Milsap RL. Pharmacokinetic principles of drug distribution in saliva. In: Malamud D, Tabak L, editors. *Saliva as a diagnostic fluid*. New York: Ann N Y Acad Sci 1993: **694**; pp. 36–47.

135 Jarczyk L, Maier H, Born IA, Scherer G, Adlkofer F. Concentration of nicotine in serum and saliva after intravenous infusion of the alkaloid and after smoking. In: Adlkofer F, Thurau K, editors. *Effects of nicotine on biological systems*. Basel: Birkhauser Verlag, 1991: pp. 51–55.

136 Rose JD, Levin ED, Benowitz NL. Saliva nicotine as an index of plasma levels in nicotine skin patch users. *Ther. Drug Monit.*, 1993: **15**; 431–35.

137 North LM, Gaudette ND, Cordiero ML, Fitchen JH, Davidson SL, Hindahl MS. Detection of cotinine in oral fluid recovered with the Orasure® collection system. In: Malamud D, Tabak L, editors. *Saliva as a diagnostic fluid*. New York: Ann N Y Acad Sci 1993: **694**; pp. 36–47.

138 Caddy B. Saliva as a specimen for drug analysis. In: Basalt RC, editor. *Advances in analytical toxicology (Volume 1)*. Foster City, CA: Biomedical Publications, 1984: 198–254.

139 Schramm W, Pomerleau OP, Pomerleau CS, Grates HB. Cotinine in an ultrafiltrate of saliva. *Prev. Med.*, 1992: **21**; 63–73.

140 Schneider NG, Jacob P, Nilsson F, Leischow SJ, Benowitz NL, Olmstead RE. Saliva cotinine levels as a function of collection method. *Addiction*, 1997: **92**; 347–51.

141 Haeckel R. Procedures for saliva sampling [Abstract]. *Eur. J. Clin. Chem. Clin. Biochem.*, 1989: **4**; 246–47.

142 Haley NJ, O'Neill IK. Collection of urine for prospective studies in passive smoking. In: O'Neill IK, Brunnemann KD, Dodet B, Hoffmann D, editors.

  *Environmental carcinogens methods of analysis and exposure measurement, (Volume 9, Passive Smoking).* Lyon: International Agency for Research on Cancer, 1987: pp. 293–97.

143 Hagan RL, Ramos JM, Jacob PM III. Increasing urinary cotinine concentrations at elevated temperature: the role of conjugated metabolites. *J. Pharm. Biomed. Anal.*, 1997: **16**; 191–97.

144 Van Vunakis H, Gjika HB, Langone JJ. Method 16 - Radioimmunoassays for nicotine and cotinine. In: O'Neill IK, Brunnemann KD, Dodet B, Hoffman D, editors. *Environmental carcinogens - methods of analysis and exposure measurement, Volume 9, Passive Smoking.* Lyon: International Agency for Research on Cancer, 1987: pp. 317–30.

145 Feyerabend C, Bryant AE. Method 15 – Determination of cotinine in physiological fluids by gas chromatography. In: O'Neill IK, Brunnemann KD, Dodet B, Hoffman D, editors. *Environmental carcinogens - methods of analysis and exposure measurement, Volume 9, Passive Smoking.* Lyon: International Agency for Research on Cancer, 1987: pp. 309–16.

146 Watts RR, Langone JJ, Knight GJ, Lewtas J. Cotinine analytical workshop report: consideration of analytical methods for determining cotinine in human body fluids as a measure of passive exposure to tobacco smoke. *Environ. Health Persp.*, 1990: **84**; 173–82.

147 Sander SC, Byrd GD. A cotinine in freeze-dried urine reference material. *J. Res. NIST*, 1989: **94**; 305-09.

148 Langone JJ, Cook G, Bjercke RJ, Lifschitz, MH. Monoclonal antibody ELISA for cotinine in saliva and urine of active and passive smokers. *J. Immunol. Methods*, 1988: **114**; 73–8.

149 Thompson SG, Barlow RD, Wald NJ, Van Vunakis H. How should urinary cotinine concentrations be adjusted for urinary creatinine concentrations? *Clin. Chim. Acta*, 1990: **187**; 289–96.

150 Boeniger MF, Lowry LK, Rosenberg J. Interpretation of urine results used to assess chemical exposure with emphasis on creatinine adjustments: a review. *Am. Ind. Hyg. Assoc. J.*, 1993: **54**; 615–27.

151 Alessio L, Berlin A, Dell'Orto A, Toffoletto F, Ghezzi I. Reliability of urinary creatinine as a parameter used to adjust values of urinary biological indicators. *Int. Arch. Occup. Environ. Health*, 1985: **55**; 99–106.

152 Peach H, Morris RW, Ellard GA, Jenner PJ. A simple, inexpensive urine test of smoking. In: Zaridze DG, Peto R, editors. *Tobacco: a major international health hazard.* Lyon: International Agency for Research on Cancer, 1986: pp. 183–93.

153 Barlow RD, Stone RB, Wald NJ, Puhakainen EVJ. The direct barbituric acid assay for nicotine metabolites in urine: a simple colorimetric test for the routine assessment of smoking status and cigarette smoke intake. *Clin. Chim. Acta*, 1987: **165**; 45–52.

154 Puhakainen EVJ, Barlow RD, Salonen JT. An automated colorimetric assay for urine nicotine metabolites: a suitable alternative to cotinine assays for the assessment of smoking status. *Clin. Chim. Acta*, 1987: **170**; 255–62.

155 Barlow RD, Thompson PA, Stone RB. Simultaneous determination of nicotine and five additional nicotine metabolites in urine of smokers using pre-column derivatization and high-performance liquid chromatography. *J. Chromatogr. B Biomed. Appl.*, 1987: **419**; 375–80.

156 Parviainen MT, Barlow RD. Assessment of exposures to environmental tobacco smoke using a high-performance liquid chromatography method for the simultaneous determination of nicotine and two of its metabolites in urine. *J. Chromatogr. B Biomed. Appl.*, 1988: **431**; 216–21.

157 Parviainen MT, Puhakainen EVJ, Laatikainen R, Savolainen K, Herranen J, Barlow RD. Nicotine metabolites in urine of smokers. *J. Chromatogr. B Biomed. Appl.*, 1990: **525**; 193–202.

158 Smith CL, Cooke M. Evaluation of diethylbarbituric acid as a spectrophotometric reagent for tobacco alkaloids. *Analyst*, 1987: **112**; 1515–18.

159 Rustemeier K, Demetriou D, Schepers G, Voncken P. High-performance liquid chromatographic determination of nicotine and its urinary metabolites via their 1,3-diethyl–2-thiobarbituric acid derivatives. *J. Chromatogr. B Biomed. Appl.*, 1993: **613**; 95–103.

160 Isaac PF, Rand MJ. Cigarette smoking and plasma levels of nicotine. *Nature*, 1972: **236**; 308–10.

161 Feyerabend C, Levitt T, Russell MAH. A rapid gas-liquid chromatographic estimation of nicotine in biological fluids. *J. Pharm. Pharmacol.*, 1975: **27**; 434–36.

162 Feyerabend C, Russell MAH. Improved gas-chromatographic method and micro-extraction technique for the measurement of nicotine in biological fluids. *J. Pharm. Pharmacol.*, 1979: **31**; 73–76.

163 Feyerabend C, Russell MAH. A rapid gas-liquid chromatographic method for the determination of cotinine and nicotine in biological fluids. *J. Pharm. Pharmacol.*, 1990: **42**; 450–52.

164 Feyerabend C, Russell MAH. Assay of nicotine in biological materials: sources of contamination and their elimination. *J. Pharm.Pharmacol.*, 1980: **32**; 178–81.

165 Beckett AH, Gorrod JW, Jenner P. The effect of smoking on nicotine metabolism in vivo in man. *J. Pharm. Pharmacol.*, 1971: **23[Supplement]**; 62S–67S.

166 Jacob III P, Benowitz NL, Yu L, Shulgin AT. Determination of nicotine-*N*-oxide by gas chromatography following thermal conversion to 2-methyl-6-(3-pyridyl)-tetrahydro-1,2-oxazine. *Anal. Chem.*, 1986: **58**; 2218–21.

167 Zhang Y, Jacob III P, Benowitz NL. Determination of nornicotine in smokers' urine by gas chromatography following reductive alkylation to *N'*-propylnornicotine. *J. Chromatogr. B Biomed. Appl.*, 1990: **525**; 349–57.

168 Dagne E, Castagnoli N. Structure of hydroxycotinine, a nicotine metabolite. *J. Med. Chem.*, 1972: **15**; 356–60.

169 Voncken P, Rustemeier K, Schepers G. Identification of *cis*-3'-hydroxycotinine as a urinary nicotine metabolite. *Xenobiotica*, 1990: **20**; 1353–56.

170 Jacob III P, Shulgin AT, Benowitz NL. Synthesis of (3'R,5'S)-*trans*-3'-hydroxycotinine, a major metabolite of nicotine – metabolic formation of 3'-hydroxycotinine in humans is highly stereoselective. *J. Med. Chem.*, 1990: **33**; 1888–91.

171 Jacob III P, Shulgin AT, Yu L, Benowitz NL. Determination of the nicotine metabolite *trans*-3'-hydroxycotinine in urine of smokers using gas chromatography with nitrogen-selective detection or selected ion monitoring. *J. Chromatogr. B Biomed. Appl.*, 1992: **583**; 145–54.

172  Benowitz NL. Pharmacokinetics and pharmacodynamics of nicotine. In: Rand MJ, Thurau K, editors. *The pharmacology of nicotine.* Oxford: IRL Press, 1988: pp. 3–18.

173  Watson ID. Rapid analysis of nicotine and cotinine in the urine of smokers by isocratic high-performance liquid chromatography. *J. Chromatogr. B Biomed. Appl.,* 1977: **143**; 203-06.

174  Maskarinec MP, Harvey RW, Caton JE. A novel method for the isolation and quantitative analysis of nicotine and cotinine in biological fluids. *J. Anal. Toxicol.,* 1978: **2**; 124–26.

175  Hariharan M, van Noord T, Greden JF. A high-performance liquid-chromatographic method for routine determination of nicotine and cotinine in plasma. *Clin. Chem.,* 1988: **34**; 724–29.

176  Hariharan M, Van Noord T. Liquid-chromatographic determination of nicotine and cotinine in urine from passive smokers: comparison with gas chromatography with a nitrogen-specific detector. *Clin. Chem.,* 1991: **37**; 1276–80.

177  Machacek DA, Jiang NS. Quantitation of cotinine in plasma and saliva by liquid chromatography. *Clin. Chem.,* 1986: **32**; 979–82.

178  Thuan NTL, Migueres ML, Roche D, Rousset G, Mahuzier G, Chretien J, Ekindjian OG. Elimination of caffeine interference in HPLC determination of urinary nicotine and cotinine. *Clin. Chem.,* 1989: **35**; 1456–59.

179  Kyerematen GA, Damiano MD, Dvorchik BH, Vesell ES. Smoking-induced changes in nicotine disposition: application of a new HPLC assay for nicotine and its metabolites. *Clin. Pharmacol. Ther.,* 1982: **32**; 769–80.

180  Shulgin AT, Jacob III P, Benowitz NL, Lau D. Identification and quantitative analysis of cotinine-*N*-oxide in human urine. *J. Chromatogr. B Biomed. Appl.,* 1987: **423**; 365–72.

181  Kyerematen GA, Taylor LH, deBethizy JD, Vesell ES. Radiometric high-performance liquid chromatographic assay for nicotine and twelve of its metabolites. *J. Chromatogr. B Biomed. Appl.,* 1987: **419**; 191–203.

182  Li Y, Castagnoli N, Gorrod JW. Is demethylcotinine $\Delta^{4',5'}$-enamine a metabolite of cotinine? Proceedings of the 1st International Meeting on Pharmacy and Pharmaceutical Sciences, Istanbul, Turkey. September 4–7, 1994.

183  Demetriou D, Rustemeier K, Voncken P, Schepers G. Two radiochromatographic high pressure liquid chromatography methods for the determination of nicotine metabolites. *Med. Sci. Res.,* 1992: **20**; 873–75.

184  Pilotti A, Enzell CR, McKennis H, Bowman ER, Dufva E, Holmstedt B. Studies on the identification of tobacco alkaloids, their mammalian metabolites and related compounds by gas chromatography – mass spectrometry. *Beiträge zur Tabakforschung,* 1976: **8**; 339–49.

185  Jones D, Curvall M, Abrahamsson L, Kazemi-Vala E, Enzell C. Quantitative analysis of plasma nicotine using selected ion monitoring at high resolution. *Biomedical Mass Spectrometry,* 1982: **9**; 539–45.

186  Gruenke, LD, Beelen TC, Craig CJ, Petrakis NL. The determination of nicotine in biological fluids at picogram levels by selected ion recording. *Anal. Biochem.,* 1979: **94**; 411–16.

187  Dow J, Hall K. Capillary column combined gas chromatography-mass spectrometry method for the estimation of nicotine in plasma by selected ion monitoring. *J. Chromatogr. A,* 1978: **153**; 521–25.

188 Skarping G, Willers S, Dalene M. Determination of cotinine in urine using glass capillary gas chromatography and selective detection, with special reference to the biological monitoring of passive smoking. *J. Chromatogr. A*, 1988: **454**; 293–301.

189 Cooper DA, Moore JM. Femtogram on-column detection of nicotine by isotope dilution gas chromatography/negative ion detection mass spectrometry. *Biological Mass Spectrometry*, 1993: **22**; 590–594.

190 Daenens P, Laruelle L, Callewaert K, deSchepper P, Galeazzi R, Van Rossum J. Determination of cotinine in biological fluids by capillary gas chromatography-mass spectrometry-selected ion monitoring. *J. Chromatogr. B Biomed. Appl.*, 1985: **342**; 79–87.

191 Jacob P, Yu L, Liang G, Shulgin AT, Benowitz NL. Gas chromatographic-mass spectrometric method for determination of anabasine, anatabine and other tobacco alkaloids in urine of smokers and smokeless tobacco users. *J. Chromatogr. B Biomed. Appl.*, 1993: **619**; 49–61.

192 Davoli E, Stramare L, Fanelli R, Diomede L, Salmona M. Rapid solid-phase extraction method for automated gas chromatographic-mass spectrometric determination of nicotine in plasma. *J. Chromatogr. B Biomed. Appl.*, 1998: **707**; 312–16.

193 Xu AS, Peng LL, Havel JA, Petersen ME, Fiene JA, Hulse JD. Determination of nicotine and cotinine in human plasma by liquid chromatography – tandem mass spectrometry with atmospheric-pressure chemical ionization interface. *J. Chromatogr. B Biomed. Appl.*, 1996: **682**; 249–57.

194 Bernert JT, Turner WE, Pirkle JL, Sosnoff CS, Akins JR, Waldrep MK, Ann Q, Covey TR, Whitfield WE, Gunter EW, Miller BB, Patterson DG, Needham LL, Hannon WH, Sampson EJ. Development and validation of sensitive method for determination of serum cotinine in smokers and non-smokers by liquid chromatography/atmospheric pressure ionization tandem mass spectrometry. *Clin. Chem.*, 1997: **43**; 2281–91.

195 Bentley MC, Abrar M, Kelk M, Cook J, Phillips K. Validation of an assay for the determination of cotinine and 3-hydroxycotinine in human saliva using automated solid-phase extraction and liquid chromatography with tandem mass spectrometric detection. *J. Chromatogr. B Biomed. Appl.*, 1999: **723**; 185–194.

196 McManus KT, deBethizy JD, Garteiz DA, Kyerematen GA, Vesell ES. A new quantitative thermospray LC-MS method for nicotine and its metabolites in biological fluids. *J. Chromatogr. Sci.*, 1990: **28**; 510–16.

197 Byrd GD, Chang KM, Greene JM, deBethizy JD. Evidence for urinary excretion of glucuronide conjugates of nicotine, cotinine, and *trans*–3′-hydroxycotinine in smokers. *Drug Metab. Dispos.*, 1992: **20**; 192–97.

198 Byrd, GD, Uhrig MS, deBethizy JD, Caldwell WS, Crooks PA, Ravard A, Riggs RM. Direct determination of cotinine-*N*-glucuronide in urine using thermospray liquid chromatography/mass spectrometry. *Biological Mass Spectrometry*, 1994: **23**; 103-07.

199 Langone JJ, Gjika HB, Van Vunakis. H. Nicotine and Its Metabolites: Radioimmunoassays for Nicotine and Cotinine. *Biochemistry*, 1973: **12**; 5025–30.

200 Haines CF, Mahajan DK, Miljkovic D, Miljkovic M, Vesell, ES. Radioimmunoassay of plasma nicotine in habituated and naive smokers. *Clin. Pharmacol. Ther.*, 1976: **16**; 1083–89.

201 Matsukura S, Sakamoto N, Seino Y, Tamada T, Matsuyama H, Muranaka H. Cotinine excretion and daily cigarette smoking in habituated smokers. *Clin. Pharmacol. Ther.*, 1979: **25**; 555–61.

202 Castro A, Prieto I. Nicotine Antibody Production: Comparison of two nicotine conjugates in different animal species. *Biochem. Biophys. Res. Commun.*, 1975: **67**; 583–89.

203 Knight GJ, Wylie P, Holman MS, Haddow, JE. Improved 125I Radioimmunoassay for cotinine by selective removal of bridge antibodies. *Clin. Chem.*, 1985: **31**; 118–21.

204 Jones SR, Amatayakul, S. Improved 125I radioimmunoassay for cotinine. *Clin. Chem.*, 1985: **31**; 1076–77.

205 Davis, RA and Stiles, MF, *Determination of nicotine and cotinine: comparison of gas chromatography and radioimmunoassay methods.* Presented at 47th Tobacco Chemists' Research Conference, Gatlinburg, TN. October 18–21, 1993.

206 Bjercke RJ, Cook C, Langone, JJ. Comparison of monoclonal and polyclonal antibodies to cotinine in nonisotopic and isotopic immunoassays. *J. Immunol. Methods*, 1987: **96**; 239–46.

207 Langone JJ, Van Vunakis H. Radioimmunoassay of nicotine, cotinine, and γ-(3-pyridyl)-γ-oxo-*N*-methylbutyramide. In: Langone JJ, Van Vunakis H, editors. *Methods in Enzymology*,Volume **84**, New York: Academic Press, 1982: pp. 628–40.

208 Anderson IGM, Proctor CJ, Husager L. Comparison of the measurement of serum cotinine levels by gas chromatography and radioimmunoassay. *Analyst*, 1991: **116**; 691–93.

209 Biber A, Scherer G, Hoepfner I, Adlkofer F, Heller W, Haddow JE, Knight, GJ. Determination of nicotine and cotinine in human serum and urine: an inter-laboratory study. *Toxicol. Lett.*, 1987: **35**; 45–52.

210 Letzel H, Fischer-Brandies A, Johnson LC, Uberla K, Biber, A. Measuring problems in estimating the exposure to passive smoking using the excretion of cotinine. *Toxicol. Lett.*, 1987: **35**; 35–44.

211 Zuccaro P, Pichini S, Altieri I, Rosa M, Pellegrini M, Pacifici, R. Interference of nicotine metabolites in cotinine determination by RIA, *Clin. Chem.*, 1997: **43**; 180–81.

212 Schepers G, Walk, RA. Cotinine determination by immunoassays may be influenced by other nicotine metabolites. *Arch. Toxicol.*, 1988: **62**; 395–7.

213 Richie JP, Leutzinger Y, Axelrad CM, Haley NJ. Contribution of *trans*-3′-hydroxycotinine and glucuronide conjugates of nicotine metabolites to the measurement of cotinine by RIA. In: Adlkofer A, Thurau K, editors. *Effects of Nicotine on Biological Systems*. Basel: Birkhäuser Verlag, 1991: pp. 77–81.

214 Byrd GD. LC-MS/MS method for profiling nicotine and its metabolites in biological fluids. [Poster] 44th American Society for Mass Spectrometry and Allied Topics, Portland, OR, May 12–16, 1996.

215 Rickert WS. Validation of measures of exposure to environmental tobacco smoke. Project report, Health Canada, Ottawa Canada, March, 1995 (Contract No. H4078–4-C815/01-SS).

216 Jenkins RA, Palausky A, Counts RW, Bayne CK, Dindal AB, Guerin, MR. Exposure to environmental tobacco smoke in sixteen cities in the United States as

determined by personal breathing zone air sampling. *J. Expo. Anal. Environ. Epidemiol.*, 1996: **6**; 473–502.

217 Benkirane S, Nicolas A, Galteau MM, Siest, G. Highly sensitive immuno-assays for the determination of cotinine in serum and saliva – comparison between RIA and an avidin-biotin ELISA. *Eur. J. Clin. Chem. Clin. Biochem.*, 1991: **29**; 405–10.

218 Bjercke RJ, Cook G, Rychlik N, Gjika HB, Van Vunakis H, Langone, JJ. Steroespecific monoclonal antibodies to nicotine and cotinine and their use in enzyme-linked immunosorbent assays. *J. Immunol. Methods*, 1986: **90**; 203–13.

219 Langone, JJ, Bjercke RJ. Idiotype-anti-idiotype hapten immunoassays: assay for cotinine. *Anal. Biochem.*, 1989: **182**; 187–92.

220 Chang, K-M, Gentry G, Davis R, Stiles M, Coggins, C. [Abstract] Determination of plasma nicotine and cotinine in rats exposed to aged and diluted cigarette sidestream smoke using an enzyme-linked immunosorbent assay (ELISA). *The Toxicologist*, 1992: **12**; 263.

221 Chang K-M, Gentry G, Davis, R, Stiles M. [Abstract] Confirmation of smoking status - determination of urinary cotinine using an enzyme-linked immunosorbent assay (ELISA). *The Toxicologist*, 1994: **14**; 397.

222 Smith CJ, McKarns SC, Davis RA, Livingston SD, Bombick BR, Avalos JT, Morgan WT, Doolittle, DJ. Human urine mutagenicity study comparing cigarettes which burn or primarily heat tobacco. *Mutat. Res.*, 1996: **361**; 1–9.

223 Byrd, GD, Davis, RA, Kazemi-Vala, E. Comparison of methods for determining nicotine and its metabolites in urine [Poster]. 1st Annual Meeting of the Society for Research on Nicotine and Tobacco; 1995 Mar 24–25; San Diego, CA.

*Chapter 15*

# Determination of nicotine and its metabolites in biological fluids: in vitro studies

Mui C. Tsai and John W. Gorrod

*Toxicology Unit, John Tabor Laboratories, The University of Essex, Wivenhoe Park, Colchester, Essex CO4 3SQ, UK*

## I. INTRODUCTION

Although much is known about nicotine metabolic pathways, most of which have been described in many reviews [1–6], the full metabolic map of nicotine is not complete and a considerable effort is still being made to fill the gaps. Some of the metabolic pathways of nicotine such as primary *C*-oxidation and *N*-oxidation are well-established, but others have only been postulated (see Chapter 3). With advances in analytical technology (see Chapters 6, 7 and 8) applied to the determination of nicotine and its metabolites, the structure of new nicotine metabolites may be elucidated or new metabolic pathways established. This chapter is intended to give some insight as to how some nicotine metabolites formed in biological fluids may be determined in vitro. From these studies, the mechanisms and enzymology involved in the metabolic pathways may be elucidated.

The analytical methods used (qualitative, semi-quantitative, quantitative), in in vitro studies on nicotine metabolism will be discussed in the present chapter. Both immunoassay and chromatographic methods, developed for the analysis of nicotine and its metabolites in biological fluids such as urine, serum, etc, have been discussed in Chapter 14. These in vivo analytical methods are also applicable to in vitro nicotine metabolic studies. For in vitro analyses, even though the concentration of nicotine metabolites present are usually in a much higher range than those detected in in vivo studies, some purification treatments are still necessary to separate the analytes from proteins, lipids and salts present in the in vitro metabolic systems. Solvent or solid phase extraction techniques are usually used to

645

isolate and concentrate the nicotine metabolites before analysis. For accurate quantitation of nicotine or its metabolites formed in vitro, internal standards (see Chapter 14), should be included during or prior to the work-up procedures.

## Nicotine

Whilst it may seem incongruous to measure nicotine during studies on nicotine metabolism, in practice this can give useful data during mass balance studies. Other situations where nicotine needs to be measured are when a metabolite is difficult to assay because of its high polarity and hence water solubility or instability during analytical procedures. Such a situation pertains with both nicotine-*N*-oxides and nicotine-$N^+$-glucuronide. In the former case the *N*-oxides can be rapidly and quantitatively reduced to the parent nicotine and nicotine-$N^+$-glucuronide can be hydrolyzed by β-glucuronidase to release the aglycone. These will be discussed later.

Both GC and HPLC are frequently used for the determination of nicotine in in vitro studies. The methods for the analysis of nicotine by GC are essentially based on those of Beckett and Triggs [7] and Beckett, Gorrod, Jenner [8,9]. Essentially nicotine is extracted from a basified sample with ether; usually this requires three extractions before the ether in a further extract contains no detectable nicotine. The extracts are bulked, the internal standard added, and evaporated to a small volume at 42°C or at room or lower temperature using a stream of nitrogen gas. Because of the volatility of nicotine it is desirable to add an acid such as either succinic or tartaric acid to the ethereal extract to convert it to a salt with lower volatility. This has been shown to prevent losses during concentration. The above procedures can be conveniently carried out in ground glass tubes with a small nipple at the base [7]. When the ether has apparently evaporated (change in opaqueness of ground glass joint), the tubes are stoppered and placed in containers of ground ice. This causes condensation of any residual ether which washes the walls of the tube and collects in the nipple. This condensate can be used directly for analysis or diluted. The internal standard which has been found to be most useful, is phenmetrazine [8,9]. GC analysis is conveniently carried out using a column packed with Chromosorb G coated with 5% potassium hydroxide and 2% carbowax 20M, and detected with a FID [7–9]. Prior to analysis the column is silanized in situ with hexamethyldisilizane after equilibration for 24 h at the operating conditions. However, improved glass capillary GC methods, coupled with NPD, have also been reported for the analysis of nicotine [10,11].

For HPLC analysis of nicotine, various analytical columns such as $C_8$-,$C_{18}$- or cyano reversed-phase columns, which are monitored at 254 nm using a UV detector have been reported ([12–14], see Table 1). The combination of HPLC and a colorimetric method using barbituric acid or diethylthio-barbituric acid can be used for the determination of nicotine and some of its metabolites formed in vitro. Prior to HPLC analysis, pre-column derivatiza-tion of the samples with diethylthiobarbituric acid and the inclusion of N'-ethylnorcotinine (as internal standard) are necessary [15,16].

Cation-exchange HPLC analysis can also be used for in vitro nicotine determination [17]. Prior to the injection of the extract into the chromato-graph, the above solvent extraction and purification techniques are essential. A cartridge packed with the same packing materials as the analytical column is usually connected in series to serve as a guard column. Recently, a sensitive and reproducible HPLC method, using a Chiral-AGP analytical column, has been developed for the determination of the individual enantiomers of nicotine [18].

## Nicotine iminium ions

Murphy [19] suggested that cotinine was formed via a $\Delta^{1'(5')}$ iminium ion intermediate (Fig. 1). He showed that the presence of cyanide in microsomal incubates of nicotine led to its incorporation to produce 5'-cyanonicotine. Furthermore, the synthesis of nicotine $\Delta^{1'(5')}$ iminium ion allowed the role of

TABLE 1

Several HPLC methods and conditions which have been employed for the separation of nicotine and some of its Phase I metabolites formed in vitro and in vivo

| Detector and column | Mobile phase system (isocratic/gradient) | Flow rate ml/min | Reference |
|---|---|---|---|
| UV (254 nm), Supelcosil LC$_8$DB | Solvent A: water/acetonitrile (96.4:3.6, v/v) containing 2 ml/l triethylamine and 0.012 M each of sodium heptanesulphonate, K$_2$HPO$_4$ and citric acid, final pH 4.7. Solvent B: water/acetonitrile (80.3:19.7, v/v) containing 2 ml/l triethylamine and 0.012 M each of sodium heptanesulphonate, K$_2$HPO$_4$ and citric acid, final pH 5.2. | 1.5 | 12 |
| UV (254 nm), Waters μ Bondapak C$_{18}$ Reverse-phase | Methanol: 0.2% phosphoric acid (40:60 v/v), buffered with triethylamine to pH 7.25. | 0.5 | 13 |
| UV (254 nm), IBM Cyano Reverse-phase | Solvent A: water/methanol/0.1M acetate buffer (pH 4.0)/acetonitrile (187.5:11:1:0.5 v/v). Solvent B: water/methanol/0.5 M acetate buffer (pH 4.0)/acetonitrile (65:29:4:2 v/v), final pH adjusted to 6.82 with triethylamine. | 1.5 | 14 |
| UV (260 nm) Nucleosil SA Cation-exchange | Sodium acetate buffer (0.2 M): methanol (70:30 v/v), and triethylamine (0.02 % v/v), adjusted to pH 4.5 with glacial acetic acid | 1.0 | 17 |

this compound to be firmly established, as this iminium ion was converted to cotinine both in vivo and in vitro [20–22].

During nicotine biotransformation (Fig. 1), the proposed $\Delta^{1'(2')}$ iminium ion and 1'-$N$-methylene iminium ion may be converted to pseudo-oxynicotine and nornicotine, respectively. Pseudo-oxynicotine was first shown to be formed from nicotine $\Delta^{1'(2')}$ iminium ion by soil bacteria [23]. These two iminium ions can also be trapped using cyanide to form the corresponding cyano-adducts [24,25].

Whilst it is clear that direct detection of nicotine $\Delta^{1'(5')}$ iminium ion can be carried out using HPLC [26–28], no methods for the detection of either the $\Delta^{1'(2')}$ iminium ion or 1'-$N$-methylene iminium ion have been described. Direct analysis of nicotine $\Delta^{1'(5')}$ iminium ion can be achieved using a Beckman Ultrasphere ODS (5 μm, 4.6 × 250 mm) column and a mobile phase consisting of 20% acetonitrile/methanol (7:3% v/v) and 80% sodium phosphate (0.03 M), pH 7.0, containing triethylamine (0.1% v/v). The separation of the iminium ion, nicotine and other nicotine metabolites was performed with isocratic elution at a flow rate of 1 ml/min, and monitored using a UV spectrometer at 260 nm [27].

Because of the inherent instability of these iminium ions, analytical methods based on the formation of isomeric cyanonicotines by the addition of cyanide to the incubation mixtures, or to post incubation extracts have been developed [26–29].

More recently [24,30,31], this method has been developed into a quantitative technique by the incorporation of [$^{14}$C]-sodium cyanide into the incubation media. This technique separates any radiolabeled isomeric cyanonicotine by solvent extraction, leaving any inorganic cyanide in the aqueous phase. The organic phase can be concentrated and counted to give total cyanide incorporation or the individual isomeric cyano-nicotines can be separated by HPLC, after extraction and concentration, to give an indication of the concentration of each individual iminium ion. This technique has been applied to nicotine and a variety of other tobacco alkaloids [32] and to nicotine metabolism by lung and various areas of the rat and mouse brain [33,34]. However, recent in vitro studies using the cyanide trapping technique emphasized that extra care is needed when interpreting data derived from incubates containing secondary amines and formaldehyde donating substrates or microsomal preparations that contained glycerol as preservative. The production of 1'-$N$-cyanomethylnornicotine in the incubates could arise from formaldehyde and nornicotine produced during incubation [35] plus cyanide and could give rise to misleading results (Fig. 2).

## Cotinine

Hucker and co-workers [36] first detected cotinine, as a primary $C$-oxidative metabolite of nicotine, using rabbit hepatic preparations. Subsequently, the

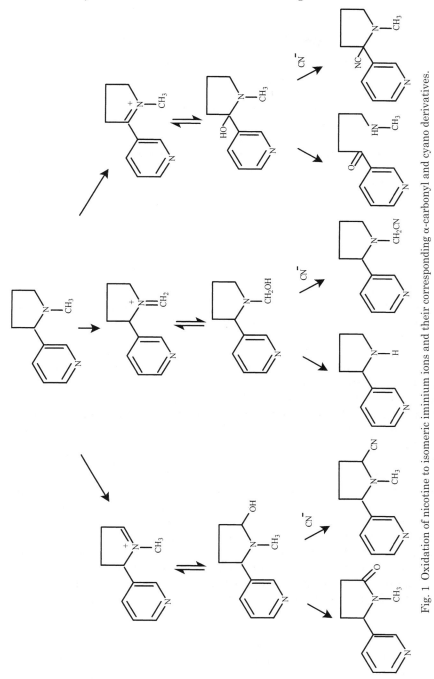

Fig. 1 Oxidation of nicotine to isomeric iminium ions and their corresponding α-carbonyl and cyano derivatives.

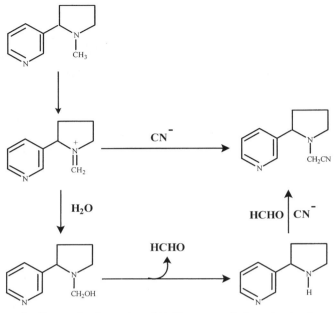

Fig. 2 Two routes leading to the formation of 1'-*N*-cyanomethylnornicotine from nicotine in the presence of cyanide.

in vitro formation of cotinine from nicotine, by rabbit [37,38], mice [39], hamster [40], and guinea pig [41] hepatic preparations, was also demonstrated.

As cotinine is soluble in organic solvents, solvent extraction is usually performed on incubates prior to analysis. First, the liver proteins are precipitated by either hydrochloric acid (0.2 M), trichloroacetic acid (5% w/v), or zinc sulphate solution (5% w/v) followed by barium chloride. After centrifugation, the supernatant is rendered alkaline (pH 9), with either sodium hydroxide, or ammonia solution (0.5 M) and extracted using organic solvents. For cotinine extraction, the solvents of choice, with good recovery, are chloroform or dichloromethane. The organic phases are collected and evaporated to dryness using a Vortex evaporator or under a stream of nitrogen gas at 30–40°C. Cotinine is less basic and less volatile than nicotine and addition of acid during concentration is usually not required. The dried extracts are reconstituted using mobile phase before analysis using various chromatographic methods.

Many of the analytical methods, used in in vitro studies, were initially developed for the simultaneous analysis of nicotine and its metabolites in biological fluids. In earlier work, qualitative paper- and thin-layer chromatography was used to detect cotinine formed from nicotine in vitro [36,38,42]. The use of spray detection agents, such as Dragendorff reagent or 1%

benzidine in ethanol (care: carcinogenic) or cyanogen bromide exposure of the developed plates, proved to be useful for the detection of many nicotine metabolites including cotinine. Subsequently, methods for the quantitative analysis of cotinine using gas chromatography (GC) or radioimmunoassay were developed [9,43]. In the former method, lignocaine was used as the internal standard for cotinine analysis. Many internal standards such as *N*-ethylnornicotine and 5-methylcotinine have been used for cotinine analysis, depending on the type of chromatographic methods used [44,45]. High-performance liquid chromatography (HPLC) is now becoming the analytical method of choice for cotinine analysis in vitro [14,46]. The HPLC chromatographic conditions listed in Table 1, can also be used for the analysis of cotinine formed in vitro.

**Nornicotine**

The *N*-demethylation of nicotine to nornicotine has been reported both in vivo [47] and in vitro [38], although other groups have failed to find this compound as a metabolite [29,48]. This ambiguity may, in part, be due to difficulties in the analysis of nornicotine due to its poor extractability into organic solvents [9]. A GC analytical method, which was developed for an in vivo study of nornicotine could also be used for in vitro studies [9]. In addition, an extraction method using sodium chloride in conjunction with either acetonitrile or dichloromethane was successful for the extraction of nornicotine and its metabolites formed in vitro (unpublished observation). Studies using hepatocytes from various species suggested that the 'intact' cells may be required to demonstrate nornicotine formation in vitro [14], although we have recently demonstrated its formation using $^{14}$C-nicotine and rabbit 10 000 g hepatic homogenates from an animal pretreated with phenobarbitone. In our analysis, a Nucleosil HPLC cation exchange column was used. The mobile phase consisted of a mixture of sodium acetate buffer (0.2 M) and methanol (70:30 v/v) and triethylamine (0.02%), with a final pH of 4.5.

The extracts were monitored at 260 nm using a UV detector, at a flow rate of 1 ml/min. Using this HPLC system, many polar nicotine metabolites such as desmethylcotinine and 3'-hydroxycotinine, together with nicotine and cotinine can also be analyzed simultaneously [14].

## II. *N*-OXIDATION

The *N*-oxidation pathway constitutes an important route of nicotine biotransformation. Nicotine-1'-*N*-oxide (NNO), the main *N*-oxide metabolite of nicotine is formed in vitro using hepatic and lung preparations from various species [27,38,49]. NNO exists as both *cis* and *trans* isomers [41,49,50].

Prior to the development of HPLC, most analyses performed on NNOs were semi-quantitative. Paper chromatography was first employed to identify these *N*-oxides formed in vitro, with the aid of synthetic standards [38,41]. Initially, two-dimensional paper chromatography, using both 'alkaline' and 'acid' solvent systems, coupled with Koenig reagents for detection were used [38]. A positive Koenig reaction is a characteristic of pyridine derivatives [51]. Subsequent paper chromatography using a descending system (*n*-butanol: *n*-propanol: 2*N* ammonia (2:1:1 v/v) was used to separate *cis* and *trans*-diastereoisomers of NNO or their reineckate salts [41]. The solvent extraction of NNOs into chlorinated organic solvents was not possible, due to their polar nature. Therefore, a solid phase extraction method, using silica as a solid phase sorbent, was used for the extraction of NNOs from biological fluids [52–55]. Using a preparative silica column (silica gel 60, 70–230 mesh), the following pre-conditioning of the column should be performed. Initially, the column is washed with methanolic ammonia, followed by aqueous acetic acid (4% v/v) prior to the application of the biological samples. Ideally, the biological samples (i.e. hepatic incubates or urine) should be acidified to pH 3.8–4.0. This elution technique was modified from the method of Park et al. [55], which was used for urinary analysis of NNO. Other sorbents such as amberlite (Dowex, XAD), $C_{18}$ or alumina can also be used.

The chromatographic properties of the complex formed by isatin and potassium dichromate oxidation of NNO has been reported [41]. Treatment of NNO with potassium dichromate produces several reaction products, which can then be separated by TLC. Furthermore, the inclusion of isatin in the reaction mixture and heating at 100°C for 15 min, produces a stable crimson complex which can be extracted into organic solvents. The NNO formed from nicotine using guinea pig liver and lung preparations have been studied using this method [41]. A gas chromatographic method for the determination of (S)-nicotine *N*-1'-oxide, involving thermal conversion to 2-methyl-6-(3-pyridyl)-tetrahydro-1,2-oxazine has been reported [52], but has not, to our knowledge, been applied to in vitro experiments.

An indirect quantitative assay for the total diastereoisomers of NNO, by reducing the *N*-oxides to nicotine with acidified titanium trichloride at room temperature, has been reported [8]. The reduced nicotine was then determined using gas–liquid chromatography [7,8] as described earlier. By using this technique in conjunction with paper chromatographic separation of the diastereoisomers of NNO, it was possible to assay each individual isomer of *N*-oxide derived from either S-(−)- or R-(+)-nicotine [49,50].

Furthermore, marked species, but not sex differences, were observed in the metabolism of R-(+)- and S-(−)-nicotine to diastereoisomeric NNOs by 10 000g hepatic preparations from rats, rabbits, mice, guinea pigs and hamsters (49). Although large inter-animal variation was observed, guinea pig was considered as the species *par excellence* in its ability to *N*-oxidize both R-(+)- and S-(−)-nicotine followed by hamster, mouse, rabbit and rat [49]. It should be remembered that N-oxidation is a reversible reaction and reduction of NNO has been observed in experimental animals and man [56,57].

With advances in HPLC technology and the use of radiolabeled nicotine, the metabolic NNOs can now be quantified directly [14,58,59]. A strong cation-exchange HPLC analytical column (Nucleosil SA 10 μm, 250×4.6 mm), together with an isocratic mobile phase consisting of a mixture of ammonium dihydrogen orthophosphate buffer (0.2 M, pH 4.3): methanol (77:23 v/v) and triethylamine (0.5% v/v), at a final pH of 4.5 have been frequently used for the analysis of in vitro nicotine metabolites including NNOs [60]. However, if the ratio of diastereoisomeric N-oxides is required, an amino:cyano (2:1) analytical column (Partisil PAC, 250×4.6 mm) and a mobile phase consisting of methanol:water (95:5% v/v) is more useful [60].

## III. FURTHER METABOLISM OF PRIMARY METABOLITES

The primary C-oxidative metabolites of nicotine i.e. cotinine and nornicotine can be further metabolized. It is now recognized that most of the further metabolites of nicotine arise via cotinine and it is estimated that more than 70% of nicotine metabolites are probably formed via this compound [61]. This includes the glucuronide conjugates of cotinine and its metabolites.

**Metabolites of cotinine**

Cotinine undergoes further N-oxidation and C-oxidation, catalysed by microsomal enzymes to produce four metabolites i.e. cotinine-N-oxide (CNO), 3′-hydroxycotinine (3-OHC), N-hydroxymethylnorcotinine (NHMC) and 5′-hydroxycotinine (5-OHC) (see Fig. 3). These metabolites derived by metabolic oxidation of cotinine and also demethylcotinine (DMC) can be quantitatively determined by the use of a cation-exchange HPLC column (Nucleosil SA 10 μm, 250×4.6 mm). The isocratic mobile phase consists of acetonitrile: ammonium dihydrogen orthophosphate (2:98% v/v) and triethylamine (0.07% v/v) with a final pH of 5.6. These four cotinine metabolites all have $\lambda_{max}$ similar to cotinine, i.e. 260 nm and are available as authentic standards; HPLC coupled with diode-array detector is very useful for in vitro studies of these metabolites.

The recoveries of CNO, 3-OHC, 5-OHC and DMC using liquid–liquid extraction (dichloromethane, chloroform, and acetonitrile) have been studied [62]. It was found that both dichloromethane and chloroform were ineffective, i.e. more than 79% of these compounds remained in the aqueous layer, although these chlorinated solvents were excellent for cotinine extraction.

Extraction with acetonitrile ($3 \times 2$ ml) in the presence of NaCl (0.5 g/3.5 ml) produced more than 50% recovery for 3-OH, 5-OHC and DMC. However, only 35% CNO was extractable using acetonitrile. The study also showed that solid–liquid extraction, i.e. with Sep-Pak $C_{18}$ cartridges produced a recovery of over 65% for all the cotinine metabolites [62]. Initially, the cartridges were activated with methanol (5 ml), followed by water (10 ml). Methanol (2 ml) was used for elution of these metabolites.

Recent in vitro studies, using HPLC, showed that cotinine was metabolized to CNO by hepatic microsomes prepared from many species. The metabolite percentages of CNO formed in mouse, rat, rabbit, dog, hamster and guinea pig hepatic microsomal incubates after 30-min incubation were 10%, 14%, 24%, 15%, 7%, 0%, respectively [62]. The formation of CNO from cotinine has previously been shown in vitro [63]. As the pKa of CNO is 3.37 (cf. cotinine: 4.44) at 15–18°C [64]; and the usual organic solvents of choice (i.e. dichloromethane and chloroform) for extraction of cotinine at physiological pH were not effective (<20%), it suggests that CNO is much more polar than cotinine. The incubation of [$^{14}$C]-cotinine with PB-induced hamster hepatic microsomes showed CNO was formed in the incubates. Unlike the production of nicotine-1'-$N$-oxide, which involves the FMO system, the formation of CNO is cytochrome P450 dependent [63].

Hydroxylation of cotinine produces 3-OHC (*cis* and *trans* isomers), NHMC and 5-OHC. The in vitro formation of 3-OHC from cotinine had been studied and analyzed using an HPLC system, although the stereoisomers were not separable [62]. 3-OHC was found to be a predominant cotinine metabolite in guinea pig hepatic preparations (78%) compared to 19% and 24% in dog and hamster liver, whereas this metabolite was not formed in mouse, rat and rabbit hepatic preparations [62].

NHMC was identified as a new primary metabolite of cotinine in vitro [65]. The metabolite was detected in cotinine incubates with hamster (<3%), guinea pig (1%) and rabbit (0.2%) hepatic microsomes in the presence of NADPH and oxygen. It was not present in dog, rat and mouse incubates [62]. The further conversion of this metabolite to norcotinine, either chemically or biochemically, was not observed in that study.

Using the above HPLC method, 5-OHC was detected as a major cotinine metabolite in incubates of hepatic preparations from rat (86%), mouse (90%), hamster (66%), dog (66%) and rabbit (76%), whereas only 21% was formed in incubates derived from guinea pig. The percentages shown for CNO, 3-OHC, NHMC and 5-OHC formation are the percentages of cotinine metabolised. Recently, 5-hydroxycotinine-$N$-oxide has been isolated as a new nicotine metabolite present in rat urine [66] but it has not yet been detected in vitro. 5-OHC (cyclic form, 80%) also exists in equilibrium with its open-chain keto-amide form, 4-(3-pyridyl)-4-oxo-$N$-methylbutyramide (20%) [67]. The transformation of this keto-amide (Fig. 4) to 4-(3-pyridyl)-4-oxo-butyric acid (POBA) involves a single-step hydrolytic reaction rather than demethylation to the primary amide followed by deamination [62]. The further metabolism of the keto acid will be discussed in a later section.

## Metabolites of nornicotine

This alkaloid which also occurs in tobacco and tobacco smoke can be converted to norcotinine in vitro [38,68]. The analysis of (S)-(−)-nornicotine incubates of rabbit liver homogenates using two-dimensional paper chromatography, in conjunction with Koenig reagent, showed four Koenig-positive zones which corresponded to parent nornicotine, norcotinine, nicotinamide and one uncharacterized compound [38]. Further in vitro studies on (S)-(−)-nornicotine showed that both norcotinine and nornicotine-$\Delta^{1'}$-nitrone were metabolites of nornicotine. Incubation of (S)-(−)-nornicotine using liver microsomes from various species showed that norcotinine was formed in the incubates of hamster, rabbit and guinea pig, but not in those of rat and dog [68]. In contrast, nornicotine-$\Delta^{1'(2')}$-nitrone was detected in incubates from all the above species. Furthermore, 1'-$N$-hydroxynornicotine (an intermediate in the formation of nornicotine-$\Delta^{1'(2')}$-nitrone) and a third uncharacterised nornicotine metabolite were also detected [68]. The HPLC analysis was performed using a cation-exchange column (Nucleosil SA) coupled to a Rapiscan multiple UV detector. The mobile phase consisted of a mixture of ammonium dihydrogen orthophosphate buffer (0.2 M, pH 4.3) and methanol (40:60 v/v) with 0.5% v/v triethylamine. The final pH of the mobile phase was adjusted to pH 4.6 using orthophosphoric acid. The retention times of nornicotine-$\Delta^{1'}$-nitrone, norcotinine, 1'-$N$-hydroxynornicotine, the uncharacterized metabolite and norcotinine were 5.5, 6.4, 7.4, 10.3 and 17.8 min, respectively.

## Norcotinine

Recently, norcotinine has been shown to be further metabolized to 4-(3-pyridyl)-4-oxobutyramide (POBAM) using rat hepatic microsomes [69]. The HPLC analysis was carried out using a Spherisorb $C_{18}$ column with a flow rate of 2 ml/min, and eluted with a mobile phase system of acetonitrile and water (15:85 v/v). Figure 3 depicts the further in vitro metabolic pathways undertaken by nornicotine, using rat hepatic microsomes.

## 4-(3-Pyridyl)-4-oxo-$N$-methylbutyramide (POMAM)

Further incubation of POMAM with rat microsomes produced 4-(3-pyridyl)-4-oxo-butyric acid (POBA) when analysed using HPLC ([69], see Fig. 3). A

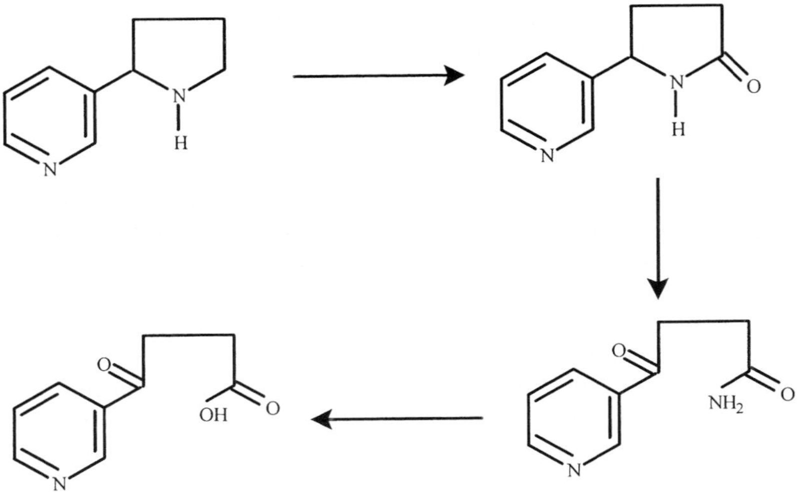

Fig. 3 In vitro biotransformation products of nornicotine using rat hepatic microsomes.

mobile phase system of acetonitrile and water (10:90 v/v) was used for the analysis [69].

## IV. PYRROLIDINE RING CLEAVAGE

The proposed metabolic pathway of POBA to 3-pyridylacetic acid is as follows: 4-(3-pyridyl)-4-oxo-butyric acid (POBA), formed from 4-(3-pyridyl)-4-oxo-N-methylbutyramide or POBAM (see above), is reduced to the corresponding alcohol, 4-(3-pyridyl)-4-hydroxybutyric acid (PHBA) which cyclizes to give 5-(3-pyridyl)-tetrahydrofuran-2-one (PTHF). PHBA is thought to lose water to produce 4-(3-pyridyl)-butenoic acid which is converted to 4-(3-pyridyl)-butyric acid (PBA). Finally, β-oxidation of this metabolite should produce 3-pyridylacetic acid (PAA) as the terminal nicotine metabolite (see Chapter 3). However, the recent identification of 3-pyridylcarbinol as a urinary metabolite of smokers suggests that 3-pyridylcarbinol, rather than 3-pyridylacetic acid, may be the terminal degradation product of nicotine [69,70].

To date, there has been only one in vitro study which claimed to show the presence of PAA; in this study nicotine was incubated with 18 500 g supernatant from rabbit liver [36]. Our in vitro studies using [2′-$^{14}$C]-nicotine incubated with PB-induced rabbit homogenates showed that POBA was the only acid nicotine metabolite produced, although the incubation conditions were different from those used earlier [72].

A gradient HPLC system (Table 2) which can separate all five proposed acid metabolites of cotinine (POBA, PHBA, PTHF, PBA and PAA) has been developed [73]. Figure 4 shows the HPLC chromatographic separation of these acid metabolites. Using [¹⁴C]-POBA as incubation substrate, these authors also failed to detect any new metabolites when hepatic microsomal preparations from rat, rabbit, guinea pig or hamster were used. However, both PHBA and PTHF were detected when POBA was used as substrate in incubates when 10 000 g and 27 000 g supernatant hepatic cell fractions of various species were used [73,74]. Clearly further studies are needed to confirm the intermediates involved in the conversion of POBA to PAA and it seems that, by analogy with fatty acid metabolism, methods may have to be

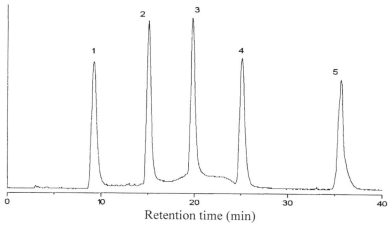

Retention time (min)

Fig. 4 Typical HPLC chromatogram of five carboxylic acid metabolites of nicotine separated using a gradient HPLC on a reverse phase Spherisorb ODS2 column. 1=3-pyridylacetic acid (PAA), 2=4-(3-pyridyl)-4-hydroxybutyric acid (PHBA), 3=4-(3-pyridyl)-4-oxo-butyric acid (POBA), 4=4-(3-pyridyl)-butyric acid (PBA), and 5=5-(3-pyridyl)-tetrahydrofuran-2-one (PTHF).

TABLE 2

Gradient HPLC system for the separation of five carboxylic acid metabolites of nicotine on a reverse phase Spherisorb ODS2 coulumn

| Time (min) | Flow rate (ml/min) | Solvent A[1] (%) | Solvent B[2] (%) |
|---|---|---|---|
| Initial | 1 | 95 | 5 |
| 3 | 1 | 95 | 5 |
| 10 | 1 | 70 | 30 |
| 35 | 1 | 30 | 70 |
| 45 | 1 | 95 | 5 |

[1] 0.1 M ammonium acetate and 0.5% triethylamine, adjusted with acetic acid to pH 4.8 (Solvent A).
[2] Solvent A : acetonitrile (10:1 v/v).

developed for the detection and quanitation of CoA derivatives of the pyridyl acids of interest.

## V. CONJUGATION REACTIONS

### Methylation

Biological methylation of nicotine and cotinine has the potential to give rise to a number of $N$-methylated metabolites such as $N'$-methylnicotinium ion, $N$-methylnicotinium ion, $N,N'$-dimethylnicotinium ion and $N$-methylcotininium ion. Crooks and coworkers have shown that cytosolic azaheterocycle-$N$-methyltransferases, purified from rabbit liver and guinea pig tissues, were responsible for the stereospecific formation of $N$-methylated metabolites of R-(+)-nicotine [75–77]. This enzyme is widely distributed in guinea pig; the order of its activity was found to be: spleen>brain>lungs>liver [76]. Methylation of nicotine has also been shown using human liver cytosolic preparations [78].

An analytical cation-exchange radio-HPLC method, coupled with UV detection, has been developed to detect and quantitate the urinary methylated nicotine metabolites in the guinea pig [79]. The analysis of $N$-methylnicotinium ion using a selective HPLC method, coupled with electrochemical detection has also been reported [80]. The UV-HPLC analytical system was the most widely used for the in vitro determination of these $N$-methylated nicotine metabolites. The UV-HPLC separation of $N$-methylcotininium ion, $N'$-methylnicotinium ion, $N$-methylnicotinium ion and $N,N'$-dimethylnicotinium ion has been achieved using a cation-exchange Partisil 10 SCX column (10 μm, 250×4.6 mm), monitored at 254 nm. The mobile phase systems used consisted of a primary buffer: sodium acetate buffer (0.3 M) and methanol (70:30 v/v), pH 4.5, followed by a secondary buffer: sodium acetate buffer (0.3 M) and methanol (70:30 v/v), containing triethylamine (1% v/v), pH 4.5 [79].

### Glucuronidation

Evidence from studies of nicotine disposition in smokers, snuff users and macaques revealed that glucuronides of nicotine, cotinine, and 3'-hydroxycotinine were major urinary metabolites [81–87]. The average percentages of glucuronide conjugates of nicotine, cotinine, and 3'-hydroxycotinine excreted, as the total dose of nicotine in man, are approximately 3%, 9%, and 23%, respectively [82]. These percentages were established using β-glucuronidase to release the aglycone (indirect hydrolysis method). Other chemical hydrolytic methods (i.e. using hot base or acid) can also be used for the indirect determination of glucuronide conjugates. The thermolabile nature of the glycosiduronic bond of these glucuronides may render their

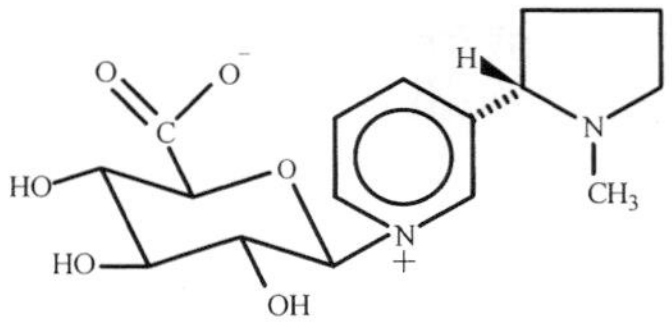

**NNG**

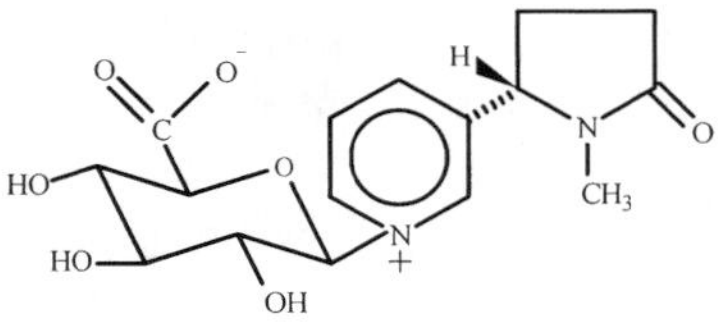

**CNG**

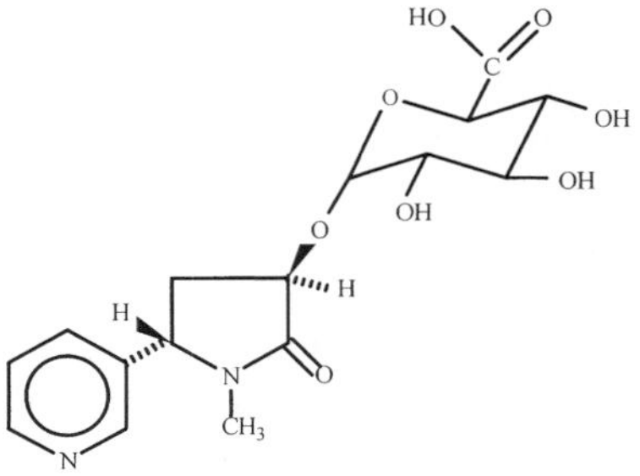

**3-OHG**

Fig. 5 The chemical structures of S-(−)-nicotine-$N^+$-1-(β)-glucuronide (NNG), S-(−)-cotinine-$N^+$-1-(β)-glucuronide (CNG), *trans*-3′-hydroxylcotinine-*O*-(β)-glucuronide (3-OHG).

analysis using GC complicated. A recent study has shown that increased urinary cotinine produced during storage is due to cotinine-$N^+$-glucuronide (CNG) hydrolysis [88]. Figure 5 shows their structures.

For more accurate determination of these glucuronides in biological fluids, or for further studies, the availability of authentic synthetic standards is desirable. The chemical synthesis and characterization of quaternary ammonium-linked glucuronides of 3′-hydroxycotinine (3-HNG) and cotinine (CNG) [89], of nicotine (NNG) [90] have been published. To date, the synthesis on *trans*-3′-hydroxycotinine-*O*-glucuronide (3-OHG) has yet to be reported. Direct determination of cotinine-$N^+$-glucuronide and nicotine-$N^+$-glucuronide in smokers urine have been reported using thermospray LC/MS [87,91]. Furthermore, these glucuronides are charged so that a cation-exchange high-performance liquid chromatographic method can be used to isolate them from smokers' urine [87,91]. The amounts of urinary NNG excreted were related to differences among smokers in their nicotine intake and metabolism, and thus large interindividual variability (>10-fold) was observed in the study [91]. Similarly, this would be expected to be true for the urinary excretion of CNG and 3-OHG in smokers.

There are only a few reports on the formation of quaternary ammonium-linked glucuronides of tertiary amine drugs by laboratory animals in vivo or in vitro. This type of glucuronide was thought to be only important for man and other higher primates. Therefore, in order to study the enzymology involved in the glucuronidation of nicotine or its metabolites, an animal model that most resembles human smokers needs to be sought. Both rat and guinea pig were found to excrete less than 0.5% of an administered nicotine dose as glucuronides of nicotine and 3'-hydroxycotinine [92].

In vitro studies on the glucuronidation of nicotine, cotinine and 3'-hydroxy-cotinine (*cis* and *trans*) have been carried out [93]. The activation of microsomal preparations with Triton X-100 is necessary because UDP-glucuronosyltransferases (UGTs) are integral membrane proteins [94]. Uridine diphospho-glucuronic acid (UDPGA), the cofactor for the UGT enzymes, is included in the incubation mixture. These studies showed that NNG was only formed when UDPGA was included in the incubates [93].

A very useful technique for glucuronidation studies is to utilize radiolabeled substrates or radiolabeled UDPGA (or both) in in vitro system. In our studies, the substrate nicotine used, (S)-( − )-[pyrrolidine-2-$^{14}$C]nicotine-di-(+)-tartrate salt, was purchased from Chemsyn Science Lab. Kansas, USA. However, for cotinine and 3-OHC studies, UDP-[U-$^{14}$C]-GA purchased from ICN Radiomedicals Ltd, UK, was used as a cofactor. In these cases, radio-HPLC is the analytical technique of choice. A Nucleosil analytical cation-exchange column (10 μm, 250×4.6 mm) was used. The mobile phase A was designed specifically for NNG detection, whereas mobile phase B was used for both CNG and 3-HNG detection. The former system consisted of a mixture of sodium acetate buffer (0.2M) and methanol (70:30 v/v) and triethylamine (0.02%), final pH 4.5; system (B) consisted of sodium acetate buffer (0.1 M), pH 4.0. The mobile phase was mixed with Quickszint flow 302 cocktail (1:2 v/v), at a flow rate of 3 ml/min for radioactivity determination. The retention times of NNG, CNG, 3-HNG and UDPGA using system A were 14.7, 5.7, 4.8, and 2.6 min respectively.

The results showed that NNG was formed when using marmoset hepatic preparations; but not those of rabbit, rat and guinea pig (Fig. 6). Further confirmation of the nature of the formed glucuronide was carried out using chemical and enzymatic hydrolytic treatments. Controls, with the inclusion of glucosaccharo-1,4-lactone (specific inhibitor of β-glucuronidase) are essential. Under the present incubation condition, the rate of NNG formation appeared almost linear up to four hours using activated marmoset hepatic microsomes (Fig. 7). The percentage of nicotine glucuronide formed accounted for approximately 6% of the nicotine substrate used [93].

Glucuronidation of cotinine or 3'-hydroxycotinine (*cis* and *trans*) were not observed in these in vitro studies [93]. Although 3-OHG was not available as a standard, we examined the chromatograms for the appearance of any new radiolabelled peak, but none was observed. The failure to observe the formation of CNG and 3-OHG suggested that different isoforms of UGT may be involved in the glucuronidation of cotinine and 3'-hydroxycotinine. Previously, human UGTs (i.e. UGT1A3 and UGT1A4) have been shown to be important isozymes for the formation of quaternary ammonium-linked glucuronides [95]. However, most of the aglycones used in the studies were aliphatic or cyclic tertiary amines. To date, the human UGTs responsible for

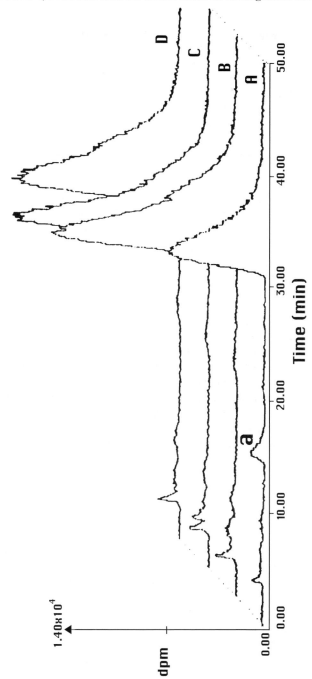

Fig. 6 Radio-HPLC chromatograms of incubates containing radiolabeled nicotine, when Triton X-100-activated microsomes of marmoset (A), rabbit (B), guinea pig (C) and rat (D) were used as enzyme source. a = S-(−)-nicotine-$N^+$-1-($\beta$)-glucuronide (NNG).

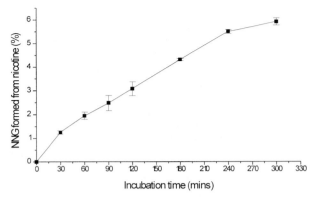

Fig. 7 In vitro rate of formation of S-(−)-nicotine-$N^+$-1-(β)-glucuronide from radiolabeled nicotine using marmoset hepatic microsomes. Results are expressed as mean±SEM, $n=3$.

the formation of quaternary ammonium-linked glucuronides of nicotine and cotinine (on the pyridine ring) have yet to be characterized.

## CONCLUSIONS

The rapid development of analytical techniques as applied to nicotine and related tobacco alkaloids has allowed the detection, characterization and quantitation of most of the intermediate metabolites involved in nicotine biodegradation. Most of the processes can now be carried out in vitro and the enzymology of the processes delineated. Future work using these techniques should enable any missing intermediates to be detected and characterized, and the factors controlling their formation and further metabolism established. This is particularly important in studies on the metabolism of tobacco alkaloids in man where the role of many intermediates in the total picture remains obscure.

## REFERENCES

1  Gorrod JW, Jenner P. The metabolism of tobacco alkaloids. *Essays Toxicol.*, 1975: **6**; 35–78.
2  Schievelbein H. Nicotine, resorption, and fate. *Pharmac. Ther.*, 1982: **18**; 233–248.
3  Nakayama H. Nicotine metabolism in mammals. *Drug Metab. Drug Interact.*, 1988: **6**; 95–122.
4  Kyerematen GA, Vesell ES. Metabolism of nicotine. *Drug Metab. Rev.*, 1991: **23**; 3–41.
5  Gorrod JW. The mammalian metabolism of nicotine: an overview. *Nicotine and Related Alkaloids: Absorption, Distribution, Metabolism and Excretion.* Gorrod JW, Wahren J. (Eds), London: Chapman & Hall, 1993; pp. 31–43.

6  Gorrod JW. Nicotine: a case study in metabolic complexity. *Eur. J. Drug Metab. Pharmacokinet.*, 1995; Special Issue: 5–9.

7  Beckett AH, Triggs EJ. Determination of nicotine and its metabolite, cotinine, in urine by gas chromatography. *Nature*, 1966: **211**; 1415–1417.

8  Beckett AH, Gorrod JW, Jenner P. The analysis of nicotine-1'-*N*-oxide in urine, in the presence of nicotine and cotinine, and its application to the study of in vivo nicotine metabolism in man. *J. Pharm. Pharmac.*, 1971: **23**; 55S–66S.

9  Beckett AH, Gorrod JW, Jenner P. A possible relation between pKa₁ and lipid solubility and the amounts excreted in urine of some tobacco alkaloids given to man. *J. Pharm. Pharmac.*, 1972: **24**; 115–120.

10 Feyerabend C, Russell MAH. Improved gas-chromatographic method and micro-extraction technique for the measurement of nicotine in biological fluids. *J. Pharm. Pharmac.*, 1979: **31**; 73–76.

11 Severson RF, McDuffie KL, Arrendale RF, Gwynn GR, Chaplin JF, Johnson AW. Rapid method for the analysis of tobacco nicotine alkaloids. *J. Chromatogr.*, 1981: **211**; 111–121.

12 Zuccaro P, Altieri I, Rosa M, Passa AR, Pichini S, Ricciarello G, et al. Determination of nicotine and four metabolites in the serum of smokers by high-performance liquid chromatography with ultraviolet detection. *J. Chromatogr.*, 1993: **621**; 257–261.

13 Saunders JA, Blume DE. Quantitation of major tobacco alkaloids by high-performance liquid chromatography. *J. Chromatogr.*, 1981: **205**; 147–154.

14 Kyerematen GA, Morgan M, Warner G, Martin LF, Vesell ES. Metabolism of nicotine by hepatocytes. *Biochem. Pharmacol.*, 1990: **40**; 1747–1756.

15 Barlow RD, Thompson PA, Stone RB. Simultaneous determination of nicotine and five additional nicotine metabolites in urine of smokers using pre-column derivatization and high-performance liquid chromatography. *J. Chromatogr. B Biomed. Appl.*, 1987: **419**; 375–380.

16 Rustemeier K, Demetriou D, Schepers G, Voncken P. High-performance liquid chromatographic determination of nicotine and its urinary metabolites via their 1,3-diethyl-2-thiobarbituric acid derivatives. *J. Chromatogr. B Biomed. Appl.*, 1993: **613**; 95–103.

17 Tsai MC, Gorrod JW. In vitro metabolism of (S)-(−)-[2'-¹⁴C]-nicotine, using various tissue preparations of marmoset. *Drug Metab. Drug Interact.*, 1999: **15**; 97–105.

18 Demetriou D, Rustemeier K, Voncken P, Schepers G. HPLC separation of the enantiomers of nicotine and nicotine-like compounds. *Chirality*, 1993: **5**; 300–302.

19 Murphy PJ. Enzymatic oxidation of nicotine to nicotine $\Delta^{1'(5')}$ iminium ion. *J. Biol. Chem.*, 1973: **248**; 2796–2800.

20 Brandange S, Lindblom L. Synthesis, structure and stability of nicotine $\Delta^{1'(5')}$ iminium ion, an intermediary metabolite of nicotine. *Acta Chem. Scand.*, 1979: **B33**; 187–191.

21 Hibberd AR, Gorrod JW. Nicotine iminium ion: a reactive intermediate in nicotine metabolism. *Adv. Exp. Med. Biol.*, 1981: **136**; 1121–1131.

22 Gorrod JW, Hibberd AR. The metabolism of nicotine $\Delta^{1'(5')}$ iminium ion in vivo and in vitro. *Eur. J. Drug Metab. Pharmacokin.*, 1982: **7**; 293–298.

23 Wada E, Yamasaki K. Degradation of nicotine by soil bacteria. *J. Am. Chem. Soc.*, 1954: **76**; 155–157.

24 Gorrod JW, Aislaitner G. The metabolism of alicyclic amines to reactive iminium ion intermediates. *Eur. J. Drug Metab. Pharmacokin.*, 1994: **19**; 209–217.

25 Nguyen T-L, Gruenke LD, Castagnoli N. Metabolic *N*-demethylation of nicotine. Trapping of a reactive iminium species with cyanide ion. *J. Med. Chem.*, 1976: **19**; 1168–1169.

26 Shigenaga MK, Trevor AJ, Castagnoli N. Metabolism-dependent covalent binding of (S)-[5-³H]nicotine to liver and lung microsomal macromolecules. *Drug Metab. Dispos.*, 1988: **16**; 397–402.

27 Williams DE, Shigenaga MK, Castagnoli N. The role of cytochrome P-450 and flavin-containing monooxygenase in the metabolism of (S)-nicotine by rabbit lung. *Drug Metab. Dispos.*, 1990: **18**; 418–428.

28 Williams DE, Ding X, Coon MJ. Rabbit nasal cytochrome P-450 Nma has high activity as a nicotine oxidase. *Biochem Biophys Res Commun* 1990: **166**; 945–952.

29 McCoy GD, Howard PC, DeMarco GJ. Characterization of hamster liver nicotine metabolism I. Relative rates of microsomal C and N oxidation. *Biochem. Pharmacol.*, 1986: **35**; 2767–2773.

30 Gorrod JW, Whittlesea CMC, Lam SP. Trapping of reactive intermediates by incorporation of ¹⁴C-sodium cyanide during microsomal oxidation. *Adv. Exp. Med. Biol.*, 1990: **283**; 657–664.

31 Whittlesea CMC, Lam SP, Gorrod JW. In vitro metabolism of some therapeutic alicyclic drugs by rabbit hepatic microsomes in the presence of ¹⁴C-sodium cyanide. *Prog. Pharmacol. Clin. Pharmacol.*, 1991: **8**; 49–58.

32 Aislaitner G, Sai Y, Gorrod JW. Preliminary evidence for the formation of reactive intermediates during the in vitro metabolism of minor tobacco alkaloids. *Med. Sci. Res.*, 1993: **21**; 819–821.

33 Aislaitner G, Bello A, Tan SC, Hutt AJ, Marriott C, Gorrod JW. Metabolism of (−)-(S)-nicotine in the isolated perfused rabbit lung. *Eur. J. Drug Metab. Pharmacokin.*, 1997: **22**; 395–402.

34 Gorrod JW, Aislaitner G, Hindmarsh J, Jenner P. Evidence for the presence of nicotine iminium ions in areas of rat brain using the cyanide trapping technique after intraperitoneal administration of nicotine. Poster 68 presented at 1st annual SNRT meeting San Diego, USA 1995.

35 Gorrod JW, Sai Y. Recognition of novel artifacts produced during the microsomal incubation of secondary alicyclic amines in the presence of cyanide. *Xenobiotica*, 1997: **27**; 389–399.

36 Hucker HB, Gillette JR, Brodie BB. Enzymatic pathway for the formation of cotinine, a major metabolite of nicotine in rabbit liver. *J. Pharmacol. Exper. Ther.*, 1960: **129**; 94–100.

37 Papadopoulos NM, Kintzios JA. Formation of metabolites from nicotine by a rabbit liver preparation. *J. Pharmacol. Exper. Ther.*, 1963: **140**; 269–277.

38 Papadopoulos NM. Formation of nornicotine and other metabolites from nicotine in vitro and in vivo. *Can. J. Biochem.*, 1964: **42**; 435–442.

39 Hanson E, Hoffmann PC, Schmiterlow CG. Metabolism of nicotine in mouse tissue slices. *Acta Physiol. Scand.*, 1964: **61**; 380–392.

40 Harke HP, Frahm B, Schlitz C, Donterwill W. Abbau von nikotin bei hamster and ratte. *Biochem. Pharmacol.*, 1970: **19**; 495–498.

41 Booth J, Boyland E. Enzymic oxidation of (−)-nicotine by guinea pig tissues in vitro. *Biochem. Pharmacol.*, 1971: **20**; 407–415.

42 Stalhandske T. The metabolism of nicotine and cotinine by a mouse liver preparation. *Acta Physiol. Scand.*, 1970: **78**; 236–248.

43  Langone JJ, Gjika HB, Van Vunakis, H. Nicotine and its metabolites. Radio-immunoassays for nicotine and cotinine. *Biochemistry.*, 1973: **12**; 5025–5030.

44  Jacob III P, Wilson M, Benowitz NL. Improved gas chromatographic method for the determination of nicotine and cotinine in biological fluids. *J. Chromatogr.*, 1981: **222**; 61–70.

45  Jacob III P, Ulgen M, Gorrod JW. Metabolism of ( − )-(S)-nicotine by guinea pig and rat brain:Identification of cotinine. *Eur. J. Drug Metab. Pharmacokin.*, 1997: **22**; 391–394.

46  Li Y, Li NY, Aislaitner G, Gorrod JW. In vitro metabolism of cotinine. *Med. Sci. Res.*, 1992: **20**; 903–904.

47  McKennis H, Turnbull LB, Schwartz SL, Tamaki E, Bowman ER. Demethylation in the metabolism of ( − )-nicotine. *J. Biol. Chem.*, 1962: **237**; 541–546.

48  McCoy GD, DeMarco GJ, Koop DR. Microsomal nicotine metabolism: A comparison of relative activities of six purified rabbit cytochrome P450 isozymes. *Biochem. Pharmacol.*, 1989: **38**; 1185–1188.

49  Jenner P, Gorrod JW, Beckett AH. Factors affecting the in vitro metabolism of R-( + )- and S-( − )-nicotine by guinea pig liver preparations. *Xenobiotica*, 1973: **3**; 563–572.

50  Testa B, Jenner P, Beckett AR, Gorrod JW. A reappraisal of the stereoselective metabolism of nicotine to nicotine–1′-*N*-oxide. *Xenobiotica*, 1976: **6**; 553–556.

51  Koenig, W. Zur Kenntnis der Pyridinfarbstoffe. *J. Prakt. Chem.*, 1904: **70**; 19–56.

52  Kyerematen, GA, Damiano MD, Dvorchik BH, Vesell, ES. Smoking-induced changes in nicotine disposition: Application of a new HPLC assay for nicotine and its metabolites. *Clin. Pharmacol. Ther.*, 1982: **32**; 769–780.

53  Jacob III P, Benowitz NL, Yu L, Shulgin AT. Determination of (S)-nicotine *N*–1′-oxide by gas chromatography following thermal conversion to 2-methyl-6-(3-pyridyl)-tetrahydro–1,2-oxazine. *Anal. Chem.*, 1986: **58**; 2218–2221.

54  Kyerematen GA, Owens GF, Chattopadhyay B, deBethizy JD, Vesell ES. Sexual dimorphism of nicotine metabolism and distribution in the rat. *Drug Metab. Dispos.*, 1988: **16**; 823–828.

55  Park SB, Jacob III P, Benowitz NL, Cashman JR. Stereoselective metabolism of (S)-( − )-nicotine in humans: Formation of *trans*-(S)-( − )-nicotine *N*-1′-oxide. *Chem. Res. Toxicol.*, 1993: **6**; 880–888.

56  Dajani RM, Gorrod JW, Beckett AH. Hepatic and extrahepatic reduction of nicotine-1′-*N*-oxide in rats. *Biochem. J.*, 1972: **130**; 88p.

57  Dajani RM, Gorrod JW, Beckett AH. In vitro hepatic and extrahepatic reduction of ( − )-nicotine–1′-*N*-oxide in rats. *Biochem. Pharmacol.*, 1975: **24**; 109–117.

58  Cashman JR, Park SB, Yang ZC, Wrighton, SA, Jacob III P, Benowitz NL. Metabolism of nicotine by human liver microsomes: Stereoselective formation of *trans*-nicotine *N*′-oxide. *Chem. Res. Toxicol.*, 1992: **5**; 639–646.

59  Demetriou D, Rustemeier K, Voncken P, Schepers G. Two radiochromatographic HPLC methods for the determination of nicotine metabolites. *Med. Sci. Res.*, 1992: **20**; 873–875.

60  Aislaitner G, Jahanshahi M, Gorrod JW. Comparative studies on the microsomal metabolism of 1′-*N*-methylanabasine and nicotine. *Med. Sci. Res.*, 1992: **20**; 891–896.

61  Benowitz NL, Jacob III P. Metabolism of nicotine to cotinine studied by a dual stable isotope method. *Clin. Pharmacol. Ther.*, 1994: **56**; 483–493.

62 Li Y, The in vitro metabolism of cotinine and related compounds. PhD thesis, University of London, 1995.

63 Hibberd AR, Gorrod JW. Comparative *N*-oxidation of nicotine and cotinine by hepatic microsomes. In: *Biological Oxidation of Nitrogen*. Gorrod JW, Damani LA. (Eds), Chichester: Ellis Horwood, 1985; pp. 246–250.

64 Li NY, Li Y, Gorrod JW. Determination of partition coefficients and ionisation constants of (S)-(−)-nicotine and certain metabolites. *Med. Sci. Res.*, 1992: **20**; 901–902.

65 Li Y, Gorrod JW. *N*-Hydroxymethylnorcotinine, a new primary in vitro metabolite of cotinine. *Xenobiotica*, 1994: **24**; 409–415.

66 Schepers G, Demetriou D, Stabbert R, Diehl B, Seeman JI. 5′-Hydroxycotinine-*N*-oxide, a new nicotine metabolite isolated from rat urine. *Xenobiotica*, 1999: **29**; 793–801.

67 Nguyen TL, Dagne E, Gruenke L, Bhargava H, Castagnoli N. The tautomeric structures of 5-hydroxycotinine, a secondary mammalian metabolite of nicotine. *J. Org. Chem.*, 1981: **46**; 758–760.

68 Aislaitner G, Li Y, Gorrod JW. In vitro metabolic studies on (−)-(S)-nornicotine. *Med. Sci. Res.*, 1992: **20**; 897–899.

69 Eldirdiri N, Ulgen M, Jacob III P, Gorrod JW. The in vitro metabolism of norcotinine and related biotransformation products by microsomal preparations. *Eur. J. Drug Metab. Pharmacokinet.*, 1997: **22**; 385–390.

70 Parviainen MT, Barlow RD. Assessment of exposure to environmental tobacco smoke using a high-performance liquid chromatography method for the simultaneous determination of nicotine and two of its metabolites in urine. *J. Chromatogr. B Biomed. Appl.*, 1988: **431**; 216–221.

71 Parviainen MT, Puhakainen EVJ, Laatikainen R, Savolainen K, Herranen J, Barlow RD. Nicotine metabolites in urine of smokers. *J. Chromatogr. B Biomed. Appl.*, 1990: **525**; 193–202.

72 Tsai MC, Sai Y, Li Y, Aislaitner G, Gorrod, JW. The Biosynthesis of [5′-¹⁴C]-cotinine and other radiolabeled nicotine metabolites. *J. Labelled Cpd. Radiopharm.*, 1999: **42**; 387–409.

73 Sai Y, Gorrod JW. Metabolic interconversion of acid metabolites of cotinine in vitro. *Drug Metab. Drug Interact.*, 1995: **12**; 93–103.

74 Sai Y. Studies on the intermediary metabolism of nicotine and related alkaloids. PhD thesis, University of London, 1996.

75 Cundy KC, Crooks PA. In vitro characteristics of guinea pig lung aromatic azaheterocycle *N*-methyltransferase. *Drug Metab. Dispos.*, 1985: **13**; 658–663.

76 Cundy KC, Godin CS, Crooks PA. Stereospecific in vitro *N*-methylation of nicotine in guinea pig tissue by an S-adenosylmethionine-dependent *N*-methyltransferase. *Biochem. Pharmacol.*, 1985: **34**; 281–284.

77 Damani LA, Shaker MS, Godin CS, Crooks PA, Ansher SS, Jakoby WB. The ability of amine *N*-methyltransferases from rabbit liver to *N*-methylate azaheterocycles. *J. Pharm. Pharmacol.*, 1986: **38**; 547–550.

78 Crooks PA, Godin CS. *N*-Methylation of nicotine enantiomers by human liver cytosol. *J. Pharm. Pharmacol.*, 1988: **40**; 153–154.

79 Cundy KC, Crooks PA. High-performance liquid chromatographic method for the determination of *N*-methylated metabolites of nicotine. *J. Chromatogr.*, 1984: **306**; 291–301.

80 Mousa S, van Loon GR, Houdi AA, Crooks PA. High-performance liquid chromatography with electrochemical detection for the determination of nicotine and *N*-methylnicotinium ion. *J. Chromatogr.*, 1985: **347**; 405–410.

81 Kyerematen GA, Morgan M, Chattopadhyay B, deBethizy JD, Vesell ES. Disposition of nicotine and eight metabolites in smokers and non-smokers: identification in smokers of two metabolites that are longer lived than cotinine. *Clin. Pharmacol. Ther.*, 1990: **48**; 641–651.

82 Curvall M, Vala EK, Englund, G. Conjugation pathways in nicotine metabolism. *Effects of Nicotine on Biological Systems*. Adlkofer F, Thurau K. (Eds), Basel: Birkhäuser Verlag, 1991; pp. 69–75.

83 Byrd GD, Chang KM, Greene JM, deBethizy JD. Evidence for urinary excretion of glucuronide conjugates of nicotine, cotinine, and *trans*-3'-hydroxycotinine in smokers. *Drug Metab. Dispos.*, 1992: **20**; 192–197.

84 Schepers G, Demetriou D, Rustemeier K, Voncken P, Diehl B. Nicotine phase 2 metabolites in human urine-structure of metabolically formed *trans*-3'-hydroxycotinine glucuronide. *Med. Sci. Res.*, 1992: **20**; 863–865.

85 Seaton M, Kyerematen GA, Morgan M, Jeszenka EV, Vesell ES. Nicotine metabolism in stumptailed macaques, *Macaca arctoides. Drug Metab. Dispos.*, 1991: **19**; 946–954.

86 Caldwell WS, Greene JM, Byrd GD, Chang KM, Uhrig MS, deBethizy JD. Characterization of the glucuronide conjugate of cotinine: a previously unidentified major metabolite of nicotine in smokers' urine. *Chem. Res. Toxicol.*, 1992: **5**; 280–285.

87 Byrd GD, Uhrig MS, deBethizy JD, Caldwell WS, Crooks PA, Ravard A, Riggs RM. Direct determination of cotinine-*N*-glucuronide in urine using thermospray liquid chromatography/mass spectrometry. *Biol. Mass Spectr.*, 1994: **23**; 103–107.

88 Hagan RL, Ramos Jr JM, Jacob III P. Increasing urinary cotinine concentrations at elevated temperatures: the role of conjugated metabolites. *J. Pharm. Biomed. Anal.*, 1997; 16: 191–197.

89 Crooks PA, Bhatti BS, Ravard A, Riggs RM, Caldwell WS. Synthesis of *N*-glucuronic acid conjugates of cotinine, *cis*-3'-hydroxycotinine and *trans*-3'-hydroxycotinine. *Med. Sci. Res.*, 1992: **20**; 881–883.

90 Seaton M, Vesell ES, Luo H, Hawes EM. Identification of radiolabeled metabolites of nicotine in rat bile. Synthesis of S-(−)-*N*-glucuronide and direct separation of nicotine-derived conjugates using high-performance liquid chromatography. *J. Chromatogr.*, 1993: **621**; 49–53.

91 Byrd GD, Caldwell WS, Crooks PA, Ravard A, Bhatti BS. Direct determination of nicotine-*N*-glucuronide in human biological samples. International Symposium on Nicotine: *Effects of Nicotine on Biological Systems II*. Montreal Canada, July 21–24, 1994. Clarke PBS, Quik M, Thurau K, Adlkofer F. (Eds), Basel: Birkhuser Verlag, 1994; Abstract P131.

92 Tsai MC, Gorrod JW. In vivo metabolism of nicotine in rat and guinea pig: evidence of urinary excretion of nicotine glucuronide conjugate in these species. *Eur. J. Drug Metab. Pharmacokinet.*, 1999; (submitted).

93 Tsai MC, Gorrod JW. Evidence for the biosynthesis of a glucuronide conjugate of (S)-(−)-nicotine, but not of (S)-(−)-cotinine or (±)-*trans*-3'-hydroxycotinine by marmoset hepatic microsomes. *Drug Metab. Drug Interact.*, 1999: **15**; (in press).

94 Mackenzie PI. The UDP glucuronosyltransferase multigene family. *Rev. Biochem. Toxicol.*, 1995: **11**; 29–72.
95 Green MD, Tephly TR. Glucuronidation of amine substrates by purified and expressed UDP-glucuronosyltransferase proteins. *Drug Metab. Dispos.*, 1998: **26**; 860–867.

*Chapter 16*

# Uses and abuses of cotinine as a marker of tobacco smoke exposure

Peter N. Lee

*P N Lee Statistics and Computing Ltd., Hamilton House, 17 Cedar Road, Sutton, Surrey SM2 5DA, UK*

## I. INTRODUCTION

Cotinine, as a biomarker of exposure to tobacco smoke, is typically measured in blood, saliva or urine. Cotinine levels in plasma and saliva are similar, but urinary levels are higher, generally by a factor of 5 to 6 [1]. Cotinine has also been detected in a wide range of other fluids, including amniotic fluid and breast milk [1], placental tissue [2], umbilical cord blood [3], semen [4], cerebrospinal fluid [5], cervical fluid [6] and even serum from a testicular biopsy [7]. Cotinine is also detectable in hair, though only at very low levels, and here nicotine is used as the marker of exposure [8].

There are seven main uses to which the measurement of cotinine has been put:

1. To provide evidence of exposure to tobacco smoke;
2. To compare exposure in smokers and non-smokers;
3. To use high levels in self-reported non-smokers to indicate that they are in fact true smokers;
4. To quantify the extent of exposure in cigarette smokers;
5. To compare exposure in different types of smokers;
6. To quantify the extent of ETS exposure in non-smokers;
7. To provide evidence of a relationship of tobacco smoke exposure with disease.

This chapter considers these uses in turn in Sections II to VIII, summarizing relevant results from the literature and commenting on the problems of interpretation that arise. Finally, some other issues relevant to the use of cotinine as a marker of tobacco smoke exposure are considered in Section IX and some general conclusions are reached in Section 1X.

## II. EVIDENCE OF EXPOSURE TO TOBACCO SMOKE

As will be seen in the following sections, there is no doubt that, on average, cotinine levels are substantially higher in smokers than in non-smokers, and are positively associated with the amount smoked by smokers and with the extent of ETS exposure in non-smokers. Although, in most cases, the presence of cotinine in body fluids arises as a result of active or passive smoking, it should be borne in mind that this need not necessarily be the case.

The most obvious alternative explanation for the presence of cotinine is exposure to other sources of nicotine, including use of smokeless tobacco products, chewing of nicotine gum, and use of nicotine patches or other aids for smoking cessation, and also occupational exposure to tobacco leaves or to nicotine products [1]. Chemical analytical error may, in theory, lead to a false reporting of the presence of cotinine, although this should not arise if appropriate methods are used (see Chapters 6, 7 and 8).

### Dietary nicotine

Nicotine has been shown to be present in measurable quantities in several foods, including tomatoes, aubergines (egg plant) and other solanaceous vegetables, tea and potatoes [9–11], and there has been some discussion in the literature regarding the possible effect that dietary intake of nicotine might have on the use of cotinine as a marker of tobacco smoke exposure. Claims that dietary nicotine might be important are not well supported. One author [12] concluded that "a person who consumed 10 cups of tea in a day, ate food comprising 1 lb of tomatoes, aubergines or peppers, and 1 lb of potatoes with their skins, might ingest nicotine equivalent to 1–2 cigarettes", overestimated the nicotine equivalent by almost two orders of magnitude, due to a confusion between dry and wet weight in the calculations. Another author [13] presented calculations demonstrating that nicotine intake from someone eating a pound of potatoes is similar to that from someone spending about half a day in a smoky room with a low concentration of ETS, and concluded that dietary intake may be of practical importance when using cotinine levels to quantify the extent of ETS exposure in non-smokers. However the ETS concentration used in these calculations, of 1 µg nicotine per m$^3$, is really quite low, 5-10 µg/m$^3$ being common and 50 µg/m$^3$ being sometimes found in bars and nightclubs [14]. In fact, though nicotine from dietary sources is potentially detectable in cotinine assays, there is no indication that it would more than minimally interfere with the use of cotinine as a quantitative marker of ETS exposure. Most researchers have argued that diet is a relatively trivial source of nicotine [15–18] compared not only to smoking but also to moderate ETS exposure. This view is supported by results from two epidemiological studies which related cotinine levels to reported dietary intake. In one large US study [19] it was estimated that the contribution of diet to serum cotinine level was less than 0.020 ng/ml, more than 10 times less than that related to ETS exposure, and 1000 times or so less than that found in smokers. Another study, in Scotland [20], found that cotinine levels in non-smokers reflected far better the nicotine in inhaled ETS than they did the nicotine in tea.

## III. RELATIVE EXPOSURE IN SMOKERS AND NON-SMOKERS

### Cotinine levels in smokers and non-smokers

A number of studies have compared average cotinine levels in self-reported smokers and non-smokers. Tables 1 (urine), 2 (serum or plasma) and 3 (saliva) summarize the findings from some of the larger studies. The tables show the population studied, the chemical analytical technique used for cotinine determination, the type of average used (mean or median) and whether non-smokers with high cotinine levels have been omitted from the averages used.

In the serum, plasma or saliva of smokers, average cotinine levels are typically in the range 200–350 ng/ml. Exceptionally they are in the range 100–150 ng/ml in pregnant women or in adolescents. In non-smokers levels are typically no more than about 1 ng/ml, provided they are calculated as medians or as means excluding non-smokers with high cotinine levels typical of smokers. Means including such individuals, e.g. in the studies of Perez-Stable [32], Pojer [33], Richmond [35] and McNeill [39], tend to be higher, about 5 ng/ml. For studies of adults, and discounting means which include non-smokers with high cotinine levels, the ratio of average cotinine for smokers compared to non-smokers typically exceeds 200. This is evident not only from the three large representative studies conducted in the UK in 1993, 1994 and 1996 [27,28,34], but also from a large representative study undertaken in the US in 1988–1991 [19]. In that study, which used state-of-the-art high-performance liquid chromatography atmospheric-pressure chemical ionization tandem mass spectrometry (LC-MS/MS) [40], non-tobacco users aged 17+ had geometric mean cotinine levels of about 0.2 ng/ml, over 500 times lower than those seen in tobacco users, which appeared to be of order 300 ng/ml from a distribution given graphically.

For cotinine in urine, typical levels in the Jarvis [22], Thompson [25] and Wald [26] studies, whether in ng/ml or in ng/mg creatinine, were about 1500 in smokers and about five in non-smokers, leading to a ratio of about 300. Again, studies which did not exclude non-smokers with high cotinine levels [21,23,24] gave much lower ratios. Two of these studies [23,24] reported cotinine levels that were exceptionally high, leading to doubts about the methodology used [41].

*Analytical Determination of Nicotine*

TABLE 1

Cotinine levels in urine (ng/ml or ng/mg creatinine)

| Study | Population | Method[1] | Units | Type of average[2] | N[3] | Cotinine level[4] | | |
|---|---|---|---|---|---|---|---|---|
| | | | | | | Non-smoker | Smoker | Ratio |
| Galanti [21] | Belgium/adults aged 18–29 | FPI/Cr | ng/mg | Mean | 2431 | 32.0 | 717 | 22.4 |
| Jarvis [22] | UK/adult outpatients | GC | ng/ml | Mean* | 190 | 4.8 | 1404 | 293 |
| Mathai [23] | UK/pregnant females | SP/Cr | ng/mg | Mean | 285 | 452 | 5980 | 13.2 |
| Matsukura [24] | Japan/adults | RIA/Cr | ng/mg | Mean | 744 | 680 | 8570 | 12.6 |
| Thompson [25] | UK/adult males | RIA/Cr | ng/mg | Median | 233 | 6.1 | 1623 | 266 |
| | | RIA/Cr | ng/mg | Mean | 233 | 13.4 | 1691 | 126 |
| Wald [26] | UK/adult males | RIA | ng/ml | Median | 453 | 4.0 | 1474 | 369 |

[1] Method: GC=gas chromatography, SP=spectrophotometric, RIA=radioimmunoassay, FPI=fluorescence polarization immunoassay, /Cr=corrected for creatinine.

[2] Type of average: * indicates that non-smokers with cotinine levels typical of smokers have been excluded from the calculation of the mean.

[3] N: number of non-smokers and smokers combined.

[4] Cotinine level: where appropriate, levels have been estimated from data by level of ETS exposure or by smoking category, or converted from nmol/ml to ng/ml.

TABLE 2

Cotinine levels in serum or plasma (ng/ml)

| Study | Population | Method[1] | Type of average[3] | N[3] | Cotinine level[4] | | |
|---|---|---|---|---|---|---|---|
| | | | | | Non-smoker | Smoker | Ratio |
| Bennett [27] | UK/adults | GC | Mean* | 4864 | 1.1 | 217 | 198 |
| Colhoun [28] | UK/adult males | GC | Mean* | 4772 | 0.9 | 232 | 255 |
| | UK/adult females | GC | Mean* | 5242 | 0.8 | 230 | 299 |
| Haddow [29] | US/adult females | RIA | Mean | 296 | 2.9 | 304 | 105 |
| | | | Median | 296 | 1.1 | 305 | 277 |
| Haddow [30] | US/pregnant females | RIA | Median | 4211 | <10 | 134 | ? |
| Jarvis [22] | UK/adult outpatients | GC | Mean* | 190 | 1.5 | 280 | 187 |
| Kemmeren [31] | Holland/adult males | GC | Mean | 253 | 1.1 | 327 | 304 |
| Perez-Stable [32] | US/Mexican adults | GC | Mean | 743 | 5.3 | 170 | 32.2 |
| Pojer [33] | Australia/adults[5] | GC | Mean | 368 | 4.4 | 335 | 76.2 |
| Prescott-Clarke [34] | UK/adult males | GC | Mean* | 5446 | 1.0 | 235 | 233 |
| | UK/adult females | GC | Mean* | 5996 | 0.8 | 222 | 266 |
| Richmond [35] | Australia/adults[6] | GC | Mean | 420 | 5.0 | 335 | 67.0 |
| Tunstall-Pedoe [20] | UK/adult males | GC | Median | 3813 | 0.7 | 250 | 368 |
| | UK/adult females | GC | Median | 3657 | 0.1 | 243 | 2430 |
| Wald [36] | UK/adult males | RIA | Mean | 304 | 0 | 276 | ∞ |

[1] Method: GC=gas chromatography, RIA=radioimmunoassay.
[2] Type of average: * indicates that non-smokers with cotinine levels typical of smokers have been excluded from the calculation of the mean.
[3] N: number of non-smokers and smokers combined.
[4] Cotinine level: where appropriate levels have been estimated from data by level of ETS exposure or by smoking category, or converted from nmol/ml to ng/ml.
[5] In smoking cessation campaign or attending clinic.
[6] Smokers in smoking cessation program.

TABLE 3
Cotinine levels in saliva (ng/ml)

| Study | Population | Method[1] | Type of average[2] | N[3] | Cotinine level[4] | | |
|---|---|---|---|---|---|---|---|
| | | | | | Non-smoker | Smoker | Ratio |
| Jarvis [22] | UK/adult outpatients | GC | Mean* | 190 | 1.7 | 314 | 185 |
| Jarvis [37] | UK/children aged 16–19 | GC | Median | 699 | 0.6 | 130 | 217 |
| Lee [17] | UK/adult females | GC | Median | 422 | 0.9 | 311 | 365 |
| | UK/adult males | GC | Median | 562 | 0.4 | 319 | 798 |
| Li [38] | US/pregnant females | RIA | Mean* | 1279 | 1.1 | 123 | 112 |
| McNeill [39] | UK/girls aged 11–16 | GC | Mean | 508 | 4.6 | 123 | 27 |

[1] Method: GC=gas chromatography, RIA=radioimmunoassay.
[2] Type of average: * indicates that non-smokers with cotinine levels typical of smokers have been excluded from the calculation of the mean.
[3] N: number of non-smokers and smokers combined.
[4] Cotinine level: where appropriate levels have been estimated from data by level of ETS exposure or by smoking category, or converted from nmol/ml to ng/ml.

## Factors affecting estimates of the ratio of cotinine levels in smokers and non-smokers

There are a number of issues that are relevant to the ratio of cotinine levels in smokers and non-smokers. One problem, as already noted, is that the ratio may be substantially underestimated if some smokers misreport their smoking and have their high cotinine levels included in the average for non-smokers. While this has little effect on the median, it biases the mean upward, sometimes substantially. Even one smoker with a serum cotinine of 300 ng/ml can double the estimate of the mean, when erroneously included within a sample of 300 non-smokers, with a mean cotinine of 1 ng/ml.

Another problem relates to analytical accuracy at low levels of cotinine. The lower the level of cotinine, the higher the percentage variability, and all methods, with the possible exception of LC-MS/MS [40], have a minimum detection limit that is higher than the level seen in a proportion of non-smokers. For values below the detection limit, there is uncertainty as to what value should be included in the calculation of the mean, and for values just above it there is considerable doubt about its reliability. While one can have moderate confidence about the reliability of the average level in smokers reported in recent studies using appropriate methods, the average level reported in non-smokers, and hence the calculated ratio of levels, is far less precise.

The effect the choice of analytical method has on the ratio of levels is well illustrated by a study [42] in which 11 laboratories from 6 countries, all experienced in performing cotinine determinations in biological fluids by radioimmunoassay and/or gas chromatography, each received 18 serum and 18 urine samples, 8 from smokers and 10 from non-smokers (2 spiked with defined amounts of nicotine and cotinine). While all the laboratories distinguished perfectly between the smokers and non-smokers, and the ranking of the samples agreed well over laboratory, there were systematic differences in absolute values between the laboratories. Notably, the ratios of urinary cotinine concentrations between active and passive smokers differed widely from laboratory to laboratory, ranging from 21 to 294.

As noted in the previous section, dietary intake of nicotine is only likely to have a small effect on cotinine levels. As its contribution will be trivial in smokers, but may be less so in non-smokers, the effect of ignoring diet will be to underestimate somewhat the ratio of cotinine levels in smokers compared to non-smokers.

## Interpretation of the ratio of cotinine levels in smokers and non-smokers

Apart from the technical problems in obtaining an accurate estimate of the ratio of cotinine levels in smokers compared to non-smokers, one must bear in mind the extent to which such a ratio can be interpreted as an index of relative exposure to tobacco smoke constituents. In the first place, cotinine is at best only a marker of recent exposure to nicotine, and is not necessarily a marker of lifetime exposure of, or exposure to, other tobacco smoke constituents. Nicotine is predominantly in the particulate phase of mainstream smoke and in the vapour phase of ETS [43] and the pattern of absorption, which depends on the pH of the smoke [14], will differ for smokers and non-smokers. It has been estimated [43] that the dose ratio between smoking and ETS exposure, while about 100–125 for nicotine, is much higher for particles, (1700–4000), and for tobacco specific nitrosamines, (3000–6000), and much lower for vapour phase constituents such as carbon monoxide, (2.7–4.2) or volatile nitrosamines, (1.5–2.5).

Even as a marker of exposure to nicotine, cotinine is only one of a number of metabolites and smokers and non-smokers may differ in their patterns of metabolism, as discussed further below.

For these reasons, one certainly cannot assume that, just because average cotinine levels in smokers may be 300 times, say, higher than those in non-smokers, it necessarily follows that exposure to those tobacco smoke constituents that are relevant to a specific disease is 300 times greater in smokers than non-smokers [44,45].

## IV. RECLASSIFYING SELF-REPORTED NON-SMOKERS AS SMOKERS

If some smokers fail to report they smoke, the prevalence of smoking in the population concerned will be underestimated. When attempting to study the relationship of ETS exposure to risk of lung cancer (or other diseases strongly related to smoking) among non-smokers, the inclusion of a small proportion of misclassified smokers in the study population can cause important bias, and create an apparent relationship of ETS exposure to risk even when no true relationship exists [46].

### Using cotinine levels to estimate misclassification rates

Failure to report previous smoking can be demonstrated from studies which have recorded smoking habits in the same individuals at different points in time [47]. Failure to report current smoking can be detected using cotinine. The rationale for regarding high cotinine levels as indicative of current smoking, rather than of extreme ETS exposure, is based on two main lines of argument. First, the distribution of cotinine values in those not reporting smoking but reporting ETS exposure is usually found to take the form of a continuous log-normal distribution with a mean a 100 times or more lower than that of smokers, together with a few very high values that do not fit the distribution at all, but are typical levels for smokers [17,19]. Second, non-smokers exposed to maximally tolerable levels of ETS in experimental chamber studies [48,49] or working in occupations with very high ETS exposure [50,51] are not found to have cotinine levels that even approach typical levels seen in smokers. The EPA has suggested that self-reported non-smokers with cotinine levels above 10% of smokers' levels can definitely be regarded as true smokers, and that cotinine levels above 30% should be regarded as indicating regular smoking [52]. However, a number of researchers use rather lower levels, e.g. 8 to 20 ng/ml in serum or saliva [47], to reclassify non-smokers as smokers. At the end of the day, cotinine levels cannot completely distinguish between non-smokers with high ETS exposure and very light active smokers. However, they can indicate levels that are certainly not due to ETS and, provided a satisfactory attempt is made to exclude other sources of nicotine such as gum or patches, are due to active smoking.

*Levels of misclassification*

Table 4 summarizes evidence on misclassification rates from over 40 studies. Three types of rate are shown:

A. percentage of self-reported non-smokers found to be smokers.
B. percentage of true current smokers (i.e. including misclassified smokers) who deny current smoking and report being non-smokers.
C. percentage of true current smokers who deny current smoking and report never having smoked.

Rate A can be calculated for all the studies (and the studies are presented in ascending order of this), but, as it depends on the underlying frequency of smokers in the population, is less valuable than rates B or C. Rate B is only available if the study was not restricted to self-reported non-smokers, while rate C additionally requires data to have been collected on past smoking.

**Factors affecting estimated misclassification rates**

As is evident from Table 4, misclassification rates vary widely between studies. There are a number of reasons for this. These include the precise choice of cut-off used, the methodology used for cotinine determination, the underlying prevalence of smoking in the population (which should only affect rate A) and failure in some studies to ask questions about all types of smoking or to consider other sources of nicotine (though many studies were very thorough in this respect). A particularly important reason is that study populations vary in willingness to admit smoking when asked.

A striking observation from Table 4 is that, in the three studies of Asian women, two conducted in Japan [74,75] and one in South-East Asian women living in the USA [76], the proportion of true smokers denying smoking (rate B) is consistently high (26.4%, 40.7% and 64.6%). Also, unlike in Western

TABLE 4

Rates of misclassification of current smoking[1]

| Study | Country[2] | Sex | Body fluid | Cut-off | among self-reported non-smokers | | % denying current smoking and reporting as | |
|---|---|---|---|---|---|---|---|---|
| | | | | | n | % (A) | Non-smokers (B) | Never smokers (C) |
| Emmons [53] | USA | M+F | Saliva | 10 ng/ml | 1 | 0.5 | | |
| Cummings [54] | USA | M+F | Urine | 90 ng/ml | 6 | 0.9 | | |
| Wald [26] | UK | M+F | Urine | 128 ng/ml | 2 | 0.9 | 0.9 | |
| Delfino [55] | Canada | M+F | Saliva | 20 ng/ml | 3 | 1.2 | | |
| Pirkle [19] | USA | M+F | Serum | 15 ng/ml | 78 | 1.3 | 2.0 | |
| Martinez [56] | USA | F | Serum | 20 ng/ml | 2 | 1.4 | 5.1 | |
| Thompson [25] | UK | M+F | Urine | 100 ng/mg | 3 | 1.6 | 5.8 | |
| English [57] | USA | F | Serum | 14 ng/mg | 42 | 1.7 | 3.4 | 2.3 |
| Haley [58] | USA | M+F | Urine | 100 ng/mg | 5 | 1.7 | | |
| Haddow [59] | USA | F | Serum | 10 ng/ml | 29 | 1.9 | | |
| Riboli [60] | Many | F | Urine | 100 ng/mg | 26 | 1.9 | | |
| Slattery II [61][3] | USA | F | Serum | 15 ng/ml | 8 | 2.1 | 4.7 | 2.3 |
| Haddow [29] | USA | F | Serum | 16 ng/ml | 5 | 2.2 | 7.2 | |
| Fontham [62] | USA | F | Urine | 100 ng/mg | 25 | 2.3 | | |
| Lee [17] | UK | M+F | Saliva | 30 ng/ml | 20 | 2.5 | 3.1 | |
| Heller II [63][4] | Germany | M+F | Serum | 15 ng/ml | 65 | 2.8 | 5.7 | 1.2 |
| Dickinson [64] | Australia | M+F | Saliva | 35 ng/ml | 10 | 2.8 | 7.7 | |
| Heller A [63][4] | Germany | M+F | Serum | 15 ng/ml | 72 | 3.0 | 5.5 | 0.8 |
| Colhoun [28] | UK | M+F | Serum | 20 ng/ml | 232 | 3.2 | 7.0 | 1.8 |
| Heller B [63][4] | Germany | M+F | Serum | 15 ng/ml | 71 | 3.2 | 7.6 | 1.8 |
| Tunstall-Pedoe [20] | UK | M+F | Serum | 20 ng/ml | 141 | 3.4 | 4.1 | |
| Bennett [27] | UK | M+F | Serum | 20 ng/ml | 121 | 3.5 | 7.3 | 2.8 |
| Luepker [65] | USA | M+F | Saliva | 20 ng/ml | 8 | 3.6 | 6.5 | 4.1 |
| Becher [66] | Germany | F | Urine | 100 ng/ml | 7 | 3.6 | | |
| Van Vunakis [67] | USA | M+F | Saliva | 25 ng/ml | 7 | 3.8 | 5.6 | |

TABLE 4 Continued

| Study | Country[2] | Sex | Body fluid | Cut-off | Current smokers among self-reported non-smokers | | % denying current smoking and reporting as | |
|---|---|---|---|---|---|---|---|---|
| | | | | | n | % (A) | Non-smokers (B) | Never smokers (C) |
| Haddow [30] | USA | F | Serum | 25 ng/ml | 112 | 3.9 | 7.7 | |
| Wagenknecht [68] | USA | M+F | Serum | 14 ng/ml | 145 | 4.2 | 8.6 | 4.3 |
| Pierce [69] | Australia | M+F | Saliva | 34 ng/ml | 31 | 5.0 | 8.1 | |
| Ogden [70] | USA | F | Saliva | 10%[5] | 45 | 5.5 | 14.1 | 6.4 |
| Coultas [71] | USA | M+F | Saliva | 20 ng/ml | 51 | 5.7 | 14.6 | 7.1 |
| Van Vunakis [67] | USA | M+F | Serum | 8 ng/ml | 11 | 5.7 | 8.0 | |
| Perez-Stable [32] | USA | M+F | Plasma | 14 ng/ml | 12 | 6.3 | 11.4 | 4.8 |
| Slattery I [61][3] | USA | M | Serum | 15 ng/ml | 7 | 6.4 | | |
| Ford [72] | N.Z. | M | Serum | 15 ng/ml | 30 | 6.6 | 26.1 | |
| Ogden [73] | USA | F | Saliva | 10 ng/ml | 8 | 7.7 | | |
| Lee [74] | Japan | F | Urine | 100 ng/mg | 28 | 8.8 | 26.4 | 20.8 |
| Akiyama [75] | Japan | F | Urine | 100 ng/mg | 11 | 9.2 | 40.7 | 36.0 |
| Wewers [76] | Asia[6] | M | Saliva | 14 ng/ml | 43 | 9.3 | 11.8 | 6.0 |
| Wewers [76] | Asia[6] | F | Saliva | 14 ng/ml | 64 | 10.9 | 64.6 | 61.6 |
| Luepker [65] | USA | M+F | Saliva | 20 ng/ml | 31 | 12.4 | 26.3 | 5.9 |
| Phillips [77] | UK | F | Saliva | 25 ng/ml | 41 | 12.5 | | |
| Slattery III [61][3] | USA | M | Serum | 15 ng/ml | 27 | 17.2 | | |
| Jarvis [22] | UK | M+F | Plasma | 14 ng/ml | 21 | 17.4 | 18.9 | |

[1] Adapted from Lee [47], with data from additional studies added.
[2] N.Z. = New Zealand.
[3] Results were reported for three different study populations; see [47].
[4] Heller A and Heller B are results from the same sample at two time points three years apart, Heller II are results from a second sample.
[5] 10% of smokers' levels.
[6] Study conducted in US but restricted to subjects from Laos, Cambodia or Vietnam.

populations, where most misclassified current smokers admit past smoking and have average cotinine levels that are substantially lower than in those who admit their current smoking, the Asian women identified as misclassified current smokers predominantly claim never to have smoked and to have cotinine levels typical of current smokers who admit it. The high misclassification rates in Asian women no doubt occur because smoking by women is socially unacceptable there.

It is well documented [78–80] that misclassification rates are likely to be particularly high in studies where patients are advised to give up smoking, patients being unwilling to admit to their doctor or to their spouse that they have failed to do what they were asked. Studies conducted in a situation where subjects had, or were likely to have had, advice to give up smoking and which reported high misclassification rates, include those by Jarvis et al. [22], Slattery et al. [61] and Ford et al. [72]. Higher than average misclassification rates were also noted in studies where non-smokers were paid to participate [77,81], concealment of smoking probably occurring for financial reasons. By contrast, screening studies, where subjects may be more ready to admit smoking, tended to produce quite low detection rates. These are only some of the factors affecting the extent to which smokers admit their habit. A comprehensive review of the evidence on misclassification of smoking as determined by cotinine gives a fuller discussion [47].

## V. EXTENT OF EXPOSURE IN CIGARETTE SMOKERS

Cotinine is often used in an attempt to quantify the extent of exposure in cigarette smokers. Table 5 summarizes evidence from various studies of smokers that have reported the correlation between cotinine level and the reported number of cigarettes per day smoked, with results shown separately for urine, serum or plasma, and saliva. Although the studies are consistent in showing a positive correlation regardless of body fluid, the magnitude of the correlation is not very strong, typically in the range 0.35–0.50, though some studies report lower values.

A positive relationship is also evident from the data shown in Table 6, showing how cotinine level varies by grouped amount smoked, with results presented from three large representative surveys of the UK population [27,28,34], a population study in Germany [93], five studies in the US – two of pregnant women [30,57] and two not [37,95,96] – and two studies in Japan [24,74]. In virtually all these studies cotinine levels increase monotonically with increasing number of cigarettes smoked.

### Do heavy smokers inhale less?

Although the relationship is monotonic, it may not be linear and there is a suggestion from some of the studies (though not all) that the increase in cotinine per cigarette becomes smaller with increasing number of cigarettes.

TABLE 5

Correlation of cotinine level with number of cigarettes per day smoked (by cigarette smokers)

| Body fluid | Study | Country | Sex | Sample size | Correlation coefficient |
|---|---|---|---|---|---|
| Urine | Bodmer [82] | UK | M+F | 38 | 0.50 |
| | Matsukura [24] | Japan | M+F | 247 | 0.15 |
| Serum/ | Adlkofer [83] | Germany | M | 200 | 0.41-0.44* |
| Plasma | Bennett [27] | UK | M+F | 1290 | 0.49 |
| | Klebanoff [84] | USA | F | 192 | 0.44 |
| | Kemmeren [31] | Holland | M | 148 | 0.44 |
| | Puntoni [85] | Italy | M+F | 120 | 0.40 |
| | Suadicani [86] | Denmark | M | 1710 | 0.27 |
| | Woodward [87] | UK | F | 1939 | 0.46 |
| | | | M | 1574 | 0.49 |
| | Worrell [88] | UK | F | 240 | 0.38 |
| | | | M | 456 | 0.38 |
| Saliva | Coultas [90] | USA | M+F | 260 | 0.52 |
| | Istvan [91] | USA | F | 2096 | 0.15 |
| | | | M | 3538 | 0.21 |
| | Pierce [69] | Australia | M+F | 353 | 0.33 |
| | Stanton [92] | New Zealand | M+F | 265 | 0.38 |

* Cotinine samples taken at three time points.

The tendency for the slope to decrease as the number of cigarettes increases is more clearly seen in the final study [97], which subdivides cigarette consumption more finely. The results of the study are claimed by the authors to explain the anomaly that "the relationship between the number of cigarettes smoked per day and the incidence of lung cancer is linear, but, from the multistage model of carcinogenesis, it should be quadratic (upwards curving)." The authors effectively argue that use of number smoked per day as an index of exposure is misleading as heavier smokers inhale less smoke from each cigarette, and that cotinine is a better and, indeed, appropriate index of exposure.

While there is evidence that cotinine levels are higher in smokers who report that they inhale than in those who report that they do not [86,88], and while it may also be true that heavy smokers do inhale less than do light smokers, the argument is not as straightforward as the authors would have it. In the first place the machine nicotine yields of cigarettes smoked by heavy and light smokers may differ. Furthermore, heavy and light smokers may differ in exposure to other lung carcinogens. A non-linear relationship of cotinine to amount smoked may arise because metabolism of nicotine to cotinine is not a simple linear function of the dose of nicotine absorbed. It is also possible that differences between heavier and lighter smokers in relation to other factors (e.g. alcohol consumption, see Section 8) might affect metabolism. The fact that the relationship between cigarette consumption

TABLE 6

Cotinine level in cigarette smokers by level of consumption

| Study (Country) | Body fluid | Sex | Sample Size | Cotinine level (ng/ml unless stated) with number per day in brackets below | | | | |
|---|---|---|---|---|---|---|---|---|
| Bennett [27] (UK) | Serum | M+F | 1270 | 115 (1-9) | 248 (10–19) | 319 (30+) | | |
| Colhoun [28] (UK) | Serum | M | 1350 | 151 | 269 | 323 | | |
| | | F | 1365 | 118 (1-9) | 253 (10–19) | 306 (20+) | | |
| Prescott-Clarke [34] (UK) | Serum | M | 1638 | 144 | 270 | 333 | | |
| | | F | 1592 | 110 (1-9) | 246 (10–19) | 308 (20+) | | |
| Heller [93] (Germany) | Serum | M | 502 | 49 | 133 | 209 | 278 | 348 |
| | | F | 385 | 46 (Occ.) | 113 (1-9) | 212 (10–19) | 278 (20–29) | 355 (30+) |
| English [57] (USA) | Serum | F** | 374 Blacks | 91 | 149 | 182 | | |
| | | | 829 Whites | 44 (1-9) | 128 (10–19) | 164 (20+) | | |
| Haddow [30] (USA) | Serum | F** | 1340 | 68 (1-7) | 119 (8–12) | 135 (13–17) | 159 (18–24) | 184 (25+) |
| Perez-Stable [94] (USA) | Serum | M | 287 | 70 | 181 | 243 | | |
| | | F | 260 | 84 (1-9) | 160 (10–19) | 257 (20+) | | |
| Sepkovic [95] (USA) | Plasma | M | 958 | 141 | 261 | 332 | 375 | 381 |
| | | F | 749 | 125 (1–10) | 363 (11–20) | 315 (21–30) | 322 (31–40) | 404 (41+) |
| Ahijevych [96] (USA) | Saliva | F | 142 | 360 (109) | 399 (10–19) | 407 (20+) | | |
| Lee [74] (Japan) | Urine* | F | 78 | 259 (1–5) | 1018 (6–10) | 1433 (11–15) | 2647 (16+) | |
| Matsukura [24] (Japan) | Urine* | M+F | 247 | 310 (1-9) | 420 (10–19) | 870 (20–29) | 1030 (30–39) | 1560 (40–49) |
| Law (97) (UK) | Serum | M | 1175+ | 45 (1-9) | 130 (10–14) | 247 (15–19) | 249 (20) | 262 (20–24) |
| | | | | 301 (25–29) | 287 (30–34) | 338 (35–39) | 343 (40+) | |

* ng/mg creatinine,     ** pregnant women     + includes never smokers.

and risk of lung cancer is less steep than predicted by the multistage model might also be due to inadequacies in this model and to random errors in reported cigarette consumption, which would tend to flatten out the observed slope.

In any case, the results of this study [97] seem inconsistent with that of another study [98] of 116 smokers, in which saliva cotinine was measured and daily cigarette consumption logged into a hand-held computer. That study reported a non-linear relationship in which cotinine levels rose rapidly up to 10 cigarettes/day, then flattened off, and then rose again above 20 cigarettes/day. In that study, saliva cotinine was also noted to be positively associated with age.

Variation in inhalation and metabolism by amount smoked and errors in reporting amount smoked may all contribute to the relationship between cotinine and daily number of cigarettes smoked not being a strong one. The correlation may also be weakened because the daily number of cigarettes smoked on average may not correspond to the actual number smoked during the days to which the cotinine value was relevant and, probably more importantly because of the considerable inter-individual variation in cotinine levels between smokers of the same number of cigarettes/day. This may arise due to differences in the cigarettes smoked, inhalation, absorption and metabolism.

## VI. COMPARING EXPOSURE FOR DIFFERENT TYPES OF SMOKERS

### Pipe and cigar smokers

Table 7 compares cotinine levels in pipe and cigar smokers with levels in cigarette smokers based on data from three large representative studies of the UK [27,28,34], five other UK studies [17,22,26,36,87] and one study in Germany [93]. In the German study, cotinine levels in both pipe-only smokers and cigar-only smokers are lower than in cigarette smokers. In the UK studies, however, the results are consistent in showing that while cigar smokers have a markedly lower cotinine level than cigarette smokers, pipe smokers have a similar level. The similarity of cotinine levels in pipe and cigarette smokers, coupled with evidence that heart disease rates are substantially higher in cigarette than in pipe smokers, was used by Wald et al. [36] to infer that nicotine is not the cause of coronary heart disease. This inference is, however, open to question. The pattern of exposure to nicotine in cigarette smokers (intermittent series of high-nicotine boli) differs sharply from that in pipe smokers (slower absorption without peaks) [99], and similar levels of cotinine need not necessarily imply similar levels of circulating nicotine [100].

Table 7 also shows that cotinine levels in smokers of cigars only or in smokers of pipes or cigars are consistently higher in those who have smoked cigarettes in the past than in those who have not done so. This is consistent with inhalation in smokers of cigars only being shallower than in smokers of cigarettes only, but with cigarette smokers switching to cigars tending to retain their inhaling habits.

### Handrolled cigarette smokers

Few studies have compared cotinine levels in smokers of handrolled cigarettes and smokers of manufactured cigarettes. Those that have done so suggest rather higher levels in handrolled cigarette smokers [17,93].

TABLE 7

Cotinine levels in smokers of pipes and cigars[1]

| Study | Country | Body fluid | Product smoked | Cigarette smoking | Sample size | Cotinine level (ng/ml) | Cotinine Level in cigarette only smokers |
|---|---|---|---|---|---|---|---|
| Bennett [27] | UK | Serum | Pipe or cigar | Ex | 118 | 151 | 234 |
| | | Serum | Pipe or cigar | Never | 71 | 49 | |
| Colhoun [28] | UK | Serum | Pipe or cigar | Ex | 227 | 136 | 263 |
| | | Serum | Pipe or cigar | Never | 112 | 57 | |
| Prescott- | UK | Serum | Pipe or cigar | Ex | 226 | 150 | 263 |
| Clarke [34] | | Serum | Pipe or cigar | Never | 147 | 56 | |
| Lee [17] | UK | Saliva | Cigar only | Non | 9 | 10 | 310 |
| Jarvis [22] | UK | Plasma | Pipe or cigar | Non | 15 | 210 | 294 |
| | | Saliva | Pipe or cigar | Non | 15 | 231 | 330 |
| | | Urine | Pipe or cigar | Non | 15 | 1185 | 1448 |
| Wald [26] | UK | Urine | Cigar only | Non | 59 | 396 | 1645 |
| | | Urine | Pipe only | Non | 42 | 1920 | |
| Wald [36] | UK | Serum | Cigar only | Ex | 55 | 149 | 306 |
| | | Serum | Cigar only | Never | 15 | 18 | |
| | | Serum | Pipe only | Ex | 33 | 415 | |
| | | Serum | Pipe only | Never | 23 | 353 | |
| Woodward [87] | UK | Serum | Cigar only | Non | 319 | 9 | 283 |
| | | Serum | Pipe only | Non | 39 | 259 | |
| | | Serum | Pipes and cigars | Non | 83 | 203 | |
| Heller [93] | Germany | Serum | Pipe only | Non | 153 | 92 | 233 |
| | | Serum | Cigar only | Non | 136 | 79 | |

[1] Data shown are for males or predominantly for males.

## Compensation by smokers of lower tar and nicotine cigarettes

Many experimental and observational studies have suggested that smokers who switch to a brand with a lower tar and nicotine level, as measured by a smoking machine under standard conditions, adjust their inhalation pattern in order to compensate [101]. A review of the evidence using three commonly used markers of intake, carboxyhaemoglobin, breath carbon monoxide and blood nicotine, suggests that though cigarette smokers switching to a lower tar and nicotine brand do compensate to some extent, such compensation is not complete, so they do inhale less carbon monoxide and nicotine [102]. More recent compensation studies have used cotinine as the marker, which seems logical if, as is commonly thought [103], nicotine determines the extent of smoke intake, given that cotinine is the major metabolite of nicotine.

Some investigators have attempted to quantify compensation by means of an index $\alpha$, which relates cotinine level C to brand nicotine yield Y by the model $C = \mu Y^{1-\alpha}$. Here $\alpha = 1$ implies complete compensation, with cotinine level independent of nicotine yield, and $\alpha=0$ implies no compensation, with cotinine level directly proportional to nicotine yield. In one such study [83], of 103 German cigarette smokers supplying blood samples on six occasions, $\alpha$ was estimated as 0.755, after adjustment for amount smoked. This is equivalent to a 50% reduction in brand nicotine yield leading to a 16% reduction in cotinine level. Similar conclusions were reached in a UK intervention trial [104] in which over 400 subjects were randomly allocated to change rapidly or slowly to a brand with a tar yield of about half that of their usual brand. Data collected at five time points included serum cotinine and, after adjustment for amount smoked, the authors estimated an average compensation index for nicotine of 0.79, with the 50% reduction in yield predicted to result in an estimated 14% reduction in cotinine.

In a later study in Germany [93], in which serum cotinine was determined in a representative study of over 4000 subjects, the compensation index was estimated as 0.70 (or a 19% reduction in cotinine for a 50% reduction in yield). Unlike the previous two studies, which looked at within-smoker changes, this looked at between-smoker changes. Such estimates may be biased downwards if more 'dedicated' smokers, who would inhale more than average under comparable circumstances, tend to choose to smoke higher yield cigarettes. In this context, it should be noted that the first German study [83] also presented compensation indices based on between-smoker differences which, averaging 0.57, were lower than the estimated value within-smoker. Though some studies, e.g. [88], have reported no significant association between cotinine level and brand nicotine yield, the literature generally suggests that though smokers compensate, they do so partly but not completely. While the evidence summarized here does not demonstrate directly that smokers compensate for nicotine specifically, and it has been suggested that titration for tar may occur [89], it is of interest to note that the UK intervention trial cited above (104) reported lower compensation indices for carbon monoxide (0.65) and for tar (0.62) than it did for nicotine (0.79).

## VII. EXTENT OF ETS EXPOSURE IN NON-SMOKERS

Evidence that ETS exposure might increase the risk of lung cancer in non-smokers was first derived from two studies conducted in 1981 [105,106] which reported that non-smoking women married to smokers had a higher risk than did non-smoking women married to non-smokers. Since then many other studies have used smoking by the husband as an index of ETS exposure in non-smoking women.

### Cotinine levels and smoking by the husband

Table 8 compares cotinine levels in non-smoking women according to their husband's smoking habits (or in some cases according to their partner's

TABLE 8

Cotinine levels in non-smokers[1] in relation to smoking by the husband

| Study | Country | Body fluid | Sample size | Cotinine level (ng/ml)[2] Husband non-smoker | Husband smoker | Ratio (Z) |
|---|---|---|---|---|---|---|
| Becher [66] | Germany | Urine* | 93 | 5.5 | 18.0 | 3.27 |
| | Poland | Urine* | 90 | 6.1 | 16.3 | 2.68 |
| Cummings [54] | USA | Urine | 225 | 6.8 | 10.5 | 1.55 |
| Heller [63][3] | Germany | Serum | 1936 | 0.36 | 0.95 | 2.66 |
| Lee [17] | UK | Saliva | 72 | 0.3 | 1.0 | 3.33 |
| Lee [74] | Japan | Urine | 264 | 18.0 | 11.5 | 0.64 |
| Ogden [73] | USA | Urine | 96 | 0.27 | 2.3 | 8.63 |
| Phillips [77][4] | UK | Saliva | 106 | 0.25 | 1.3 | 5.20 |
| | UK | Saliva | 101 | 0.25 | 0.90 | 3.60 |
| Pirkle [19][5] | USA | Serum | 2672 | 0.20 | 0.77 | 3.85 |
| Rebagliato [107] | Spain | Saliva | 698 | 0.73 | 1.28 | 1.75 |
| Riboli [60][5] | Many | Urine* | 1322 | 3.23 | 9.26 | 2.87 |
| Wald [108] | USA | Urine | 121 | 8.5 | 25.2 | 2.96 |
| Willers [109] | Sweden | Urine | 34 | 1.4 | 6.6 | 4.71 |

[1] Self-reported non-smokers with high cotinine levels indicative of smoking have generally been removed from the analyses presented.
[2] ng/mg creatinine where indicated by an asterisk in the body fluid column.
[3] Data are for partner smoked and are for sexes combined.
[4] Results are for pre-monitoring (line 1) and post-monitoring (line 2) cotinine levels, and are for smoking by the partner.
[5] Data are for any household exposure.

habits or to smoking in the household). It can be seen that, with the striking exception of the single study in Japan [74], where cotinine levels were non-significantly lower if the husband smoked, the studies consistently show higher cotinine levels in the partner if the husband smoked.

## Z ratios

A number of meta-analyses of the evidence relating husband's smoking to lung cancer, (e.g. [14,52,110,111]), have attempted to correct overall estimates of relative risk for the background ETS exposure of wives of non-smoking husbands. The method used assumes that the relationship between risk of lung cancer and ETS exposure is linear, and that the ratio of ETS exposure for women married to a smoker and for women married to a non-smoker is validly estimated by their relative cotinine exposure $Z$. Given these assumptions, the meta-analysis risk estimates for women married to a smoker, relative to women married to a non-smoker ($R$) can be converted into an estimate, relative to totally unexposed women, ($R'$), by using the formula

$$R' = 1 + Z(R-1)/(Z-R)$$

while the risk for women married to a non-smoker relative to totally unexposed women, $R''$, can be estimated using the formula

$$R'' = 1 + (R-1)/(Z-R).$$

For example, if $Z=3$ and $R=1.25$, the relative risk estimates adjusted for background become $R'=1.43$ (married to a smoker) and $R''=1.14$ (married to a non-smoker).

While it is reasonable to assume that the higher cotinine levels seen in non-smoking women married to a smoker indicate they receive greater ETS exposure, there are a number of problems with this background adjustment procedure [112]. In the first place, as can be seen from Table 8, there is considerable variability between studies in the estimated Z ratio. Thus, while one review [110] used the value of three obtained from an early UK study [108], another review [52] used a much lower value of 1.55, based on a single US study [54]. For the same R value of 1.25, the differing Z values have a large effect on the estimates of $R'$ (1.43 for $Z=3$, 2.29 for $Z=1.55$), with a consequently large effect on the estimated risk of lung cancer attributable to ETS.

Variations between studies in the estimated Z ratio may arise for a number of reasons. These include sampling error, analytical error at low cotinine levels (which will depend on the method used) and differences between the study populations in the extent to which men smoke and to which husbands smoke in the presence of the wife. The differing relative importance, in the differing study populations, of other sources of nicotine exposure (including other sources of ETS exposure, nicotine replacement products and diet) would also affect the Z ratio. Though marriage to a smoker may be a reasonable marker of increased exposure one must have distinct reservations about the described use of Z ratios, particularly when the populations investigated in the lung cancer studies differ from those used to estimate Z.

## Cotinine levels and workplace ETS exposure

Table 9 shows that cotinine levels in non-smokers are also increased in relation to workplace ETS exposure. However, exposed/unexposed ratios tend to be lower than for spousal smoking (Table 8) suggesting that, on average, marriage to a smoker is a more important source of ETS exposure than is working with a smoker. This conclusion is consistent with that of a number of studies which show higher cotinine levels in those with ETS exposure only at home than in those exposed only at work. For example, the IARC multicountry study [60] reported urinary cotinine levels of, respectively, 2.7, 4.8, 9.0 and 10.0 ng/mg creatinine for non-smoking women with reported ETS exposure neither at home nor at work, at work only, at home only, and both at home and at work. Similarly, in a large US study of both

687

TABLE 9

Cotinine levels in non-smokers[1] in relation to workplace ETS exposure

| | | | | Cotinine level (ng/ml)[2] | | |
| | | | | Workplace ETS exposure | | |
| Study | Country | Body fluid | Sample size | No | Yes | Ratio |
| --- | --- | --- | --- | --- | --- | --- |
| Becher [66] | Germany | Urine* | 93 | 8.6 | 8.8 | 1.03 |
| | Poland | Urine* | 90 | 8.1 | 13.4 | 1.65 |
| Matsukura [24] | Japan | Urine* | 277 | 0.22 | 0.72 | 3.27 |
| Ogden [73] | USA | Urine | 96 | 0.64 | 0.93 | 1.45 |
| Pirkle [19] | USA | Serum | 2672 | 0.23 | 0.46 | 2.00 |
| Rebagliato [107] | Spain | Saliva | 698 | 0.85 | 1.3 | 1.53 |
| Riboli [60] | Many | Urine* | 1322 | 5.0 | 6.7 | 1.35 |
| Sepkovic [95] | USA | Urine* | 267 | 4.5 | 7.1 | 1.58 |

[1] Self-reported non-smokers with high cotinine levels indicative of smoking have generally been removed from the analyses presented.
[2] Except where indicated by an asterisk in the body fluid column.

sexes [19] serum cotinine levels in non-tobacco users were, respectively, 0.132, 0.381, 0.651 and 0.926 ng/ml for the same four groups. While clearly some individual non-smokers are exposed to more ETS at work than at home, these and additional data (see [113]) make it clear that the claim by OSHA (114) that the workplace is "the overwhelming exposure location" has little foundation in fact. Similarly, the approximately 2-fold higher cotinine levels associated with at home rather than at work ETS exposure cited above [19,60] suggest strongly that a theoretically derived estimate of an approximately 4-fold higher tar exposure associated with at work ETS exposure [115] (which has been used for risk estimation [116]) is likely to be misleading.

### Cotinine levels and extent of ETS exposure

Quite a large number of studies have related cotinine levels in adults to questionnaire-based assessments of the extent of ETS exposure, either at home, at work or overall. Generally, a clear dose-response relationship can be seen, as, for example, in the three large representative UK studies [27,28,34] where serum cotinine in non-smokers was related to the number of cigarettes reported to be smoked per day by other household members (Table 10). Even in the highest categories, cotinine levels are typically two orders of magnitude less than those seen in smokers.

### Cotinine levels in children and parental smoking

Table 11 summarizes results for a number of studies which have compared cotinine levels in children according to smoking by the parents. In some

TABLE 10

Serum cotinine level in non-smokers[1] in relation to extent of household ETS exposure in three large representative UK studies

| Study | Sex | Sample size | Cotinine level (ng/ml)[2] | | | |
|---|---|---|---|---|---|---|
| | | | Passive smoke exposure[3] | | | |
| | | | None | Light | Moderate | Heavy |
| Bennett [27] | M+F | 3385 | 0.9 | 1.3 | 1.8 | 2.8 |
| Calhoun [28] | M | 3083 | 0.7 | 1.5 | 1.9 | 2.6 |
| | F | 3877 | 0.6 | 1.1 | 1.4 | 2.4 |
| Prescott- | M | 3435 | 0.8 | 1.3 | 1.9 | 2.9 |
| Clarke [34] | F | 4404 | 0.7 | 0.9 | 1.4 | 2.0 |

[1] Excluding those using nicotine replacement products, and self-reported non-smokers with cotinine levels of 20 ng/ml.
[2] Mean cotinine levels are shown; median levels are typically 30-50% lower but still show a clear dose-relationship.
[3] Based on total number reported to be smoked per day by other members of the household; light=less than 10; moderate=10, less than 20; heavy=20 or more.

TABLE 11

Cotinine levels in infants and non-smoking children in relation to their parents' smoking habits

| Study | Country | Age[2] | Body fluid | Sample size | Cotinine level (ng/ml)[2] | | | |
|---|---|---|---|---|---|---|---|---|
| | | | | | Smoking by parents | | | |
| | | | | | Neither | Father[3] only | Mother only | Both |
| Bakoula [117] | Greece | 1-5 | Urine* | 501 | 23.7 | 57.5 | 70.1 | 114.0 |
| Bono [118] | Italy | 14 | Urine* | 387 | 7.1 | 9.5** | 12.4 | 16.3 |
| Cook [119] | UK | 5-7 | Saliva | 2721 | 0.33 | 1.17 | 2.20 | 4.05 |
| Jarvis [120] | UK | 11-16 | Saliva | 569 | 0.44 | 1.31 | 1.95 | 3.38 |
| Jarvis [121] | UK | 11-16 | Saliva | 330 | 1.56 | 2.57 | 3.75 | 6.12 |
| Marbury [122] | USA | <2 | Urine* | 48 | 10 | 320 | 622 | 1650 |
| Preston [123] | USA | 2-11 | Urine | 175 | 5.9 | 19.7 | 41.8 | 31.1 |
| Reese [124] | Australia | 3.8 | Urine* | 443 | 31 | 43 | 39 | 74 |
| Weaver [125] | USA | 3.8 | Urine* | 79 | 45 | 80** | 89 | 128 |
| Willers [126] | Sweden | 7.5 | Urine | 77 | 2.0 | 5.8 | 12.0 | 9.4 |
| Willers [109] | Sweden | 8.9 | Urine* | 46 | 1.7 | 6.1 | 8.0 | 13.0 |

[1] ng/mg creatinine where indicated by an asterisk in the body fluid column.
[2] Range, mean or median.
[3] Other household member where indicated by a double asterisk.

studies the father is replaced by another household member in these analyses. Though the actual levels (particularly in urine) vary markedly from study to study, the studies are reasonably consistent in showing that cotinine increases with the number of parents who smoke. The increase seen is greater if the mother smokes than if the father smokes, consistent with children spending more time in the presence of their mother than their father.

A strong association between cotinine levels in children and ETS exposure at home was also seen in a large representative US study [19]. Serum cotinine levels in children who did not use tobacco were 0.119 ng/ml at age 4–11 and 0.113 ng/ml at age 12–16 if there was no reported ETS exposure at home, but were, respectively, 1.14 ng/ml and 0.808 ng/ml if there was such exposure. Some studies (e.g. [117,127–129]) have reported a clear relationship between cotinine levels in children and the number of cigarettes smoked per day by the parents.

## Cotinine levels in non-smoking children as an index of equivalent active smoking

In one study cited in Table 11 [120] the salivary cotinine level in 11–16 year old non-smoking children with a mother who smoked was estimated as 1.95 ng/ml. Noting that the active smoking of 20 cigarettes per day was associated with a salivary cotinine level of 300 ng/ml, the authors estimated that the dose of nicotine received from mother's smoking was equivalent to the active smoking of $(1.95 \times 20 \times 365)/300 = 47.45$ cigarettes per year or about 50 cigarettes per year. Similarly they estimated that the dose of nicotine received from father's smoking, where the cotinine level was 1.31 ng/ml, was equivalent to about 30 cigarettes per year, and that from both parents smoking [3.38 ng/ml] was equivalent to about 80 cigarettes per year. The authors went on to argue that "this unsolicited burden may be prolonged throughout childhood and poses a definite risk to health". There are various reasons why the conclusions are open to question. In the first place, all the salivary cotinine in children whose parents both smoke is assumed to be due to their smoking, which is inconsistent with the value of 0.44 ng/ml seen in children neither of whose parents smoke. Also, because ETS and mainstream smoke are dissimilar in their composition, properties and modes of intake, comparisons on the basis of cigarette equivalents are misleading (see also Section 3). Even ignoring these points, the conclusion that this poses a definite risk to health seems dubious. Epidemiological studies have not indicated any excess risk of disease in smokers of even 80 cigarettes per year (0.2 per day) [130] and other researchers estimating rather similar uptakes of nicotine by non-smokers have regarded them as 'negligible' [48].

## VIII. EVIDENCE OF A RELATIONSHIP OF TOBACCO SMOKE TO DISEASE

### Cotinine as a marker in studies of diseases in infants and children

Cotinine has been quite widely used as a marker in studies relating ETS exposure to various diseases or conditions in infants and children. Most commonly, these studies concern aspects of respiratory morbidity, such as asthma, lung function or the presence of various symptoms [124,126,131–145]. Otitis media [146–148], sudden infant death syndrome [149,150], Down's syndrome [151], and even child development [152,153] have also been investigated using cotinine. It is beyond the scope of this paper to summarize the relevant findings. However, some points are worth noting.

First, relatively few of the studies allow one to distinguish whether cotinine or reported indices of ETS exposure correlate better with the health effects. In those that did, the evidence was not clear. Two studies [132,133] found that cotinine did not have significant predictive power over and above reported indices, one study [135] found, at least for some of the diseases studied, that it did, while another study [139] found associations with reported ETS exposure but not with cotinine.

Second, measurements of cotinine do not avoid the problems of distinguishing between potential pre- and post-natal effects on health of maternal smoking. Most smokers who smoke after pregnancy also smoke during pregnancy. Particularly in infants and young children, correlation of health indices with cotinine levels may indirectly reflect effects of smoking during pregnancy and need not necessarily indicate effects of ETS exposure.

Third, there is a theoretical possibility that metabolism of nicotine to cotinine may be affected by the presence of some diseases and also perhaps that it may be genetically linked to susceptibility to some diseases. As far as the present author is aware, these possibilities have not been studied.

### Cotinine as a marker in studies of birthweight

A number of studies [23,30,38,57,59,154–158] have related birthweight of infants to cotinine levels measured in the mother during pregnancy. Some of these studies found that, after adjustment for reported number of cigarettes smoked by the mother, there was a clear association between increasing cotinine level and reduced birthweight [23,30,57,154,157,158]. One large study [30] presented a two-way table of birthweight by grouped cotinine level and by grouped reported number of cigarettes per day, which clearly showed that cotinine was a better predictor of birthweight. This may be partly because cotinine is intrinsically a better marker of relevant exposure in this context and partly because mothers' reports of amount smoked may be

691

inaccurate, especially in a situation where they are likely to have been advised to give up or cut down.

Further research questions in this area include the specific time point in pregnancy at which cotinine correlates most strongly with final birthweight, and whether pregnancy itself, or complications of pregnancy, affects the metabolism of nicotine.

## Cotinine as a marker in studies of diseases in adults

Studies of health effects in adults have only rarely used cotinine as a marker of the extent of smoking or of ETS exposure. Five particular studies merit comment.

The first [159] examined the relation of biochemical, physical examination, and depression assessments to self-reported cigarette consumption and serum cotinine in a sample of 743 Mexican-Americans. After controlling for sex, age and education, serum cotinine was a significant predictor of haematocrit, haemoglobin, red blood cells, white blood cells, blood lead and diastolic blood pressure. Self-reported cigarette consumption was only significant for mean corpuscular volume. The authors concluded that "serum cotinine may be a better method of quantifying risks from cigarette use in epidemiological studies." The value of cotinine must surely depend, however, on the disease in question, in particular whether its incidence is affected by nicotine exposure or by other tobacco smoke constituents and whether the disease is acute or chronic, cotinine being likely to be better correlated with recent smoking than with past smoking.

In the second study [160], conducted in Scotland, the relationship between self-reported ETS exposure, serum cotinine and evidence of respiratory or coronary disease was studied in 786 men and 1492 women who reported never having smoked tobacco and who had serum cotinine concentrations below 17.5 ng/ml. It is interesting to note that, though self-reported ETS exposure showed strong, statistically significant, dose-response relationships with various respiratory symptoms, these relationships were weak or absent for serum cotinine. The authors suggest that the presence of respiratory disease might bias self-reports of the extent of ETS exposure, in this study classified rather subjectively as 'none at all', 'a little', 'some' or 'a lot'. Using the more objective marker, cotinine, avoided such bias.

Virtually all the epidemiological studies relating ETS exposure to the incidence of lung cancer among non-smokers have relied on self- or proxy-report to obtain information on smoking habits and ETS exposure.

One exception is a large US case-control study [62] which merely used urinary cotinine levels in an attempt to exclude misclassified non-smokers. Using it as an index of ETS exposure was not attempted, presumably as cotinine measured after diagnosis would not be a reliable marker of lifetime ETS exposure. It is also true, however, that many lung cancer patients give up smoking at or around the time of diagnosis, and cotinine levels determined from a sample of urine taken days or weeks later would therefore still not reliably have excluded all misclassified smokers.

The other ETS/lung cancer study to use cotinine [161], involved the collection and freezing of urine samples from over 20 000 Dutch women. Over the following years, 92 of the women were identified as having contracted lung cancer and the urine samples of these women and of 305 controls were analysed for cotinine. Women identified as smokers, based on cotinine, had a significant 8-fold higher risk of lung cancer than did non-smokers, but the authors concluded that "urinary cotinine did not seem to be more predictive of lung cancer risk than self-reported cigarette consumption." Among women identified as never smokers, with no self-report of past or present smoking and no evidence of current smoking from their cotinine level, lung cancer risks were over 2-fold higher in those with a cotinine/creatinine ratio of between 9.2 and 23.4 ng/mg or between 23.4 and 100 ng/mg than in those with a ratio of less than 9.2 ng/mg. However, the differences seen were not statistically significant, the analysis being based on only 23 lung cancer cases. The study indicates the enormous effort that would be required to use cotinine as a marker of ETS exposure in a lung cancer prospective study, and obtain enough deaths for reliable inferences to be drawn.

The final study of adults using cotinine as a marker that merits particular attention, also prospective, is a study of 3270 Danish men aged 53–74 [162] who reported their previous and current tobacco habits, including chewing tobacco and snuff, and were identified as misclassified smokers based on a serum cotinine level of 100 ng/ml. Based on deaths over the next 8 years, the authors found that smokers who denied smoking had a risk of heart disease that was 4.0 times higher (95% confidence interval 1.8–9.1) than that of smokers who admitted to smoking. This finding is consistent with denial of smoking being substantially greater in smokers advised to give up by their doctors [78], smokers with a first non-fatal myocardial infarction and at increased risk of a second fatal one being particularly likely to receive such advice. While this study did not provide data on risk of disease in non-smokers in relation to cotinine level, it is of particular importance for the interpretation of the evidence on possible effects of ETS on chronic diseases, as it suggests that standard methods for adjustment for bias due to misclassification of smoking habits may underestimate bias by failing to take into account the possibility that misclassification rates may be higher in those subjects who are more at risk of disease (163).

## Cotinine as a predictor of risk of chronic diseases

It can be seen from the above that there is very little evidence on which to judge how well a single cotinine measurement is likely to predict risk of subsequent death from chronic diseases such as lung cancer, heart disease or chronic bronchitis. By taking into account aspects of inhalation, it is possible that it might be a better predictor of risk than number of cigarettes per day in people who continue to smoke at a relevant consistent rate. However, since it is only a marker of recent smoking, it may prove a worse predictor than a lifetime smoking history. Of course, it is possible to measure cotinine at more than one time point.

## Cotinine as a marker of exposure to disease risk factors other than smoking

One problem in interpreting epidemiological studies of potential effects of smoking and of ETS exposure lies in the possibility of confounding by other risk factors. While it is well known that, on average, smokers drink more alcohol and coffee than non-smokers and tend, in many countries, to be of lower social class, the full extent to which smoking, and ETS exposure, are associated with increased exposure to a variety of lifestyle risk factors is perhaps not so well realised. A recent study [164] of a representative sample of over 9000 British adults compared the distribution of 33 lifestyle risk factors generally considered to be associated with adverse health in current smokers, ex-smokers, never smokers living with a smoker and other never smokers. Of the 33 risk factors, 27 showed a significantly higher prevalence in heavy smokers than in never smokers and only two showed a significantly lower prevalence. For many risk factors, prevalence increased with amount smoked, decreased with time since smoking cessation, and was increased in never smokers living with a smoker. These results strongly suggest that some of the observed excess risk of disease in smokers and passive smokers is due to their poorer lifestyle in other respects.

These results were based on self-reports of smoking habits or ETS exposure. That using cotinine as a marker of tobacco smoke exposure does not eliminate the possibility of confounding is illustrated by some previously unpublished work I have carried out with my colleagues Jan Hamling and John Fry (fuller details of which are available on request). This was based on data from two UK studies. The first (Health and Lifestyle Survey follow up=HALS2) was based on reinterviews of the same population used for the study referred to above [164], in which data on saliva cotinine, self-reported smoking habits and ETS exposure, and 27 of the 33 risk factors originally studied were available on 3668 adults. The second (Health Survey of England 1993=HSE93) was a survey of 4813 adults, in which data on serum cotinine, self-reported smoking habits and ETS exposure, and many identical or similar risk factors were available.

## Relationship of cotinine levels in never smokers to other risk factors (HALS2 and HSE93 studies)

Table 12 shows how, in never smokers, age and sex adjusted prevalence of many of the risk factors varies markedly by cotinine level. The table shows

TABLE 12

Prevalence of lifestyle health risk factors in never smokers[a] classified by cotinine level

| Risk factor[c] | Saliva cotinine level (ng/ml)[b] | | | | | HALS2 Trend | HSE93 Trend |
|---|---|---|---|---|---|---|---|
| | 0.0-0.5 | 0.6-1.0 | 1.1-2.0 | 2.1-5.0 | 5.1-30.0 | | |
| | %[d] | %[d] | %[d] | %[d] | %[d] | p[e] | p[e] |
| Father dead | 61.1 | 59.8 | 64.6 | 61.4 | 67.7 | NS | NS |
| Mother dead | 44.6 | 41.5 | 45.7 | 46.7 | 59.5 | + | + |
| No educational qualifications | 28.2 | 30.1 | 39.0 | 38.7 | 56.7 | +++ | +++ |
| Separated, divorced or widowed | 12.5 | 15.0 | 18.4 | 18.5 | 17.5 | + | NS |
| Not in paid employment | 38.2 | 32.9 | 34.3 | 34.1 | 40.2 | NS | NS |
| Low social class | 37.3 | 41.6 | 53.3 | 54.1 | 63.1 | +++ | +++ |
| High alcohol consumption | 18.2 | 23.5 | 23.2 | 35.7 | 29.3 | ++ | +++ |
| High bread consumption | 45.5 | 42.0 | 47.0 | 49.6 | 52.2 | + | NS |
| Low fruit consumption | 26.3 | 22.8 | 31.8 | 30.2 | 39.5 | ++ | +++ |
| Low vegetable consumption | 50.1 | 48.1 | 51.0 | 49.4 | 56.6 | NS | |
| Low vegetable/salad consumption[f] | 27.6 | 27.1 | 27.8 | 32.9 | 39.7 | | +++ |
| Low salad consumption | 49.3 | 46.0 | 46.6 | 49.2 | 52.5 | NS | |
| High sweet food consumption | 59.8 | 58.9 | 55.0 | 49.7 | 48.1 | − − | − |
| High salt consumption in food[f] | 37.4 | 39.0 | 40.3 | 38.3 | 47.4 | | ++ |
| No use of low fat/PU spread | 21.7 | 20.5 | 20.1 | 27.7 | 38.6 | ++ | NS |
| High body mass index | 51.1 | 55.9 | 58.1 | 62.7 | 64.6 | +++ | ++ |
| Underweight | 2.3 | 1.0 | 1.1 | 1.3 | 2.8 | NS | NS |
| 'Risky' occupation | 21.9 | 21.2 | 26.5 | 33.2 | 44.3 | +++ | |
| Low income | 45.2 | 44.0 | 49.2 | 53.8 | 57.6 | +++ | |
| Do nothing to keep healthy | 28.3 | 29.6 | 28.6 | 35.8 | 43.8 | ++ | |
| Low activity[f] | 13.1 | 15.8 | 14.6 | 12.9 | 15.3 | | NS |
| Little sleep | 37.7 | 38.5 | 36.5 | 46.5 | 42.9 | NS | |
| Long time before first meal | 14.7 | 17.4 | 19.0 | 21.1 | 15.9 | ++ | |
| High fried food consumption | 15.3 | 20.1 | 20.5 | 24.6 | 29.8 | +++ | |
| Low breakfast cereal consumption | 25.1 | 28.4 | 30.3 | 29.1 | 39.0 | +++ | |
| Sugar in tea or coffee | 35.6 | 35.5 | 39.8 | 36.8 | 51.7 | + | |
| High tea consumption | 18.5 | 21.0 | 17.8 | 20.0 | 22.2 | NS | |
| High coffee consumption | 7.2 | 5.3 | 8.5 | 6.7 | 7.1 | NS | |
| High neuroticism | 39.8 | 44.6 | 54.0 | 48.2 | 42.2 | (+) | |
| High extroversion | 32.0 | 42.5 | 43.1 | 49.2 | 56.1 | +++ | |
| Low control at work[f] | 17.7 | 20.4 | 24.1 | 28.8 | 32.2 | | +++ |
| High pace index[f] | 31.6 | 39.6 | 37.1 | 33.7 | 33.8 | | NS |
| Number of subjects (HALS2) | 511 | 278 | 241 | 221 | 93 | | |
| Number of subjects (HSE93)[g] | 353 | 485 | 390 | 263 | 194 | | |

[a] Excluding never smokers with no data on social class or living with a smoker and with either saliva cotinine above 30 ng/ml (HALS2) or with serum cotinine above 20 ng/ml (HSE93).
[b] Where data for the risk factor are available from HALS2, prevalences shown are for HALS2 data. Otherwise prevalences shown are for HSE93 with categories, based on serum cotinine, of 0.0–0.2, 0.3–0.5, 0.6–1.0, 1.1–2.0 and 2.1–20.0 ng/ml.
[c] A detailed definition of the risk factors and the high/low cut offs used is available from the author.
[d] Percentages are adjusted for sex and age (<30, 30–39, 40–49, 50–59, 60–69, 70+) to the overall population of never smokers considered.
[e] Significance codes: +++, − − − $p<0.001$, ++, − − $p<0.01$, +, − $p<0.05$, (+), (-) $p<0.1$, NS not significant $p \geqslant 0.1$. Plus signs indicate increasing trend, minus signs decreasing trend.
[f] Only data for HSE93 available.
[g] See note b for categories used.

results of trend tests for both surveys, where available, but only shows prevalence by cotinine level for one of the surveys (HALS2 if available) as patterns were generally found to be quite similar in the two surveys. Never smokers with cotinine levels consistent with active smoking are excluded from the analyses.

The results clearly show that though a high cotinine level in a never smoker is a marker of increased ETS exposure, it is also a marker of many other things. Thus, it is clearly associated with poorer education, lower income, lower social class, with working in a job with increased potential exposure to carcinogens, with higher alcohol consumption, with higher extroversion scores, with a higher body mass index and with higher fried food consumption, and with lower consumption of fruit, breakfast cereal and sweet foods. Additional analysis (data not shown) showed that prevalence of risk factors was generally more strongly associated with cotinine level than with the index "living with a smoker".

### Relationship of cotinine levels in cigarette smokers to other risk factors (HALS2 and HSE93 studies)

Table 13 compares prevalences in never smokers and in current cigarette smokers classified by cotinine level. Based on trends including the never smoking group, which to some extent reflect the current smoker/never smoker difference, the table shows that increased smoking (as quantified by cotinine) is associated with a highly significant ($p < 0.001$) increase in prevalence of virtually all the risk factors studied, the notable exceptions being that increased smoking is associated with a lower body mass index and with lower sweet food consumption.

The significance of trends in prevalence by cotinine level in analyses restricted to current cigarette smokers was also calculated, though is not shown in Table 13. For many of the risk factors, including no educational qualifications, low social class, low fruit consumption, low salad consumption, high salt consumption, no use of low fat/polyunsaturated spread, 'risky' occupation, do nothing to keep healthy, long time before first meal, high fried food consumption, low breakfast cereal consumption, taking sugar in tea or coffee, high tea consumption, high coffee consumption, the trend seen remained significant at least at $p < 0.05$. However for others they were not apparent, indicating the major difference in prevalence is between current smokers and never smokers. Interestingly, however, in both studies, though frequency of high alcohol consumption was clearly commoner in current than in never smokers, it was (in both studies) significantly negatively related to cotinine level in current smokers.

Further analyses (data not shown) showed that while, for some risk factors, prevalence, in smokers, was more strongly associated with cotinine level, for others it was more strongly associated with number of cigarettes

TABLE 13

Prevalence of lifestyle health risk factors in current cigarette smokers classified by cotinine level and in never smokers[a]

| Risk factor[d] | Never smoked[a] | Current cigarette smoker | | | | HALS2 Trend[c] | HSE93 Trend[c] |
|---|---|---|---|---|---|---|---|
| | | Saliva cotinine level (ng/ml)[b] | | | | | |
| | | <200 | 200.1–350 | 350.1–450 | >450 | | |
| | %[e] | %[e] | %[e] | %[e] | %[e] | p[f] | p[f] |
| Father dead | 60.6 | 60.1 | 63.7 | 65.9 | 62.5 | (+) | + |
| Mother dead | 43.9 | 41.2 | 48.6 | 49.6 | 48.3 | ++ | NS |
| No educational qualifications | 32.7 | 45.0 | 52.2 | 58.5 | 54.5 | +++ | +++ |
| Separated, divorced or widowed | 14.1 | 19.7 | 19.7 | 23.9 | 22.2 | +++ | +++ |
| Not in paid employment | 34.5 | 41.4 | 41.5 | 40.5 | 43.6 | +++ | +++ |
| Low social class | 45.3 | 61.0 | 65.7 | 65.5 | 72.9 | +++ | +++ |
| High alcohol consumption | 24.4 | 43.7 | 42.5 | 35.4 | 30.7 | +++ | +++ |
| High bread consumption | 47.6 | 46.9 | 53.5 | 50.6 | 52.1 | +++ | NS |
| Low fruit consumption | 28.7 | 49.6 | 51.5 | 60.1 | 59.8 | +++ | +++ |
| Low vegetable consumption | 51.5 | 51.6 | 51.2 | 55.1 | 58.3 | + | |
| Low vegetable/salad consumption[g] | 30.5 | 39.4 | 39.1 | 41.2 | 43.4 | | +++ |
| Low salad consumption | 48.9 | 52.6 | 61.4 | 66.0 | 71.5 | +++ | |
| High sweet food consumption | 55.6 | 38.6 | 36.7 | 33.2 | 38.4 | - - - | - - - |
| High salt consumption in food[g] | 40.1 | 55.0 | 51.4 | 61.0 | 56.0 | | +++ |
| No use of low fat/PU spread | 22.4 | 34.0 | 32.5 | 35.2 | 47.3 | +++ | +++ |
| High body mass index | 55.3 | 48.5 | 56.3 | 47.3 | 48.1 | - - - | - - - |
| Underweight | 1.9 | 2.7 | 5.1 | 4.7 | 5.0 | +++ | +++ |
| 'Risky' occupation | 26.3 | 31.7 | 38.9 | 36.6 | 46.6 | +++ | |
| Low income | 46.1 | 61.4 | 58.1 | 60.9 | +++ | | |
| Do nothing to keep healthy | 30.9 | 39.2 | 40.7 | 39.8 | 48.1 | +++ | |
| Low activity[g] | 13.8 | 10.9 | 20.3 | 15.0 | 11.0 | +++ | +++ |
| Little sleep | 38.6 | 46.6 | 46.4 | 47.6 | 47.7 | +++ | |
| Long time before first meal | 17.4 | 23.9 | 40.1 | 48.5 | 52.3 | +++ | |
| High fried food consumption | 20.1 | 30.5 | 33.6 | 33.2 | 40.4 | +++ | |
| Low breakfast cereal consumption | 28.8 | 43.1 | 50.4 | 54.9 | 54.9 | +++ | |
| Sugar in tea or coffee | 37.8 | 50.9 | 53.7 | 56.4 | 66.0 | +++ | |
| High tea consumption | 19.2 | 28.7 | 39.6 | 44.4 | 50.5 | +++ | |
| High coffee consumption | 6.6 | 17.3 | 20.2 | 27.5 | 27.7 | +++ | |
| High neuroticism | 43.1 | 55.9 | 53.1 | 47.7 | 49.0 | +++ | |
| High extroversion | 40.6 | 56.3 | 53.3 | 56.2 | 53.1 | +++ | |
| Low control at work[g] | 23.0 | 14.3 | 28.5 | 28.4 | 23.5 | | ++ |
| High pace index[g] | 35.6 | 39.6 | 29.3 | 36.2 | 23.3 | | NS |
| Number of subjects (HALS2) | 1344 | 229 | 340 | 235 | 259 | | |
| Number of subjects (HSE93)[h] | 1685 | 278 | 414 | 317 | 262 | | |

[a] Excluding never smokers with no data on social class or living with a smoker and with either saliva cotinine above 30 ng/ml (HALS2) or with serum cotinine above 20 ng/ml (HSE93).
[b] Where data for the risk factor are available from HALS2, prevalences shown are for HALS2 data. Otherwise prevalences shown are for HSE93 with categories, based on serum cotinine, of <100, 100.1–250, 250.1–350, >350 ng/ml.
[c] Trend including never smokers.
[d] A detailed definition of the risk factors and the high/low cut offs used is available from the author.
[e] Percentages are adjusted for sex and age (<30, 30–39, 40–49, 50–59 and 60–69, 70+ for males or 60+ for females) to the overall population considered.
[f] Significance codes: +++, − − − $p<0.001$, ++, − − $p<0.01$, +, − $p<0.05$, (+), (−) $p<0.1$, NS not significant $p \geqslant 0.1$. Plus signs indicate increasing trend, minus signs decreasing trend.
[g] Only data for HSE93 available.
[h] See note b for categories used.

697

smoked. This may reflect the fact that, in smokers, cotinine is an indicator not only of how many cigarettes are smoked, but also of how much cotinine is obtained per cigarette. The depth to which people inhale cigarettes and the extent to which they metabolize nicotine to cotinine may vary according to various factors.

### Effect of alcohol consumption on cotinine levels (HALS2 and HSE93 studies)

The data relating to alcohol consumption from the two surveys are particularly interesting in this respect. While the frequency of high alcohol consumption in smokers clearly increases with increasing amount smoked [164], it clearly decreases with increasing cotinine level (see Table 13). In other words, increasing alcohol consumption is associated with decreased cotinine per cigarette smoked. This may be due to heavy drinking being associated with reduced inhalation, but other possible explanations include alcohol having an effect on the method used to determine cotinine or on the metabolism of nicotine. An effect of alcohol on nicotine metabolism is supported by the results of a pharmacokinetic study in rats pretreated with ethanol [165]. It is interesting to note that if alcohol affects nicotine metabolism in humans, adjustment for alcohol consumption may be necessary in studies using cotinine as an index of exposure to active smoking or ETS.

### Effect of body mass index on cotinine levels in smokers (HALS2 and HSE93 studies)

Body mass index was another risk factor found in our analyses to be positively related to number of cigarettes smoked and negatively related to cotinine. In a previous report of a similar finding [91] nicotine metabolism, energy intake and measurement issues were considered as possible explanations. These findings are consistent with those of another study [166] which reported increased salivary cotinine concentrations in smokers during rapid weight loss following participation in a protein-sparing modified fast.

### Relationship of cotinine to factors other than smoking – some evidence from studies other than HALS2 and HSE93

A large number of other studies have reported data relating cotinine levels in adults or children to factors other than smoking. The evidence seems to be quite consistent that increased cotinine levels are associated with low social class or poor education [17,119,167–170], Black race [19,57,152,168,170–173], household crowding or small room size [19,128,169] and alcohol and drug consumption [21]. Cotinine levels in male and female smokers of the same number of cigarettes appear, however, to be similar (see Table 6). Pregnancy was not associated with decreased cotinine levels in women when measured at the time of diagnosis [174].

Consistent with evidence from earlier studies (175,176), a recent study (177) reported levels of urinary cotinine in the infants of nursing mothers who smoked were markedly higher if they were breast-fed than if they were bottle-fed. The authors suggested breast-feeding, rather than direct inhalation of ETS, is their primary determinant of cotinine.

## IX. OTHER ISSUES

### Correlation between multiple cotinine measurements taken at different time points

Cotinine levels are determined by smoking and ETS exposure in the last few days, yet are used as markers in studies of diseases which may reflect exposure to tobacco smoke constituents over months or years. If cotinine levels vary markedly over time, due to variations in exposure, metabolism, etc., a single cotinine measurement may substantially underestimate any true relationship of smoking or ETS exposure with disease. If, on the other hand, cotinine levels are quite consistent over time, a single measurement may be more relevant.

A few studies have investigated the consistency of cotinine measurements taken in the same subjects at different points of time. In one such study, of 169 girls aged 11–16 in which salivary cotinine was measured in 1985 and again in 1986 [121], there was a correlation of 0.75 between the two measurements. Another study, of 111 children aged 5–7 [135], reported a correlation of 0.81 between measurements taken at 6-month intervals. In a third study, in which plasma cotinine in 148 smokers was measured on two occasions separated in time by 14 weeks [31], the correlation was 0.81.

One paper [83], reported results from two studies of cigarette smokers, one of 200 men providing blood on 3 occasions, the other of 51 men and 52 women providing blood on 6 occasions. In both studies, standard deviations of blood cotinine within smokers, of about 63 ng/ml, were 2 to 3 times less than standard deviations between smokers.

In a study of 20 US non-smokers measurements of urinary cotinine on 10 days over a 10 week period [178] the highest values recorded for a subject were typically about twice as high as their mean value.

In a recent study, an attempt was made to collect urine for cotinine determination every two months from birth to two years of age in 99 infants [179]. When individual values for an infant were compared to the average of the multiple measurements, about a third were in error by more than 100 ng/mg. Classifying infants as having low or high exposure depending on whether or not 50% of their readings were less than 32 ng/mg, 84% were correctly classified by their individual values.

In a detailed study of ETS exposure in 91 asthmatic children [180], the two month test-retest correlation coefficient was in the range 0.60 to 0.88 for nearly all the indices studied, including parental smoking, cigarette butts and ETS as determined by an air monitor. However, for urinary cotinine, it was only 0.04, and the authors considered it unreliable.

In a further study [181], levels of salivary cotinine in two smokers were recorded from samples collected daily over a one month period. The maximum values coincided with weekends or holidays when the individual's smoking behaviour was not recorded.

Generally, though not always, these results suggest that a single cotinine determination may be useful in studies of tobacco smoke exposure and disease. However, multiple measurements will clearly add power to detect any true associations of average cotinine level with disease.

## Correlation between cotinine measurements taken in different body fluids

In some studies, cotinine has been measured in more than one body fluid. Correlations have, for example, been reported between levels in saliva and serum, $r=0.84$ [67], urine and blood, $r=0.72$ [182], and saliva and urine, $r=0.92$ [183]. Although these correlations are relatively high, they are far from perfect, and suggest that the value of cotinine, as a marker of tobacco smoke exposure, may vary in relation to the body fluid in which it is measured.

## Correlation between cotinine levels and air nicotine levels

A number of studies of children and non-smoking adults have reported the correlation between cotinine levels, as measured in urine or saliva, and nicotine levels, as measured from air samples taken by personal monitors or passive monitors placed in the room [77,122,178,184–192]. While positive correlations are always reported, they are far from perfect, never exceeding more than about 0.8 and typically averaging around 0.5.

The lack of perfect correlation will partly be due to errors in determining cotinine levels and air nicotine levels, and partly because the former, but not the latter, will be affected by inter-individual differences in inhalation and nicotine metabolism.

## Metabolism of nicotine and half-life of cotinine

In a study in which six healthy volunteers were infused over four days with doses of nicotine [193] varying up to 480 μg/kg/day, cotinine concentrations in plasma were found to be linearly and directly related to nicotine intake. The mean regression line was cotinine concentration (μg/l)$=0.783\times$nicotine (μg/kg/day).

The half-life of cotinine has been reported variously as 12 h [194], 17 h [18], 16–19 h [195], 19 h [196], 29 h [197] and 55 h [184]. For a half-life of 18 h, an average smoker would have to have stopped smoking for about four days before his cotinine level declined to that typical of an ETS-exposed non-smoker and for about a week before his cotinine level became undetectable using modern analytical methods.

In one study in which plasma, saliva and urinary cotinine concentrations were determined in eight smokers who stopped smoking [198], plasma and urinary levels dropped markedly over a five-and-a-half day period, by a factor of 40 or more. In contrast, salivary cotinine levels only dropped by a factor of two over this period. However, a subsequent study failed to replicate this finding, reporting no tendency for the half-life in saliva to be longer than in plasma or urine [195].

The published evidence appears to be somewhat conflicting regarding whether smokers metabolize nicotine to cotinine differently from non-smokers. While one study reported nicotine clearance was significantly slower in smokers [199], some other studies [58,200,201] have reported a longer half-life in non-smokers. However, two of these studies [200,201] used radiolabelled racemic nicotine, which is a mixture of two forms of nicotine, only one of which is present in tobacco. Furthermore, other studies [18,195,202,203] have reported that the pharmacokinetics of nicotine and/or cotinine are similar in smokers and non-smokers.

One study [173] compared nicotine metabolism and intake in black and white smokers and explained the higher levels of cotinine per cigarette smoked by blacks in terms of their slower clearance of cotinine and higher intake of nicotine per cigarette.

A possible limitation of cotinine as a marker of nicotine exposure is highlighted by the report [204] of a woman who was found to have the expected plasma levels of nicotine but unusually low plasma levels of cotinine, both when smoking cigarettes and while receiving transdermal nicotine. The subject converted only 9% of nicotine to cotinine, compared to 72% for control subjects. The frequency of such deficient C-oxidation of nicotine to cotinine is not known, but it may help to explain why some smokers have low cotinine levels [17,19], it being generally regarded as implausible that adults would claim to smoke when they did not in fact do so.

## Storage of samples

In one study [205], the ratio of cotinine as determined by: (a) radio-immunoassay; and (b) gas chromatography, was measured both in fresh serum samples and in samples stored for more than four years. The ratios in the two sets of samples were similar for serum from smokers but were much less consistent in the stored samples for serum from non-smokers. The authors suggested that aging may affect the determination when dealing with low levels of cotinine.

In another study [206], plasma left at room temperature for 12 days had cotinine concentrations which correlated highly [$r=0.99$] with those in duplicate samples frozen immediately. There was no evidence of a shift in values or an increase in variability due to delayed freezing.

In a third study [207], two saliva specimens were taken simultaneously from 10 smokers and 10 non-smokers. One was frozen immediately and the other sent through the local US mail and then subjected to additional physical agitation and heat before being frozen. No cotinine was detectable in any of the non-smokers' specimens, and there was good agreement [$r=0.88$] between the two specimens for the smokers. The authors concluded that the mailing of saliva specimens for cotinine analysis was practical. Similar conclusions were reached in another study [208], which compared saliva samples, one of which was immediately stored on ice and transported to the laboratory, the other being mailed immediately by the subject to the laboratory.

In a further study [209], cotinine was measured in 1988 in urine collected in 1976–77, at a time when data on smoking status were also collected. Cotinine measurements showed a clear separation of smokers and non-smokers (specificity 95%, sensitivity 100%), suggesting that cotinine retained its ability to discriminate, even after 10 years of storage.

A conclusion from these data seems to be that one can satisfactorily use stored samples for cotinine analysis. However a study [210] in which urine, taken from smokers, was stored for 30 days at temperatures of 5, 25, 40, 50 or 60°C shows that this conclusion does not hold true for storage at high temperature. While cotinine levels remained relatively constant at 5°C they increased with time at higher temperatures, doubling over the 30 day period at 60°C. This is presumably due to release of cotinine from its glucuronide conjugate.

## X. SOME GENERAL CONCLUSIONS

Cotinine is the main metabolite of nicotine, the principal alkaloid of the tobacco plant. Since 1966, when a reasonably reliable chemical method for its determination was first described [211], techniques for its determination in a wide range of body fluids have been developed (see Chapter 8). As described in the previous sections of this paper, cotinine determinations have been made in many human populations for many purposes.

There are a number of advantages in using cotinine as a marker of tobacco smoke exposure, which are summarized below.

*Objectivity*. Most epidemiological studies relating smoking or ETS exposure to the incidence of various diseases have relied upon data reported by the subject or another family member. Such data are unreliable [47] and errors are clearly not random. The well documented tendency [78–80] for patients advised to give up smoking to be particularly likely to deny later that they have failed to do so, is one illustration of this. Use of an objective marker of exposure, such as cotinine, avoids such recall bias, and indeed in one study [162] was able to demonstrate a high subsequent risk of coronary death in misclassified smokers. The results of a Scottish study [160] which demonstrated an association of respiratory symptoms with self-reported ETS exposure but not with cotinine are consistent with use of an objective marker avoiding the bias which arises because subjects with symptoms tend to perceive their ETS exposure as more severe.

*Sensitivity.* Cotinine levels in humans can be detected over more than a thousand-fold range. Not only can they clearly distinguish smokers from non-smokers (providing care is taken to exclude other sources of nicotine such as chewing tobacco, nicotine gum and nicotine patches), but, at the lower end of the detectable range, can discriminate non-smokers with high, medium or low ETS exposure.

*Practical uniqueness to tobacco.* It has been argued [45] that an appropriate marker of exposure to tobacco smoke must be unique to tobacco. From an epidemiologist's viewpoint, such an argument is not correct; the ideal marker is one that measures exposure to the actual agent that causes the disease under consideration, whether or not the agent is present only in tobacco smoke. However, from the point of view of distinguishing smokers from non-smokers, or of quantifying relative exposure of different smokers or relative ETS exposure of different non-smokers, use of a marker that is not affected in a major way by non-tobacco sources is an advantage. As discussed in section 2, nicotine is present in diet [9–11], but not at levels that are likely to mimic more than moderate ETS exposure at most [15–20].

*Index of inhalation.* While some epidemiological studies do include a question concerning inhalation by smokers, these data are regarded as rather soft, as the concept is difficult to quantify and smokers have little detailed knowledge of how far into their lungs their smoke goes. An attractive feature of cotinine, as a marker of tobacco smoke exposure, is that it attempts to measure uptake rather than available dose, as would be measured by number of cigarettes smoked per day. In theory, two smokers of 20 cigarettes per day could smoke very differently, not only in terms of inhalation but also in terms of puffing characteristics, so that they have very different relevant exposures to the target tissue.

There are also a number of limitations in using cotinine.

*Marker only of recent exposure.* Though cotinine has a much longer half-life than nicotine, and is therefore not nearly so dependent on when the last cigarette was smoked or when the last heavy ETS exposure occurred, the fact that its half-life in serum or urine is only about a day or so [18,194–197] means that it is only a marker of quite recent exposure. A cotinine measurement taken on a Friday will hardly reflect exposures on the previous weekend and may differ from that taken on a Monday, especially if the subject smokes more heavily at home than at work or goes out to smoky bars on a Friday or Saturday night. Though measurement of nicotine in hair may allow detection of exposure over a period of weeks or months [8], nicotine-based markers can seemingly not be used as reliable markers of lifetime exposure, unless they are taken at regular intervals.

*Marker only of nicotine exposure.* If comparisons are made between non-smokers with no, light or heavy ETS exposure, the relative increases in marker levels associated with heavy and light exposure would be expected to be similar regardless of the marker used. In this context, the fact that cotinine is only strictly an index of exposure to nicotine, need not matter much, as a doubling of the cotinine increase associated with heavy rather than light exposure might reasonably be taken to indicate that exposure to all constituents of ETS is approximately doubled. A similar argument might be applied to increases in markers in smokers associated with different numbers of cigarettes smoked, though perhaps less confidently as for some chemicals saturation might occur, with a doubling of cigarettes smoked not associated with a doubling of marker levels.

However, the fact that nicotine is predominantly in the particulate phase in mainstream smoke and in the vapour phase in ETS, and the fact that inhalation of tobacco smoke by smokers is very different from normal breathing by non-smokers implies that the exposure of smokers compared to non-smokers as determined by a nicotine-based marker, such as cotinine, may be very different from markers based on other tobacco smoke constituents. Certainly cotinine cannot be used to determine relative retention of particles in smokers and non-smokers.

*Analytical error at low levels.* The inaccuracy of cotinine determinations is proportionately much greater at low levels than at high levels. Indeed, even with the most sensitive methods, some non-smokers have levels below the detectable limit, and for less sensitive methods, such as some radio-immunoassay techniques, there are problems due to non-specificity [42].

*Index of metabolism.* Cotinine is only one metabolite of nicotine and the speed and extent to which cotinine is metabolised to nicotine may vary for a number of reasons. Inasmuch as the intent of using cotinine is to measure exposure to tobacco smoke, the report of a women who metabolised very little nicotine to cotinine [204] may be important, especially if she is shown not to be a one-off case. The evidence regarding factors, genetic or environmental, which may affect metabolism to cotinine is very limited, and more research is needed. If, for example, it were shown that certain diseases affected metabolism, there would be obvious problems in using cotinine as a marker of tobacco smoke exposure in studies investigating the relationship of smoking as ETS exposure to these diseases.

*Marker of exposure to other risk factors.* Just as self-reported smoking and ETS exposure are correlated with a wide range of other disease factors [164], the same is true for cotinine (Tables 12 and 13). This indicates that while use of cotinine as a marker in studies attempting to relate tobacco smoke exposure to disease can avoid recall bias, it does not avoid potential problems of confounding. Indeed, if some of the associations between cotinine and other risk factors are due to effects of the risk factors on metabolism of nicotine to cotinine, the potential for bias due to uncontrolled confounding is increased. In this context, the observation that cotinine per cigarette smoked decreases with increasing alcohol consumption and with increasing body mass index is of relevance, as it raises the possibility that alcohol and body mass index may affect metabolism.

Bearing in mind these advantages and limitations and the evidence summarized in this chapter, it is possible to come to a conclusion concerning the usefulness of cotinine as a marker in the various contexts in which it has been used.

Provided that the possibility of other sources of nicotine such as chewing tobacco, gum and patches can be excluded, it is clear that cotinine is very useful for identifying true current smokers. It is perhaps slightly less useful for identifying true current non-smokers. This is because some smokers smoke or inhale so little that their cotinine levels become indistinguishable from those of non-smokers with high ETS exposure, and because smokers vary in the rate at which they metabolise nicotine to cotinine, with some metabolising very little.

Cotinine is also of value in categorising non-smokers in terms of light, moderate or heavy ETS exposure, though the possibility cannot be excluded that dietary nicotine may affect the categorisation in a few subjects.

Cotinine is also of value in demonstrating that, in most populations at least, the home is on average a greater source of ETS exposure than is the workplace (see Section 7).

Cotinine may also be of value in categorising smokers in terms of extent of exposure to tobacco and hence in terms of relative propensity to smoking-associated diseases. Whether cotinine is of more value than a reported smoking history is likely to depend on a number of factors, including number of cotinine measurements taken, relevance of nicotine to the disease in question, relevance of inhalation to the disease in question, and whether the disease is an acute or chronic one.

Cotinine is also of some value in indicating that smokers switching to a brand of cigarettes with a lower yield of nicotine as determined by a smoking machine partly, but by no means completely 'compensate' for the reduced yield by increasing their inhalation. However, it does not directly demonstrate that the compensation is specifically for nicotine.

While cotinine can discriminate between levels of ETS exposure in non-smokers, it is unreliable in estimating the actual (Z-) ratio of exposure in non-smokers married to a smoker and married to a non-smoker, for reasons discussed in more detail in Section 7, which include sampling error, analytical error at low cotinine levels and the contribution of diet.

Cotinine is clearly extremely unreliable for estimating the relative tobacco smoke exposure of smokers and non-smokers (and for expressing ETS exposure in terms of cigarette equivalents [120]). The problems include differences between inhalation by smokers and normal breathing by non-smokers, differences in the phase distribution of nicotine in mainstream smoke and ETS, analytical error at the low levels of cotinine seen in non-smokers and the possibility that smoking itself affects metabolism of nicotine. The relative nicotine exposure of smokers and non-smokers, even if it could be accurately measured by cotinine, may differ markedly from the relative exposure to other constituents.

Cotinine is also somewhat unreliable when trying to demonstrate that heavy smokers inhale less [97], as discussed in Section 5, and in drawing inferences regarding the lack of effect of nicotine on heart disease based on the similarity of cotinine levels in pipe and cigarette smokers [36], as discussed in Section 6.

Overall, the general conclusion is that cotinine is a marker of tobacco smoke exposure which is very useful in some contexts, but far less useful in others. Care should therefore be taken to take its limitations into account.

## AKNOWLEDGEMENTS

I thank my colleagues Jan Hamling and John Fry for their agreement to present previously unpublished data. I also thank the Health and Lifestyle Survey at the University of Cambridge and the ESRC Data Archive at the University of Essex for permission to use the data for the Health and Lifestyle Survey. The data from the Health Survey of England are crown copyright and are also made available through the ESRC Data Archive. I also thank Mrs P J Wassell and Mrs D P Morris for their expert typing of this manuscript, Dr F J C Roe for helpful comments on an earlier draft, and British American Tobacco Co for providing financial support. I alone bear the responsibility for all opinions expressed in this paper.

## REFERENCES

1 California Environmental Protection Agency (EPA). *Health effects of exposure to environmental tobacco smoke.* USA: CaliEPA Office of Environmental Health Hazard Assessment; 1997.

2 Luck W, Nau H, Hensen R, Steldinger R. Extent of nicotine and cotinine transfer to the human fetus, placenta and amniotic fluid of smoking mothers. *Develop. Pharmacol. Therap.*, 1985: **8**; 384–395.

3 Ueda Y, Morikawa H, Funakoshi T, Kobayashi A, Yamasaki A, Takeuchi K, et al. Estimation of passive smoking during pregnancy by cotinine measurement and its effect on fetal growth. *Acta Obstet. Gynaecol. Jap.*, 1989: **41**; 454–460.

4 Pacifici R, Altieri I, Gandini L, Lenzi A, Passa AR, Pichini S, et al. Environmental tobacco smoke: nicotine and cotinine concentration in semen. *Environ. Res.*, 1995: **68**; 69–72.

5 Paulson GW, Olson BL. Can smoking be detected from cerebrospinal fluid? [Letter]. *Clin. Neuropharmacol.*, 1995: **18**; 375–376.

6 Poppe WA, Peeters R, Daenens P, Ide PS, Van Assche FA. Tobacco smoking and the uterine cervix: cotinine in blood, urine and cervical fluid. *Gynecol. Obstet. Invest.*, 1995: **39**; 110–114.

7 Sofikitis N, Miyagawa I, Zavos P, Monga M, Sikka SC, Toda T, et al. Effects of smoking on testicular function, semen quality, and sperm fertilizing capacity [Abstract]. *J. Androl.*, 1995: **16(Sup)**; 31.

8 Nilsen T, Zahlsen K, Nilsen OG. Uptake of nicotine in hair during controlled environmental air exposure to nicotine vapour: evidence for a major contribution of environmental nicotine to the overall nicotine found in hair from smokers and non-smokers. *Pharmacol. Toxicol.*, 1994: **75**; 136–142.

9 Dawson RF, Solt ML, Christman DR. Nicotine and its botanical sources. *Ann. NY Acad. Sci.*, 1960: **90**; 7–12.

10 Castro A, Monji N. Dietary nicotine and its significance in studies on tobacco smoking. *Biochem. Arch.*, 1986: **2**; 91–97.

11 Sheen SJ. Detection of nicotine in foods and plant materials. *J. Fd. Sci.*, 1988: **53**; 1572–1573.

12 Idle JR. Titrating exposure to tobacco smoke using cotinine – a minefield of misunderstandings. *J. Clin. Epidemiol.*, 1990: **43**; 313–317.

13 Domino EF, Hornbach E, Demana T. The nicotine content of common vegetables [Letter]. *N. Engl. J. Med.*, 1993: **329**; 437.

14 Committee on Passive Smoking, Board on Environmental Studies and Toxicology, National Research Council. *Environmental tobacco smoke. Measuring exposures and assessing health effects.* Washington D.C.: National Academy Press; 1986.

15 Jarvis MJ. Dietary nicotine: ... unless subjects eat 90 kg tomatoes a day [Letter]. *BMJ*, 1994: **308**; 62.

16 Repace JL. Dietary nicotine - Won't mislead on passive smoking... [Letter]. *BMJ*, 1994: **308**; 61–62.

17 Lee PN. Passive smoking and lung cancer association: a result of bias? *Human Toxicol.*, 1987: **6**; 517–524.

18 Benowitz NL. Cotinine as a biomarker of environmental tobacco smoke exposure. *Epidemiol. Rev.*, 1996: **18**; 188–204.

19 Pirkle JL, Flegal KM, Bernert JT, Brody DJ, Etzel RA, Maurer KR. Exposure of the US population to environmental tobacco smoke. The Third National Health and Nutrition Examination Survey, 1988 to 1991. *JAMA*, 1996: **275**; 1233–1240.

20 Tunstall-Pedoe H, Woodward M, Brown CA. Tea drinking, passive smoking, smoking deception and serum cotinine in the Scottish Heart Health Study. *J. Clin. Epidemiol.*, 1991: **44**; 1411–1414.

21 Galanti LM, Manigart PE, Dubois P. Tobacco smoking and alcohol and drug consumption in a large, young healthy population. *Arch. Environ. Hlth.*, 1998: **53**; 156–160.

22 Jarvis MJ, Tunstall-Pedoe H, Feyerabend C, Vesey C, Saloojee Y. Comparison of tests used to distinguish smokers from non-smokers. *Am. J. Publ. Hlth.*, 1987: **77**; 1435–1438.

23 Mathai M, Skinner A, Lawton K, Weindling AM. Maternal smoking, urinary cotinine levels and birth-weight. *Aust. NZ J. Obstet. Gynaecol.*, 1990: **30**; 33–36.

24 Matsukura S, Taminato T, Kitano N, Seino Y, Hamada H, Uchihashi M, et al. Effects of environmental tobacco smoke on urinary cotinine excretion in non-smokers. Evidence for passive smoking. *N. Engl. J. Med.*, 1984: **311**; 828–832.

25 Thompson SG, Stone R, Nanchahal K, Wald NJ. Relation of urinary cotinine concentrations to cigarette smoking and to exposure to other people's smoke. *Thorax*, 1990: **45**; 356–361.

26 Wald NJ, Boreham J, Bailey A, Ritchie C, Haddow JE, Knight G. Urinary cotinine as marker of breathing other people's tobacco smoke [Letter]. *Lancet*, 1984: **1**; 230–231.

27 Bennett N, Dodd T, Flatley J, Freeth S, Bolling K. *Health Survey for England 1993.* London: HMSO; 1995. Series HS No. 3.

28 Colhoun H, Prescott-Clarke P, editors. *Health Survey for England 1994.* London: HMSO; 1996.

29 Haddow JE, Palomaki GE, Knight GJ. Use of serum cotinine to assess the accuracy of self reported non-smoking [Letter]. *BMJ*, 1986: **293**; 1306.

30 Haddow JE, Knight GT, Palomaki GE, Kloza EM, Wald NJ. Cigarette consumption and serum cotinine in relation to birthweight. *Br J. Obstet. Gynaecol.*, 1987: **94**; 678–681.

31 Kemmeren JM, van Poppel G, Verhoef P, Jarvis MJ. Plasma cotinine: stability in smokers and validation of self-reported smoke exposure in non-smokers. *Environ. Res.*, 1994: **66**; 235–243.

32 Perez-Stable EJ, Marin G, Marin BV, Benowitz NL. Misclassification of smoking status by self-reported cigarette consumption. *Am. Rev. Respira. Dis.*, 1992: **145**; 53–57.

33 Pojer R, Whitfield JB, Poulos V, Eckhard IF, Richmond R, Hensley WJ. Carboxyhemoglobin, cotinine, and thiocyanate assay compared for distinguishing smokers from non-smokers. *Clin. Chem.*, 1984: **30**; 1377–1380.

34 Prescott-Clarke P, Primatesta P, editors. *Health Survey for England '96.* London: HMSO; 1998.

35 Richmond R, Webster I. Blood cotinine, carboxyhaemoglobin, and thiocyanate concentrations and cigarette consumption. *BMJ*, 1986: **293**; 1280.

36 Wald NJ, Idle M, Boreham J, Bailey A, van Vunakis H. Serum cotinine levels in pipe smokers: evidence against nicotine as cause of coronary heart disease. *Lancet*, 1981: **2**; 775–777.

37 Jarvis MJ, McNeill AD, Russell MAH. Factors determining exposure to passive smoking in young adults living at home: quantitative analysis using saliva cotinine concentrations. *Int. J. Epidemiol.*, 1991: **20**; 126–131.

38 Li CQ, Windsor RA, Perkins L, Goldenberg RL, Lowe JB. The impact on infant birth weight and gestational age of cotinine-validated smoking reduction during pregnancy. *JAMA*, 1993: **269**; 1519–1524.

39 McNeill AD, Jarvis MJ, West R, Russell MAH, Bryant A. Saliva cotinine as an indicator of cigarette smoking in adolescents. *Brit. J. Addict.*, 1987: **82**; 1355–1360.

40 Bernert JT Jr, Sosnoff C, Turner WE, Akins J, Waldrep MK, Whitfield W, et al. Development of a rapid and sensitive method for serum cotinine analysis as a marker of exposure to environmental tobacco smoke [Abstract]. *Clin. Chem.*, 1994: **40**; 1075.

41 Adlkofer F, Scherer G, von Hees U. Passive smoking. *N. Engl. J. Med.*, 1985: **312**; 719–720.

42 Biber A, Scherer G, Hoepfner I, Adlkofer F, Heller W-D, Haddow JE, et al. Determination of nicotine and cotinine in human serum and urine: an inter-laboratory study. *Toxicol. Lett.*, 1987: **35**; 45–52.

43 Scherer G, von Maltzan C, von Meyerinck L, Westphal K, Adlkofer F. Biomonitoring after controlled exposure to environmental tobacco smoke (ETS). *Exp. Path.*, 1989: **37**; 158–163.

44 Bayard S, Jinot J, Flatman G. Environmental tobacco smoke and lung cancer: uncertainties in the population estimates but not in the causal association – a rejoinder to Gross. *Environmetrics*, 1995: **6**; 413–418.

45 Roberfroid MB. Correlating exposure to environmental tobacco smoke exposure with increased incidence of lung cancer in non smokers: is cotinine a valid marker? *Cancer J.*, 1994: **7**; 108–114.

46 Lee PN. *Environmental tobacco smoke and mortality*. Basle: Karger; 1992.

47 Lee PN, Forey BA. Misclassification of smoking habits as determined by cotinine or by repeated self-report – a summary of evidence from 42 studies. *J. Smoking-Relat. Disord.*, 1995: **6**; 109–129.

48 Hugod C, Hawkins LH, Astrup P. Exposure of passive smokers to tobacco smoke constituents. *Int. Arch. Occup. Environ. Hlth*, 1978: **42**; 21–29.

49 Curvall M, Kazemi-Vala E, Enzell CR. Inhaled amount of tobacco smoke during passive smoking. In: Seifert B, Esdorn H, Fischer M, Rüden H, Wegner J, editors. *Indoor Air '87. Proceedings of the 4th International Conference on Indoor Air Quality and Climate, Berlin (West), 17–21 August 1987. Volume 2.* pp. 57–60.

50 Bergman TA, Johnson DL, Boatright DT, Smallwood KG, Rando RJ. Occupational exposure of non-smoking nightclub musicians to environmental tobacco smoke. *Am. Ind. Hyg. Assoc. J.*, 1996: **57**; 746–752.

51 Jarvis MJ, Foulds J, Feyerabend C. Exposure to passive smoking among bar staff. *Brit. J. Addict.*, 1992: **87**; 111–113.

52 US Environmental Protection Agency. *Smoking and tobacco control, Monograph 4: Respiratory health effects of passive smoking: lung cancer and other disorders.* USA: National Institutes of Health; 1993.

53 Emmons KM, Abrams DB, Marshall R, Marcus BH, Kane M, Novotny TE, et al. An evaluation of the relationship between self-report and biochemical measures of environmental tobacco smoke exposure. *Prev. Med.*, 1994: **23**; 35–39.

54 Cummings KM, Markello SJ, Mahoney M, Bhargava AK, McElroy PD, Marshall JR. Measurement of current exposure to environmental tobacco smoke. *Arch. Environ. Hlth.*, 1990: **45**; 74–79.

55 Delfino RJ, Ernst P, Jaakkola M, Solomon S, Becklake MR. Questionnaire assessments of recent exposure to environmental tobacco smoke in relation to salivary cotinine. *Eur. Respir. J.*, 1993: **6**; 1104–1108.

56 Martinez FD, Wright AL, Taussig LM, the Group Health Medical Associates. The effect of paternal smoking on the birthweight of newborns whose mothers did not smoke. *Am. J. Publ. Hlth.*, 1994: **84**; 1489–1491.

57 English PB, Eskenazi B, Christianson RE. Black-white differences in serum cotinine levels among pregnant women and subsequent effects on infant birthweight. *Am. J. Publ. Hlth.*, 1994: **84**; 1439–1443.

58 Haley NJ, Colosimo SG, Axelrad CM, Harris R, Sepkovic DW. Biochemical validation of self-reported exposure to environmental tobacco smoke. *Environ. Res.*, 1989: **49**; 127–135.

59 Haddow JE, Knight GJ, Palomaki GE, McCarthy JE. Second-trimester serum cotinine levels in non-smokers in relation to birth weight. *Am. J. Obstet. Gynecol.*, 1988: **159**; 481–484.

60 Riboli E, Preston-Martin S, Saracci R, Haley NJ, Trichopoulos D, Becher H, et al. Exposure of non-smoking women to environmental tobacco smoke: a 10 country collaborative study. *Cancer Causes Control*, 1990: **1**; 243–252.

61 Slattery ML, Hunt SC, French TK, Ford MH, Williams RR. Validity of cigarette smoking habits in three epidemiologic studies in Utah. *Prev. Med.*, 1989: **18**; 11–19.

62 Fontham ETH, Correa P, Reynolds P, Wu-Williams A, Buffler PA, Greenberg RS, et al. Environmental tobacco smoke and lung cancer in non-smoking women. A multicenter study. *JAMA*, 1994: **271**; 1752–1759.

63 Heller W-D, Sennewald E, Gostomzyk J-G, Scherer G, Adlkofer F. Validation of ETS exposure in a representative population in Southern Germany. *Indoor Air '93. Proceedings of the 6th International Conference on Indoor Air Quality and Climate, Helsinki, Finland July 4–8, 1993. Volume 3.*, 1993. p. 361–365.

64 Dickinson JA, Rostas JAP, Nichols B, Ebeling W, Dickeson J, Matthews D, et al. *The use of saliva cotinine to validate smoking self-report in primary care.* Unpublished; 1988. Provided as a personal communication by Prof. R.W. Gibberd, University of Newcastle, New South Wales, Australia.

65 Luepker RV, Pallonen UE, Murray DM, Pirie PL. Validity of telephone surveys in assessing cigarette smoking in young adults. *Am. J. Publ. Hlth.*, 1989: **79**; 202–204.

66 Becher H, Zatonski W, Jockel K-H. Passive smoking in Germany and Poland: comparison of exposure levels, sources of exposure, validity and perception. *Epidemiology*, 1992: **3**; 509–514.

67 Van Vunakis H, Tashkin DP, Rigas B, Simmons M, Gjika HB, Clark VA. Relative sensitivity and specificity of salivary and serum cotinine in identifying tobacco-smoking status of self-reported non-smokers and smokers of tobacco and/or marijuana. *Arch. Environ. Hlth.*, 1989: **44**; 53–58.

68 Wagenknecht LE, Burke GL, Perkins LL, Haley NJ, Friedman GD. Misclassification of smoking status in the CARDIA study: a comparison of self-report with serum cotinine levels. *Am. J. Publ. Hlth.*, 1992: **82**; 33–36.

69 Pierce JP, Dwyer T, DiGiusto E, Carpenter T, Hannam C, Amin A, et al. Cotinine validation of self-reported smoking in commercially run community surveys. *J. Chron. Dis.*, 1987: **40**; 689–695.

70 Ogden MW, Morgan WT, Heavner DL, Davis RA, Steichen TJ. National incidence of smoking and misclassification among the US married female population. *J. Clin. Epidemiol.*, 1997: **50**; 253–263.

71 Coultas DB, Howard CA, Peake GT, Skipper BJ, Samet JM. Salivary cotinine levels and involuntary tobacco smoke exposure in children and adults in New Mexico. *Am. Rev. Respirat. Dis.*, 1987: **136**; 305–309.

72 Ford RPK, Tappin DMN, Schluter PJ, Wild CJ. Smoking during pregnancy: how reliable are maternal self reports in New Zealand? *J. Epidemiol. Commun. Hlth.*, 1997: **51**; 246–251.

73 Ogden MW, Davis RA, Maiolo KC, Stiles MF, Heavner DL, Hege RB, et al. Multiple measures of personal ETS exposure in a population-based survey of non-smoking women in Columbus, Ohio. *Indoor Air '93. Proceedings of the 6th International Conference on Indoor Air Quality and Climate, Helsinki, Finland July 4–8, 1993. Volume 1.* 1993. p. 523–528.

74 Lee PN. 'Marriage to a smoker' may not be a valid marker of exposure in studies relating environmental tobacco smoke to risk of lung cancer in Japanese non-smoking women. *Int. Arch. Occup. Environ. Hlth*, 1995: **67**; 287–294.

75 Akiyama Y, Ohkawa Y, Matsuki H, Yokoyama H, Kawano Y, Kasuga H, et al. Misclassification of smoking status: comparison of questionnaire data and urinary cotinine analysis. In: Leslie GB, Leslie KJ, Huang J, Qin Y, editors. *Proceedings of the second International Conference on Indoor Air Quality in Asia.* Switzerland: Indoor Air International; 1994. p. 319–323.

76 Wewers ME, Dhatt RK, Moeschberger ML, Guthrie RM, Kuun P, Chen MS. Misclassification of smoking status among Southeast Asian adult immigrants. *Am. J. Respir. Crit. Care Med.*, 1995: **152**; 1917–1921.

77 Phillips K, Howard DA, Browne D, Lewsley JM. Assessment of personal exposures to environmental tobacco smoke in British non-smokers. *Environ. Internat.*, 1994: **20**; 693–712.

78 Lee PN. Misclassification of smoking habits and passive smoking. A review of the evidence. *International Archives of Occupational and Environmental Health Supplement.* Heidelberg: Springer-Verlag; 1988: 114pp.

79 Wilcox RG, Hughes J, Roland J. Verification of smoking history in patients after infarction using urinary nicotine and cotinine measurements. *BMJ*, 1979: **2**; 1026–1028.

80 Woodward M, Tunstall-Pedoe H. Biochemical evidence of persistent heavy smoking after a coronary diagnosis despite self-reported reduction: analysis from the Scottish Heart Health Study. *Europ. Hlth. J.*, 1992: **13**; 160–165.

81 Apseloff G, Ashton HM, Friedman H, Gerber N. The importance of measuring cotinine levels to identify smokers in clinical trials. *Clin. Pharmacol. Ther.*, 1994: **56**; 460–462.

82 Bodmer CW, MacFarlane IA, Flavell HJ, Wallymahmed M, Calverley PMA. How accurate is the smoking history in newly diagnosed diabetic patients? *Diabetes Res. Clin. Pract.*, 1990: **10**; 215–220.

83 Adlkofer F, Scherer G, Biber A, Heller W-D, Lee PN, Schievelbein H. Consistency of nicotine intake in smokers of cigarettes with varying nicotine yields. In: Wald N, Froggatt P, editors. *Nicotine, smoking and the low tar programme.* Oxford: Oxford University Press; 1989. p. 116–131.

84 Klebanoff MA, Levine RJ, Clemens JD, DerSimonian R, Wilkins DG. Serum cotinine concentration and self-reported smoking during pregnancy. *Am. J. Epidemiol.*, 1998: **148**; 259–262.

85 Puntoni R, Fornai E, Maggiorelli F, Di Pede F, Corlando A, Kuylenstierna F, et al. Assessment of smoking habit in relation to nicotine and cotinine plasma values in smokers [Abstract]. *Eur. Respir. J.*, 1991: **4**; 5035.

86 Suadicani P, Hein HO, Gyntelberg F. Serum validated tobacco use and social inequalities in risk of ischaemic heart disease. *Int. J. Epidemiol.*, 1994: **23**; 293–300.

87 Woodward M, Tunstall-Pedoe H, Smith WCS, Tavendale R. Smoking characteristics and inhalation biochemistry in the Scottish population. *J. Clin. Epidemiol.*, 1991: **44**; 1405–1410.

88 Worrell PC, Edwards R, Powsell JT. Smoking markers as a reflection of smoking habit. *J. Smoking-Relat. Disord.*, 1995: **6**; 89–97.

89 Hasenfratz M, Baldinger B, Battig K. Nicotine or tar titration in cigarette smoking behavior? *Psychopharmacology (Berl.)*, 1993: **112**; 253–258.

90 Coultas DB, Stidley CA, Samet JM. Cigarette yields of tar and nicotine and markers of exposure to tobacco smoke. *Am. Rev. Respirat. Dis.*, 1993: **148**; 435–440.

91 Istvan JA, Nides MA, Buist AS, Greene P, Voelker H. Salivary cotinine, frequency of cigarette smoking, and body mass index: findings at baseline in the Lung Health Study. *Am. J. Epidemiol.*, 1994: **139**; 628–636.

92 Stanton WR, McClelland M, Elwood C, Ferry D, Silva PA. Prevalence, reliability and bias of adolescents' reports of smoking and quitting. *Addiction*, 1996: **91**; 1705–1714.

93 Heller W-D, Scherer G, Sennewald E, Adlkofer F. Misclassification of smoking in a follow-up population study in southern Germany. *J. Clin. Epidemiol.*, 1998: **51**; 211–218.

94 Perez-Stable EJ, Marin BV, Marin G, Brody DJ, Benowitz NL. Apparent underreporting of cigarette consumption among Mexican American smokers. *Am. J. Publ. Hlth.*, 1990: **80**; 1057–1061.

95 Sepkovic DW, Axelrad CM, Colosimo SG, Haley NJ. *Measuring tobacco smoke exposure: clinical applications and passive smoking. Presentation at the 80th Annual Meeting of APCA New York, New York June 21–26, 1987.* New York: APCA; 1987. (APCA = Association dedicated to Air Pollution Control and Hazardous Waste Management)

96 Ahijevych KL, Wewers ME. Patterns of cigarette consumption and cotinine levels among African American women smokers. *Am. J. Respir. Crit. Care Med.*, 1994: **150**; 1229–1233.

97 Law MR, Morris JK, Wald NJ. The dose-response relationship between cigarette consumption, biochemical markers and risk of lung cancer. *Br. J. Cancer*, 1997: **75**; 1690–1693.

98 Swan GE, Habina K, Means B, Jobe JB, Esposito JL. Saliva cotinine and recent smoking – evidence for a nonlinear relationship. *Publ. Hlth. Rep.*, 1993: **108**; 779–783.

99 Saloojee Y, Cole PV. Nicotine, carbon monoxide, and heart disease [Letter]. *Lancet*, 1981: **2**; 1044.

100 Jarvis MJ, Russell MAH. Nicotine, carbon monoxide, and heart disease [Letter]. *Lancet*, 1981: **2**; 1045.

101 International Agency for Research on Cancer. *IARC Monographs on the evaluation of the carcinogenic risk of chemicals to humans. Volume 38. Tobacco Smoking.* Lyon, France: IARC; 1986.

102 Stephen A, Frost C, Thompson S, Wald N. Estimating the extent of compensatory smoking. In: Wald N, Froggatt P, editors. *Nicotine, smoking and the low tar programme.* Oxford: Oxford University Press; 1989. p. 100–115.

103 McMorrow MJ, Foxx RM. Nicotine's role in smoking: an analysis of nicotine regulation. *Psychol. Bull.*, 1983: **93**; 302–327.

104 Frost C, Fullerton FM, Stephen AM, Stone R, Nicolaides-Bouman A, Densem J, et al. The tar reduction study: randomised trial of the effect of cigarette tar yield reduction on compensatory smoking. *Thorax*, 1995: **50**; 1038–1043.

105 Hirayama T. Non-smoking wives of heavy smokers have a higher risk of lung cancer: a study from Japan. *BMJ*, 1981: **282**; 183–185.

106 Trichopoulos D, Kalandidi A, Sparros L, MacMahon B. Lung cancer and passive smoking. *Int. J. Cancer*, 1981: **27**; 1–4.

107 Rebagliato M, Bolumar F, Florey Cdu V. Assessment of exposure to environmental tobacco smoke in non-smoking pregnant women in different environments of daily living. *Am. J. Epidemiol.*, 1995: **142**; 525–530.

108 Wald NJ, Ritchie C. Validation of studies on lung cancer in non-smokers married to smokers [Letter]. *Lancet*, 1984: **1**; 1067.

109 Willers S, Attewell R, Bensryd I, Schutz A, Skarping G, Vahter M. Exposure to environmental tobacco smoke in the household and urinary cotinine excretion, heavy metals retention, and lung function. *Arch. Environ. Hlth.*, 1992: **47**; 357–363.

110 Wald NJ, Nanchahal K, Thompson SG, Cuckle HS. Does breathing other people's tobacco smoke cause lung cancer? *BMJ*, 1986: **293**; 1217–1222.

111 Hackshaw AK, Law MR, Wald NJ. The accumulated evidence on lung cancer and environmental tobacco smoke. *BMJ*, 1997: **315**; 980–988.

112 Layard MW. The background adjustment in risk assessment of environmental tobacco smoke and lung cancer. *Environ. Internat.*, 1992: **18**; 453–461.

113 Lee PN. *Environmental tobacco smoke, a commentary on some of the evidence provided by OSHA in support of their proposed rules.* Washington DC: Occupational Safety and Health Administration; 1994. OSHA docket 9–47513.

114 Occupational safety and health administration. Indoor air quality, proposed rule. *Fed. Reg.*, 1994: **59**; 15968–16039.

115 Repace JL, Lowrey AH. A quantitative estimate of non-smokers' lung cancer risk from passive smoking. *Environ. Internat.*, 1985: **11**; 3–22.

116 Kawachi I, Pearce NE, Jackson RT. Deaths from lung cancer and ischaemic heart disease due to passive smoking in New Zealand. *NZ Med. J.*, 1989: **102**; 337–340.

117 Bakoula CG, Kafritsa YJ, Kavadias GD, Haley NJ, Matsaniotis NS. Factors modifying exposure to environmental tobacco smoke in children (Athens, Greece). *Cancer Causes Control*, 1997: **8**; 73–76.

118 Bono R, Arossa W, Russo R, Scursatone E, Gilli G. Passive exposure to environmental tobacco smoke in 434 14-year-old schoolchildren. *Indoor Air '93. Proceedings of the 6th International Conference on Indoor Air Quality and Climate, Helsinki, Finland July 4–8, 1993. Volume 1.* 1993. p. 663–668.

119 Cook DG, Whincup PH, Jarvis MJ, Strachan DP, Papacosta O, Bryant A. Passive exposure to tobacco smoke in children aged 5–7 years: individual, family, and community factors. *BMJ*, 1994: **308**; 384–389.

120 Jarvis MJ, Russell MAH, Feyerabend C, Reiser JR, Morgan M, Gammage P, et al. Passive exposure to tobacco smoke: saliva cotinine concentrations in a representative population sample of non-smoking school children. *BMJ*, 1985: **291**; 927–929.

121 Jarvis MJ, McNell AD, Russell MAH, West RJ, Bryant A, Feyerabend C. Passive smoking in adolescents: one-year stability of exposure in the home [Letter]. *Lancet*, 1987: **1**; 1324–1325.

122 Marbury MC, Hammond SK, Haley NJ. Measuring exposure to environmental tobacco smoke in studies of acute health effects. *Am. J. Epidemiol.*, 1993: **137**; 1089–1097.

123 Preston AM, Ramos LJ, Calderon C, Sahai H. Exposure of Puerto Rican children to environmental tobacco smoke. *Prev. Med.*, 1997: **26**; 1–7.

124 Reese AC, James IR, Landau LI, Lesouef PN. Relationship between urinary cotinine level and diagnosis in children admitted to hospital. *Am. Rev. Respirat. Dis.*, 1992: **146**; 66–70.

125 Weaver VM, Davoli CT, Murphy SE, Sunyer J, Heller PJ, Colosimo SG, et al. Environmental tobacco smoke exposure in inner-city children. *Cancer Epidemiol. Biomarkers Prev*, 1996: **5**; 135–137.

126 Willers S, Svenonius E, Skarping G. Passive smoking and childhood asthma. Urinary cotinine levels in children with asthma and in referents. *Allergy*, 1991: **46**; 330–334.

127 Dell'Orco V, Forastiere F, Agabiti N, Corbo GM, Pistelli R, Pacifici R, et al. Household and community determinants of exposure to involuntary smoking: a study of urinary cotinine in children and adolescents. *Am. J. Epidemiol.*, 1995: **142**; 419–427.

128 Forastière F, Agabiti N, Dell'orco V, Pistelli R, Corbo GM, Brancato G, et al. Questionnaire data as predictors of urinary cotinine levels among non-smoking adolescents. *Arch. Environ. Hlth.*, 1993: **48**; 230–234.

129 Woodward A, Grgurinovich N, Ryan P. Breast feeding and smoking hygiene: Major influences on cotinine in urine of smoker's infants. *J. Epidemiol. Commun. Hlth.*, 1986: **40**; 309–315.

130 Lee PN. Passive exposure to tobacco smoke [Letter]. *BMJ*, 1985: **291**; 1646.

131 Bakoula CG, Kafrista YJ, Kavadias GD, Lazopoulou DD, Theodoridou MC, Maravelias KP, et al. Objective passive-smoking indicators and respiratory morbidity in young children. *Lancet*, 1995: **346**; 280–281.

132 Casale R, Colantonio D, Cialente M, Colorizio V, Barnabei R, Pasqualetti P. Impaired pulmonary function in schoolchildren exposed to passive smoking: detection by questionnaire and urinary cotinine levels. *Respiration*, 1991: **58**; 198–203.

133 Chilmonczyk BA, Salmun LM, Megathlin KN, Neveux LM, Palomaki GE, Knight GJ, et al. Association between exposure to environmental tobacco smoke and exacerbations of asthma in children. *N. Engl. J. Med.*, 1993: **328**; 1665–1669.

134 Clark SJ, Warner JO, Dean TP. Passive smoking amongst asthmatic children. Questionnaire or objective assessment? *Clin. Exp. Allergy*, 1994: **24**; 276–280.

135 Cook DG, Whincup PH, Papacosta O, Strachan DP, Jarvis MJ, Bryant A. Relation of passive smoking as assessed by salivary cotinine concentration and questionnaire to spirometric indices in children. *Thorax*, 1993: **48**; 14–20.

136 Duff AL, Pomeranz ES, Gelber LE, Price GW, Farris H, Hayden FG, et al. Risk factors for acute wheezing in infants and children: viruses, passive smoke, and IgE antibodies to inhalant allergens. *Pediatrics*, 1993: **92**; 535–540.

137 Ehrlich R, Kattan M, Godbold J, Saltzberg DS, Grimm KT, Landrigan PJ, et al. Childhood asthma and passive smoking: Urinary cotinine as a biomarker of exposure. *Am. Rev. Respirat. Dis.*, 1992: **145**; 594–599.

138 Henderson FW, Henry MM, Ivins SS, Morris R, Neebe EC, Leu S-Y, et al. Correlates of recurrent wheezing in school-age children. *Am. J. Respir. Crit. Care Med.*, 1995: **151**; 1786–1793.

139 Margolis PA, Keyes LL, Greenberg RA, Bauman KE, La Vange LM. Urinary cotinine and parent history (questionnaire) as indicators of passive smoking and predictors of lower respiratory illness in infants. *Pediatr. Pulmonol*, 1997: **23**; 417–423.

140 Ogborn CJ, Duggan AK, DeAngelis C. Urinary cotinine as a measure of passive smoke exposure in asthmatic children. *Clin. Pediatr.*, 1994: **33**; 220–226.

141 Rylander E, Pershagen G, Eriksson M, Bermann G. Parental smoking, urinary cotinine, and wheezing bronchitis in children. *Epidemiology*, 1995: **6**; 289–293.

142 Strachan DP, Jarvis MJ, Feyerabend C. The relationship of salivary cotinine to respiratory symptoms, spirometry, and exercise-induced bronchospasm in seven-year-old children. *Am. Rev. Respirat. Dis.*, 1990: **142**; 147–151.

143 Tager IB, Hanrahan JP, Tosteson TD, Castile RG, Brown RW, Weiss ST, et al. Lung function, pre- and post-natal smoke exposure, and wheezing in the first year of life. *Am. Rev. Respirat. Dis.*, 1993: **147**; 811–817.

144 Tager IB, Ngo L, Hanrahan JP. Maternal smoking during pregnancy. Effects on lung function during the first 18 months of life. *Am. J. Respir. Crit. Care Med.*, 1995: **152**; 977–983.

145 Winkelstein ML, Tarzian A, Wood RA. Parental smoking behavior and passive smoke exposure in children with asthma. *Ann. Allergy Asthma Immunol.*, 1997: **78**; 419–423.

146 Adair-Bischoff CE, Sauve RS. Environmental tobacco smoke and middle ear disease in preschool-age children. *Arch. Pediatr. Adolesc. Med.*, 1998: **152**; 127–133.

147 Etzel RA, Pattishall EN, Haley NJ, Fletcher RH, Henderson FW. Passive smoking and middle ear effusion among children in day care. *Pediatrics*, 1992: **90**; 228–232.

148 Strachan DP, Jarvis MJ, Feyerabend C. Passive smoking, salivary cotinine concentrations, and middle ear effusion in 7 year old children. *BMJ*, 1989: **298**; 1549–1552.

149 McMartin KI, Klein J, Platt MS, Koren G. The effect of tobacco smoking on nicotine and cotinine levels in lung tissue of SIDS and non-SIDS cases [Abstract]. *Clin. Pharmacol. Ther.*, 1996: **59**; 145.

150 Milerad J, Rajs J, Gidlung E. Nicotine and cotinine levels in pericardial fluid in victims of SIDS. *Acta Paediatr.*, 1994: **83**; 59–62.

151 Cuckle HS, Alberman E, Wald NJ, Royston P, Knight G. Maternal smoking habits and Down's syndrome. *Prenat. Diagn.*, 1990: **10**; 561–567.

152 Eskenazi B, Bergmann JJ. Passive and active maternal smoking during pregnancy, as measured by serum cotinine, and postnatal smoke exposure. I. Effects on physical growth at age 5 years. *Am. J. Epidemiol.*, 1995: **142**; S10–18.

153 Eskenazi B, Trupin LS. Passive and active maternal smoking during pregnancy, as measured by serum cotinine, and postnatal smoke exposure. II. Effects on neurodevelopment at age 5 years. *Am. J. Epidemiol.*, 1995: **142**; S19–29.

154 Bardy AH, Seppälä T, Lillsunde P, Kataja JM, Koskela P, Pikkarainen J, et al. Objectively measured tobacco exposure during pregnancy: neonatal effects and relation to maternal smoking. *Br. J. Obstet. Gynaecol.*, 1993: **100**; 721–726.

155 Haddow JE, Knight GJ, Kloza EM, Palomaki GE, Wald NJ. Cotinine-assisted intervention in pregnancy to reduce smoking and low birthweight delivery. *Br. J. Obstet. Gynaecol.*, 1991: **98**; 859–865.

156 Rebagliato M, Florey Cdu V, Bolumar F. Exposure to environmental tobacco smoke in non-smoking pregnant women in relation to birth weight. *Am. J. Epidemiol.*, 1995: **142**; 531–537.

157 Peacock JL, Cook DG, Carey IM, Jarvis MJ, Bryant AE, Anderson HR. Maternal cotinine level during pregnancy and birthweight for gestational age. *Int. J. Epidemiol.*, 1998: **27**; 647–656.

158 Secker-Walker, R.H., Vacek PM, Flynn BS, Mead PB. Estimated gains in birth weight associated with reductions in smoking during pregnancy. *J. Reprod. Med.*, 1998: **43**; 967–974.

159 Perez-Stable EJ, Benowitz NL, Marin G. Is serum cotinine a better measure of cigarette smoking than self-report? *Prev. Med.*, 1995: **24**; 272–279.

160 Tunstall-Pedoe H, Brown CA, Woodward M, Tavendale R. Passive smoking by self report and serum cotinine and the prevalence of respiratory and coronary heart disease in the Scottish heart health study. *J. Epidemiol. Commun. Hlth.*, 1995: **49**; 139–143.

161 de Waard F, Kemmeren JM, van Ginkel LA, Stolker AAM. Urinary cotinine and lung cancer risk in a female cohort. *Br. J. Cancer*, 1995: **72**; 784–787.

162 Suadicani P, Hein HO, Gyntelberg F. Mortality and morbidity of potentially misclassified smokers. *Int. J. Epidemiol.*, 1997: **26**; 321–327.

163 Lee PN, Forey BA. Misclassification of smoking habits as a source of bias in the study of environmental tobacco smoke and lung cancer. *Stat. Med.*, 1996: **15**; 581–605.

164 Thornton A, Lee P, Fry J. Differences between smokers, ex-smokers, passive smokers and non-smokers. *J. Clin. Epidemiol.*, 1994: **47**; 1143–1162.

165 Adir J, Wildfeuer W, Miller RP. Effect of ethanol pretreatment on the pharmacokinetics of nicotine in rats. *J. Pharmacol. Exp. Ther.*, 1980: **212**; 274–279.

166 Niaura R, Clark MM, Raciti MA, Pera V, Abrams DB. Increased saliva cotinine concentrations in smokers during rapid weight loss. *J. Consult. Clin. Psychol.*, 1992: **60**; 985–987.

167 Spiekerkötter EF, Henschen M, Pracht T, Forster J, Frischer T, Kuehr J. Relationship between urinary cotinine, passive smoke exposure and respiratory health in children [Abstract]. *Pediatr. Res.*, 1994: **36**; 216.

168 Wagenknecht LE, Manolio TA, Sidney S, Burke GL, Haley NJ. Environmental tobacco smoke exposure as determined by cotinine in black and white young adults: The CARDIA Study. *Environ. Res.*, 1993: **63**; 39–46.

169 Henschen M, Frischer T, Pracht T, Spiekerkötter E, Karmaus W, Meinert R, et al. The internal dose of passive smoking at home depends on the size of the dwelling. *Environ. Res.*, 1997: **72**; 65–71.

170 Wagenknecht LE, Cutter GR, Haley NJ, Sidney S, Manolio TA, Hughes GH, et al. Racial differences in serum cotinine levels among smokers in the Coronary Artery Risk Development in (young) Adults study. *Am. J. Publ. Hlth.*, 1990: **80**; 1053–1056.

171 Caraballo RS, Giovino GA, Pechacek TF, Mowery PD, Richter PA, Strauss WJ, et al. Racial and ethnic differences in serum cotinine levels of cigarette smokers. Third National Health and Nutrition Examination Survey, 1988–1991. *JAMA*, 1998: **280**; 135–139.

172 Knight JM, Eliopoulos C, Klein J, Greenwald M, Koren G. Passive smoking in children. Racial differences in systemic exposure to cotinine by hair and urine analysis. *Chest*, 1996: **109**; 446–450.

173 Pérez-Stable EJ, Herrera B, Jacob P III, Benowitz NL. Nicotine metabolism and intake in black and white smokers. *JAMA*, 1998: **280**; 152–156.

174 Foundas M, Hawkrigg NC, Smith SMS, Devadason SG, Le Souef PN. Urinary cotinine levels in early pregnancy. *Aust NZ J. Obstet. Gynaecol.*, 1997: **37**; 383–386.

175 Labrecque M, Marcoux S, Weber J-P, Fabia J, Ferron L. Feeding and urine cotinine values in babies whose mothers smoke. *Pediatrics*, 1989: **83**; 93–97.

176 Schwartz-Bickenbach D, Schulte-Hobein B, Abt S, Plum C, Nau H. Smoking and passive smoking during pregnancy and early infancy: effects on birth weight, lactation period, and cotinine concentrations in mother's milk and infant's urine. *Toxicol. Lett.*, 1987: **35**; 73–81.

177 Mascola MA, Van Vunakis H, Tager IB, Speizer FE, Hanrahan JP. Exposure of young infants to environmental tobacco smoke: breast-feeding among smoking mothers. *Am. J. Publ. Hlth.*, 1998: **88**; 893–896.

178 Coultas DB, Samet JM, McCarthy JF, Spengler JD. Variability of measures of exposure to environmental tobacco smoke in the home. *Am. Rev. Respirat. Dis.*, 1990: **142**; 602–606.

179 Peterson EL, Johnson CC, Ownby DR. Use of urinary cotinine and questionnaires in the evaluation of infant exposure to tobacco smoke in epidemiologic studies. *J. Clin. Epidemiol.*, 1997: **50**; 917–923.

180 Emerson JA, Hovell MF, Meltzer SB, Zakarian JM, Hofstetter CR, Wahlgren DR, et al. The accuracy of environmental tobacco smoke exposure measures among asthmatic children. *J. Clin. Epidemiol.*, 1995: **48**; 1251–1259.

181 Haley NJ, Axelrad CM, Tilton KA. Validation of self-reported smoking behavior: biochemical analyses of cotinine and thiocyanate. *Am. J. Publ. Hlth.*, 1983: **73**; 1204–1207.

182 Vine MF, Hulka BS, Margolin BH, Truong YK, Hu P-C, Schramm MM, et al. Cotinine concentrations in semen, urine and blood of smokers and non-smokers. *Am. J. Publ. Hlth.*, 1993: **83**; 1335–1338.

183 Waage H, Silsand T, Urdal P, Langård S. Discrimination of smoking status by thiocyanate and cotinine in serum, and carbon monoxide in expired air. *Int. J. Epidemiol.*, 1992: **21**; 488–493.

184 Collier AM, Goldstein GM, Shrewsbury RP, Davis SM, Koch GG, Zhang C-A, et al. Cotinine elimination and its use as a biomarker in young children involuntarily exposed to environmental tobacco smoke. *Indoor Environ.*, 1994: **3**; 353–359.

185 Coultas DB, Samet JM, McCarthy JF, Spengler JD. A personal monitoring study to assess workplace exposure to environmental tobacco smoke. *Am. J. Publ. Hlth.*, 1990: **80**; 988–990.

186 Proctor CJ, Warren ND, Bevan MAJ, Baker-Rogers J. A comparison of methods of assessing exposure to environmental tobacco smoke in non-smoking British women. *Environ. Internat.*, 1991: **17**; 287–297.

187 Phillips K, Bentley MC, Howard DA, Alván G. Assessment of air quality in Stockholm by personal monitoring of non-smokers for respirable suspended particles and environmental tobacco smoke. *Scand. J. Work Environ. Hlth.*, 1996: **22(Suppl 1)**; 24pp.

188 Phillips K, Howard DA, Bentley MC, Alván G. Assessment of air quality in Turin by personal monitoring of nonsmokers for respirable suspended particles and environmental tobacco smoke. *Environ. Internat.*, 1997: **23**; 851–871.

189 Phillips K, Bentley MC, Howard DA, Alván G, Huici A. Assessment of air quality in Barcelona by personal monitoring of nonsmokers for respirable suspended particles and environmental tobacco smoke. *Environ. Internat.*, 1997: **23**; 173–196.

190 Henderson FW, Morris R, Reid HF, Hu PC. Serum and urine cotinine as quantitative measures of passive tobacco smoke exposure in young children. In: Seifert B, Esdorn H, Fischer M, Rüden H, Wegner J, editors. *Indoor Air '87. Proceedings of the 4th International Conference on Indoor Air Quality and Climate, Berlin (West), 17–21 August 1987. Volume 2.* 1987. p. 18–21.

191 Chan C-C, Chen S-C, Wang J-D. Relationship between indoor nicotine concentrations, time-activity data, and urine cotinine-creatinine ratios in evaluating children's exposure to environmental tobacco smoke. *Arch. Environ. Hlth.*, 1995: **50**; 230–234.

192 O'Connor TZ, Holford TR, Leaderer BP, Hammond SK, Bracken MB. Measurement of exposure to environmental tobacco smoke in pregnant women. *Am. J. Epidemiol.*, 1995: **142**; 1315–1321.

193 Galeazzi RL, Daenens P, Gugger M. Steady-state concentrations of cotinine as a measure of nicotine-intake by smokers. *Eur. J. Clin. Pharmacol.*, 1985: **28**; 301–304.

194 Zhou S, Wei Y, Li S. Evaluation of passive smoking under chamber test. *Indoor Air '93. Proceedings of the 6th International Conference on Indoor Air Quality and Climate, Helsinki, Finland July 4–8, 1993. Volume 1.* 1993. p. 705–710.

195 Jarvis MJ, Russell MAH, Benowitz NL, Feyerabend C. Elimination of cotinine from body fluids: implications for noninvasive measurement to tobacco smoke exposure. *Am. J. Publ. Hlth.*, 1988: **78**; 696–698.

196 Willers S, Skarping G, Dalene M, Skerfving S. Urinary cotinine in children and adults during and after semiexperimental exposure to environmental tobacco smoke. *Arch. Environ. Hlth,*, 1995: **50**; 130–138.

197 Goldstein GM, Collier A, Etzel R, Lewtas J, Haley N. Elimination of urinary cotinine in children exposed to known levels of side-stream cigarette smoke. In: Seifert B, Esdorn H, Fischer M, Rüden H, Wegner J, editors. *Indoor Air '87. Proceedings of the 4th International Conference on Indoor Air Quality and Climate,*

*Berlin (West), 17–21 August 1987. Volume 2.* Berlin: Institute for Water, Soil and Air Hygiene; 1987. p. 61–67.

198 Sepkovic DW, Haley NJ. Biomedical applications of cotinine quantitation in smoking related research. *Am. J. Publ. Hlth.*, 1985: **75**; 663–665.

199 Benowitz NL, Jacob P III. Nicotine and cotinine elimination pharmacokinetics in smokers and non-smokers. *Clin. Pharmacol. Ther.*, 1993: **53**; 316–323.

200 Kyerematen GA, Damiano MD, Dvorchik BH, Vesell ES. Smoking-induced changes in nicotine disposition: application of a new HPLC assay for nicotine and its metabolites. *Clin. Pharmacol. Ther.*, 1982: **32**; 769–780.

201 Kyerematen GA, Morgan ML, Chattopadhyay B, deBethizy JD, Vessell ES. Disposition of nicotine and eight metabolites in smokers and non-smokers: identification in smokers of two metabolites that are longer lived than cotinine. *Clin. Pharmacol. Ther.*, 1990: **48**; 641–651.

202 De Schepper PJ, Van Hecken A, Daenens P, Rossum JM. Kinetics of cotinine after oral and intravenous administration to man. *Eur. J. Clin. Pharmacol.*, 1987: **31**; 583–588.

203 Curvall M, Vala EK, Enzell CR, Wahren J. Simulation and evaluation of nicotine intake during passive smoking: cotinine measurements in body fluids of non-smokers given intravenous infusions of nicotine. *Clin. Pharmacol. Ther.*, 1990: **47**; 42–49.

204 Benowitz NL, Jacob P III, Sachs DPL. Deficient C-oxidation of nicotine. *Clin. Pharmacol. Ther.*, 1995: **57**; 590–594.

205 Anderson IGM, Proctor CJ. Comparison of the measurement of serum cotinine levels by gas chromatography and radioimmunoassay. *Analyst*, 1991: **116**; 691–3.

206 Foulds J, Feyerabend C, Stapleton J, Jarvis MJ, Russell MAH. Stability of nicotine and cotinine in unfrozen plasma. *J. Smoking-Relat. Disor.*, 1994: **5**; 41–44.

207 Greeley DA, Valois RF, Bernstein DA. Stability of salivary cotinine sent through the U.S. mail for verification of smoking status. *Addict. Behav.*, 1992: **17**; 291–296.

208 Murray DM, McBride C, Lindquist R, Belcher JD. Sensitivity and specificity of saliva thiocyanate and cotinine for cigarette smoking: a comparison of two collection methods. *Addict. Behav.*, 1991: **16**; 161–166.

209 Riboli E, Haley NJ, De Waard F, Saracci R. Validity of urinary biomarkers of exposure to tobacco smoke following prolonged storage. *Int. J. Epidemiol.*, 1995: **24**; 354–358.

210 Hagan RL, Ramos JM Jnr, Jacob P III. Increasing urinary cotinine concentrations at elevated temperatures: the role of conjugated metabolites. *J. Pharm. Biomed. Anal.*, 1997: **16**; 191–197.

211 Beckett AH, Triggs EJ. Determination of nicotine and its metabolite, cotinine, in urine by gas chromatography. *Nature*, 1966: **211**; 1415–1417.

# Index

Note: Figures and Tables are indicated (in this index) by *italic page numbers*.

Acronyms:

CE = capillary electrophoresis;
ELISA = enzyme-linked immunosorbent assay;
ETS = environmental tobacco smoke;
GC = gas chromatography;
HPLC = high-performance liquid chromatography;
LC = liquid chromatography;
MS = mass spectrometry;
NMR = nuclear magnetic resonance (spectroscopy).
RIA = radioimmunoassay.